T5-DGG-360

Methods in Enzymology

Volume X

OXIDATION AND PHOSPHORYLATION

METHODS IN ENZYMOLOGY

EDITORS-IN-CHIEF

Sidney P. Colowick Nathan O. Kaplan

Methods in Enzymology

Volume X

Oxidation and Phosphorylation

EDITED BY

Ronald W. Estabrook

DEPARTMENT OF BIOPHYSICS AND PHYSICAL BIOCHEMISTRY
JOHNSON RESEARCH FOUNDATION
UNIVERSITY OF PENNSYLVANIA SCHOOL OF MEDICINE
PHILADELPHIA, PENNSYLVANIA

and

Maynard E. Pullman

DEPARTMENT OF BIOCHEMISTRY
THE PUBLIC HEALTH RESEARCH INSTITUTE OF THE CITY OF NEW YORK, INC.
NEW YORK, NEW YORK

1967

ACADEMIC PRESS New York and London

ACADEMIC PRESS INC.
111 Fifth Avenue, New York, New York 10003

United Kingdom Edition published by
ACADEMIC PRESS INC. (LONDON) LTD.
Berkeley Square House, London W.1

Library of Congress Catalog Card Number: 54-9110

PRINTED IN THE UNITED STATES OF AMERICA

Contributors to Volume X

Article numbers are shown in parentheses following the names of contributors
Affiliations listed are current

K. AHMED (119), *Division of Metabolic Research, The Chicago Medical School, Institute for Medical Research, Chicago, Illinois*

DAVID W. ALLMANN (71, 72a), *Institute for Enzyme Research, The University of Wisconsin, Madison, Wisconsin*

ELISABETH BACHMANN (72a), *Department of Biology, Sterling College, Tuscaloosa, Alabama*

WALTER X. BALCAVAGE (26), *Department of Physiological Chemistry, The Johns Hopkins University School of Medicine, Baltimore, Maryland*

R. E. BASFORD (16), *Department of Biochemistry, University of Pittsburgh School of Medicine, Pittsburgh, Pennsylvania*

ANDERS BERGSTRAND (72c), *Department of Pathology II, Karolinska Institutet, Stockholm, Sweden*

ROBERT E. BEYER (34, 81), *Biophysics Research Division, Institute of Science and Technology and Department of Zoology, University of Michigan, Ann Arbor, Michigan*

L. L. BIEBER (117), *Molecular Biology Institute and Department of Chemistry, University of California, Los Angeles, California*

PAUL V. BLAIR (12, 38), *Institute for Enzyme Research, The University of Wisconsin, Madison, Wisconsin*

WALTER D. BONNER, JR. (24), *Johnson Research Foundation, University of Pennsylvania, Philadelphia, Pennsylvania*

P. D. BOYER (10, 117), *Molecular Biology Institute and Department of Chemistry, University of California, Los Angeles, California*

ARNOLD F. BRODIE (30), *Department of Microbiology, University of Southern California School of Medicine, Los Angeles, California*

PHILIP E. BRUMBY (73), *Department of Biological Chemistry, University of Michigan, Ann Arbor, Michigan*

DONNA M. BRYAN (10), *Molecular Biology Institute and Department of Chemistry, University of California, Los Angeles, California*

RONALD A. BUTOW (18), *Department of Chemistry, Frick Chemical Laboratory, Princeton University, Princeton, New Jersey*

ERNESTO CARAFOLI (114), *Institute of General Pathology, University of Modena, Modena, Italy*

B. CHANCE (99), *Johnson Research Foundation, University of Pennsylvania, Philadelphia, Pennsylvania*

T. E. CONOVER (85), *Department of Contractile Proteins, Institute for Muscle Diseases, Inc., New York, New York*

DAVID Y. COOPER (65, 96), *Harrison Department of Surgical Research, University of Pennsylvania School of Medicine, Philadelphia, Pennsylvania*

F. CRANE (48), *Department of Biological Sciences, Purdue University, Lafayette, Indiana*

RICHARD S. CRIDDLE (102), *Department of Biochemistry and Biophysics, University of California, Davis, California*

JOHN R. CRONIN (55), *Department of Biochemistry, Yale University, New Haven, Connecticut*

M. ANNE DANNENBERG (84a), *Department of Biochemistry and Molecular Biology, Cornell University, Ithaca, New York*

JOHN TERRANCE DAVIS (21), *Department of Pathology, University of Pennsyl-*

vania School of Medicine, Philadelphia, Pennsylvania

STARKEY D. DAVIS (106), *Department of Pediatrics, University of Washington School of Medicine, Seattle, Washington*

CHRISTIAN DE DUVE (2), *The Rockefeller University, New York, New York*

SAMIR S. DEEB (66), *Biochemistry Division, Department of Chemistry, University of Illinois, Urbana, Illinois*

THOMAS M. DEVLIN (20), *Research Laboratories, Merck Sharp & Dohme, Rahway, New Jersey*

W. B. ELLIOTT (32), *State University of New York at Buffalo School of Medicine, Buffalo, New York*

LARS ERNSTER (5, 14, 56, 72c, 87, 92a, 112, 113), *Wenner-Gren Institute, University of Stockholm, Stockholm, Sweden*

RONALD W. ESTABROOK (7, 65, 74, 109, 110), *Department of Biophysics and Physical Biochemistry, Johnson Research Foundation, University of Pennsylvania School of Medicine, Philadelphia, Pennsylvania*

S. P. FELTON (77), *Division of Biochemistry, Scripps Clinic Research Foundation, La Jolla, California*

JUNE M. FESSENDEN (35, 84a, 84b, 107), *Department of Biochemistry and Molecular Biology, Cornell University, Ithaca, New York*

BECCA FLEISCHER (70), *Department of Molecular Biology, Vanderbilt University, Nashville, Tennessee*

SIDNEY FLEISCHER (69, 70), *Department of Molecular Biology, Vanderbilt University, Nashville, Tennessee*

MAURILLE J. FOURNIER, JR. (115), *Department of Biology, State University of New York at Stony Brook, Stony Brook, New York*

RENE FRENKEL (74), *Johnson Research Foundation, University of Pennsylvania, Philadelphia, Pennsylvania*

WILHELM R. FRISELL (55), *Department of Biochemistry, University of Colorado School of Medicine, Denver, Colorado*

ROBERT P. GLAZE (86), *Department of Biochemistry, University of Alabama, Birmingham, Alabama*

JEANINE GONZE (11, 17), *Johnson Research Foundation, University of Pennsylvania, Philadelphia, Pennsylvania*

J. W. GREENAWALT (27), *Department of Physiological Chemistry, The Johns Hopkins University School of Medicine, Baltimore, Maryland*

CHARLES T. GREGG (33), *Los Alamos Scientific Laboratory, University of California, Los Alamos, New Mexico*

HELMUT GREIM (109), *Institute for Toxicology, University of Tübingen, Tübingen, Germany*

H. GUTFREUND (110), *Department of Biochemistry, University of Bristol, Bristol, England*

D. W. HAAS (32), *Department of Cardio-Respiratory Diseases, Walter Reed Army Institute of Research, Washington D. C.*

LOWELL P. HAGER (66, 67), *Biochemistry Division, Department of Chemistry and Chemical Engineering, University of Illinois, Urbana, Illinois*

D. O. HALL (27), *Department of Botany, Kings College, University of London, Strand, London, England*

Y. HATEFI (41, 43), *Scripps Institute and Research Foundation, La Jolla, California*

HANS W. HELDT (75), *Physiologisch-Chemisches Institut der Universität Marburg, Marburg, Germany*

ROBERT L. HOWARD (52), *Laboratory for Respiratory Enzymology, Oregon State University, Corvallis, Oregon*

F. M. HUENNEKENS (77), *Division of Biochemistry, Scripps Clinic Research Foundation, La Jolla, California*

F. EDMUND HUNTER, JR. (105b), *Department of Pharmacology, Washington University School of Medicine, St. Louis, Missouri*

SHINJI ISHIKAWA (31), *Biochemistry Division, Institute of Brain Research and Department of Chemistry, Faculty of*

Medicine, University of Tokyo, Bunkyo-Ku, Tokyo, Japan

EIJI ITAGAKI (67), *Biochemistry Division, Department of Chemistry and Chemical Engineering, University of Illinois, Urbana, Illinois*

E. E. JACOBS (6), *Biophysics Laboratory, Stanford University, Stanford, California*

DIANE JOHNSON (15), *History of Science Department, The University of Wisconsin, Madison, Wisconsin*

J. D. JUDAH (119), *Division of Metabolic Research, The Chicago Medical School, Institute for Medical Research, Chicago, Illinois*

YASUO KAGAWA (79, 83), *Department of Chemistry, Institute of Infectious Diseases, University of Tokyo, Minato-ku, Tokyo, Japan*

HENRY KAMIN (92), *Radioisotope Service, Veterans Administration Hospital, and Department of Biochemistry, Duke University, Durham, North Carolina*

NATHAN O. KAPLAN (57), *Graduate Department of Biochemistry, Brandeis University, Waltham, Massachusetts*

TSOO E. KING (37, 40, 52, 58, 98), *Laboratory for Respiratory Enzymology, Oregon State University School of Science, Corvallis, Oregon*

M. KLINGENBERG (1, 75, 104), *Physiologisch-Chemisches Institut der Universität Marburg, Marburg, Germany*

KRYSTYNA C. KOPACZYK (47), *Institute for Enzyme Research, The University of Wisconsin, Madison Wisconsin*

BO KUYLENSTIERNA (72c), *Wenner-Gren Institute, University of Stockholm, Stockholm, Sweden*

HENRY LARDY (15), *Institute for Enzyme Research, The University of Wisconsin, Madison, Wisconsin*

ALBERT LAUWERS (71), *Pharmaceutisch Institute, Ryksuniversiteit Ghent, Ghent, Belgium*

CHUAN-PU LEE (5, 87, 112, 113), *Department of Biophysics and Physical Biochemistry, Johnson Research Foundation, University of Pennsylvania, Philadelphia, Pennsylvania*

ALBERT L. LEHNINGER (114), *Department of Physiological Chemistry, The Johns Hopkins University School of Medicine, Baltimore, Maryland*

GIORGIO LENAZ (71, 78), *Institute for Enzyme Research, The University of Wisconsin, Madison, Wisconsin*

ROBERT L. LESTER (50), *Department of Biochemistry, University of Kentucky, Lexington, Kentucky*

COSMO G. MACKENZIE (55), *Department of Biochemistry, University of Colorado School of Medicine, Denver, Colorado*

BRUCE MACKLER (46, 49, 53, 88, 106), *Department of Pediatrics, University of Washington School of Medicine, Seattle, Washington*

D. H. MACLENNAN (78), *Institute for Enzyme Research, The University of Wisconsin, Madison, Wisconsin*

PABITRA K. MAITRA (74), *Molecular Biology Unit, Tata Institute of Fundamental Research, Bombay, India*

E. MARGOLIASH (61), *Abbott Laboratories, Scientific Divisions, North Chicago, Illinois*

VINCENT MASSEY (73), *Department of Biological Chemistry, University of Michigan, Ann Arbor, Michigan*

BETTIE SUE SILER MASTERS (92), *Duke University Medical Center, Durham, North Carolina*

JAMES R. MATTOON (26), *Department of Physiological Chemistry, The Johns Hopkins University School of Medicine, Baltimore, Maryland*

THOMAS D. MEHL (106), *Department of Pediatrics, University of Washington School of Medicine, Seattle, Washington*

GLADYS C. MONROY (80), *Department of Biochemistry, The Public Health Research Institute of the City of New York, Inc., New York, New York*

ROY O. MORRIS (98), *Laboratory of Respiratory Enzymology, Oregon State University, Corvallis, Oregon*

WALTER L. NELSON (18), *Department of Biochemistry, Cornell University, Ithaca, New York*

M. NISHIMURA (99), *Johnson Research Foundation, University of Pennsylvania, Philadelphia, Pennsylvania*

KERSTIN NORDENBRAND (14, 92a), *Department of Biochemistry, Wenner-Gren Institute, University of Stockholm, Stockholm, Sweden*

TSUNEO OMURA (65, 90), *Institute for Protein Research, Osaka University, Osaka, Japan*

LESTER PACKER (105a), *Department of Physiology, University of California at Berkeley, Berkeley, California*

GRAHAM PALMER (93, 94), *Institute of Science and Technology, Biophysics Research Division, University of Michigan, Ann Arbor, Michigan*

H. G. PANDIT-HOVENKAMP (29), *Laboratory of Physiological Chemistry, University of Amsterdam, Amsterdam, The Netherlands*

D. F. PARSONS (72b, 101), *Department of Biophysics, Roswell Park Memorial Institute, Buffalo, New York*

HARVEY S. PENEFSKY (82, 108), *Department of Biochemistry and Molecular Biology, Cornell University, Ithaca, New York*

JOHN T. PENNISTON (51), *Institute for Enzyme Research, The University of Wisconsin, Madison, Wisconsin*

ERICH PFAFF (104), *Physiologisch-Chemisches Institut der Universität Marburg, Marburg, Germany*

RICHARD L. PHARO (54), *Department of Bioenergetics, Institute of Biological and Medical Sciences, Retina Foundation, Boston, Massachusetts*

GIFFORD B. PINCHOT (28), *Department of Biology and McCollum-Pratt Institute, The Johns Hopkins University, Baltimore, Maryland*

M. POE (110), *Johnson Research Foundation, University of Pennsylvania, Philadelphia, Pennsylvania*

R. L. POST (116, 118), *Department of Physiology, Vanderbilt University School of Medicine, Nashville, Tennessee*

BERTON C. PRESSMAN (111), *Johnson Research Foundation, University of Pennsylvania, Philadelphia, Pennsylvania*

MAYNARD E. PULLMAN (9, 80), *Department of Biochemistry, The Public Health Research Institute of the City of New York, Inc., New York, New York*

EFRAIM RACKER (35, 84a, 84b, 107), *Department of Biochemistry and Molecular Biology, Cornell University, Ithaca, New York*

N. APPAJI RAO (77), *Division of Biochemistry, Scripps Clinic Research Foundation, La Jolla, California*

E. R. REDFEARN (68), *Department of Biochemistry, University of Leicester, Leicester, England*

HERBERT REMMER (109), *Institute for Toxicology, University of Tübingen, Tübingen, Germany*

J. S. RIESKE (41, 42, 43, 44, 62, 63, 64, 76), *Department of Biochemistry, Ohio State University, Columbus, Ohio*

OTTO ROSENTHAL (96), *Department of Surgery, Pennsylvania Hospital, Philadelphia, Pennsylvania*

GEORGE ROUSSER (69), *Department of Biochemistry, City of Hope Medical Center, Medical Research Institute, Duarte, California*

D. R. SANADI (6, 54), *Department of Bioenergetics, Institute of Biological and Medical Sciences, Retina Foundation, Boston, Massachusetts*

E. SANDERS (65), *Johnson Research Foundation, University of Pennsylvania, Philadelphia, Pennsylvania*

RYO SATO (90), *Institute for Protein Research, Osaka University, Osaka, Japan*

LEONARD A. SAUER (19), *Medical Research, The Rockefeller University, New York, New York*

GOTTFRIED SCHATZ (4, 36), *Institute of*

Biochemistry, University of Vienna, Vienna, Austria

JOHN B. SCHENKMAN (109), *Johnson Research Foundation, University of Pennsylvania, Philadelphia, Pennsylvania*

FREDERICK J. SCHINDLER (97), *Johnson Research Foundation, University of Pennsylvania, Philadelphia, Pennsylvania*

A. K. SEN (116, 118), *Department of Physiology, Vanderbilt University School of Medicine, Nashville, Tennessee*

FRED SHERMAN (95), *Department of Radiation Biology and Biophysics, University of Rochester Medical School, Rochester, New York*

MELVIN V. SIMPSON (115), *Department of Biology, State University of New York at Stony Brook, Stony Brook, New York*

DOROTHY M. SKINNER (115), *Department of Biology, Oak Ridge National Laboratory, Oak Ridge, Tennessee*

E. C. SLATER (3, 8), *Laboratory of Biochemistry, University of Amsterdam, Amsterdam, The Netherlands*

ARCHIE L. SMITH (13), *Division of Biochemistry, Sloan Kettering Institute, New York, New York*

E. E. SMITH (105b), *Department of Pharmacology, Washington University School of Medicine, St. Louis, Missouri*

LOUIS A. SORDAHL (54), *Department of Pharmacology, Baylor University School of Medicine, Houston, Texas*

GIAN LUIGI SOTTOCASA (5, 72c), *Istituto di Chimica Biologica, University of Trieste, Trieste, Italy*

CLINTON D. STONER (103), *Department of Biochemistry, Ohio State University, Columbus, Ohio*

PHILIPP STRITTMATTER (89, 91), *Department of Biological Chemistry, Washington University School of Medicine, St. Louis, Missouri*

S. C. STUART (100), *Department of Biochemistry, University of Toronto, Toronto, Ontario, Canada*

KUNI TAKAYAMA (103), *Division of Biochemistry, Veterans Administration Hospital, Madison, Wisconsin*

HOWARD D. TISDALE (39, 62, 63), *Division of Molecular Biology, Veterans' Administration Hospital, San Francisco, California*

D. D. TYLER (11, 17), *Royal Veterinary College, London, England*

ALEXANDER TZAGOLOFF (71), *Institute for Enzyme Research, The University of Wisconsin, Madison, Wisconsin*

SIMON G. VAN DEN BERGH (22, 114a), *Laboratory of Biochemistry, University of Amsterdam, Amsterdam, The Netherlands*

CHARLES L. WADKINS (86), *Department of Biochemistry, University of Arkansas, Little Rock, Arkansas*

O. F. WALASEK (61), *Scientific Divisions, Abbott Laboratories, North Chicago, Illinois*

O. C. WALLIS (27), *Department of Microbiology, Queen Elizabeth College, London, England*

RALPH J. WEDGWOOD (106), *Department of Pediatrics, University of Washington School of Medicine, Seattle, Washington*

DAVID C. WHARTON (45), *Biochemistry Section, Cornell University, Ithaca, New York*

CHARLES H. WILLIAMS, JR. (92), *Veterans Administration Hospital, and Department of Biological Chemistry, University of Michigan, Ann Arbor, Michigan*

G. R. WILLIAMS (72b, 100), *Department of Biochemistry, University of Toronto, Toronto, Ontario, Canada*

JOHN R. WILLIAMSON (74), *Johnson Research Foundation, University of Pennsylvania, Philadelphia, Pennsylvania*

J. T. WISKICH (23), *Department of Botany, University of Adelaide, Adelaide, South Australia*

RAY WU (19), *Department of Biochemistry and Molecular Biology, Cornell University, Ithaca, New York*

TAKASHI YONETANI (59, 60), *Johnson Research Foundation, University of Pennsylvania, Philadelphia, Pennsylvania*

H. ZALKIN (85), *Department of Biochemistry, Purdue University, Lafayette, Indiana*

I. ZELITCH (25), *Department of Biochemistry, Connecticut Agricultural Experiment Station, New Haven, Connecticut*

D. ZEIGLER (42), *Clayton Foundation Biochemical Institute, University of Texas, Austin, Texas*

Preface

Though important advances have been made in the last 15 years toward an understanding of the mechanism of energy-coupled respiration, the details of this major mitochondrial function remain largely unknown. The nature and volume of the experimental work on mitochondria reported in the past few years reflect the fact that not only has the enzymologist broadened his approach to this problem, but that this field has attracted from other areas of biology increasing numbers of investigators including electron microscopists, cell physiologists, and geneticists. The diversity of approaches employed by investigators interested in electron transport, oxidative phosphorylation, and other aspects of mitochondrial physiology indicated that this was an appropriate time to compile a methodological reference for this area of biochemistry.

The aim of this volume is to provide not only the specialist and the advanced student but also investigators from other areas of research with a *single authoritative* source for the vast and often difficult to retrieve and evaluate methodology associated with mitochondrial research. The accounts of procedures, written by leading investigators personally experienced with the methods, are intended to be sufficiently detailed and comprehensive to insure reproducibility without having to refer to original papers. Alternate methods are mentioned, and when possible an effort has been made to evaluate the more commonly used procedures.

The scope of coverage ranges from the commonly used techniques of measuring P:O ratios and the isolation of mitochondria to more specialized techniques, such as the application of electron microscopy to the study of mitochondria or the measurement of heat exchange occurring during electron transport. A chapter was devoted to the application of inhibitors and uncouplers to the study of electron transport and oxidative phosphorylation and includes the source, specificity and optimal concentrations required. While the major portion of the volume is devoted to mitochondrial respiratory enzymes and associated factors for oxidative phosphorylation, a number of newer procedures related to the properties and the purification of microsomal as well as bacterial respiratory pigments have been included. Related methods and techniques which have been included in previous volumes of this series have not been repeated here unless it was felt that recent developments represented significant improvements over the earlier described methods. Nevertheless references to these articles appear at the beginning of each section of this volume where they would normally have appeared.

We would like to express our gratitude to all of the contributors for their cooperation in making this volume possible. We are grateful to

Drs. B. Chance, D. E. Green, H. Lardy, A. L. Lehninger, E. Racker, and E. C. Slater for their advice and constructive criticism in the organization of the original outline for this volume. We also wish to thank Mr. Philip Blackwood, Mrs. Edith Casper, and Mrs. Phyllis Pullman for the excellent secretarial assistance they provided. Finally, we wish to acknowledge the warm cooperation and friendly patience of the staff of Academic Press.

RONALD W. ESTABROOK
MAYNARD E. PULLMAN

April, 1966

Table of Contents

Section I. General Techniques for the Measurement of Oxidative and Phosphorylation Reactions

Section II. Preparations, Properties, and Conditions for Assay of Mitochondria

Section III. Preparations, Properties, and Conditions for Assay of Submitochondrial Particles

Section IV. Purified Respiratory Chain Oxidation-Reduction Components

Section V. Isolation and Determination of Other Mitochondrial Constituents

Section VI. Isolation and Assay of Coupling Factors for Oxidative Phosphorylation

Section VII. Microsomal Electron Transport

Section VIII. Special Assays and Techniques

Section IX. Measurement of Energy-Linked and Associated Reactions

METHODS IN ENZYMOLOGY

EDITED BY

Sidney P. Colowick and Nathan O. Kaplan

VANDERBILT UNIVERSITY
SCHOOL OF MEDICINE
NASHVILLE, TENNESSEE

GRADUATE DEPARTMENT OF
BIOCHEMISTRY
BRANDEIS UNIVERSITY
WALTHAM, MASSACHUSETTS

METHODS IN ENZYMOLOGY

EDITORS-IN-CHIEF

Sidney P. Colowick   Nathan O. Kaplan

VOLUME VIII

Complex Carbohydrates
Edited by ELIZABETH F. NEUFELD AND VICTOR GINSBURG

VOLUME IX

Carbohydrate Metabolism
Edited by WILLIS A. WOOD

VOLUME X

Oxidation and Phosphorylation
Edited by RONALD W. ESTABROOK AND MAYNARD E. PULLMAN

In Preparation

Enzyme Structure
Edited by C. H. W. Hirs

Nucleic Acids, Parts A and B
Edited by LAWRENCE GROSSMAN AND KIVIE MOLDAVE

Fast Reactions
Edited by KENNETH KUSTIN

Citric Acid Cycle
Edited by J. M. LOWENSTEIN

Lipid Metabolism
Edited by J. M. LOWENSTEIN AND R. B. CLAYTON

Photosynthesis
Edited by A. SAN PIETRO

Methods in Enzymology

Volume X

OXIDATION AND PHOSPHORYLATION

Section I

General Techniques for the Measurement of Oxidative and Phosphorylation Reactions

PREVIOUSLY PUBLISHED ARTICLES FROM METHODS IN ENZYMOLOGY RELATED TO SECTION I

Vol. I [3]. Fractionation of Cell Components of Animal Tissues. George H. Hogeboom.

Vol. II [101]. Coupling of Phosphorylation with Oxidation. F. Edmund Hunter, Jr.

Vol. IV [13]. Use of Artificial Electron Acceptors in the Study of Dehydrogenases. J. H. Quastel.

Vol. IV [37]. O^{18} Containing Phosphate and Related Compounds. M. Cohn.

Vol. VIII [10]. Assay of Inorganic Phosphate, Total Phosphate, and Phosphatases. Bruce N. Ames.

[1] Enzyme Profiles in Mitochondria

By MARTIN KLINGENBERG

The content in mitochondria of various enzymes can be described in an enzyme profile. The profile summarizes the data in a graph which is particularly designed for comparing the equipment of the mitochondria from various organs.

In most cases the enzyme profile corresponds to an "enzyme activity pattern."[1-3] The measure for the content is in this case the activity of the enzymes as found in a crude extract or sediment of the mitochondria. They include the dehydrogenases, phosphate transferases, transaminases, etc. A molarity pattern can be obtained for enzymes with fixed prosthetic groups which can be distinguished by direct measurement in the crude extract or by a more specific extraction procedure. In this way are obtained the molar content of the cytochromes on the basis of their specific absorption[4-6] and—approximately—the molar content of two flavin-specific dehydrogenases, such as the DPNH and succinate dehydrogenases, by measuring the content of FMN or of peptide-bound FAD.[7] In principle, the enzyme activity can be converted to a molar content by the turnover number.[7] In this case, however, the kinetic properties of the enzyme in the crude extract must agree with those of the isolated enzyme.

The enzyme activity, in particular of regulatory enzymes, might be modified in the crude extract. Among the mitochondrial enzymes this concerns, for example, glutamate dehydrogenase and DPN-specific isocitrate dehydrogenase, the activity of which can be influenced by various modifiers, possibly present in the extract. Furthermore, the enzyme activity test can be perturbed by other enzymes in a crude extract,

[1] T. Bücher and M. Klingenberg, *Angew. Chem.* **70**, 552 (1958).
[2] A. Delbrück, E. Zebe, and T. Bücher, *Biochem. Z.* **331**, 273 (1959).
[3] D. Pette, M. Klingenberg, and T. Bücher, *Biochem. Biophys. Res. Commun.* **7**, 425 (1962).
[4] B. Chance and G. R. Williams, *Advan. Enzymol.* **17**, 56 (1956).
[5] R. Estabrook and A. Holowinsky, *J. Biophys. Biochem. Cytol.* **9**, 19 (1961).
[6] M. Klingenberg, Die funktionelle Biochemie der Mitochondrien. *GDNA Sympo. Funktionelle und Morphologische Organisation der Zelle, Rottach-Egern, 1961*, p. 69. Springer, Heidelberg, 1962.
[7] See M. Klingenberg, *Ergebn. Physiol. Biol. Chem. Exptl. Pharmakol.* **55**, 129 (1964).

such as by hydrolases, isomerases, which may compete for the substrate. Careful control of these perturbations must be assured.[8]

It can be assumed however, that in the majority of cases the activity of an enzyme in a crude extract is the same as that of the purified enzyme and the activity measurements give a meaningful result on the molar content of the enzyme. This prerequisite must also be fulfilled for comparing enzyme activity profiles between mitochondria from various sources.

Preparation of Mitochondria

Although the procedures for preparation of mitochondria are described in another chapter, some remarks are necessary here. In the majority of cases the enzyme profile of the isolated mitochondria should reflect that of the mitochondria in the tissue. Criteria for the intactness of the mitochondria based on oxidative phosphorylation (respiratory control, P:O ratio, etc.) do not reflect necessarily an intact equipment with enzymes. The isolation procedure must avoid the loss of proteins easily leaking from the mitochondria. This concerns in particular some phosphate transferases and cytochrome *c*.[9,10] The mitochondria must be isolated in isotonic or hypotonic media with low ionic strength. Slightly buffered solutions of sucrose give the best results.

The mechanical procedure for the release of mitochondria from the tissue should be as gentle as possible to preserve the intactness of the mitochondria. Even for example "heavy mitochondria" from beef heart are not any more sufficiently intact for obtaining a representative enzyme profile, since they have already lost part of the cytochrome *c* and creatine kinase. A digestion of the tissue by bacterial proteinase preceding the homogenization must be very well controlled in order to avoid breakdown of some enzymes.

Extraction

The following procedure is for the preparation of a crude extract containing all enzymes which can be extracted by a buffer at neutral pH. The mitochondria are suspended in 0.1 *M* phosphate buffer, pH 7.2 at a concentration of about 2–10 mg protein per milliliter; 10 m*M* reduced glutathione and—for the preservation of DPN-specific isocitrate dehydrogenase—1 m*M* ADP are added. For the disintegration the suspension

[8] See discussion in T. Bücher, W. Luh, and D. Pette, "Hoppe-Seyler/Thierfelder Handbuch der Physiologisch- und Pathologisch-Chemischen Analyse," Vol. VI, p. 293. Springer, Heidelberg, 1964.

[9] H. Jacobs, Dissertation, Univ. of Marburg, 1965.

[10] K. Schwalbach and M. Klingenberg, unpublished.

is carefully kept near 0° by ice or ice–salt mixture during the following mechanical disintegration. The mitochondria are broken by an appropriate rotatory disintegrator such as Ultraturrax, which has a fingerlike protruding rod. The disintegration may also be accomplished by sonification. Here again most practical is the finger type, such as the Branson Sonifier. Disintegration by Ultraturrax may need 2–3 minutes; with sonifier, 1–2 minutes at high power output. The disintegration with the sonifier must be interrupted after 20-second intervals for cooling of the suspension.

For separation of soluble enzymes from the particulate material the suspension is centrifuged at about 100,000 *g* for 15–30 minutes when obtained by means of a rotatory disintegrator, and up to 60 minutes when prepared by sonification.

The supernatant can be used as a crude extract for the enzyme assays. The sediment is washed again by resuspending in phosphate buffer of the same volume in a tight-fitting glass homogenizer. After centrifugation, the sediment is again resuspended and can be used for measurements of the particle-bound enzyme activities.

The particle-bound DPNH oxidase activity must be carefully removed by the centrifugation since it disturbs the enzyme tests. For this purpose the centrifugation must be very thorough after sonication, when the particles are particularly finely dispersed. Residual DPNH oxidase may be inhibited by the addition of KCN to the enzyme test mixture. KCN can, however, be inhibitory to some dehydrogenases, as for example to DPN-specific isocitrate dehydrogenase.

Fractional Extraction

Part of the enzyme activity may be dissociated from the mitochondria without mechanical disintegration by extraction with buffers. This extract may contain for some enzymes the largest portion of the total activity. In this extraction procedure, three fractions are obtained: (1) easily dissociable by nonmechanical treatment; (2) neutral pH dissociable by mechanical treatment; (3) and structurally bound.

The mitochondria are incubated in 0.1 *M* phosphate buffer, at comparatively high dilution, at less than 2 mg protein per milliliter. After 30 minutes' incubation at 0° the suspension is centrifuged at 60,000 *g* for 20 minutes. The supernatant contains the easily dissociable portion of the enzyme activities. The sediment is resuspended in a phosphate buffer at a somewhat higher concentration and now mechanically disintegrated. The subsequent separation follows that described above for the preparation of the total soluble extract. Between extract steps 1 and 2 the sediment is washed once by resuspension in the extraction medium and centrifugation. After extraction step 2 the sediment is extracted once again by resuspension, sonication, and centrifugation.

A survey of the extractability of various enzymes is given in the table.

EXTRACTABILITY OF MITOCHONDRIAL ENZYMES

Solubility and treatment	Enzyme types
Easily soluble,[a] incubation with 0.1 M phosphate at pH 7.2	Phosphate transferases (e.g., adenylate kinase, creatine kinase), cytochrome c
Soluble, mechanical breakage	DPN, TPN-linked dehydrogenases
	Transaminases
	Flavoprotein (lipoate dehydrogenase)
Difficultly soluble, prolonged sonic treatment	Flavoproteins (succinate dehydrogenase[b])
	ATPase
Insoluble	Cytochromes (cytochromes b, c_1, a, a_3)
	Flavoproteins (DPNH dehydrogenase)

[a] Soluble, defined as: to become soluble without the application of detergents, lipophilic solvents, or lipases.

[b] Alkaline treatment.

Evaluation of Enzyme Profiles

The enzyme activity is mostly referred to the total protein content of the mitochondria. A graphic plot with a logarithmic ordinate has been designed to facilitate the comparison of the activation of various enzymes from various sources.[2] The logarithmic plot has the advantage of coping easily with the large range of activity. Furthermore, in this plot constant ratios of activity correspond to constant distances of the ordinate when different types of mitochondria are compared, even at different levels of activity. The constant ratios may reflect functional relationships between various enzymes.

For this reason, in an enzyme pattern the enzyme activities may be referred to a key enzyme instead of to the protein content. For example, enzyme activity of mitochondria might be referred to the molar content of cytochrome a as illustrated in Fig. 1. In this case the enzyme activity has the meaning of a turnover of cytochrome a when multiplied by 2 in order to account for the one electron-accepting capacity of cytochrome a. In enzyme profiles related to cytochrome a, for example, the enzyme activity of malate dehydrogenase has been found to be constant in a wide variety of mitochondria (cf. Fig. 1).[3] Enzymes having a constant ratio of their activity to each other have been defined as constant-proportion groups.[11]

[11] T. Bücher, *Boll. Soc. Ital. Biol. Sper.* **36**, 1509 (1960).

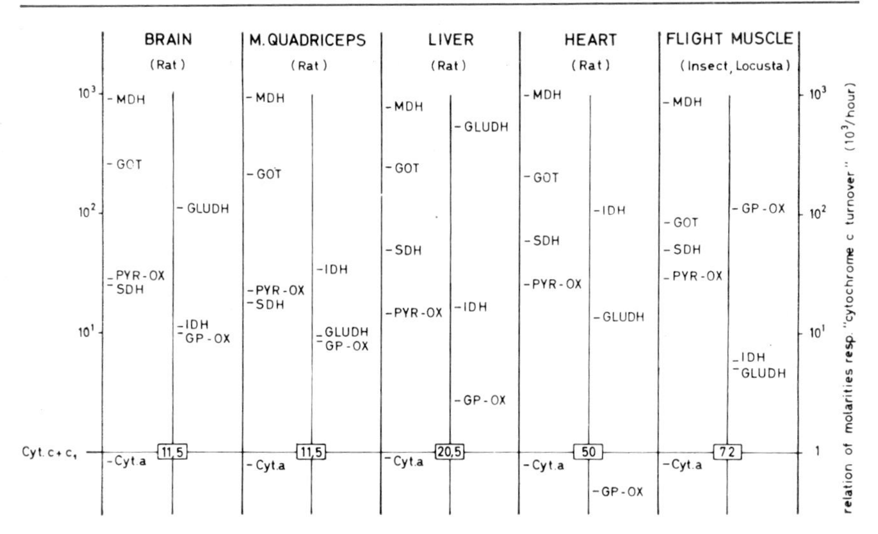

FIG. 1. Enzyme activity profile for mitochondria from various animal sources. The enzyme activities are referred to the content of cytochrome *c* ("turnover" of cytochrome *c*), i.e. dividing the activities by the content of cytochrome *c* and multiplying by 2. Abbreviations: MDH, malate dehydrogenase; GludH, glutamate dehydrogenase; SDH, succinate dehydrogenase; GOT, glutamate oxaloacetate transaminase; IDH, TPN-specific isocitrate dehydrogenase; PYR-Ox, pyruvate oxidase; GP-Ox, glycerol-1-phosphate oxidase. [From D. Pette, M. Klingenberg, and Th. Bücher, *Biochem. Biophys. Res. Commun.* **7**, 425 (1962).]

One ultimate goal of studying enzyme activities in mitochondria is elucidation of the molecular composition of mitochondria with enzymes. With the exception of the cytochromes, this evaluation is not unequivocal in most cases. However, the data merely on the enzyme activity distribution contain valuable information when viewed in comparison of mitochondria from various organs.

[2] Criteria of Homogeneity and Purity of Mitochondria

By CHRISTIAN DE DUVE

Mitochondria vs. Mitochondrial Fractions

Mitochondria are intracellular organelles characterized by certain well defined structural and biochemical properties. They are most easily recognized morphologically by the presence of two membranes, of which the inner one shows infoldings or cristae. They are the specific bearers of the main phosphorylating electron transport chain and can be detected and assayed by their ability to carry out characteristic reactions of this system, for instance the oxidation of cytochrome *c* or of succinate.

Biochemists engaged in the study of mitochondria usually work on crude mitochondrial fractions isolated from disrupted tissue preparations by means of centrifugal techniques designed for the collection of subcellular elements showing a sedimentation coefficient in 0.25 *M* sucrose comprised between about 10^5 and 5×10^3 S. Such fractions are frequently referred to in the biochemical literature under the name "mitochondria."

It is clearly undesirable to have the same word carry different meanings depending on the area of specialization of the investigator. In this chapter, the word "mitochondria" will be used in its narrow and specific cytological sense; subcellular fractions containing mitochondria as their main component will be referred to as "mitochondrial fractions."

This distinction is introduced, not for the purpose of splitting hairs, but to convey the fundamental notion that there are many more things than mitochondria in mitochondrial fractions. Not to confuse the object of an investigation with its purpose, it is necessary, in every piece of work directed toward the understanding of mitochondria, but performed on mitochondrial fractions, to raise and, if possible, to answer the following questions:

1. What is the likelihood that the results obtained express properties of contaminants rather than of the mitochondria themselves?

2. If the observed properties are truly mitochondrial, what is the likelihood that they have been modified or affected in any way by the presence of contaminants?

Such questions are not easily answered, but this is no reason for ignoring them. It is the purpose of this section to enumerate some of the contaminants which may be encountered in mitochondrial fractions and to describe briefly the technical procedures whereby they can be recognized and, in some cases, eliminated. Readers are referred to previous reviews for more detailed information.[1-3b]

Methods

Morphological Methods

Examination in the electron microscope is useful to verify that mitochondria do indeed form the main component of a fraction and to assess

[1] C. de Duve, R. Wattiaux, and P. Baudhuin, *Advan. Enzymol.* **24,** 291 (1962).
[2] C. de Duve, *J. Theoret. Biol.* **6,** 33 (1964).
[3] C. de Duve, *Harvey Lectures Ser.* **59,** 49 (1965).
[3a] C. de Duve and R. Wattiaux, *Ann. Rev. Physiol.* **28,** 435 (1966).
[3b] C. de Duve and P. Baudhuin, *Physiol. Rev.* **46,** 323 (1966).

their degree of morphological preservation. As a means of detecting and evaluating contaminants, this method has to contend with almost insuperable problems of sampling, identification, and resolution.

A sample of 1 mg dry weight contains approximately 10^{10} particles of mitochondrial size. If the composition of this sample is to be assessed accurately in a reasonable time, perfect random distribution of the particles throughout the sample is an essential prerequisite. This condition is not met in the currently used techniques relying on centrifugation for packing the particles. A promising method using Millipore filtration has been recently worked out.[4] Granting random sampling, one is still left with the problem of identifying each particle within a given field. Every worker who was looked at mitochondrial fractions in the electron microscope knows that this is not possible. The terms "damaged mitochondria" or "mitochondrial debris" are standard euphemisms that cover a multitude of sins. Finally, morphological detection becomes increasingly hazardous the smaller the size of the contaminant. But in biochemical work, dimension is hardly a good criterion to gauge the importance of an impurity.

To these intrinsic difficulties may be added a practical one. Few biochemists are familiar with electron microscopy and probably even fewer have colleagues in morphology departments who are ready to devote the necessary time and effort to an ungrateful and relatively unrewarding task.

Biochemical Methods

Increasing knowledge of the enzymatic composition of nonmitochondrial elements likely to contaminate mitochondrial fractions has provided a number of relatively simple techniques that can serve to evaluate the composition of such fractions and to monitor purification attempts. These tests become particularly powerful when used in conjunction with fractionation techniques of high resolution. In our experience, a considerable amount of information can be obtained by subfractionating the particles by density equilibration in a sucrose gradient and analyzing the subfractions for their relative content in various marker enzymes. The resolution varies somewhat according to the origin of the material; it may sometimes be improved by subjecting the animals to a preliminary treatment which alters the physical properties of some particles. Even when quantitative separation is not possible, relevant questions can often be answered by comparing the distribution of the enzyme or component under study against those of the marker enzymes. It may some-

[4] P. Baudhuin, P. Evrard, and J. Berthet, to be published.

times be necessary to use other gradients besides sucrose gradients. The principles of these techniques have been discussed elaborately elsewhere.[2,3,5-8]

It is essential in such biochemical controls that the analyses be performed on the original homogenate and on the discarded fractions, as well as on the mitochondrial fraction itself, so that a balance sheet can be drawn and the amount found in the mitochondrial fraction related to the total content of the tissue. This is important in two ways. When a marker enzyme known to be associated with a nonmitochondrial type of particle is assayed in order to ascertain the degree of contamination of the fraction by this particle, for instance acid phosphatase for lysosomes or catalase for peroxisomes, it must be remembered that the enzyme measured is actually representative of a whole group of enzymes present in the fraction in approximately the same relative proportion. The relative value is therefore much more informative than the absolute activity of the particular enzyme chosen as control. On the other hand, when the true localization of an enzyme or other trace component is not known, the relative amount found in the fraction may help to evaluate its significance and serve to guide further experiments.

Contaminants of Mitochondrial Fractions

The nature of the contaminants which may be encountered in mitochondrial fractions obviously depends on the biological material from which they are isolated. In the following list, the occurrence of each type of contaminant in various kinds of cells will be mentioned in so far as it is known.

Whole Cells and Cell Fragments

Except for erythrocytes, which are easily recognized, few cells are small enough to escape sedimentation with the nuclear fraction. On the other hand, the possibility that pinched-off cell fragments may come down as whole units with the mitochondria cannot be lightly discounted. A striking example of this phenomenon is provided by the work of Gray and Whittaker,[9] who found nerve endings in mitochondrial fractions

[5] C. de Duve, J. Berthet, and H. Beaufay, *Progr. Biophys. Biophys. Chem.* **9**, 325 (1959).

[6] H. Beaufay and J. Berthet, *Biochem. Soc. Symp. (Cambridge, Engl.)* **23**, 66 (1963).

[7] H. Beaufay, D. S. Bendall, P. Baudhuin, R. Wattiaux, and C. de Duve, *Biochem. J.* **73**, 628 (1959).

[8] H. Beaufay, P. Jacques, P. Baudhuin, O. Z. Sellinger, J. Berthet, and C. de Duve, *Biochem. J.* **92**, 184 (1964).

[9] E. G. Gray and V. P. Whittaker, *J. Anat.* **96**, 79 (1962).

from guinea pig brain and were able to separate them from mitochondria by density gradient centrifugation. This finding probably explains the presence of glycolytic enzymes in so-called "brain mitochondria," the object of a long and fruitless controversy.

Nuclei and Nuclear Fragments

It is probably impossible to homogenize even a soft tissue like liver without damaging at least 10% of the nuclei. It is also known that particulate components of relatively high sedimentation coefficient, such as chromosomes, chromatin threads, and nucleoli, can be separated from disrupted nuclei. Even intact nuclei may contaminate mitochondrial fractions if decantation of the cytoplasmic extract has been performed too generously. They are then seen as a grayish sediment, usually associated with red blood cells, at the bottom of the mitochondrial pellet.

With the present interest in mitochondrial DNA, these facts deserve careful consideration. Here typically is a problem which can be solved satisfactorily only with techniques of high enough resolution to allow the demonstration of a close and unique correlation between DNA and reliable mitochondrial markers. This is particularly true since, as will be mentioned below, phagocytized DNA could also contaminate mitochondrial fractions.

As a possible test for nucleoli, one may mention NAD pyrophosphorylase, an enzyme which has been localized in the nuclei in liver[10] and has been found to be associated with the nucleoli in starfish oocytes.[11]

Cytoplasmic Granules

Secretion Granules and Golgi Vesicles. Granules containing secretory products have been isolated from a large number of exocrine and endocrine glands. As shown, for instance, by the work of Palade and coworkers[12] on the pancreas, such granules may comprise cisternae derived from the rough endoplasmic reticulum, smooth-walled vesicles presumably related to the Golgi apparatus, and mature storage granules, at least when the secreted material is of protein nature. The size range of secretory granules usually encompasses or overlaps that of mitochondria, leading to heavy cross-contamination when simple centrifugal procedures are used. In a number of cases, better separations have been achieved by means of specially worked out schemes of differential centrifugation or by density gradient centrifugation.

[10] G. H. Hogeboom and W. C. Schneider, *J. Biol. Chem.* **197,** 611 (1952).

[11] E. Baltus, *Biochim. Biophys. Acta* **15,** 263 (1954).

[12] G. E. Palade, P. Siekevitz, and L. G. Caro, *Ciba Found. Symp. Exocrine Pancreas* p. 23 (1962).

In most instances, secretory products can be measured by chemical or enzymatic determination or by some kind of bioassay which can serve for their determination in mitochondrial fractions. Exceptionally, an immunoassay may be necessary, as, for instance, for the evaluation of granules containing newly synthesized plasma proteins in mitochondrial fractions from liver.[13]

Lysosomes. Lysosomes appear to be almost ubiquitous, at least in animal cells, and are usually extremely difficult to separate from mitochondria. This is unfortunate since they contain a whole spectrum of hydrolytic enzymes acting on a large variety of substrates; it is true that these enzymes display their optimum activity at an acid pH, but many still show significant activity in the neutral range. As will be mentioned below, lysosomes also contain a variety of substances at various stages of digestion as well as undigestible residues. They are particularly abundant in liver, kidney, spleen, leukocytes, and macrophages; they occur in smaller amounts in brain and in muscle. However, the latter tissues also contain fewer mitochondria on a weight basis and should not be expected to yield cleaner mitochondria. For instance, beef heart sarcosome preparations, which have been widely used for mitochondrial studies, actually have higher specific activities of some lysosomal enzymes than rat liver mitochondrial fractions.[14] One way to reduce contamination by lysosomes is to subject the animal to a treatment which alters the sedimentation coefficient or the density of these particles. Injection of egg-white,[15] Triton WR-1339,[16] dextran,[17] or sucrose[18] has been found to produce changes that can be exploited for this purpose. These treatments do not alter the physical characteristics of the mitochondria, but it is not known whether they affect their functional properties.

The contamination by lysosomes can be ascertained by assaying the preparation for one or preferably more typical acid hydrolases. Convenient enzymes are acid phosphatase, β-glucuronidase, β-galactosidase, cathepsin D, and acid deoxyribonuclease. As already pointed out, it is essential that the results of such assays be related to the total activity of the tissue so that the contamination by lysosomes may be evaluated on

[13] T. Peters, Jr., *J. Biol. Chem.* **237,** 1181, 1186 (1962).
[14] N. Stagni, G. L. Sottocasa, R. Cremese, and B. de Bernard, *Ital. J. Biochem.* **10,** 519 (1961).
[15] W. Straus, *J. Biophys. Biochem. Cytol.* **3,** 933 (1957).
[16] R. Wattiaux, M. Wibo, and P. Baudhuin, *Ciba Found. Symp. Lysosomes* p. 176 (1963).
[17] P. Baudhuin, H. Beaufay, and C. de Duve, *J. Cell Biol.* **26,** 219 (1965).
[18] R. Wattiaux, S. Wattiaux-De Coninck, M.-J. Rutgeerts, and P. Tulkens, *Nature* **203,** 757 (1964).

a percentage basis. It is also important to remember that the value arrived at in this manner is representative of numerous enzymes in addition to the ones actually assayed.

Phagosomes. This is the name given by Straus[19] to pinocytic or phagocytic vesicles containing exogenous material engulfed by the cell. Experiments performed with a variety of materials have shown that such vesicles frequently reach a size which causes them to sediment together with the mitochondria. When first formed, phagosomes differ from lysosomes in that they do not contain any acid hydrolases, but later they fuse with these particles and become indistinguishable from them. The important point concerning phagosomes (and lysosomes) is that we really have no idea of the nature of the materials which are normally taken up in such particles. They are known to segregate foreign substances of various kinds, cell debris, viruses, bacteria as well as normal plasma components, and therefore represent a possible site of location of any constituent occurring in trace amounts in mitochondrial fractions, including such important substances as nucleic acids, enzymes, hormones, vitamins, metals, or drugs. This point can be checked by some of the techniques mentioned above for the selective separation of lysosomes.

Autophagic Vacuoles or Cytolysomes. These structures represent fragments of the cell's own cytoplasm, usually including one or more mitochondria, segregated by cellular autophagy and in the process of being digested by lysosomal enzymes. They may be considered a special type of lysosome containing endogenous cell components in a more or less degenerate state. They are usually too few in number to deserve serious consideration as contaminants of mitochondrial fractions, except under certain circumstances, such as starvation, treatment with antimetabolites or with glucagon, which stimulate their formation. Nothing is known so far concerning their centrifugal behavior.

Peroxisomes or Microbodies. These particles are a particularly obnoxious contaminant of mitochondrial fractions since their function is an essentially oxidative one. They contain large amounts of catalase, which probably acts mainly in its peroxidatic capacity, as well as a number of hydrogen peroxide-producing oxidases acting on uric acid, D-amino acids, L-amino acids, and L-α-hydroxy acids, including L-lactate. Typical peroxisomes have been identified so far in mammalian liver and kidney, as well as in *Tetrahymena pyriformis.*[3b,8,17] In these three materials, the oxidases show significant differences in specificity and relative activity, but the evidence strongly suggests that they, as well as catalase, belong entirely to peroxisomes and do not occur in mitochondria. The study of

[19] W. Straus, *J. Biophys. Biochem. Cytol.* **4,** 541 (1958).

these particles has only recently started and it is quite possible that they may contain other enzymatic activities that have not yet been recognized. The literature contains several examples of "*in vitro*" interactions between peroxisome and mitochondrial components. Such interactions are likely to be artifacts and careful investigations are obviously needed if interferences of this sort are to be detected and avoided in the future.

Peroxisomes can be largely separated from mitochondria by isopycnic centrifugation in a sucrose gradient, in which they exhibit a higher equilibrium density.[3b,8,17] They can be recognized by their constitutive enzymes. In the particular case of liver, it may be useful to carry out assays for urate oxidase as well as for one of the other enzymes, for instance catalase. Urate oxidase is attached to the insoluble core of the peroxisomes, whereas catalase is soluble and escapes from the particles when they are injured. Therefore the content in catalase of the fraction is a measure of intact peroxisomes; the excess of urate oxidase over catalase gives an estimate of isolated cores. This test is not applicable to all preparations; for instance, the peroxisomes of rat kidney and of *Tetrahymena* are devoid of urate oxidase.

Melanosomes. Pigmented cells contain melanin-forming particles that are the sites of location of tyrosine oxidase and DOPA oxidase. The melanosomes appear to be different from mitochondria and can be partly separated from them by density gradient centrifugation.[20]

Particulate Glycogen. Owing to its high density, particulate glycogen represents a significant contaminant of mitochondrial fractions isolated from the liver of fed animals. Several enzymes involved in the synthesis and breakdown of glycogen are largely associated with these particles.[21] This type of contamination can be considerably reduced by the use of fasted animals.

Ribosomes. Even when associated into polysomes, ribosomes are too small to sediment in significant amounts under the forces used to separate mitochondria. One or two washings should reduce their level to a very low value unless they have the ability to attach themselves to large components by adsorption or agglutination. This point is not known. On the other hand, ribosomes combined with ergastoplasmic cisternae and other microsomal components may very well contaminate mitochondrial fractions in highly significant amounts. Some idea of this contamination can be obtained by assaying for an enzyme located in the microsomal membranes (see below), but this may lead to an underestimation since

[20] M. Seiji and T. B. Fitzpatrick, *J. Biochem.* **49,** 700 (1961).
[21] L. F. Leloir and S. H. Goldemberg, *J. Biol. Chem.* **235,** 919 (1960).

the more rapidly sedimenting membrane components are likely to be the richest in ribosomes. This problem is of obvious importance in relation to recent studies on the RNA content and capacity for protein synthesis of isolated mitochondria.

Membrane Components

Cell Membranes. The fate of cell membranes in centrifugal fractionation is not known, except in the case of liver where large membrane fragments, probably derived from the bile canaliculi, occur in the nuclear fraction, from which they can be extracted.[22] It is probable that smaller fragments come down with the mitochondrial and the microsomal fractions. Cell membranes are believed to contain the ouabain-sensitive Na^+/K^+-activated ATPase. However, the complex nature of the ATPase activity of mitochondria is likely to invalidate the assay of this enzyme as a test for cell membrane material.

According to cytochemical staining observations, the plasma membrane possesses the ability to split ADP and CDP, a property which it shares with the Golgi apparatus, but not with the endoplasmic reticulum.[23] In kidney and intestine, alkaline phosphatase can be used to detect fragments of the brush borders.

Golgi Membranes. It is generally assumed that the elements of the Golgi apparatus are fragmented by homogenization and recovered mostly with the microsomal fraction. Since mitochondrial preparations are generally contaminated by microsomes, they presumably include also elements of the Golgi system. According to cytochemical data, these could be recognized by their nucleoside-diphosphatase activity (with ADP or CDP as substrate) and, possibly in a more specific manner, by their ability to split thiamine pyrophosphate.[24]

Endoplasmic Reticulum. In liver and kidney preparations (and in guinea pig intestine), fragments of the endoplasmic reticulum can be detected by their glucose 6-phosphatase activity. In tissues where this enzyme is absent, the assay for nucleoside-diphosphatase with GDP, UDP, or IDP as substrate may possibly provide an adequate test, which, however, is not expected to distinguish between components of the endoplasmic reticulum and other membrane elements.[23,24] As already mentioned, tests of this sort may, by comparison with the RNA:enzyme ratio of the microsomal fraction, provide a rough estimate of the amount

[22] D. M. Neville, Jr., *J. Biophys. Biochem. Cytol.* **8**, 413 (1960).
[23] A. B. Novikoff, E. Essner, S. Goldfischer, and M. Heus, "The Interpretation of Ultrastructure," *Symp. Intern. Soc. Cell Biol.* **1**, 149 (1962).
[24] A. B. Novikoff and M. Heus, *J. Biol. Chem.* **238**, 710 (1963).

of ribosomal RNA likely to be present as a contaminant in a mitochondrial fraction.

Sarcotubules. These elements are likely contaminants of mitochondrial preparations isolated from striated muscles. They can be recognized by their relaxing activity.[25]

Myelin Sheaths. Fragments of myelin sheaths contaminate mitochondrial preparations from brain and nerve tissue. Their low density makes it possible to separate them by flotation in 0.8 *M* sucrose.[9]

Fibrous Material

Connective Tissue Fibers. The presence of collagen fragments is not likely to affect the properties of mitochondrial fractions in any significant manner. If necessary, they can be detected by measurements of hydroxyproline.

Myofibrils. These are a likely contaminant of mitochondrial fractions isolated from muscle tissue.[26] Their ATPase activity differs from those of mitochondria and of sarcotubules, but selective assays based on these differences have not been worked out.

Neurofilaments. To our knowledge, no specific biochemical test is available for the determination of neurofibrils which occur as contaminants in mitochondrial preparations from brain and nerve tissue.[9]

Soluble Material

When, as is the case for a number of dehydrogenases, an enzyme occurs both in the mitochondrial fraction and in the final supernatant, one has to distinguish between three possibilities: (1) the enzyme is mitochondrial and its presence in the supernatant is due to leakage from damaged mitochondria; (2) the enzyme belongs to the cell sap and its presence in the mitochondrial fraction is due to an adsorption artifact; (3) the enzyme belongs truly both to the mitochondria and to the cell sap in the intact cell.

Adsorption artifacts can often be recognized by elution attempts with solutions of high ionic strength. On the other hand, latency measurements provide a very useful test of the binding of an enzyme. In general, enzymes present in the matrix of mitochondria are restricted in their ability to act on external substrates by the permeability properties of the mitochondrial membrane; disruption of the mitochondria by a procedure which does not denature the enzyme then results in a considerable

[25] E. Andersson-Cedergren and U. Muscatello, *Proc. 1st Intern. Pharmacol. Meeting, London, 1961,* Vol. 5, p. 71. Pergamon Press, Oxford, 1963.

[26] S. V. Perry and T. C. Grey, *Biochem. J.* **64**, 184 (1956).

increase in activity. The lysosomal hydrolases and the peroxisome catalase also display latency, but these particles differ from the mitochondria and from each other in their sensitivity to some disrupting procedures.[3b, 27] Thus latency studies can even serve to distinguish between mitochondrial and nonmitochondrial enzymes. When an enzyme has been identified as being truly mitochondrial, it may be desirable to find out whether the activity present in the supernatant fraction could have originated from injured mitochondria. Comparison with an enzyme known to be entirely localized in the mitochondria, for instance glutamate dehydrogenase, may be used for this purpose.

Other Components

Although long enough to provide food for thought, the preceding list is probably not exhaustive. No mention has been made, for instance, of centrioles, of spindle fibers and their presumed precursors, the cytoplasmic microtubules, nor of the other structures of unknown nature and function that can be seen in electron micrographs of cells. We must accept the limitations of our ignorance and keep in mind the possibility that new organelles with specific biochemical, enzymatic, and metabolic properties still remain to be discovered. Some of them could be present as contaminants in mitochondrial fractions.

Homogeneity of Mitochondria

Having discussed the lack of homogeneity of mitochondrial fractions, we may now raise the question of the homogeneity of the mitochondria themselves. This important problem has been the subject of numerous publications, but usually the distinction has not been clearly made between the mitochondria and the subcellular fraction in which they are contained. We can only attempt to answer this question largely in terms of our own experience with rat liver preparations.

Liver mitochondria are derived primarily from parenchymal cells. This simplifies the problem since we may, in first approximation, neglect cell-specific differences in the composition of mitochondria. There remains the question whether individual mitochondria from a given cell type differ significantly from each other either within the same cell or according to the histological situation of their host cell. We will discuss this question only in so far as it is accessible to biochemical analysis and pertinent to biochemical work on isolated mitochondria.

Mitochondria obviously differ in size; they also differ in density, at

[27] D. S. Bendall and C. de Duve, *Biochem. J.* **74,** 444 (1960).

least as they are present in homogenates and subcellular fractions. This heterogeneity is responsible for the fact that they do not all sediment exactly at the same rate in a given centrifugal field nor occupy exactly the same equilibrium position after isopycnic centrifugation in a density gradient. When expressed in terms of mitochondrial mass, frequency distribution curves of either sedimentation coefficient or equilibrium density are reasonably symmetrical and unimodal. The bimodal density distribution curves that can be obtained under certain conditions have been shown to be an artifact.[8] Thus, mitochondria behave very much like any other biological population by exhibiting a certain variability of the parameters that determine their physical properties. However, it is not known to what extent the observed differences in density reflect true differences in chemical composition or are the result of more or less accidental variations in the size of their water spaces due to small permeability changes. The density and chemical composition of mitochondria can be altered experimentally,[28,29] but whether differential changes of this sort can take place in the living animal under normal conditions has not been established.

It seems to us significant that no evidence of a clear-cut enzymatic heterogeneity has yet been obtained on rat liver mitochondria. In all experiments in which the distribution of two or more mitochondrial enzymes has been studied by means of high resolution techniques throughout the whole mitochondrial population, the observed distribution patterns have been found to coincide with each other and with that of the total proteins within the limits set by the experimental error and by the known interferences by contaminants. The impression derived from this work is that hepatic mitochondria are relatively homogeneous, at least with respect to the number of molecules of each enzyme species per unit weight of total protein. This opinion is advanced very tentatively since only a small number of enzymes has been studied in this manner so far. However, it is consistent with current ideas on the existence of stoichiometric relationships among the various constituents of the mitochondrial membranes, and it is supported by the fact that every enzyme whose distribution has been found to deviate significantly from that of a mitochondrial enzyme has invariably turned out to belong to a contaminant.

[28] J. W. Greenawalt, C. S. Rossi, and A. L. Lehninger, *J. Cell Biol.* **23,** 21 (1964).
[29] D. J. Luck, *J. Cell Biol.* **24,** 445 (1965).

[3] Manometric Methods and Phosphate Determination

By E. C. SLATER

Introduction

One of the characteristic properties of mitochondria is their ability to catalyze the synthesis of ATP from ADP and inorganic phosphate, coupled with the oxidation by oxygen of a number of oxidizable substrates, each of which is activated by a specific dehydrogenase. The yield of oxidative phosphorylation is generally expressed by the P:O ratio, i.e., the ratio of the number of moles ADP phosphorylated (to ATP) to the number of gram atoms of oxygen consumed.

The first requirement for a useful measurement is that conditions are chosen such that there is a clear stoichiometric relationship between the amounts of oxygen and substrate consumed. For example, (i) in the presence of *catalytic* amounts of a Krebs-cycle intermediate, the oxidation of pyruvate may be written

$$CH_3—CO—COOH + 5O \rightarrow 3CO_2 + 2H_2O$$

(ii) glutamate is oxidized by mammalian mitochondria almost exclusively according to the equation

$$HOOC—CH_2—CH_2—CH(NH_2)—COOH + 3O \rightarrow HOOC—CH_2—CH(NH_2)—COOH + CO_2 + H_2O$$

(iii) in the presence of malonate, α-oxoglutarate is very largely oxidized to succinate

$$HOOC—CH_2—CH_2—CO—COOH + O \rightarrow HOOC—CH_2—CH_2—COOH + CO_2$$

On the other hand, with liver mitochondria oxidizing succinate, in the concentrations very often employed, the relationship between oxygen uptake and the amount of the products of the one-step oxidation of succinate (fumarate + malate) are such that measurements of the yield of oxidative phosphorylation in this reaction have little value.[1] The measurements become meaningful only by adding sufficient succinate[1] or by adding rotenone to prevent the oxidation of malate.

The oxidative phosphorylation reaction may be written

$$m \text{ substrate} + n \text{ O} + p \text{ ADP} + p \text{ P}_i \rightarrow m \text{ product} + p \text{ ATP} \quad (1)$$

Although, for calculation of the P:O ratio, measurements of n and p suffice, many mistakes would be avoided if it became established practice

[1] P. Greengard, K. Minnaert, E. C. Slater, and I. Betel, *Biochem. J.* 73, 637 (1959).

to measure *m* as well. Most substrates and products can nowadays be readily determined, for example by specific enzymatic methods.[2]

Measurement of Oxygen Uptake

The oxygen uptake may be measured either manometrically or polarographically. This chapter will be confined to the first method, since the latter is considered elsewhere in this volume [7].

If properly used, and under correctly chosen conditions, the manometer is still the most accurate instrument for measuring the oxygen uptake. It must be remembered, however, that it is essentially an instrument for measuring the *rate* of O_2 uptake during steady state conditions. It responds slowly to changes in the steady state.

Since many excellent manuals (e.g. those of Dixon[3] and of Umbreit *et al.*[4]) describing the use of manometers exist, only certain aspects will be touched upon here. It might be mentioned, however, that recent publications in the scientific literature have revealed that some of the precautions which need to be taken in using manometers are not as widely known as they used to be.

For some experiments, all that is required is the measurement of the respiratory activity. More usually, however, it is necessary that the phosphorylation be measured simultaneously. This poses extra problems, since it becomes necessary to measure the amount of oxygen consumed during the period in which the phosphorylation is measured. The author prefers differential manometers, in which both arms of the manometer are joined to flasks, to the now more commonly used constant-volume Warburg manometer, in which one arm is open to the atmosphere. Small flasks of the Warburg type (total volume about 5–7.5 ml) and small capillaries (0.4–0.6 mm diameter) are used. The use of the small capillaries, which is essential in order to obtain sufficiently low manometer constants, and a paraffin manometer fluid [for example, a petroleum ether fraction (sp. gr. 0.7862 at 22.8°) colored with Sudan III (0.3 mg/ml)] allows very accurate measurements of the height of the manometer fluid (estimates to 0.1 mm of manometer fluid may be made). Manometer constants are calculated as described by van Dorp and Slater.[5]

[2] H. U. Bergmeyer, "Methoden der enzymatischen Analyse." Verlag Chemie, Weinheim, 1962.

[3] M. Dixon, "Manometric Methods as Applied to the Measurement of Cell Respiration and Other Processes." Cambridge Univ. Press, London and New York, 1943.

[4] W. W. Umbreit, R. H. Burris, and J. F. Stauffer, "Manometric Techniques and Tissue Metabolism." Burgess, Minneapolis, Minnesota, 1949.

[5] A. van Dorp and E. C. Slater, *Biochim. Biophys. Acta* **33**, 559 (1959).

The standard procedure used in the author's laboratory for more than 10 years[6] is as follows: the various components of the reaction mixture,[7] in a total volume of 0.9 ml, are pipetted into the manometer flask standing on the laboratory bench (at about 20–22°). Routinely the last addition is hexokinase (0.01–0.02 ml), and $MgCl_2$ is added just before the hexokinase.[8] Sufficient 10% (w/v) KOH (usually 0.07–0.1 ml) is pipetted into the center well, the inside rim of which is lightly smeared with Vaseline, to make it about three-quarters full. At zero time, 0.1 ml of mitochondrial suspension (at about 0°) is added to the main flask and a fluted Whatman filter paper is placed in the center well so that it projects 2–3 mm above the center well. The flask is then attached to the manometer (already provided with the compensating flask containing 1.0 ml 0.1 *N* H_2SO_4). The manometer cone is greased with Vaseline.[9] The manometer is then placed in a constant-temperature bath (usually at 25°). The time between adding the mitochondria and placing the manometer in the bath is usually about 0.5–1 minute. After about 3 minutes the flasks are firmly bound to the manometers and the taps are closed. The first reading is taken between 5 and 10 minutes after placing the manometer in the bath. Readings are then taken at intervals of 2–5 minutes. The total reaction period is usually 20–30 minutes. One minute before the end of the reaction, the flask is removed from the manometer, the paper in the center well is removed, and the reaction is stopped by the addition of 0.15 ml of 40% (w/v) trichloroacetic acid.

The O_2 uptake between zero time and the first reading, and between the last reading and the end of the reaction, are determined by extrapolation.

Figure 1 described two records with two concentrations of mitochondria. The first point on each curve represents the first manometer reading. The oxygen uptake between zero time and the time of this reading was obtained by extrapolation, and the value obtained was added to all readings. Curve *1* illustrates the precision with which even very small oxygen uptakes can be measured with differential manometers and

[6] E. C. Slater and F. A. Holton, *Biochem. J.* **55,** 530 (1953).

[7] Since the composition of the reaction mixture depends upon the type of mitochondria, it is not possible to go into this important question here. It might be mentioned, however, that although the P:O ratio of rat heart mitochondria is not greatly affected by considerable variations in the reaction medium, at least 0.14 *M* sucrose and low phosphate are essential to obtain satisfactory ratios with beef heart mitochondria.

[8] If $MgCl_2$ is added to concentrated solutions containing phosphate, ADP, and fluoride, a precipitate is formed which does not readily dissolve on addition of water.

[9] If the bath temperature is 37°, lanolin is used instead of Vaseline.

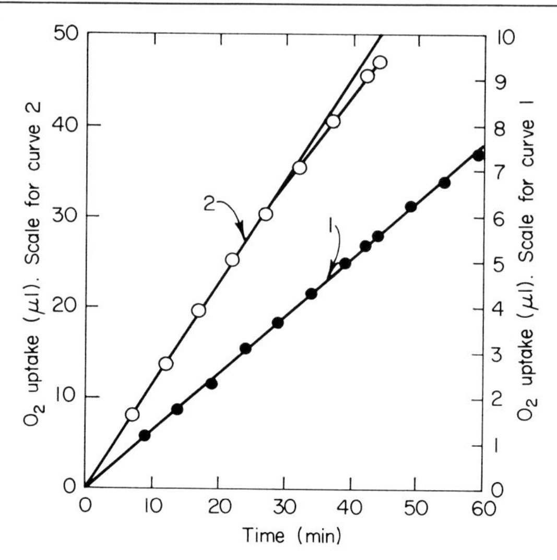

FIG. 1. Course of oxidation of α-oxoglutarate by heart mitochondria in a reaction mixture containing 30 m*M* phosphate, 10 m*M* malonate, 40 m*M* NaF, 20 m*M* glucose, 1 m*M* EDTA, 0.6 m*M* ADP, 0.6 m*M* AMP, 5 m*M* $MgCl_2$, 5 m*M* α-oxoglutarate, and 50 μ*M* cytochrome *c*. Curve *1*, 0.12 mg protein per milliliter; curve *2*, 0.60 mg protein per milliliter. Two different preparations were used. Note different scales of the ordinates. See text for further details. Reproduced from E. C. Slater and F. A. Holton [*Biochem. J.* **55**, 530 (1953)] with kind permission of the publisher.

small flasks (constant of manometer = 0.47 μl O_2 per millimeter manometer fluid).

The procedure described above was adopted only after a thorough examination of the possible errors. The manometer readings in the first few minutes after the mitochondria were added, if they could be assumed correctly to indicate the consumption of oxygen by the reaction mixture,

TABLE I
RELATIONSHIP BETWEEN OXYGEN UPTAKE, α-OXOGLUTARATE DISAPPEARANCE AND Δ ESTERIFIED PHOSPHATE AT DIFFERENT TIMES[a,b]

Time (min)	O_2 uptake[c] (μatoms)	−Δα-oxoglutarate (micromoles)	ΔEstimated P (micromoles)	P:O
15	1.57	1.58	4.31	2.75
30	3.13	3.19	8.46	2.71
45	4.31	4.35	11.78	2.73

[a] Taken from E. C. Slater and F. A. Holton [*Biochem. J.* **55**, 530 (1953)].
[b] Replicate flasks, identical with those whose O_2 uptake is shown in curve *2* (Fig. 1) were removed at different times and treated with trichloroacetic acid.
[c] O_2 uptake in first 7 minutes obtained by extrapolation.

would indicate that there is an initial lag in the oxidation of the α-ketoglutarate. The determinations summarized in Table I show that there is no lag and illustrate the validity of the extrapolation procedure, in which the initial manometer readings are disregarded.

Several factors may contribute to the initial lag in the manometer readings:

1. The contents of the reaction flask are at a temperature a little below that of the reference flask. Some time is required before the gas phase in the reaction flask reaches the same temperature as that of the reference flask. This time is probably quite short in the experimental procedure described above. This factor can, however, be a serious source of error when thick-walled flasks are cooled in ice before adding the mitochondria (see Fig. 3 of Slater and Holton).[10]

2. The manometer records the change of pressure in the gas phase, but the reaction takes place in the liquid phase. The steady state concentration of oxygen in the liquid phase depends upon the rate of diffusion of oxygen from the gas phase and the rate of consumption of oxygen. It will always be less than the concentration of oxygen in equilibrium with air, how much less depending upon the relative rates of these two processes. The rate of diffusion depends upon the geometry of the flask and on the rate of shaking. A finite time is necessary to set up the new steady state. If instead of the "extrapolation" procedure recommended above, the O_2 uptake is calculated from the change of pressure in the manometer after adding the mitochondria, this uptake will be underestimated by an amount equal to volume times $\Delta[O_2]$, where $\Delta[O_2]$ is the difference between the concentration of O_2 in equilibrium with air and that in the steady state.

The importance of this error has been emphasized in recent publications by Haslam,[11] Brierley,[12] and Lenaz and Beyer.[13] Haslam[11] showed that if the oxygen uptake exceeded 0.2–0.3 μatom/minute in his apparatus, there was a measurable uptake of gas after addition of trichloroacetic acid to stop the reaction. This amount of O_2 equals that required to bring the solution back to equilibrium with air. When the rate of oxygen uptake exceeded about 0.5 μatom/minute, the uptake of gas after addition of trichloroacetic acid equaled the amount which dissolves in the volume of reaction mixtures used; in other words, the reaction mixture must have been completely anaerobic at the end of the reaction. This was checked directly by polarographic measurements of O_2 concen-

[10] E. C. Slater and F. A. Holton, *Biochem. J.* **56,** 28 (1954).

[11] J. M. Haslam, *Biochim. Biophys. Acta* **105,** 184 (1965).

[12] G. P. Brierley, *Biochem. Biophys. Res. Commun.* **19,** 500 (1965).

[13] G. Lenaz and R. E. Beyer, *J. Biol. Chem.* **240,** 3653 (1965).

tration. Under these circumstances, the rate of O_2 uptake is completely dependent upon the rate of diffusion of oxygen from the gas phase into the liquid and is unrelated to the capabilities for oxygen consumption by the mitochondria. Attention was drawn to this point by Dixon and Tunnicliffe[14] as long ago as 1923. The simplest way of determining the rate of diffusion of oxygen into the solution is to measure the oxygen consumption of various concentrations of a mitochondrial suspension (or yeast suspension). When the curve relating rate of O_2 consumption to concentration of respiring particles shows a sharp break (X in Fig. 2),

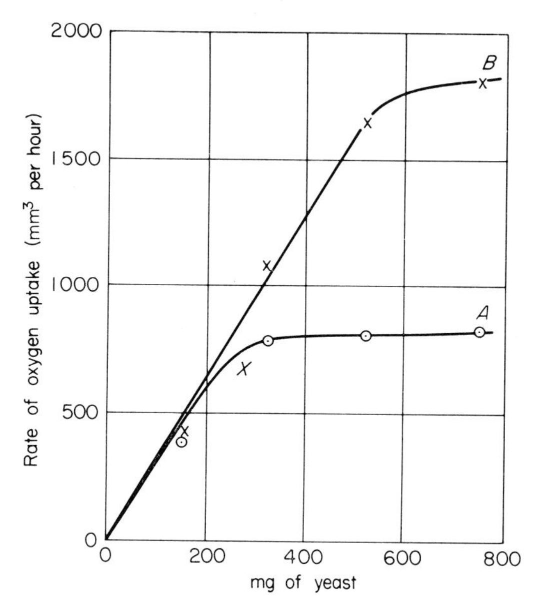

FIG. 2. Dependence of rate of O_2 uptake of yeast suspension on the concentration of yeast. The amounts of yeast given were suspended in 3 ml phosphate buffer. The experiments were carried out at room temperature with the flasks filled with air. Curve A, manometers were shaken at 102 oscillations per minute; curve B, at 138 oscillations per minute. Reproduced from M. Dixon, "Manometric Methods as Applied to the Measurement of Cell Respiration and Other Processes." Cambridge Univ. Press, London and New York, 1943, with kind permission of author and publisher.

diffusion has become rate limiting. This has also been discussed by Roughton[15] and Myers and Matsen.[16]

Clearly the O_2 uptake will be underestimated even by the extrapola-

[14] M. Dixon and H. E. Tunnicliffe, *Proc. Roy. Soc.* **B94,** 266 (1923).
[15] F. J. W. Roughton, *J. Biol. Chem.* **141,** 129 (1941).
[16] J. Myers and F. A. Matsen, *Arch. Biochem. Biophys.* **55,** 373 (1955).

tion method if conditions are such that the rate of O_2 uptake is governed by the rate of diffusion of oxygen into the liquid phase. For every apparatus, experiments similar to that illustrated in Fig. 2 should be carried out, and conditions should be chosen so that the rate of O_2 uptake is well below the rate of diffusion.

3. Time is also required to establish a new steady state concentration of CO_2 in the gas phase (cf. Dixon[3] and Myers and Matsen[16]), the time depending upon the rate of formation of CO_2 and on the efficiency of its absorption. Slater and Holton[6] emphasized that care must be taken to provide a sufficient surface of filter paper, wet with KOH, for the efficient absorption of CO_2 formed during the oxidation of α-oxoglutarate to succinate (respiratory quotient = 2). Lenaz and Beyer[13] have recently come to the same conclusion. The seriousness of insufficient absorption of CO_2 as a source of error in measuring the O_2 uptake, when certain procedures are followed, has been emphasized by Haslam.[11]

4. Artifacts can be caused by movement of the flask on the manometer joint when a solution is added from the side-arm by tipping the manometer.

Because of all these factors, the author considers it absolutely essential that the manometer be used only to measure the steady state rate of O_2 consumption, so that the total O_2 consumption can be calculated only when there is a constant rate of O_2 consumption for at least 15 minutes after addition of mitochondria. If this is not the case (e.g., with succinate in the presence of dinitrophenol), accurate measurements of O_2 uptake are not possible with a manometer, and polarographic methods should be used.

Examples of incorrectly high values of the P:O ratio which can be obtained by incorrect manometric procedures have been given by Haslam,[11] Lenaz and Beyer,[13] Brierley,[12] and Koivusalo *et al.*[17] It is unfortunate that errors in measuring P:O ratios usually lead to high values.

Measurement of Phosphorylation

ADP (or ATP), glucose, and yeast hexokinase (EC 2.7.1.1) are generally added to the reaction mixture used for measurement of oxidative phosphorylation. The ATP formed in Eq. 1 reacts with glucose, thus

$$p\ \text{ATP} + p\ \text{Glc} \rightarrow p\ \text{ADP} + p\ \text{HMP} \tag{2}$$

and the sum reaction is

$$m\ \text{substrate} + n\ \text{O} + p\ \text{Glc} + p\ \text{P}_i \rightarrow m\ \text{product} + p\ \text{HMP} \tag{3}$$

[17] M. Koivusalo, R. D. Currie, and E. C. Slater, *Biochem. Z.* **344**, 221 (1966).

where HMP stands for hexose monophosphate, usually consisting of the equilibrium mixture of glucose 6-phosphate and fructose 6-phosphate formed by glucose phosphate isomerase (EC 5.3.1.9) present in the hexokinase. Sufficient hexokinase should be added to compete succesfully for the ATP formed with any extramitochondrial ATPases which might be present.

The adenine nucleotides act as catalysts and so do not appear in the sum reaction (Eq. 3). However, they take part in a side reaction (Eq. 4) catalyzed by adenylate kinase (EC 2.7.4.3), which may have to be taken into account.

$$2\ \text{ADP} \rightleftharpoons \text{AMP} + \text{ATP} \tag{4}$$

The value of n may be determined by measuring either the disappearance of P_i or the formation of HMP.

Disappearance of P_i. Owing to its simplicity, this is the method which is mostly used, but it is inaccurate when only small amounts of P_i disappear, and it suffers from the disadvantage that considerations of analytical accuracy determine the amount of phosphate that can be added to the reaction mixture.

The reaction mixture after addition of trichloroacetic acid is diluted with 2 ml of 5% trichloroacetic acid and cooled to 0°. The precipitated protein is removed by centrifugation, and the P_i content of a sample of the suitably diluted supernatant is determined by the molybdenum blue method. The procedure described by Sumner[18] has been found convenient. The P_i content of a "zero-time control," i.e., in which trichloroacetic acid is added before the mitochondria, is determined in the same way. The difference between the two values for P_i gives the amount that is taken up.

Formation of HMP. The HMP formed may be measured enzymatically or, when $^{32}P_i$ is used in the reaction mixture, radioisotopically.

(i) ENZYMATIC. Two enzymatic methods may be used. The HMP may be estimated with the help of glucose 6-phosphate dehydrogenase (EC 1.1.1.49) and excess $NADP^+$.[19] With this method, the reaction should be stopped with $HClO_4$ rather than with trichloroacetic acid, since trichloroacetate strongly inhibits glucose 6-phosphate dehydrogenase. $HClO_4$ (0.1 ml of 35%, w/v) is added to the manometer flask. After the precipitated protein has been centrifuged off, the supernatant is neutralized with KOH, the mixture is kept in ice for approximately 30 minutes, and the precipitated $KClO_4$ is removed by centrifugation. Sufficient

[18] J. B. Sumner, *Science* **100**, 413 (1944).
[19] G. B. Pinchot, *J. Biol. Chem.* **205**, 65 (1953).

glucose 6-phosphate isomerase, probably in the glucose 6-phosphate dehydrogenase, should be present in the assay. Reduction of $NADP^+$ is measured spectrophotometrically at 340 mμ in a reaction mixture containing 50 mM Tris-HCl buffer (pH 7.4), 10 mM $MgCl_2$, 2 mM EDTA, and approximately 1 micromole $NADP^+$ in a final volume of 2.5 ml. The reaction is started by addition to each cuvette of 0.02 IUB unit of glucose 6-phosphate dehydrogenase diluted in a solution of serum albumin (1 mg/ml).

The second method, introduced by Slater,[20,21] depends upon the reaction

$$\text{HMP} + \sim\text{P} + 2\ \text{NADH} + 2\ \text{H}^+ \rightarrow 2\ \text{glycerol-}P + 2\ \text{NAD}^+ \quad (5)$$

(where ~P indicates the two terminal phosphate groups of ATP, and the terminal group of ADP), catalyzed by the enzymes present in partially purified rabbit muscle extracts. When excess ~P is added, the oxidation of NADH (measured spectrophotometrically at 340 mμ) is a measure of the amount of HMP present; when excess HMP is present, the amount of NADH oxidized is a measure of the ~P present.

Traces of hexokinase interfere seriously with the determinations. Since treatment with trichloroacetic acid does not completely inactivate the hexokinase, it is most important that a compact protein precipitate be obtained after centrifugation and that the supernatant sample be free from any contaminating protein. When the amount of mitochondrial protein is small (less than 0.5 mg) it is advisable to add some serum albumin (about 1 mg) in order to increase the volume of the precipitate.

After the addition of the 40% trichloroacetic acid to the reaction mixture and removal of residual KOH from the center well, 2 ml 5% trichloroacetic acid is added and the mixture is transferred to small plastic centrifuge tubes. After centrifugation at 20,000 g for 10 minutes, 2 ml of the supernatant is carefully pipetted into 0.2 ml 0.5 M phosphate buffer (pH 6.8), and sufficient 1 N KOH (usually about 0.6 ml) to bring the pH to about 6.8. This solution should be kept at 4° (not frozen) until analyzed. Analyses should preferably be carried out within 2 days.

The procedure now described for the HMP determination is suitable for a spectrophotometer cuvette holder carrying 4 cuvettes. The reference cell is provided with 0.25 ml 0.25 M Tris-HCl buffer (pH 7.4) and 2.2 ml water. The other cuvettes are provided with 1 ml of a mixture containing 0.25 micromole ATP, 7 micromoles $MgCl_2$, 0.25 micromole NADH, 30 micromoles Tris-HCl buffer (pH 7.4), and 2.5 micromoles EDTA. A sample (containing about 0.05 micromole HMP) of a suitably diluted

[20] E. C. Slater, *Biochem. J.* **53**, 157 (1953).
[21] E. C. Slater, *Biochem. J.* **53**, 521 (1953).

neutralized supernatant, prepared as described in the above paragraph, is added together with sufficient water to make the volume 2.5 ml. One cuvette should contain only 1.5 ml water in addition to the 1 ml ATP mixture. After thorough mixing, the absorbances at 340 mμ of the three cuvettes with reference to the reference cuvette are determined. Immediately (zero time), 0.1 ml of enzyme mixture containing approximately 400 μg of rabbit muscle fraction A[20] and 500 μg of rabbit muscle fraction B[20] is added to each cell, the solutions are well mixed, and readings of each cell are taken each minute for 10–15 minutes. The readings are plotted against time on graph paper and extrapolated to zero time. From the difference between the initial and the extrapolated absorbances, the amount of NADH oxidized by HMP may be calculated (remember that 1 mole HMP oxidizes 2 moles NADH). This value should be corrected for the small $\Delta A_{340\ m\mu}$ obtained in the cuvette containing only ATP mixture and water. The initial rapid decline in $A_{340\ m\mu}$ in the cuvettes containing HMP should be completed within 5 minutes. If this is not the case, the enzyme preparation is insufficiently active and should be discarded.

The ~P content is determined similarly, with the exception that glucose 6-phosphate (0.6 micromole) replaces the ATP. The points reach the extrapolated line rather more slowly in this assay.

Owing to inhibition of the rabbit muscle enzymes by trichloroacetate, the sample of the neutralized supernatant should not exceed 0.3 ml in either assay. Both determinations are very precise. With a good spectrophotometer, the accuracy is limited by the accuracy with which the samples are pipetted.

Since the enzyme preparations also contain lactate and malate dehydrogenases, any pyruvate or oxalacetate present will be determined as HMP and also as ~P. This is most easily corrected for by carrying out an additional determination in which phosphohexokinase (fraction B) and Mg^{++} are omitted from the reaction mixture and 10 m*M* EDTA is added to inhibit residual phosphohexokinase activity.

When either enzymatic method is used, a correction is necessary for the amount of HMP formed by reaction with glucose of ATP present initially in the reaction mixture, or formed from ADP by the adenylate kinase reaction (Eq. 4). The amount of ATP initially present is most easily determined by allowing the hexokinase reaction to proceed to completion before adding the trichloroacetic acid in the "zero time" control. The adenylate kinase reaction may be allowed for in two ways (cf. Slater[22]).

[22] E. C. Slater, *Biochem. J.* **59**, 392 (1955).

Method 1: A separate control is run in which substrate is omitted, and 1 mM cyanide is added to prevent oxidation of endogenous substrate.[21]

Method 2: $\sim$P determinations are made as well as HMP on the "experimental" flasks and on the "zero-time controls." Since neither the hexokinase nor the adenylate kinase reaction leads to a change in $\sim P + HMP$, $\Delta{\sim}P + \Delta HMP$ (giving a negative sign to disappearance) measures the oxidative phosphorylation (see Table II, for example[23]).

TABLE II
EXAMPLE OF CALCULATION OF Δ ESTERIFIED P FROM MEASUREMENTS OF HMP AND $\sim$P[a]

Determination	Time (min) 0	15
$\sim$P	5.63[b]	1.07
$\Delta{\sim}$P	—	−4.56
HMP	0.04	7.58
ΔHMP	—	7.54
$\Delta HMP + \Delta{\sim}P$	—	2.98

[a] Taken from M. Koivusalo and E. C. Slater [*Biochem. Z.* **342,** 246 (1965)].
[b] Values are stated as micromoles.

Method 1 can cause a slight underestimation of the amount of phosphorylation in short-time experiments, since it does not take account of the steady state concentration of ATP.[6] Which method is used will depend to a large extent on the particular conditions.

(ii) ISOTOPIC. Ernster *et al.*[24] and Nielsen and Lehninger[25] measure the amount of esterified P by determining the amount of $^{32}P_i$ incorporated into the organic phosphate, separated from P_i by conversion of the latter into phosphomolybdate and extraction into benzene–isobutanol. The organic phosphate remains behind in the water layer.

Measurement of Substrate Disappearance or Product Formation

Methods will be found in Vol. III and in Bergmeyer.[2]

[23] M. Koivusalo and E. C. Slater, *Biochem. Z.* **342,** 246 (1965).
[24] L. Ernster, R. Zetterström, and O. Lindberg, *Acta Chem. Scand.* **4,** 942 (1950).
[25] S. O. Nielsen and A. L. Lehninger, *J. Biol. Chem.* **215,** 555 (1955).

[4] The Measurement of Oxidative Phosphorylation in the NADH–Cytochrome *b* Segment of the Mitochondrial Respiratory Chain

By GOTTFRIED SCHATZ[1]

The Coenzyme Q_1 Assay

Principle. The assay is based on the measurement of ATP formation coupled to the NADH-linked reduction of exogenous coenzyme Q_1 (CoQ_1). It has been shown to be applicable to submitochondrial particles from beef heart and from rat liver.

Reagents

Potassium phosphate buffer, 0.2 *M*, pH 7.4, containing purified $^{32}P_i$ at a final specific radioactivity of 0.5 to 1.5×10^5 cpm/per micromole of P_i. Purification of commercially available $^{32}P_i$ (essentially according to Conover *et al.*[1a]) is carried out as follows: The sample (up to 20 mC of $^{32}P_i$) is evaporated to dryness in a Pyrex test tube, moistened with a few drops of a 10% $Mg(NO_3)_2$—solution in 95% ethanol and heated until the evolution of brownish fumes has almost ceased. This operation is repeated once. The residue is dissolved in a few ml of 3 *N* HCl and heated for 30 minutes at 100°, followed by evaporating most of the HCl over an open flame. To this hydrolyzed solution are then added 200 micromoles $MgCl_2$, 200 micromoles P_i, and 15 ml of 5 *N* NH_4OH. After mixing, the formed precipitate is collected by centrifugation, washed once with 5 *N* NH_4OH by centrifugation and dissolved in a minimal volume of 1 *N* HCl containing 20 m*M* $MgCl_2$. This procedure is repeated twice by again adding NH_4OH and centrifuging off the precipitate. The recrystallized sample is then treated twice with Dowex 50 (H^+ form) in order to remove Mg^{++} ions. Careful purification of the $^{32}P_i$ is essential for the assay described here.

KCN, 40 m*M*, in 0.1 *M* Tris-sulfate buffer pH 7.4 (final pH 8.2), prepared fresh every day and kept at 0°.

CoQ_1, 0.4 m*M*, in 0.25 *M* sucrose. This solution is prepared fresh every day by slowly pipetting a 80 m*M* solution of CoQ_1 in 95% ethanol into 200 volumes of rapidly stirred 0.25 *M* sucrose at

[1] This manuscript was prepared while the author was at the Department of Biochemistry, The Public Health Research Institute of the City of New York, Inc., New York.

[1a] T. E. Conover, R. L. Prairie, and E. Racker, *J. Biol. Chem.* **238,** 2831 (1963).

room temperature. It is important to obtain a solution essentially free of turbidity. The concentration of CoQ_1 in the stock solution is assayed spectrophotometrically by measuring either the extinction at 275 mμ ($\epsilon_{mM} = 14.3$) or the extinction decrease at this wavelength upon addition of KBH_4 ($\epsilon_{mM} = 12.1$).

NADH, 2–5 m*M*, in 0.01 *M* unneutralized Tris, prepared fresh and kept at 0°. The concentration of NADH is measured spectrophotometrically at 340 mμ ($\epsilon_{mM} = 6.22$).

Stock solution A, containing in a volume of 0.1 ml: 2.0 micromoles $MgCl_2$, 0.48 micromole EDTA, 5.0 micromole Tris-sulfate pH 7.4, 32 micromoles glucose, 1.0 micromole ATP, 10 units hexokinase and 2.0 mg bovine serum albumin. This solution is stable for several months if stored at —20° or below. The bovine serum albumin should be purified by dissolving the crystalline commercial product in 0.01 *M* Tris-sulfate pH 7.4 and dialyzing overnight against 200 volumes of the same buffer in the cold. The hexokinase used is purified by dialyzing a suspension of the crystalline enzyme in ammonium sulfate overnight against 200–500 volume of 5 m*M* EDTA pH 7.4 containing 1% glucose in the cold.

Submitochondrial particles, adjusted to a protein concentration of approximately 2–8 mg/ml with 0.25 *M* sucrose containing 20 m*M* Tris-sulfate pH 7.4

Trichloroacetic acid, 50%

Sucrose, 0.25 *M*

Perchloric acid, 10 *N*

Ammonium molybdate, 5 *N*

Isobutanol:benzene 1:1 (v/v)

Water-saturated isobutanol

Assay Procedure for Submitochondrial Particles from Beef Heart. To a 1 ml silica spectrophotometer cell are added 0.1 ml stock solution A, 0.05 ml 0.2 *M* radioactive phosphate buffer pH 7.4, 0.04 ml 40 m*M* KCN solution, 0.03 ml 0.4 m*M* CoQ_1 solution in 0.25 *M* sucrose and 0.25 *M* sucrose to give a final volume of 1.0 ml after the addition of NADH and particles. A second cuvette, representing the control, receives the same components except the CoQ_1. After the contents of both cuvettes have been mixed, an equal amount of submitochondrial particles (50–200 μg protein) and, finally, 0.08 micromole NADH are added to both, followed by rapid mixing. Oxidation of NADH by CoQ_1 is then measured at 340 mμ. Since initially both cuvettes exhibit approximately equal extinction at this wavelength, the measurement is most conveniently carried out by placing the cuvette containing the CoQ_1 in the reference beam of a recording spectrophotometer. However, simpler instruments such as the Beckman DU spectrophotometer can also be used. As soon as no further change in the extinction at 340 mμ occurs, the contents of both cuvettes are mixed with 0.1 ml 50% trichloroacetic acid and centrifuged. The supernatant solutions are then extracted with molybdic acid and isobutanol–benzene as described by Lindberg and Ernster[2] except that 0.3 mg glucose-6-P (sodium salt) is added as carrier and that an additional extraction with water-saturated isobutanol and then with ether is performed in order to remove the last traces of $^{32}P_i$. The amount of glucose-6-^{32}P formed in the CoQ_1-free control is subtracted from that formed in the complete reaction mixture. This correction usually amounts to less than 10%. The glucose-6-^{32}P:NADH ratio is taken as a measure of phosphorylation efficiency. Under these conditions the amount of glucose-6-^{32}P formed corresponds closely to the amount of P_i taken up.[3]

Assay Procedure for Submitochondrial Particles from Rat Liver. The assay is identical to that described above for beef heart particles except that only 0.12 ml 0.4 m*M* CoQ_1 and 0.03 micromole NADH are added to the reaction mixture and that the specific radioactivity of the $^{32}P_i$ is increased to 2 to 5×10^5 cpm per micromole P_i.

Comments. Under the assay conditions given above the rate of NADH oxidation with CoQ_1 as acceptor is 90–95% inhibited by rotenone and is approximately 1.1–2.3 times faster than the rate with oxygen as the acceptor. It is important to use exactly the concentrations of CoQ_1 given above since at higher concentrations CoQ_1 is also reduced via a nonphosphorylating pathway which is insensitive to rotenone.[3]

Other Methods for Measuring Oxidative Phosphorylation at the First Site

Several other assay procedures for measuring ATP formation in the NADH-cytochrome *b* segment have been proposed. These are either based on the difference between the P:O ratios obtained with NADH and succinate as substrate[4] or on the phosphorylation which was claimed to acompany the NADH-linked reduction of artificial electron acceptors such as ferricyanide,[5] phenazine methosulfate,[6] or fumarate.[7,8] A com-

[2] O. Lindberg and L. Ernster, *Methods Biochem. Anal.* **3**, 1956.

[3] G. Schatz and E. Racker, *J. Biol. Chem.* **241**, 1429 (1966).

[4] D. E. Green, R. E. Beyer, M. Hansen, A. L. Smith, and G. Webster, *Federation Proc.* **22**, 1460 (1963).

[5] G. Webster, *J. Biol. Chem.* **240**, 1365 (1965).

[6] A. L. Smith and M. Hansen, *Biochem. Biophys. Res. Commun.* **8**, 136 (1962).

[7] D. R. Sanadi and A. L. Fluharty, *Biochemistry* **2**, 523 (1963).

[8] D. W. Haas, *Biochim. Biophys. Acta* **92**, 433 (1964).

parative reinvestigation of these various methods revealed, however, that none of them was satisfactory if applied to submitochondrial particles from beef heart[3]: The first method relies on a difference between two ratios and is usually too inaccurate to be of any real value. Concerning the methods employing artificial electron acceptors it was found that with submitochondrial particles from beef heart oxidative phosphorylation is obtained only with fumarate, but not with either ferricyanide (0.1 or 1.0 m*M*) or phenazine methosulfate (0.02 m*M*) as acceptor. Although these findings seemed to point to the fumarate assay as a useful method, it too was found to be unsuitable for accurate measurements. This was due to the slow reaction rate (2–6% of that of the NADH oxidase reaction) and the high rate of ATP formation in the fumarate-free controls. The fumarate assay of Sanadi and Fluharty[7] and Haas[8], however, may be of great value if oxidative phosphorylation in the NADH-cytochrome *b* segment is studied in submitochondrial particles from cells whose NADH-oxidase chain is insensitive to rotenone, such as *Saccharomyces cerevisiae*. In these cases it is very difficult to ascertain the point of interaction of CoQ_1 with the respiratory chain. This predicament is not encountered with the fumarate assay since it can be relatively safely assumed that fumarate will always be reduced by succinate dehydrogenase.

[5] Use of Artificial Electron Acceptors for Abbreviated Phosphorylating Electron Transport: Flavin-Cytochrome c

By CHUAN-PU LEE,[1] GIAN LUIGI SOTTOCASA, and LARS ERNSTER

It is well established that the second of the three energy-coupling sites of the respiratory chain, coupling site II, is located on the path of electrons from cytochrome *b* to cytochrome *c*. Below, two systems are described which are suited for the study of the phosphorylation originating from coupling site II: the succinate–ferricyanide system, which specifically involves the phosphorylation at coupling site II; and the "TMPD shunt," which involves a selective bypass over this site.

The Succinate–Ferricyanide System

Ferricyanide is known[1a,2] as the most suitable artificial electron

[1] This manuscript was prepared while the author was at the Wenner-Gren Institute, University of Stockholm, Stockholm, Sweden.

[1a] J. H. Copenhaver, Jr. and H. A. Lardy, *J. Biol. Chem.* **195**, 225 (1952).

[2] B. C. Pressman, *Biochim. Biophys. Acta* **17**, 273 (1955).

acceptor for phosphorylating electron transport in the cytochrome *c* region of the respiratory chain. It can be used in connection with both pyridine nucleotide-linked substrates and succinate. In the former case, the electron transport pathway is sensitive to amytal and rotenone and involves coupling sites I and II of the respiratory chain. In the latter case, the system is insensitive to amytal and rotenone, and involves only coupling site II. In fact, the succinate–ferricyanide system is presently the only known specific system for study of coupling site II.

The assay (Table I) involves incubation of freshly prepared mito-

TABLE I
PHOSPHORYLATION COUPLED TO THE OXIDATION OF SUCCINATE BY FERRICYANIDE[a]

Substrate	$K_3Fe(CN)_6$ reduced (micromoles)	P_i esterified (micromoles)	P:2e^-
Succinate	5.06	2.50	0.99
Succinate + AA	0.36	0	0

[a] The reaction mixture contained, in a final volume of 1 ml: 10 mM succinate, 1 mM KCN, 13.2 mM potassium phosphate, pH 7.5, 2 mM ATP, 50 mM glucose, 50 K.M. units of yeast hexokinase, 180 mM sucrose, 50 mM tris-acetate buffer, pH 7.5, and 3 mg rat liver mitochondrial protein. When indicated, 1 μg antimycin A (AA) dissolved in 0.002 ml ethanol was added. Samples were preincubated for 5 minutes at 30°, and the reaction was started by the addition of 0.1 ml 54 mM $K_3Fe(CN)_6$. Incubation time, 6 minutes.

chondria, e.g., from rat liver, in a buffered, isotonic medium, in the presence of succinate, ferricyanide, KCN (to block cytochrome oxidase), P_i, ADP, hexokinase, and glucose. After the incubation, the sample is fixed with $HClO_4$, and the amount of ferricyanide is determined with an aliquot of the extract spectrophotometrically at 420 mμ; the molar extinction coefficient of ferricyanide is 1×10^3 cm^{-1}. Esterification of P_i is determined with another aliquot of the extract, using, for example, the isotope distribution method.[3] The phosphorylating efficiency is expressed as the ratio of moles of P_i esterified to 2 moles of ferricyanide reduced (P:2e^- ratio). The generally accepted maximal P:2e^- ratio for the succinate–ferricyanide system is 1.

The assay may also be performed by monitoring the ferricyanide reduction spectrophotometrically during the incubation.[4] The P:2e^- ratio may be estimated by determining the increase in ferricyanide reduction

[3] O. Lindberg and L. Ernster, *Methods Biochem. Anal.* 3, 1 (1955).
[4] R. W. Estabrook, *J. Biol. Chem.* **236**, 3051 (1961).

upon the addition of a limiting amount of ADP, in a way analogous to the estimation of ADP:O ratio with the aid of the polarographic method as described by Chance and Williams.[5]

Errors in the estimation of the P:2e^- ratio with the above system may arise from a direct interaction of succinate dehydrogenase with ferricyanide. The extent of this error can be determined by the use of antimycin A. In intact mitochondria, antimycin A virtually completely blocks the reduction of ferricyanide by succinate (and by NAD^+-linked substrates), a phenomenon indicating that the succinate dehydrogenase (and the NADH dehydrogenase) are inaccessible to ferricyanide. To observe the antimycin A inhibition, however, it is important (1) that antimycin A is mixed with the mitochondria prior to the addition of ferricyanide, since otherwise ferricyanide may modify and inactivate antimycin A[6]; and (2) that the mitochondria are intact. Damaged mitochondria, such as those exposed to aging, hypotonicity, detergents, etc., as well as submitochondrial particles, show no sensitivity of the ferricyanide reduction to antimycin A, with either succinate or NAD^+-linked substrates; in the latter case, the sensitivity of the ferricyanide reduction to amytal and rotenone is likewise lost.[7] Apparently, the inaccessibility of the succinate and NADH dehydrogenases to ferricyanide is confined to the intact mitochondria. As a consequence, the use of ferricyanide as an artificial electron acceptor for abbreviated phosphorylating electron transport is limited to intact mitochondria.

Bypass of Coupling Site II

The system is based on the principle[8] that electron transport between cytochrome *b* and cytochrome *c* (or c_1) is blocked by antimycin A, and a bypass over this span is established by means of an artificial electron mediator, *N*,*N*,*N*′,*N*′-tetramethyl-*p*-phenylenediamine (TMPD). A suitable assay system is as follows: To an oxygen electrode vessel are added rat liver mitochondria in an isotonic, buffered medium containing ATP, glucose, hexokinase, and a relatively high concentration of Mg^{++} (which is required in order to minimize the uncoupling effect of TMPD[9]). Oxygen uptake is initiated by the addition of 3.3 m*M* either glutamate (or another pyridine nucleotide-linked substrate), or succinate. Four traces

[5] B. Chance and G. R. Williams, *J. Biol. Chem.* **217,** 383 (1955).

[6] P. Walter and H. A. Lardy, *Biochemistry* **3,** 812 (1964).

[7] G. L. Sottocasa, unpublished results.

[8] C. P. Lee, K. Nordenbrand, and L. Ernster, *Proc. Intern. Symp. Oxidases Related Redox Systems, Amherst, Mass., 1964* p. 960. Wiley, New York, 1965.

[9] J. H. Park, B. P. Heriwether, C. R. Park, and L. Specter, *Federation Proc.* **16,** 97 (1957).

are run for each substrate: one without any further addition, a second with 0.3 mM TMPD, a third with 0.5 μg antimycin A, and a fourth with 0.3 mM TMPD + 0.5 μg antimycin A included in the medium. An additional series of traces may be run with ascorbate as the electron donor to cytochrome c via TMPD, as first described by Jacobs.[10] Respiration is recorded until about 70% of the oxygen in the medium has been consumed, after which the samples are fixed and phosphate uptake is determined. Results of a typical experiment are shown in Table II. The

TABLE II
EFFECT OF ANTIMYCIN A AND TMPD ON RESPIRATION AND PHOSPHORYLATION[a,b]

Substrate	TMPD (0.3 mM)	Antimycin A (0.5 μg)	Oxygen (mμatoms/min)	P:O
Glutamate	−	−	300	2.75
	+	−	400	2.08
	−	+	28	—
	+	+	290	1.48
Succinate	−	−	380	1.78
	+	−	490	1.29
	−	+	25	—
	+	+	254	0.73
Ascorbate	−	−	14	—
	+	−	442	0.78
	−	+	16	—
	+	+	475	0.78

[a] From C. P. Lee, K. Nordenbrand, and L. Ernster, *Proc. Intern. Symp. Oxidases Related Redox Systems, Amherst, Mass., 1964* p. 960. Wiley, New York, 1965.

[b] The reaction mixture consisted of 125 mM KCl, 25 mM Tris-HCl buffer, pH 7.5, 2 mM P_i, 8 mM $MgCl_2$, 1 mM ATP, 4.8 mM glucose, 50 K.M. units of yeast hexokinase; and rat liver mitochondria containing 4 mg protein. Other additions as indicated: 0.3 mM TMPD; 0.5 μg antimycin A. The substrate was either 3.3 mM glutamate, or 3.3 mM succinate, or 1.5 mM ascorbate. Final volume, 3 ml; temperature 30°. Oxygen consumption was measured with a Clark oxygen electrode, and the esterification of P_i was determined by the isotope distribution method. [O. Lindberg and L. Ernster, *Methods Biochem. Anal.* **3,** 1 (1955).]

data illustrate the bypass of the antimycin A sensitive site of the respiratory chain by means of TMPD, with a simultaneous lowering of the P:O ratio by approximately one-third with glutamate and one-half with succinate as substrate. In the absence of antimycin A, TMPD causes an increase in oxygen uptake with these substrates, and a corresponding decrease of the P:O ratio. The results are consistent with the conclusion

[10] E. E. Jacobs, *Biochem. Biophys. Res. Commun.* **3,** 536 (1960).

that the TMPD shunt circumvents coupling site II of the respiratory chain. When antimycin A is present, only the shunt is operating, and the P:O ratio is lowered by 1 unit; when antimycin A is absent, respiration proceeds by both the normal pathway and the TMPD shunt, this resulting in an increase in oxygen uptake and a decrease in P:O ratio. With ascorbate + TMPD, the respiration is, as expected, insensitive to antimycin A, and the P:O ratio is below 1. A P:O ratio in the latter system slightly higher than 1 may be observed occasionally and has been interpreted[11] as being indicative of the occurrence of more than one phosphorylation in the region of cytochrome $c \rightarrow O_2$. Alternatively, however, it may be due to phosphorylation connected with the oxidation of endogenous NAD^+-linked substrate, mediated by the TMPD shunt.[8]

The TMPD-induced respiration of the antimycin A blocked system is inhibited by KCN. It is also inhibited by malonate in the case of succinate as substrate, and by amytal and rotenone in the case of NAD^+-linked substrates.

Submitochondrial particles can be assayed under conditions similar to those described for intact mitochondria. Succinate or NADH may be used as substrates. A difference from the intact mitochondrial system is that amytal and rotenone do not inhibit NADH oxidation in the presence of TMPD, which oxidizes added NADH nonenzymatically.

The oxidation of reduced cytochrome b upon the addition of TMPD to the antimycin A blocked respiratory chain can be demonstrated spectrophotometrically.

Figure 1 summarizes in a schematic form the electron transfer pathway involved in the TMPD shunt.

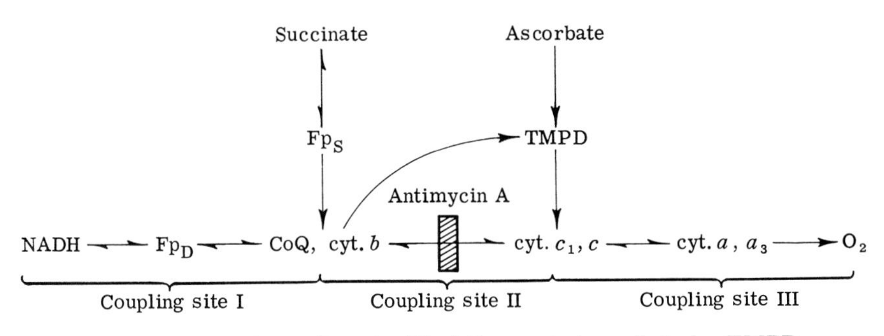

FIG. 1. Bypass of coupling site II of the respiratory chain by TMPD.

[11] J. L. Howland, *Biochim. Biophys. Acta* **77**, 419 (1963).

[6] Assay of Oxidative Phosphorylation at the Cytochrome Oxidase Region (Site III)

By D. R. SANADI[1] and E. E. JACOBS

In early studies directed toward the localization of the phosphorylation sites in the respiratory chain, ascorbate and exogenous cytochrome *c* were used to feed electrons in the segment of the respiratory chain between cytochrome *c* and oxygen.[1a-3] These, and subsequent related studies,[4-9] clearly established that the P:O coupled to the oxidation of ferrocytochrome *c* in mitochondrial preparations was close to 1.0. More recently, renewed interest in the assay of phosphorylation in this terminal segment of the respiratory chain has arisen in order to establish the locus of the phosphorylation site at which different coupling factors exert their effect. The most commonly used method at the present time utilizes the dye, *N,N,N′,N′*-tetramethyl-*p*-phenylenediamine (TMPD) as the artificial electron carrier between ascorbate and the cytochrome *c* region of the respiratory chain.[10]

Assay Method

Principle. The TMPD is kept largely in the reduced state by ascorbate, and oxidized by the cytochrome oxidase system. The exact site of entry of the electrons into the respiratory chain is not established, but is on the oxygen side of the antimycin-binding site. The resulting oxygen uptake is measured in Warburg manometers. Phosphate esterification, which occurs coincident with oxidation, is measured by the decrease in inorganic phosphate.

Reagents

Tris buffer: Tris base is neutralized with acetic acid to a final pH of 7.5 and diluted to 0.1 *M*

[1] This manuscript was prepared while the author was at the Gerontology Branch, National Heart Institute, National Institutes of Health, Bethesda, Maryland, and The Baltimore City Hospitals, Baltimore, Maryland.

[1a] M. Friedkin and A. L. Lehninger, *J. Biol. Chem.* **178**, 611 (1949).

[2] J. D. Judah, *Biochem. J.* **49**, 271 (1951).

[3] J. H. Copenhaver, Jr. and H. A. Lardy, *J. Biol. Chem.* **195**, 225 (1952).

[4] G. F. Maley and H. A. Lardy, *J. Biol. Chem.* **210**, 903 (1954).

[5] A. L. Lehninger, H. Mansoor, and H. C. Sudduth, *J. Biol. Chem.* **210**, 911 (1954).

[6] S. Nielsen and A. L. Lehninger, *J. Biol. Chem.* **215**, 555 (1955).

[7] E. C. Slater, *Nature* **174**, 1143 (1954).

[8] W. C. Hulsmann and E. C. Slater, *Nature* **180**, 372 (1952).

[9] E. E. Jacobs and D. R. Sanadi, *Biochim. Biophys. Acta* **38**, 12 (1960).

[10] E. E. Jacobs, *Biochem. Biophys. Res. Commun.* **3**, 536 (1960).

Phosphate buffer: 0.1 M sodium phosphate at pH 7.5

Ascorbate: ascorbic acid is dissolved in water and freshly neutralized to pH 6.0–6.5

Magnesium chloride, 0.1 M

ATP, 15 mM, pH 7.5

Glucose, 0.5 M

Hexokinase. Type III hexokinase, supplied by Sigma Chemical Co., is quite adequate for the purpose. It is dissolved in 0.1 M glucose to a concentration of 10 mg/ml.

TMPD. This compound is generally available as the dihydrochloride. It is recrystallized from ethanol and dissolved in water to a concentration of 15 mM. The TMPD solution seems to change color on standing. This apparently does not affect the activity, since we have successfully used samples stored at —20° for several weeks.

Rotenone. Rotenone, supplied by K & K Laboratories, is recrystallized from trichloroethylene[11] before use. It is dissolved in ethanol to 1 mM and subsequently diluted before use to a concentration of 3 μM. Rotenone is susceptible to destruction by light and oxygen. The stock solution should be made up at least once a week.

Bovine serum albumin. 1% crystalline albumin solution is neutralized and dialyzed against 10 mM Tris acetate, pH 7.5

KOH, 5 M, for use in the central well of Warburg flasks.

Procedure. The following components are added to the main compartment of the Warburg vessel: 0.6 ml Tris-acetate, 0.3 ml phosphate, 0.25 ml $MgCl_2$, 0.2 ml ascorbate, 50 μl TMPD, 20 μl rotenone, 0.2 ml bovine serum albumin, and 7–9 mg protein rat liver mitochondria or 1.5–2.0 mg beef heart submitochondrial particles. The side arm of the Warburg vessel receives 50 μl ATP, 0.2 ml glucose, and 50 μl hexokinase. The central well should contain 0.2 ml KOH with a piece of filter paper. After temperature equilibration at 30° for 6 minutes in the usual manner, the stopcock of the manometer is closed, and the side arm contents are tipped into the main compartment. The reaction is measured for 15–20 minutes; during this time 5–10 μatoms of oxygen are consumed at 30°. The reaction is terminated by the addition of trichloroacetic acid to a final concentration of 5%. The precipitated protein is centrifuged off, and the inorganic phosphate in the supernatant fluid is determined by the Fiske-SubbaRow[12] method. Under these conditions, P:O values of 0.8

[11] G. Buchi, L. Crombie, P. J. Godin, J. S. Kaltenbroon, K. S. Siddalingaiah, and D. A. Whiting, *J. Chem. Soc.* 2843 (1961).

[12] C. H. Fiske and Y. SubbaRow, *J. Biol. Chem.* **66**, 375 (1925).

to near 1.0 and 0.5–0.7 are obtained with rat liver mitochondria and beef heart submitochondrial particles, respectively.

Addition of serum albumin or Tris buffer is not critical in the assay. The former does improve the P:O with mitochondria that are not entirely fresh.

The polarographic technique has been used for measuring P:O with ascorbate-TMPD.[13] However, this method is not fully satisfactory since the respiratory control ratio, even with intact rat liver mitochondria, is quite low. However, the polarographic method of determination of oxygen consumption may be combined with the phosphate esterification assay. Since the dissolved oxygen is rather low, it is necessary to use ^{32}P-labeled phosphate in order to obtain reliable measurements of phosphorylation. It is desirable to purify the ^{32}P obtained from the Oak Ridge National Laboratory by chromatography on DEAE-cellulose.[14] The reaction may be carried out in the same medium as above, in a closed vessel using the Clark electrode. The $^{32}P_i$ is separated from the esterified ^{32}P by reverse phase partition chromatography[15] or by extraction of the phosphomolybdate with isobutanol.[16]

Comments

If endogenous substrates are oxidized concomitantly with the oxidation of TMPD, P:O values greater than 1 would be encountered. Several respiratory inhibitors have been used to suppress the oxidation of endogenous substrates, including arsenite,[17] antimycin,[18] and rotenone.[19] Arsenite inhibits α-ketoglutarate oxidation, but does not significantly affect other oxidations. Although antimycin is a suitable respiratory inhibitor for many purposes, its use in the presence of dyes like TMPD is not entirely satisfactory, since endogenous substrates can reduce TMPD without the electrons going through the antimycin site. For example, NADH produced from fatty acids may be oxidized by TMPD via the NADH dehydrogenase system allowing phosphorylation at site I. This additional phosphorylation would contribute to the apparent P:O with ascorbate-TMPD. The most suitable inhibitor to suppress oxidation of endogenous substrates appears to be rotenone.[19]

[13] L. Packer and E. E. Jacobs, *Biochim. Biophys. Acta* **57**, 371 (1962).
[14] T. E. Andreoli, K.-W. Lam, and D. R. Sanadi, *J. Biol. Chem.* **240**, 2644 (1965).
[15] B. Hagihara and H. A. Lardy, *J. Biol. Chem.* **235**, 889 (1960).
[16] I. A. Rose and S. Ochoa, *J. Biol. Chem.* **220**, 307 (1956).
[17] J. L. Howland, *Biochim. Biophys. Acta* **77**, 419 (1963).
[18] J. Ramirez and A. Mujica, *Biochim. Biophys. Acta* **86**, 1 (1964).
[19] D. D. Tyler, R. W. Estabrook, and D. R. Sanadi, *Biochem. Biophys. Res. Commun.* **18**, 264 (1965).

Several artificial electron carriers may be suitable in place of TMPD. If cytochrome *c* is the electron carrier, the use of fluoride or EDTA or higher pH of the reaction medium is desirable for obtaining maximal P:O values.[9] Complex anions like silicomolybdate and phosphotungstate[9] have also been used successfully. A higher concentration of magnesium chloride, at least 7.5 m*M*, appears to be necessary for maximal P:O.[9]

The recent report of Lee and co-workers[20] may provide another satisfactory assay for phosphorylation at the cytochrome oxidase site. They have observed that succinate oxidation, which is inhibited by antimycin, may be restored by the addition of TMPD to mitochondria. Presumably, succinate dehydrogenase is capable of reducing TMPD; this establishes a bypass around the antimycin block so that site II phosphorylation does not occur. The reduced TMPD is then oxidized via cytochrome oxidase, and only site III phosphorylation is observed under these conditions.

[20] C. P. Lee, K. Nordenbrand, and L. Ernster, *Proc. Intern. Symp. Oxidases Related Redox Systems, Amherst, Mass., 1964.* Wiley, New York, 1965.

[7] Mitochondrial Respiratory Control and the Polarographic Measurement of ADP : O Ratios[1]

By RONALD W. ESTABROOK

The convenience and simplicity of the polarographic "oxygen electrode" technique for measuring rapid changes in the rate of oxygen utilization by cellular and subcellular systems is now leading to its more general application in many laboratories. The types and design of oxygen electrodes vary, depending on the investigator's ingenuity and the specific requirements of the system under investigation. Stationary electrodes,[2] rotating electrodes,[3-5] or vibrating electrodes,[6,7] with or without special coatings[2,3] or membranes,[8-10] have been employed. For most routine uses,

[1] See also W. W. Kielley, Vol. VI [33].
[2] P. W. Davies and F. J. Brink, *Rev. Sci. Instr.* **13**, 524 (1942).
[3] B. Hagihara, *Biochim. Biophys. Acta* **46**, 134 (1961).
[4] L. Packer, *Res. Rept. Walter Reed Army Inst. Res.* **143**, (1957).
[5] I. S. Longmuir, *Biochem. J.* **57**, 81 (1954).
[6] B. Chance, *Science* **120**, 767 (1954).
[7] B. Chance, *Harvey Lectures Ser.* **49**, 145 (1955).
[8] L. C. Clark, Jr., *Trans. Am. Soc. Artificial Internal Organs* **2**, 41 (1956).
[9] L. C. Clark, Jr., R. Wolf, D. Granger, and Z. Taylor, *J. Appl. Physiol.* **6**, 189 (1953).
[10] D. O. Voss, J. C. Cowles, and M. Bacila, *Anal. Biochem.* **6**, 211 (1963).

comparable results can be obtained with any of the various types of oxygen electrode construction.

Although the polarographic method of measuring changes in oxygen concentration of photosynthetic systems,[11] yeast cells,[12] and nerve[13] had been studied in the 1940's, the application of the oxygen electrode technique to a study of mitochondrial respiration and oxidative phosphorylation was first reported by Chance and Williams[14] in 1955. Earlier studies by Lardy and Wellman[15] had demonstrated the dependence of mitochondrial oxygen utilization on the availability of inorganic phosphate or ADP. The concept of "respiratory control," introduced by these studies, now serves as a foundation for much of our understanding of oxidative phosphorylation.

Apparatus

The principle of the oxygen electrode has been summarized in a number of recent reviews.[16-18] In brief, the apparatus consists of a platinum or gold wire sealed in glass or plastic as the cathode with a calomel electrode connected via a KCl agar bridge as reference anode. The calomel electrode has frequently been replaced by a silver wire immersed in a chloride-containing solution. When a voltage is imposed across the two electrodes immersed in an oxygen-containing solution, with the platinum electrode negative relative to the reference electrode, oxygen undergoes an electrolytic reduction. The relationship between the steady state current and the polarization potential has been published in a number of papers.[2,17] When current is plotted as a function of polarizing voltage, a plateau region is observed between 0.5 and 0.8 volt. With a polarization voltage of −0.6 volt, current is directly proportional[2] to the oxygen concentration of the solution. The current generated is generally measured with a galvanometer or with a suitable amplifier and recorder combination. Figure 1 illustrates the physical arrangement[19,20] of the reac-

[11] L. R. Blinks and R. K. Skow, *Proc. Natl. Acad. Sci. U.S.* **24**, 420 (1938).

[12] R. J. Winzler, *J. Cellular Comp. Physiol.* **17**, 263 (1941).

[13] D. W. Bronk, F. Brink, Jr., and P. W. Davies, *Am. J. Physiol.* **133**, 224 (1941).

[14] B. Chance and G. R. Williams, *Nature* **175**, 1120 (1955).

[15] H. Lardy, and H. Wellman, *J. Biol. Chem.* **195**, 215 (1952).

[16] B. Chance, *Federation Proc.* **16**, 671 (1957).

[17] P. W. Davies, *in* "Physical Techniques in Biological Research," Vol. IV, "Special Methods" (W. L. Nastuk, ed.), p. 137. Academic Press, New York, 1962.

[18] I. S. Longmuir, *Advan. Polarog. Proc. 2nd Intern. Congr., Cambridge, Engl., 1959* Vol. 3, p. 1011. Pergamon, London, 1960.

[19] J. B. Chappell, *in* "Biological Structure and Function" (T. W. Goodwin and O. Lindberg, eds.), Vol. II, p. 71. Academic Press, New York, 1961.

[20] W. W. Kielley and J. R. Bronk, *J. Biol. Chem.* **230**, 521 (1958).

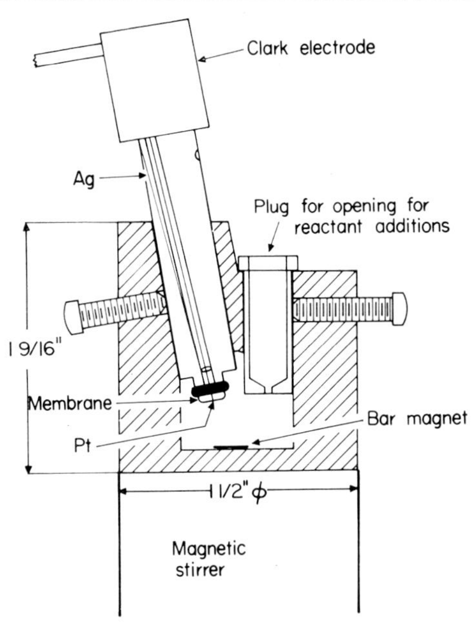

FIG. 1. Reaction vessel for measurement of oxygen utilization with the oxygen electrode.

tion chamber, magnetic stirrer, and Clark type[8,9] membrane-coated oxygen electrode routinely used in the author's laboratory. The reaction vessel can be constructed of glass or Plexiglas, with or without a water temperature jacket. It is imperative, however, that closed vessels are constructed so that air bubbles are not trapped and back diffusion of oxygen is reduced to a minimum. The use of a magnetic stirrer with a glass- or plastic-encased "flea" permits continual mixing of the reaction medium and facilitates establishment of the equilibrium between the oxygen dissolved in solution and the gas diffusing through a polyethylene membrane of the oxygen electrode. The apparatus is equilibrated with air-saturated buffer of a constant temperature, and reactants are added through the small opening at the top of the reaction vessel. The vessel is cleaned after each experiment by suction aspiration of the fluid and repeated water rinsing of the interior of the chamber. The oxygen electrode is not sealed into the Plexiglas vessel, but is held only by a press-fit permitting its easy removal for changing the membrane. A second type of reaction vessel, which is popular, has been described in detail by Hagihara.[3] This system employs a rotating collodion-coated platinum electrode and a KCl bridge connected to a calomel electrode. The design of the vibrating oxygen electrode for use with spectrophotometric studies is illustrated in a paper by Chance.[6]

Electrode

Electrodes can be constructed with ease in the laboratory or they can be purchased commercially. A very simple and inexpensive design for a membrane coated electrode has recently been described by Kahn.[21] The basic steps in the construction of an electrode involve the sealing of a thin platinum wire (0.0079 inch in diameter) in soft-glass tubing of about 1–2 mm outside diameter. The tip of the electrode is carefully ground with fine sandpaper and then further polished with a mixture of fine Carborundum suspended in light machine oil. The success of the platinum-glass seal should be inspected with a magnifying lens to ensure that no cracks or air bubbles are present at the seal. The platinum electrode can then be used directly by mounting it on a vibrator or a rotator with a suitable anode reference electrode, or it can be sealed in epoxy with a silver anode. In the latter case the excess epoxy resin can be removed by gentle sandpaper grinding, using care not to destroy the platinum-glass seal. Electrodes of varying sizes can be prepared, depending on the skill of the investigator. For membrane coated electrodes, a small drop of 1 *M* KCl or saturated KCl is placed on a piece of polyethylene or Teflon film (about 0.25 to 1 mil thickness) and the membrane is held in place on the electrode with a ring of polyethylene tubing.

Electronics

The polarizing voltage source and sensitivity control adjustment are obtained using a 1.5 volt flashlight battery, three potentiometers, a voltmeter, and an on-off switch. A schematic representation for one type of arrangement is illustrated in Fig. 2. Current is measured as the voltage change across resistor R_1, using a suitable millivolt amplifier, and then recorded with a strip chart recorder. The potentiometer R_1 for sensitivity control should permit use of the 10 mV scale of the amplifier. A voltage change of greater than 10 mV should be avoided to reduce modification of the polarizing voltage. The polarizing voltage is established by the variable resistor R_2. In addition the circuit includes a variable resistor R_3 for a zero offset when measuring small changes in oxygen uptake.

Calibration

The oxygen electrode apparatus can be calibrated in a number of ways. An initial estimate of oxygen content of the buffer system em-

[21] J. S. Kahn, *Anal. Biochem.* **9**, 389 (1964).

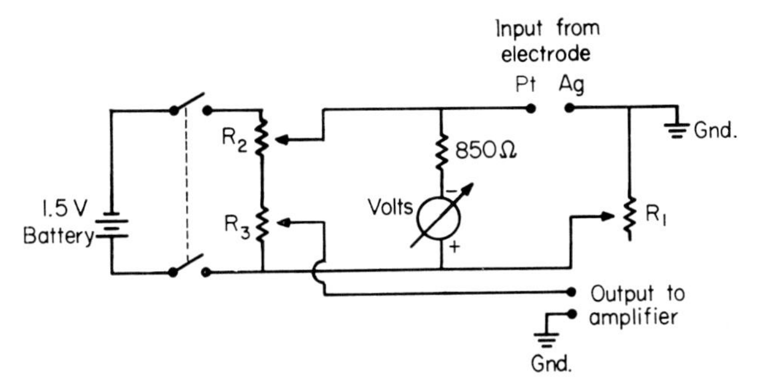

FIG. 2. Schematic representation of the polarizing voltage and sensitivity control for an oxygen electrode. The variable resistor R_1 for sensitivity control equals 50,000 ohms; R_2 for establishing the proper polarizing voltage equals 2500 ohms; and R_3 for zero offset control is of 100 ohms.

ployed can be calculated from the Handbook of Chemistry and Physics,[22] This rough estimate of oxygen content is illustrated diagrammatically in Fig. 3. A more accurate calibration of oxygen content can be obtained

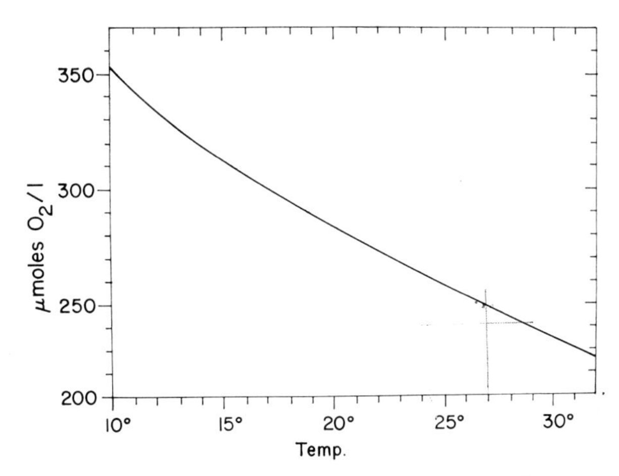

FIG. 3. Relationship between oxygen content of air-saturated water and temperature. Calculated from the "Handbook of Chemistry and Physics," 33rd ed., 1951–1952, p. 1481. Chemical Rubber Publ. Co., Cleveland, Ohio.

by gas equilibration with various nitrogen–oxygen mixtures. The relationship between current generated and oxygen content should be linear. In the absence of known gas mixtures, a convenient and rapid means of accurately determining the oxygen content of the reaction medium is

[22] "Handbook of Chemistry and Physics," 33rd ed., 1951–1952, p. 1481. Chemical Rubber Publ. Co., Cleveland, Ohio.

by employing submitochondrial particles, such as heart muscle ETP, with limiting concentrations of DPNH. The high affinity of the heart muscle preparation for DPNH permits a stoichiometric titration[23] of oxygen content. Using spectrophotometrically standardized DPNH, one can add limiting concentrations of DPNH and determine directly the change in current occurring on complete oxidation of the DPNH. In this manner a direct calibration can be obtained. In addition, experiments of this type permit the estimation of back diffusion of oxygen into the reaction medium. If sufficient DPNH is added to heart muscle particles suspended in suitable buffer to utilize 50% of the oxygen in the reaction vessel, the drift of the current observed after complete oxidation of the DPNH can then be directly measured and used to establish the rate of back diffusion of oxygen into the reaction vessel.

Determination of ADP:O Ratios

When "tightly coupled" mitochondria are suspended in an isotonic buffer, using an apparatus of the type described above, a slow rate of oxygen uptake is measured in the presence of substrate and absence of ADP. As illustrated in Fig. 4, addition of ADP causes an immediate

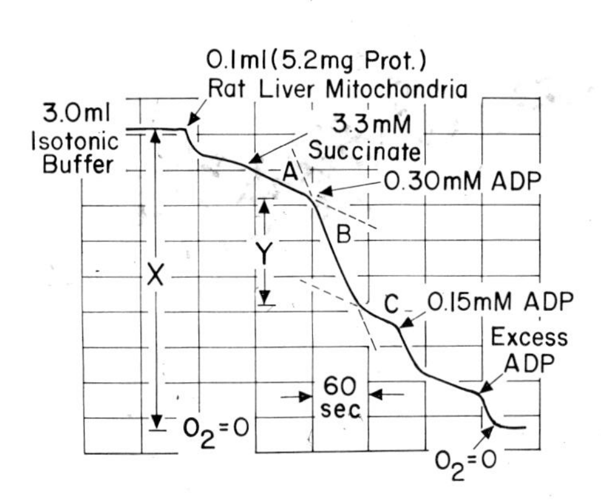

FIG. 4. Respiratory control and ADP:O ratio determination. Rat liver mitochondria was diluted with an isotonic buffer containing 0.225 *M* sucrose, 10 m*M* potassium phosphate, pH 7.4, 5 m*M* $MgCl_2$, 20 m*M* KCl, and 20 m*M* triethanolamine buffer, pH 7.4. Succinate and ADP were added as indicated. The calculation of ADP:O and respiratory control is described in the text.

increase in the rate of oxygen utilization. The duration of the increased rate of oxygen uptake is dependent on the concentration of ADP added to the reaction mixture. As described by Chance and Williams[14] and

[23] R. W. Estabrook and B. Mackler, *J. Biol. Chem.* **229**, 1091 (1957).

examined in detail by Hagihara,[3] the concentration of oxygen utilized is proportional to the amount of ADP phosphorylated to ATP. An ADP:O ratio (equivalent to a P:O ratio) can be directly calculated from oxygen electrode tracings of the type presented in Fig. 4 as follows:

(a) The distance Y is determined by extending the slopes of lines A, B, and C, and measuring the number of recorder divisions (or mm) from the intersect of curves A-B and curves B-C. This is a measure of the extent of ADP-stimulated respiration (state 3).

(b) The recorder deflection X expresses the total oxygen content of the reaction medium (cf. Fig. 3 or above section on calibration).

(c) (Oxygen content of medium/X units) $\times$ Y units $\times$ ml reaction medium $\times$ 2 = microatoms oxygen utilized.

(d) Micromoles ADP added/microatoms oxygen utilized = ADP:O ratio.

For the example illustrated in Fig. 4, the medium contains 240 millimicromoles O_2 per milliliter and $X = 86$ divisions. Y equals 30 divisions; therefore 520 mμatoms of oxygen is utilized during the activated (or state 3) respiration. A 0.01-ml aliquot of 90 mM ADP was added at the point indicated, i.e., 900 millimicromoles ADP. The ADP:O ratio is therefore 900:520 or 1.7.

Frequently, the degree of "respiratory control" is of interest. Chance[24] has defined the "respiratory control ratio" as "the respiratory rate in the presence of added ADP to the rate obtained *following* its (ADP) expenditure." In the example illustrated in Fig. 4, the respiratory control ratio (RCR) is rate B/rate C or 5.9.

Concluding Remarks

The above-described method for measuring oxygen utilization is rapid, convenient, and simple. Even so, a number of precautions must be observed. Uncoated platinum electrodes are easily poisoned by reactive chemicals such as cyanide and iodide. Membrane-coated electrodes frequently are sluggish in their response, delays as great as 15 seconds being often encountered. Special precautions must be used when studying the influence of inhibitors dissolved in organic solvents. Membrane-coated electrodes are often affected by organic solvents giving artifactual responses. In addition, Plexiglas (plastic) reaction vessels have an avidity for inhibitors such as antimycin, oligomycin, etc., and frequently the carry-over of inhibitor from one experiment to the next can give spurious results.

[24] B. Chance, *Ciba Found. Symp. Regulation Cell Metabolism,* p. 91. Little, Brown, Boston, Massachusetts, 1959.

[8] Application of Inhibitors and Uncouplers for a Study of Oxidative Phosphorylation

By E. C. Slater

In this chapter no attempt will be made exhaustively to treat the large number of inhibitors and uncouplers of oxidative phosphorylation that have been described. The discussion will be confined to those inhibitors and uncouplers that are particularly useful in the study of oxidative phosphorylation.

All oxidative phosphorylation reactions may be described by Eq. (1)

$$AH_2 + B + ADP + P_i \rightleftharpoons A + BH_2 + ATP + H_2O \quad (1)$$

The various inhibitors and uncouplers may be classified on the basis of the chemical-coupling theory of oxidative phosphorylation which may be formulated:

$$AH_2 + B + C \rightleftharpoons A{\sim}C + BH_2 \quad (2)$$
$$A{\sim}C + ADP + P_i \rightleftharpoons A + C + ATP + H_2O \quad (3)$$

The sum of Eqs. (2) and (3) is Eq. (1).

Respiratory-chain inhibitors inhibit the reaction given by Eq. (2); inhibitors of oxidative phosphorylation inhibit the reaction given by Eq. (3) or, by combining with $A \sim C$, inhibit both the reaction given by Eq. (2) and that given by Eq. (3), whereas uncouplers permit the oxidation of AH_2 by B to proceed without phosphorylation [Eq. (4)].

$$AH_2 + B \rightleftharpoons A + BH_2 \quad (4)$$

Uncouplers induce ATPase activity.

Respiratory-Chain Inhibitors

Inhibitors of Cytochrome *c* Oxidase

Cyanide. Cyanide is a powerful reversible inhibitor of cytochrome *c* oxidase, but it should be used with great caution. The NADH oxidase of the Keilin and Hartree heart muscle preparation was inhibited by 93.9% by 0.33 m*M*, 96.9% by 0.5 m*M*, 98.3% by 1 m*M*, 98.9% by 3.33 m*M*, and 99.6% by 10 m*M* cyanide.[1] For many purposes, 1 m*M* cyanide is sufficient. However, allowance should always be made in a suitable control experiment for the "leak" past the cyanide block.

Cyanide is most conveniently used in the form of KCN neutralized

[1] E. C. Slater, *Biochem. J.* **46**, 484 (1950).

immediately before use. Owing to its volatility, HCN quickly disappears from dilute solutions left open to the atmosphere. It will also distill over from the manometer flask into the KOH in the center well unless this also contains the correct concentration of cyanide (for further details, see Umbreit *et al.*[2]). Cyanide is ineffective as an inhibitor in the presence of keto acids owing to removal of the cyanide by cyanhydrin formation, a comparatively slow reaction.[3]

If oxidative phosphorylation is to be measured, the concentration of cyanide should not exceed 1 m*M* since higher concentrations inhibit or uncouple.[3]

If cyanide is present together with ferricytochrome *c*, the slow reaction between the two should be taken into account, since the ferricytochrome *c*-cyanide complex is not reduced by cytochrome *c* reductase systems.[3-5] Either the cyanide or the ferricytochrome *c* should be added to the reaction mixture immediately before the start of the reaction.

Sulfide. Sulfide is also a powerful inhibitor of cytochrome *c* oxidase. The NADH oxidase system of the Keilin and Hartree heart muscle preparation was inhibited by 96.3% by 0.1 m*M* and by 99% by 10 m*M*.[1] Many of the precautions to be taken with cyanide apply also to sulfide. However, sulfide does not react with keto acids. It cannot be used in experiments in which ferricytochrome *c* is used as acceptor since it rapidly reduces the ferri compound.

Azide, hydroxylamine, and CO, also inhibitors of cytochrome *c* oxidase, are not recommended because of incomplete inhibition.

Inhibitors of the Respiratory Chain between Cytochromes *b* and c_1

Antimycin. Since only one antimycin is used in the literature, the term antimycin A may be dropped. It is a mixture of at least 4 components, designated as A_1-A_4.[6] Antimycin may be purchased from the Wisconsin Alumni Research Foundation, University of Wisconsin, or from Kyowa Fermentation Industries, Ltd., Tokyo, Japan. Since this compound is bound very firmly to mitochondria or mitochondrial fragments, its concentration is best expressed as micrograms or micromoles per milligram protein. The molecular weight of antimycin A_1 is 548 and of A_2 is 520. For calculation of molar concentrations, the former molecular

[2] W. W. Umbreit, R. H. Burris, and J. F. Stauffer, "Manometric Techniques and Tissue Metabolism." Burgess, Minneapolis, Minnesota, 1951.

[3] E. C. Slater, *Biochem. J.* **59**, 392 (1955).

[4] V. R. Potter, *J. Biol. Chem.* **137**, 13 (1941).

[5] B. L. Horecker and A. Kornberg, *J. Biol. Chem.* **165**, 11 (1946).

[6] E. E. van Tamelen, J. P. Dickie, M. E. Loomans, R. S. Dewey, and F. M. Strong, *J. Am. Chem. Soc.* **83**, 1639 (1961).

weight is generally used and, for the sake of uniformity, it is recommended that this be continued. Antimycin is usually added as a small amount of an ethanolic solution, but for accurate titration a solution in serum albumin is preferable. This is prepared by adding 1 volume of an ethanolic solution to 99 volumes of 0.1% serum albumin in phosphate buffer. The concentration in the stock ethanolic solution may be determined spectrophotometrically ($\epsilon_{mM} = 4.8$ at 320 mμ).[7] The stock solution may be kept indefinitely at —20°. The control should contain the same amount of ethanol (and serum albumin) as the experimental.

The curve relating degree of inhibition to amount of antimycin has a sigmoid form. The amount of antimycin required for maximal inhibition is much greater for heart than for liver mitochondria.[8] From Estabrook's[8] data it can be calculated that about 0.07 micromole antimycin per gram mitochondrial protein is necessary. Maximal inhibition by antimycin often requires about 1 minute for full development. As a routine 0.2 micromole (about 110 μg) per gram mitochondrial protein for liver and 0.8 micromole (about 440 μg/g) for heart may be used. High concentrations of antimycin should be avoided, not only because of the expense, but also because antimycin, in higher concentrations, uncouples.[9,10] Haas[10] found uncoupling with digitonin fragments of heart mitochondria with a concentration of antimycin exceeding 0.8 micromole per gram protein. It is advisable to determine in a preliminary experiment the most suitable concentration of antimycin.

Inhibition of respiration by antimycin is incomplete. The activity of the antimycin-resistant pathway is largely independent of substrate (about 20 nanoatoms per milligram protein per minute for heart mitochondria).[11] The antimycin-resistant respiration is not coupled to phosphorylation. It is not affected by dinitrophenol, oligomycin, or Amytal, but it is inhibited by cyanide.

Care is necessary in using antimycin in the presence of ferricyanide, since the latter causes the destruction of the antimycin, in the absence of mitochondria.[12]

[7] F. M. Strong, J. P. Dickie, M. E. Loomans, E. E. van Tamelen, and R. S. Dewey, *J. Am. Chem. Soc.* **82,** 1513 (1960).

[8] R. W. Estabrook, *Biochim. Biophys. Acta* **60,** 236 (1962).

[9] H. Löw and I. Vallin, *Biochim. Biophys. Acta* **69,** 361 (1963).

[10] D. W. Haas, *Biochim. Biophys. Acta* **92,** 433 (1964).

[11] E. C. Slater, *Symp. Intracellular Respiration: Phosphorylating and Non-Phosphorylating Oxidation Reactions, Proc. 5th Intern. Congr. Biochem., Moscow, 1961* **5,** 325. Pergamon, London, 1963.

[12] P. Walter and H. A. Lardy, *Biochemistry* **3,** 812 (1964).

n-Heptylquinoline N-oxide (HQNO). HQNO (mol. wt. 241) can be used as an alternative inhibitor to antimycin.[13] Unlike antimycin, it is also effective with bacterial preparations. Howland[14] found that about 40 μg (0.166 micromole) HQNO per milligram protein were necessary completely to inhibit the oxidation of succinate by rat liver mitochondria. Higher concentrations are uncoupling.[10] Inhibition by HQNO may be reversed by dinitrophenol.[15]

Amytal. Amytal (5-ethyl-5-isoamylbarbituric acid) is readily available. In concentrations of 1.8 mM it maximally inhibits the oxidation of NAD-linked substrates without inhibiting the oxidation of succinate.[16] Indeed, the oxidation of succinate is often stimulated, probably owing to a decrease in the concentration of oxalacetate.[17] Inhibition by Amytal of the respiration of NAD-linked substrates is incomplete. The Amytal-resistant respiration has properties similar to those of the antimycin-resistant respiration; 1.8 mM Amytal causes some decline in the P:O ratio.[17]

Amytal is most conveniently used as an ethanolic solution.

Rotenone. Rotenone (mol. wt. 394) is available from S. B. Penick and Company, New York. It acts like Amytal but in much lower concentrations.[18–20] An amount of rotenone equivalent to about 30 nanomoles per gram protein is sufficient to inhibit the oxidation of NAD-linked substrates by rat liver mitochondria.[19] Excess rotenone does not cause a lowering of the P:O ratio.[19] It is usually added as an ethanolic solution.

Specific Inhibitors of Succinate Oxidation

Malonate. The degree of inhibition by malonate depends upon K_m: K_i and the succinate concentration. The K_i is 18 μM, the K_m depends upon the preparation. Up to 10 mM malonate may be used without any effect on the phosphorylation. This concentration is sufficient completely to inhibit the oxidation of succinate formed from α-oxoglutarate by rat liver mitochondria, but there is some "through" oxidation with rat heart mitochondria.[21]

[13] J. W. Lightbown and F. L. Jackson, *Biochem. J.* **63**, 130 (1956).
[14] J. L. Howland, *Biochim. Biophys. Acta* **77**, 419 (1963).
[15] J. L. Howland, *Biochim. Biophys. Acta* **77**, 659 (1963).
[16] O. Jalling, O. Lindberg, and L. Ernster, *Acta Chem. Scand.* **9**, 198 (1955).
[17] P. Greengard, K. Minnaert, E. C. Slater, and I. Betel, *Biochem. J.* **73**, 637 (1959).
[18] P. E. Lindahl and K. E. Öberg, *Exptl. Cell Res.* **23**, 228 (1961).
[19] L. Ernster, G. Dallner, and G. F. Azzone, *J. Biol. Chem.* **238**, 1124 (1963).
[20] J. Burgos and E. R. Redfearn, *Biochim. Biophys. Acta* **110**, 475 (1965).
[21] E. C. Slater and F. A. Holton, *Biochem. J.* **56**, 28 (1954).

Specific Inhibition of α-Oxoglutarate Oxidation

Arsenite. A concentration of 1 m*M* arsenite is sufficient to block the oxidation of α-oxoglutarate, after a short time lag. This amount of arsenite causes some decline of the P:O ratio.[22,23]

Inhibitors of Oxidative Phosphorylation

Non-Site-Specific

Oligomycin. Oligomycin is widely used to inhibit the synthesis of ATP by respiratory-chain oxidative phosphorylation without inhibiting the initial conservation of energy [Eq. (2)]. Oligomycin preparations (which may be obtained from the Wisconsin Alumni Research Foundation) are a mixture of three structurally related compounds (A, B, and C) of molecular weights 425, 397, and 479, respectively.[24-26] The relative proportions of the three compounds can be determined by paper chromatography.[25] Rutamycin (Lilly compound A 272)[27] is a related compound of molecular weight 439. Oligomycin C is slightly less active than the other three in inhibiting respiration.[26] All are equally active in inhibiting the ATP-P_i exchange.[26] Oligomycin C is much less active than the others in inhibiting the ATPase induced by dinitrophenol and other uncouplers.[26]

Oligomycin may be stored in ethanolic solution. Oligomycin A was found to lose its activity on storage in 50% ethanol for 4 months at 5°. The other oligomycins were stable in this solvent. All oligomycins are stable in absolute ethanol.[26]

Oligomycin inhibits coupled respiration, probably by acting on the reaction given by Eq. (3), but has no effect on uncoupled respiration.[28,29] It also has no effect on substrate-linked phosphorylation.[30] There is some discrepancy in the literature about the amount of oligo-

[22] A. Fluharty and D. R. Sanadi, *Proc. Natl. Acad. Sci. U.S.* **46,** 608 (1960).

[23] E. C. Slater and J. M. Tager, *Biochim. Biophys. Acta* **77,** 276 (1963).

[24] S. Masamune, J. M. Sehgel, E. E. van Tamelen, F. M. Strong, and W. H. Peterson, *J. Am. Chem. Soc.* **80,** 6092 (1958).

[25] J. Visser, D. E. Weinauer, R. C. Davis, W. H. Peterson, W. Nazarewicz, and H. Ordway, *J. Biochem. Microbiol. Technol. Eng.* **2,** 31 (1960).

[26] H. A. Lardy, P. Witonsky, and D. Johnson, *Biochemistry* **4,** 552 (1965).

[27] R. Q. Thompson, M. M. Hoehn, and C. E. Higgins, "Antimicrobial Agents and Chemotherapy," p. 474. Am. Soc. Microbiologists, Detroit, Michigan, 1961.

[28] H. A. Lardy, D. Johnson, and W. C. McMurray, *Arch. Biochem. Biophys.* **78,** 587 (1958).

[29] F. Huijing and E. C. Slater, *J. Biochem.* (*Tokyo*) **49,** 493 (1961).

[30] J. B. Chappell and G. D. Greville, *Nature* **190,** 502 (1961).

mycin required for maximal inhibition of coupled respiration. Ernster *et al.*[19] found 0.2 micromole per gram sufficient for rat liver mitochondria, but Lardy *et al.*[31] find 0.4 micromole necessary. Much larger amounts are required for submitochondrial particles from heart. Here 1 micromole per gram protein stimulates the phosphorylation and higher concentrations inhibit.[32]

Oligomycin inhibits the dinitrophenol-induced ATPase of mitochondria and also the ATPase of mitochondrial fragments. Since it is without effect on the ATPase of microsomes and, except in very high concentrations,[33] of cell membranes, oligomycin may be used to measure the degree of contamination of mitochondrial fragments by these extraneous ATPases.

Aurovertin. Aurovertin (mol. wt. 490) is not available commercially. It is used in ethanolic solution (ϵ_{mM}, 42.7 at 367.5 mμ).[34] It acts rather like oligomycin and is approximately equally effective on a weight basis.[31,35] However, it is much less effective than oligomycin in inhibiting ATPase induced by uncoupling agents. Lenaz[36] has reported that 0.4 micromole aurovertin per gram protein inhibits oxidative phosphorylation at all sites.

Site I-Specific Inhibitors

Amytal. Since inhibition by Amytal is partially relieved by dinitrophenol,[37] this compound may be classed as an inhibitor of oxidative phosphorylation, but since it strongly inhibits nonphosphorylating preparations of the respiratory chain, it has been placed under the inhibitors of the respiratory chain.

Alkylguanidines. Galegine (4-methyl-3-butenylguanidine) is available from the California Corporation for Biochemical Research. Other alkylguanidines are not available commercially but are easily prepared. These compounds are much more effective at site I than at other phosphorylation sites.[10,38-41] The effectiveness increases with increasing size

[31] H. A. Lardy, J. L. Connelly, and D. Johnson, *Biochemistry* **3**, 1961 (1964).
[32] C. P. Lee and L. Ernster, *Biochem. Biophys. Res. Commun.* **18**, 523 (1965).
[33] H. E. M. van Groningen and E. C. Slater, *Biochim. Biophys. Acta* **73**, 527 (1963).
[34] C. L. Baldwin, L. C. Weaver, R. M. Brooker, T. N. Jacobsen, C. E. Osborne, and H. A. Nash, *Lloydia* **27**, 88 (1964).
[35] J. L. Connelly and H. A. Lardy, *Biochemistry* **3**, 1969 (1964).
[36] G. Lenaz, *Biochem. Biophys. Res. Commun.* **21**, 170 (1965).
[37] B. Chance and G. Hollunger, *J. Biol. Chem.* **238**, 418 (1963).
[38] B. C. Pressman, *J. Biol. Chem.* **238**, 401 (1963).
[39] B. Chance and G. Hollunger, *J. Biol. Chem.* **238**, 432 (1963).
[40] J. B. Chappell, *J. Biol. Chem.* **238**, 410 (1963).
[41] R. J. Guillory and E. C. Slater, *Biochim. Biophys. Acta* **105**, 221 (1965).

of an alkyl substituent.[38] Unlike oligomycin, the alkylguanidines probably act on the reaction given by Eq. (2). An amount of hexylguanidine equal to 500 micromoles per gram mitochondrial protein has no effect on the oxidation of succinate by rat liver mitochondria in the presence of Amytal, whereas 20 micromoles per gram mitochondrial protein was sufficient maximally to inhibit glutamate oxidation, when added before the ADP.[41] The inhibition is slow to develop when the inhibitor is added in the presence of ADP.[40,42]

Site II-Specific Inhibitors

HQNO. Like Amytal this compound may also be regarded as an inhibitor of oxidative phosphorylation, but for the reasons given for Amytal it is placed under the respiratory inhibitors.

Phenylethylbiguanide. Phenylethylbiguanide is available commercially as the oral hypoglycemic agent DBI. It is relatively specific for site II.[10,40,42] Like the alkylguanidines, and unlike oligomycin, phenylethylbiguanide probably acts on the reaction given by Eq. (2).

Site III Inhibitors

Synthalin (Decamethylenediguanidine). Synthalin is readily available commercially. It has no effect on site I phosphorylation,[10] and inhibits site III at a concentration of about 30 micromoles per gram mitochondrial protein.[41] It is not known whether it has any effect on site II. In slightly higher concentrations (50% inhibition with 40 micromoles per gram mitochondrial protein), Synthalin inhibits the ATP-ADP exchange reaction, like oligomycin.[41] The other guanidines have little effect on this reaction.

Inhibitor of Phosphorylation of Added ADP

Atractyloside. Atractyloside (also known as potassium atractylate) inhibits the phosphorylation of added ADP[43,44] without having any effect on that of the endogenous ADP of the mitochondria.[45-47]

Atractyloside (mol. wt. 803), which is not available commercially, is easily soluble in water. The aqueous solution is stable if kept frozen. A

[42] B. C. Pressman, *in* "Energy-Linked Functions in Mitochondria" (B. Chance, ed.), p. 181. Academic Press, New York, 1963.

[43] A. Bruni and A. R. Contessa, *Nature* **191,** 818 (1961).

[44] A. Bruni, S. Luciana, and A. R. Contessa, *Nature* **201,** 1219 (1964).

[45] A. Kemp, Jr. and E. C. Slater, *Biochim. Biophys. Acta* **92,** 178 (1964).

[46] H. W. Heldt, H. Jacobs, and M. Klingenberg, *Biochem. Biophys. Res. Commun.* **18,** 174 (1965).

[47] J. B. Chappell and A. R. Crofts, *Biochem. J.* **95,** 707 (1965).

concentration of 0.25 micromole per gram protein is sufficient for complete inhibition of the phosphorylation of 1 m*M* added ADP. The inhibition by atractyloside is competitive with respect to the ADP.[44]

Uncouplers

Dinitrophenols

2,4-Dinitrophenol and 2,6-Dinitrophenol. The most commonly used uncoupler is 2,4-dinitrophenol, which is readily available commercially. Aqueous solutions containing up to 10 m*M* can be prepared. It affects all phosphorylation sites in the respiratory chain, but has no effect on the substrate-linked phosphorylation. The degree of uncoupling depends on the pH. 2,6-Dinitrophenol acts very similarly to 2,4-dinitrophenol. The concentrations of 2,6-dinitrophenol, at different pH's, inducing the maximal ATPase and maximal stimulation of respiration in the absence of ADP and phosphate[48,49] are listed in the table. In higher concentrations, the ATPase and respiration are inhibited.

4-Isooctyl-2,6-dinitrophenol. Alkyl-substituted dinitrophenols, such as 4-isooctyl-2,6-dinitrophenol, are not available commercially, but are

CONCENTRATIONS OF 2,6-DINITROPHENOL AND 4-ISOOCTYL-2,6-DINITROPHENOL NECESSARY FOR OPTIMAL ATPASE AND RESPIRATORY ACTIVITY[a,b]

	Optimal concentration (μM)					
	2,6-Dinitrophenol			4-Isooctyl-2,6-dinitrophenol		
Reaction	pH 6.0	pH 7.0	pH 8.0	pH 6.0	pH 7.0	pH 8.0
ATPase	60	210	760	1.3	2.0	3.9
Oxidation of succinate[c,d]	2.7	27	270	0.16	1.2	10
Oxidation of pyruvate + malate[c]	3.7	11	38	—	0.67[e]	0.67[e]
Oxidation of β-hydroxybutyrate	3.8	12	40	0.41	1.3	5

[a] Compiled from H. C. Hemker [*Biochim. Biophys. Acta* **63,** 46 (1962); *ibid.*, **81,** 1 (1964)].

[b] Except where indicated, these concentrations are independent of the concentration of the mitochondria.

[c] In absence of ADP and phosphate.

[d] In presence of 1.2 m*M* Amytal.

[e] Per milligram of mitochondrial protein per milliliter [H. C. Hemker, *Biochim. Biophys. Acta* **63,** 46 (1962)].

[48] H. C. Hemker, *Biochim. Biophys. Acta* **63,** 46 (1962).

[49] H. C. Hemker, *Biochim. Biophys. Acta* **81,** 1 (1964).

readily synthesized.[48] This compound acts like 2,4-dinitrophenol, but is effective in much lower concentrations (see the table). Thus, in contrast to 2,4-dinitrophenol, it is effective in concentrations which give little color to the solution. This can be an advantage for spectrophotometric measurements.

Dicoumarol. Dicoumarol [3,3′-methylenebis(4-hydroxycoumarin)] is readily available commercially. A concentration of 50 micromoles per liter is sufficient for complete uncoupling of rat liver mitochondria. Its action may be completely reversed by the subsequent addition of serum albumin.[47] It is active on all phosphorylating sites of the respiratory chain and has no effect on the substrate-linked phosphorylation.

Carbonyl Cyanide *m*-Chlorophenylhydrazone (m-Cl-CCP)

m-Cl-CCP (mol. wt. 204.5) is not available commercially. Aqueous solutions at pH 7.4 may be kept for a few days, but deterioration has been noted on long storage.[50] Complete uncoupling was obtained with a variety of mitochondria with 1.6 μM m-Cl-CCP (0.3–3.0 mg mitochondrial protein).[50] 1,2- and 1,3-Aminothiols protect against uncoupling when added before the uncoupler.[50] A related compound, *p*-trifluoromethoxyphenylhydrazone, exhibits uncoupling effects at concentrations down to $10^{-8} M$, and completely uncouples mouse liver mitochondria at $10^{-7} M$ (2.5–6.2 mg protein per milliliter).[51]

Arsenate

Arsenate readily uncouples substrate-linked phosphorylations[52,53] but is relatively ineffective against respiratory-chain phosphorylation.[54] There is virtually no arsenate-induced ATPase.[54] In the absence of added phosphate, 40 mM arsenate stimulates respiration to the extent of about 40% of that maximally induced by 2,4-dinitrophenol. The addition of ADP further stimulates the respiration (cf. Estabrook and Itada[55]). A small amount of phosphate inhibits the arsenate-stimulated respiration (50% inhibition with 0.5 mM phosphate of respiration stimulated by 40 mM arsenate).

Arsenate is particularly useful in combination with dinitrophenol in order to uncouple all phosphorylation steps. Experiments with arsenate

[50] P. G. Heytler, *Biochemistry* **2**, 357 (1963).

[51] P. G. Heytler and W. W. Prichard, *Biochem. Biophys. Res. Commun.* **7**, 272 (1962).

[52] D. M. Needham and R. K. Pillai, *Biochem. J.* **31**, 1837 (1937).

[53] D. R. Sanadi, D. M. Gibson, P. Ayengar, and L. Ouellet, *Biochim. Biophys. Acta* **13**, 146 (1954).

[54] H. F. Ter Welle and E. C. Slater, *Biochim. Biophys. Acta* **89**, 385 (1964).

[55] R. W. Estabrook and N. Itada, *Federation Proc.* **21**, 55 (1962).

should be carried out in the presence of 1 mM EDTA in order to avoid structural damage to the mitochondria.[54]

Other Uncouplers

In general, all compounds which promote the utilization of $A \sim C$ other than by the reaction given by Eq. (3) may be regarded as uncouplers. Thus, gramicidin and valinomycin which promote the utilization of intermediates of oxidative phosphorylation for the uptake of alkali metal ions are uncouplers in the presence of these ions (see this volume [111]). Indeed, uncouplers such as 2,4-dinitrophenol may act by essentially the same mechanism, the ion in this case being H^+ or OH^-. Similarly, the addition of α-oxoglutarate + NH_3 to rat liver mitochondria has an uncoupling action,[56] since it promotes the energy-utilizing reversed-electron transport (see this volume [113]).

[56] J. M. Tager and E. C. Slater, *Biochim. Biophys. Acta* **77,** 246 (1963).

[9] Measurement of ATPase, ^{14}C-ADP-ATP, and ^{32}P$_i$-ATP Exchange Reactions

By Maynard E. Pullman

General Principle

The general principles and description of the assay of these reactions have been described in a previous volume.[1] Alternative or modified procedures will be described here for both exchange reactions either because, compared to the previously described procedures, the methods described here were found by the author to be more rapid and convenient or more reliable and reproducible.

Assay Method for ATPase[1]

The previously described procedure has been modified in one respect. The concentrations of ATP and Mg^{++} have been increased to 3 mM to ensure maximal activity. This procedure is applicable to mitochondria, submitochondrial particles, and the soluble enzyme.

Assay for the ADP-ATP Exchange Reaction

The incubation condition for the assay of this reaction are similar to those described by Wadkins and Lehninger.[2] The nucleotides are separated by chromatography on Dowex 1-Cl columns.[3]

[1] See Vol. VI. ATPase [34]; exchange reactions [32].

[2] C. L. Wadkins and A. L. Lehninger, *J. Biol. Chem.* **233,** 1589 (1958).

[3] H. Zalkin, M. E. Pullman, and E. Racker, *J. Biol. Chem.* **240,** 4011 (1965).

Procedure. Each reaction mixture contains 6 mM ATP, 2 mM ^{14}C-ADP (40,000–130,000 cpm), 10 mM Tris-SO_4, pH 7.0, and 1.0 mg of mitochondrial protein in a total volume of 1.0 ml. The incubation is carried out in test tubes with gentle shaking for 10 minutes at 30°. The reaction is terminated with 2 ml of water at 0°. The entire reaction mixture is poured onto a 1 × 3 cm Dowex 1-Cl-2% cross-linked[4] column. Each tube is rinsed twice with 5 ml of H_2O and the rinses are added to the column. The nucleotides are eluted at a flow rate of 0.8–1.5 ml/minute according to the following schedule: 13 ml water, 15 ml of 0.01 N HCl (AMP), 10 ml of 0.01 N HCl, 5 ml of 0.04 N HCl, 20 ml of 0.04 N HCl (ADP), 10 ml of 0.04 N HCl, 10 ml of 0.1 N HCl, and 22.5 ml of 0.1 N HCl (ATP).

A millimolar extinction coefficient of 14.5 at 260 mμ is used for the spectrophotometric determination of the concentration of the nucleotide fractions.

Aliquots of the fractions are plated on planchets containing 0.1 ml of 0.1% agar and a few drops of ethanol to facilitate uniform plating. The solution is covered by a lens paper disk, and the planchets are placed in a 250° oven to dry. Counting of the planchets is carried out in a thin window, low background, gas flow counter.

The amount of ^{14}C-ATP formed is calculated by dividing the total radioactivity of the ATP fraction isolated from the column by the initial specific activity of the ^{14}C-ADP, i.e.

$$\text{Micromoles } ^{14}\text{C-ATP formed} = \frac{\text{total cpm (ATP)}}{\text{cpm (ADP)/micromoles ADP}}$$

Comments. Mitochondria and submitochondrial particles contain many known enzymes that are capable of catalyzing an ATP-ADP exchange reaction.[2] With intact mitochondria, in the absence of added Mg^{++} (see below), it is generally assumed that only the exchange activity associated with oxidative phosphorylation is sensitive to uncoupling agents. Therefore, with intact mitochondria, an uncoupling agent such as 2,4-dinitrophenol may be employed in a control experiment to distinguish the relevant exchange.

Over 90% of the exchange activity of intact mitochondria, measured in the absence of added Mg^{++}, is sensitive to dinitrophenol, while less then 50% of the total exchange activity measured in the presence of added Mg^{++} is sensitive to dinitrophenol.[5] Thus the omission of Mg^{++} provides another means of distinguishing the exchange reaction involved

[4] Bio-Rad AG1-X2. Prior to use, each column used is washed with 3 N HCl until the ultraviolet absorbance at 260 mμ has disappeared (15–20 ml). The columns are then rinsed with water until the water effluent is neutral to phenol red.

[5] C. L. Wadkins and A. L. Lehninger, *J. Biol. Chem.* **238**, 2555 (1963).

in oxidative phosphorylation. With most, but not all, preparations of submitochondrial particles, the relevant as well as the nonspecific exchange reactions requires added Mg^{++} which results in over a tenfold increase in the nonspecific background activity. Under these circumstances, even the presence of uncouplers does not help in detecting the relevant reaction. No satisfactory method has yet been described that determines unequivocally the contribution of other exchange enzymes to the incorporation of ^{14}C-ADP into ATP under these conditions.

Assay of the $^{32}P_i$-ATP Exchange

The incubation conditions and the estimation of the $^{32}P_i$ content of ATP are similar to those described by Conover *et al.*[6]

Each reaction mixture contains in a final volume of 1.0 ml: 10 m*M* Mg^{++}, 10 m*M* ATP, 20 m*M* P_i, pH 7.4, containing 40,000–80,000 cpm of recrystallized $^{32}P_i$[7] and 0.7–1.5 mg of mitochondrial or submitochondrial protein. The incubation is carried out in test tubes with gentle shaking for 10 minutes at 30°. The reaction is stopped by the addition of 0.1 ml of 35% perchloric acid, and an aliquot (0.1 ml) of the deproteinized reaction mixture is used for analysis of $^{32}P_i$-labeled ATP by the method described below. With small amounts of sample, i.e., containing less than 0.2 mg of protein, it is not necessary to deproteinize, and an aliquot of the reaction mixture may be added directly to 1.25 *N* perchloric acid (see below).

Under conditions where more than 20% of the ATP present is hydrolyzed by ATPase, it is important for reasons discussed later, to determine the specific radioactivity of the remaining ATP, rather than to rely on the total counts found in the aqueous fraction. In these circumstances, a separate 0.1-ml aliquot of the reaction mixture is immersed in a Dry Ice–acetone bath and subsequently heated at 100° for 2 minutes; it is used for determining the ATP remaining.[8]

Measurement of ^{32}P-ATP

Reagents

Perchloric acid, 1.25 *N*
Ammonium molybdate, 5%
Isobutanol–benzene 1:1 (v/v)
Isobutanol, saturated with water

[6] T. E. Conover, R. L. Prairie, and E. Racker, *J. Biol. Chem.* **238,** 2831 (1963).
[7] See this volume [4].
[8] ATP is measured at 340 mμ by the reduction of NADP in the presence of Mg^{++}, hexokinase, glucose, and glucose 6-P dehydrogenase [cf. A. Kornberg, *J. Biol. Chem.* **182,** 770 (1950)].

Procedure. To 16 × 150 mm tubes containing 4.0 ml of perchloric acid and 5 ml isobutanol-benzene is added 0.1–0.2 ml of the reaction mixture. Immediately 1.0 ml of ammonium molybdate is added, and the tube is shaken vigorously for 10 seconds.[9] After all the tubes have been shaken, the shaking procedure is repeated again for another 10 seconds. The two phases separate in about 1–2 minutes. The upper organic phase is removed by aspiration with a water pump, equipped with a trap. The lower aqueous phase is reextracted with 5.0 ml of isobutanol. After separation of the two phases, the upper phase is removed by aspiration and a second extraction with 5 ml isobutanol is carried out. The sides of the tube are washed and the last traces of isobutanol are removed by shaking with 1.0–2.0 ml ethyl ether, which is also removed by aspiration. The aqueous phase should be clear. The radioactivity of the sample is determined by counting 1.0 ml of the extracted aqueous phase.

The amount of $^{32}P_i$ incorporated into ATP is calculated by dividing the *total* radioactivity of the aqueous phase by the specific radioactivity of the inorganic phosphate of the original reaction mixture.

Comments. Though it is generally assumed that the $^{32}P_i$ recovered in the aqueous phase, under the described conditions, represents $AT^{32}P$, it should be remembered that all organic phosphate esters as well as inorganic pyrophosphate will be found in this fraction.

Substances which stimulate or inhibit the hydrolysis of ATP will affect the total counts found in the aqueous phase, by either removing or sparing the $^{32}P_i$ incorporated into ATP. Apparent effects on the exchange reaction may therefore be simply reflections of increased or decreased ATPase activity. This pitfall may be largely circumvented by measuring the ATP remaining at the end of the reaction and expressing the exchange reaction in terms of the specific radioactivity of the ATP.[10]

[9] A convenient and efficient mixing device is the Vortex Jr. mixer, manufactured by Scientific Industries, Inc., Queens Village, New York.

[10] M. E. Pullman and G. Monroy, *J. Biol. Chem.* **238**, 3762 (1963).

[10] The Application of ^{18}O Methods to Oxidative Phosphorylation

By P. D. Boyer and Donna M. Bryan

^{18}O techniques as used in studies of oxidative phosphorylation usually are directed toward measurement of the extent of the $P_i \rightleftharpoons HOH$ or $ATP \rightleftharpoons HOH$ exchange reactions, the extent of water oxygen incorporated into P_i accompanying ATP hydrolysis, or the source of oxygen in reaction products. The primary analytical procedure usually used is measurement of ^{18}O content of P_i isolated from the reaction mixture or from the hydrolysis of the appropriate phosphate compound. The procedure described herein involves isolation of the P_i as KH_2PO_4, and conversion of the phosphate oxygens to CO_2 for mass spectrometry by heating with guanidine hydrochloride.[1] The procedure is also applicable to measurement of the ^{18}O content of HOH.

Other useful procedures for determination of ^{18}O in P_i include unaided equilibrium with water[2,3] or equilibration with water hastened by a hot Pt wire,[4,5] electric discharge,[6] or sulfite.[5] Direct conversion of phosphate oxygen to gases suitable for mass spectrometry include heating $Ba_3(PO_4)_2$ with C to yield CO,[7] heating of KH_2PO_4 with $Hg(CN)_2$ to yield CO_2,[8] heating of Ag_3PO_4 to give O_2,[9] and treatment of $BiPO_4$ with BrF_3 to give O_2.[10]

A recent method for determination of ^{18}O in milligram quantities of HOH is based on oxidation with BrF_5 to release O_2, followed by conversion of O_2 to CO_2 by hot carbon.[11] The fluorine oxidation procedures give quantitative conversion of oxygen of P_i or HOH to O_2, which in turn may be converted quantitatively to CO_2. Such quantitative procedures are preferable when highest accuracy is necessary. In procedures where conversion is not quantitative, small isotope effects may be encountered. These are not important in many biochemical experiments where measurement of isotopic ratios to a precision of $\pm 1\%$ usually suffices.

Vacuum Train, Mass Spectrometer and Accessories

A vacuum train capable of readily evacuating 30–100 ml reaction vessels to less than 1 μ of pressure within several minutes is desirable. A schematic diagram is shown in Fig. 1. A suitable train may be assembled with a Welch Duo Seal forepump (Model 1402), connected through a 3–4 inch diameter trap immersed in Dry Ice-acetone, to a Consolidated Vacuum Corporation glass oil (Model GF-20) two-stage diffusion pump.

[1] P. D. Boyer, D. J. Graves, C. H. Suelter, and M. E. Dempsey, *Anal. Chem.* **33**, 1906 (1961).
[2] M. Cohn, *J. Biol. Chem.* **201**, 735 (1953).
[3] S. Epstein and T. K. Mayeda, *Geochim. Cosmochim. Acta* **4**, 213 (1953).
[4] I. Dostrovsky and F. S. Klein, *Anal. Chem.* **24**, 414 (1952).
[5] W. H. Harrison, P. D. Boyer, and A. B. Falcone, *J. Biol. Chem.* **215**, 303 (1955).
[6] A. B. Falcone, *Anal. Biochem.* **2**, 147 (1961).
[7] M. Cohn and G. R. Drysdale, *J. Biol. Chem.* **216**, 831, (1955).
[8] F. R. Williams and L. P. Hager, *Science* **128**, 1434 (1958).
[9] M. Anbar, M. Halmann, and B. Silver, *Anal. Chem.* **32**, 841 (1960).
[10] A. P. Tudge, *Geochem. Cosmochim. Acta* **18**, 81 (1960).
[11] J. R. O'Neil and S. Epstein, *J. Geophys. Res.* **71**, 4955 (1966).

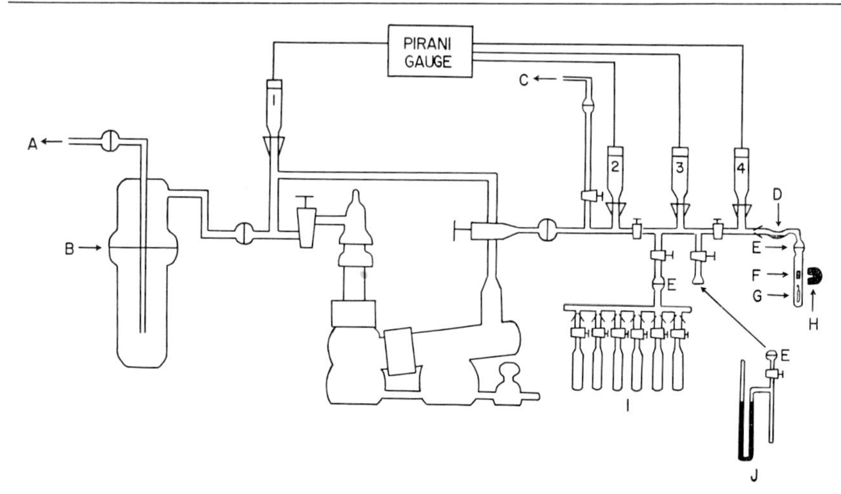

FIG. 1. Vacuum train for isotopic gas analyses: *A,* to forepump; *B,* trap with O-ring seal; *C,* to McLeod gauge; *D,* H_2SO_4 trap; *E,* O-ring seal ball joints; *F,* iron bar; *G,* break-seal tube; *H,* magnet; *I,* collection bulbs; *J,* Manometer.

A diffusion pump bypass line allows large volumes of air to be removed without passing through the hot pump vapors. Small volumes of air, e.g., up to about 50 ml, may be passed through the hot silicon oil vapors of the diffusion pump without turning the pump heater off. A four-stage Consolidated Pirani vacuum gauge (type GP-140) allows monitoring of gas at various points in the train. A McLeod gauge (Consolidated Type GM-100A) gives a convenient means of calibrating the vacuum gauges, but is not essential for most work. A calibrated manometer is useful to measure CO_2 yields.

For gas collection, tubes approximately 12 cm $\times$ 1.8 cm fitted with 2-mm hollow plug vacuum stopcocks and 10/30 inner joints are used. These may be conveniently immersed in liquid nitrogen for CO_2 collection. Prior to use for CO_2 collection, the bulb is thoroughly degassed by evacuation for several hours or preferably overnight.

A mass spectrometer capable of measuring the 46/44 ratio of 1 micromole of CO_2 to $\pm 1\%$ suffices for most work. The reader is referred to Vol. IV [20 and 37], for additional information on mass spectrometry and ^{18}O measurement.

Preparation of CO_2 Standards

It is essential to have several CO_2 standards containing ^{18}O in the range likely to be encountered in the experimental work. Although an absolute standard is desirable, all that is necessary is a means of accurately comparing the relative amounts of ^{18}O in the starting material and isolated product. The stated ^{18}O content of the water supplied may be taken as a convenient approximation for calculations.

Standards may be prepared by equilibrating water with CO_2 by shaking for several days[2]; addition of certain salts hastens the equilibration.[5] Use of bicarbonate as a source of CO_2 is convenient as it makes arrangement for addition of known amounts of gaseous CO_2 unnecessary. Use of Na_2CO_3 and KH_2PO_4 gives a convenient bicarbonate-phosphate source, and the phosphate provides buffer capacity allowing release of CO_2 with decreased conversion of HCO_3^- to CO_3^{2-} and loss of CO_2 yield.

Principle. A bicarbonate-phosphate solution in $H^{18}OH$ of known content is shaken in a bulb from which most of the air has been removed for equilibration of the oxygens of H_2O, CO_2, and HCO_3^-. Water and CO_2 are solidified by liquid N_2, air is removed, and the equilibrated CO_2 is released by warming to Dry-Ice-acetone temperature.

Reagents

- $H^{18}OH$; obtainable as 1.5 atom % excess or higher from Yeda Development Corp., Rehovoth, Israel
- Na_2CO_3, KH_2PO_4
- Dry Ice, acetone, and liquid N_2

Procedure. Dilutions of the $H^{18}OH$ are prepared with suitable volumetric pipettes to give water containing known relative amounts of excess ^{18}O, e.g., 1.5, 0.80, 0.40, 0.20, 0.10, and 0.05 atom % excess. A smaller or larger series of standards will suffice depending upon work contemplated. A small amount of Dry Ice may be used as a source of nonisotopic CO_2. If mass spectrometry of greater precision than $\pm 1\%$ is available, and if higher accuracy is desirable, gravimetric procedures should be used for the preparation of standards.

Convenient bulbs for equilibration are made by attaching a 100 ml round-bottom flask to an about 5 cm length of 8 mm O.D. tubing. In each dry flask is placed 2 millimoles (212 mg) of Na_2CO_3 and 5 millimoles (544 mg) of KH_2PO_4. Then 20 ml of the appropriate $H^{18}OH$ is added, and the flask is joined, by use of an oxygen flame, to about 5 cm of 8 mm O.D. tubing fitted with a precision-ground stopcock (greased with vacuum grease, and with a spring-clip to maintain tension) and an 18/9 inner joint for connection to the vacuum train. A total length of about 12 cm from stopcock to bulb allows convenient immersion in Dewar flasks. When standards are exhausted, the connecting tubing is scratched and snapped, and the bulbs and stopcocks are reused.

The bulbs are cooled in Dry Ice–acetone to freeze water, then in liquid N_2, and evacuated to less than roughly 1000 μ or so. The thawed bulbs are then shaken for 48–72 hours at 24–26°. They are then stored for withdrawal of CO_2. Standing without shaking for longer periods would probably suffice, but has not been tested.

To collect a portion of the standard, the bulb is exposed to Dry Ice–acetone until the water freezes, then to liquid N_2 to freeze the CO_2. Air is removed on the vacuum line. If much air is present, the sample is thawed and refrozen and air is again removed. The bulb is then warmed in Dry Ice–acetone, and the desired portion of the CO_2 released is collected in a regular sample tube by use of liquid N_2. The standard is reused until exhausted, with removal of any small amounts of air from leaking, degassing of grease, etc., as required by freezing in liquid N_2 and evacuation. If a large air leak develops, the standard is discarded.

The $NaHCO_3$-phosphate solution is saved for later collection of additional standard CO_2. Several collections are possible before the CO_2-HCO_2 reservoir of the solution is exhausted. By adding solid citric acid to the frozen solution and then thawing, a final yield of CO_2 will be obtained.

For calculation of the amount of ^{18}O in the CO_2, account must be taken of the equilibrium distribution of ^{18}O between water and CO_2. At 25°, the CO_2-H_2O fractionation factor, α, has been determined as equal 1.0407.[11] Thus $\alpha \times (^{18}O/^{16}O)_{HOH} = (^{18}O/^{16}O)_{CO_2}$, where $^{18}O/^{16}O$ is the ratio of the two isotopes in H_2O and CO_2, respectively. The ratio $^{18}O/^{16}O$ in HOH $= F/(1 - F)$, where F equals the atom fraction ^{18}O in HOH or $F = {}^{18}O/(^{18}O + {}^{16}O)$. For preparation of standards as outlined above, dilution of the total oxygen pool of water by oxygen from bicarbonate may be neglected, and thus F = initial total atom % $^{18}O/100$. For nonisotopic Lake Michigan water, $F = 0.00198$.[12] The atom % ^{18}O in the CO_2 equilibrated with the water standards is given by

$$\text{atom \% } {}^{18}O \text{ in } CO_2 = 100 \times \frac{(^{18}O/^{16}O)_{CO_2}}{1 + (^{18}O/^{16}O)_{CO_2}}$$

The expected mass 46/44 ratio equals $2 \times (^{18}O/^{16}O)_{CO_2}$ because each CO_2 molecule contains 2 oxygens.

Preparation of P_i-^{18}O

Principle. Water is exchanged with P_i by heating with KH_2PO_4, as described by Cohn and Drysdale.[7]

Procedure. A 2.8 M solution of KH_2PO_4 is placed in a sealed Pyrex tube. A higher ratio of P_i to HOH is advisable if expensive highly

[12] M. Dole, *Chem. Rev.* **51**, 263 (1952).

enriched water is used. The tube, placed inside an appropriate iron jacket (e.g., an ordinary 1-inch plumbing pipe 6-inch nipple with caps), is heated at 120° for about 8 days. In the authors' experience, exchange is not complete at this time, but longer heating gives only small additional ^{18}O incorporation.

If desired for higher ^{18}O samples, the water is recovered by lyophilization. The KH_2PO_4 is dissolved in a minimum quantity of hot 50% ethanol, any insoluble material (polyphosphates?) is removed by filtration while hot or by careful settling and decanting, and the KH_2PO_4 is crystallized by cooling. For highest purity, crystallization is repeated. The sample is dried under vacuum.

Measurement of ^{18}O in P_i or in HOH

Principle. The oxygens of KH_2PO_4 or of water are converted to CO_2 by heating with guanidine hydrochloride, ammonia and nitrogenous compounds are removed by exposure to H_2SO_4, and the CO_2 is collected for mass spectrometry.[1]

Reagents

Guanidine hydrochloride, recrystallized from water and dried under vacuum

H_2SO_4, Concentrated

Procedure. A sample of 2–10 micromoles of dry KH_2PO_4 and roughly 5 mg guanidine hydrochloride is placed in an approximately 0.8×12 cm glass tube which has been sealed at one end. A long narrow strip of folded weighing paper aids sample transfer. The tube is heated about 1½ inches above the bottom in an oxygen flame, and constricted to roughly about 0.5 mm or less by pulling with 2–3 inches elongation. The tube is connected to the vacuum train by rubber vacuum tubing lightly lubricated with vacuum grease. Air is removed to a pressure of less than 100 μ, a soft flame is applied to the constriction, and, as melting occurs, the tube is removed with a bending motion and heating to make a fragile break-seal. The tube while held in tweezers is heated in a soft flame (e.g., an air- and oxygen-free natural gas flame about 1 inch high) for about 0.5–1 minute, sufficient for all visible reaction to cease. Heating is usually commenced slightly above the solids in the tube. Alternatively, the samples may be heated for 6–12 hours in an oven at 260°–280°.

After removing excess carbon with a tissue, the warm tube is placed inside a heated outer tube (hot but not uncomfortable to touch) as indicated in Fig. 1. If the sample tube is allowed to cool, it is rewarmed before placing in the outer tube. A small volume, e.g., about 0.5–1 ml of

H_2SO_4 is previously placed in the indentation in the system, with aid of a long-tipped transfer dropper. One charge of H_2SO_4 may be used for up to 10–12 samples. Warming helps volatilize carbamates and shorten transfer time. With longer transfer times, unwarmed samples may be used. The system is evacuated to less than 50 μ, and the break-seal is broken with aid of the bar and the magnet. Reaction is usually visible at the surface of the H_2SO_4 as ammonia is adsorbed. The pressure gauge gives indication of the CO_2 yield. The outer tube is successively immersed in water, Dry Ice–acetone, and then liquid N_2. When CO_2 freezes as indicated by a pressure drop, any residual air and other gases are removed by evacuation to less than 5 μ. The CO_2 is volatilized by immersing the outside tube in Dry Ice–acetone. The gas is then collected in a previously evacuated sample tube for mass spectrometry.

The procedure for measurement of ^{18}O in water is similar to the above, but with the use of roughly 1–5 μl of water added by a small, thin capillary. For evacuation and making the break-seal, the sample is immersed in liquid N_2 briefly just as the evacuation is begun. The break-seal is made while the sample remains frozen. Water samples require more heating than phosphate samples, and the oven heating at 260° is convenient.

Measurement of $P_i \leftrightarrow$ HOH and ATP $\leftrightarrow$ HOH Exchanges

Procedures are described here for typical incubation with liver mitochondria, with isolation of the P_i and P_i from ATP for measurement of oxygen incorporation from $H^{18}OH$. If desired, the exchanges could also be measured by loss of ^{18}O from P_i and/or ATP. Use of highly labeled P_i or ATP would allow the use of smaller samples with appropriate addition of carrier P_i. A recent publication gives references to various contributions of ^{18}O methodology to oxidative phosphorylation.[13]

Principle. Mitochondria are incubated with P_i, ATP, substrate, and $H^{18}OH$ for oxygen exchange to occur. P_i is isolated and converted to KH_2PO_4. ATP is adsorbed on charcoal, and P_i is liberated by acid hydrolysis and converted to KH_2PO_4. Extent of exchange is calculated from measurement of the ^{18}O in the KH_2PO_4.

Reagents

Rat liver mitochondria in 0.25 M sucrose

P_i, ATP, $MgCl_2$, ATP, sucrose, and Tris solutions for convenient

[13] P. D. Boyer, *in* "Biological Oxidations" (T. P. Singer, ed.), in press. Wiley, New York, 1966.

preparation of reaction medium described below. For maximum addition of $H^{18}OH$, reaction mixtures without mitochondria may be lyophilized.

$(NH_4)_2MoO_4$, 0.08 *M*. Dissolve 60.7 g of $[(NH_4)_2]_6Mo_7O_{24} \cdot 4\ H_2O$ in 500 ml of 0.01 *N* H_2SO_4.

$(C_2H_5)_3N$, 0.8 *M*, pH 5. Dissolve 8.1 g of redistilled triethylamine in 30 ml H_2O. Adjust the pH with 6 *N* HCl first until pH 7–8 is reached, then with 0.1 *N* HCl to pH 5. The triethylamine will not all dissolve until the pH is almost 5. Make up to 100 ml with water.

H_2SO_4, 1 *N*

1:1 NH_4OH and 1:3 NH_4OH:HOH, v:v

Magnesia mixture. Dissolve 55 g of $MgCl_2 \cdot 6\ H_2O$ and 100 g NH_4Cl in 500–600 ml of water, add 100 ml of 15 *M* NH_4OH, and make the solution up to 1 liter. The final mixture is filtered if turbid.

Dowex 50-X8(H^+). Recycle twice, free from fines, and wash shortly before using.

0.1 *N* CO_2-low KOH. Prepare from solid KOH and freshly boiled water, and store in a plastic aspirator bottle fitted with a soda-lime tube.

Bromocresol green, 0.1%, aqueous, pH 3.8–5.4, yellow-blue

Acid-washed charcoal. Norit A charcoal is heated for 5 minutes with 1 *N* HCl in a boiling H_2O bath, filtered on a Büchner funnel, washed with water, and dried in an oven.

KCl, 0.1 *N*

Procedure. A 2-ml reaction mixture is prepared containing 5 m*M* succinate, 5 m*M* P_i, 5 m*M* ATP, 100 m*M* Tris, 50 m*M* sucrose, and rat liver mitochondria (equivalent to about 20 mg of mitochondrial protein) at a final pH of 7.4. Water of the incubation medium contains about 1 atom % excess ^{18}O. After addition of the mitochondria, the mixture is incubated for 5 minutes at 30°, then 2 ml of cold 0.6 *M* $HClO_4$ is added, and the sample is chilled. Protein is removed by centrifugation, and the supernatant is decanted. The $HClO_4$ extract is mixed with 50 mg of charcoal and filtered on a small Hirsch funnel. The filtrate and a small (about 0.5 ml) water wash are collected in a 12-ml thick-walled conical centrifuge tube. If ATP-^{18}O measurement is to be made, the charcoal is washed with about 3 ml of 0.1 *N* KCl, then with about 1 ml of water, and treated as described later.

For P_i isolation, to the filtrate is added 0.25 ml of the molybdate solution. (Samples 0.3 to 1 *N* in perchloric or sulfuric acid are satisfactory. For other experimental conditions, if more than 5 micromoles of P_i is present, 0.25 ml of molybdate is added for each 5 micromoles.) The sample is mixed and 0.05 ml of the triethylamine solution is added (or 0.05 ml for each 5 micromoles if more than 5 micromoles of P_i is present); the sample is again mixed and is allowed to stand near 0° for 1–2 hours. The precipitate is collected by centrifugation and washed by suspension and centrifugation with 5 ml of 1.0 *N* H_2SO_4. A fine glass rod serves to break up the pellets.

The well drained precipitate is dissolved in a minimum of 1:1 NH_4OH (about 1 ml for 5 micromoles), and made to a volume of about 3 ml. The P_i is precipitated by addition of 0.1 ml of magnesia mixture per about 4 micromoles of P_i present, with standing in a refrigerator overnight.

The precipitate is collected by centrifugation, the supernatant is carefully decanted, and the sample is drained well. Any adhering liquid near the mouth is removed with tissue. Care must be taken to not lose easily suspended precipitate particles. The sample is washed once by suspension and centrifugation with about 5 ml of cold 1:3 NH_4OH. If the samples are visibly contaminated, they may be reprecipitated as $MgNH_4PO_4$, or put through both the molybdate and magnesia precipitations again.

The well drained sample, with liquid removed by tissue, is dried in an oven at 100° for 10–15 minutes to remove NH_3. A Dowex-50 resin column about 1.5 cm high is prepared in an about 6 mm I.D. by 75 mm glass medicine dropper plugged with a small amount of glass wool. The column is conveniently suspended over a test tube through a rubber disk cut from a rubber stopper with a suitable center hole. The freshly prepared column is washed with 1–2 ml of water. The dried sample is dissolved with a minimum amount of freshly washed resin and 0.2–0.3 ml of water. After the column has been suspended above a 5-ml beaker, a small excess of resin is added to the sample, and the slurry is transferred to the column by a long-tipped dropper fitted with a rubber bulb. Most of the residual resin and sample are transferred to the column with several rinses of about 0.3 ml of water. The eluate and washings are placed in an oven at 105–110° or a vacuum oven at about 60° and concentrated to near-dryness.

About 0.2 ml of water and 10 μl of 0.1% bromocresol green solution are added. The sample is titrated to pH of about 4.1 with the KOH solution while observing the inclined beaker over a white, lighted surface. A 0.1 ml long-tipped pipette graduated in 0.01 ml is used, and comparison is made with an appropriate standard at pH 4.1 with bromocresol green added. Slight overtitration is allowable, and addition of HCl to compensate for extra KOH is without effect as small amounts of KCl do not interfere in the analyses. The amount of KOH used should correspond approximately to the amount of P_i expected (0.01 ml of 0.1 N KOH is equivalent to 1 micromole of P_i).

The sample is transferred to a numbered break-seal tube with the aid of a long-tipped dropper, and the beaker is rinsed with about 0.1 ml of water. The sample is evaporated in a vacuum oven at 60°. (Evaporation at 60° gives less than 1% exchange of the oxygens. Evaporation at 100–110° without vacuum gives 5–10% exchange of KH_2PO_4 with water.) Alternatively, the samples may be lyophilized. Several samples may be conveniently lyophilized in a larger container. When samples are visibly dry, about 5 mg of powdered guanidine hydrochloride is added to each. Samples are then dried on the high-vacuum line until the pressure gauge indicates no more gas (water) is coming off when the stopcock leading to the diffusion pump is closed. A heating time of 30–60 minutes is usually more than sufficient. The tubes are sealed immediately after drying or stored in a desiccator. Analyses are performed as described above for determination of ^{18}O in KH_2PO_4.

When ^{18}O content of ATP is to be determined, the washed charcoal through which the perchloric acid extract was passed is heated for 30 minutes with 2 ml of 2 N H_2SO_4 in a boiling water bath. Less than 1% exchange of P_i with HOH occurs under these conditions. The sample is cooled and filtered through a small Hirsch funnel, fitted with moistened retentive filter paper, directly into a 12-ml centrifuge tube. The test tube used for heating and the charcoal are washed about 3 times with 1 ml of water. P_i is isolated from the filtrate by the molybdate and magnesia procedures and prepared for analysis as described above.

Comments. With samples of 3–5 micromoles, contamination by other sources of oxygen has usually been found to be negligible. The heating of such large samples with guanidine HCl in the presence of air, not evacuated as described above, has given little error. With samples of 0.2–1.0 micromole, small amounts of nonisotopic oxygen (present as water, as impurities, or from slight exchange with glass or air) may reduce observed 46/44 ratios slightly. Useful checks for presence of impurities and validation of analyses can be made by isolation of nonisotopic P_i and of P_i of known ^{18}O content from simulated reaction mixtures.

Procedures other than drying for final isolation of the KH_2PO_4 may be used. For highest accuracy and with large amounts of P_i, crystallization may be advisable. Alternatively, KH_2PO_4 may be precipitated from solution with ethanol[2] or with dioxane.[5] This may have the advantage of removing traces of organic impurities, but losses are encountered with small samples.

Calculations. The atom fraction of ^{18}O in the CO_2, F, may be calculated from the observed 46/44 ratio, R, by the relation

$$F_{CO_2} = \frac{R}{2 + R}$$

The atom % excess is given by the relation

$$(F_{\text{sample}} - F_{\text{nonisotopic control}}) \times 100 = \text{atom \% excess } ^{18}\text{O in sample}$$

The nonisotopic value should be determined with the P_i used in the experiment and should be close to 0.20 atom %. In preference to calculation from the 46/44 ratio, values of atom % excess ^{18}O may be conveniently read from an empirical graph of plots of atom % excess in CO_2 standards against observed mass spectrometer readings. This procedure, with standards read at the same time as samples, obviates variations due to changes in mass spectrometer behavior.

The fraction exchange in the P_i is given by the relation

$$F_{ex} = \frac{\text{atom \% excess } ^{18}\text{O in P}_i}{\text{atom \% excess } ^{18}\text{O in HOH}}$$

For the ATP, if only the 3 oxygens of the terminal phosphoryl are involved (usually adenylate kinase obviates this simple assumption) the fraction exchange of these 3 oxygens is equal to

$$\frac{8/3 \times \text{atom \% excess } ^{18}\text{O in P}_i \text{ from hydrolysis of ATP}}{\text{atom \% excess in HOH}}$$

The total millimolarity of water exchanging with P_i, $P_i \rightleftharpoons HOH$, if the fraction exchange is low, is given by $F_{ex} \times 4 \times mM\ P_i$. For the ATP, the millimolarity for the ATP $\rightleftharpoons$ HOH, if only the terminal phosphoryl is considered, is $F_{ex} \times 3 \times mM$ ATP. If the fraction exchange is large, correction must be made for replacement of ^{18}O already in P_i by ^{18}O of HOH, by the general relation

$$\text{Amount of reaction} = \frac{-(A)(B)}{(A) + (B)} \ln(1 - F_{ex})$$

where A and B are the pool sizes of exchanging reactants. With HOH in great excess to P_i, the relation reduces to

$$mM \text{ of HOH} \leftrightarrow P_i \text{ exchange} = 4 \times mM \text{ of } P_i \times \ln \frac{1}{1 - F}$$

Expression of results as millimolarity or other units of P_i exchanged is not recommended because of confusion as to whether exchange of 1 or 4 oxygens of the P_i molecule is meant. The exchanges may be expressed in terms of micromoles or microatoms. The millimolarity of the exchange is equal to the micromoles of HOH exchanging with P_i per ml of reaction medium, or the microatoms of oxygen of P_i exchanging per ml.

The above calculations assume the ATP and P_i pool sizes remain constant during incubation. If relatively small changes in pool size occur, the exchanges can be approximated using the average pool size.

Estimation of the oxygen incorporation by the above procedures does not give information as to whether the exchange was directly with water, or whether other intermediates or routes are involved.

Acknowledgment

Supported in part by contract AT(11-1)-34, Project No. 102, of the U.S. Atomic Energy Commission.

Section II

Preparations, Properties, and Conditions for Assay of Mitochondria

PREVIOUSLY PUBLISHED ARTICLES FROM METHODS IN ENZYMOLOGY RELATED TO SECTION II

Vol. I [4]. Preparation of Mitochondria from Plants. Bernard Axelrod.
Vol. VI [35]. Oxidative Phosphorylation Systems: Microbial. Arnold F. Brodie.

[11] The Preparation of Heart Mitochondria from Laboratory Animals

By D. D. TYLER[1] and JEANINE GONZE

Methods for the isolation of heart mitochondria from the rat,[1a-4] cat,[1a] guinea pig,[4] rabbit,[5] and pigeon[6] have been described, and one method[1a] has been applied to the heart muscle of several laboratory animals.[7] In the recommended procedure to be described, which is a modification of the method of Chance and Hagihara,[6] a gentle, complete homogenization of the tough muscle tissue is obtained after brief digestion with bacterial proteinase.

Method

Reagents

MSE medium (0.225 *M* mannitol, 0.075 *M* sucrose, and 0.05 m*M* EDTA, pH 7.4)
Unneutralized Tris buffer, 1 *M*
Crystalline bacterial proteinase (Nagarse)

Procedure. PREPARATION OF THE HOMOGENATE. All containers are surrounded with ice and water, and the centrifuge is set to run at 1°. Two rat hearts (or one pigeon heart) are washed in cold MSE medium and finely chopped with a pair of scissors. Rapid chilling and washing of the freshly removed hearts is essential to obtain a stable preparation of mitochondria with low haemoglobin content. Between four and six washes may be required to remove adhering blood, about 40 ml of MSE being used for each wash. The last volume of washing fluid, which should be clear and colorless, is decanted as completely as possible from the chopped tissue. A solution of 10 mg of proteinase in 5 ml of MSE and 0.05 ml of Tris buffer is added to the tissue, and the mixture is transferred to

[1] This manuscript was prepared while the author was at the Johnson Research Foundation, University of Pennsylvania, Philadelphia, Pennsylvania.
[1a] K. W. Cleland and E. C. Slater, *Biochem. J.* **53**, 547 (1953).
[2] F. A. Holton, W. C. Hulsmann, D. K. Myers, and E. C. Slater, *Biochem. J.* **67**, 579 (1957).
[3] C. M. Montgomery and J. L. Webb, *J. Biol. Chem.* **221**, 347 (1956).
[4] G. F. Maley and G. W. E. Plaut, *J. Biol. Chem.* **205**, 297 (1953).
[5] J. W. Harman and M. Feigelson, *Exptl. Cell Res.* **3**, 47, 58 (1952).
[6] B. Chance and B. Hagihara, *Proc. 5th Intern. Congr. Biochem., Moscow, 1961* Vol. 5, p. 3. Pergamon Press, London.
[7] L. Packer, *Exptl. Cell Res.* **15**, 551 (1958).

a glass homogenizer fitted with a Teflon pestle. Three passes of the pestle are made within 30 seconds, and the homogenate is diluted to about 40 ml with more fresh MSE. Homogenization of the diluted preparation is then continued until a smooth uniform homogenate is obtained (usually 30–60 seconds). The tissue must not be exposed to the action of the concentrated proteinase solution for more than 30 seconds, since prolonged exposure results in a low yield of mitochondria of inferior quality. The pH of the homogenate is measured (pH indicator paper is convenient) and adjusted if necessary to pH 7.2.

CENTRIFUGATION OF THE HOMOGENATE. The homogenate is diluted to 80 ml with MSE, divided equally into two centrifuge tubes, and centrifuged at 8000 *g* for 10 minutes. The supernatant, which contains the proteinase, is decanted and discarded, and the whole pellet is rehomogenized and resuspended to 80 ml with MSE. After a second centrifugation at 700 *g* for 10 minutes, the supernatant is divided into two clean centrifuge tubes; the pellet, which contains nuclei, red cells, and debris, is discarded. A third centrifugation at 8000 *g* for 10 minutes yields a brown pellet (mitochondrial fraction), covered by a loosely packed fluffy layer, and a supernatant which is discarded. The tubes are shaken gently with a small volume of MSE to remove the fluffy layer as completely as possible from the tightly packed brown pellet. The mitochondrial fraction is gently broken up with a glass rod (avoiding the resuspension of the spot of red cells at the bottom of the tubes), rehomogenized, and diluted to 40 ml. The mitochondria are washed twice (fourth and fifth centrifugation) by repeating the procedure used during and after the third centrifugation.

The pellet obtained after the fifth centrifugation is resuspended in fresh MSE with a small glass homogenizer, to yield 1.5 ml of a uniform suspension of mitochondria containing about 30 mg biuret protein/ml.

Properties

Many properties of pigeon heart mitochondria are described in a single paper.[6] Typical values obtained from the polarographic assay of oxidative phosphorylation in rat and pigeon heart mitochondria are presented in Table I. Rat heart mitochondria prepared by the proteinase method are superior to those isolated from conventional or quartz sand homogenates when judged by the following criteria: Proteinase preparations show (1) higher P:O ratios and respiratory control ratios, especially with succinate as substrate; (2) greater stability of respiratory control and oxidase activities during storage at 1°; (3) low magnesium-activated ATPase activity compared to that induced by dinitrophenol, with both activities powerfully inhibited by oligomycin (Table II). Mitochondria isolated according to an earlier preparative method[1a] contained nearly equal magnesium and dinitrophenol activated ATPase activities,[2] with the magnesium activated ATPase inhibited only 50% by oligomycin.[8]

TABLE I
OXIDATIVE PHOSPHORYLATION IN HEART MITOCHONDRIA[a]

	Rat			Pigeon		
Substrate	O_2 uptake	RCR	P:O	O_2 uptake	RCR	P:O
Glutamate + malate	5.5	6.2	2.8	4.9	6.5	2.9
β-Hydroxybutyrate	4.4	5.0	2.6	0.3	—	—
Pyruvate + malate	7.3	5.3	2.6	4.9	4.2	2.7
Ascorbate + 50 μ*M* TMPD	3.7	1.6	1.1	5.5	1.6	0.9
Palmityl carnitine	7.3	6.8	2.5	5.0	6.5	2.4
Succinate + rotenone	7.6	3.2	1.8	6.3	2.8	1.7
DPNH	1.0	1.0	—	1.2	1.0	—

[a] Measurements were made polarographically in a 3-ml reaction mixture containing mannitol (0.225 *M*), sucrose (0.070 *M*), EDTA (1 m*M*), K phosphate buffer (0.010 *M*, pH 7.2) and 1.5–3.0 mg protein. Sucrose was replaced by KCl (0.035 *M*) when ascorbate-TMPD was used as substrate. Substrates were added at 10 m*M* except for palmityl carnitine (20 μ*M*). Rates of oxygen uptake, expressed as micromoles oxygen consumed per milligram of protein per hour, were measured after the addition of ADP (0.1–0.4 m*M*). Average values observed with 3 rat and 3 pigeon preparations are given. RCR, respiratory control ratio. Temperature, 22°.

TABLE II
COMPARISON OF THE ATPASE ACTIVITIES OF RAT HEART AND PIGEON HEART MITOCHONDRIA AT pH 7.4[a]

DNP (m*M*)	$MgCl_2$ (m*M*)	Oligomycin (μg/ml)	ATPase activity (micromoles P/mg protein/hour) Rat heart	Pigeon heart
0	0	0	0.6	0.4
0.1	0	0	51.8	37.7
0	3	0	11.6	6.7
0.1	3	0	38.4	27.5
0.1	0	4	0.3	0.2
0	3	4	0.3	0.2

[a] The hydrolysis of ATP was assayed by the method of Holton *et al.* [F. A. Holton, W. C. Hulsmann, D. K. Myers, and E. C. Slater, *Biochem. J.* **67,** 579 (1957)], using a 5-minute incubation period. Average values observed with 3 rat and 3 pigeon preparations are given.

[8] D. D. Tyler, unpublished observations (1963).

[12] The Large-Scale Preparation and Properties of Heart Mitochondria from Slaughterhouse Material

By Paul V. Blair

Principle

Methods for the preparation of mitochondria from the organs of laboratory animals, based upon homogenization and differential centrifugation,[1] have been modified to meet the needs of the investigator who is primarily interested in the isolation of enzymes and multienzymatic complexes originating in mitochondria. Fresh tissues (e.g., beef heart) are obtained from the slaughterhouse and are sliced immediately into several chunks before they are packed in ice.

General Procedure for the Preparation of Beef Heart Mitochondria

Fat, connective tissue, and blood vessels are trimmed from fresh chilled beef hearts before the meat is diced and passed through a precooled electric meat grinder. To 400 g of ground heart is added 1200 ml of a solution which is 0.25 *M* in sucrose and 0.01 *M* in phosphate. The pH is continuously maintained between 7.2 and 7.4 with KOH as the meat is homogenized in a high speed blender of large capacity.[2,3] The blender is operated for 30–45 seconds at full speed. The homogenate (8 batches of blendate, representing 9–10 beef hearts) is immediately centrifuged for 10 minutes at 1900 rpm (1600 *g*) in a refrigerated International Serum Centrifuge (Model 13L) designed to handle 13 liters of fluid. Other types of centrifuges may be used provided the same centrifugal force is generated.[4] The cherry red supernatant suspension containing the mitochondria is decanted through a double layer of cheesecloth (the sediment and intermediate fluffy layer are discarded). Further purification will depend in part upon the intended experimental use of the isolated mitochondria, but sucrose solutions are usually the medium of choice for diluting the suspension. However, some procedures and isolation tech-

[1] See small-scale preparation of mitochondria from laboratory animals, see this volume [11].

[2] A 0.5 horsepower motor is fitted with a shaft 25 cm long and a 3-pointed blade which rotates at 18,000 rpm without load (Handrow blender; rheostat setting at 120).

[3] An Ultra-Turrax TP 18/2 (Janke and Kunkel KG., Staufer in Brisgare, West Germany), which is a stainless steel tube slit radially with a head having two rotating knives generating a speed of 24,000 rpm.

[4] A. Löw and I. Vallin, *Biochim. Biophys. Acta* **69**, 361 (1963).

niques are more definitive if the suspensions of mitochondria and other subcellular fragments are diluted with isotonic KCI.

Mitochondria Isolated in Sucrose.[5] Two liters of 0.25 *M* sucrose solution are added to 8 liters of the supernatant fluid (the yield from 13 liters of original suspension) described above. This suspension is passed through a refrigerated Sharples supercentrifuge (Model T-1P) at 50,000 rpm (62,000 *g*) with a nozzle (3/32 inch) allowing the passage of 10 liters of fluid in 7 minutes. The mitochondria (110–125 ml), which have been sedimented in the bowl of the centrifuge, are scraped out in the cold room (0°). These mitochondria are blended with 400 ml of 0.25 *M* sucrose for 10–15 seconds at low speed, or until they form a smooth paste. The paste from one bowl is brought to a final volume of 2 liters with 0.25 *M* sucrose. The suspension, consisting mainly of mitochondria, is again passed through the Sharples centrifuge at a rate of 10 liters in 20 minutes. The resulting mitochondria are generally satisfactory for use in fractionation procedures; but for the study of energy-linked functions of mitochondria it is desirable to split them into heavy beef heart mitochondria (HBHM) and light beef heart mitochondria (LBHM).[6]

Separation of Mitochondria into Heavy and Light Fractions. The residue from the Sharples centrifuge bowl is suspended in 8 volumes of 0.25 *M* sucrose which is 0.01 *M* with respect to Tris, and the mixture is adjusted to pH 7.8 with HCl. The suspension is centrifuged for 10 minutes at 17,000 rpm in the No. 30 rotor of the Spinco preparative ultracentrifuge. The sediment is usually composed of two distinct layers (more are visible on occasion) with a relatively dark area at the bottom of the pellet. The faster sedimenting layer of mitochondria is designated "heavy" particles and the slower sedimenting layer is designated "light" particles. The light beef heart mitochondria LBHM are sloughed off and separated from the heavy fraction by means of a stirring rod. The heavy beef heart mitochondria HBHM are resuspended in the sucrose-Tris solution as given in the above procedure; after gentle homogenization, the light-heavy separation is again effected. The procedure is repeated until LBHM are no longer detectable in the heavy fraction. Usually, the third separation yields a heavy fraction free of light particles. Essentially the same procedure may be followed to obtain a light fraction devoid of HBHM. Almost equal amounts of HBHM and LBHM are isolated from the mitochondrial paste. The differential sedimentation properties of light and heavy particles are accentuated by proper adjustment of the pH of the medium, but very little by the nature of the buffer. At low pH values,

[5] F. L. Crane, J. L. Glenn, and D. E. Green, *Biochim. Biophys. Acta* **22**, 475 (1956).
[6] Y. Hatefi and R. L. Lester, *Biochim. Biophys. Acta* **27**, 83 (1958).

the separation of particle types is poor; high pH values effect good separation but damage the enzymatic activity of both types of mitochondria.

Mitochondria Prepared in KCl. Eight liters of 0.12 *M* KCl are added to every 2 liters of original supernatant fluid from the first centrifugation in the serum centrifuge (Model 13L). After it is mixed the suspension is centrifuged in the Sharples at a rate of 10 liters in 7 minutes. The residue in the bowl of the centrifuge is removed as described above and is suspended in 0.12 *M* KCl to the desired dilution. The separation into light and heavy mitochondria can also be carried out in this medium provided the pH is appropriately adjusted to 7.8 and an extended centrifugation time is used (30,000 rpm, No. 30 rotor for 20 minutes), but the energy-linked functions are impaired, perhaps owing to loss of cytochrome *c*. All the above procedures are carried out at approximately 0°.

Assay and Properties

In general the methods used for assaying mitochondrial preparations involve the measurement of respiratory phosphorylation. Krebs cycle acids are used as primary electron donors, and oxygen uptake is measured either in the Warburg apparatus[7] or by the polarographic method.[8] In the Warburg manometer oxygen consumption is measured at 30°; during the initial 5–10 minute period of thermal equilibration, the oxygen uptake is assumed to be linear and to be equivalent to that measured in the following experimental interval. However, this commonly used assumption is not valid, and appropriate corrections are required, especially if short incubation periods are used. Three to 8 mg of mitochondrial protein is added to a Warburg flask containing in a volume of 3 ml, 20–80 micromoles of phosphate (pH 7.2), 5 micromoles of ATP, 0.1 mg of hexokinase, 5 micromoles of $MgCl_2$, 100 micromoles of glucose, 750 micromoles of sucrose, 20 micromoles of pyruvate, and 20 micromoles of malate. Other substrates such as those oxidized by DPN-linked reactions may be substituted for pyruvate plus malate with an expected phosphorylation efficiency (P:O) of 3 or greater. The oxidation of succinate may also be used with a phosphorylative efficiency of 2 or greater. Phosphate uptake is measured by comparing parallel reactions which are conducted for 0 and 30 minutes respectively.[7] Rates of oxidation are expressed as microatoms of oxygen consumed per milligram of protein per minute. Rates of phosphorylation are expressed as micromoles of phosphate esterified

[7] For discussion, see this volume [3].
[8] For discussion, see this volume [7].

per milligram of protein per minute. The efficiency of phosphorylation is expressed as the P:O ratio. The oxidation rate of HBHM with the mixture of pyruvate plus malate as substrate is about 0.20, and with succinate about 0.10 microatoms of oxygen per milligram of protein per minute with P:O ratios near 3 and 2, respectively. HBHM may be stored at —20° for several months without appreciable reduction in the P:O ratio generated in the presence of pyruvate plus malate, or similar DPN-linked oxidative systems. The succinic oxidase system shows an increase in oxidation rate, and a concomitant decrease in coupled phosphorylation, which leads to a definite loss in phosphorylation efficiency. The oxidation rate of LBHM is about 0.1 with pyruvate plus malate, and 0.2 with succinate. The P:O ratios are approximately 2.7 and 0.4, respectively.

The respiratory control (ratio of oxygen uptake in the presence of ADP plus phosphate to that before introducing ADP and phosphate) of HBHM is usually greater than 5 for pyruvate plus malate. These mitochondria appear to be "healthy" when examined under the electron microscope. HBHM are capable of carrying out the other energy-linked reactions such as transhydrogenation (upon preparation of submitochondrial particles from them), ion accumulation, swelling-shrinking, and "reversal of electron flow" which are normally associated with intact mitochondria. Heart mitochondria are more stable than those isolated from soft tissues (liver, kidney, and brain) and as a consequence have been used extensively for the preparation of submitochondrial particles.

[13] Preparation, Properties, and Conditions for Assay of Mitochondria: Slaughterhouse Material, Small-Scale

By Archie L. Smith

Heart muscle has many advantages over other mammalian tissues as a source of mitochondria. The mitochondria derived from this tissue are stable with respect to oxidation and phosphorylation when stored at 0° for up to a week, and for up to a year at —15° to —20°. Although the technical difficulties of liberating mitochondria from heart muscle presents some problems, these are offset by the stability and purity of the mitochondria that are obtained in this manner. Slaughterhouse material provides an abundance of tissue for the isolation of intact mitochondria. Beef heart is the tissue of choice, but this procedure can be applied to pig heart as well. Because of the fibrous nature of heart muscle, homogenization of the tissue is the major problem in the isolation of intact mitochondria. Three procedures will be described for the isolation of mitochondria from slaughterhouse material. These procedures are modifications of methods described previously.[1-4]

Special Equipment

A glass-Teflon homogenizer is used to comminute the heart mince. The Teflon pestle (A. H. Thomas Co., size C) was reduced in size so that it would be loose fitting in the glass homogenizer vessel. The pestle was turned down on a lathe so that the diameter of the pestle was reduced from 0.993 inch to 0.983 inch, to provide a loose fit in the glass homogenizer vessel. The clearance between the pestle and the vessel was 0.016 inch.

A heavy duty, 3/8 horsepower, industrial drill (Sears, Roebuck), which is mounted on a drill stand, is used to turn the pestle during homogenization. The revolutions per minute of the drill can be regulated by a variable power transformer (Powerstat).

Preparation of Mitochondria

The three procedures for the isolation of beef heart mitochondria differ only in the manner in which the heart mince is homogenized. The fractionation of the mitochondria and washing steps are the same in the three procedures.

Procedure 1. One or two beef hearts are obtained from the slaughterhouse within 1–2 hours after the animal is slaughtered. The hearts are placed in cracked ice to ensure cooling of the tissue for transport to the laboratory. All subsequent procedures are carried out at 2–4°. Fat and connective tissue are trimmed from the heart, and the tissue is cut into approximately 5-cm cubes. Three hundred grams of these cubes are passed through a meat grinder which is maintained at 4°. The plate holes of the meat grinder are 4–5 mm. The resulting mince is placed in 400 ml of 0.25 M sucrose (Mallinckrodt) and 0.01 M Tris-Cl, pH 7.8. Since this mixture may be at pH levels (5.5 or less) that are injurious to the mitochondria, it is necessary to adjust the pH to 7.5 ± 0.1 as rapidly as possible. This is accomplished by the addition of 6 M KOH or 2 M Tris (pH 10.8, unneutralized). The neutralized ground heart mince is placed in a double layer of cheesecloth and is squeezed free of the sucrose solution.

Two hundred grams of this ground, neutralized heart tissue is sus-

[1] F. L. Crane, J. L. Glenn, and D. E. Green, *Biochim. Biophys. Acta* **22,** 475 (1956).
[2] Y. Hatefi and R. L. Lester, *Biochim. Biophys. Acta* **27,** 83 (1958).
[3] Y. Hatefi, P. Jurtshuk, and A. G. Haavik, *Arch. Biochem. Biophys.* **94,** 148 (1961).
[4] D. E. Green and D. M. Ziegler, Vol. VI [58].

pended in 400 ml of 0.25 *M* sucrose, 0.01 *M* Tris-Cl pH 7.8, 1 m*M* Tris-succinate, and 0.2 m*M* EDTA (this solution is used throughout the isolation procedure and is referred to as the sucrose solution). Fifty milliliters of the suspension is placed in a glass homogenizing vessel. The loose-fitting pestle driven by the heavy-duty drill at 1400 rpm is inserted into the vessel for one pass of 10 seconds and two subsequent passes of 5 seconds each. The remainder of the suspension is similarly homogenized. The final homogenate is readjusted to pH 7.8 with the addition of 1 *M* KOH or 2 *M* unneutralized Tris.

The homogenate is centrifuged for 20 minutes at 1200 *g* (International Centrifuge PR-2, rotor 284 at a setting of 2000 rpm or Lourdes centrifuge, rotor 3RA at a setting of 2400 rpm) to remove unruptured muscle tissue and nuclei. The supernatant solution is carefully decanted so that the loosely packed fluffy layer is not disturbed. The supernatant solution is filtered through two layers of cheesecloth to remove the lipid granules. The pH of the suspension is readjusted to 7.8 with 1 *M* KOH or 2.0 *M* unneutralized Tris, and the suspension is centrifuged for 15 minutes in the 30 rotor of the Spinco model L centrifuge at 26,000 *g* (17,000 rpm). The resulting pellet usually consists of three distinct layers: (1) a light, loosely packed, buff-colored layer (light beef heart mitochondria, LBHM[2]), (2) a dark brown layer (heavy beef heart mitochondria, HBHM[2]), and (3) a tiny brown-black button at the bottom of the tube. The top, buff-colored layer consists of damaged mitochondria and is discarded. This is achieved by decanting about 25 ml of the supernatant solution and gently shaking the remaining sucrose solution in the centrifuge tube. This procedure dislodges the loosely packed damaged mitochondria, and the mixture is decanted and discarded. A portion of the LBHM adheres to the wall of the centrifuge tube and may be removed with the aid of a glass stirring rod. The dark brown layer of heavy beef heart mitochondria is then dislodged by means of a glass stirring rod, mixed with 10 ml of the sucrose solution, and decanted, leaving behind the brown-black pellet. This mitochondrial suspension is homogenized in a tight-fitting glass–Teflon homogenizer (clearance of 0.006 inch) with two passes, each of 5 seconds, at 1400 rpm. The volume of the homogenate is adjusted to 180 ml with the sucrose solution, and the pH of the suspension is adjusted to 7.8. The suspension is centrifuged as above at 26,000 *g* for 15 minutes. The resulting pellet now consists of relatively small top and bottom layers, and only the middle dark brown layer is collected, as described. The mitochondrial suspension is homogenized, and the volume is adjusted to 60 ml with the sucrose solution. The pH of the suspension is adjusted to 7.8, and the suspension is centrifuged at 26,000 *g* for 15 minutes. Again the dark brown middle layer is collected, suspended in a small volume of the sucrose solution, and homogenized; the protein concentration is adjusted by dilution with the sucrose solution to 20–40 mg protein per milliliter. The protein determination is done by the biuret procedure.[5] The average yield of intact HBHM is approximately 1 mg protein per gram of starting mince.

Procedure 2. This method of preparation involves the use of a proteolytic enzyme to aid in the homogenization of the heart mince. This method is similar to the procedures described elsewhere[3,6] with some modifications. After 200 g of the ground heart mince is washed with 0.25 *M* sucrose, as previously described, it is suspended in 400 ml sucrose solution containing 50–55 mg of the proteolytic enzyme Nagarse (Teikoku Chemical Industry Co., Ltd., 7, Itachibori, 1-Chome Nishi-Ku, Osaka, Japan; U.S. Distributors: Biddle Sawer Corp., 20 Vesey Street, New York, New York). The suspension is stirred slowly for 20 minutes; the pH is maintained between 7.2 and 7.4 at the temperature of 0–2°. After the proteolytic digestion, the suspension is homogenized with the loose-fitting homogenizer and the mitochondria are isolated as described in procedure 1. This procedure has the advantage of producing a somewhat higher yield of mitochondria than the first procedure. Mitochondria prepared by this procedure manifest higher respiratory control ratios.[6]

Procedure 3. This procedure leads to the formation of a large proportion of damaged mitochondria (LBHM) but has the advantage that large amounts of material can be worked up at one time. This method uses a Waring blendor to homogenize the heart mince. Two hundred grams of the ground heart mince is placed in 400 ml of the sucrose solution in the Waring blendor container, and 1 ml of 6 *M* KOH or 3 ml of 2 *M* unneutralized Tris is added to the mixture. The blendor is operated at high speed for 15 seconds, 1 ml KOH or 3 ml of Tris is then added, and the homogenization is continued for another 5 seconds. The pH of the homogenate is now adjusted to 7.8 with either KOH or Tris. From this point on the isolation of the mitochondria is the same as described in procedure 1.

Assay Procedure

The determination of P:O ratios is based on procedures described previously.[2,7] Oxygen consumption by the mitochondria is followed by

[5] A. G. Gornall, C. J. Bardawill, and M. M. David, *J. Biol. Chem.* **177,** 751 (1949).

[6] B. Chance and B. Hagihara, *Proc. 5th Intern. Congr. Biochem., Moscow, 1961* Vol. 5, p. 3. Pergamon, London, 1963.

[7] A. L. Smith and M. Hansen, *Biochem. Biophys. Res. Commun.* **15,** 431 (1964).

conventional manometric techniques.[8] Phosphorylation is followed by measuring inorganic phosphate concentration before and after the incubation period; the phosphate concentration was measured by the procedure of Martin and Doty[9] as described by Lindberg and Ernster.[10] The reaction medium used for phosphorylation assays with HBHM contains in 3.0 ml: 750 micromoles of sucrose, 5–15 micromoles of both sodium pyruvate and Tris-malate or 25 micromoles of Tris-succinate, 100 micromoles of glucose, 15 micromoles of $MgCl_2$, 35 micromoles of potassium phosphate pH 7.4, 0.3 micromoles of EDTA, 15 mg of crystalline bovine serum albumin, 1–4 mg of HBHM. The reaction mixture is placed in the main compartment of the Warburg vessel; in the side arm of the vessel, 5 micromoles of ADP (Sigma) and 50–100 μg of crystalline hexokinase (Worthington, 430 units/mg). After a 5-minute thermal equilibration period, the phosphorylation reaction is initiated by the addition of ADP and hexokinase from the side arm of the vessel; the reaction times range from 20 to 30 minutes, and the reaction is terminated by the addition of 1.0 ml of 1.5 *M* perchloric acid. All reagents are pipetted into the Warburg vessel at room temperature and the reaction is run at 30.0°.

Comments

The mitochondria prepared by the above procedures exhibit respiratory control for up to 5 days when they are maintained on ice. The respiratory control ratios as measured manometrically are 8–15 with pyruvate as substrate whereas the control ratios are 3–5 when measured polarographically. The P:O ratios approach 3 and higher for pyruvate plus malate, with oxidation rates of 0.2–0.3 μatoms O_2 per minute per milligram protein. The P:O ratios with succinate as substrate are well above 2, and the oxidation rates are in the range of 0.05–0.07 μatoms O_2 per minute per milligram of protein.

The quality of the distilled water used for the isolation media and for the assay reagents must be of the highest quality. Mitochondria prepared and assayed with glass-distilled (1X) water results in P:O ratios well below 3 for pyruvate and below 2 for succinate. Mitochondria that are prepared and assayed with distilled water that had been passed through two ion exchange beds (Illco Way Exchanger, Research Model, Illinois Water Treatment Co., Rockford, Illinois) consistently exhibit P:O

[8] W. W. Umbreit, R. H. Burris, J. F. Stauffer, "Manometric Techniques" Burgess, Minneapolis, Minnesota, 1959.

[9] J. B. Martin and D. M. Doty, *Anal. Chem.* **21,** 965 (1949).

[10] O. Lindberg and L. Ernster, *Methods Biochem. Anal.* **3,** 1 (1956).

ratios greater than 3 for pyruvate, β-hydroxybutarate, α-ketoglutarate and greater than 2 for succinate. The Illco Way Exchanger reduces the conductivity of distilled water to well below 0.1 part per million. The nature of the inhibitor of phosphorylation found in distilled water is not known, but it can be easily removed by passage of the distilled water through the exchange resin.

[14] Skeletal Muscle Mitochondria

By LARS ERNSTER and KERSTIN NORDENBRAND

The bulk of the muscle tissue consists of myofibrils, and the mitochondria as a rule constitute a much smaller portion of the total mass of the tissue than in other organs. It is quite natural, therefore, that early interest in studies with isolated muscle mitochondria was confined to muscles with a relatively high content of mitochondria, such as the flight-muscle of insects (cf. this volume [22]) or the heart muscle of vertebrates (cf. this volume [11]). Among skeletal muscles, again, the first successful procedures for the isolation of mitochondria were developed with one particularly rich in mitochondria, namely, pigeon breast muscle.[1-4] Reliable procedures are now available also for the isolation of skeletal muscle mitochondria from mammalian organisms, primarily human[5,6] and rat.[4,7,8] A comprehensive review on muscle mitochondria has recently been published by Klingenberg.[9]

Preparation

A crucial difference in the preparation of mitochondria from skeletal muscle as compared with other tissues is that sucrose, or in general, nonelectrolytes, cannot be used with advantage as the homogenizing medium. Preparations made in sucrose medium not only may be of

[1] A. Kitiyakara and J. W. Harman, *J. Exptl. Med.* **97,** 553 (1953). See also J. W. Harman and U. H. Osborne, *J. Exptl. Med.* **98,** 81 (1953).
[2] J. B. Chappell and S. V. Perry, *Biochem. J.* **55,** 586 (1953).
[3] J. B. Chappell and S. V. Perry, *Nature* **173,** 1094 (1954).
[4] G. F. Azzone and E. Carafoli, *Exptl. Cell Res.* **21,** 447 (1960). See also G. F. Azzone, E. Carafoli, and U. Muscatello, *Exptl. Cell Res.* **21,** 456 (1960).
[5] L. Ernster, D. Ikkos, and R. Luft, *Nature* **184,** 1851 (1959).
[6] G. F. Azzone, O. Eeg-Olofsson, L. Ernster, R. Luft, and G. Szabolcsi, *Exptl. Cell Res.* **22,** 415 (1961).
[7] M. Klingenberg and P. Schollmeyer, *Biochem. Z.* **333,** 335 (1960); **335,** 231 (1961).
[8] R. Hedman, *Exptl. Cell Res.* **38,** 1 (1965).
[9] M. Klingenberg, *Ergeb. Physiol. Biol. Chem. Exptl. Pharmakol.* **55,** 131 (1964).

inferior quality (poor phosphorylating efficiency, high ATPase activity[3]) but also the yield of mitochondria is very low.[8] The problem of quality originates mainly from the relatively high content of Ca^{++} in muscle tissue which adsorbs to the mitochondria during homogenization. This complication, as first shown by Slater and Cleland[10] with heart muscle, can be eliminated by including a suitable concentration of EDTA in the sucrose medium. The problem of the yield arises from the fact that skeletal muscle, when homogenized in a nonelectrolyte medium, often assumes a gelatinous consistency which makes it difficult to obtain a sufficient disintegration of the myofibrils. Chappell and Perry[3] have overcome this difficulty by devising a medium which maintains the myofibrils in such a physical state that homogenization and subsequent differential centrifugation can easily be performed, and the resulting mitochondrial preparation is satisfactory both quantitatively and qualitatively. The medium, hereinafter referred to as the Chappell-Perry medium, consists of 0.1 *M* KCl, 0.05 *M* Tris-HCl buffer, pH 7.4, 0.001 *M* Na-ATP, 0.005 *M* $MgSO_4$, and 0.001 *M* EDTA. This medium has been used successfully for the preparation of mitochondria from pigeon breast muscle[3] as well as from human[6] and rat[4] skeletal muscle.

Procedure. The excised muscle specimen is blotted with filter paper, freed from fat and connective tissue, quickly weighed, and immersed into ice cold 0.15 *M* KCl. The tissue is cut with scissors into small pieces, and rinsed with several portions of 0.15 *M* KCl. Cutting is continued until a fine mince is obtained; alternatively, a mincer (of garlic-press type) may be used. The minced tissue is rinsed with Chappell-Perry medium and suspended in about 1 volume of the same medium. Homogenization is carried out with a relatively loosely fitting, all-glass Potter-Elvehjem homogenizer for 1–2 minutes. The homogenization and all following operations are performed at 0–2°. The homogenate is diluted with Chappell-Perry medium to a volume of 10 times the initial weight of the muscle and centrifuged at 600–650 *g* for 5–10 minutes. The supernatant fraction is decanted into a new tube and recentrifuged as before, in order to remove residual myofibrils. The resulting supernatant is decanted and centrifuged, either at 8000 *g* for 10 minutes in the case of pigeon breast muscle, or at 14,000 *g* for 10 minutes in the case of human and rat skeletal muscle. The mitochondrial pellets are resuspended in Chappell-Perry medium and recentrifuged as above; the washing may be repeated once. The washing medium is discarded, then the surface of the tightly packed pellet may be rinsed with 0.15 *M* KCl or 0.25 *M* sucrose in order to remove the remainder of the medium. The mitochon-

[10] E. C. Slater and K. W. Cleland, *Nature* **170**, 118 (1952).

drial pellet is finally suspended in 0.15 *M* KCl or 0.25 *M* sucrose to contain 6–10 mg of mitochondrial protein per milliliter of suspension. The protein content of the suspension is determined by the biuret method. The yield is approximately 5 mg of mitochondrial protein per gram of pigeon breast muscle, or 2–3 mg of mitochondrial protein per gram of human or rat skeletal muscle. In the case of rat skeletal muscle, the mitochondria may also be sedimented at 3500 *g* (rather than 14,000 *g*) for 10 minutes; the yield of mitochondria is only 50% of that at 14000 *g*, but the preparation is considerably purer.[8]

Preparation procedures slightly different from the one described above have been devised by Azzone and Carafoli[4] for pigeon breast muscle, and by Klingenberg and Schollmeyer[7] for rat skeletal muscle.

Properties

Like mitochondria from most animal tissues, skeletal muscle mitochondria contain the respiratory chain catalysts, cytochromes *a*, *c*, and *b*, ubiquinone, flavins, and pyridine nucleotides. The contents of the individual cytochromes and pyridine nucleotides and of ubiquinone in mitochondria from pigeon breast and rat leg muscle are indicated in Table I. Characteristic of skeletal muscle mitochondria (as of those from

TABLE I
REDOX COMPONENTS OF THE RESPIRATORY CHAIN IN MITOCHONDRIA FROM RAT LEG AND PIGEON BREAST MUSCLE[a,b]

Component	Rat leg	Pigeon breast
Cytochrome *a*	0.35	0.70
Cytochromes $c + c_1$	0.46	0.86
Cytochrome *b*	0.40	0.56
Ubiquinone	5.2	6.9
NAD	7.6	11.0
NADP	1.2	1.8

[a] From M. Klingenberg, *Ergeb. Physiol. Biol. Chem. Exptl. Pharmakol.* **55**, 131 (1964).
[b] Values stated as micromoles per gram of protein.

heart muscle) is a high content of ubiquinone and a low content of $NADP^+$; this picture is the converse to that found with liver mitochondria. The high content of cytochromes in mitochondria from pigeon breast muscle, which is similar to that found in heart, is a rather unique feature among skeletal muscles, not shared by, for example, rat leg.

The respiratory activities of mitochondria from pigeon breast and human and rat skeletal muscle with various substrates are summarized in Table II. Skeletal muscle mitochondria also resemble mitochondria

TABLE II
RESPIRATORY ACTIVITY OF SKELETAL MUSCLE MITOCHONDRIA WITH VARIOUS SUBSTRATES[a]

Substrate	Respiration (μatoms oxygen/hr/mg mitochondrial protein)[b]		
	Pigeon breast muscle	Human skeletal muscle	Rat skeletal muscle
Pyruvate + malate	2.4	6.1	12.0
α-Ketoglutarate	4.5	3.4	10.0
Succinate	4.0	0.9	3.0
Succinate (+ amytal or rotenone)	—	2.6	11.3
Citrate	0.2	0.7	1.5
Glutamate	2.5	2.9	15.0
Glycerol 1-phosphate	0.3	1.6	7.6
Butyrate	0	—	—
Butyrylcarnitine	5.3	—	—
Palmitate	0	—	—
Palmitylcarnitine	3.9	—	—

[a] Data were compiled from the following sources:
G. F. Azzone and E. Carafoli, *Exptl. Cell Res.* **21,** 447 (1960).
G. F. Azzone, E. Carafoli, and U. Muscatello, *Exptl. Cell Res.* **21,** 456 (1960).
G. F. Azzone, O. Eeg-Olofsson, L. Ernster, R. Luft, and G. Szabolcsi, *Exptl. Cell Res.* **22,** 415 (1961).
C. Bode and M. Klingenberg, *Biochem. Z.* **341,** 271 (1965).
R. Hedman, *Exptl. Cell Res.* **38,** 1 (1965).
M. Klingenberg, *Ergeb. Physiol. Biol. Chem. Exptl. Pharmakol.* **55,** 131 (1964).
[b] The values for pigeon breast muscle and rat skeletal muscle refer to 25°, and those for human skeletal muscle to 30°.

from other tissues in their capacity to catalyze the oxidation of various Krebs cycle metabolites. Pyruvate plus malate and α-ketoglutarate are oxidized at high rates. Succinate also is oxidized at a high rate, provided that oxalacetate formation is prevented, for example by the addition of amytal or rotenone. Added citrate (and isocitrate) are oxidized only at a very low rate, probably because of a limited permeability of the mitochondria to these substrates.[4] There is evidence that endogenous isocitrate (formed in the Krebs cycle) is oxidized at a high rate, and that the oxidation proceeds primarily via the NAD^+-linked isocitrate dehydrogenase.[9] The activity of this enzyme, which requires ADP as an activator, exceeds in skeletal muscle mitochondria the activity of $NADP^+$-linked isocitrate dehydrogenase.

Glutamate is oxidized in skeletal muscle mitochondria only in small part by way of glutamate dehydrogenase.[9] These mitochondria contain a potent glutamate–oxalacetate transaminase, which is the major pathway of glutamate catabolism. Human and rat skeletal muscle mitochondria catalyze the oxidation of glycerol 1-phosphate at an appreciable rate, whereas with pigeon breast muscle mitochondria, the rate of glycerol 1-phosphate oxidation is only marginal. Skeletal muscle mitochondria are unable to oxidize free fatty acids, but catalyze a high rate of oxidation of fatty acyl carnitines.[11] They also oxidize acetoacetate, after activation by way of succinyl∼CoA, i.e., provided that the latter (generated, for example, by the oxidation of α-ketoglutarate) is available.[12] With all substrates, added cytochrome *c* may stimulate respiration,[8] probably because of some loss of endogenous cytochrome *c* from the mitochondria during the preparation in the salt medium used.

Exogenous NADH is not oxidized by skeletal muscle mitochondria in the absence of added cytochrome *c*. When cytochrome *c* is added, exogenous NADH is oxidized at a high rate in the case of rat, and a lower but still substantial rate in the case of human skeletal muscle mitochondria.[13] In both cases, the oxidation is to a large extent sensitive to antimycin A, amytal, and rotenone (which is in contrast to the situation found with rat liver mitochondria[14] but similar to that found with mitochondria from bull sperm[15]). Oxidation of exogenous NADH in rat skeletal muscle mitochondria can also be activated by the addition of NAD^+-linked glycerol 1-phosphate dehydrogenase and a catalytic amount of glycerol 1-phosphate,[16] i.e., by establishing a "glycerol 1-phosphate cycle."

The phosphorylating efficiency of skeletal muscle mitochondria is similar to that observed with mitochondria from other tissues. The P:O ratio approaches 4 with α-ketoglutarate (in the presence of malonate), 3 with other NAD^+-linked substrates, and 2 with succinate and glycerol 1-phosphate. Added cytochrome *c*, which stimulates respiration, lowers the P:O ratio, especially in the case of succinate and glycerol 1-phosphate.[8] The cytochrome *c*-dependent oxidation of exogenous NADH is accompanied by little or no phosphorylation.[13]

[11] C. Bode and M. Klingenberg, *Biochem. Z.* **341,** 271 (1965).

[12] E. M. Suranyi and L. Ernster, *Proc. 1st Meeting Federation European Biochem. Soc., London, 1964,* p. 69. Academic Press, New York, 1964.

[13] R. Hedman, E. M. Suranyi, R. Luft, and L. Ernster, *Biochem. Biophys. Res. Commun.* **8,** 314 (1962).

[14] A. L. Lehninger, *Harvey Lectures Ser.* **49,** 174 (1955).

[15] H. Mohri, T. Mohri, and L. Ernster, *Exptl. Cell Res.* **38,** 217 (1965).

[16] E. M. Suranyi, R. Hedman, R. Luft, and L. Ernster, *Acta Chem. Scand.* **17,** 877 (1963).

Freshly prepared skeletal muscle mitochondria exhibit a high degree of respiratory control. The respiratory control ratio (ratio of respiratory rates in "state 3" and "state 4,"[17] i.e., with and without P_i and/or ADP) is 5–20 with NAD^+-linked substrates. It is lower with succinate, but can be increased by the addition of ATP[7]; the effect of ATP consists partly of a depression of the state 4 respiration (by a reversal of the electron flow through the respiratory chain; see further below), and partly of a stimulation of the state 3 respiration (by relieving oxalacetate inhibition of succinate dehydrogenase). With glycerol 1-phosphate as substrate, ATP has a similar effect on the state 4 respiration, but depresses state 3 respiration as well, owing to an inhibition of glycerol 1-phosphate dehydrogenase.[7] Bovine serum albumin lowers the state 4 respiration with either succinate or glycerol 1-phosphate, (probably by removing an endogenous uncoupler) and thus improves the respiratory control with both substrates.[7]

Energy-linked reduction of endogenous NAD^+ can be demonstrated with skeletal muscle mitochondria using either succinate or glycerol 1-phosphate as substrate.[7] Addition of bovine serum albumin or ATP enhances both the rate and extent of NAD^+ reduction, with a simultaneous increase in the steady-state level of reduced cytochrome *b* and a decrease in the steady-state level of reduced cytochrome *c*; these changes become more striking, and are also accompanied by a decrease of the steady-state level of reduced cytochrome *a*, if the respiration is partially inhibited by the addition of a low concentration of azide.

The effects of thyroidectomy and experimentally induced hypothyroidism on the function and morphology of rat skeletal muscle mitochondria have been studied by Tata *et al.*[18] Thyroidectomy leads to a decrease, and thyroid hormone treatment to an increase, of the respiratory and phosphorylating activities of rat skeletal muscle mitochondria, with unaltered respiratory control and P:O ratios. On the other hand, both thyroidectomy and thyroid hormone treatment result in a striking enlargement of the individual size and total mass of skeletal-muscle mitochondria, as revealed by electron microscopy of tissue sections.[19] A detailed study of the changes in mitochondrial enzyme and coenzyme levels following thyroid hormone treatment of rats has been carried out

[17] B. Chance and G. R. Williams, *Advan. Enzymol.* **17**, 65 (1956).

[18] J. R. Tata, L. Ernster, and O. Lindberg, *Nature* **193**, 1058 (1962). See also J. R. Tata, L. Ernster, O. Lindberg, E. Arrhenius, S. Pedersen, and R. Hedman, *Biochem. J.* **86**, 408 (1963).

[19] R. Gustafsson, J. R. Tata, O. Lindberg, and L. Ernster, *J. Cell Biol.* **26**, 555 (1965).

by Kadenbach.[20] Ernster and associates[5,21] have investigated the biochemical properties of skeletal muscle mitochondria from hypermetabolic human subjects, in particular a clinical case of extremely severe hypermetabolism of nonthyroid origin (basal metabolic rate of the order of 200% above the normal). They have found that the skeletal muscle mitochondria of this patient are in the "loosely coupled" state, i.e., lack respiratory control but can still exhibit a substantial extent of phosphorylating efficiency.

Assay

Respiration of isolated skeletal muscle mitochondria can be measured by the conventional manometric or polarographic methods. A suitable incubation mixture for the manometric assay is as follows: 3–5 mg of mitochondrial protein in 0.5 ml of 0.15 *M* KCl or 0.25 *M* sucrose is added to a Warburg vessel which contains 1.5 ml of a medium consisting of 50 m*M* KCl, 25 m*M* Tris-HCl buffer, pH 7.5, 25 m*M* K-phosphate buffer, *p*H 7.5, 0.2% bovine serum albumin, 4–8 m*M* $MgCl_2$, 10 m*M* substrate, 1 m*M* ATP, 30 m*M* glucose, 100 Kunitz-MacDonald units of hexokinase, and, if desired, 0.01–0.02 m*M* cytochrome *c*. The reaction volume may be reduced to 1 ml, containing one-half the amount of mitochondria, if small (5–6 ml) Warburg vessels are used; this may be an advantage when working with limited amounts of material, such as human muscle biopsy specimens. Incubation is made at 30°, with 5 minutes of thermoequilibration, after which manometer readings are taken at 5-minute intervals. For determination of phosphate uptake, the incubation is terminated after 20–30 minutes by the addition of 0.2 ml 5 *M* H_2SO_4, and P_i is estimated either colorimetrically or by the ^{32}P distribution procedure.[22] In the calculation of the P:O ratio, the oxygen uptake is extrapolated for the thermoequilibration period.

In the polarographic assay, a reaction mixture containing mitochondria, sucrose, KCl, and Tris-HCl buffer, in concentrations as indicated above, is placed in the chamber of an O_2 electrode. The final volume may be 1–3 ml, depending on the size of the chamber. After adjustment of the recorder, 5–10 m*M* substrate, 5–10 mM P_i, and 0.2–0.4 m*M* ADP are added, in the order indicated, at 2–3 minute intervals; the additions are made in volumes of 5–20 μl. With freshly prepared mitochondria, the respiratory rate is low after the addition of substrate and P_i; it increases

[20] B. Kadenbach, Thesis, University of Marburg, 1964; *Biochem. Z.* **344,** 49 (1966).

[21] R. Luft, D. Ikkos, G. Palmieri, L. Ernster, and B. Afzelius, *J. Clin. Invest.* **41,** 1776 (1962). See also L. Ernster, and R. Luft, *Exptl. Cell Res.* **32,** 26 (1963).

[22] O. Lindberg and L. Ernster, *Methods Biochem. Anal.* **3,** 1 (1955).

substantially (5–20-fold in the case of NAD^+-linked substrates) upon the addition of ADP (state 4–state 3 transition). When all the ADP added has been converted into ATP, the respiratory rate will decline (state 3–state 4 transition) and will increase again upon the repeated addition of ADP. The ADP:O (=P:O) ratio can be calculated by dividing the number of micromoles of ADP added by the number of microatoms of oxygen consumed during state 3.[23]

It is sometimes found that, although ADP greatly stimulates oxygen uptake when added to muscle mitochondria respiring in the presence of substrate and P_i, the respiratory rate declines only little when the added ADP would be expected to be converted into ATP; this lack of state 3–state 4 transition may be accentuated by the presence of Mg^{++}. These findings have been interpreted[24] in terms of a deficient respiratory control and an "uncoupling" effect of Mg^{++}. However, more recent work[6,8] has provided evidence that this phenomenon may be explained by a contamination of such muscle-mitochondrial preparations with a Mg^{++}-stimulated ATPase, originating from myofibrils, sarcotubular membranes, or submitochondrial particles, which causes a continuous breakdown of ATP, thereby maintaining a steady supply of ADP. Under these circumstances, clearly, the ADP:O ratio derived from polarographic records is not an accurate measure of the phosphorylating efficiency of the mitochondria. For the same reason, the respiratory control ratio in such mitochondrial preparations can be deduced more adequately from the extent of stimulation of respiration by ADP (state 4–state 3 transition) than from the extent of respiratory decline upon the expected exhaustion of ADP (state 3–state 4 transition).

2,4-Dinitrophenol, 10^{-4} *M*, uncouples oxidative phosphorylation and stimulates the ATPase activity of skeletal muscle mitochondria, but may also depress the respiration occurring in the presence of P_i and ADP.[6] In the presence of 10^{-5} *M* dinitrophenol, respiration is maximal in both the presence and absence of P_i and ADP, while the P:O ratio is only slightly diminished; i.e., oxidative phosphorylation is "loosely coupled." Oligomycin abolishes the stimulation of respiration by P_i and ADP,[21] but not that by dinitrophenol, and it inhibits the dinitrophenol-stimulated ATPase,[8] all in a fashion similar to that of mitochondria from other tissues.[25]

ATPase activity of skeletal muscle mitochondria can be assayed with

[23] B. Chance and G. R. Williams, *J. Biol. Chem.* **217,** 383 (1955).

[24] B. Chance and M. Baltscheffsky, *Biochem. J.* **68,** 283 (1958). See also B. Chance, *Ciba Found. Symp. Regulation Cell Metabolism,* 1959, p. 91. Little, Brown, Boston, Massachusetts, 1959.

[25] L. Ernster and C. P. Lee, *Ann. Rev. Biochem.* **33,** 729 (1964).

a reaction mixture containing 0.5–1 mg of mitochondrial protein, 5 mM ATP, 50 mM Tris-HCl buffer, pH 7.5, and 100 mM KCl, in a final volume of 1 ml. $MgCl_2$ (5 mM), dinitrophenol (0.1 mM), and oligomycin (2 μg/mg protein) may be added, according to the aim of the experiment. Incubation is made at 30° for 5 and/or 10 minutes. The samples are fixed with 1 ml of cold 1 M perchloric acid, and P_i liberated is determined colorimetrically in 1 ml of the perchloric acid extract.

Energy-linked reduction of endogenous NAD^+ in skeletal muscle mitochondria is assayed essentially as described in this volume [112]. Details regarding assay systems used particularly for skeletal muscle mitochondria are described by Klingenberg and Schollmeyer.[7]

[15] Isolation of Liver or Kidney Mitochondria

By Diane Johnson and Henry Lardy

Principle. The selected tissue is disrupted by homogenization in cold isotonic sucrose. Differential centrifugation is then employed to separate the mitochondria from cell debris, red blood cells, nuclei, microsomes, and soluble components.[1]

Reagents and Equipment

- Sucrose, 0.25 M, CO_2-free, or 0.2 M mannitol
- International refrigerated centrifuge, Model PR-2, with multispeed attachment and head No. 296
- Potter-Elvehjem homogenizers with either glass or Teflon pestles. The pestles should be 0.006–0.008 inch smaller in diameter than the inside bore of the homogenizer tube.
- Stirring motor with shaft rotation of 600 rpm

All glassware used for preparation of mitochondria should be routinely cleaned by immersion in a hot sulfuric–nitric acid bath. It should be thoroughly rinsed with tap water, distilled water and finally with water that has been deionized and distilled from an all-glass or quartz apparatus.

All solutions that come in contact with the mitochondria must be made with highest purity reagents and water. Suitable water can be prepared by distilling from a glass or stainless steel vessel, passing

[1] W. C. Schneider, *J. Biol. Chem.* **176**, 259 (1948).

through a mixed-bed ion exchange resin, and then distilling from an all-quartz still. On storage, such water will absorb CO_2 from the atmosphere and should therefore be boiled before use or may be neutralized with a small amount of tris(hydroxymethyl)aminomethane, triethylamine, or alkali metal hydroxide. Alternatively, distilled water may be purified sufficiently by passing through a suitable mixed-bed ion exchange resin on the all-plastic container such as that supplied by Continental H_2O, Chicago, Illinois.

Preparation of Liver Homogenate. The animal is stunned and decapitated. Profuse bleeding is encouraged with flowing cold water. The liver is removed, blotted, and immediately placed in sucrose solution at 0°. All subsequent operations are carried out at 0° (vessels are kept immersed in chipped ice). Ten grams of liver is homogenized in 25–50 ml of cold 0.25 M sucrose by passing the homogenizer tube up and down past the rotating pestle. The tube is immersed in ice during this procedure. As little as 3 g or as much as 10 g at a time may be disrupted depending on the size of the homogenizer. Caution must be exercised to avoid pulling a vacuum in the tube. Homogenization is to be discontinued when liver tissue is no longer discernible. Excessive homogenization damages mitochondria and should be avoided. The homogenate is adjusted to a volume of 80 ml with cold sucrose.

Isolation of Liver Mitochondria (M_w). The homogenate is distributed into Lusteroid centrifuge cups and centrifuged at 600 g for 10 minutes. The supernatant fraction is decanted and saved. The pellets, containing cells, tissue fragments, and some mitochondria, are washed once with a total volume of 20 ml of sucrose. The pellets may be dispersed by using the side of a stirring rod against the wall of the cup or by hand-operating the homogenizer. The resuspended material is centrifuged at 600 g for 10 minutes. The supernatant fractions are combined. The pellets are discarded.

This washing contributes not only to the yield of the final mitochondrial preparation, but also to its integrity, apparently by permitting the recovery of the larger mitochondria. The combined supernatants are centrifuged at 15,000 g for 5 minutes. The resultant supernatant is discarded along with any lightly packed pink microsomes. Lightly packed tan mitochondria are retained. The pellets are resuspended as the nuclear pellet was, quantitatively collected in one cup, centrifuged at 15,000 g for 5 minutes, and washed twice with 15–20 ml of sucrose. During each decanting operation any question is decided in favor of the purity of the mitochondria-containing fraction at the expense of complete mitochondrial recovery.

It is imperative that the isolation be completed without any delay.

If there must be lapse of time between the preparation of the mitochondria and their use, the M_w are stored as the pellet, in ice.

Isolation of Kidney M_w. Killing of the animal and removal of the tissue is described above. The kidney capsule is removed by gently squeezing the kidney through the thumb and forefinger. The kidney is then cut sagittally. The medullary portion is removed and discarded. Mitochondria are then prepared from the cortex following the method described above for liver, with the exception that the mitochondrial pellet need be washed only once.

Testing the Quality of Mitochondria. The respiration of tightly coupled mitochondria is strikingly dependent on the presence of P_i and a phosphate acceptor system.[2] ADP is the immediate phosphoryl acceptor and may be generated from ATP by creatine and creatine-kinase or glucose and hexokinase, or it may be added in stoichiometric amounts. The phosphoryl acceptor system should enhance oxygen consumption of liver mitochondria 4- to 10-fold above the rate without the acceptor when pyridine nucleotide-linked substrates are provided. The oxidation of succinate, fatty acids, and α-glycerophosphate is usually enhanced 2- to 5-fold by the phosphoryl acceptor. Enhancement of respiratory rate by phosphoryl acceptor is a far more sensitive test of mitochondrial integrity than is efficiency of oxidative phosphorylation, for relatively high P:O ratios may be measured with mitochondria that show only little or no respiratory control.

Typical rates of oxygen uptake by liver mitochondria with various substrates in the presence of phosphoryl acceptor are: succinate, 310–410; glutamate, 280–375; α-ketoglutarate, 230–360; β-hydroxybutyrate, 160–260; all expressed as microliters of oxygen × milligrams mitochondrial nitrogen^{-1} × hours^{-1} at 30°.

[2] H. A. Lardy and H. Wellman, *J. Biol. Chem.* **195**, 215 (1952).

[16] Preparation and Properties of Brain Mitochondria

By R. E. Basford

The separation of mitochondria from other particulate portions of the cell is more difficult using brain tissue as the source rather than heart or liver, owing mainly to the high myelin content of brain. The preparation of brain mitochondria[1] described below has been developed

[1] W. L. Stahl, J. C. Smith, L. M. Napolitano, and R. E. Basford, *J. Cell Biol.* **19**, 293 (1961).

for the isolation of mitochondria from bovine brain. A synthetic anticoagulant, polyethylene sulfonate (PES), average molecular weight, 5900 (Lloyd Bros., Cincinnati, Ohio) is added to a sucrose–EDTA medium to minimize agglutination of mitochondria with other subcellular particles. Heparin (sodium salt) may be substituted for PES. Separation of a relatively pure mitochondrial fraction is also facilitated by increasing the density of the medium by the addition of Ficoll, a high molecular weight polysaccharide (Pharmacia, Rochester, Minnesota).

The procedure may also be used for the preparation of mitochondria from rat or rabbit brain; however, the simpler procedure described by Løvtrup and Zelander[2] also yields a mitochondrial preparation with quite satisfactory enzymatic and morphological properties.

Assay Methods

Oxidative capacity is determined manometrically or polarographically. The composition of the manometric assay medium is: KH_2PO_4, 25 micromoles; ATP, 1.0 micromole; glucose, 12 micromoles; hexokinase (Sigma Chemical Company, St. Louis, Missouri), Type III, 0.5 mg or crystalline, free of ammonium sulfate, 23–30 KM units; $MgCl_2$, 20 micromoles; dialyzed serum albumin, 10 mg; sucrose, 360 micromoles; substrate, 50–60 micromoles; mitochondrial protein, 3–5 mg (biuret assay)[3]; final volume, 2.5 ml, pH 7.4. Optimal rates of oxidation are usually obtained without the addition of DPN or cytochrome *c*.

Phosphorylative efficiency in manometric determinations is calculated from the disappearance of inorganic phosphate as determined according to Dryer *et al.*[4] or by the appearance of glucose-6-P as determined with glucose-6-P dehydrogenase.[5]

Polarographic determination of oxidative phosphorylation is carried out according to Chance and Williams[6] using the assay medium of Voss *et al.*[7]: mannitol, 600 micromoles; Tris, 20 micromoles; KCl, 20 micromoles; phosphate, 20 micromoles; substrate, 25 micromoles; mitochondrial protein, 2.5–4.5 mg; final volume, 2.0 ml, pH 7.4. The P:O ratios are calculated as ADP:O from the oxygen uptake in microatoms during the active state of respiration and the micromole of ADP added. The ADP:O ratios may be confirmed by the determination of

[2] S. Løvtrup and T. Zelander, *Exptl. Cell Res.* **27**, 468 (1962).

[3] A. G. Gornall, C. J. Bardawill, and M. M. David, *J. Biol. Chem.* **225**, 177 (1957).

[4] R. L. Dryer, A. R. Tammes, and J. L. Routh, *J. Biol. Chem.* **225**, 177 (1957).

[5] See Vol. III [19].

[6] B. Chance and G. R. Williams, *Nature* **175**, 1120 (1955).

[7] D. O. Voss, A. P. Campello, and M. Bacila, *Biochem. Biophys. Res. Commun.* **4**, 48 (1961).

the amount of ATP formed from ADP using the luciferin-luciferase assay.[8]

Reagents

Medium A: 0.4 *M* sucrose, 0.001 *M* EDTA, 0.02% PES or heparin. Adjust to pH 6.8–7.0 with KOH.

Medium F: Medium A to which Ficoll is added to a final concentration of 8%.

Isolation Procedure

Undamaged calf or beef brains are obtained from exsanguinated animals at a slaughterhouse. The brains are removed within 5–10 minutes after the death of the animals. Most slaughterhouses are equipped with an apparatus for removal of the brain from the cranium without damage. The brains are placed immediately into a polyethylene bag containing ice cold Medium A and stored in ice for transport to the laboratory. The lapse of time between death of the animals and the arrival of the specimens at the laboratory should be no more than 30 minutes.

In a 5° cold room, the cerebral hemispheres are removed from the brains and the meninges are removed with forceps. The gray matter is scraped from the cortices with a dull spatula or spoonula, leaving behind as much of the white matter as possible without spending undue time for careful dissection. Two brains yield about 100 g of wet tissue. The tissue is homogenized (ten passes of the pestle) in a 200 ml capacity Teflon and glass homogenizer, 0.004–0.006 inch clearance (Kontes Glass Company, Vineland, New Jersey), using 2 ml of medium A per gram of wet tissue.

The homogenized tissue is adjusted to pH 6.9–7.0 with a few drops of 2 *M* Tris, pH 10.8, 1 mg of Nagarse[9] per gram of tissue is added, and the mixture is stirred at 0–4° for 15 minutes on a magnetic stirrer. The pH of the suspension is maintained throughout the digestion by the addition of more Tris, if necessary. The suspension is diluted with Medium A (20 ml per gram of original tissue), transferred to polyethylene "catchup bottles," and centrifuged at 184 *g* for 20 minutes[10] followed (without transfer of supernatant) by a centrifugation at 1153 *g* for 20 minutes in the International PR-2 centrifuge, rotor No. 845.

[8] B. L. Strehler and J. R. Totter, *Arch. Biochem. Biophys.* **40,** 28 (1952).

[9] B. Chance and B. Hagihara, *Proc. 5th Intern. Congr. Biochem. Moscow, 1961* Vol. 5, p. 3, Pergamon, London, 1963. Nagarse, crystalline bacterial proteinase, may be purchased from Enzyme Development Co., 64 Wall Street, New York, New York 10005.

[10] All centrifugal forces are given as the maximal gravitational force, g_{max}.

After the initial low speed centrifugation, the residue, R_1, is discarded and the supernatant, S_1, is transferred to cellulose acetate tubes for the Spinco Model L or Sorvall RC-2 centrifuge (No. 30 or SS 34 rotor, respectively) and centrifuged at 12,000 g for 15 minutes, to yield a crude mitochondrial pellet, R_2, and a supernatant, S_2, which is discarded. Fraction R_2 is transferred to a Teflon–glass homogenizer with Medium F (6 ml per gram of original tissue), gently homogenized, and centrifuged at 12,000 g for 30 minutes. The resulting supernatant fluid, S_3, is decanted from the brown pellet, R_3, which is usually visibly free of contaminating white fluffy material.[11]

The mitochondrial fraction, R_3, is washed by homogenization in 4 ml of Medium A per gram of original tissue and centrifugation at 12,000 g for 15 minutes, to yield the final mitochondrial fraction, R_4, which is homogenized in the desired volume of Medium A.

The yield of mitochondrial protein is between 100 and 140 mg per 50 g wet weight of original tissue. The percentage yield diminishes slightly as the amount of starting material is increased.

In the preparation of the tissue for homogenization, if the entire cerebrum is used without removal of as much of the white matter as possible, a much higher percentage of "fluff" is obtained in fraction R_2 and more time is required for resuspension and centrifugation in Medium F than is spent in gross dissection, to obtain a comparable degree of purity.

Properties

Oxidative and Phosphorylative Ability. Representative data on respiratory rate, P:O ratio, and respiratory control ratio are shown in the table for several substrates as measured both manometrically and polarographically. No phosphorylation can be demonstrated polarographically in the absence of bovine serum albumin. The addition of DNP prior to the addition of substrate or ADP produces no effect on the rate of oxygen utilization. Stimulation of the rate by DNP can be demonstrated after the addition of substrate and ADP.

Contamination and Morphology. Slight contamination of the mitochondrial preparation by nerve-ending particles with entrapped mitochondria and synaptic vesicles and by myelin fragments can be seen in electron micrographs of thin sections of mitochondrial pellet.[1] The mitochondria have a "zebra" appearance indicating some probably morphological damage. If an oxidizable substrate is included in the isola-

[11] If R_3 is not free of fluffy material, it may be resuspended and homogenized in Medium F and centrifuged at 12,000 g for 30 minutes, to yield fractions R'_3 and S'_3.

OXIDATIVE AND PHOSPHORYLATIVE ABILITY OF BOVINE BRAIN MITOCHONDRIA[a]

Method of assay	Temp. (°C)	Substrate	Additions	Number of assays	Rate[b] (μA O/min/mg protein)		P:O[b]	Respiratory control ratio
Manometric	30	Pyruvate	Malate, 3 μmoles	8	0.103 ± 0.034		2.94 ± 0.42	3–10
Manometric	30	Glutamate	—	3	0.078 ± 0.022		3.08 ± 0.47	3–5
Manometric	30	α-Ketoglutarate	—	3	0.057 ± 0.022		2.47 ± 0.66	4–5
Manometric	30	Succinate	Amytal, 2.5 μmoles	5	0.165 ± 0.022		1.93 ± 0.17	3–4
					+ADP	−ADP		
Polarographic	25	Pyruvate	Malate, 1.5 μmoles	5	0.070	0.016	2.53	4–5
Polarographic	25	Succinate	Seconal, 2 μmoles	6	0.111	0.053	1.49	2
Polarographic	25	Succinate	Seconal, 2 μmoles + 2nd add'n ADP	2	0.089	0.024	1.93	3–4
Polarographic	25	α-Glycerophosphate	—	2	0.062	0.041	1.39	1–2
Polarographic	25	α-Glycerophosphate	$MgCl_2$, 20 μmoles	3	0.068	0.027	1.67	2–3
Polarographic	25	α-Glycerophosphate	EDTA, 4 μmoles	1	0.057	0.049	0	1
Polarographic	25	α-Glycerophosphate	EDTA, 4 μmoles + $MgCl_2$, 20 μmoles + 2nd add'n ADP	2	0.043	0.012	2.39	2

[a] The data are taken from Tables I and II of W. L. Stahl, J. C. Smith, L. M. Napolitano, and R. E. Basford [*J. Cell Biol.* **19,** 293 (1963)] reprinted by permission of The Rockefeller Institute Press.

[b] Plus-minus standard deviation, where indicated.

tion medium, the morphology as shown in electron micrographs more closely resembles that of mitochondria in thin sections of cortex; however, the preparations have a much higher degree of contamination by other subcellular particles.

Reliable estimations of contamination by enzymatic means cannot be made because of the use of Nagarse in the preparative procedure. Mechanical homogenization (ground glass homogenizer)[1] may be used in place of "proteolytic homogenization" by Nagarse; in this case, the preparations are somewhat less free of contamination and show lower respiratory rates. Estimation of contamination in the latter type of preparation by cholinesterase activity and by enzymes of the glycolytic cycle indicate 3–4% and 0.2–4%, respectively.[1,12]

Stability. The phosphorylative ability of the mitochondrial preparation is not stable to freezing (—20°) and thawing but is stable (decrease in P:O, 0–5%) after storage in ice for 24 hours. After storage in ice for 48–72 hours, the P:O ratios diminish 25%, and after 5 days no phosphorylation can be demonstrated.

[12] D. S. Beattie, H. R. Sloan, and R. E. Basford, *J. Cell Biol.* **19**, 309 (1963).

[17] The Preparation of Thyroid and Thymus Mitochondria

By D. D. TYLER[1] and JEANINE GONZE

Thyroid Mitochondria

Isolation Method

Preparations of phosphorylating mitochondria have been isolated from sheep,[1a,2] human, and beef[3] thyroid glands. The recommended isolation procedure described below, adapted from the proteinase method of Chance and Hagihara,[4] has been developed during a study[5] of the influence of the isolation medium on the properties of thyroid mitochondria.

Reagents

MSE medium: 0.225 *M* mannitol, 0.075 *M* sucrose, and 0.05 m*M* EDTA, pH 7.4

[1] See footnote 1, page 75.
[1a] R. W. Turkington and B. Nordwind, *J. Clin. Invest.* **41**, 1725 (1952).
[2] J. E. Dumont and L. J. DeGroot, personal communications (1965).
[3] J. Gonze and D. D. Tyler, *Biochem. Biophys. Res. Commun.* **19**, 67 (1965).
[4] B. Chance and B. Hagihara, *Proc. 5th Intern. Congr. Biochem., Moscow, 1961* Vol. 5, p. 3. Pergamon, London, 1963.
[5] J. Gonze and D. D. Tyler, unpublished observations (1964).

Unneutralized Tris buffer, 1 *M*
Crystalline bacterial proteinase (Nagarse)

Procedure. PREPARATION OF THE HOMOGENATE. All containers are surrounded with ice and water, and the centrifuge is set to run at 1°. Fresh beef thyroid glands are collected in cold MSE medium at the slaughterhouse and transferred to the laboratory as quickly as possible. About 120 g of thyroid, trimmed free of fat and connective tissue, are chopped finely with scissors and washed several times with fresh volumes (about 100 ml) of MSE. The last volume of washing fluid, which should be clear and colorless, is decanted as completely as possible from the chopped tissue. The tissue is resuspended in fresh MSE (final volume, 300 ml), and 0.25 ml of Tris buffer and a solution of 100 mg of proteinase in 10 ml of MSE are stirred in. The suspension is homogenized in a glass homogenizer, first with a loose-fitting Teflon pestle and then with a tightly fitting pestle. The homogenate is passed through two layers of cheesecloth, adjusted if necessary to pH 7.2, and divided equally into eight centrifuge tubes.

CENTRIFUGATION OF THE HOMOGENATE. After the first centrifugation at 8000 *g* for 10 minutes, the supernatant, which contains the proteinase, is discarded and the pellets are rehomogenized and diluted to 300 ml with MSE. A second centrifugation at 700 *g* for 10 minutes yields a supernatant, which is transferred into clean centrifuge tubes; the pellets, which contain nuclei, red cells, and debris, are discarded. A third centrifugation at 8000 *g* for 10 minutes yields a brown pellet (mitochondrial fraction), covered by a gray loosely packed layer, and a supernatant. The supernatant is discarded together with any freely flowing material from the pellet, and the mitochondrial fraction is gently broken up with a glass rod, avoiding the resuspension of any red cells present at the bottom of the tube. The suspension thus obtained is homogenized and diluted to 80 ml with MSE and divided equally into two tubes. The mitochondria are washed twice (fourth and fifth centrifugation) by repeating the procedure used during and after the third centrifugation.

The pellet obtained after the fifth centrifugation is resuspended in fresh MSE with a small glass homogenizer, to yield 3 ml of a uniform suspension of mitochondria containing 20–30 mg protein per milliliter.

Properties

Thyroid mitochondria exhibit many of the enzymatic activities characteristic of mitochondria isolated from other mammalian tissues (e.g., liver and heart). Several metabolities, including citric acid cycle intermediates, can serve as substrates for oxidative phosphorylation, yielding P:O ratios between 2 and 3 for DPN-linked oxidations and between 1 and 2 with succinate.[1a,3] Both ADP and AMP can serve as phosphate acceptor, since the mitochondria contain an active adenylate kinase.[3] The respiratory chain contains cytochromes b, c, c_1, a, and a_3 (0.2 millimicromole cytochrome a per milligram of protein), flavoproteins, and both diphospho- and triphospho- pyridine nucleotides (about 5 millimicromoles pyridine nucleotide per milligram of protein).[6] Several of the standard mitochondrial electron and energy transfer inhibitors induce their usual effects on the respiratory chain phosphorylation system.[3]

Thymus Mitochondria

Isolation Method

The preparative method is similar to that described above for the thyroid gland, except that the Tris buffer and proteinase are omitted. The thymus glands, trimmed free of fat and connective tissue, are passed through a tissue mincer before homogenization.

Properties

Thymus mitochondria contain the usual mammalian phosphorylating respiratory chain system, with cytochromes b, c, c_1, a, and a_3, flavoproteins, and pyridine nucleotides present. Succinate supports the highest rate of oxygen uptake (P:O = 1.6), and both succinate and glutamate plus malate oxidation (P:O = 2.4) exhibit a respiratory control ratio of about 3.0. Treatment with antimycin, rotenone, amytal, cyanide, oligomycin, or uncoupling agents induces the expected effects on the rate of oxidation of added substrates.[5]

[6] J. Gonze and D. D. Tyler, unpublished observations (1965).

[18] Guinea Pig Mammary Gland Mitochondria

By WALTER L. NELSON and RONALD A. BUTOW[1]

General Procedure

Principle. The preparation of phosphorylating mitochondria from guinea pig mammary gland tissue is carried out by disruption of the cells with the aid of all-glass homogenizing vessels and subsequent isolation of the mitochondria by differential centrifugation.

[1] This manuscript was prepared while the author was at the Department of Biochemistry, The Public Health Research Institute of the City of New York, Inc., New York.

Reagents

Sucrose, 0.25 *M*
EDTA, 5×10^{-3} *M*

Procedure. The guinea pig is sacrificed by a sharp blow at the base of the skull followed by decapitation. After thorough bleeding, the area of skin containing the mammary gland is removed and the glands are dissected away from the skin with scissors. Most of the superficial adipose and connective tissue is removed and the glands are placed in cold 0.25 *M* sucrose containing 5×10^{-3} *M* EDTA adjusted to pH 7.4 with potassium hydroxide.

Preparation of Mitochondria

All operations are carried out at 1°–2°. An all-glass homogenizer of the Potter-Elvehjem type with a loose-fitting and snug-fitting pestle is used for preparation of homogenates. ("Loose fitting" is defined as a fit in which the vessel will readily slide away from the pestle whereas a snug-fitting pestle will cause enough friction so that the vessel will slide slowly down the pestle.)

Step 1. Weighed amounts of tissue are minced with scissors and homogenized (1 g tissue + 9 ml sucrose–EDTA) with the loose-fitting pestle until obvious pieces have been suspended. The snug-fitting pestle is used to complete the cell distintegration by further homogenization for 1–2 minutes.

Step 2. Filter the homogenate through two layers of cheesecloth to remove tissue fibers and fatty material.

Step 3. Centrifuge the filtrate at 800 *g* for 10 minutes to remove unbroken cells and cell debris. Carefully remove the supernatant solution, with the aid of a pipette, from beneath the top fatty layer.

Step 4. Centrifuge the supernatant solution from step 3 for 20 minutes at 9000 *g*. Carefully remove the supernatant solution and wipe the adhering fat from the tube with adsorbent tissue. Suspend the residue (mitochondrial fraction) in the sucrose-EDTA solution by stirring gently and then aspirating with a pipette.

Step 5. Centrifuge the mitochondrial suspension for 20 minutes at 9000 *g*. Remove the supernatant solution, wipe adhering fat from the tube and resuspend the mitochondrial pellet in sucrose-EDTA as in step 4 so that 1 ml of suspension is equivalent to 3–4 g of fresh tissue.

[19] Preparation and Assay of Phosphorylating Mitochondria from Ascites Tumor Cells

By RAY WU[1] and LEONARD A. SAUER

A number of methods have been described for the preparation of mitochondrial fractions from mouse ascites tumor cells.[1a-11] This seems to be a result of the fact that ascites tumor cells are very resistant to homogenization. However, even with the variety of homogenization methods that have been used, the literature shows that mitochondria capable of oxidative phosphorylation, but with various degrees of efficiency, may be isolated from ascites cells. An adequate description of each published method would be beyond the scope of this section; therefore, we have chosen to describe in detail two quite different procedures with which we have had considerable experience. These methods involve homogenization by grinding with glass beads (R.W.) or with the Dounce homogenizer (L.A.S.). Both methods yield mitochondrial fractions with high P:O and respiratory control ratios.

Materials

Tumor Transplantation and Harvesting. The procedures used are the same as previously described.[4,10]

Reagents. Crystalline hexokinase (140 units/mg) was purchased from Boehringer Mannheim Co., New York. The enzyme was dialyzed overnight against 100 volumes of 50 m*M* triethanolamine buffer (pH 7.4) containing 1 m*M* glucose, and kept frozen at —20°. Succinate was pur-

[1] This manuscript was prepared while the author was at the Department of Biochemistry, The Public Health Research Institute of the City of New York, Inc., New York.

[1a] E. Kun, P. Talalay, and H. G. Williams-Ashman, *Cancer Res.* **11**, 855 (1951).

[2] O. Lindberg, M. Ljunggren, L. Ernster, and L. Révész, *Exptl. Cell Res.* **4**, 243 (1953).

[3] C. E. Wenner and S. Weinhouse, *Cancer Res.* **13**, 21 (1953).

[4] R. Wu and E. Racker, *J. Biol. Chem.* **234**, 1029, 1036 (1959).

[5] B. Chance and B. Hess, *J. Biol. Chem.* **234**, 2413 (1959).

[6] P. Emmelot, C. J. Bos, P. J. Brombacher, and J. F. Hampe, *Brit. J. Cancer* **13**, 348 (1959).

[7] W. Luehrs, G. Bacigalupo, B. Kadenbach, and E. Heise, *Experientia* **15**, 376 (1959).

[8] P. Borst, *J. Biophys. Biochem. Cytol.* **7**, 381 (1960).

[9] A. O. Hawtrey and M. H. Silk, *Biochem. J.* **74**, 21 (1960).

[10] L. A. Sauer, A. P. Martin, and E. Stotz, *Cancer Res.* **20**, 251 (1960).

[11] L. A. Sauer, A. P. Martin, and E. Stotz, *Cancer Res.* **22**, 632 (1962).

chased from the same company or from Eastman Kodak Co. All other reagents used were analytical grade.

Procedure for Measuring Oxidative Phosphorylation

Both the principle and the procedure for this assay have been previously described.[12]

Oxygen uptake is determined manometrically. Each Warburg vessel (15 ml capacity) contained the following in a total volume of 2 ml: substrate for respiration, 10 mM; potassium phosphate (pH 7.4), 15 mM; ATP, 2 mM; $MgCl_2$, 5 mM; triethanolamine (or Tris-HCl) buffer (pH 7.4), 15 mM; DPN, 0.5 mM; glucose, 20 mM; KCl (recrystallized from 2 mM EDTA solution), 25 mM; mannitol or sucrose, 60 mM; EDTA, 1 mM; dialyzed crystalline hexokinase, 2.5 units per vessel (or Sigma Type III, 0.6 mg per vessel); and tumor mitochondria (1.2–4 mg of protein). In the center well are placed 0.15 ml of 5 N KOH and a roll of filter paper. The vessels are equilibrated at 30° for 5 minutes before the substrate is tipped in from the side arm of the vessel. The oxygen uptake is measured for the next 30 or 40 minutes. Usually 4–8 microatoms of oxygen and 6–20 micromoles of P_i are taken up. The reaction is stopped by removing the vessel from the bath and is immediately deproteinized with 2 ml of 10% trichloroacetic acid. After centrifugation, 0.05–0.1 ml aliquots of the supernatant solution are removed for P_i determination.[13]

Procedure for Measuring Oxygen Uptake and ADP Respiratory Control Ratio

Oxygen consumption was measured with the use of a Clark oxygen electrode.[14] The reaction vessel contained the following in a total volume of 1.0 ml (or 6.5 ml): substrate for respiration, 5 or 10 mM; potassium phosphate (pH 7.4), 5 mM; $MgCl_2$, 10 mM; triethanolamine buffer (pH 7.4), 30 mM; KCl, 25 mM; serum albumin (dialyzed against 50 mM triethanolamine buffer for 2 days), 0.3 or 1%; KF, 5 mM; EDTA, 1 mM; mannitol (or sucrose), 50 mM; and tumor mitochondria, 0.7 to 1.4 mg of protein/ml (or 5–8 mg dry wt/6.5 ml). After 2 minutes at 25°,

[12] See Vol. II [101]; also see this volume [3].

[13] K. Lohmann and L. Jendrassik, *Biochem. Z.* **178**, 419 (1926). Modified by using 5% (instead of 2.5%) ammonium molybdate in 5 N H_2SO_4. See W. D. Harris and P. Popat, *J. Am. Oil Chem. Soc.* **31**, 124 (1954) as modified by T. Conover, Thesis, Univ. of Rochester, 1959.

[14] See Vol. VI [33]. The electrode used by one of us (R.W.) was modified by mounting the electrode in a horizontal position near the bottom of the reaction vessel (1 ml capacity), so that reagents can be conveniently added from the top of the reaction vessel. This modification was made by Dr. M. E. Pullman and Mr. M. Kandrach and is gratefully acknowledged.

ADP (final concentration 0.1–0.4 mM) was added and the oxygen consumption was recorded for approximately 10 more minutes. The respiratory control ratio (defined as the rate of respiration in the presence of ADP divided by the rate obtained following the expenditure of ADP)[15,16] was then calculated.

Procedure for the Preparation of Ascites Tumor Mitochondria

Method A. Grinding of the Tumor Cells with Mortar and Pestle. The Ehrlich ascites tumor cells are obtained by harvesting the ascites fluids from 4–6 mice (7–9 days tumor age). The fluids are immediately diluted in 2 volumes of ice-cold mannitol[17] (0.25 M) solution containing EDTA (2 mM), pH 7.5, and the cells are centrifuged at 600 g for 5 minutes. The packed cells are washed twice by resuspending the cells in approximately 10 volumes of the same solution and centrifuging as above. For homogenization, the packed tumor cells (approximately 5 ml) are mixed with 1 ml of the mannitol-EDTA solution and 0.1 ml of 1 M triethanolamine buffer (pH 7.4). The mixture is poured into a precooled mortar (7 cm diameter); 3 g of washed glass beads[18] are added, and the mixture is ground forcefully with a pestle for 6 minutes at 4°. Then 25 ml of mannitol-EDTA solution is added to the mortar and the mixture is ground for 20 seconds to give an even suspension. The mixture is poured into a 40 ml polyethylene tube and centrifuged at 600 g for 10 minutes. The supernatant solution is decanted and stored at 4°. The pellet is mixed with 2 ml of the mannitol-EDTA solution, 0.1 ml of 1 M triethanolamine buffer, 2 g of washed glass beads; the mixture is ground for another 4 minutes. Then 25 ml of mannitol-EDTA solution is added and the mixture is centrifuged as above. The supernatant solution from the second batch is combined with that from the first, and the combined solution is divided among 2 polyethylene tubes and centrifuged at 9000 g for 15 minutes in the 9 RA head of the Lourdes refrigerated centrifuge. The supernatant solution and "fluffy layer" are decanted and discarded. The mitochondrial pellets are washed twice after their suspension in 25 ml of mannitol-EDTA solution with a Teflon pestle inserted directly into the centrifuge tube. The final pellets are suspended in 2 ml of the

[15] B. Chance and M. Baltscheffsky, *Biochem. J.* **68**, 283 (1958).

[16] B. Chance, *Ciba Found. Symp. Regulation Cell Metabolism* p. 91. Little, Brown, Boston, Massachusetts, 1959.

[17] Sucrose (0.25 M) solution containing 1 or 2 mM EDTA may be used in place of mannitol-EDTA solution. All further steps were carried out at 4°.

[18] Superbrite glass beads (0.1 mm diameter, type 130-5005, from Minnesota Mining and Manufacturing Company, St. Paul, Minnesota) were washed by soaking in 2 volumes of 5 N HCl overnight and washed with 10 volumes of deionized water about 6 times until the acidity of the washes approached pH 5.

mannitol-EDTA solution and evenly dispersed with the use of a Teflon pestle.

Method B. Dounce Homogenizer Method.[11,19] The ascitic fluid from 5 or 6 mice (8–11 days tumor age) is removed by syringe and transferred to two 40-ml graduated centrifuge tubes packed in ice. All further procedures are carried out in the cold room. The ascitic fluid is centrifuged for 5 minutes at 900 *g*, and the supernatant fluid is decanted and discarded. The packed cell volume should be about 10–15 ml of bloodless cells.[20] The packed cells are washed twice with 30-ml portions of 0.25 *M* sucrose that contains 0.001 *M* disodium EDTA, pH 4–5 (solution S). The twice washed tumor cells are then resuspended with 40 ml of solution S, to which has been added 0.1 ml of 0.1 *M* citric acid, and transferred to a Dounce homogenizer (40-ml size) packed in ice. For support the homogenizer is inserted into a hole in a large rubber stopper.[19] Solution S is added until fluid fills the cylinder and lower portion of the reservoir of the homogenizer. Homogenization is accomplished by 10–15 vigorous, directly vertical passes of the pestle. The homogenate should be examined microscopically for unbroken cells and more passes made if necessary.[21] The resultant homogenate (approximately 60 ml, depending upon the size of the homogenizer used) is divided between two 40-ml centrifuge tubes and centrifuged at 900 *g* for 20 minutes. The top two-thirds of the supernatant solutions obtained are decanted and stored in ice. The sediments are combined and resuspended to 60 ml with solution

[19] A. L. Dounce, R. F. Witter, K. J. Monty, S. Pate, and M. A. Cottone, *J. Biophys. and Biochem. Cytol.* **1**, 139 (1955). Dounce homogenizers may be purchased from Blaessig Glass Specialties, Rochester, New York, or from Kontes Glass, Vineland, New Jersey. The clearance found most useful for the fractionation of ascites cells is 0.0005 inch. This corresponds to the tight-fitting pestle as supplied by Blaessig or the loose-fitting pestle as supplied by Kontes.

[20] Ascites cell samples that are grossly bloody should not be used. Samples that are only slightly bloody are suitable for fractionation after removal of the erythrocytes (RBC). This may be accomplished in two ways. Low speed centrifugations (125 *g* for 4 minutes) in isotonic sucrose will sediment most of the tumor cells while leaving the RBC in the supernatant fluid. The RBC are then decanted and the procedure is repeated until the number of RBC are reduced to acceptable levels. Erythrocytes may also be removed by "differential lysis." In this procedure the contaminated ascites cells are suspended to 4 or 5 times their volume with cold 1:4 diluted Ringer solution and centrifuged for 5 minutes at 960 *g*. The swollen, intact ascites cells will sediment and the supernatant fluid which contains the hemoglobin is decanted. The ascites cells may then be washed in isotonic media. The latter method is recommended since there is little or no apparent damage to the tumor cells and the removal of RBC is quick and nearly complete.

[21] If on microscopic examination the cell breakage is less than 60% after at least 20 passes, it is likely that homogenization is either not vigorous enough or that the clearance of the homogenizer is too large or too small.

S; the homogenization and centrifugation (900 *g* for 20 minutes) are repeated. During the second homogenization, extremely vigorous passes are not necessary. The yield from the second homogenization is usually small. The upper portions of the supernatant solutions obtained from the centrifugation of the second homogenate are combined with those obtained from the first, and the mixture is distributed among 8 No. 40 tubes for the Spinco centrifuge and centrifuged for 15 minutes at 8000 *g*. Any refrigerated centrifuge capable of developing this centrifugal force is suitable. The supernatant solutions and fluffy layers are decanted and discarded. Each sediment is resuspended with 6 ml of solution S by inserting a small motor-drive Teflon pestle (rotating at slow speed) directly into each centrifuge tube. The suspended sediments are again centrifuged at 8000 *g* for 15 minutes, and then this washing procedure is repeated. The twice washed sediments are resuspended to 10 ml (the mitochondrial fraction), which contains about 12–15 mg dry weight per milliliter (8–10 mg protein). The mitochondrial fraction represents a 7–10% recovery of the total dry weight of the homogenate.

Properties of the Ascites Tumor Mitochondria

The P:O ratio, the respiratory control ratio, the yield, and the specific activity (millimicromoles of P_i esterified per minute per milligram of protein) of mitochondria prepared by the two methods are shown in the table.

OXIDATIVE PHOSPHORYLATION WITH EHRLICH ASCITES TUMOR MITOCHONDRIA

Mitochondria prepared by	Yield	Substrate	P:O ratio	Specific activity	ADP respiratory control ratio
Grinding	26[a]	Succinate	1.8	270	5.6
		α-Ketoglutarate	2.4	260	5.3
		Pyruvate + malate	2.4	260	5.0
Dounce homogenization	85[b]	Succinate	1.6	210	4.5
		α-Ketoglutarate	2.4	160	5.7
		Pyruvate + malate	2.2	160	4.0

[a] Milligrams of mitochondrial protein obtained per gram of cell protein.

[b] The values given for the manometric assays represent averages of duplicate determinations for a single preparation. The ADP respiratory control ratio values, which were obtained polarographically, represent averages for 3 (pyruvate + malate), 4 (succinate), or 12 (α-ketoglutarate) preparations. Considerable variation among preparations is to be expected, for example, for the latter 12 preparations the range for the respiratory control ratios was 2.5–9.5. This variation has been noted by others [e.g., P. Borst, *J. Biophys. Biochem. Cytol.* **7,** 381 (1960)].

Comments

Grinding of ascites tumor cells with glass beads in the mortar and pestle (Method A) can be applied to small amounts of cells, e.g., 1 ml of packed cells (140 mg of cell protein), if necessary. When more than 5 ml of packed cells is used, either a larger mortar can be used or the grinding procedure can be operated with 5 ml of packed cells in each run. Homogenization of tumor cells in a Dounce homogenizer (method B) is more applicable to experiments where larger amounts of mitochondria are needed. Mitochondria prepared by these two methods give high P:O ratio and respiratory control ratio, comparable to mitochondria prepared from other tissue sources.

[20] Neoplastic Tissue Mitochondria

By THOMAS M. DEVLIN

The multiplicity of solid tumor types, originating from a variety of normal tissues, makes it impossible to generalize about the isolation, properties, and activities of mitochondria from neoplasms. Even with one tumor type, variations in the amount of connective tissue, degree of necrosis, contamination by normal tissue, and lability of the mitochondria requires that each tumor line be considered as a different entity, with preparative and assay procedures tailored to the individual malignancy.

Homogenization and Isolation Medium

With some tumors, isotonic sucrose (0.25 M) suffices for the isolation of active mitochondria comparable to those from normal tissues.[1-3] This is particularly the case for soft tumors, such as hepatomas, where disruption of the cells occurs under mild homogenization.[3] Higher sucrose concentrations (e.g., 0.35 M) have been employed to yield mitochondria with latent ATPase from Murphy-Sturm rat lymphosarcoma.[4] The use of mannitol, polyvinylpyrrolidone, or raffinose has not been extensively tested, but may yield better preparations from some tumors.[5] Emmelot

[1] R. K. Kielley, *Cancer Res.* **12,** 124 (1952).

[2] A. C. Aisenberg, *Cancer Res.* **21,** 295 (1961); *Cancer Res.* **21,** 304 (1961).

[3] G. E. Boxer and T. M. Devlin, *Science* **134,** 1495 (1961). See also T. M. Devlin and M. P. Pruss, *Proc. Am. Assoc. Cancer Res.* **3,** 315 (1962).

[4] M. Blecher and A. White, *J. Biol. Chem.* **235,** 3404 (1960).

[5] P. Emmelot, C. J. Bos, P. J. Brombacher, and J. F. Hampe, *Brit. J. Cancer* **13,** 348 (1959).

et al.[5] have reported that inclusion of 0.001 M EDTA, pH 7.4, in 0.25 M sucrose yields preparations from some but not all tumors with low ATPase and high phosphorylating activity. It is generally advisable to include EDTA if the tissue is difficult to homogenize (Walker-256 carcinosarcoma, fibrosarcomas, and mammary carcinomas) or the mitochondria are extremely sensitive to aging (rapidly growing hepatomas). The inclusion of nicotinamide (0.02 M) is beneficial in protecting the mitochondrial NAD from destruction by NADase[5,6] during isolation and assay. Isolated mitochondria from some tumors contain high NADase activity[7] presumably due to contamination by microsomes.[8] The low phosphorylating capacities of preparations from some neoplasms (e.g., Novikoff and Dunning hepatomas) can be stimulated by serum albumin in the incubation system.[9] In the author's laboratory, similar P:O ratios were obtained if the albumin (10 mg/ml) was present in the isolation medium or added to the assay system.

Isolation of Mitochondria

The choice of experimental tumor depends on the objectives of the experiment and the availability of the tumor. Some mouse and rat tumors are in general use (e.g., Morris hepatomas, Novikoff hepatoma, Walker-256, Sarcoma-180, adenocarcinoma-755, Murphy-Sturm lymphosarcoma, etc.) and are obtained easily. Many tumors are, however, only available in the laboratory where the tumor originated. The availability of the series of transplanted Morris hepatomas,[10] which vary in their degree of dedifferentiation and rate of growth, permit comparative studies of the activities of mitochondria from normal and neoplastic liver. The methods of isolation employed are variations of that described by Schneider and Hogeboom.[11] All steps are performed at 0–4°. After excising the tumor, adjoining normal tissue and gross necrotic areas are removed, and the tumor is rinsed in cold homogenization medium to remove liquefied necrotic contamination and blood. In comparing yields of mitochondria from different tumors, histological examination is necessary to monitor the degree of contamination by normal tissue and to detect any major change in cell type. With tumors transplanted many

[6] L. A. Sauer, A. P. Martin, and E. Stotz, *Cancer Res.* **22**, 632 (1962).

[7] P. Emmelot and P. J. Brombacher, *Biochim. Biophys. Acta* **22**, 487 (1956).

[8] P. Emmelot, *Exptl. Cell Res.* **13**, 601 (1957).

[9] T. M. Devlin and M. P. Pruss, *Federation Proc.* **17**, 211 (1958). Also T. M. Devlin, unpublished observations.

[10] H. P. Morris, *Advan. Cancer Res.* **9**, 232 (1965).

[11] W. C. Schneider and G. H. Hogeboom, *J. Biol. Chem.* **183**, 123 (1950). See also G. H. Hogeboom, Vol. I [3].

times and growing intraperitoneally this is not a stringent requirement. The tumor is blotted dry with filter paper and weighed.

The tissue is disrupted with a Potter-Elvehjem type homogenizer, employing a Teflon pestle rotated at 2000 rpm; attempts to use other types of mechanical tissue homogenizers have been unsuccessful. After dicing the tissue, homogenization is performed using a ratio of 1 g tumor to 6–9 ml of medium. The minimum number of passes (10–20) is preferable. With soft tumors this should give nearly 100% cell disruption, but with tissues containing large amounts of connective tissue and stroma, the homogenization should be interrupted for recooling or preferably the undisrupted tissue should be rehomogenized in fresh medium. The homogenate, diluted to 10% (w/v), is filtered through two layers of cheesecloth to remove pieces of intact tissue and fibrous material. Nuclei and cell debris are sedimented by centrifugation at 600–700 *g* for 10 minutes. Extreme care should be employed in decanting the supernatant to exclude cell debris and intact erythrocytes. For maximum yields of mitochondria, the pellet can be resuspended (and rehomogenized, if necessary) and recentrifuged at 600 *g*; the supernatants are combined. In this laboratory, this step is omitted because it lengthens the time for isolation and the increase in the yield of mitochondria is usually small.

Mitochondria can be sedimented at forces of 8000–12,000 *g* for 10 minutes; the higher centrifugal forces are often necessary because mitochondria from some tumors are smaller than those of normal tissues, but the possibility of contamination and damage to the mitochondria is also increased. Extreme care is employed in decantation or aspiration to completely remove the buffy coat. A lipid layer frequently observed at the top of the tube, should be completely removed by aspiration and wiping the inner wall of the tube with gauze. This lipid, if it contaminates subsequent steps, can lead to mitochondrial preparations with uncoupled oxidative phosphorylation, due to the presence of free fatty acids.[12]

The mitochondrial pellet is resuspended with the use of a cold finger (thin test tube containing ice) by first making a smooth paste followed by the slow addition of medium. With this technique, clumps of agglutinated mitochondria are rarely observed.

The mitochondria are washed at least two times, employing the same centrifugal speeds as before and with the same volume as the original homogenate. This author has found that mitochondria washed only once are badly contaminated with soluble enzymes and microsomes, and their use leads to erroneous interpretations of the activities assayed. In many instances, three times washed mitochondria have been employed with

[12] P. Emmelot, *Cancer Res.* **22**, 38 (1962).

excellent results. The total time of the isolation procedure should be kept to a minimum, each step being executed without delay; the mitochondria should be used immediately. Because of the varying yields from tumors, the final suspension of mitochondria must be adjusted to fit the needs of the experiment; suspensions containing the equivalent of 2.0–3.0 g of the original wet weight of tissue per milliliter usually yields a suspension containing 1–3 mg protein nitrogen per milliliter.

Assay Systems Employing Neoplastic Tissue Mitochondria

The test systems for measurement of mitochondrial activities (e.g., respiration, phosphorylation, ATPase, swelling, and contraction) are essentially those employed for normal tissue mitochondria.[13] In contrast to liver mitochondria, exogenous cytochrome *c* ($3 \times 10^{-5} M$) and NAD (1.0 m*M*) are often required for maximum rates of respiration, particularly where long-term incubations (manometric studies) are performed.[1,2,14,15] Because of the higher ATPase activity of some neoplastic mitochondria, added fluoride (0.01 *M*) is often an absolute requirement for measurement of phosphorylation. Serum albumin (5–10 mg/ml) in the incubation medium also is employed to obtain maximum P:O ratios and latent ATPase activity with some tumors.[5,9] Nicotinamide (0.01–0.05 *M*) is beneficial if the preparations are contaminated with NADase. In short-term experiments (polarographic measurement of respiration) these various cofactors are often unnecessary, since their addition is usually required to offset the effect of aging that occurs during long incubations.

Properties of Neoplastic Mitochondria

The properties and activities of mitochondria from malignant in comparison to normal tissues have been reviewed by Aisenberg,[16] and the reader is referred to this text for a complete description. It is now generally considered that most of the differences are quantitative rather than qualitative and that some of the alterations, such as higher ATPase and lower pyridine nucleotide content, may be due to changes occurring during isolation.[16] The one general characteristic of neoplastic compared to normal tissue mitochondria is apparently their greater lability during isolation and storage.[1]

Isolated mitochondria from a variety of neoplastic tissues manifest

[13] See this volume [3, 9, and 105b].

[14] H. G. Williams-Ashman and E. P. Kennedy, *Cancer Res.* **12**, 415 (1952).

[15] C. E. Wenner and S. Weinhouse, *Cancer Res.* **13**, 21 (1952).

[16] A. C. Aisenberg, "The Glycolysis and Respiration of Tumors." Academic Press, New York, 1961.

respiratory rates and yield P:O ratios nearly equivalent to rat liver mitochondria.[2,3,5] The relative intactness of mitochondria from a series of rat hepatomas has been demonstrated, in the absence of added cofactors, as measured by their inability to oxidize exogenous NADH, low ATPase and high values of respiratory control.[3] Some degree of respiratory control for neoplastic tissue mitochondria has also been reported by Aisenberg[2] when measured manometrically, and values approaching three to five have been observed[3] when assayed polarographically. The difficulties in obtaining high P:O ratios in earlier studies[14] as well as the inability of some tumors to oxidize fatty acids[5] can in most cases be attributed to their high ATPase activity. Phosphorylation in all cases is abolished by 2,4-dinitrophenol. It is of interest, that with some tumors manifesting phosphorylation, such as the Morris hepatoma 3683 and 3924, 2,4-dinitrophenol does not stimulate a significant ATPase.[3] This observation suggests the potential usefulness of mitochondria from neoplastic tissues in elucidating mechanisms sometimes obscured in normal tissues.

Mitochondria from some malignant tissues, however, have been observed to yield very low P:O ratios, whereas other activities have been normal, suggesting some impairment in phosphorylating capacity.[9] Whether this is due to some subtle damage to the mitochondria during isolation or is actually a manifestation of the state of oxidative phosphorylation in the cell has yet to be determined.

[21] A Technique for the Isolation of Mitochondria from Bovine Lymphocytes

By JOHN TERRANCE DAVIS

In connection with a study in progress at this laboratory, it was necessary to develop a technique for isolating intact mitochondria from bovine lymphocytes. As it is almost impossible to obtain a relatively pure suspension of lymphocytes from peripheral blood, the lymph node was utilized as a source of lymphocytes for mitochondrial study.

The procedure is essentially a modification of standard cellular fractionation procedures,[1] and consists of two steps: (1) the preparation of a suspension of lymphocytes from the nodes, and (2) the isolation of mitochondria from the suspension by standard techniques.

[1] See G. H. Hogeboom, Vol. I, p. 16.

Materials

Solutions. Two different solutions are utilized in this procedure, and their compositions are presented in the table. The "Transport" medium is simply an isotonic sugar solution used to transport excised nodes from the slaughter house to the laboratory. The "Preparation" medium is an isotonic sugar solution containing bovine serum albumin (BSA) and ethylenediaminetetraacetic acid (EDTA). BSA is an inert protein which acts to stabilize the mitochondria. In addition to possibly binding toxic substances ("toxic" to oxidative phosphorylation) released when the cell is disrupted, there is evidently a second stabilizing action of BSA, as has been shown in the preparation of plant mitochondria where BSA is also of value.[2,3] EDTA is used as a chelating agent to bind any ions which may be in the solution as contaminant. In combination with BSA, only a very small amount is necessary. Both media may be frozen if desired.

COMPOSITION OF SOLUTIONS USED IN ISOLATION OF MITOCHONDRIA FROM BOVINE LYMPHOCYTES

Medium	Substance	Conc.
Transport	Sucrose	0.75 *M*
	Mannitol	0.23 *M*
Preparation	Sucrose	0.75 *M*
	Mannitol	0.23 *M*
	EDTA	5×10^{-5} *M*
	BSA	0.5%
	PO_4 buffer (pH 7.2)	0.01 *M*

Methods

Acquisition of Nodes. Normal nodes obtained at the abattoir or diseased nodes obtained at autopsy should be removed from the freshly killed animal as quickly as possible. Intact nodes are placed directly into a 1000-ml beaker containing 500 ml of cold "Transport" medium in an ice bucket. About 150 g of nodal tissue is required to yield approximately 50 mg of mitochondrial protein. Probably no more than 2 hours should elapse from the obtaining of the nodes until the isolation procedure begins.

[2] G. D. Thorneberry, *Plant Physiol.* **36**, 302 (1961).
[3] J. T. Wiskich and W. D. Bonner, *Plant Physiol.* **38**, 594 (1963).

Preparation of Cell Suspension. Nodes are trimmed free of surrounding fat and cut into small pieces of approximately 1 cubic centimeter in volume. The capsule need not be trimmed off these pieces, but any grossly necrotic material should be excluded. Five to seven of these pieces are pressed in a small, prechilled, hand-operated tissue press, the bottom of which is perforated with holes 2 mm in diameter. The size of the holes is such that the thick, interlacing stroma of the node is held back while the cells and clumps of cells pass through freely. The material expressed is directly placed into a flask containing 75 ml of chilled "Preparation" medium. The pieces are stirred and repressed until no more material is expressable from the pulp remaining. When the cells expressed make a thick paste on the bottom of the tissue press, a spatula is employed to scrape them into the "Preparation" medium. This entire procedure is repeated until all the nodes have been so treated. The flask is then agitated briefly to break up any remaining clumps, and the contents are filtered through a single layer of cheesecloth into a second prechilled flask.

Microscopic examination of the resultant suspension reveals a relatively pure suspension of lymphocytes with some red cells, monocytes, macrophages, granulocytes, and cellular debris as contaminants. The degree of purity of the lymphocyte suspension will depend on the physiological or pathological state of the node and the blood picture in general.

Isolation of Mitochondria from Cell Suspension. The cell suspension is homogenized in aliquots using a 50-ml homogenizer chamber with a serrated-tip Teflon pestle. A moderately slow speed is utilized as the Teflon core makes 5 passes through the suspension. The homogenate is spun down in a refrigerated centrifuge (4°) for 10 minutes at 755 *g*. The supernatant is collected and the precipitate, which contains nuclear matter and cell debris, is discarded. The supernatant is again centrifuged at 755 *g* for 10 minutes to sediment any additional debris. The precipitate is discarded, and the supernatant is centrifuged for 15 minutes at 7710 *g*. This will bring down the mitochondria; the supernatant containing the microsomal fraction is discarded. Resuspension is accomplished in preparation medium by vigorous pipetting with a capillary pipette. The mitochondria are reprecipitated by centrifuging at 7710 *g* for 15 minutes. The supernatant is discarded; the resultant pellet of mitochondria may be rewashed as necessary, depending on the requirements of the particular experiment.

The final preparation is made by resuspending the mitochondria in 0.5–2.5 ml of "Preparation" medium, depending on the protein concentration or oxygen consumption desired.

Results

The mitochondria obtained are well coupled, showing marked stimulation by adenosinediphosphate (ADP) and dinitrophenol (DNP) in the presence of suitable substrates. In a typical preparation studied with the oxygen electrode, using a concentration of malate (with glutamate and malonate) which will yield a maximal velocity of oxygen consumption, and using 0.1 ml of mitochondrial preparation, ADP will increase oxygen consumption from 6 mμmoles O_2 per 3 ml per minute to 60 mμmoles per 3 ml per minute. Succinate-supported rates of respiration are even greater for the same amount of mitochondrial protein. The preparation is stable for at least 3 hours after preparation.

Acknowledgment

The author wishes to acknowledge his indebtedness to Drs. David Tyler, John Furth, Peter Nowell, and Ronald Estabrook for their help in the development of this procedure.

[22] Insect Mitochondria

By Simon G. Van den Bergh

Introduction

Although the isolation of mitochondria has been described from a variety of insect tissues during various stages of insect development, we shall in this chapter confine ourselves to mitochondria isolated from the flight muscle of adult insects.

On the basis of their metabolic fuel for flight activity, winged insects can be roughly divided into two groups. Insects capable of long-lasting flight, like locusts and butterflies, use fat as major substrate for flight-muscle activity, whereas insects capable of only short flights, like flies and bees, use carbohydrate as their main source of flight energy. Accordingly, flight-muscle mitochondria, isolated from the two groups of insects, have partly different properties. As examples of the two groups of mitochondrial preparations, house fly and locust mitochondria are described below. Reasonably good preparations from honey bee,[1] blow

[1] D. D. Hoskins, V. H. Cheldelin, and R. W. Newburgh, *J. Gen. Physiol.* **39**, 705 (1956).

fly,[2] mosquito,[3] cockroach,[4] and Colorado potato beetle[5] have also been described.

General Properties of Insect Mitochondria

Characteristic properties of all insect mitochondria are their low stability, their exceptionally high respiratory and phosphorylative activity with their physiological substrates, their relatively poor rate of oxidation of Krebs-cycle intermediates and the low P:O ratios accompanying these slow oxidations. (By addition of 1–2% bovine serum albumin to the reaction medium, these P:O ratios can be raised to values comparable to those of mammalian mitochondria.)

The phosphorylating respiratory chain of insect mitochondria strongly resembles that of mammalian mitochondria,[6] and it is affected by the same inhibitors and uncouplers in comparable concentrations. Insect ubiquinone is a mixture of three homologs with eight, nine, and ten isoprenoid units in the side chain. Insect mitochondria contain 3–4 millimicromoles of NAD per milligram of protein and little, if any, NADP (less than 0.2 millimicromole per milligram of protein).

Like other types of muscle mitochondria, insect mitochondria exhibit a considerable ATPase activity in the presence of Mg^{++} ions. Reversed electron transport can be demonstrated by the α-glycerophosphate-linked reduction of NAD.[7] All insect mitochondria incorporate labeled amino acids into mitochondrial proteins, accumulate divalent cations, and show the normal pattern of swelling and shrinkage phenomena, but only mitochondria from fat-utilizing insects can incorporate labeled acetate and malonate into long-chain fatty acids.

The markedly lower stability of insect mitochondria as compared with other animal mitochondria requires special precautions: (1) Both the isolation and the reaction medium should contain at least 1 m*M* EDTA. (2) The number of manipulations during the isolation procedure, the total duration of this procedure, and the time between preparing and testing the mitochondria should be kept to a minimum. (3) The initial tissue extract should be as concentrated as possible. (4) All operations should be carried out as close to 0° as possible. (5) All reactions should be started by addition of the mitochondria to an otherwise complete reaction medium.

[2] M. I. Watanabe and C. M. Williams, *J. Gen. Physiol.* **34,** 675 (1951).
[3] O. Gonda, A. Traub, and Y. Avi-Dor, *Biochem. J.* **67,** 487 (1957).
[4] D. G. Cochran, *Biochim. Biophys. Acta* **78,** 393 (1963).
[5] D. Stegwee and A. R. Van Kammen-Wertheim, *J. Insect. Physiol.* **8,** 117 (1962).
[6] R. W. Estabrook and B. Sacktor, *Arch. Biochem. Biophys.* **76,** 532 (1958).
[7] M. Klingenberg and T. Bücher, *Biochem. Z.* **334,** 1 (1961).

House Fly (*Musca domestica*) Mitochondria

Isolation Method

Animals. Although the various activities of the mitochondria are independent of the adult age of the experimental animals, the yield of mitochondrial protein increases about threefold during the first week of adult life. It is, therefore, advantageous to use flies older than 9 days after emergence for the preparation of mitochondria. No differences have been found between mitochondria isolated from male and female house flies.

Isolation medium. 0.154 *M* KCl, 1 m*M* EDTA, adjusted at 0° to pH 7.4 with KOH.

Isolation procedure.[8] Two hundred flies are immobilized by cooling below 4° for 30 minutes and placed in the cold room on a glass plate resting on ice. The heads and abdomens are cut off with the help of a small scalpel and tweezers. The remainder of the flies, consisting of the thorax with attached legs and wings, is placed in a chilled mortar, 5 ml of isolation medium is added, and the mash is gently pounded (without any grinding action) until a smooth brei results. This brei is filtered by suction through two layers of muslin (previously saturated with isolation medium and placed in a small Büchner funnel) into a Thunberg tube immersed in an ice bath. The filtrate is centrifuged at 4° for 3 minutes at 150 *g* to remove pieces of wings and scales and muscle fibrils, if present. The slight, grayish-black powderlike sediment is discarded, and the supernatant is again centrifuged for 8 minutes at 3000 *g*. After the second centrifugation, the supernatant is decanted and any fat that might stick to the wall of the tube is removed with a paper tissue. The mitochondrial pellet is rinsed carefully, without disturbing the pellet, with two portions of isolation medium and then suspended in 4 ml of isolation medium with the help of a small Potter-Elvehjem homogenizer.

Yield. This procedure gives mitochondrial suspensions containing 6–8 mg of protein per milliliter, representing a yield of 52–70%.

Properties of Mitochondria

The respiratory and phosphorylative activities of house fly mitochondria are summarized in Table I. Only α-glycerophosphate and pyruvate—the end products of glycolysis in the insect flight muscle and, therefore, the physiological substrates for the mitochondria—are oxidized at appreciable rates. The low rate of oxidation of Krebs-cycle inter-

[8] This procedure is based on the technique described by Watanabe and Williams.[2]

TABLE I
PROPERTIES OF HOUSE FLY MITOCHONDRIA[a]

Substrate	Respiratory rate (mμmoles O_2/ minute per mg protein)	P:O ratio	Respiratory control index[b]
α-Glycerophosphate	500[c]	1.3	1.7
Pyruvate	504	2.4	9.8
Succinate	16	0.4	1.0
Glutamate	16	0.7	1.0
α-Oxoglutarate	11	0.8	1.0

[a] Values given are means of a great number of manometric experiments at pH 7.5 and 25°.

[b] Ratio of respiratory rates before and after addition of phosphate acceptor to an acceptorless system.

[c] This rate can be increased up to 820 by addition of cytochrome *c* to the reaction medium. Phosphate esterification is not enhanced by addition of cytochrome *c* and consequently the P:O ratio is markedly lowered. Pyruvate oxidation is unaffected by addition of cytochrome *c*.

mediates and glutamate is caused by the very limited permeability of the mitochondria toward these substrates.[9] It is particularly noteworthy that the mitochondria can oxidize pyruvate at optimal rates for periods up to 1 hour in the absence of added Krebs-cycle intermediates.

Respiratory control by omission of inorganic phosphate can also be demonstrated.[9] The mitochondria are completely devoid of endogenous substrate. Neither free fatty acids nor their carnitine esters can be oxidized by house fly mitochondria.

The mitochondria have a considerable ATPase activity in the presence of Mg^{++} ions, which is 93–96% inhibited by oligomycin. The ATPase activity is stimulated about sixfold by addition of 0.1 m*M* 2,4-dinitrophenol.

Locust (*Locusta migratoria*) Mitochondria

Isolation Method

Animals. Optimal results are obtained with mitochondria isolated from locusts 7–17 days after the final molt.[10] At this age locusts show the greatest tendency for migratory flights.

[9] S. G. Van den Bergh and E. C. Slater, *Biochem. J.* **82,** 362 (1962). See also S. G. Van den Bergh, *Biochem. J.* **93,** 128 (1964).

[10] The optimal age may vary with different rearing conditions. In the author's laboratory locusts are grown at 34° and 65% relative humidity with 16 hours of artificial light per day.

Isolation Medium. 0.154 *M* KCl, 1 m*M* EDTA, adjusted at 0° to pH 7.4 with KOH.

Isolation procedure.[11] In the cold room, twelve animals are killed by twisting their heads through 90° in both directions. This should be done carefully so as not to disconnect the head from the gut. The abdomen is severed with a pair of scissors, and the head is pulled off so that the attached gut is now removed from the thorax. The thorax is cut open ventrally by inserting one point of the scissors in the hole from which the gut has been removed. The thorax is opened and pinned to a small dissecting table of cork or solid paraffin. The fat body is wiped off with paper tissue. Small scissors with curved points are used to cut loose the flight muscles, first laterally and then in the middle; the muscles are immediately transferred to a small beaker containing 20 ml of ice-cold isolation medium.

The muscles are ground for about 30 seconds at low speed in a chilled Potter-Elvehjem homogenizer. The homogenate is filtered through a small bag of finely woven nylon cloth. The contents of the bag are washed with another 10 ml of isolation medium and gently squeezed. The combined filtrates are subsequently centrifuged for 4 minutes at 150 *g* in a cooled centrifuge. The supernatant fluid is carefully decanted and centrifuged for 8 minutes at 3000 *g*. After the second centrifugation the supernatant is discarded and fatty deposits on the wall of the tube are removed with paper tissue. The precipitate of mitochondria is resuspended in 12 ml of isolation medium merely by stirring with a glass rod, and the suspension is again centrifuged for 8 minutes at 3000 *g*. The mitochondrial precipitate is finally resuspended in 2 ml of isolation medium with the help of a small Potter-Elvehjem homogenizer.

Yield. This procedure yields mitochondrial suspensions containing 6–8 mg of protein per milliliter.

Properties of Mitochondria

The respiratory and phosphorylative activities of locust mitochondria are summarized in Table II. Noteworthy differences from house fly mitochondria are in the rate of respiration and phosphorylation with glutamate as substrate, in the fact that locust mitochondria can oxidize pyruvate only for very short periods in the absence of added Krebs-cycle intermediates, in their ability to oxidize carnitine esters of fatty acids (but not free fatty acids),[12] and in the substantial amount of endogenous substrate present in locust mitochondria.

[11] This procedure is based on the technique described by M. Klingenberg and T. Bücher, *Biochem. Z.* **331**, 312 (1959).

[12] C. Bode and M. Klingenberg, *Biochim. Biophys. Acta* **84**, 93 (1964).

TABLE II
PROPERTIES OF LOCUST MITOCHONDRIA[a]

Substrate	Respiratory rate ($m\mu$moles O_2/minute per mg protein)	P:O ratio
α-Glycerophosphate	140[b]	0.9
Pyruvate + malate	238	2.4
Succinate	49	0.5
Glutamate	170	2.2
α-Oxoglutarate	74	1.2
Palmitoyl-carnitine[c]	179	2.1

[a] Values given are means of a great number of manometric experiments at pH 7.5 and 25°.

[b] This rate can be doubled by addition of cytochrome *c* to the reaction medium. See, however, footnote *c* to Table I.

[c] Bovine serum albumin, 2%, was added. Carnitine esters of long-chain fatty acids are potent detergents; at concentrations at which they are used as substrates in manometric experiments, they completely lyse the mitochondria within a few seconds unless bovine serum albumin is added to the reaction medium.

Respiratory control by omission of phosphate or phosphate acceptor can be demonstrated with pyruvate plus malate and glutamate as substrates. With α-glycerophosphate as a substrate, respiratory control is weaker and less reproducible.

[23] Avocado and Apple Mitochondria

By J. T. WISKICH

General Considerations

The preparation requires that the tissue be disrupted in isotonic sucrose under conditions which will give reasonable yields and yet not disrupt the mitochondria. Acidity, tannin formation, and oxidation products lead to inactivation of mitochondria and release of fatty acids can cause uncoupling of oxidative phosphorylation. These difficulties are usually overcome by adding alkaline buffers, polyvinylpyrollidone, cysteine (or ascorbate), and bovine serum albumin to the isolating medium. To date, only soluble polyvinylpyrollidone has been used, but the use of an insoluble cross-linked form[1] may be more advantageous.

[1] W. D. Loomis, and J. Battaile, *Plant Physiol.* **39,** Suppl., xxi (1964).

The activities of mitochondria isolated from these fruits tend to display seasonal variations[2,3] which are particularly associated with the climacteric pattern of respiration. These variations are possibly not inherent but a reflection of the "ease" or "difficulty" in isolating the desired mitochondria. Apart from problems associated with the presence of inhibiting-substances in the tissue there is sometimes a need for strong shearing forces to disrupt the tissue. These forces are also detrimental to the mitochondria.

The property of respiratory control[4] is regarded here as a biochemical basis of intactness.

Preparation

The media, utensils, and tissue are prechilled before use. All operations are performed at 0–1°. The practice of continually monitoring pH and adding alkali is preferred to either adding a strong buffer or allowing the disruption of the tissue to neutralize an extremely high original pH of the medium.

Avocado

Materials

Kitchen-type grater
Muslin
Mortar and pestle
Potter-Elvehjem type homogenizer
Narrow-range indicator paper, 1 *N* KOH

Reagents

Isolating medium: 0.4 *M* sucrose containing 5 m*M* EDTA pH 7.0, 6 m*M* $MgCl_2$, and 4 m*M* cysteine
Washing medium: 0.4 *M* sucrose containing 6 m*M* $MgCl_2$.

Procedure. The tissue (100 g) is grated and added to 300 ml of isolating medium, ground with sand,[5] while the pH is maintained between 7.0 and 7.5 by dropwise addition of KOH. The brei is squeezed through muslin; the filtrate, diluted with 200 ml of the washing medium, is sub-

[2] J. T. Wiskich, R. E. Young, and J. B. Biale, *Plant Physiol.* **39,** 312 (1964).
[3] A. C. Hulme and J. D. Jones, *in* "Enzyme Chemistry of Phenolic Compounds" (J. B. Pridham, ed.), p. 97. Pergamon, London, 1963. See also A. C. Hulme, J. D. Jones, and L. S. C. Wooltorton, *Proc. Roy. Soc.* **B158,** 514 (1963).
[4] B. Chance and M. Baltscheffsky, *Biochem. J.* **68,** 283 (1958).
[5] Acid-washed sand, 40–100 mesh.

jected to the following centrifugings: (a) 1500 *g* for 15 minutes—discard precipitate; (b) 10,000 *g* for 15 minutes—discard supernatant; (c) suspend pellets in 200 ml of washing medium with the aid of the homogenizer; 2500 *g* for 10 minutes—discard precipitate; (d) repeat b; (e) repeat c; (f) repeat d; (g) suspend pellets in 3 ml of washing medium. Any lipid layers formed during centrifugation are removed by suction. Normal yield is 3–4 mg mitochondrial nitrogen per 100 g avocado tissue.

Comment. This procedure has consistently yielded a highly active preparation from climacteric-peak fruit.[2] However, on standing for some time in an ice-bucket the suspension may tend to lose activity and to aggregate; this loss of activity may be partially overcome by the inclusion of bovine serum albumin (1%). Aggregation is a problem during isolation and is partially counteracted by the repeated low speed centrifugings, and the use of a large volume of washing medium containing $MgCl_2$. A modification[6] of the above procedure permits the isolation of active mitochondria from fruit at earlier stages of ripeness, but the yield is low.

Apple

Materials

Kitchen-type grater
Blendor
Muslin
Potter-Elvehjem type homogenizer
Narrow-range indicator paper (pH 6.4–8.0)

Reagents

Isolating medium: 0.4 *M* sucrose containing 10 m*M* EDTA, 1% bovine serum albumin (fraction V powder), 1% polyvinylpyrollidone (mol. wt. 30,000), 0.05% cysteine

Procedure. Apple cortex (200 g) is grated into the isolating medium (300 ml) while the pH is continually maintained at approximately 7.4 by addition of 1 *M* Tris from a burette. The mixture is blended by stepwise increases in voltage (0–70 volts). Blending is continued for 20 seconds, and the brei is filtered through muslin. The filtrate is subjected to the following centrifugings: (a) 1500 *g* for 15 minutes—discard precipitate; (b) 10,000 *g* for 15 minutes—discard supernatant; (c) suspend pellets in 100 ml of 0.4 *M* sucrose with the aid of a homogenizer; 10,000 *g*

[6] G. E. Hobson, C. Lance, R. E. Young, and J. B. Biale, *Nature* **209**, 1242 (1966).

for 15 minutes; (d) repeat c; (e) suspend pellets in 2 ml of 0.4 *M* sucrose. Normal yield is 0.5–1.0 mg of mitochondrial nitrogen per 100 g apple tissue.

Comments. The procedure has been used successfully with late-preclimacteric fruit, but the yields were low. Severe blending caused complete loss of activity, and the use of a mortar and pestle brought no improvement. Mitochondria have also been isolated from apple peel,[7] which was previously infiltrated with buffer. Maintenance of an extremely high pH[8] is not necessary for activity.

Assay

Oxygen consumption can be measured either manometrically (see this volume [3]) or polarographically (see this volume [7]), for which a coated electrode is preferred because of the extraneous matter sometimes contained in these preparations. Partial reactions of the electron-transfer chain can be measured spectrophotometrically (see this volume [4]). A standard reaction mixture used successfully for both oxidative and phosphorylative studies contains 0.25 *M* sucrose, 10 m*M* potassium phosphate buffer (pH 7.2), 10 m*M* Tris-HCl buffer (pH 7.2), 5 m*M* $MgCl_2$. Under some conditions the inclusion of 0.5 m*M* EDTA or of 1% bovine serum albumin is beneficial. Oxidative phosphorylation can be measured with a hexokinase-glucose trap[2] (see Vol. II [101]). Total nitrogen can be determined by acid digestion and titration.[9] The biuret or Nessler methods are not always successful with isolated plant mitochondria.

Properties

The mitochondria usually require the addition of cofactors for maximum oxidative activity. Cytochrome *c*, NAD, and thiamine pyrophosphate are the most common additives required. With oxygen electrode studies of succinate oxidation the first addition of ADP produces a sluggish response[2] which can be alleviated by preincubation with ATP[10] for a few minutes. This has been attributed to oxalacetate inhibition of succinate oxidation.[11] Both the above preparations are green, presumably because of the presence of chloroplast fragments, but there is no evidence

[7] A. C. Hulme, J. D. Jones, and L. S. C. Wooltorton, *Phytochemistry* **3**, 173 (1964). See also J. D. Jones, A. C. Hulme, and L. S. C. Wooltorton, *Phytochemistry* **3**, 201 (1964).

[8] M. Lieberman, *Plant Physiol.* **35**, 796 (1960).

[9] H. A. McKenzie and H. S. Wallace, *Australian J. Chem.* **7**, 55 (1954).

[10] J. T. Wiskich and W. D. Bonner, *Plant Physiol.* **38**, 594 (1963).

[11] M. Avron (Abramsky) and J. B. Biale, *J. Biol. Chem.* **225**, 699 (1957).

to suggest interference due to this contamination. Some of these fragments can be removed by narrowing the differential of centrifugal force or by centrifuging through a layer of 1 *M* sucrose. The preparations are free of endogenous oxidizable substrates as judged by polarographic experiments.

Discussion

Avocado. The operation of the complete tricarboxylic acid cycle in isolated mitochondria has been established.[12] The P:O ratios obtained from polarographic studies approach the expected maxima, but these have not yet been confirmed by direct analysis.[2] Respiratory control ratios for the oxidation of succinate, malate, and α-ketoglutarate (which is severely inhibited by the substrate-level phosphorylation) have been reported.[2,3] Malate oxidation proceeds at a gradually decreasing rate, but this can be prevented by adding glutamate which is not oxidized.[2] No information is available on the cytochrome components.

Apple. The difficulties of isolating the mitochondria and of using a reference on which to base the results are discussed elsewhere.[3,7] True respiratory control ratios have not yet been reported, but those obtained do not exceed 2. However, P:O ratios approaching the expected maxima have been reported, when polyvinylpyrollidone and bovine serum albumin are used in the isolation medium.[7] The presence of a *b* type cytochrome and of cytochrome a_3 has been detected.[8]

[12] M. Avron (Abramsky) and J. B. Biale, *Plant Physiol.* **32**, 100 (1957).

[24] A General Method for The Preparation of Plant Mitochondria

By WALTER D. BONNER, JR.

General Principles

The general principles for the isolation of mitochondria from higher plants remain basically those outlined in 1955 by Axelrod.[1] Since that time, however, it has become possible to define precisely what properties isolated mitochondria must possess if they have been isolated in a reasonably intact condition. These properties include: (a) good respiratory control by ADP concentration. This control must include not only acceleration of substrate oxidation rate by ADP addition, but also *inhibition* of oxidation rate when the ADP concentration in the reaction

[1] B. Axelrod, Vol. I, p. 19.

medium reaches zero; (b) no influence of externally added DPN on the rate of oxidation of DPN-linked substrates and no influence of added cytochrome *c* on any substrate oxidation rate; (c) retention of respiratory control for some hours following isolation; (d) retention of mitochondrial fine structure and minimal contamination by nonmitochondrial structures as determined by electron microscopic examination.

For various reasons the methods outlined by Axelrod fail to produce mitochondrial suspensions that meet the criteria listed above. The problems that have been overcome to meet these criteria include: (a) devising suitable extracting and washing media; (b) devising suitable methods for the disruption of plant cell walls; (c) defining the centrifugal forces required to sediment the mitochondria.

The method outlined below, when followed carefully, will produce mitochondrial suspensions that fulfill adequately the above-listed requirements. The method is inefficient in the sense that only a small percentage of the total mitochondrial population present in a plant tissue is recovered. However, at the present state of the art one can proceed best as outlined below. Large quantities of tissue are used to isolate small yields of very high quality mitochondria. It is anticipated that the efficiency of the isolation will improve in the years ahead.

Choice of Tissue

While it is possible to prepare mitochondria from selected tissues of all higher plants, it remains obvious that some kinds of tissues and organs are eminently more suitable for experimental material than are others. Those plants whose tissues contain rigid, highly lignified cell walls should be avoided for routine use, as should tissues which release a large amount of vacuolar acid on cell rupture. In general, mitochondrial preparations that are free of green chloroplast fragments are desirable; hence, the choice of starting material is limited to those tissues which do not contain chlorophyll. Such considerations have led to the widespread use of two sources of starting material, viz; chlorophyll-free organs that are obtainable from local markets and young, etiolated (dark-grown) seedlings. Widely used plant tissues that are obtainable from local markets include sweet potatoes (*Ipomea batatas*), white potatoes (*Solanum tuberosum*), cauliflower buds (*Brassica oleracea batrytis*), cabbage heads (*Brassica oleracea*), Jerusalem artichoke (*Helianthus tuberosus*), beet root (*Beta vulgaris*), summer squash (*Cucurbita pepo*), and carrots (*Daucus carota*). When one considers the complexities of modern day agriculture and subsequent marketing it is indeed remarkable that it is possible to isolate good mitochondria from market purchased vegetables. The fact remains, however, that it is not possible to prepare consistently good mitochondria over a period of time from produce purchased at a local market. The variation in mitochondrial preparations is due not only to variations in the "freshness" of the material, but also to marked differences between preparations of mitochondria isolated from storage tubers. These differences depend on the stage of dormancy of the tuber. Thus, mitochondria isolated from freshly harvested white potato tubers (which are in deep dormancy) are markedly different in many respects (concentration of pyridine nucleotide, respiratory control, etc.) from those isolated from tubers that are breaking dormancy. Obviously local markets sell potatoes that range all the way from deep dormancy to tubers that are actively sprouting. Therefore it is strongly recommended that the investigator use young etiolated seedlings, grown under controlled environmental conditions. Such material is always fresh and of the same physiological age, two criteria that are important in the standardization of mitochondrial preparations. If one has a source of freshly harvested produce, white potatoes and sweet potatoes are excellent material for mitochondrial investigations but, in contrast to etiolated seedlings, the use of such material is seasonal.

Methods for Growing Etiolated Seedlings

There are a variety of seedlings that grow rapidly in the dark. Those that are most easily grown include any one of a wide variety of beans. Mung beans (*Phaseolus aureus*) and black valentine beans (*Phaseolus vulgaris*) have proved to be reliable sources of tissue. Mung beans are ready for harvest 5 days after planting, black valentine beans require 7–8 days. One need not restrict oneself to these seedlings, however; most dicotyledonous seedlings can be used.

The seeds that are used must be fresh and of high percentage germination. Seeds should be soaked in 0.5 NaOCl for 10 minutes and then in water for 4–6 hours. The soaked seeds are then germinated and grown either in moist vermiculite or, preferably, on perforated aluminum holders that are suspended over water-filled trays. The seedlings grow well at 28° and 50% relative humidity. White light must be rigidly excluded from the germinating and growing seedlings, low intensity green light has no harmful physiological effects on them. Harvesting of the seedlings can be performed in white light. Only etiolated stems (hypocotyls) are used for mitochondrial preparation; the roots, cotyledons, and leaves must be removed (Fig. 1).

Preparation of the Mitochondria

The method is outlined in the flow diagram (p. 130). As indicated, preparation of tissue prior to grinding is important. High starch-contain-

FIG. 1. Separating the roots and leaves from mung bean hypocotyls.

ing objects, e.g., potatoes, should be stored overnight at 4° in order to reduce the number of large starch granules. Large bulky tissues, such as potatoes, sweet potatoes, turnips, squash, must be peeled, washed, and then cut into cubes 2–3 cm on a side. Cauliflower heads are similarly cut into small pieces. In the case of dark-grown seedlings the chilled tissue is cut into sections 2–3 cm long; this is most easily accomplished with a sharp knife and a wooden butcher's block.

As indicated, the grinding medium consists of a cold mixture of mannitol (0.3 *M*), EDTA (1 m*M*), bovine serum albumin (0.1%), and cysteine (0.05%). The cysteine should be added immediately before using the medium. All operations must be carried out in the cold, 0–4°. Mannitol is preferable to sucrose in the grinding medium because it is easier to separate mitochondria from starch in mannitol. Actually, the concentration of mannitol is not critical, but 0.3 *M* has proved excellent for a very wide variety of plant tissues. The addition of phosphate buffer to

Grinding medium	Wash medium
0.3 *M* mannitol 1.0 m*M* EDTA 0.1% BSA 0.05% cysteine pH 7.2	0.3 *M* mannitol 1.0 m*M* EDTA 0.1% BSA pH 7.2
Massive tissues	**Etiolated seedlings**
Peel, wash, chill, cut in 2 - 3 cm cubes. 1 part tissue : 2 parts grinding medium. Grind in electric salad maker. Adjust pH to 7.2	Cut chilled hypocotyls into sections 2 - 3 cm long; suspend in grinding medium. Hand grind 100 - 200 ml aliquots for 30 seconds. Adjust pH to 7.2
Brei squeezed through muslin (calico)	Brei squeezed through muslin (calico)

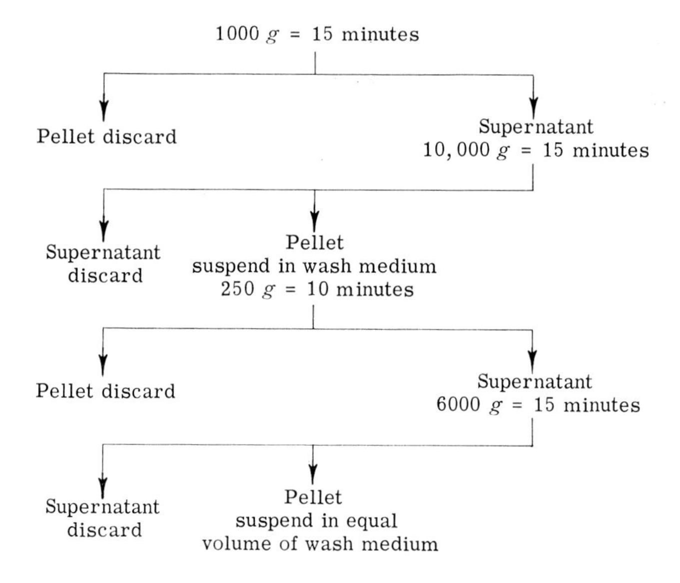

the media, which has been recommended in the past,[1,2] is to be avoided. Not only does phosphate extract cytochrome *c*, but it also leads to mitochondrial rupture. The presence of a mild reducing agent in the grinding medium is required in order to minimize the concentration of oxidizing compounds, e.g., quinones, which are formed on cell rupture.

Bovine serum albumin (BSA) has the capacity of binding a variety of substances. Why BSA is so beneficial to the mitochondrial preparation is

[2] A. Millerd, J. Bonner, B. Axelrod, and R. Bandursky, *Proc. Natl. Acad. Sci. U.S.* **37**, 855 (1951).

not known, but its presence in the grinding medium is required for the isolation of good mitochondria from most plant tissues.[3-7]

There is a difference in the grinding procedure employed for etiolated seedlings and massive tissues. The cut-up seedling stem sections are suspended in grinding medium (tissue to medium ratio = 1:2) and then 100–200 ml aliquots are hand-ground for 30 seconds in a cold mortar. The resulting brei is squeezed through muslin (calico) and centrifuged as outlined in the flow diagram. During the grinding procedure it is important to monitor the pH, using narrow range indicator paper, and to maintain the pH at 7.2 by the dropwise addition of KOH. The importance of pH control is discussed by Wiskich and Bonner.[6]

Mitochondria can be liberated from massive tissues through the use of an electric salad maker. [Such an instrument, the "Moulinex Electric Professional Salad Maker," is obtainable from Varco Inc., 91 Broadway, Jersey City 6, New Jersey or Varco Inc., 11 Rue Jules-Ferry, Bagnolet (Seine), France.] The grater drum of this salad maker does an excellent job of rupturing plant cells in tissues of potatoes, sweet potatoes, squash, cauliflower, etc.; however, it does not work well with the etiolated seedlings. The cut-up tissue is pushed through the salad maker together with a continuous stream of grinding medium. Again, the tissue to grinding medium ratio should be 1:2. In order to allow a continuous stream of grinding medium to pass through the grater, holes are drilled in the lower end of the plunger. Again, it is very important to monitor and maintain the pH of the ground tissue at 7.2. Before centrifugation, the brei is squeezed through muslin (calico).

As the flow diagram shows, after an initial low speed centrifugation, the mitochondria are sedimented at 10,000 g and, after washing, resedimented at 6000 g. The supernatant fluid is most conveniently removed from the sedimented mitochondria by aspiration. It is very important to remove all the fluffy material which layers on top of the mitochondrial pellet and to separate carefully the mitochondrial pellet from the starch immediately below it.

An indication of the amount of tissue required to produce a reasonable amount of high-quality mitochondria (approximately 1.5 ml mitochondria which contain on the order of 150 mg of mitochondrial protein) can be produced from the following amounts of tissues: (a) 2.5 kg of 5-day old

[3] C. A. Price and K. V. Thimann, *Plant Physiol.* **29,** 113 (1954).
[4] F. L. Crane, *Plant Physiol.* **32,** 619 (1957).
[5] G. D. Throneberry, *Plant Physiol.* **36,** 302 (1961).
[6] J. T. Wiskich and W. D. Bonner, *Plant Physiol.* **38,** 594 (1963).
[7] C. Lance, G. E. Hobson, R. E. Young, and J. B. Biale, *Plant Physiol.* **40,** 1116 (1965).

mung bean hypocotyls; (b) 8 kg of white potatoes; (c) 5 kg of Jerusalem artichoke.

Assay of the Mitochondria

The mitochondrial preparations are assayed in a pH 7.2 reaction medium consisting of 0.3 M mannitol, 10 mM potassium phosphate buffer, 10 mM KCl, and 5 mM $MgCl_2$. The mitochondria are suspended, most conveniently in the reaction medium in the cuvette of an "oxygen electrode." The mitochondria should actively oxidize succinate, malate, and DPNH; the apparent Michaelis constants for these three substrates, as determined with mung bean mitochondria, are respectively 0.4, 3.0, and 0.07 mM. α-Ketoglutarate is oxidized by the plant mitochondria, but this oxidation requires the addition of thiamine-pyrophosphate. Citrate is oxidized at rates much lower than the above-mentioned substrates.

With succinate as substrate in the first state 4, the "substrate rate" is much slower than subsequent state 4's; the inhibited "substrate rate" can be eliminated by pretreatment of the mitochondrial suspension with a small amount (50 μM) of ATP. Examples of succinate, malate, and DPNH oxidations by mitochondria isolated from white potatoes and mung bean hypocotyls are shown in Fig. 2.

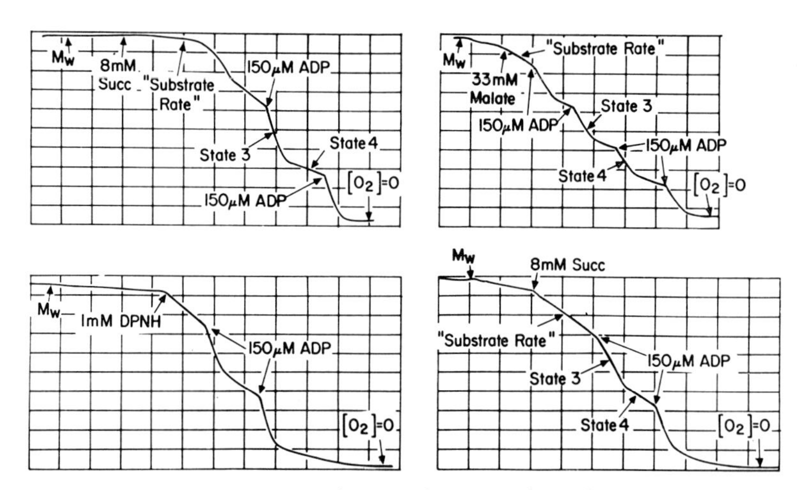

FIG. 2. Oxidation of various substrates by mung bean hypocotyl mitochondria and white potato mitochondria. The trace in the lower right corner is with white potato mitochondria, the other three traces are with mung bean mitochondria. Conditions are described in the text.

Comments

A general method for the isolation of intact mitochondria from higher plants has been detailed. The important features of the method are: (a) control of temperature (0–4°) and pH (7.0–7.2); (b) avoidance of violent methods for the disruption of the tissue; (c) sedimentation of the washed mitochondria at 6000 *g*; and (d) careful separation of the mitochondrial fraction from other materials that sediment with the mitochondria.

[25] Mitochondria from Spinach Leaves

By I. ZELITCH

Ohmura[1] first showed that mitochondria from spinach carried out oxidative phosphorylation. The following modified method produces mitochondria that give high P:O ratios with substrates of the citric acid cycle and DPNH.[2] Other studies on mitochondria isolated from green leaves have been described for tobacco.[3]

Preparation

Spinach was purchased locally and kept at 5° in a plastic bag, sometimes for as long as several weeks, until used. Such storage had no effect on the activity of the isolated mitochondria. The younger pale-green leaves were taken, the midribs were excised, and a weighed quantity of leaf blade was washed in water and chilled on ice before being ground in the cold room.

Grinding was done in a cold mortar with washed sand, with the addition of two volumes of grinding medium of the following composition: 0.45 *M* sucrose, 0.05 *M* mannitol-borate buffer at pH 7.2, 0.03 *M* potassium citrate, 0.01 *M* EDTA at pH 7.5, and 0.05 *M* tris(hydroxymethyl)aminomethane chloride (Tris) buffer at pH 8.3. The ground tissue, final pH of the suspension 7.2, was squeezed through two layers of cheesecloth and then centrifuged for 5 minutes at 600 *g* at 5°. The residue, containing chloroplasts and other heavy cell fragments, was discarded, and the supernatant fluid was centrifuged for 20 minutes at 10,000 *g*. The residue, containing the mitochondria, was gently suspended in a washing medium (1 ml per gram of leaf) composed as follows: 0.3 *M* sucrose, 2×10^{-4} *M* EDTA, and 0.05 *M* Tris buffer at pH 7.5. The suspension was centrifuged for 20 minutes at 10,000 *g*. The mitochondria were finally taken up in the washing medium (1 ml per 1.8 g of leaf) for

[1] T. Ohmura, *Arch. Biochem. Biophys.* **57**, 187 (1955).
[2] I. Zelitch and G. A. Barber, *Plant Physiol.* **35**, 205 (1960).
[3] W. S. Pierpoint, *Biochem. J.* **71**, 518 (1959). See also W. S. Pierpoint, *Biochem. J.* **82**, 143 (1962).

use in the experiments. Usually 0.9 ml of the suspension was placed in each Warburg vessel. This is equivalent to 1.6 g of original leaf tissue, and contained 0.5–0.8 mg of protein N, as measured by nesslerization after Kjeldahl digestion of the particles which had been washed with trichloroacetic acid. The preparation was green in color, doubtless because of the presence of material originating in the chloroplasts.

Oxidative Phosphorylation

Oxygen uptake was measured by conventional techniques in 15-ml Warburg vessels in an air atmosphere at 30°. Substrate, cofactors, and the mitochondria suspended in the washing medium were placed in the main compartment of chilled vessels. The center well contained KOH, and to the side arm glucose and an excess of hexokinase were added. A typical reaction mixture contained the following: sucrose, 300 micromoles; $MgSO_4$, 10 micromoles; substrate, 20 micromoles; potassium phosphate buffer at pH 7.0, 37 micromoles; yeast coenzyme concentrate, 1 mg; ATP, 2 micromoles; mitochondria in 0.9 ml of washing medium; and water to make the final volume 2.0 ml. In the side arm were glucose, 50 micromoles, and yeast hexokinase, 0.2 mg.

After temperature equilibration for 5 minutes, at zero time, the contents of the side arm were tipped into the main compartment. At the end of the reaction period, usually 25 minutes, the vessels were quickly transferred to an ice bath and 1 ml of 3% perchloric acid was added to stop enzymatic reaction. After centrifugation of the reaction mixture, the orthophosphate remaining in the supernatant fluid was determined.[4] Good agreement was obtained when the disappearance of orthophosphate was compared with the net increase in glucose 6-phosphate. In typical experiments,[2] the following P:O ratios were obtained: succinate, 2.1; fumarate, 2.8; L-malate, 2.6; pyruvate (in presence of a catalytic amount of malate), 2.7; DPNH, 2.5; glycolate, 0.1–0.3.

Inhibitors

The mitochondrial preparations described also oxidize glycolate, but there is negligible phosphorylation accompanying oxidation of this substrate. A competitive inhibitor of glycolate oxidase, α-hydroxy-2-pyridinemethanesulfonic acid, at $5 \times 10^{-4}\,M$, inhibited the oxidation of glycolate 68%, but had no effect on the oxidation or phosphorylation with succinate. Azide, at $1 \times 10^{-3}\,M$, inhibited succinate oxidation 67% and diminished phosphorylation, whereas oxygen uptake with glycolate

[4] K. Lohmann and L. Jendrassik, *Biochem. Z.* **178**, 419 (1926).

was enhanced. Thus the pathway of electron transport by the mitochondria is undoubtedly through the cytochrome system, except with glycolate when a flavoprotein oxidase is involved.

[26] Yeast Mitochondria and Submitochondrial Particles

By JAMES R. MATTOON and WALTER X. BALCAVAGE

General Considerations

Mitochondria and submitochondrial particles have been prepared from several species of yeast, including *Candida* (*Torulopsis*) *utilis*,[1] *Saccharomyces cerevisiae*,[2,3] and closely related species such as *S. carlsbergensis*.[4] The foremost experimental obstacle in the preparation of intact yeast mitochondria is the need for efficiently breaking the refractory cell wall without extensively damaging the liberated mitochondria. Two approaches to this problem have been used: (1) Cells are ruptured mechanically and mitochondrial fragments are separated from more intact mitochondria by differential centrifugation and careful separation of supernatants and loosely packed (fluffy) layers (submitochondrial particles) from firmer pellets (mitochondria). (2) Yeast cells are subjected to enzymatic cell wall digestion in order to form osmotically sensitive spheroplasts (protoplasts), from which mitochondria are liberated by osmotic shock and mild homogenization.

The main disadvantages of the mechanical rupture method are poor yields, the need for empirical selection of optimum breakage conditions, and the highly subjective process of fluffy layer separation. The enzymatic method, while giving good yields of reasonably intact mitochondria, requires careful control of enzyme source and/or digestion conditions to avoid premature lysis of spheroplasts. Since the most commonly used source of digestion enzymes, gut juice from the snail *Helix pomatia*, contains materials detrimental to mitochondrial activity (even after enzyme fractionation), careful washing of spheroplasts is required. The most critical problem encountered in the spheroplast method is the variation of cell wall sensitivity to digesting enzymes with different strains of yeast and with physiological state of the yeast. Stationary phase cells must be subjected to successive digestions in the

[1] A. W. Linnane, E. Vitols, and P. Nowland, *J. Cell Biol.* **13**, 345 (1962).
[2] E. A. Duell, S. Inoue, and M. F. Utter, *J. Bacteriol.* **88**, 1762 (1964).
[3] E. Vitols and A. W. Linnane, *J. Biochem. Biophys. Cytol.* **9**, 701 (1961).
[4] T. Ohnishi and B. Hagihara, *J. Biochem.* (*Tokyo*) **55**, 584 (1964).

presence of thiol compounds,[2] while a single digestion is adequate for cells harvested during logarithmic growth. The need for a high ratio of snail gut juice to yeast makes the procedure costly. Other enzyme preparations, rich in glucanases, which have been used to prepare yeast spheroplasts, have not yet been used extensively for mitochondrial preparation.

The primary advantage of the enzymatic method is that it provides a high yield of mitochondria with a level of physiological "intactness" (as measured by control of respiration by phosphate-acceptor ADP)[5] which often exceeds that obtained with mitochondria liberated by mechanical means. The main advantages of the mechanical procedure are its rapidity and the lack of dependence on physiological condition or strain of yeast.

Several mechanical devices have been employed for breaking yeast, including various shakers,[2] the French pressure cell, blendors with overhead drive,[3] and certain types of mills. All these devices under appropriate conditions have yielded yeast mitochondria or submitochondrial particles capable of oxidative phosphorylation, but in only a few instances has acceptor control of respiration been demonstrated with such preparations. The procedures given below employ a colloid mill; one method yields mitochondria with acceptor control.

Yeast Culture Conditions

Effects of Carbon Source and Aeration. Yeasts of the genus *Saccharomyces* (including most bakers' yeast) are facultative aerobes and are subject to severe catabolite repression of mitochondrial development and aerobic respiration when grown on a fermentable carbon source, particularly glucose. This repressive effect is magnified when aeration is inadequate, and the yeast utilizes its fermentative capacity to provide the bulk of its energy. It is therefore desirable to grow yeast on nonfermentable carbon sources such as ethanol, lactic acid, or glycerol under highly aerobic conditions. If it is necessary to use fermentable carbon sources—for example, when certain mutant strains are grown—special care must be exercised to ensure adequate aeration and agitation. Fermentable carbon sources, such as raffinose or melibiose, which are slowly utilized may be used[6] or glucose may be fed continuously at growth-limiting concentrations under conditions where oxygen supply is

[5] Acceptor control ratio is defined as respiratory rate after ADP addition: respiratory rate before ADP addition, i.e., state 3 rate/second state 4 rate. Rates are measured polarographically.

[6] C. Reilly and F. Sherman, *Biochim. Biophys. Acta* **95**, 640 (1965).

nonlimiting.[7] Nonlimiting oxygen supply is usually not attained in many common laboratory procedures for yeast growth. However, reasonably good aeration can be obtained with Erlenmeyer flasks filled to 10% or less of capacity and incubated on a rotary shaker. Reciprocating shakers are less efficient, while laboratory fermentors are superior to most flask arrangements.

Effects of Physiological State. The physiological state of the yeast also appears to influence the condition of the mitochondria in the cells. Cells grown on nonfermentable carbon sources exhibit high respiratory capacity (Q_{O_2}) throughout logarithmic growth, while cells grown on glucose show a high Q_{O_2} only near the end of logarithmic growth when ethanol derived from glucose may in fact be the main source of energy.[8] The relative levels of nitrogen source and growth factor are undoubtedly of importance, but these parameters have not been thoroughly studied. The pH of the medium should be about 4.5 initially. It is recommended that Q_{O_2} be measured prior to mitochondrial preparation, in order to ascertain the physiological state of the starting material.

Preparation of Yeast for Mitochondrial Isolation. Commercial bakers' yeast may be conveniently utilized for preparation of yeast mitochondria. Fresh yeast is preferable, but yeast cake 3–5 days old has been used. Commercial yeast is commonly grown on fermentable sugar in the form of molasses which is slowly fed under conditions of efficient aeration. However, to ensure a highly derepressed state[5] before preparation of mitochondria, it may be desirable to condition washed cells by allowing them to grow 1–3 hours in a medium containing 1% yeast extract, 2% casein hydrolyzate (peptone) and 3% (v/v) ethanol; pH 4.5 (YPE medium). About 60 g cells (pressed cake), suspended in 500 ml of YPE medium in a 2-liter Erlenmeyer flask, are incubated for 2 hours at 30° on a rotary shaker operated at 250 rpm. This treatment is sufficient to raise the Q_O (μl oxygen consumed per hour per milligram dry weight at 25°) by 10–20%, but brief enough to avoid significant overgrowth by bacteria which are almost always present in relatively small numbers in commercial bakers' yeast. Although these conditions are not optimal for aeration, they avoid large volumes. No major differences have been observed between mitochondria prepared from derepressed bakers' yeast and *S. cerevisiae* (our strain D-261) grown aseptically in a fermentor on YPE medium.

Pure *S. cerevisiae* is grown on YPE medium in 10-liter batches in

[7] W. D. Maxon and M. J. Johnson, *Ind. Eng. Chem.* **45**, 2554 (1953).

[8] B. Ephrussi, P. P. Slonimski, Y. Yotsuyanagi, and J. Tavlitzki, *Compt. Rend. Trav. Lab. Carlsberg. Ser. Phys.* **26**, 87 (1956).

14-liter fermentor jars on a New Brunswick Microferm Model MF 314 fermentor drive assembly at 30°. Air is bubbled through the medium at 11 liters/minute by means of a sparger, and the impeller is operated at 650–700 rpm. When 10 g (wet weight) of cells are used for inoculum, yeast harvested in the early stationary phase after 19–20 hours of growth displays a high Q_{O_2}. Harvested cells are washed 2 or 3 times with distilled water.

Preparation of Phosphorylating Mitochondria and Submitochondrial Particles

Reagents and Equipment

MTEB medium: (final concentrations) 0.25 *M* mannitol, 0.02 *M* Tris, 10^{-3} *M* EDTA, 0.2% crystalline bovine serum albumin (Armour Pharmaceutical Co., Kankakee, Illinois); final pH, 7.4.

Glass beads: 0.2 mm (No. 4285-M20, Arthur H. Thomas Co., Philadelphia, Pennsylvania). Beads are soaked in *aqua regia* for 24 hours and exhaustively washed.

Colloid mill: MV-6-3 Micro-Mill (Gifford-Wood Co., Hudson, New York) cooled by circulating ethanol at —5°. This model has a minimum operating volume of 150 ml, a maximum of 1 gallon. Grinding is accomplished by a conical rotor-stator arrangement with adjustable gap. A smaller capacity mill (25–75 ml) of somewhat different design is available, but it has not been tested by the authors.

Centrifuges: Lourdes Model L or Model SL with VRA rotor and Servall Model SS-1. Polypropylene bottles and cups are used.

Glass homogenizer: piston type tissue grinder, 4.5 ml operating capacity, 25 mm abrasive interface (Arthur H. Thomas Co., No. 4288g).

Procedure. Forty grams wet weight (10 g dry weight) of water-washed cells are rinsed once with a small volume of MTEB medium, centrifuged, and resuspended in 150 ml of fresh medium. Fifty ml (70 g) of glass beads is added, and the pH of the entire suspension is adjusted to 7.4. The suspension is chilled to 0° and homogenized in the colloid mill which has been precooled 3–5 minutes by operating at low speed at a gap setting of 60 (0.060 inch clearance between rotor and stator) with about 200 ml of distilled water in the mill hopper. Care must be taken to avoid freezing. With the mill still running at low speed the water is quickly drained, and the cold cell-bead suspension is immediately poured into the hopper. The powerstat of the mill is then set at 85 (120 volts), the gap is closed to 35, and the yeast is homogenized with recycling for 30 seconds. The gap is then opened to 60 and the mill speed reduced. The homogenate is drained and a second similar suspension is homogenized. This process is repeated as desired. After the last homogenization, the mill is quickly rinsed with 50 ml of cold MTEB medium by operating the mill at top speed (gap 60). All subsequent operations are carried out at 5° or lower. The pH, which is usually slightly greater than 7.0, is immediately adjusted to 7.1 and the beads are separated by decantation. Beads are washed with 100 ml of medium, and the wash is added to the homogenate.

The homogenate is centrifuged for 10 minutes at 2600 rpm (780 *g*)[9] in a Lourdes centrifuge in order to remove intact cells, debris, and nuclei. The resulting supernatant is carefully decanted to avoid disturbing any loosely packed layer. Mitochondria and mitochondrial membrane fragments are collected by centrifugation of the decanted supernatant for 20 minutes at 10,000 rpm (8800 *g*) in the Servall centrifuge. All subsequent steps employ this centrifuge. The pH of the resulting supernatant is about 6.7. The mitochondrial pellets are very gently homogenized in a glass homogenizer, suspended in a final volume of 20 ml with MTEB medium, and further fractionated by centrifugation for 15 minutes at 3800 rpm (1400 *g*). The pellet contains remaining cells, nuclei, and debris, possibly some yeast glycogen and some mitochondria. This fraction is discarded since it exhibits high endogenous respiration, presumably due to intact cells. Final separation of reasonably intact mitochondria from the bulk of the phosphorylating mitochondrial membrane fragments is accomplished by centrifuging the supernatant from the preceding step for 20 minutes at 8300 rpm (6800 *g*). The supernatant and fluffy layer are carefully decanted from the mitochondrial pellet or drawn off with a pipette.

The yield at this point is about 25–30 mg mitochondrial protein per 120 g (3 batches) of wet weight yeast. The fractionated mitochondria from the pellet catalyze high rates of oxidative phosphorylation with both pyruvate (+ malate sparker), representing soluble, NAD-linked matrix enzymes, and succinate, representing membrane-bound enzymes. Submitochondrial particles may be isolated from the supernatant (plus fluffy layer) by centrifugation at 10,000 *g*. These particles display significant oxidative phosphorylation only with substrates linked to membrane-bound enzymes, e.g., succinate and NADH. An increased

[9] Relative centrifugal force (*g*) values are calculated from revolutions per minute and an operating radius measured from the centrifuge axis to a point located at the center of the volume of homogenate contained in the centrifuge cup.

fraction of the mitochondria are converted to these submitochondrial particles if longer homogenization times or smaller gaps are used. A convenient index of the level of "intactness" of the mitochondrial fraction is the phosphate uptake obtained with pyruvate + malate relative to that obtained with succinate.

Mitochondria prepared by the above method give P:O ratios determined manometrically, of 1.0–1.6 with succinate, 1.5–2.0 with pyruvate + malate, and 1.6–2.0 with ethanol. Ethanol usually gives P:O ratios that are 20–40 per cent higher than those given by pyruvate + malate. Tetramethyl-*p*-phenylenediamine (TMPD) reduced by glutathione gives P:O ratios of about 0.4–0.6, but this value is reduced to 0.2–0.3 when antimycin A is included. Succinate oxidation is stimulated 2- to 4-fold by addition of a small increment of ADP, but the respiratory rate does not return to the ADP-limited (state 4)[5] rate. Oxidation of NAD-linked substrates is only partially (about 20–25%) inhibited by amobarbital and rotenone. These results indicate that the above preparation yields mitochondria which are sufficiently intact to retain matrix enzymes and some NAD and which carry out phosphorylation at two or three sites in the electron transport chain. However, the mitochondria have been damaged so that phosphorylation efficiency, particularly at site one (if present) and at the cytochrome *c*-oxygen site, is considerably below the currently accepted value of 1.0 phosphorylation per site. Furthermore, these mitochondria appear to be "loosely coupled" because of their behavior when ADP is added to stimulate respiration.

Preparation of Mitochondria Exhibiting Acceptor Control

While mitochondria prepared by the above procedure are useful for manometric studies, it is often desirable to have more "physiologically intact" mitochondria which can be conveniently studied by polarographic means. With appropriate modification of the above procedure such mitochondria can be prepared in somewhat smaller yields.

Reagents

Homogenization medium: (final concentrations) 0.3 *M* mannitol; 10^{-4} *M* EDTA; 0.4% bovine serum albumin, fraction V (Calbiochem, Los Angeles, California)[10]; the pH is adjusted to 7.0 with 1.0 *N* NaOH.

[10] Although use of crystalline albumin instead of fraction V was found to yield mitochondria with higher P:O values in the former procedure, in this procedure there was virtually no difference, possibly because a higher level of albumin was used.

Wash medium: (final concentrations) 0.3 *M* mannitol; 0.2% bovine serum albumin, fraction V; the pH is adjusted to 7.0 with 1.0 *N* NaOH.

Procedure. Mitochondrial preparations using the colloid mill represent a heterogeneous population of particles with varying degrees of damage. In this method the percentage of highly intact mitochondria in a population is increased by using milder homogenization conditions, and a more critical selection process permits the isolation from this population of mitochondria which exhibit acceptor control. Three hundred to 500 g (wet weight) of cells are suspended in 1 liter of homogenization medium, cooled to 0–5°, and placed in the precooled colloid mill. The gap is opened to the widest setting (76) and the powerstat is set at 15 to give minimum speed. Then 500 ml (700 g) of 0.2 mm glass beads are slowly added to the mill. A combination pH electrode positioned in the mill is used to monitor the pH, which is maintained at 7.0 during homogenization by incremental additions of 1.0 *N* NaOH. After initial pH adjustment the gap setting is decreased to 50 and the speed increased by setting the powerstat at 45.[11] Under these conditions adequate cell breakage occurs within 4 minutes. During this period the temperature rises 1–2° and 5–10 ml of 1.0 *N* NaOH must be added to maintain neutrality.

During differential centrifugation whole cells and debris are removed by a 10-minute centrifugation at 3200 rpm (1250 *g*), and mitochondria are collected by centrifuging the resulting supernatant fraction for 15 minutes at 7500 rpm (6400 *g*). The preceding large-volume centrifugations are carried out in the Lourdes centrifuge, and the succeeding steps are performed in the Servall centrifuge. The mitochondrial pellets obtained above are evenly dispersed in 3–5 ml of wash medium by repeatedly drawing up the mixture of medium and mitochondria into a pipette. The resulting thick suspensions are pooled in a 50-ml centrifuge tube and the volume is adjusted to 35–40 ml with wash medium. This suspension is centrifuged at 3500 rpm (750 *g*) for 7 minutes, and the resulting pellet is discarded. The supernatant is recentrifuged for 15 minutes at 7700 rpm (5200 *g*). The pellet from this step contains a voluminous, tan fluffy layer which is completely removed by aspiration from the lower darker layer of "intact" mitochondria. These mitochondria are suspended in a minimal volume (0.2–0.4 ml) of wash medium. About 8–12 mg of mitochondrial protein is obtained from 400 g (wet weight) of yeast.

Mitochondria isolated from bakers' yeast and from strain D-261 by

[11] These settings were found to be optimal for the particular colloid mill employed in the authors' laboratory.

this procedure give ADP:O ratios of 1.2–1.8 with succinate, and acceptor control ratios[5] of 1.5–2.8 with the same substrate. These mitochondria also exhibit acceptor control with D-lactate, ethanol, and α-ketoglutarate as substrates.

Assay for Oxidative Phosphorylation

Assay methods will be outlined briefly since they are very similar to methods used elsewhere in this volume. It is important to emphasize that the pH values of assay media were 6.7 or 6.5.

Oxygen uptake was measured by classical manometric procedures in 20-ml Warburg vessels for 30 minutes at 25°. Reaction mixtures contained in a total volume of 2.0 ml: 20 micromoles substrate (2 micromoles L-malate was used as sparker for pyruvate), 2 micromoles KH_2PO_4, 2 micromoles ADP, 20 micromoles glucose, 150 Kunitz-MacDonald units of hexokinase, 10 micromoles $MgCl_2$, 10 mg crystalline bovine serum albumin, 60 micromoles Tris, 2 micromoles EDTA, and 0.4–2.5 mg of mitochondrial protein. Final pH was 6.7. Phosphate uptake was measured by the method of Gomori.[12]

For polarographic determination of acceptor control a Clark oxygen electrode was used. A semiclosed system, 2 ml reaction chamber (in which atmospheric pressure is maintained and in which back-diffusion of air is minimal) was used with the power supply and recorder described by Kielley.[13] The composition and final concentration of the reaction medium was as follows: 0.3 *M* mannitol, 0.04 *M* KCl, 0.01 *M* sodium phosphate buffer, pH 6.5. Oxygen uptake rates were determined after successive additions of mitochondria (about 0.40 mg protein in 20 μl), substrate (25 μl of 1.0 *M* substrate), and ADP (5 μl of 0.05 *M*) to 1.95 ml of reaction medium.

[12] G. Gomori, *J. Lab. Clin. Med.* **27**, 955 (1942).
[13] See Vol. VI [33].

[27] Preparation and Properties of *Neurospora* Mitochondria

By J. W. Greenawalt, D. O. Hall,[1] and O. C. Wallis[1]

Principles

The preparation of structurally and functionally intact mitochondria from *Neurospora* requires a method of breaking or removing the resistant, heteropolysaccharide-containing walls of these cells without extensively

[1] This manuscript was prepared while the authors were at the Department of Physiological Chemistry, The Johns Hopkins University, School of Medicine, Baltimore, Maryland.

damaging the relatively fragile mitochondria. Not all cell types of *Neurospora* are equally susceptible to a particular method of disruption; some methods which are useful in breaking the long vegetative hyphae are ineffective in rupturing the smaller, more spherical conidia. Mitochondria have been isolated from hyphae ground with sand in a mortar and pestle[1a]; however, this procedure has limited applicability since only small quantities of hyphal cells can be broken easily in this way. Mitochondria have also been prepared from conidia and hyphae converted enzymatically to "protoplasts."[2] The extremely long incubation time required to produce "protoplasts" (10–15 hours) makes this treatment unsuitable for many studies. Mitochondria have been obtained on a large scale, however, from conidia, germinating conidia, and hyphae of *Neurospora* by high speed homogenization in the presence of glass beads.[3] Also, extremely clean mitochondrial preparations have been isolated from *Neurospora* conidia and hyphae after brief (1 hour) enzymatic digestion of the cell walls followed by gentle homogenization.[4] The high concentration of degradative enzyme required and the low yields of mitochondria restrict the application of this method to small preparations. These latter two methods are described below and although they have been used by the authors only with wild-type *N. crassa,* strain SY7A, and the cholineless mutant, 34486, there is no reason to believe that they cannot be applied successfully to other strains of *Neurospora* and to other fungi.

Production of Conidia and Growth of *N. crassa*

For biochemical studies it is convenient to have a supply of conidia readily available as a source from which mitochondria can be prepared directly or for use as inocula for the growth of germinating conidia or hyphae. The formation of mature conida by *N. crassa* requires 7–8 days. Conidia harvested and suspended in distilled water remain viable with no change in the rate or percentage of germination for at least 7 days if stored under aseptic conditions at 4°.

Wild-type, Strain SY7A. Wild-type conidia are grown and harvested essentially as described by Wainwright.[5] A single, large conidiation flask yields slightly less than 200 ml of suspension containing ca. 2×10^8 conidia/ml. The quantity of conidia produced can be altered by varying the number of conidiation flasks seeded; the concentration of conidia may be readily changed by centrifuging the original suspension (500 *g*

[1a] D. J. L. Luck, *J. Cell Biol.* **16,** 483 (1963).
[2] B. Weiss, *J. Gen. Microbiol.* **39,** 85 (1965).
[3] D. O. Hall and J. W. Greenawalt, *Biochem. Biophys. Res. Commun.* **17,** 565 (1964).
[4] O. C. Wallis and J. W. Greenawalt, in preparation.
[5] S. D. Wainwright, *Can. J. Biochem. Physiol.* **37,** 527 (1959).

for 5 minutes) and resuspending the pellet of conidia in the volume of distilled water required to give the desired concentration.

To grow germinating conidia and hyphae 50 ml of the conidial suspension (ca. 1×10^{10} conidia) are used to inoculate a 2-liter Erlenmeyer flask containing 500 ml of Vogel's[6] complete, liquid medium plus 3 drops of silicone antifoam; cultures are grown at 30° on a rotary shaker (ca. 265 rpm). Cells can be harvested at any desired time during the growth of the fungus; however, the rate of germination and growth of *Neurospora* will vary with the size of inoculum and conditions of growth.

Choline-Requiring Mutant, Strain 34486 (chol-1). Conidia of this mutant strain of *N. crassa* have been obtained in large quantities by using Vogel's[6] minimal medium supplemented with 100 μg choline chloride per milliliter, 1% sucrose, 1% glycerol, and 2% agar as conidiation medium. The yield of conidia is about the same as that recovered with the wild-type strain (ca. 4×10^{10} conidia per conidiation flask). Hyphae are grown from a conidial inoculum (ca. 7×10^{5} conidia per milliliter) in Vogel's[6] minimal medium supplemented with 2% sucrose and 100 μg choline chloride per milliliter.

Preparation of Mitochondria

Physical Disruption of Cell Walls. All operations are carried out at 0–4°. Wild-type conidia, germinating conidia, or hyphae are harvested and washed by centrifugation at 500 *g* for 5 minutes and resuspended in 500 ml of a preparation medium consisting of 0.25 *M* sucrose, 0.005 *M* EDTA (pH 7.0), and 0.15% crystalline bovine serum albumin (BSA). About 800 ml of the original conidial suspension (ca. 2×10^{8} conidia per milliliter), 4 liters of growth medium containing germinating conidia (3½ hours old), or 1 liter of 1–2-day-old hyphae yield enough mitochondria for biochemical experiments. To aliquots (250 ml) of the conidia or hyphae in preparation medium are added 500 g of acid-washed glass beads (0.2 mm) and 4 drops of silicone antifoam; the mixture of beads and cells are poured into a prechilled Eppenbach Micro-Mill, Model MV-6-3, (Gifford-Wood Co., Hudson, New York) and ground at maximum speed (10,000 rpm) for 1 minute at a gap setting of 1/30,000 inch. The ratio of liquid to beads and the time of homogenization must be critically controlled for adequate cell breakage and subsequent separation of functional mitochondria. The gap setting of the mill may require periodic adjustment to ensure reproducible results. It is estimated that about 75% of *Neurospora* hyphae are broken under these conditions. Broken cells and beads are spun out of the mill at low speed with the

[6] H. J. Vogel, *Microbial Genet. Bull.* **13**, 42 (1956).

gap completely open into a large beaker and the cell contents in the supernatant fluid are decanted. The beads are washed twice to remove trapped cellular components and the washes are added to the decanted liquid.

Enzymatic Degradation of Cell Walls. Mitochondria have been prepared from conidia and hyphae of wild-type *Neurospora crassa* and of the cholineless mutant, strain 34486 (chol-1), after partial digestion of the cell walls by a method described for the formation of spheroplasts from yeast cells.[7] About 4×10^{10} conidia or hyphae (ca. 400 mg of protein) grown as described above are washed and resuspended in 20 ml of reaction medium containing sorbitol, 0.63 *M*; citric acid-K_2HPO_4 buffer (pH 5.8), 0.1 *M* with respect to citric acid; 2-mercaptoethylamine ·HCl, 0.03 *M*; EDTA, 4×10^{-4} *M*; and 2.0 ml (ca. 200,000 units of glucuronidase) of the crude snail gut enzyme, Glusulase (Endo Laboratories, Inc.) or 0.2 g of Helicase (Industrie Biologique Francaise, S.A.). Reaction mixtures are incubated for 1 hour with gentle shaking at 30°. All subsequent steps are carried out at 0–4°.

The cells are centrifuged (500 *g*, 5 minutes) from the digestion medium and washed twice with 100 ml of cold, 0.9 *M* sorbitol. The pellets are resuspended in 40 ml of cold preparation medium consisting of sucrose, EDTA, and BSA (see above) and are homogenized in 20 ml aliquots with a Ten Broeck ground-glass homogenizer or a glass homogenizer with a Teflon pestle. Six up and down cycles are made with a pestle attached to a mechanically rotating shaft. Homogenates, combined and diluted with preparation medium to a volume of 100 ml, are centrifuged (ca. 800 *g*, 10 minutes) to remove unbroken cells and large cell fragments.

Separation of Mitochondria. Mitochondrial fractions from cells disrupted by either of the two methods described above are collected by differential centrifugation between 1500 *g* (10 minutes) and 8000 *g* (30 minutes). Most of the small fragments of cell walls, especially prevalent in preparations isolated from cells homogenized in the presence of glass beads, can be removed by repeating this centrifugation procedure. The mitochondrial fraction is finally resuspended in 3–5 ml of preparation medium to give a final concentration of 20–30 mg of protein per milliliter. All resuspensions are done with Ten Broeck ground-glass homogenizers. Preparations can be completed in about 2½–3 hours.

Fragments of cell wall material comprise the major recognizable contaminant in mitochondrial pellets observed in the electron microscope. Further purification of the mitochondrial fractions can be obtained by

[7] E. A. Duell, S. Inoue, and M. F. Utter, *J. Bacteriol.* **88**, 1762 (1964).

density gradient centrifugation. To a 4.6 ml gradient (20–65%, w/v; 0.58–1.9 *M* sucrose)[1a,8] is added 0.5–0.9 ml of the mitochondrial suspension; after centrifugation (40,000 rpm, 1 hour), fractions are collected through a perforation in the bottom of the tube. Gradients are made and fractions collected with the aid of a Buchler Densigrad apparatus.

Properties of *Neurospora* Mitochondria

Phosphorylation coupled to the oxidation of succinate has been obtained with mitochondria from *Neurospora* conidia, germinating conidia, and hyphae homogenized with glass beads[3]; P:O ratios from 0.7 to 1.5 have been measured. To obtain these P:O ratios Mg^{++}, ADP, P_i, and hexokinase are required; sucrose and BSA are also necessary, but failure to add these compounds only slightly decreases the P:O ratios since they are present in the preparation medium added with the mitochondria. The common uncouplers of phosphorylation and inhibitors of respiration in animal mitochondria, such as 2,4-dinitrophenol, antimycin A, oligomycin, cyanide, and atractyloside, also inhibit these reactions in *Neurospora* mitochondria. Mitochondria from 2-day-old hyphae couple phosphorylation to the oxidation of 7 different substrates; oxidation of β-hydroxybutyrate has not been detected in any preparations of *Neurospora* mitochondria tested.

It should be emphasized that the biochemical properties of the isolated mitochondria may vary with the physiological age of the cells from which they are prepared. Oxidation of pyruvate (malate), citrate, isocitrate, and α-ketoglutarate is barely measurable in mitochondria from conidia, is increased in mitochondria from germinating conidia ($3\frac{1}{2}$ hours old) and is highest in mitochondria from hyphae 1–2 days old.[9] Mitochondria isolated from *Neurospora* at all three stages of growth contain cytochromes *b*, *c*, and *a*; present also are activities commonly associated with oxidative phosphorylation in animal mitochondria, such as ATP-ADP exchange, ATP-P_i exchange, and Ca^{++} accumulation.[9] Electron microscope studies of mitochondria prepared from cells by high speed homogenization with glass beads indicate that this isolation procedure causes considerable damage to mitochondrial structures. This conclusion is supported by the high Mg^{++}-activated ATPase activity and low respiratory control measurements observed with mitochondria prepared by this method.

Mitochondria prepared from *Neurospora* cells with walls enzymatically digested as described above are more intact than mitochondria prepared by homogenization with glass beads. The mitochondrial band

[8] R. J. Britten and R. B. Roberts, *Science* **131,** 32 (1960).

obtained on density gradient centrifugation contains over 80% of the total cytochrome oxidase activity of the preparation, and observation of this band in the electron microscope shows an extremely clean, homogeneous preparation of mitochondria. The respiratory control of these mitochondria is higher than that of the mitochondria obtained by homogenization with glass beads.[4]

[28] Microbial Phosphorylating Preparations: *Alcaligenes faecalis*

By GIFFORD B. PINCHOT

This organism is used because extracts fail to metabolize glucose 6-phosphate, making it possible to trap ATP with hexokinase without glycolytic ATP formation. Strain 8750 was obtained from the American Type Culture Collection in 1951, and a stock has been maintained lyophilized in sterile milk since that time. For daily use the strain is carried on 2% nutrient agar slants, but after repeated serial passage the organism ceases to perform satisfactorily. Strain 8750 obtained recently from the American Type Culture Collection has also been unsatisfactory.

Preparative Methods

Crude Extract. The organisms are grown in liquid media containing 2% trypticase (Baltimore Biological Labs) and 0.5% yeast extract (Difco) fortified with 0.1 mg of calcium pantothenate per liter and the following salts (reported in grams per liter, final concentration): KH_2PO_4 0.2, K_2HPO 0.8, NaCl 0.2, $MgSo_4 \cdot 7\ H_2O$ 0.1, $CaSO_4 \cdot 2\ H_2O$ 0.1, $Na_2MoO_4 \cdot 2\ H_2O$ 0.0038, $FeSO_4 \cdot 7\ H_2O$ 0.0225, and citric acid 0.15.[1] The pH is adjusted to 8.0.

One slant is used to inoculate a liter of medium, and flasks are grown at 37° for 18 hours with shaking. These cultures are harvested by centrifugation, or 1 liter is used to inoculate 100 liters of medium containing 1 ml of Dow Antifoam A on an FS1-130 Fermentor Stand (New Brunswick Scientific Co. Inc.). The culture is incubated at 37° with aeration (1.5 cubic feet/minute) and stirring until the optical density at 660 mμ reaches approximately 3.500. Refrigerated Sharples centrifuges are used for harvesting, and the wet weight yield approximates 5 g/liter. The bacteria are washed in 10 liter of cold 0.9% NaCl and then in 10

[1] D. Burk and M. Lineweaver, *J. Bacteriol* **19**, 389 (1930).

liters of cold distilled water. A Waring blendor run at very slow speed is used to suspend them. After washing they are resuspended in water in the ratio of 150 mg wet weight per milliliter, and disrupted in 50-ml batches in a Raytheon 250-watt 10 kc sonic oscillator at maximum amperage and cavitation for 5 minutes at 0°. The suspension is centrifuged for 15 minutes at 25,000 *g*, the pellet is discarded, and the supernatant solution, referred to as *crude extract,* is frozen. It remains active for months in the deep freeze. It contains 10–20 mg of protein per milliliter, and oxidizes 2–5 micromoles of DPNH per minute per milliliter in 0.1 *M* phosphate buffer pH 7.4, calculated from the rate of disappearance during the first minute with an initial reading of approximately 0.550 at 340 mμ.

Fractionation of Crude Extract. The crude extract, which catalyzes oxidative phosphorylation, can be fractionated into a particulate DPNH oxidase, a soluble protein, and a polynucleotide, all of which are required for phosphorylation. Details of this fractionation are given elsewhere.[2,3]

The isolation and purification procedure for the polynucleotide have been described.[3,4] The active component can be identified as an oligoribonucleotide containing adenine, guanine, and uridine in a molar ratio of 2:1:1.[4] The polynucleotide in conjunction with Mg^{++} can be shown to bind the coupling enzyme to the particles.[5]

Magnesium or Phosphorylating Particles. Active phosphorylating particles are centrifuged from crude extract which has been incubated with 0.04 *M* magnesium chloride for 20 minutes at 0°. Centrifugation is at 100,000 *g* for 30 minutes at 0°. The pellet is resuspended in the original volume of 0.01 *M* KCl-0.04 *M* $MgCl_2$ with the aid of a Lucite plunger made to fit the tubes of the No. 40 Spinco head accurately for the bottom 15–20 mm. Above this the plunger shaft is smaller, allowing dispersal of the pellet with minimal frothing by repeated passage between plunger and tube. The pellet is recovered and resuspended in 0.1–0.2 volume of 0.01 *M* KCl. This is magnesium or phosphorylating particle. Crude extracts made from bacteria grown in 1-liter batches have produced better phosphorylating particles than crudes made from large batches, perhaps because of the antifoam in the latter.

Oxidative Phosphorylation Assay

The reaction volume is 1 ml containing 50 micromoles of glycylglycine or Tris buffer, 1 micromole of potassium phosphate buffer pH 7.4, in-

[2] G. B. Pinchot, *J. Biol. Chem.* **205,** 65 (1953).

[3] G. B. Pinchot, *J. Biol. Chem.* **229, 1,** 11, 25 (1957).

[4] S. Shibko and G. B. Pinchot, *Arch. Biochem. Biophys.* **94,** 257 (1961).

[5] S. Shibko and G. B. Pinchot, *Arch. Biochem. Biophys.* **93,** 140 (1961).

organic pyrophosphate-free ^{32}P (E. R. Squibb & Co.) with approximately 1×10^6 cpm, 6 micromoles of $MgCl_2$, 25 micromoles of glucose, 0.1 micromole of ADP or ATP (Sigma) crystalline hexokinase (Sigma), and 1 micromole of DPNH (Sigma) as substrate, or an equal amount of DPN as control. Either crude extract 0.2 ml, or Mg particles 0.05–0.10 ml, or oxidase particles 0.10 ml, crude coupling enzyme 0.20 ml, and polynucleotide equivalent to 0.5 micromole ADP are used as enzyme complex.

The enzymes are incubated for 4 minutes with P_i, buffer, $MgCl_2$, and ADP, and the reaction is started by the addition of hexokinase, $^{32}P_i$, glucose, and substrate, then stopped after 1–10 minutes by adding 0.5 ml of 10% trichloroacetic acid. Incorporation of $^{32}P_i$ into the organic phase is determined by the method of Berenblum and Chain[6] modified to use 3 isobutanol-benzene extractions and one ether extraction to remove inorganic phosphate. This method extracts 99.95% of the inorganic ^{32}P.[5] Aliquots are counted using a gas flow end window counter, timing 1000 counts. If ATP is determined rather than ^{32}P, the reaction is stopped by boiling and an aliquot is assayed for G-6-P using G-6-P dehydrogenase (Boehringer) and TPN. Care is taken that 6-phosphogluconic dehydrogenase is inactive in the assay system. The difference in phosphorylation between the DPNH vessel and the DPN control is reported. Oxygen consumption is not measured routinely but the P:O ratios range from 0.1 to 0.3.[2] Rarely, higher values have been found. Typical values for oxidative phosphorylation in crude extracts are in the range of 20–40 millimicromoles $^{32}P_i$ incorporation per milligram protein in 10 minutes.[7] Phosphorylating particles catalyze 50–70 millimicromoles per milligram of protein. Particles from the large cultures give lower values.

Formation of Intermediate

When phosphorylating particles are allowed to oxidize DPNH in the absence of P_i and ADP, proteins are released into solution, and the particles when recovered by centrifugation have lost most of their ability to phosphorylate. The supernatant solution (S^{DPNH}) has the ability to esterify ^{32}P-labeled P_i, and to form ATP when incubated with P_i and ADP. S^{DPN} has comparatively little activity.[8] The experiments are carried out as follows: Phosphorylating particles containing 10–15 mg of protein are incubated in a total volume of 2.5 ml at 24° for 3–5 minutes in a mixture containing 12 micromoles of $MgCl_2$, 5.9 micromoles of KCl, and 3 micromoles of DPNH or of DPN in the control vessel. The pH is uncontrolled or maintained at 8.5–9.5 with Tris buffer. The tubes are

[6] J. Berenblum and E. Chain, *Biochem. J.* **32,** 295 (1938).
[7] O. Warburg and W. Christian, *Biochem. Z.* **310,** 384 (1941).
[8] G. B. Pinchot, *Proc. Natl. Acad. Sci. U.S.* **46,** 929 (1960).

chilled and centrifuged at 100,000 g for 30 minutes, and the supernatant solutions are decanted, lyophilized, and taken up in water to 1:10 the original volume and dialyzed for 18–24 hours against 0.005 M Tris buffer pH 7.4. This material is studied for $^{32}P_i$ incorporation and ATP formation.

Assay of Intermediate

$^{32}P_i$ incorporation is studied with the same assay as particles except that no substrate (DPNH) is added and the P_i concentration is 0.5 micromoles, and the assay time may be up to 60 minutes. Hexokinase and glucose are generally absent and ADP is the acceptor. The control is "P_i discharged" intermediate in which the intermediate is incubated with cold P_i and $MgCl_2$ for 5–15 minutes before addition of $^{32}P_i$ and ADP. P_i alone apparently forms an unstable compound, and S^{DPNH} is regarded with suspicion if more than 50% of the $^{32}P_i$ esterification remains after P_i discharge.

ATP Assay

This assay is done in the reaction mixture detailed above with the addition of 25 micromoles of glucose and excess crystalline hexokinase (Sigma). No ^{32}P is added. The reaction is either stopped by boiling or is not stopped and aliquots are added to a 1-ml assay system containing glycylglycine buffer 25 micromoles, 0.2 micromole of TPN (Sigma) and enough crystalline glucose-6-P dehydrogenase so that the reaction comes to completion in 1–3 minutes. Significant amounts of 6-phosphogluconic dehydrogenase are not present. The intermediate preparations contain traces of myokinase, and in order to differentiate between ATP formed by this enzyme from that formed from the intermediate, the increment in ATP produced in a vessel containing phosphate over a control containing no phosphate is reported. An added control is to leave out ADP. This value is close to zero. ATP assay has also been done with the firefly assay system[8] of McElroy.[9]

Purification of Intermediate

The most satisfactory method to date for purification has been to subject lyophilized S^{DPNH} to starch block electrophoresis by the method of Smith[10] using a tray 38 × 10 × 1.3–5.0 cm deep. Up to 36 ml of S^{DPNH} is lyophilized and taken up in the minimal amount of water and dialyzed

[9] W. D. McElroy, *Harvey Lectures Ser.* **51**, 240 (1957).

[10] I. Smith, "Chromatographic and Electrophoretic Techniques," Volume II. Heinemann, London, 1960.

against 0.03 *M* borate buffer pH 8.48. A tray of starch suspended in the same buffer is prepared, and a transverse central section (not extending to the edges of the tray) is cut out, mixed with S^{DPNH}, and poured back into the center of the block. The block is developed in the cold room at a potential of 300–900 volts. Protein peaks are identified under ultraviolet light. A major peak moves toward the cathode and another toward the anode. These are excised, eluted with water, and tested for $^{32}P_i$ incorporation and ATP formation. As can be seen in the table, the protein

$^{32}P_i$ UPTAKE AND ATP FORMATION BY S^{DPNH} FRACTIONS[a]

			Millimicromoles phosphorylation			
			1 Hour		2 Hours	
Expt.	Fractions tested	Assay	P_i	No P_i	P_i	No P_i
1	Anodic	$^{32}P_i$	19.0	—	44.6	—
2	Cathodic	$^{32}P_i$	1.2	—	1.8	—
3	Anodic	ATP	0.8	0.7	0.5	0.5
4	Cathodic	ATP	40.3	—	55.3	—
Sum. Expt. 3 + 4	—	ATP	41.1	—	55.8	—
5	Cathodic + anodic	ATP	59.5	37.6	99.8	60.5
5 − (Sum Expt. 3 + 4)	—	ATP	18.4	—	44.0	—
5 (P_i − No P_i)	—	ATP	21.9	—	39.3	—

[a] $^{32}P_i$ assay as described, no hexokinase or glucose, 0.5 micromole ADP as acceptor. ATP assay 0.2 micromole ADP, 25 micromoles of glucose, and 40 μg crystalline hexokinase run with and without 0.5 micromoles of P_i in a 1 ml reaction containing 6.0 micromoles of $MgCl_2$. Aliquots, 0.5 ml, of reaction mixture assayed for G-6-P with G-6-P dehydrogenase and 0.2 μmoles TPN. Anodic fraction contained 55 μg of protein, and cathodic 71 μg.

moving toward the anode takes up P_i, but does not make ATP unless the other cathodic protein is added. The amount of P_i taken up agrees well with the phosphate-stimulated ATP formation and with the increment in ATP formed by both fractions together over the sum formed by the fractions separately.

Nature of the Intermediate

It has previously been shown that the intermediate contains a coupling enzyme since it restores oxidative phosphorylation to particles made defective by preincubation with DPNH.[8] The compound bound to the enzyme has been identified as DPN by splitting off the DPN either with P_i in the discharge reaction or with the complete ATP-forming reaction, then precipitating the protein and identifying DPN in the supernatant by paper chromatography or by reduction with alcohol dehydrogenase. ^{14}C-labeled DPNH was used to form the intermediate and found to travel with the anode protein intermediate. In addition it was possible to label the protein with ^{14}C-labeled DPN in an ATP-requiring reverse reaction. Details of these experiments are reported elsewhere.[11]

[11] G. B. Pinchot and M. Hormanski, *Proc. Natl. Acad. Sci. U.S.* **48,** 1970 (1962).

[29] Microbial Phosphorylating Preparations: *Azotobacter*

By H. G. Pandit-Hovenkamp

Preparations

Principle. A cell-free extract of *Azotobacter vinelandii* is prepared, which contains particles capable of catalyzing the oxidation of NADH and other substrates, and a soluble factor which is required for coupled phosphorylation. The factor is bound to the particles in the presence of a suitably high concentration of salt and released when the salt concentration is decreased. This release is more complete in the presence of low concentrations (1–2 m*M*) of EDTA or phosphate.

Growth of Bacteria and Preparation of Cell-Free Extract. Azotobacter vinelandii (strain ATCC 478) is grown at 30° on Burk's nitrogen-free medium,[1] containing 3.8 mg Na_2MoO_4 and 20 g sucrose per liter, previously sterilized for 15 minutes at 120°. The cells are incubated for 24 hours on a rotary shaker with 100 ml of the medium. The contents of the incubating flask are transferred to 4 liters of the medium, contained in a 5-liter bottle, and the mixture is aerated sufficiently to prevent accumulation of acids, but not too vigorously (1 liter of air per minute). The bacteria are harvested after 48 hours at which stage the color of the suspension should be brown. The cells are washed four times with deionized water (3×1000 ml; 1×100 ml). The last centrifugation, which is carried out for 10 minutes at 20,000 *g*, yields 2–4 g of wet cells per liter of the medium.

The cells are suspended in ice-cold deionized water (6 ml/g wet weight of cells) and treated in 10-ml portions for 5 minutes with sonic vibration generated by a 60-Watt MSE magnetostrictor oscillator oper-

[1] D. Burk and H. Lineweaver, *J. Bacteriol.* **19,** 389 (1930).

ating at 20 kc/sec.[1a] The temperature of all the preparations is kept between 0° and 2°. The suspension is centrifuged for 10 minutes at 3000 *g*, then for 20 minutes at 16,000 *g*. The cell-free extract, which contains about 10 mg protein per milliliter, is stored at —16°. Repeated freezing and thawing should be avoided.

Although cell-free extracts may also be prepared by grinding with Alumina or glass powder,[2] the yield of phosphorylating particles in these preparations is much lower.

Preparation of Phosphorylating Particles. This is similar to the method described by Shibko and Pinchot[3] for *Alcaligenes*.

The cell-free extract is incubated for 20 minutes in the presence of 0.1 *M* KCl–2 m*M* phosphate buffer (pH 7.0) and centrifuged for 30 minutes at 140,000 *g* in a Spinco ultracentrifuge.[4] The sediment is washed with the same volume of 0.1 *M* KCl–2 m*M* phosphate, centrifuged at the same speed, and suspended in 0.08 *M* KCl (30–40% of the original volume of the extract). The yield of the particles is 0.8–2 mg of protein per milliliter of cell-free extract.

The KCl can be replaced by 0.1 *M* NaCl, NH_4Cl, or Tris-chloride buffer (pH 7.0), by 0.05 *M* phosphate buffer (pH 7.0), or by 0.01 *M* $MgCl_2$, $MnCl_2$, or $CaCl_2$; in the presence of the bivalent cations, however, Tris-chloride instead of phosphate should be used as buffer. KCl–Tris-chloride buffer is recommended if particles with a low phosphate content are required.

Preparation of Nonphosphorylating Particles for Assay of the Soluble Factor. The extract is incubated with KCl and phosphate as described for the phosphorylating particles and centrifuged for 30 minutes at 55,000 *g*. The sediment is washed in 2 m*M* phosphate buffer (pH 7.0), centrifuged at the same speed and suspended in deionized water (25% of the original volume of the extract). These particles should be made freshly for each assay of the soluble factor and used immediately after preparation. The phosphate buffer may be replaced by 2 m*M* EDTA (pH 7.0).

Preparation of Soluble Factor. Phosphorylating particles are prepared as described, except that the centrifugations are carried out at 80,000 *g*. The particles are suspended in deionized water (one-fifth of the original volume of the cell-free extract) and dialyzed for 16–20 hours against

[1a] The conditions required to obtain optimal yield with maximal P:O ratios should be established empirically for each oscillator.

[2] A. Tissières, H. G. Hovenkamp, and E. C. Slater, *Biochim. Biophys. Acta* **25**, 336 (1957).

[3] S. Shibko and G. B. Pinchot, *Arch. Biochem. Biophys.* **93**, 140 (1961).

[4] Centrifugal forces apply to the bottom of the tube.

a solution containing either 2 mM Tris-chloride (pH 7.4) with 1 mM EDTA or 2 mM phosphate buffer (pH 7.4). The dialysis solution is changed once. The particles are removed by 30 minutes centrifugation at 140,000 g. The supernatant is centrifuged three times at 140,000 g[5] for 30, 60, and 120 minutes, respectively, and each time carefully pipetted from a gelatinous yellow precipitate which contains some residual NADH oxidase. After the last centrifugation the NADH oxidase activity at pH 7.4 and 22° should not exceed 0.01 micromole/min/ml.

The clear supernatant, which is colorless or very light yellow, contains 10–15% of the protein content of the particles. It is stored at —16°.

Assay of Soluble Factor

Principle. The soluble factor is incubated for 60 minutes with nonphosphorylating particles in the presence of 25 mM $MgCl_2$ and 2 mM Tris-chloride buffer (pH 7.0). Oxidative phosphorylation with NADH as a substrate may be either determined with the mixture as such or with the sediment after centrifugation at 140,000 g. Under the conditions of the assay the soluble factor alone does not catalyze the formation of glucose 6-phosphate. When various amounts of soluble factor are tested,

ASSAY OF SOLUBLE FACTOR OF OXIDATIVE PHOSPHORYLATION

Incubation mixture component	Protein conc. (mg/ml)	Tube number 1	2	3	4	5	6
Nonphosphorylating particles	4.6	0.1[a]	0.1	0.1	0.1	0.1	0.1
Soluble factor	0.53	—	0.05	0.1	0.15	—	—
Soluble factor	1.6	—	—	—	—	0.1	0.15
Water		0.10	0.07	0.035	—	0.08	0.07
0.05 M Tris-chloride buffer (pH 7.0)		0.02	0.02	0.02	0.02	0.02	0.02
$MgCl_2$ 0.1 M		0.075	0.08	0.085	0.09	0.10	0.11
Total volume		0.295	0.32	0.34	0.36	0.40	0.45
Addition after 60 min: 0.025 M $MgCl_2$		0.15	0.13	0.11	0.09	0.05	0.00
Assay of oxidative phosphorylation with 0.1 ml of incubation mixture							
P:NADH ratio		0.01	0.03	0.05	0.15	0.24	0.25
Ratio soluble factor to particles (mg protein)		0	0.06	0.12	0.18	0.34	0.52

[a] Milliliters of incubation mixture component added.

[5] The speed after dialysis should be higher than before, but both may be lowered (to 80,000 g and 30,000 g, respectively) if this is more convenient.

the incubation volume should be adjusted to keep the protein concentration approximately constant in order to avoid differences in stability of the particles due to variations in the protein concentration.

Assay. An example of an assay is given in the table.

The order of addition is important; incubation of particles with $MgCl_2$ and Tris buffer before addition of the soluble factor may result in lower phosphorylating activities. Addition of $MgCl_2$ without Tris buffer lowers the pH. The total protein concentration should preferably be above 1 mg/ml; the activation of the particles will be less in more dilute preparations.

As the P:NADH ratio obtained is dependent upon both the amount of soluble factor added and the quality of the particle preparation used, different preparations of the soluble factor are best compared with the same particle preparation at the same time. If this is not possible, activities may be roughly compared by estimation of the ratio of factor protein to particle protein needed for maximal stimulation. In most cases this ratio is about 0.3.

Assay of Oxidative Phosphorylation

Standard manometric procedures may be used,[6] but when isolated particles are used, 0.1% serum albumin should be added to stabilize the enzymes. Since the adenylate kinase (EC 2.7.4.3) and ATPase (EC 3.6.1.4) content of the particles is low, NaF may be omitted. The NADPH oxidase activity of the particles is too high to permit use of the spectrophotometric method described by Pinchot (cf. Brodie[6]).

A method used for the estimation of phosphorylation with NADH as a substrate by the particles is described below.

Principle. A measured quantity of NADH is oxidized completely by phosphorylating particles in a medium containing orthophosphate, ADP, $MgCl_2$, hexokinase, and glucose. The glucose 6-phosphate formed is measured and corrected for a control containing NAD^+ instead of NADH.

Procedure. The reaction is carried out at pH 7.4 in a spectrophotometric cuvette which contains 2 ml of the following medium: 0.5 mM ADP, 1 mM EDTA, 30 mM orthophosphate, 5 mM $MgCl_2$, 5 mM glucose, 2 International Units of hexokinase, and 0.7 mM NADH. All solutions except the hexokinase and NADH are mixed before being pipetted into the reaction vessel. The NADH solution is prepared freshly for each experiment, and the concentration is measured enzymatically.[7]

[6] See Vol. VI [35].
[7] See Vol. III [128].

The reaction is started with 0.1 ml of the particle suspension (containing 0.1–0.4 mg of protein), and the absorbance at 340 mμ is measured against a reference cuvette where NADH is replaced by the same concentration of NAD^+. The reaction is completed when the absorbance at 340 mμ is constant (2–6 minutes); since the NADH concentration is measured separately, the initial readings are not important. The contents of both cuvettes are transferred to centrifuge tubes which are stoppered with glass marbles, immersed in boiling water for 20 minutes, cooled, and centrifuged. The glucose 6-phosphate is estimated in the clear supernatant.[8] The P:NADH ratio is the difference in glucose 6-phosphate content of the two tubes, divided by the amount of NADH added.

The glucose 6-phosphate content of the control (due to slight adenylate kinase activity) is generally between 0.01 and 0.03 micromole.

If cell-free extract is used instead of the particles, NaF should be included in the reaction mixture to inhibit the adenylate kinase reaction and deproteinization should be carried out with $HClO_4$, which is removed as the potassium salt, to avoid the risk of side reactions during the heating.

Properties of the Preparations

Both phosphorylating and nonphosphorylating particles catalyze the oxidation of succinate, malate, lactate, NADPH, and NADH; the phosphorylating particles can, in addition, catalyze phosphorylation of ADP to ATP coupled to these oxidations. The P:NADH ratios obtained with phosphorylating particles made from fresh extract vary between 0.25 and 0.5. This ratio gradually declines when the extract is stored at —16°; when it is below 0.3 the extract should not be used to make nonphosphorylating particles, though it may still give a satisfactory preparation of the soluble factor. The phosphorylating enzymes are inactivated if the extract is frozen after the addition of $MgCl_2$ (cf. Brodie[9]) or KCl; there is little loss of activity when the extract is kept 1 night at 0° with or without buffered $MgCl_2$. The stability of the particles is dependent upon the protein concentration; with 1 mg/ml or less, inactivation of phosphorylating enzymes may occur by shaking at 25°–30°, which may result in erroneously low P:O ratios in manometric experiments. Addition of serum albumin stabilizes the enzymes. Considerable activity of the soluble factor is retained after 3 months' storage[10] at —16° or 1 night at 22°, but at 100° the activity is lost within 5 minutes.

[8] E. C. Slater, *Biochem. J.* **53**, 157 (1953).

[9] A. F. Brodie, Vol. V [4].

[10] Longer periods were not tried.

The ATPase activity of the phosphorylating particles is low (0.025 micromole ATP per minute per milligram of protein at pH 7.4) in contrast to the high rate of phosphorylation of ADP (0.7–1.1 micromole per minute per milligram of protein with NADH as a substrate at pH 7.4). The ATPase activity of the soluble factor is still lower (0.004 micromole per minute per milligram of protein at pH 7.4).

Compared with mammalian preparations, the oxidation and phosphorylation in *Azotobacter* is less sensitive toward most inhibitors or uncouplers. The following uncouplers of mammalian oxidative phosphorylation cause at the concentration mentioned a 30–50% decrease of the P:NADH ratio obtained with 0.2–0.3 mg of phosphorylating particles: 4-isooctyl-2,6-dinitrophenol (10^{-5} *M*); carbonyl cyanide *m*-chlorophenylhydrazone (1.4×10^{-6} *M*); pentachlorophenol (2.5×10^{-5} *M*); dicoumarol (5×10^{-5} *M*); chlorpromazine (5×10^{-5} *M*); 2-heptyl-4-hydroxyquinoline-*N*-oxide (3×10^{-5} *M*); menadione (1.2×10^{-5} *M*). In several cases lowering of the P:NADH ratio was noticed only when the particles were incubated for some minutes with the reaction medium containing the uncoupler, before the reaction was started with the addition of NADH. No decrease of P:NADH ratio was observed with oligomycin (30 μg/mg of protein) or dinitrophenol (10^{-4} *M*). The NADH oxidase is insensitive to antimycin (0.9 μg per milligram of protein), oligomycin (5 μg per milligram of protein) or Amytal (1.8 m*M*).

[30] Microbial Phosphorylating Preparations: *Mycobacterium*

By ARNOLD F. BRODIE

Introduction

A number of systems from both aerobic and facultative microorganisms have been described which carry out oxidative phosphorylation.[1–9]

[1] G. B. Pinchot, *J. Biol. Chem.* **225,** 65 (1953).
[2] A. F. Brodie and C. T. Gray, *Biochim. Biophys. Acta* **17,** 146 (1955).
[3] P. M. Nossal, D. B. Keech, and D. J. Morton, *Biochim. Biophys. Acta* **22,** 412 (1956).
[4] L. A. Rose and S. Ochoa, *J. Biol. Chem.* **230,** 307 (1956).
[5] P. E. Hartman, A. F. Brodie, and C. T. Gray, *J. Bacteriol.* **74,** 319 (1957).
[6] A. Tissières, H. G. Hovenkamp, and E. C. Slater, *Biochim. Biophys. Acta* **25,** 336 (1957).
[7] S. Ishikawa and A. L. Lehninger, *J. Biol. Chem.* **237,** 2401 (1962).
[8] E. R. Kashket and A. F. Brodie, *Biochim. Biophys. Acta* **78,** 52 (1963).
[9] T. Yamanaka, A. Ota, and K. Okunuki, *J. Biochem.* (*Tokyo*) **51,** 253 (1962).

In general, the systems of bacterial origin differ from intact mammalian mitochondria in that they have lower P:O ratios and fail to exhibit respiratory control.[10] Nevertheless, studies of the effect of uncoupling agents, respiratory inhibitors, and investigation of the sites of phosphorylation[10] indicate that the differences between mammalian and bacterial coupling processes are not at the mechanistic level but instead appear to be a reflection of the differences in structural organization. P:O ratios greater than one have been obtained with a cell-free system from *Corynebacterium creatinovorans* and *Mycobacterium phlei*.[11] The respiratory chains in *M. phlei* are similar to those of mammalian mitochondria. They differ in quinone composition. In addition, this bacterial system contains an additional respiratory chain for malate oxidation (malate-vitamin K reductase).[12]

Fractionation of Particles and Supernatant Fluid

The system from *M. phlei* which couples phosphorylation to oxidation has been resolved into two components, a particulate and supernatant fraction.[13] Both fractions are required for restoration of oxidative phosphorylation. The particulate fractions contains the bulk of the respiratory carriers[12,14] whereas the soluble fraction contains oxidative factors, coupling factor proteins, nonphosphorylative electron transport "bypass" enzymes and the bulk of the enzymes which carry out the exchange reactions associated with oxidative phosphorylation.[15]

The preparation of bacterial particles has been described in an earlier volume of this series (Vol. V [4]). *M. phlei* cells (ATCC 354) are disrupted by sonic oscillation, and the cellular debris is removed by centrifugation at low speed. The crude extract obtained following removal of the cellular debris is capable of coupling phosphorylation to oxidation without further addition of cofactors. P:O ratios greater than one are obtained with the crude extract and most substrate.[11] The components of oxidative phosphorylation are resolved into particles and soluble coupling factors by centrifugation of the crude extract at 144,000 g for 90 minutes. The particles are washed twice with 0.15 M KCl containing 0.01 M $MgCl_2$ and 0.01 M Tris buffer (pH 7.4) in order to remove contaminating coupling proteins. The washing procedure also results in a loss of NAD^+ which must be added back for restoration of oxidation of NAD^+-linked

[10] A. Asano and A. F. Brodie, *J. Biol. Chem.* **240,** 4002 (1965).
[11] A. F. Brodie and C. T. Gray, *J. Biol. Chem.* **219,** 853 (1956).
[12] A. Asano and A. F. Brodie, *J. Biol. Chem.* **239,** 4280 (1964).
[13] A. F. Brodie, *J. Biol. Chem.* **234,** 398 (1959).
[14] A. F. Brodie and C. T. Gray, *Science* **125,** 534 (1957).
[15] A. Asano and A. F. Brodie, *Biochem. Biophys. Res. Commun.* **13,** 416 (1963).

substrates.[16] The particulate fraction can be stored after quick freezing in an acetone–dry ice mixture but coupled activity decreases rapidly. The rate and characteristics of cytochrome *b* reduction are considerably altered by prolonged storage of the particles. In contrast, the crude extract is more stable and can be kept for at least 2 weeks with only a slight loss of activity after quick freezing.

Fractionation of the Supernatant Factors

A method for obtaining 16-fold purification of the supernatant factors has been presented in a previous volume of this series (Vol. VI [35]). The soluble material obtained following removal of the particles is used as the starting material. Two fractions, AS I and AS II, containing components for restoration of coupled activity with different segments of the respiratory chain are resolved by fractionation of the supernatant fluid with ammonium sulfate.[13]

Step I. The protein obtained following precipitation between 0 and 30% ammonium sulfate saturation (referred to as AS I) is necessary for restoration of phosphorylation at site III. Site III phosphorylation is measured by the method of Jacobs[17] utilizing ascorbate and tetramethyl phenylenediamine as the electron donor and 2-nonylhydroxyquinoline *N*-oxide (15 μg) as a respiratory block at the cytochrome *b* level of oxidation. The precipitate obtained between 0 and 30% ammonium sulfate is dissolved in a minimal amount of distilled water, and the pH is adjusted to 7.2–7.4 with Tris buffer before use. Further purification of fraction AS I is achieved by the addition of 10% streptomycin solution added dropwise until the ratio of 280:260 absorption is minimized. The precipitate formed on streptomycin addition is removed and discarded. The supernatant is dialyzed against distilled water in the cold overnight and fractionated with ammonium sulfate (0–30% saturation). The precipitate is removed and dissolved in distilled water and the pH is adjusted to 7.2–7.4.

Step II. The bulk of the supernatant components are found in fraction AS II, which is obtained following 30–60% ammonium sulfate fractionation of the crude supernatant fraction. This fraction contains an oxidative component necessary for restoration of NAD^+-linked respiration and malate-vitamin K reductase as well as coupling factors for succinoxidase NAD^+-linked oxidation and for the malate-vitamin K reductase pathway. The preliminary fractionation steps for separating these components are described in a previous volume of this series (Vol. V [4]). The am-

[16] A. Asano and A. F. Brodie, unpublished observations.

[17] E. E. Jacobs, *Biochem. Biophys. Res. Commun.* **3**, 536 (1960).

monium sulfate or protamine fractionated material is further purified by column chromatography on DEAE-cellulose.[15] The dialyzed AS II fraction buffered with 0.005 M Tris at pH 7.2 is absorbed on a DEAE-cellulose column (1.6×13 cm) which has been previously washed with 0.005 M Tris-HCl buffer (pH 8.2). The column is washed with 10 ml of 0.01 M Tris buffer (pH 8.2) and developed with a chloride gradient. The nonlinear elution system consists of a mixing chamber containing 250 ml of 0.01 M Tris buffer (pH 8.2) and a reservoir containing 250 ml of 0.35 N KCl in the same buffer. Fractions are collected in 15-ml aliquots. The factors necessary for restoration of activity with the

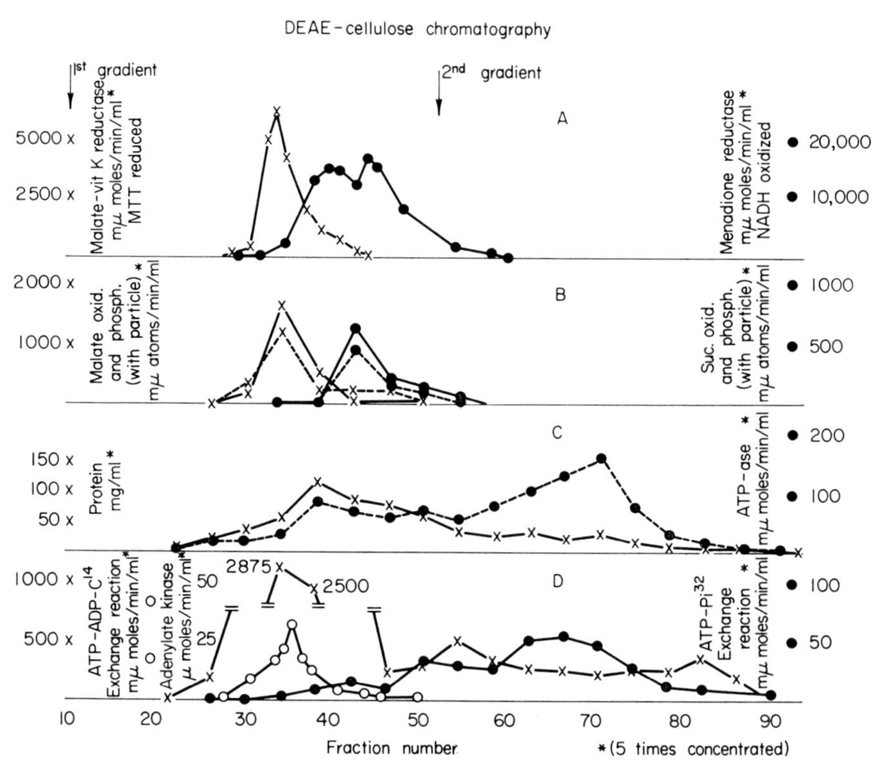

FIG. 1. Distribution of various activities following DEAE-cellulose fractionation of the supernatant fraction. Curve A: Malate-vitamin K reductase activity (x-x), menadione reductase (●-●). Curve B: Distribution of factors necessary for restoration of oxidative phosphorylation with malate (x-x, oxidation; x----x, phosphorylation) or succinate (●-●, oxidation; ●----●, phosphorylation) as electron donor. Curve C: Distribution of protein (x-x) and ATPase activity (●----●). Curve D: Distribution of the ADP-^{14}C-ATP exchange reaction (x-x), the ATP-$^{32}P_i$ exchange reaction (●-●), and adenylate kinase activity (○-○). From Asano and Brodie, *Biochem. Biophys. Res. Commun.* **13**, 416 (1963).

malate-vitamin K reductase pathway are obtained first and can be separated from those required for restoration of the malate (NAD^+-linked) and succinate pathways (Fig. 1). The distribution of protein, one of the nonphosphorylative bypass enzymes (menadione reductase), ATPase, and exchange enzymes is shown in this figure.

Nonphosphorylative Bypass Enzymes

The soluble enzymes which mediate the oxidation of NADH and NADPH are referred to as "bypass" enzymes since they transfer electrons directly to oxygen and thus bypass the particulate chain entirely, or reenter the particulate chain at the cytochrome *c* region and thus bypass the electron carriers between NADH and cytochrome *c*.[18] These enzymes contribute to the low P:O ratios observed with NADH as the electron donor. They can be separated from the factors necessary for restoration of malate oxidation by chromatography on DEAE-cellulose by the procedure described above.

Quinone-Dependent System

An association of a naphthoquinone with the electron transport chain and coupled phosphorylation in *M. phlei* has been demonstrated with the quinone-depleted system.[19-21] The natural quinone (vitamin K_9H) is primarily localized in the particulate fraction.[22] The structure of the natural quinone from *M. phlei* (K_9H) has been elucidated[23] and differs from the simple isoprenolog scheme in that one of the 9 isoprenoid units of the side chain is reduced. The chemical properties of the quinones have made it possible to disassociate then from the respiratory chain without disruption of the orientation of the enzymes and coenzymes necessary for electron transport and coupled activity.[19,21] The quinone-depleted system is obtained by exposure of the particulate and supernatant fractions to light at 360 mμ. The procedure is described in a previous volume of this series (Vol. VI [36]).

Photooxidation with Methylene Blue

Inactivation of ATPase, the $^{32}P_i$-ATP exchange reaction and adenylate kinase activity of the supernatant fraction can be accomplished by

[18] A. Asano and A. F. Brodie, *Biochem. Biophys. Res. Commun.* **19,** 121 (1965).
[19] A. F. Brodie and J. Ballantine, *J. Biol. Chem.* **235,** 226 (1960).
[20] A. F. Brodie and J. Ballantine, *J. Biol. Chem.* **235,** 232 (1960).
[21] A. F. Brodie, *Federation Proc.* **20,** 995 (1961).
[22] A. F. Brodie, B. R. Davis, and F. L. Fieser, *J. Am. Chem. Soc.* **80,** 6454 (1958).
[23] P. H. Gale, B. H. Arison, N. R. Trenner, A. C. Page, K. Folkers, and A. F. Brodie, *Biochemistry* **2,** 200 (1963).

photooxidation with visible light and methylene blue.[24] This technique results in a loss of histidine and methionine residues in the light-treated preparation.[25] Methylene blue is added to a final concentration of 0.5 mg/ml supernatant fluid, and the pH is adjusted to 8.2. The material is irradiated at 3° with visible light at 30 cm by a spotlight for 4 hours. The irradiated supernatant material is fractionated with ammonium sulfate, and the precipitate obtained between 30 and 60% saturation removed and dialyzed against distilled water at 3° for 48 hours. The dialyzed material is treated with 1–2% charcoal to remove the methylene blue.

Enzyme Assays

1. *Oxidative Phosphorylation*

The crude cell-free extract (15–20 mg of protein) or fractionated components (particles 10–15 mg of protein), and fractionated supernatant fluid (5–10 mg of protein) are incubated with a suitable electron donor, inorganic orthophosphate, and a phosphate acceptor system. Oxidative phosphorylation can be measured by the manometric procedure, spectrophotometric assay,[26] or polarographic techniques. The components of the system and procedures have been described in volume VI [35] of this series.

2. *Phosphorylation Coupled to the One-Step Oxidation of Substrates*

One-step oxidation of substrates used by the three major respiratory chains of *M. phlei* is obtained by selecting appropriate conditions (see the table). Conditions for the one-step oxidation of succinate depend on obtaining particles depleted of NAD^+. Thus, the oxidation of succinate with the depleted particles and supernatant fraction containing fumarase results in the accumulation of malate.[10] NAD^+ depletion is obtained by repeated washing of the particulate fraction.

The one-step oxidation of NAD^+-linked substrate is studied with ethanol, β-hydroxybutyrate, or L-malate as electron donors. The oxidation products of these substrates, acetaldehyde, acetoacetate and oxalacetate are not further metabolized by the resolved *M. phlei* system. The ammonium sulfate fractionated supernatant fluid contains oxalacetate

[24] J. W. Adelson, A. Asano, and A. F. Brodie, *Proc. Natl. Acad. Sci. U.S.* **51**, 402 (1964).

[25] L. S. Weil, S. James, and A. Buchert, *Arch. Biochem. Biophys.* **46**, 266 (1953).

[26] G. B. Pinchot, *J. Biol. Chem.* **229**, 11 (1957).

CONDITIONS FOR ONE-STEP OXIDATION OF SUBSTRATES IN *Mycobacterium phlei*

Substrate	Pathway	End product	Conditions
Malate	NAD^+-linked	Oxalacetate	Elicited by the addition of NAD^+ to washed particles and DEAE-fractionated supernatant fraction containing malic dehydrogenase[a]
Malate	FAD-linked	Oxalacetate	Elicited by the addition of FAD to well washed particles and DEAE-fractionated supernatant fraction containing malate-vitamin K reductase activity[b]
β-Hydroxybutyrate	NAD^+-linked	Acetoacetate	Elicited by the addition of NAD^+ to particles plus AS II supernatant fraction. Acetoacetate is not further metabolized
Succinate	Succinoxidase pathway	Malate[c]	Some activity, (20–40%) observed with well washed particles. Addition of DEAE-cellulose fractionated supernatant, containing factors necessary for coupled activity with succinate, stimulates activity to level observed with crude extract
Ethanol	NAD^+-linked	Acetaldehyde	Elicited by the addition of NAD^+ to particles and AS II supernatant fraction. Acetaldehyde is not further metabolized

[a] A. Asano and A. F. Brodie, *Biochem. Biophys. Res. Commun.* **13,** 416 (1963).

[b] The use of this ammonium-sulfate fractionated supernatant AS II fraction instead of DEAE-fractionated enzyme results in the accumulation of pyruvate as the end product. The AS II fraction contains Mg^{++}-activated oxalacetate decarboxylase, which is removed by fractionation on DEAE-cellulose.

[c] Malate accumulation due to NAD^+ depletion on washing of particles. Supernatant contains fumarase activity.

decarboxylase which gives rise to pyruvate. This enzyme is removed from the components necessary for malate oxidation following chromatography on DEAE-cellulose. This procedure is essential for obtaining the one-step oxidation of malate via the malate-vitamin K reductase pathway.[27]

[27] A. Asano and A. F. Brodie, *Biochim. Biophys. Res. Commun.* **13,** 423 (1963).

3. *Measurement of Phosphorylation Coupled to Segments of the Respiratory Chain*

a. SITE I

Phosphorylation at site I is measured with the fractionated system (AS II) lacking coupling factors for phosphorylation between cytochrome *c* and oxygen. The particulate and supernatant fraction are irradiated with light at 360 mμ in order to remove the natural naphthoquinone and supplemented with dihydrophytyl vitamin K_1 or lapachol, quinones that do not restore phosphorylation.[21,28,29] Phosphorylation is obtained by utilizing the particulate-bound NAD^+ (i.e., substrates such as malate or β-hydroxybutyrate). In contrast, phosphorylation is not observed with added NADH. Oxidation via the particulate-NAD^+ pathway is sensitive to Amytal (10^{-3} *M*).

The span between NAD^+ and cytochrome *c* is measured spectrophotometrically under anaerobic conditions with mammalian cytochrome *c* or thiazolyl blue tetrazolium (MTT) as the electron acceptor.

Reagents

- $MgCl_2$ 15 micromoles
- EDTA, 15 micromoles
- KF, 25 micromoles
- ADP, 5.0 micromoles
- FAD, 25 millimicromoles
- Orthophosphate, 20 micromoles containing $^{32}P_i$ (200,000–300,000 cpm)
- Mammalian cytochrome *c*, 200 millimicromoles
- MTT, 240 millimicromoles
- Tris-HCl buffer pH 7.4, 100 micromoles
- A mixture of trichloroacetic acid (10%) containing 0.05% phosphoric acid
- Activated charcoal (Darco G-60)
- Perchloric acid (10%)

Procedure. The cytochrome *c* assay system consists of washed particles (5–10 mg protein), fractionated supernatant AS II (3–5 mg protein), 15 micromoles of $MgCl_2$, 100 micromoles of Tris-HCl buffer (pH

[28] A. F. Brodie and J. Adelson, *Science* **149**, 265 (1965).

[29] A. F. Brodie and P. J. Russell, Jr., *Proc. 5th Intern. Congr. Biochem. Moscow, 1961* Vol. V, p. 89. Pergamon, London, 1963.

7.4), 3 micromoles EDTA, 0.1 micromole of NAD^+, 20 micromoles of orthophosphate, 200,000–300,000 cpm of carrier-free $^{32}P_i$, 50 millimicromoles of KF, 200 millimicromoles of cytochrome *c*, 5.0 micromoles of ADP, and water to a final volume of 3.2 ml. The reaction is carried out *in vacuo* in a Thunberg type cuvette with a 1-cm light path. The reaction is started by the addition of 100 micromoles of L-malate or DL-β-hydroxybutyrate and the rate of cytochrome *c* followed spectrophotometrically at 550 mμ. The reaction is stopped by the addition of either 1 ml of 10% trichloroacetic acid containing 0.05% phosphoric acid or by the addition of 1 ml of 10% perchloric acid. The precipitated protein is removed and ATP (^{32}P) is assayed by the method of Nielson and Lehninger[30] or by absorption on charcoal. In the latter procedure a 2% suspension of activated charcoal (1.0 ml) is added to the trichloroacetic acid supernatant and the suspension is kept on ice for 10 minutes with occasional shaking. The charcoal suspension is then filtered over a Millipore filter (0.45 μ pore size) and washed twice with 5% trichloroacetic acid containing 0.05% phosphoric acid. This wash is followed by three washes with water containing 5% ethanol. The Millipore filter containing the washed charcoal is transferred to a planchette, 0.15 ml of a 2% starch solution added and the planchette dried overnight at room temperature. The radioactivity is measured and corrections made for self absorption in the calculation of $^{32}P_i$ incorporation. The P:2*e* ratio is calculated from the amount of cytochrome *c* reduced and amount of $AT^{32}P$ formed. Isolation of $AT^{32}P$ by Dowex 1-X2 (chloride) or glucose 6-phosphate (^{32}P) isolation by Dowex 1-formate can be used to measure the extent of phosphate incorporation. In the latter case hexokinase (3 mg) and glucose (20 micromoles) are added to the incubation mixture and the reaction is stopped with perchloric acid.

The span from succinate or malate-vitamin K reductase to cytochrome *c* is measured with a similar system. With succinate, NAD^+ is omitted from the system and the reaction is started with the addition of 100 micromoles of succinate. The span between malate-vitamin K reductase and cytochrome *c* is measured by using either purified malate-vitamin K reductase or AS II and 25 millimicromoles of FAD. $AT^{32}P$ formation is measured by the procedures described above.

The MTT assay system is similar to that used in the cytochrome *c* assay except that 240 mμmoles of MTT are used to replace cytochrome *c*. The reaction is carried out anaerobically in a 1 cm light path Thunberg cuvette and followed spectrophotometrically at 565 mμ.

[30] S. D. Nielson and A. L. Lehninger, *J. Biol. Chem.* **215,** 555 (1955).

b. Site III

Phosphorylation associated with the electron transport span between cytochrome *c* and oxygen can be followed manometrically with the crude extract or with the particles and dialyzed crude supernatant material. Fraction ASI can be used instead of the crude supernatant fraction; however, this fraction appears to be less stable. Evidence for a third site can be obtained with crude extract, but precaution must be taken to eliminate the endogenous activity which tends to obscure the phosphorylation coupled to ascorbate oxidation.

Reagents

Ascorbate, 10^{-2} *M*
Tetramethylphenylenediamine (TMPD), 10^{-6} *M*
2-Nonyl hydroxyquinoline N-oxide, 15 μg
ADP, 2.5 micromoles
$MgCl_2$, 15 micromoles
KF, 25 μmoles
Glucose, 20 micromoles
Hexokinase, 3 mg

Procedure. The Warburg vessels contain crude extract (15–20 mg protein), 2.5 micromoles of ADP, 15 micromoles of $MgCl_2$, 25 micromoles of KF, 20 micromoles of glucose, 3 mg hexokinase, 22 micromoles of ascorbate, and water to a final volume of 1.5 ml. The reaction is started by the addition of TMPD (10^{-6} *M*) and carried out at 30° for 10 or 15 minutes. The reaction is stopped by the addition of 10% trichloroacetic acid and phosphate determined by the method of Fiske and SubbaRow.[31]

4. *Assay for Coupling Factors*

The restoration of phosphorylation coupled to the oxidation of succinate or NAD^+-linked substrates requires the addition of soluble coupling factors found in fractions AS I and AS II. The manometric or spectrophotometric techniques can be used to assay for coupling factors.[13] Differences in phosphate esterification over control systems lacking coupling factors are determined by the colorimetric method of Fiske and SubbaRow or by determining $AT^{32}P$ formation. The assay system contains washed particles, substrate, 15 micromoles of $MgCl_2$, 25 micromoles of KF, 10 micromoles of phosphate, and a phosphate acceptor

[31] C. H. Fiske and Y. SubbaRow, *J. Biol. Chem.* **66**, 375 (1925).

system. The fraction containing coupling factors is added to the particles before addition of substrate.

5. *ATPase and Adenylate Kinase Activity*

ATPase activity is measured by following the liberation of inorganic phosphate from ATP whereas the adenylate kinase activity is assayed spectrophotometrically (340 mμ) following the addition of hexokinase, TPN, and glucose 6-phosphate dehydrogenase. Both ATPase and adenylate kinase require the addition of Mg^{++} ions for optimal activity.

6. *$^{32}P_i$-ATP and ADP-^{14}C-ATP Exchange Reactions*

The exchange reactions are assayed at 30° by the method of Cooper and Kulka[32] after chromatographic separation of the nucleotides on Dowex 1-X2 (chloride form).[32,33]

7. *Malate-Vitamin K Reductase*

This enzyme is required with particles for restoration of coupled activity of the malate-FAD-linked pathway. In the absence of particles it can be measured by following the reduction of thiazole blue tetrazolium spectrophotometrically at 565 mμ. FAD and vitamin K_1 in phospholipid micelles are required as cofactors for the dye assay system.[34]

Reagents

Thiazolyl blue tetrazolium (MTT), 0.024 *M*
Tris-HCl buffer, 0.1 *M*, pH 7.4
KCl, 0.04 *M*
FAD, 0.0125 *M*
L-Malate, 0.05 *M*
Vitamin K_1, 0.89 micromole, in phospholipid micelles prepared by the method of Fleischer and Klouwen.[35]

Procedure. The assay system consists of 0.24 micromole of MTT, 100 micromoles of Tris buffer (pH 7.4), 40 micromoles of KCl, 0.89 micromole of K_1 in phospholipid micelles, 12.5 micromoles of FAD, 0.05–0.1 ml of enzyme containing 0.5–3.5 mg of protein and water to a final volume of 1.5 ml. The reaction is started by the addition of 50 micromoles of L-malate and followed spectrophotometrically at 565 mμ. The millimolar

[32] C. Cooper and R. G. Kulka, *J. Biol. Chem.* **236**, 2351 (1961).
[33] W. E. Cohn, see Vol. III, p. 867.
[34] A. Asano, T. Kaneshiro, and A. F. Brodie, *J. Biol. Chem.* **240**, 895 (1965).
[35] S. Fleischer and H. Klouwen, *Biochem. Biophys. Res. Commun.* **5**, 378 (1961).

extinction coefficient of enzymatically reduced MTT at 565 mμ is 15.0 mM^{-1} cm^{-1}.

Properties of the *M. phlei* Phosphorylating System Respiratory Chains

The system from *M. phlei* contains three major respiratory chains which are capable of coupling phosphorylation to oxidation; the succinate, NAD^+-linked, and malate-vitamin K reductase pathways.[12] These pathways are distinguished by their response to added cofactors,[36] by their response to purified supernatant factors,[13,15] by analysis of the electron transport carriers,[12,14] and from the effects of different respiratory inhibitors and uncoupling agents on the different pathways.[10,12,14]

The particulate fraction contains bound NAD^+, flavins, a naphthoquinone (vitamin K_9H) and cytochromes b, c_1, c, a, and a_3. The ratio of cytochromes $a + a_3$, b, $c + c_1$, flavin, and vitamin K_9H was found to be 1.0, 0.68, 2.3, 2.5, and 43.7. The NAD^+-linked pathway of the particles is similar to that of mammalian tissues with the exception of the naphthoquinone. The quinone in *M. phlei* functions on the NAD^+-pathway between the flavoprotein and cytochrome b. The soluble malate-vitamin K reductase requires FAD and converges with the NAD^+-particulate chain at the naphthoquinone level. The succinoxidase pathway is found in the particles and contains a flavoprotein, metal, and unidentified light-sensitive factor. The succinate chain converges with the NAD^+-linked chain at the cytochrome b level of oxidation. The electron transfer sequence of the terminal respiratory chain of this microorganism, like the mitochondrial system, flows from cytochrome b to c to $a + a_3$ to oxygen.

Respiration with succinate as an electron donor is inhibited 85% by Atebrin (3×10^{-3} M), 98% by Dicumarol (10^{-3} M), 98% by NHQNO (2 μg), 60% by TTB (10^{-3} M), 95% by KCN (3×10^{-3} M), and 85% by pCMB (7×10^{-4} M). Inhibition of malate oxidation was 75% by Atebrin, 90% by dicoumarol (10^{-4} M), 80% by NHQNO (2 μg/mg protein), 95% by KCN (3×10^{-3} M), 75% by pentachlorophenol (10^{-4} M), 100% by Tween 80 (0.13%), and 100% by BRIJ-35 (0.33%). The *M. phlei* system is insensitive to antimycin A.

The phosphorylation associated with the three major respiratory chains is sensitive to a wide variety of uncoupling agents. Some differences in sensitivity to uncoupling agents are observed with the different respiratory chains.[10] Dinitrophenol (7×10^{-4} M) is effective with the succinate and NAD^+-linked chains. M-Cl-CCP (10^{-5} M) uncouples the phosphorylation associated with malate-vitamin K reductase and NAD^+-

[36] A. Asano and A. F. Brodie, unpublished observations, 1965.

linked pathways, but inhibits oxidation with succinate. Hydroxalamine ($5 \times 10^{-4} M$) uncouples all three respiratory chains but is less effective with succinate.[10] Dicoumarol ($10^{-4} M$) inhibits oxidation of the malate and NAD^+-linked pathways and uncouples phosphorylation with all three pathways. Galegine ($5 \times 10^{-3} M$) is effective against the NAD^+-linked chain.[10] In addition to the uncoupling agents listed above, the crude extract was found to be sensitive to gramicidin ($8 \times 10^{-6} M$), thyroxine ($8 \times 10^{-5} M$), lapachol ($5 \times 10^{-4} M$), methylene blue ($8 \times 10^{-5} M$), and usnic acid ($10^{-4} M$).[11] The *M. phlei* system is insensitive to oligomycin.

Comments

Bacterial systems offer an additional tool for the study of the mechanism of phosphate-bond energy generation. Systems which can be fractionated and reconstituted facilitate analysis of the essential components. The differences in respiratory pathways and ease of obtaining respiratory mutants may provide a means of obtaining respiratory segments unencumbered by the complete chain. The low P:O ratios observed with bacterial systems appears to be a reflection of the harsh disruptive methods used to obtain cell-free extracts and in part is due to the presence of nonphosphorylative electron transport bypass reactions. To some extent these bypass enzymes can be removed from the supernatant fluid by fractionation.

[31] Microbial Phosphorylating Preparations: *Micrococcus lysodeikticus*

By SHINJI ISHIKAWA

Protoplast membrane fragments of *Micrococcus lysodeikticus* prepared in the presence of 5 mM Mg^{++} can couple NADH or malate oxidation to the phosphorylation of ADP.[1] This system may be fractionated into three discrete components: nonphosphorylating respiratory chain enzymes bound to protoplast membrane fragments, a heat-labile coupling factor (cF), and a heat-stable coupling factor (Ft).[1,2]

Fractionation Procedures

Growth of Cells. Micrococcus lysodeikticus (strain 4698) is transferred from a slant to a nutrient broth (200 ml) containing 2% glucose,

[1] S. Ishikawa and A. L. Lehninger, *J. Biol. Chem.* **237,** 2401 (1962).
[2] S. Yamashita and S. Ishikawa, *J. Biochem.* (*Tokyo*) **57,** 232 (1965).

0.4% beef extract, 0.3% yeast extract, and 0.25% (v/v) salt mixture solution which contains the following components: 4% $MgSO_4$, 0.2% $FeSO_4$, 0.6% $MnSO_4$, 0.2% NaCl, 0.02 *M* potassium phosphate, pH 9.0. One-twentieth volume of the overnight culture at 37° is transferred to fresh broth, and the cells are allowed to grow in the log phase for 5 hours, or to the stationary phase. Either type of culture suffices as starting material. The cells are collected by centrifugation and washed twice with 50 ml of a medium containing 0.25 *M* sucrose, 0.02 *M* Tris-Cl, pH 7.40, and 5 m*M* $MgCl_2$ for each 200 ml of the culture. The washed cells are suspended in a solution of 1 *M* sucrose, 0.02 *M* Tris-Cl, pH 7.40, and 5 m*M* $MgCl_2$ so that 0.02 ml of the suspension added to 5 ml of 1 *M* sucrose yields an absorbancy of 0.5 in a 10 mm cuvette at 520 mμ. Such suspensions can be stored at —20° almost indefinitely.

Step 1. Phosphorylating Membrane Fragments.[1] Protoplasts of *M. lysodeikticus* are obtained by adding 3.5 mg of crystalline muramidase (lysozyme) to every 10 ml of the cell suspension, followed by incubation for 15 minutes at 37°. The protoplasts are recovered by centrifugation at 30,000 rpm for 40 minutes. The protoplast pellets are suspended in 30 ml of 0.25 *M* sucrose–Tris–Mg mixture for every 10 ml of original suspension in an ice bath with stirring, to induce lysis. After 20 minutes, the flocculant suspension is subjected to sonic vibration at 10 kc for 10 minutes at 2°. The shockate is centrifuged at 9000 *g* for 20 minutes to eliminate unbroken cells and large particles. The supernatant fluid is then centrifuged at 68,000 *g* for 70 minutes to yield a yellow-brown pellet, which is suspended in 0.25 *M* sucrose-Tris-Mg mixture to a protein concentration of 10 mg/ml. This preparation can couple NADH oxidation to phosphorylation at a P:O around 0.3. The supernatant fraction from the last centrifugation is saved as starting material for the preparation of the heat-stable coupling factor.

Step 2. Nonphosphorylating Membrane Fragments.[1] The pellets from the above centrifugation are suspended in 5 ml of redistilled water for each 10 ml of the original suspension of cells and the suspension is allowed to stand in an ice bath for 30 minutes. The suspension is repeatedly sucked up into a Pasteur pipette and discharged until a homogeneous appearance is attained. The suspension of membrane fragments is centrifuged at 105,000 *g* for 70 minutes. The supernatant fraction which contains both cF and Ft, is carefully pipetted off and the pellet is suspended again in redistilled water to give a protein concentration of 10 mg/ml or more. The nonphosphorylating NADH oxidase bound to membrane fragments so obtained can be stored at —20° without loss of activity. A low P:O ratio in the reconstituted system will result from repeated freezing and thawing or from standing at 37° for more than 30 minutes. The use of phosphate instead of Tris buffer in any step of the preparation will give unstable membrane fragments.

Step 3. Heat-Labile Coupling Factor (cF).[1,3] To the supernatant fraction from the last centrifugation of step 2, is added 1 *M* potassium phosphate, pH 7.40, to bring the phosphate concentration to 0.01 *M*. To this solution, solid ammonium sulfate is added to achieve 45% of saturation. After 30 minutes of stirring, the precipitate is collected by centrifugation and dissolved in 0.01 *M* potassium phosphate, pH 7.40, to give a protein concentration of 5 mg/ml. The solution is dialyzed for 24 hours against the same buffer. The crude cF in the dialysis tube is recovered, clarified by centrifugation, and stored at —20°. The cF preparation is largely free of Ft. This preparation is stable for more than one year.

Step 4. Heat-Stable Coupling Factor (Ft).[2,3] Although the starting supernatant solution of step 3 contains some Ft, (near saturating quantity as compared with cF), the bulk of Ft is present in the last supernatant fraction of step 1. This fraction is heated at 85° for 10 minutes in a boiling water bath and rapidly cooled in ice. Denatured protein is separated by centrifugation and discarded. To the yellow supernatant fluid, cold 50% trichloroacetic acid is added at 0° to give a final concentration of 5%. The precipitate formed is collected by centrifugation at 15,000 *g* for 15 minutes, and the supernatant fluid is discarded. The precipitate is suspended in 15 ml of water for each 100 ml of the starting supernatant fraction of this step, and the pH is adjusted to 8.0 with 0.5 *N* ammonium hydroxide. One milligram of crystalline pancreatic RNase is added to every 60 mg of protein and the solution is incubated at 37° for an hour. Trichloroacetic acid is added to terminate the digestion, to a final concentration of 5%. After the action of ribonuclease, the absorbancy at 260 mμ of the preparation decreases to one-fifth without any loss of the coupling activity. The collected precipitate is again dissolved in water, and the pH of the solution is adjusted to 7.40 with 0.05 *N* ammonium hydroxide. The crude Ft preparation obtained is dialyzed against 10 m*M* phosphate buffer pH 6.0 for 2 hours, and is brought to 5 mg/ml in protein concentration. A 50 ml portion of the dialyzed solution is placed on a DEAE-cellulose column (2.9 $\times$ 22 cm) previously equilibrated with 0.01 *M* potassium phosphate (pH 6.0) containing 0.2 *M* KCl. The concentration of KCl is then linearly increased with 400 ml of 0.2 *M* KCl in 0.01 *M* phosphate (pH 6.0) in the mixing chamber and 400 ml of 0.6 *M* KCl in 0.01 *M* phosphate (pH 6.0) in a reservoir. Ft activity is eluted from the column

[3] S. Yamashita and S. Ishikawa, in preparation.

when the KCl concentration is between 0.40 and 0.45 *M* KCl. Over 20-fold purification is achieved by column chromatography; the protein peak coincides with the activity peak in the elution diagram. The specific activity can be further increased by repeated chromatography as described above. Purified Ft is free of nucleic acids, adenylate kinase, and ATPase activity. The digestion of Ft with proteinase (trypsin or subtilisin) added in 1:20 ratio to protein rapidly destroys the activity.

Assay of Oxidative Phosphorylation

Although conventional manometric assay of oxygen consumption and inorganic phosphate uptake is needed for certain experiments, the assay of coupling factor activities can be done without measuring oxygen uptake, since the nonphosphorylating membrane fragments obtained according to the procedure described above invariably gives constant specific activity of the NADH oxidase, which is not altered by the presence of the coupling factors.

Reagents

Stock solution, sufficient for 1000 assays, contains the following components in 100 ml: 15 mmoles of potassium phosphate, pH 7.40 containing ^{32}P, specific activity 3×10^5 cpm per micromole phosphate, 1 millimole of ATP, 20 millimoles of glucose, 2500 I.U. of yeast hexokinase, 2 g of bovine serum albumin.

$MgCl_2$, 0.3 *M*

NADH and NAD, 0.01 *M*, pH 7.4

Nonphosphorylating membrane fragments, cF and Ft

Trichloroacetic acid, 20%

Procedure. Of the stock solution described above, 0.1 ml is added to each reaction tube, followed by 100 μg of nonphosphorylating membrane fragments and varying amounts of cF and Ft and water to make the volume to 0.85 ml. The tubes are equilibrated in a water bath at 28° with shaking, and 0.05 ml of 0.30 *M* $MgCl_2$ is added. After 5 minutes 0.10 ml of NADH or NAD is added to start the reaction, and the incubation is continued 20 more minutes; the oxidation of the added NADH is complete in about 10 minutes. Five-tenths milliliter of cold 20% trichloroacetic acid is added to stop the reaction, and aliquots from the supernatant fluid, after removal of protein by centrifugation, are taken for analysis of esterified ^{32}P.[4] The small amounts of ^{32}P esterified in the presence of NAD largely due to contaminated polynucleotide phosphoryl-

[4] S. O. Nielsen and A. L. Lehninger, *J. Biol. Chem.* **215**, 555 (1955).

ase, ranging up to 10% of that with NADH as a substrate, is subtracted from the latter to obtain the net oxidative phosphorylation. The phosphorylation is dependent on both cF and Ft, as illustrated in Fig. 1. Preincubation of all three components in the presence of Mg^{++} is necessary for the highest P:O ratios. Even if coupling factors are present in a constant ratio to membrane fragments in the assay mixture, the P:O ratio increases almost proportionally to the concentration of membrane fragments. For this reason, a series of experiments is required in which a fixed amount of membrane fragments is employed. The specific activity of a coupling factor will be calculated as the reciprocal amount of the coupling factor protein necessary for half-maximal P:O ratio, in the presence of an excess of the other coupling factor.

Other Properties of the System

The Respiratory Chain.[5] The location of the respiratory chain in the protoplasmic membrane has been shown.[6] Nonphosphorylating membrane fragments contain cytochromes of types *b*, *c*, and *a*,[1,7] as well as vitamin K with 45 carbon atoms in the side chain,[5] but no ubiquinone. The degree of unsaturation of the vitamin K_{45} is yet to be determined. The respiratory chain catalyzes oxidation of NADH (0.8–1.1 μatoms O per minute per milligram of protein) and malate (0.2–0.3 μatom o per minute per milligram of protein) in the presence of 10 m*M* phosphate, 5 m*M* $MgCl_2$, and 0.1% bovine serum albumin, at pH 7.40 at 23°. Ascorbate plus tetramethylparaphenylenediamine (TMPD) is also oxidized slowly (0.16 μatom O per minute per milligram of protein). Irradiation of the respiratory particles by near ultraviolet light (15 minutes) inactivates NADH oxidation up to 95%. Malate and NADH oxidation is relatively insensitive to HCN (50% inhibition at 2 m*M*). On the contrary, ascorbate oxidation in the presence of TMPD is highly sensitive to HCN (50% inhibition at 0.1 m*M*). From these findings and spectrophotometric data, the conclusion is reached that NADH oxidation involves cytochrome *b*, but does not need the participation of cytochromes of types *c* and *a*,[8] which are by-passed. Both NADH and ascorbate oxidation have high affinity toward oxygen ($K_m \leqq 10^{-5}$ *M*).

Phosphorylation. Phosphorylation is completely dependent on the presence of vitamin K. Among the externally added naphthoquinones, $K_{2(20)}$ (0.1 m*M*) added as a dispersion in pentaalanyldodecylamide

[5] M. Fujita, S. Ishikawa, and N. Shimazono, *J. Biochem.* (*Tokyo*) **59,** 104 (1966).

[6] V. I. Biryuzova, M. A. Lyukoyanova, N. S. Gelman, and A. I. Oparin, *Dokl. Akad. Nauk. SSSR,* **156,** 198 (1964).

[7] F. N. Jackson and V. D. Lawton, *Biochim. Biophys. Acta* **35,** 76 (1959).

[8] S. Ishikawa, M. Fujita, and S. Yamashita, unpublished results.

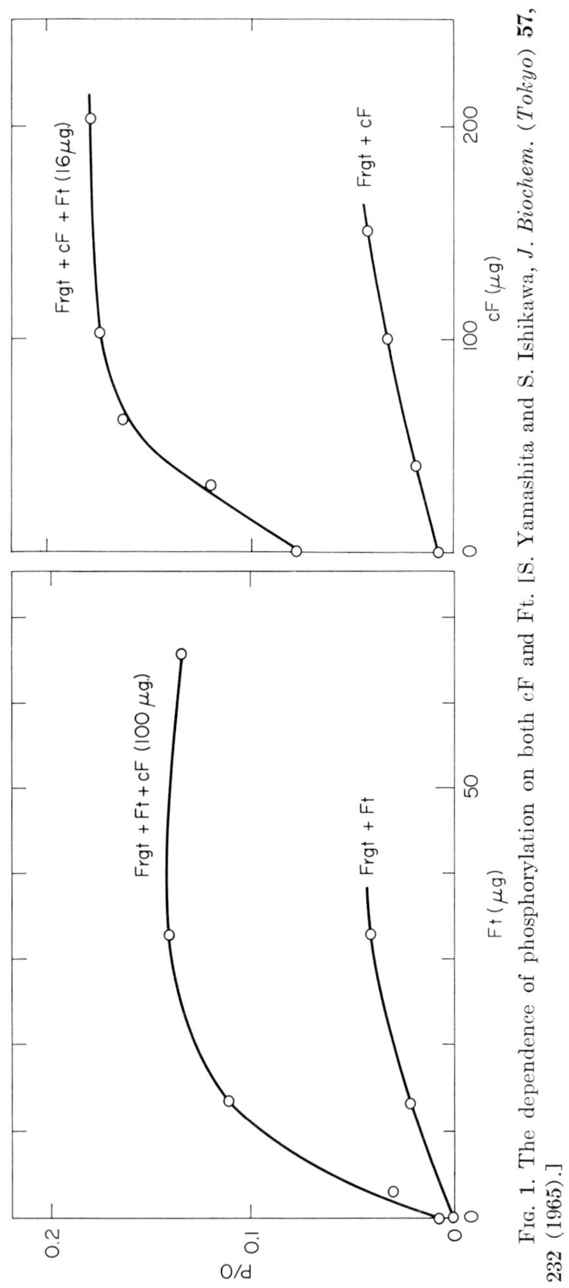

FIG. 1. The dependence of phosphorylation on both cF and Ft. [S. Yamashita and S. Ishikawa, *J. Biochem.* (*Tokyo*) **57,** 232 (1965).]

(final concentration, 0.003%) is the most effective.[5] Optimal concentration of Mg^{++} is high (20 mM); most of the Mg^{++} can be replaced by spermine or spermidine, both at 12 mM concentrations, if the system contains a minimal concentration of Mg^{++} (0.5 mM).[1] Phosphorylation is not observed when ascorbate plus TMPD is used as a substrate. Pentachlorophenol (0.5 mM), menadione (0.5 mM), and arsenate (twice the concentration of phosphate) uncouple phosphorylation, but dinitrophenol (1 mM) and oligomycin (10 μg/ml) have no action.[1] Potassium chloride at concentrations higher than 0.2 M and Tris-Cl higher than 0.05 M will uncouple phosphorylation.[8] Although the P:O ratio obtained according to the procedure described does not usually exceed 0.5, inclusion of 0.3 M sucrose in the reaction mixture substantially increases the P:O ratio.

Partial Reactions. Membrane fragments alone do not catalyze the partial reactions. The cF after ammonium sulfate fractionation exhibits Mg^{++}-dependent ATPase activity which develops in the presence of DNA.[9] This ATPase activity is inhibited by pentachlorophenol (0.5 mM) and polyethylene sulfonate (1 μM). DNA added at concentrations equivalent to 10 μM deoxymononucleotide suffices to bring about the maximal activation of the ATPase activity (50–80 millimicromoles P_i per minute per milligram of protein). cF also catalyzes an ATP-ADP exchange reaction in the presence of 5 mM $MgCl_2$.[1] The relation of these reactions to phosphate transfer reactions during oxidative phosphorylation has not been clarified.

[9] S. Ishikawa, S. Yamashita, S. Araki, and N. Shimazono, *J. Biochem.* (*Tokyo*) **57**, 235 (1965).

Section III

Preparations, Properties, and Conditions for Assay of Submitochondrial Particles

Previously Published Articles from Methods in Enzymology Related to Section III

Vol. II [125]. Triphosphopyridine and Diphosphopyridine Nucleotide Oxidases. Erwin Haas.

Vol. II [130]. Cytochromes *a*, a_1, a_2, and a_3. Lucile Smith.

Vol. II [131]. Cytochrome *b* (Mammals). Elmer Stotz.

Vol. VI [32]. Preparation and Assay of Phosphorylating Submitochondrial Particles. C. L. Wadkins and A. L. Lehninger.

Vol. VI [33]. Preparation and Assay of Phosphorylating Submitochondrial Particles: Sonicated Mitochondria. W. Wayne Kielley.

Vol. VI [34]. Preparation and Assay of Phosphorylating Submitochondrial Systems: Mechanically Ruptured Mitochondria. Maynard E. Pullman and Harvey S. Penefsky.

Vol. VI [57]. Lipid-Dependent DPNH Cytochrome *c* Reductase from Mammalian Skeletal and Heart Muscle. Alvin Nason and Frank D. Vasington.

Vol. VI [58]. Electron Transport Particles. D. E. Green and D. M. Ziegler.

[32] Preparation and Properties of Digitonin Particles from Beef Heart[1]

By W. B. ELLIOTT and D. W. HAAS

Mitochondria, isolated on a large scale from beef heart, are treated with digitonin to produce fragments, with phosphorylating ability and respiratory control, which can be stored in liquid nitrogen for several weeks without significant change in properties.

Preparative Methods

Reagents and Equipment

KOH,[2] 4 *N*
Sucrose, 0.25 *M*
Sucrose, 0.25 *M*, containing 0.05 *M* Tris buffer and 0.001 *M*, EDTA adjusted to pH 7.4
Digitonin, 1%, was prepared by adding 1.0 g of digitonin (only Calbiochem Grade A has been satisfactory) to 100 ml of distilled water and heating on a hot plate with stirring until all the digitonin was in solution. This reagent should be prepared while the homogenate is being centrifuged, then held at 2° until needed. Since it is supersaturated, it may precipitate if held too long before use.
Servall RC2 centrifuge[3] with GSA and SS34 angle heads and KSB continual flow accessory
Spinco Model L ultracentrifuge[4] with 30 rotor
Waring blendor, 1 gallon capacity[5]
Dounce homogenizer[6]

Preparation of Mitochondria. Beef hearts are opened and placed in

[1] The present method has been described briefly [D. W. Haas and W. B. Elliott, *J. Biol. Chem.* **238,** 1132 (1963)] and was developed by modification of the procedures of C. Cooper and A. L. Lehninger [*J. Biol. Chem.* **219,** 489 (1956)] and T. M. Devlin and A. L. Lehninger [*J. Biol. Chem.* **233,** 1586 (1958)].

[2] All reagents are made up in redistilled water, and all equipment and solutions used are precooled in ice or a cold room.

[3] Ivan Sorvall, Inc. Norwalk, Connecticut.

[4] Spinco Division, Beckman Instruments Co., Belmont, California.

[5] Waring Products Corporation, Winsted, Connecticut.

[6] Kontes Glass Co., Vineland, New Jersey.

cracked ice as soon as removed from the carcass. As soon as possible, the hearts are freed of connective tissue and visible fat and cut into about 1 cm^3 chunks. Up to 2400 g of the chunks of muscle are homogenized in the Waring blendor using 800 g muscle and 3 liters of 0.25 M sucrose–Tris–EDTA per batch. Homogenization is for approximately 1 second each at low and medium speeds followed by 20 seconds at 19,000 rpm, 1-minute cooling period (to prevent excessive heating of the blendor bearing), and 20 seconds more to 19,000 rpm. While the homogenate is still in the blendor, the pH is adjusted by adding 9 ml of 4 N KOH per 800 g tissue. The homogenate is then centrifuged for 10 minutes at 3500 rpm (250-ml polyethylene bottles) in the GSA rotor of the Servall RCII centrifuge at 0°. The supernatant (about 10 liters) is filtered through four layers of cheesecloth (suspended in a funnel but not touching the inner surface of the funnel), to remove floating lipid, into 6-liter Erlenmeyer flasks placed in ice. The supernatant is transferred to 4-liter aspirator bottle, also packed in ice (in a polyethylene tank with boss for the tube to pass out through) and attached to the KSB continual flow attachment of the Servall RCII centrifuge in which the tubes have been filled with 2°–4° 0.25 M sucrose and the SS34 rotor precooled. Centrifugation is at 37,000 g at +1° with a flow rate of 150 ml per minute. On opening the centrifuge tubes, the lipid layer is carefully removed and the upper inside area of the tubes is wiped with Kimwipes.[7] The sedimented particles are suspended in 300 ml of 0.25 M sucrose by use of a Dounce homogenizer with loose-fitting pestle and centrifuged at 13,000 rpm for 25 minutes at −2° in precooled 50-ml polypropylene tubes. If a significant layer of fluffy "light mitochondria" is present, it should be decanted or rinsed off before the dense mitochondrial layer is suspended. Suspension and centrifugation are repeated twice more, increasing the speed to 15,000 rpm for the last two centrifugations. The washings remove microsomes, hemoglobin, and myoglobin.

Submitochondrial Fragments. The washed mitochondria are suspended by use of the loose-fitting pestle of the Dounce homogenizer in 1% aqueous digitonin to give a protein concentration of about 100 mg/ml (usually 50 ml of 1% aqueous digitonin per 300 g ground heart). The suspension is allowed to stand for 30 minutes at 0° in an ice bath and then centrifuged in 50-ml polypropylene tubes at 37,000 g for 15 minutes at −2° in the Servall RCII centrifuge (or 5 minutes at 25,000 rpm in the 30 rotor of the Spinco Model L ultracentrifuge). The supernatant including any loose precipitate is transferred to Lusteroid tubes, placed in the precooled 30 rotor, and centrifuged at 105,000 g for 40 minutes in

[7] Kimberly-Clark Corporation, Neenah, Wisconsin.

the Spinco Model L ultracentrifuge.[8] The resulting pellet is suspended by use of the Dounce homogenizer, to give about 20–30 mg protein per milliliter, in 0.25 M sucrose containing 0.01 M Tris (pH 7.4), and is then used immediately or is transferred to 10-ml screw-topped Lusteroid centrifuge tube,[3] frozen 1 hour in a deep-freeze (or in the cold vapor above liquid nitrogen in a Dewar or liquid nitrogen refrigerator), and placed in the liquid nitrogen refrigerator.

Properties

The digitonin fragments have virtually no endogenous respiration, but take up oxygen rapidly in the presence of β-hydroxybutyrate, succinate, or NADH and ascorbate + TMPD. Only slight oxygen uptake has been obtained with other Krebs cycle intermediates. Phosphate to oxygen ratios up to 2.5 (freshly prepared) or 2.3 (stored) have been obtained with β-hydroxybutyrate and 1.8 with succinate. Respiratory control ratios (oxygen uptake in presence of substrate and ADP: oxygen uptake in presence of substrate) of 3.1–3.7 (β-hydroxybutyrate), 3.4–4.2 (NADH), and 1.2–1.4 (succinate) have been obtained in better preparations.[1] The digitonin fragments also carry on reversed electron transfer[9] and have energy-linked transhydrogenase activity.[9] An ADP:2e ratio of about 0.7 has been obtained for first phosphorylation site as measured in the NADH → fumarate reaction.[10] Occasional preparations have shown negative respiratory control ratios (i.e., have shown reverse acceptor control). In the better preparations, the cytochrome concentrations are about threefold higher than in the mitochondria from which the fragments were isolated.[1]

[8] The supernatant from this centrifugation may be centrifuged at 180,000 g for 2 hours in the 50 rotor in the Spinco Model L ultracentrifuge to obtain particles which will oxidize reduced cytochrome c or NADH without significant phosphorylation.

[9] D. W. Haas, *Biochim. Biophys. Acta* **82**, 200 (1964).

[10] D. W. Haas, *Biochim. Biophys. Acta* **92**, 433 (1964).

[33] Preparation and Assay of Phosphorylating Submitochondrial Particles: Particles from Rat Liver Prepared by Drastic Sonication

By CHARLES T. GREGG

Sonication of rat liver mitochondria in a hypotonic medium for up to 60 minutes yields particles which carry out oxidative phosphorylation

(P:O ratio approximately 1.0), the 2,4-dinitrophenol-stimulated ATPase (DNP-ATPase) reaction, and the ATP-P_i and ATP-ADP exchange reactions. The particles are obtained in good yield and possess a specific respiratory activity comparable to that of other submitochondrial preparations. These preparations are noteworthy because of their stability to freezing and thawing, the peculiar behavior of the partial reactions, and the fact that they show very pronounced reverse acceptor control of respiration.[1-3]

Preparation of the Particles

Mitochondria are prepared from 20–30 g of rat liver,[4] washed three times in 0.25 *M* sucrose, then suspended in 25 ml of 0.01 *M* Tris-HCl buffer (pH 7.4). Two disruption procedures were used; both gave identical results.

Procedure I. The mitochondrial suspension is treated in the Raytheon 9-kc, 60-watt sonic oscillator for four 15-minute periods, separated by 1- or 2-minute intervals to reduce sample heating. The sample chamber is cooled with a flow of ice water to keep the sample temperature between 0° and 5°. The oscillator plate voltage is set to a maximum, and the frequency control is adjusted to give a minimum on the plate voltage meter, thus yielding maximum power transfer. The oscillator is checked frequently (as described by the manufacturer) for its ability to deliver at least 4.4 watts of acoustic power under these conditions.

Procedure II. With the large probe of the MSE 20-kc, 60-watt sonic oscillator, a single treatment time of 15 minutes is used. The sample tube is immersed in ethanol precooled to —20°. The temperature of the ethanol is kept between —5° and —10° during sonication by the occasional addition of solid carbon dioxide, and the ethanol bath is vigorously stirred with a magnetic stirrer; the sample temperature remains between 0° and 5°. The frequency control is set to give maximum plate current. The acoustic power delivered should raise the temperature of 25 ml of water from 20° to 35° in 2 minutes in the absence of external cooling.

The treated suspension from either procedure is centrifuged at 25,000 *g* for 20 minutes in the Spinco model L ultracentrifuge, and the turbid supernatant suspension is decanted and recentrifuged at 144,000 *g* for 30 minutes. The supernatant solution is discarded, and the firmly

[1] C. T. Gregg, *Biochim. Biophys. Acta* **74**, 573 (1963).

[2] A. L. Lehninger and C. T. Gregg, *Biochim. Biophys. Acta* **78**, 12 (1963).

[3] C. T. Gregg and A. L. Lehninger, *Biochim. Biophys. Acta* **78**, 27 (1963).

[4] W. C. Schneider, *in* "Manometric Techniques" (W. W. Umbreit, R. Burris, and J. E. Stauffer, eds.), p. 189. Burgess, Minneapolis, Minnesota, 1957.

packed pellet is suspended in cold water or 0.25 *M* sucrose by gentle homogenization. The yield is 15–20% of the protein of the starting material. In our hands these procedures yield extremely uniform results.

Assay for Oxidative Phosphorylation

Measurement of Respiration

WITH THE OXYGEN ELECTRODE

Reaction Medium. The complete system contains 10 m*M* substrate, 10 m*M* Tris-HCl buffer (pH 7.4), 0.2% bovine serum albumin, 3.0 m*M* $MgCl_2$, 3.0 m*M* glucose, 300 Kunitz-MacDonald units of hexokinase, 2.4 m*M* ADP, 15 m*M* potassium phosphate (about 10^6 cpm of ^{32}P), about 5.0 mg particle protein (measured by a biuret method),[5] and air-saturated water to make 2.0 ml. When β-hydroxybutyrate is the substrate, NAD^+ (0.2 m*M*) is added.

Procedure. The vessels are incubated at 25°, and respiration is measured with the oxygen electrode essentially as described by Kielley,[6] until about 60% of the dissolved oxygen has been consumed, usually 1–4 minutes.

BY DETERMINATION OF ACETOACETATE WHEN β-HYDROXYBUTYRATE IS THE SUBSTRATE

Reaction Medium. The system is the same as for the oxygen electrode measurements except that each vessel contains 0.2 mg protein.

Procedure. All the assays described here, except the oxygen electrode measurements, are carried out in 20-ml beakers in a Dubnoff shaker at 25°; sonic particles are added last to start the reactions. Assays are made on the filtrates after removing protein precipitated by trichloroacetic acid. The formation of acetoacetate is measured by Walker's[7] method as described before,[8] after stopping the reaction with 4.0 ml of 0.25 *M* trichloroacetic acid.

Measurement of Phosphate Uptake

Incorporation of labeled P_i into ATP is determined by the isobutanol–benzene extraction method.[8]

[5] A. G. Gornall, C. J. Bardawill, and M. M. David, *J. Biol. Chem.* **177,** 751 (1949).
[6] See Vol. VI [33].
[7] P. J. Walker, *Biochem. J.* **58,** 699 (1954).
[8] See Vol. VI [32].

Assay for ATPase Activity

Reaction Medium. Each vessel contains 10 mM ATP, 10 mM Tris-HCl (pH 7.4), 1.0 mg particle protein, and water to make 2.0 ml.

Procedure. The reaction is run for 10 minutes, and the P_i liberated from ATP is determined by the Fiske-SubbaRow[9] method, except that Elon (*p*-methylaminophenol sulfate) is used as the reducing agent. The reaction is stopped with 0.2 ml of 4.0 M trichloroacetic acid.

Assay for ATP-P_i Exchange

Reaction Medium. Each vessel contains 6.0 mM ATP (pH 6.5), 10 mM potassium phosphate (pH 6.5, about 10^6 cpm of ^{32}P), 1.0 mM $MgCl_2$, 0.4% serum albumin, 1.0 mg particle protein, and water to make 1.0 ml.

Procedure. Reaction time is 20 minutes. The radioactivity of the $AT^{32}P$ formed is determined as described earlier,[8] after stopping the reaction with 1.0 ml of 0.8 M trichloroacetic acid.

Assay for ATP-ADP Exchange

Reaction Medium. Each vessel contains 10 mM ATP, 10 mM ADP-^{14}C (about 3×10^4 cpm), 1.0 mM $MgCl_2$, 1.2 mg particle protein, and water to make 1.0 ml (pH 6.8).

Procedure. After a 10-minute incubation the incorporation of radioactivity into ATP is determined as before,[8] except that the nucleotides are separated by high-voltage paper electrophoresis in citrate buffer (0.05 M, pH 4.5).

Properties of the Particles

Oxidation, Phosphorylation. The particles oxidize NADH, β-hydroxybutyrate, succinate, and choline at relative rates of 6:4:2:1 with P:O ratios of almost exactly 1.0. Malate, fumarate, citrate, pyruvate, and α-ketoglutarate are not oxidized. Succinate and ATP reduce neither NAD^+ nor acetoacetate. Failure to observe these ATP-driven reductions is consistent with other data, indicating that phosphorylation occurs only at the cytochrome *b* level (site 2) in these particles. There is no oxidation or phosphorylation without added substrate, and no phosphorylation without added ADP and P_i. NAD^+ is an absolute requirement for β-hydroxybutyrate oxidation; the requirement for Mg^{++} for phosphorylation is frequently absolute. Neither spermine nor spermidine can replace Mg^{++}.[10] Oxidative phosphorylation is completely suppressed by un-

[9] See Vol. III [115].

[10] See this volume [31].

coupling agents, although higher concentrations are required than in intact mitochondria. Calcium ion does not uncouple, and these particles are incapable of active uptake of calcium. Phosphorylation is completely inhibited by antimycin A, Amytal, and NaCN, but the inhibition of β-hydroxybutyrate oxidation is incomplete.

Stability. The particles are remarkably stable toward freezing and thawing; if stored at $-20°$ as a concentrated suspension in 0.25 M sucrose and suspended in fresh medium before assay, 85–90% of the original oxidative phosphorylation activity is retained. This activity remains through at least 5 cycles of freezing and thawing or in frozen storage for at least 3 weeks.

Partial Reactions. The DNP-ATPase and the ATP-ADP and ATP-P_i exchange reactions occur in the sonic particles. The rates of the three reactions relative to that of oxidative phosphorylation are approximately 4.0, 0.4, and 0.01, respectively, when assayed in the presence of 1.0 mM $MgCl_2$. ATPase activity is inhibited by azide, and the activity elicited by DNP and Mg^{++} together is completely inhibited by oligomycin. DNP-ATPase is strongly inhibited by reduction of the respiratory carriers. Unexpectedly, NaCN greatly stimulates DNP-ATPase. The activity stimulated by NaCN is reduced by added substrate, suggesting that ATPase associated with site 3 may be activated by reducing the carriers. DNP completely inhibits the ATP-P_i exchange and reduces the ATP-ADP exchange about 50%. Oligomycin behaves similarly toward the ATP-P_i exchange but has no effect on the ATP-ADP exchange. The ATP-P_i exchange is insensitive to antimycin A or Amytal.

Stimulation and Inhibition of Respiration. The stimulation of respiration by ADP and P_i in the sonic particles is composed of two opposing effects: a stimulation of respiration by P_i, and an inhibition by ADP (reverse acceptor control).[2,3] The latter effect is highly specific for ADP, whereas the stimulation by P_i can be mimicked by EDTA, PP_i, GSH, or serum albumin. Both effects are seen with all oxidizable substrates, and both primarily involve phosphorylating, inhibitor-sensitive electron transport. The basis for these striking results is not yet understood, although a working hypothesis has been formulated.[3]

The correlation between the ability to stimulate respiration in the absence of ADP and P_i and to stimulate ATPase activity[2] is not seen in the sonic particles. EDTA stimulates respiration but inhibits ATPase activity completely. ADP inhibits respiration but not ATPase activity, whereas the effect of Mg^{++} is just the opposite of that of ADP. The action of uncoupling agents is similarly anomalous; dicoumarol, DNP, and gramicidin all inhibit acceptor-less respiration. DNP stimulates ATPase activity, whereas dicoumarol and gramicidin actually inhibit ATPase activity.

[34] Preparation, Properties, and Conditions for Assay of Phosphorylating Electron Transport Particles (ETPH) and Its Variations

By ROBERT E. BEYER

Assay Method

Principle. The activity of submitochondrial particles prepared by sonic treatment of beef heart mitochondria is frequently related to phosphorylative efficiency, or the ratio of micromoles of P_i esterified to microatoms of oxygen utilized (P:O). Therefore, the assay method usually involves the determination of oxygen consumed during the assay either by the Warburg technique or by a polarographic method and the determination of phosphate either disappearing from the P_i pool, or appearing in an esterified form. Since most submitochondrial phosphorylating particles catalyze an effective ATPase, it is important that ATP formed during the assay be transferred to a metabolically more stable form. This is accomplished by including hexokinase and glucose which transfers the γ-phosphate of ATP to glucose to form glucose 6-phosphate. Since submitochondrial particles from beef heart mitochondria may be used to study phosphorylations occurring at individual phosphorylation sites (cf. Beyer[1]), the procedure for such assays is provided below.

Reagents. Unless otherwise stated, all reagent solutions are made in water.

(a) Sucrose, 1 *M*
(b) Glucose, 0.2 *M*
(c) Potassium phosphate, 0.1 *M*, pH 7.5
(d) $MgCl_2$, 0.1 *M*
(e) ADP, 0.1 *M*, pH 6.8
(f) Hexokinase (EC 2.7.1.1, ATP: D-hexose 6-phosphotransferase), 5 mg protein/ml
(g) Alcohol dehydrogenase (EC 1.1.1.1, Alcohol: NAD oxidoreductase), 10 mg protein/ml
(h) NAD, 0.01 *M*
(i) Semicarbazide hydrochloride, 0.2 *M*, pH 7.5
(j) Ethanol, 1.0 *M*

[1] R. E. Beyer, *Biochem. Biophys. Res. Commun.* **16,** 460 (1964).

(k) Antimycin a, 0.2 mg/ml in 95% ethanol
(l) Potassium ascorbate, 0.1 M, pH 7.2
(m) Phenazine methosulfate, 0.1 mM
(n) Potassium succinate, 0.25 M, pH 7.5
(o) Submitochondrial particle, 20 mg protein/ml
(p) 20% potassium hydroxide

Reagents (a), (b), (c), (d), (e), (h), (i), (j), (k), (n), and (p) require no comment. Reagent (f) is acquired[2] as a 1× crystallized suspension in 70% saturated ammonium sulfate. It is prepared by dialysis against a solution 0.2 M in glucose and 0.05 M in Tris-acetate, pH 7.5, and stored at —20°. Reagent (g) is obtained[2] as a 2× crystallized, dry powder, dissolved in 0.05 M Tris-acetate, pH 7.5, and stored at —20°. Reagents (l) and (m) are prepared the day of assay, and reagent (m) is kept in low actinic glassware.

Manometric Assay Procedure. The reagents and the Warburg vessels are kept at 0° until incubated. Two-tenths milliliter of reagent (p) is placed into the center well of 15–20 ml Warburg vessels having one side arm, and a roll of filter paper, frayed at the top, is inserted in the center well so that 3–5 mm extend out of the center well. Then 0.75 ml of (a), 0.5 ml of (b), 0.05 ml of (d), 0.05 ml of (e), and 0.02 ml of (f) are added to the main compartment. The volume of (c) to be added depends on the particle to be assayed, being 0.5 ml for ETPH (Mg^{++}, Mn^{++}), 0.35 ml for ETPH (Mg^{++}), and 0.2 ml for ETPH (EDTA I and II) and ETPH (urea).

When all three phosphorylation sites are to be assayed, NADH is generated by adding 0.03 ml of (g), 0.04 ml of (h), and 0.09 ml of (i) to the main compartment and 0.06 ml of (j) to the side arm. When phosphorylation sites II and III are to be assayed, 0.1 ml of (n) is added to the side arm, and when site III alone is to be assayed, 0.005 ml of (k) is added to the main compartment and 0.2 ml of (l) and 0.03 ml of (m) are added to the side arm. The total fluid volume of the Warburg vessel is adjusted to 3.15 with cold water, the side arm is stoppered, 0.05 ml of (o) is added to the main compartment and the vessels are applied to the manometers. The shaking mechanism is turned on, and the flasks are placed in the Warburg bath at 10-second intervals and allowed to approach temperature equilibration for 5 minutes. At this time the fluid contents are mixed, again at 10-second intervals. At 4 minutes after the initiation of the reaction all stopcocks are closed and readings are recorded at 5, 10, 15, and 20 minutes. The reaction is

[2] Worthington Biochemical Corporation, Freehold, New Jersey.

terminated at 20 minutes by the addition of 2 ml of 1.5 M perchloric acid. Adequate blanks such as "zero" time controls and "no-substrate" controls are included. Precipitated protein is removed by centrifugation. A 0.5 ml sample of the supernatant is assayed for P_i by the colorimetric method of Martin and Doty[3] as described by Lindberg and Ernster.[4] Interference in color development and partition caused by ascorbate is alleviated by the addition of 0.1 ml of 0.03 M KIO_3 to the phosphate mixing tube prior to the addition of the reaction mixture. Oxygen consumption is extrapolated to zero time from the readings obtained between 5 and 10 minutes.

Polarographic Assay Procedure. The procedure described below has been used with an oscillating platinum electrode and a 2-ml reaction cell,[5] but it may be adapted easily to larger reaction cells and the stationary Clark electrode.[6] Reagents are the same as listed above except that $^{32}P_i$ is added to the phosphate solution to a specific activity of 10,000 cpm per micromole. The radiophosphate is obtained (Nuclear Division, Union Carbide Corporation, Oak Ridge, Tennessee 37831) as carrier-free $H_3{}^{32}PO_4$ in dilute HCl. It should be boiled prior to use to remove pyrophosphates which may be present. For the radiometric assay, the same concentration of P_i is employed regardless of the particle used.

The following reagents are pipetted into test tubes contained in shaved ice: 0.75 ml of (a), 0.6 ml of (b), 0.2 ml of (c), 0.05 ml of (d), 0.05 ml of (e), and 0.02 ml of (f). When NADH is to be generated as substrate, 0.03 ml of (g), 0.04 ml of (h), and 0.08 ml of (i) are added. Water is added to a volume of 2.95 ml, and the mixture is incubated for 5 minutes at 30° with frequent swirling. Then 0.05 ml of (o) is added, the contents are again mixed, and 2 ml is transferred to the electrode chamber. The remainder is returned to the 30° water bath. A blank rate of O_2 consumption is recorded, and 0.01 ml of (j) is added to the chamber and the contents mixed. When the oxygen in solution is depleted, an aliquot of 0.2 ml is transferred to a test tube containing 4.3 ml of silicotungstic acid, and the $^{32}P_i$ and glucose-6-^{32}P are separated according to Lindberg and Ernster.[4] After separation of the two phases, the top layer is aspirated off and the bottom layer containing the glucose 6-phosphate is extracted twice with 5-ml aliquots of water-saturated isobutanol. Of the mixture not incubated with substrate, 0.2 ml is treated identically and serves as a control for each incubation condition.

[3] J. B. Martin and D. M. Doty, *Anal. Chem.* **21**, 965 (1949).
[4] O. Lindberg and L. Ernster, *Methods Biochem. Anal.* **3**, 1 (1956).
[5] Oxygraph, Gibson Medical Electronics, Middleton, Wisconsin.
[6] Yellow Springs Instrument Co., Yellow Springs, Ohio.

The same procedure is followed for the oxidation of succinate or reduced cytochrome *c* except that (g), (h), (i), and (j) are omitted and 0.05 ml of (n) is added to the electrode cell for succinate oxidation and 0.005 ml of (k) and 0.2 ml of (l) are added to the test tube and 0.03 ml of (m) is added to the cell for reduced cytochrome *c* oxidation. The rate of oxygen consumption in the latter assay in the absence of phenazine methosulfate is negligible.

Preparation of Particles

All manipulations are carried out at ice bucket temperature unless so indicated. The Spinco Model L Ultracentrifuge is used for all centrifugations.

ETPH (Mg^{++}, Mn^{++}). The preparation of this particle has been described by Hansen and Smith.[7] Suspension of pellets following centrifugation refers to gentle homogenization by way of several passes through a Teflon pestle tissue grinder.[8] HBHM are prepared as described in this volume [12], the "light-heavy split"[9] being accomplished in a solution 0.25 *M* in sucrose, 0.01 *M* in Tris-HCl, pH 7.8, 1 m*M* in ATP, 1 m*M* in $MgCl_2$, and 1 m*M* in potassium succinate. A solution containing these five components at the concentrations given above and at pH 7.8 is abbreviated as STAMS. The HBHM are suspended in STAMS at a concentration of 40 mg protein per milliliter and maintained at —20° overnight. The suspension of HBHM is thawed immediately prior to the preparation of ETPH (Mg^{++}, Mn^{++}), adjusted to a concentration of 30 mg of protein per milliliter by the addition of STAMS; the pH is adjusted to 7.8 with 1 *N* KOH, and a "light-heavy split" is accomplished. The resulting HBHM are suspended to a protein concentration of 30 mg/ml in a solution 0.25 *M* in sucrose, 0.01 *M* in Tris-HCl (pH 7.5), 1 m*M* in ATP, 1 mM in potassium succinate, 5 m*M* in $MgCl_2$, and 10 m*M* in $MnCl_2$; the suspension is adjusted to pH 7.5 with the glass electrode using either 1 *N* HCl or KOH. Aliquots (20 ml) are placed in a jacketed 30-ml beaker, through which is circulating ethylene glycol at —10°, and subjected to 20 kc sonic irradiation for 30 seconds from a Sonifier[10] at a power output between 7 and 8 amps. The 0.5-inch step-horn tip is immersed to within 10 mm of the bottom of the beaker.

[7] M. Hansen and A. L. Smith, *Biochim. Biophys. Acta* **81**, 214 (1964).

[8] We generally use No. 4288-B Tissue Grinders, size C, Arthur H. Thomas Co., for large amounts of particle and either the same number and supplier, size A, or No. K-88548, Tissue Grinder, Size B, Kontes Glass Co., Vineland, New Jersey, for smaller amounts of particle.

[9] Y. Hatefi and R. L. Lester, *Biochim. Biophys. Acta* **27**, 83 (1958).

[10] Branson Ultrasonics Corporation, Model LS-75.

The pH is adjusted to 7.5, and the suspension is centrifuged 14×10^4 g-min.[11] (20,000 rpm, 7 minutes, No. 40 rotor). The supernatant solution is decanted, care being taken to exclude the fluffy layer, and centrifuged 3.8×10^6 g-min (50,000 rpm, 25 minutes). The surface of the tightly packed reddish-brown pellet is rinsed with a few milliliters of a solution 0.25 M in sucrose and 0.01 M in Tris-HCl, pH 7.5, and suspended to a protein concentration of at least 20 mg protein per milliliter in more of the sucrose-Tris solution immediately above if the particles are to be assayed immediately, or in a preserving mix[12] which is 0.25 M in sucrose, 0.01 M in Tris-HCl, pH 7.5, 5 mM in $MgCl_2$, 2 mM in GSH, 2 mM in ATP, and 4 mM in succinate.

ETPH (Mg^{++}). The preparation of this particle has been described by Linnane and Ziegler[13] and appears below with minor modifications. HBHM are prepared as described in this volume [12], the "light-heavy split" being performed in a solution 0.25 M in sucrose and 0.01 M in Tris-HCl, pH 7.8. The pH of the suspension is checked immediately prior to centrifugation. The resulting HBHM are suspended in fresh sucrose–Tris-HCl, pH 7.8, solution listed above to a concentration of 40 mg protein per milliliter and maintained at —20° overnight. The ETPH (Mg^{++}) may be prepared without freezing, but yields are lower than when the freezing step is included. Upon thawing, 0.25 M sucrose–0.01 M Tris-HCl, pH 7.5, is added to adjust the protein concentration to 20 mg/ml, $MgCl_2$ is added to 15 mM, and ATP is added to 1 mM. The pH is adjusted to pH 7.3 and the suspension is treated, with sonic irradiation, in 20-ml aliquots contained in a jacketed 30-ml beaker through which ethylene glycol (—10°) is circulated. A Branson probe-type Sonifier, model LS-75 is used at an output of 7–8 amps. The suspension is centrifuged 8.9×10^4 g-min (No. 40 rotor, 15,000 rpm, 6 minutes), and the supernatant suspension is decanted so as to exclude the loosely packed fluff above a hard-packed pellet. The decanted suspension is centrifuged 4.2×10^6 g-min (No. 40 rotor, 40,000 rpm, 40 minutes). The pellet is suspended in one-fourth the volume of the original supernatant of a solution 0.25 M in sucrose, 0.01 M in Tris-HCl, pH 7.5, 0.01 M in $MgCl_2$, and 1 mM in ATP. The suspension is again centrifuged 4.2×10^6 g-min (No. 40 rotor, 40,000 rpm, 40 minutes) and the pellet suspended in 0.25 M sucrose–0.01 M Tris-HCl, pH 7.5, if it is to be assayed immediately, or in the preserving mix of Linnane and Titchener[12] if the particle is to be stored for any length of time.

[11] All centrifugal forces are calculated on the basis of average radius.

[12] A. W. Linnane and E. B. Titchener, *Biochim. Biophys. Acta* **38,** 469 (1960).

[13] A. W. Linnane and D. M. Ziegler, *Biochim. Biophys. Acta* **29,** 630 (1958).

ETPH (EDTA-1). The preparation of this submitochondrial particle is similar to that described by Linnane and Titchener[12] except that it is more rapid and the resulting particle synthesizes ATP more efficiently than the modified ETPH [ETPH (EDTA-2)] of Linnane and Titchener.[12] HBHM are prepared as described in this volume [12], the "light-heavy split" being performed in a solution 0.25 *M* in sucrose and 0.01 *M* in Tris-HCl, pH 7.8. The HBHM are suspended to a protein concentration of 10 mg/ml in a solution 0.25 *M* in sucrose, 0.01 *M* in Tris-HCl, pH 7.5, and 1 m*M* in EDTA and centrifuged 1.4×10^5 *g*-min (No. 30 rotor, 15,000 rpm, 7 minutes). The sediment is suspended to 10 mg of protein per ml in a solution 0.25 *M* in sucrose and 0.01 *M* in Tris-HCl, pH 7.5, and sedimented again at 1.4×10^5 *g*-min. The sediment is suspended to a protein concentration of 20 mg/ml in a solution 0.25 *M* in sucrose and 0.01 *M* in Tris-HCl, pH 7.5, and stored at —20° overnight. Upon thawing the suspension is made 1 m*M* with respect to EDTA, the pH is adjusted to 7.5, and the suspension is irradiated with 20 kc sound for 60 seconds as described above in the sections on ETPH (Mg^{++}, Mn^{++}) and ETPH (Mg^{++}). The suspension is centrifuged 2.6×10^5 *g*-min (No. 40 rotor, 20,000 rpm, 10 minutes), and the carefully decanted supernatant solution is centrifuged again for 4.2×10^6 *g*-min (No. 40 rotor, 40,000 rpm, 40 minutes). The resulting sediment is washed twice by suspending the particles to 10 mg protein per milliliter in a solution 0.25 *M* in sucrose and 0.01 *M* in Tris-HCl, pH 7.5, and centrifuging at 4.2×10^6 *g*-min. The final sediment is suspended to 20 mg of protein per milliliter in a solution 0.25 *M* in sucrose and 0.01 *M* in Tris-HCl, pH 7.5, for immediate use or the preserving mix[12] listed under the above section on the preparation of ETPH (Mg^{++}, Mn^{++}) for storage.

ETPH (EDTA-2). The preparation of ETPH (EDTA-2) has been described by Linnane and Titchener.[12] This particle has been referred to as modified ETPH[12] and as METPH.[1] HBHM are prepared as described in this volume [12], the "light-heavy split" being carried out after centrifugation from a solution 0.25 *M* in sucrose and 0.01 *M* in Tris-HCl, pH 7.8. The HBHM are suspended to a concentration of 10 mg protein per milliliter in a solution 0.25 *M* in sucrose, 0.01 *M* in Tris-HCl, pH 7.5, and 1 m*M* EDTA, and sedimented again at 1.4×10^5 *g*-min. The precipitate is suspended in the 0.25 *M* sucrose–0.01 *M* Tris-HCl, pH 7.6, solution to 20 mg of protein per milliliter and stored at —20° for 1 week, during which time the preparation is subjected to at least 5 freeze-thaw cycles. Following each thaw the pH is adjusted to 7.5. Immediately prior to sonic treatment the solution is made 2 m*M* with respect to EDTA, the pH is adjusted to 7.5, and the suspension is subjected to 20 kc sonic irradiation for 60 seconds as reported in the section on the preparation of ETPH (Mg^{++}, Mn^{++}) above. The pH of the suspension is adjusted to 7.5 and the preparation is centrifuged at 2.6×10^5 *g*-min (No. 40 rotor, 20,000 rpm, 10 minutes). The supernatant suspension is decanted carefully and centrifuged at 4.2×10^6 *g*-min (No. 40 rotor, 40,000 rpm, 40 minutes). The pellet is suspended in one-half the original volume of a solution 0.25 *M* in sucrose, 0.01 *M* in Tris-HCl, pH 7.5, and 1 m*M* in EDTA and is sedimented at 4.2×10^6 *g*-min. This procedure of suspension and sedimentation is repeated 2 times in a solution 0.25 *M* in sucrose and 0.01 *M* in Tris-HCl, pH 7.5, excluding EDTA, and the final pellet is suspended in 0.25 *M* sucrose–0.01 *M* Tris-HCl, pH 7.5, for immediate use, or in the preserving mixture recorded in the section on ETPH (Mg^{++}, Mn^{++}) above for storage.

ETPH (*Urea*). The preparation of an ETPH (urea) particle has been described by Sanadi and his colleagues,[14,15] and the procedure given below is a modification thereof. ETPH (Mg^{++}, Mn^{++}) is prepared as described above, the final pellet being suspended in 0.25 *M* sucrose–0.01 *M* Tris-HCl, pH 7.5, to 40 mg particle protein per milliliter. To the suspension is added ATP, urea, and Tris-HCl so that the final concentrations are: ETPH (Mg^{++}, Mn^{++}), 5 mg protein/ml; ATP, 1 m*M*; urea, 1 *M*; sucrose, 0.63 *M*; Tris-HCl, 8.8 m*M*. The following recipe is used: To 2 ml of ETPH (Mg^{++}, Mn^{++}) at 40 mg protein/ml is added 0.8 ml of 10 *M* urea, 0.16 ml of 0.1 *M* ATP (pH 7.5), and the suspension is made to 8 ml with 0.01 *M* Tris-HCl, pH 7.5. The pH is controlled with a glass electrode and adjusted to 7.5. The suspension is stirred with a Teflon-covered magnetic bar and magnetic stirrer for 60 minutes at 0°. The period of urea treatment should be controlled according to the requirements of the experiment and the properties of the mitochondrial system. The author has observed variability between different starting ETPH (Mg^{++}, Mn^{++}) preparations, but not a great deal of variability in properties when the same batch of ETPH (Mg^{++}, Mn^{++}) is employed. The suspension is diluted to 2 mg protein per milliliter with 0.25 *M* sucrose–0.01 *M* Tris-HCl, pH 7.5, and centrifuged at 3×10^6 *g*-min (No. 50 rotor, 50,000 rpm, 20 minutes). The supernatant solution is discarded, and the pellet is suspended in a solution 0.25 *M* in sucrose and 0.01 *M* in Tris-HCl, pH 7.5, to 2 mg of particle protein per milliliter. The suspension is centrifuged again for 3×10^6 *g*-min, the surface of the pellet is rinsed with 0.25 *M* sucrose–0.01 *M* Tris-HCl, pH 7.5, and suspended in

[14] D. R. Sanadi, T. E. Andreoli, and K. W. Lam, *Biochem. Biophys. Res. Commun.* **17**, 582 (1964).

[15] T. E. Andreoli, K. W. Lam, and D. R. Sanadi, *J. Biol. Chem.* **240**, 2644 (1965).

either the sucrose–Tris-HCl solution to a concentration of 20 mg particle protein per milliliter for immediate use, or in the preserving mix listed in the section on the preparation of ETPH (Mg^{++}, Mn^{++}) above for storage at $-20°$.

Properties

Several remarkable properties distinguish submitochondrial particles prepared by sonic treatment from the parent particle, HBHM. The most obvious, of course, is size. The intact mitochondrion is easily seen in fresh suspension under the phase contrast microscope at a magnification of about 1200, while the submitochondrial particles are barely discernible under the same conditions and are at the limit of resolution of the light microscope. Another difference is in the specificity of the nucleoside diphosphate which will accept activated P_i during oxidative phosphorylation. Löw *et al.*[16] have shown that whereas intact beef heart mitochondria are specific for ADP as a phosphate acceptor during oxidative phosphorylation, various nucleoside diphosphates, in the approximate descending order of efficiency of ADP > IDP > GDP > UDP > CDP, may accept P_i to form the corresponding nucleoside triphosphate in submitochondrial particles such as ETPH (Mg^{++}), and presumably in the other "sonic-particles." Still another property of the sonic submitochondrial particle, as distinguished from the HBHM, is its ability to reduce

P:O Ratios of Various ETPH Preparations

Preparation	P:O ratio		
	NADH	Succinate	Reduced Cyt. *c*
ETPH (Mg^{++}, Mn^{++})	2.5–2.9	1.6–1.8	0.3–0.7
ETPH (Mg^{++})	1.4–1.8	0.8–1.2	0.3–0.5
ETPH (EDTA-1)	0.5–0.7	0.2–0.4	0.05–0.2
ETPH (EDTA-2)	0.3–0.4	0.1–0.2	0–0.05
ETPH (urea)[a]	0.3–0.4	0.2–0.3	0–0.05

[a] As indicated by Andreoli *et al.* [T. E. Andreoli, K. W. Lam, and D. R. Sanadi, *J. Biol. Chem.* **240,** 2644 (1965)], the ability of ETPH (urea) to catalyze reactions such as the energy-linked succinate-NAD reductase is variable and may be somewhat controlled by the length of time to which the ETPH (Mg^{++}, Mn^{++}) is exposed to urea during the preparation of the particle. It should be stressed that the procedures for the preparation of the submitochondrial particles may require local adjustment according to the type and quality of the slaughterhouse material available and the method of preparing the starting HBHM employed.

[16] H. Löw, I. Vallin, and B. Alm, *in* "Energy-Linked Functions of Mitochondria" (B. Chance, ed.), p. 5. Academic Press, New York, 1963.

exogenous NAD through, presumably, a reversal of the reactions of oxidative phosphorylation.[17] The submitochondrial particle also has the advantage of being able to be produced with varying, and within limits predictable, phosphorylative capacities. This may be seen in the table, in which the P:O ratios of the various particles described above are tabulated, the energy at the three phosphorylation sites being generated and assayed as described in the assay section above.

[17] H. Löw, H. Krueger, and D. M. Ziegler, *Biochem. Biophys. Res. Commun.* **5**, 231 (1961).

[35] A-Particles[1] and P-Particles[2]

By JUNE M. FESSENDEN[2a] and EFRAIM RACKER[2a]

$$P_i + ADP \xrightarrow[\text{coupling factors}]{\text{Apo-particles}} ATP$$

A partial resolution of oxidative phosphorylation in submitochondrial particles can be achieved by a variety of physical, chemical, and enzymatic degradation procedures. Such apo-particles catalyze oxidation of DPNH or succinate with little accompanying ATP formation. The oxidation can be coupled to phosphorylation by the addition of specific proteins (coupling factors) isolated from mitochondria. Among the most useful and reproducible preparations of apo-particles are the A-particles and P-particles which are stimulated by coupling factor 1,[3] coupling factor 2,[4] coupling factor 3[5] and coupling factor 4[5a] (F_1, F_2, F_3, and F_4, respectively).

Assay Method

Principle. Deficient particles are used which catalyze oxidation but negligible phosphorylation. On addition of multiple coupling factors to these particles, a marked stimulation of phosphorylation is observed.

Reagents

Potassium phosphate, 0.3 *M*, pH 7.4
ATP, 0.1 *M*, pH 7.4

[1] J. M. Fessenden and E. Racker, *J. Biol. Chem.* **241**, 2483 (1966).
[2] T. E. Conover, R. L. Prairie, and E. Racker, *J. Biol. Chem.* **238**, 2831 (1963).
[2a] This manuscript was prepared while the authors were at the Department of Biochemistry, The Public Health Research Institute of the City of New York, Inc., New York.
[3] See this volume [82].
[4] See this volume [84a].
[5] See this volume [84b].
[5a] See this volume [85].

$MgSO_4$, 0.1 *M*
Sodium succinate, 0.5 *M*, pH 7.4
F_1 (1 mg/ml)[3,6]
F_2 (1 mg/ml)[4] in 0.3 *M* potassium phosphate pH 7.4
F_3 (10 mg/ml)[5] or
F_4 (10 mg/ml).[5a] The 12–13% ammonium sulfate fraction is dialyzed 50 hours against 200 volumes of 0.25 *M* sucrose–0.01 *M* Tris–5 m*M* EDTA, pH 10.7.
Deficient particles (25 mg/ml)
Hexokinase (Boehringer) dialyzed overnight against 400 volumes of 5 m*M* EDTA, pH 7.4–1% glucose (500 units/ml)
Tris-SO_4, 1.0 *M*, pH 8.0
$MgSO_4$, 1.0 *M*
ATP, 0.2 *M*, pH 7.4
Glucose, 1.0 *M*
EDTA, 0.24 *M*, pH 7.4
Sucrose, 1.0 *M*
Bovine serum albumin, 20% (crystalline, Armour Co.) dialyzed overnight against 100 volumes of 0.01 *M* Tris-SO_4, pH 7.4
Recrystallized $^{32}P_i$ (2×10^6 cpm/ml)[7]
Trichloroacetic acid, 50%

Preparation of Stock Solution. A mixture of 0.2 ml of hexokinase, 0.05 ml of Tris-SO_4, 0.02 ml of 1.0 *M* $MgSO_4$, 0.05 ml of 0.2 *M* ATP, 0.32 ml of glucose, 0.02 ml of EDTA, 0.05 ml of bovine serum albumin, and 0.29 ml of sucrose is prepared and stored at —20°.

Procedure. To the main compartment of two Warburg vessels 0.02 ml of 0.1 *M* ATP, 0.02 ml of 0.1 *M* $MgSO_4$, and 0.02 ml of deficient particles are pipetted. To one of the vessels 0.1 ml potassium phosphate, and to the other vessel 0.04 ml of F_1, 0.1 ml of F_2 (which contains potassium phosphate), and 0.02 ml of F_4 or 0.03 ml of F_3 are added in a volume of 0.38 ml. After incubation at 23° for 5 minutes, the vessels are put in ice and 0.05 ml of the stock solution, 0.02 ml of succinate, and 0.05 ml of $^{32}P_i$ are added. The side arm of the vessel contains 0.05 ml of 50% trichloroacetic acid. The vessels are placed on manometers and equilibrated for 5½ minutes at 30°. The stopcock is then closed, and the first reading is taken at 6 minutes. Subsequent readings are taken at 6-minute intervals

[6] Protein determinations for F_1 are measured spectrophotometrically [O. Warburg and W. Christian, *Biochem. Z.* **310,** 384 (1941)], and for all others by the biuret procedure [(E. E. Jacobs, M. Jacob, D. R. Sanadi, and L. B. Bradley, *J. Biol. Chem.* **223,** 147 (1956)].

[7] See this volume [4].

for 24 or 30 minutes. The initial oxygen uptake for the first 6 minutes is determined by extrapolation of the observed rate. The reaction is terminated by tipping in the contents of the side arm. The precipitate is centrifuged out, and 0.1 ml of the supernatant solution is extracted with isobutanol–benzene,[8] and 1.0 ml of the water layer is plated, dried, and counted in a Nuclear-Chicago flow counter.

Preparation of Ammonia Particles (A-Particles)

A 2.5-g sample of heavy layer beef heart mitochondria[9] is suspended in a final volume of 120 ml containing 0.15 ml of 0.5 *M* EDTA, pH 7.4, and 22 ml of 0.25 *M* sucrose. The pH of the solution is adjusted to 9.2 with a freshly diluted solution of 1 *N* ammonium hydroxide. The suspension is exposed to sonic oscillation in 30-ml batches for 2 minutes in a Raytheon sonic oscillator (250 watts, 10 kc) cooled by flowing ice water. The suspension is centrifuged at 26,000 *g* for 10 minutes, and the pellet is discarded. The supernatant fluid is recentrifuged at 104,000 *g* for 60 minutes, and the supernatant solution is discarded. The pellet is homogenized in 80 ml of 0.25 *M* sucrose–1 m*M* EDTA, pH 7.5, and centrifuged at 104,000 *g* for 30 minutes. The pellet is again homogenized with 80 ml of 0.25 *M* sucrose and centrifuged as above. The final pellet is homogenized in 0.25 *M* sucrose at a protein concentration of about 25 mg/ml and stored at —55°.

Preparation of Phosphatide Suspension

Five grams of asolectin[2] are homogenized in 50 ml of a solution containing 0.25 *M* sucrose, 0.01 *M* Tris-SO_4, pH 7.4, 0.5 m*M* EDTA, pH 7.4, and 0.01 *M* thioglycerol, then exposed to sonic oscillation for 30 minutes in a Raytheon (250 watts, 10 kc) sonic oscillator cooled by flowing ice water. The suspension is dialyzed 12–18 hours against several changes of 10 volumes of the above buffer. The dialyzed suspension is centrifuged at 104,000 *g* for 60 minutes, and the pellet is discarded. The supernatant solution is flushed with nitrogen and stored in the refrigerator. This 10% asolectin preparation contains 3.25 mg of esterified phosphate per milliliter. It should be used as fresh as possible, since bacteria grow in this solution even at 2°.

Preparation of P-Particles

One gram of heavy layer beef heart mitochondria[9] is mixed with 0.4 ml of 0.5 *M* EDTA, pH 7.4, 10 ml of 1.0 *M* sucrose, and 8 ml of asolectin in a final volume of 40 ml. This suspension is exposed to sonic

[8] See this volume [9].
[9] See Vol. VI [58].

oscillation in 20-ml batches in a Raytheon (250 watts, 10 kc) sonic oscillator for 2 minutes. The suspension is centrifuged at 26,000 *g* for 10 minutes. The pellet is discarded, and the supernatant fluid is recentrifuged at 104,000 *g* for 60 minutes. The supernatant solution is discarded, and the pellet is homogenized with 36 ml of 0.25 *M* sucrose–1 m*M* EDTA–1% asolectin. The suspension is centrifuged at 104,000 *g* for 60 minutes, and the supernatant solution is discarded. The pellet is homogenized in 36 ml 0.25 *M* sucrose–1 m*M* EDTA and centrifuged for 60 minutes at 104,000 *g*. The supernatant solution is again discarded, and the pellet is homogenized in 0.25 *M* sucrose to a final protein concentration of about 25 mg/ml and stored at −55°.

[36] Stable Phosphorylating Submitochondrial Particles from Bakers' Yeast

By GOTTFRIED SCHATZ[1]

Preparation[1a]

Packed, commercially grown (National Yeast Corp., Belleville, New Jersey) bakers' yeast cells, 681 g (1.5 pounds), are washed three times with distilled water and once with 0.25 *M* mannitol–20 m*M* Tris-SO_4, pH 7.4,–1 m*M* EDTA (MTE medium). For each washing step, the cells are suspended in the washing medium to a final volume of approximately 3 liters and are reisolated by centrifugation (15 minutes at 1300 *g*). The packed pellet of washed cells is then stored at +4° overnight. From this point on all operations are carried out in the cold. For homogenization, the cells are evenly suspended in MTE medium to a final volume of 960 ml, and 20–28 ml aliquots (see below) are shaken for 20 seconds with exactly 50 g of glass beads (diameter 0.45–0.50 mm; Bronwill Scientific Division, Rochester, New York) at 4000 cycles per minute in the Braun MSK Mechanical Cell Homogenizer (Bronwill Scientific Division), using the 75-ml glass homogenizing flasks. During the shaking, the temperature within the flasks is maintained below +4° by cooling with liquid carbon dioxide. In order to obtain satisfactory (25–40%) cell breakage it is of great importance to fill the flasks to not more than two-thirds of capacity. The amount of cell suspension added has to be adjusted accordingly. The contents of each flask are shaken, then transferred to a chilled beaker. The flask is rinsed with approximately 20 ml of MTE medium, and the washing is added to the already homogenized cell suspension. The flask

[1] See footnote 1, page 30.

[1a] G. Schatz and E. Racker, *Biochem. Biophys. Res. Commun.* **22**, 579 (1966).

is then refilled with glass beads and fresh cell suspension for the next homogenization. If carried out by two operators, homogenization of 681 g commercially available yeast cake requires approximately 1 hour. When all the cell suspension has been homogenized, the turbid fluid is decanted from the glass beads. These are rinsed several times with MTE, and the washings are combined with the homogenized suspension to give a final volume of 1900 ml. This suspension is then freed from residual glass beads, unbroken cells, and large debris by two successive centrifugations at 2000 *g* for 10 minutes (Lourdes VRA rotor, 3500 rpm). The sediments (including all loosely packed material) are discarded. The supernatant from the second centrifugation at 2000 *g* is referred to as "homogenate." Its pH is 6.2–6.4 and is not further adjusted. All further centrifugation steps are carried out in the Spinco Model L preparative ultracentrifuge, using the appropriate stainless steel tubes filled to maximal capacity. Supernatants and loosely packed particles are always poured off by inverting the centrifuge tubes for exactly 3 seconds. The homogenate is centrifuged for 15 minutes at 10,000 rpm in the No. 21 rotor. The supernatant and the loosely packed particles are discarded. The sediment, which contains 55–75% of the DPNH- and the succinate oxidase activity of the homogenate, is resuspended in 300 ml MTE by repeated gentle aspiration with a 10-ml pipette (tip width approximately 1 mm) and is rehomogenized in a Potter-Elvehjem homogenizer by applying one up and down stroke with a loosely fitting Teflon pestle rotating at 800 rpm. The resuspended particles are centrifuged for 15 minutes at 11,000 rpm in the No. 30 rotor. The supernatant and the loosely packed particles are discarded. The sediment is rehomogenized in 60 ml MTE as described above and recentrifuged for 15 minutes at 15,000 rpm in the No. 40 rotor. The supernatant and the loosely packed particles are again discarded. The sediment represents the "heavy" yeast mitochondria. They are suspended in approximately 10 ml of 10 m*M* Tris-SO_4, pH 7.4. The protein concentration of the suspension is adjusted to 20 mg/ml. The mitochondria are disintegrated by exposure to ultrasonic oscillation in a 20 kc ultrasonic disintegrator manufactured by Measuring and Scientific Equipment, Ltd. (Great Britain), using the small probe. Aliquots (8 ml) of the mitochondrial suspension are placed into a double-walled cylindrical stainless steel cup (volume 27 ml; diameter 25 mm) cooled by circulating ice water. After the instrument has been adjusted to minimal power output, the cup is slowly raised toward the probe until a sudden increase in the noise level indicates that the tip of the probe has just penetrated the surface of the mitochondrial suspension. The instrument is then immediately tuned to maximal power output (1.1–1.5 A at 115 volts) and sonication is carried out for 40 seconds. The sonicated particle suspension is centrifuged for 15 minutes at 17,000 rpm in the No. 40 rotor. The sediment is discarded and the turbid supernatant is centrifuged for 30 minutes at 50,000 rpm in the No. 50 rotor. The yellowish, still slightly turbid supernatant is discarded, as the small amount of unsedimented particles exhibits very low phosphorylative capacity. The sediment from this final centrifugation step consists

TABLE I

FLOW DIAGRAM SUMMARIZING THE FINAL STEPS AND THE YIELDS IN THE PREPARATION OF PHOSPHORYLATING SUBMITOCHONDRIAL PARTICLES FROM BAKERS' YEAST

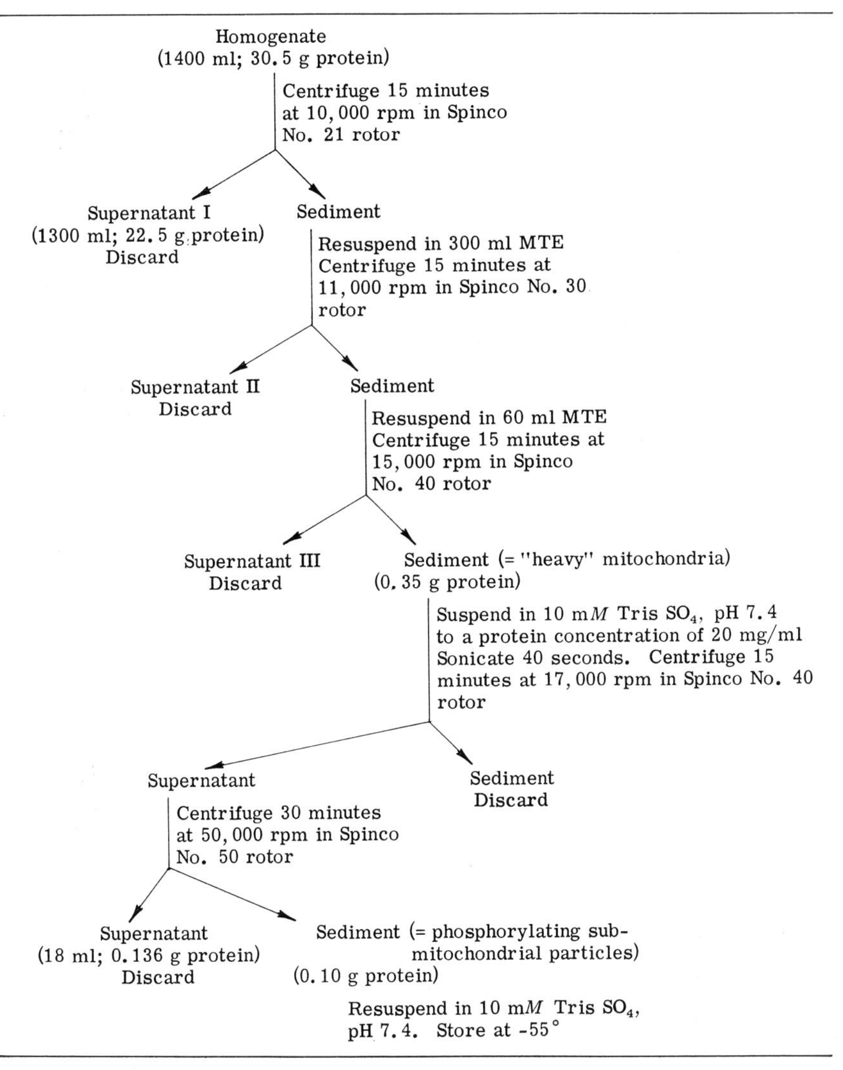

of phosphorylating submitochondrial particles. They are suspended in 2–4 ml of 10 mM Tris-SO_4, pH 7.4, to a final protein concentration of 30–40 mg/ml and are stored in aliquots at —55°. The final steps of the preparative procedure are summarized in the flow diagram of Table I.

Yield

The procedure described here yields 0.35–1.0 g (protein basis) of mitochondria, which upon sonication yield 80–150 mg of phosphorylating submitochondrial particles. Protein was measured by the biuret procedure in the presence of 0.66% deoxycholate.[2]

Properties[1a, 3]

The oxidation rates and P:O ratios with different substrates observed with mitochondria and submitochondrial particles are listed in Table II. Identical P:O ratios are always obtained with flavin-linked substrates (succinate or α-glycerophosphate) and DPN-linked substrates (pyruvate-malate or DPN^+, alcohol dehydrogenase, and ethanol). Similar observations with isolated yeast mitochondria have been reported by Vitols and Linnane[4] and Ohnishi and Hagihara.[5] The measurements of the individual phosphorylation sites (Table II) indicate that the mitochondria as well as the submitochondrial particles as isolated by the present procedure lack a functional first phosphorylation site. The submitochondrial particles also appear to be unable to catalyze the ATP-dependent reduction of DPN^+ by succinate as well as an ATP-dependent transhydrogenation between DPNH and TPN^+. Their DPNH-cytochrome *b* segment also lacks the characteristic electron paramagnetic resonance signal at $g = 1.94$ observed in mammalian mitochondria. The DPNH oxidase system is insensitive to rotenone, as observed also by Ohnishi *et al.*[5] The phosphorylative capacity of the mitochondria slowly declines upon storage at —55°. In contrast, the submitochondrial particles are stable at —55° for at least 3 weeks, if freezing and thawing are avoided. They form a single and well defined band upon equilibrium centrifugation in a sucrose density gradient. In the electron microscope they appear as vesicles with diameters between 0.07 and 0.20 μ after staining with phosphotungstate. No contamination with unbroken mitochondria (diameter 0.6–1.2 μ) can be seen. Phosphorylating "light" mitochondria, isolated from supernatants II and III (Table I) by cen-

[2] E. E. Jacobs, M. Jacob, D. R. Sanadi, and L. B. Bradley, *J. Biol. Chem.* **223,** 147 (1956).

[3] G. Schatz, E. Racker, D. D. Tyler, J. Gonze, and R. W. Estabrook, *Biochem. Biophys. Res. Commun.* **22,** 585 (1966).

[4] E. Vitols and A. Linnane, *J. Biochem. Biophys. Cytol.* **9,** 701 (1961).

[5] T. Ohnishi, K. Kawaguchi, and B. Hagihara, *J. Biol. Chem.* **241,** 1797 (1966).

TABLE II
OXIDATION RATES AND P:O RATIOS WITH VARIOUS SUBSTRATES AND ACCEPTORS OBSERVED WITH MITOCHONDRIA AND SUBMITOCHONDRIAL PARTICLES FROM BAKERS' YEAST[a]

Reaction measured	Mitochondria		Submitochondrial particles	
	Specific activity[b]	P:O	Specific activity[b]	P:O
Succinate → O_2	0.20	1.15	0.29	0.55
α-Glycerophosphate → O_2	0.28	1.02	0.37	0.45
Ethanol → O_2	0.16	1.20	0.15[c]	0.38[c]
Pyruvate-malate → O_2	0.046	1.20	0.00[d]	—
Site I (DPNH → fumarate)	0.004	0.00	0.007	0.00
(DPNH → CoQ_1)	3.14	0.00	4.57	0.00
Site II	0.16	0.36	0.22	0.18
Site III	0.32	0.70	0.44	0.20

[a] All measurements were carried out at 22–24° in the presence of 2.0 micromoles $MgCl_2$, 0.48 micromole EDTA, 5.0 micromoles Tris-SO_4, pH 7.4, 32 micromoles glucose, 1.0 micromole ATP, 30 units dialyzed hexokinase, 2.0 mg dialyzed bovine serum albumin, 16 micromoles $^{32}P_i$, pH 7.4 ($0.8–2.0 \times 10^5$ cpm/micromole) and substrates, acceptors, inhibitors, and particles as indicated. The final volume was always adjusted to 1.03 ml by the addition of 0.25 *M* sucrose. The following concentration of substrates was used: succinate, 10 m*M*; α-glycerophosphate, 6 m*M*; pyruvate and malate, 1 and 7 m*M*; ethanol, 15 m*M*. Oxidative phosphorylation with oxygen as acceptor was assayed polarographically at a particle concentration of 620 μg/ml. Oxidative phosphorylation at site I was assayed spectrophotometrically at 340 mμ, either with CoQ_1 as acceptor in the presence of 100 μg of particles as described elsewhere in this volume [4], or by employing the reduction of fumarate (4 m*M*) by DPNH (0.21 m*M*) [see D. R. Sanadi and A. L. Fluharty, *Biochemistry* **2,** 523 (1963)] in the presence of 1.6 m*M* KCN and 518 μg of particles. Phosphorylation in these systems was corrected for that observed in the absence of fumarate and coenzyme Q_1, respectively. Site II was measured by assaying the phosphorylation coupled to the reduction of ferricyanide (0.8 m*M*) by succinate (10 m*M*) in the presence of 1.6 m*M* KCN. The concentration of particles was 280 μg/ml. Reduction of ferricyanide was followed spectrophotometrically at 420 mμ. In calculating P:2*e* ratios in this system, two corrections were applied: The amount of ferricyanide reduced was corrected for that reduced in a reaction insensitive to 1 μg antimycin per milligram of particle protein, as this antimycin-insensitive reduction was nonphosphorylating. This first correction amounted to 15–19%. Moreover, the amount of ATP formed in the absence of ferricyanide was subtracted from that formed in its presence. This second correction amounted to 3–7%. Site III was measured polarographically in the presence of 300 μg of particles, 2 μg of antimycin, 0.1 micromole of tetramethyldiamino-*p*-phenylenediamine and 2 micromoles ascorbate pH 7.4. Incorporation of $^{32}P_i$ into glucose-6-P was determined as described elsewhere in this volume [4]. Hexokinase was assayed as described by M. E. Pullman, H. S. Penefsky, A. Datta, and E. Racker [*J. Biol. Chem.* **235,** 3322 (1960)].

[b] Micromoles of substrate oxidized per minute per milligram of protein.

[c] In the presence of 0.2 micromole DPN^+ and 300 μg yeast alcohol dehydrogenase.

[d] In the presence of 0.2 micromole DPN^+.

trifugation at 35,000 *g* for 15 minutes differ from the submitochondrial particles described here in that they are considerably less stable upon storage and are not homogeneous as revealed by density gradient centrifugation.

[37] The Keilin-Hartree Heart Muscle Preparation[1]

By TSOO E. KING

The heart muscle preparation has been extensively used in the study of intracellular respiration and as a starting material for large-scale isolations of respiratory components. It is a particulate suspension of physically disintegrated mitochondrial membrane. The manipulations for the preparation are simple and the yield is high.

Assay Methods

Principle. A number of criteria may be used for the determination of the activity. Among them, oxidation of succinate or NADH by molecular oxygen is useful and informational. The following describe a manometric method for succinate oxidase and a polarographic procedure for NADH oxidase.

Reagents

Phosphate buffer, 0.2 *M* (the Sørensen type), pH 7.8
Succinic acid, 0.6 *M* in water, adjusted to pH 7.8 with NaOH
NADH, 0.02 *M* in water, to be used within 4 hours
Cytochrome *c*, 0.0009 *M* in water

Manometric Method for Succinate Oxidation. In the main compartment of a Warburg flask are placed 0.1 ml cytochrome *c*, 1.5 ml phosphate buffer, and an amount of the preparation to give 10–80 μl oxygen uptake per 10 minutes. Water is added up to 2.8 ml. In the side arm, 0.2 ml succinate is pipetted.

Three or four levels of the preparation are tested. Care must be taken because oxygen diffusion may become limiting when the reaction

[1] Voluminous literature is available. Readers are referred to authoritative reviews by D. Keilin and E. F. Hartree, [*Nature* **176**, 200 (1955)] ["Cellular Respiration and Cytochromes." Cambridge Univ. Press, London and New York, 1966] and also others.[2-4]

[2] B. Chance and G. R. Williams, *Advan. Enzymol.* **17**, 65 (1956).

[3] E. C. Slater, *Advan. Enzymol.* **20**, 147 (1958).

[4] T. E. King, *Advan. Enzymol.* **28**, 155 (1966).

is fast. It is advisable to select those Warburg flasks with large diameters at the bottom of the main compartment.

After temperature equilibration at 37° for 8 minutes, the succinate is tipped into the main compartment. Readings are taken for 1 hour at 10-minute intervals. The oxygen-uptake is linear with time; occasionally a very slight deviation from linearity occurs after the first 30 minutes.

Polarographic Method for NADH. We usually use an Oxygraph (oxygen electrode) manufactured by Gilson Medical Electronics. The manufacturer's manual should be consulted for calibration of the instrument. Other types of polarographic apparatus for oxygen determination are equally usable.

In the reaction vessel are pipetted 1.2 ml phosphate buffer, 0.08 ml cytochrome *c*, and an amount of the heart muscle preparation equivalent to 8–40 μM of oxygen uptake per minute. Water is added to a volume of 2.35 ml. A recording for 1–2 minutes is made to establish the base line. Then 0.05 ml NADH is stirred in, and the recording continues. After a lag period of less than 1 or 2 minutes,[5] the oxygen consumption is then linear with time until the dissolved oxygen is almost exhausted.

Several levels of the heart muscle preparation should be tested. Any temperature, between 23° and 37°, may be used for the reaction. However, consideration of the high temperature coefficient[6] in the oxidation of NADH must be taken into account (see Properties).

Calculation. The observed amounts of the oxygen consumed are plotted as a function of the concentration of the heart muscle preparation in terms of protein. The linear portion of the curve is used for calculation.

In the manometric method, the oxygen uptake in microliters can be converted to mM (millimolar) oxygen reacted by using a factor of 0.0149 or to mM succinate oxidized by 0.0298. In the polarographic method, it suffices to use the values 0.25 mM for oxygen concentration (i.e., solubility) as the air-saturated solution at room temperature (25°) although the method depends on the activity of the dissolved oxygen. The activity of oxygen varies with a number of factors such as the electrolyte concentration in the system. A factor of 67.1 is used for converting mM oxygen reduced to microliters.

The specific activity may be expressed as micromoles of substrate oxidized per minute per milligram of protein or as Q_{O_2}. The latter term refers to microliters of oxygen consumed per hour per milligram of protein.

[5] S. Minakami, F. J. Schindler, and R. W. Estabrook, *J. Biol. Chem.* **239**, 2042 (1964).

[6] E. C. Slater, *Biochem. J.* **46**, 484 (1950).

Determination of Protein. The protein in the heart muscle preparation is estimated either as the fat-free dry weight or by the biuret method. By the latter, the heart muscle preparation must be first "solubilized." An aliquot of the heart muscle preparation, diluted 10 times with water, is mixed with 0.2 ml of 5% sodium deoxycholate in a total volume of 1 ml. The suspension is usually clarified within 5 minutes. Then the conventional biuret method may be applied.[7]

The fat-free dry weight is determined according to Slater.[8] In a weighed, glass centrifuge tube are placed 1 ml of the heart muscle preparation, and 4 ml of water; these are mixed well. Then 1 ml of 20% trichloroacetic acid is added. The flocculent precipitate is collected by centrifuging the mixture at about 3000 g for 5 minutes, washed once with 5 ml of 50% ethanol, and once with 5 ml of 96% ethanol by centrifugation. The residue is then dried to constant weight at 100°.

The result from the biuret method is usually about 25% lower than that by the fat-free weight. We ordinarily use the Slater method.

Preparations[4, 9]

Keilin[10] originally employed the method of the preparation for his studies of intracellular respiration. Later Keilin and Hartree[11] "standardized" the preparation. Subsequently due mainly to the availability of high-speed centrifuges and adaptations for special purposes, some variances in manipulations have been introduced.

Method 1

This is the adaptation[9] of the original method with some steps similar to those used by Slater[8] and by Tsou.[12] The procedure includes washing of heart mince, extraction by phosphate, grinding with sand, and differential centrifugation. The preparation thus obtained is somewhat deficient in cytochrome *c* but free of hemoglobin and free of or extremely low in myoglobin.

Step 1. Preliminary Preparation of Tissue. Heart tissue immersed in ice water is brought from a slaughterhouse to the laboratory as soon as possible. It is cleaned of fat and connective tissues and either used immediately or kept frozen at —25°. The frozen tissue should be used within 3–4 weeks.

[7] A. G. Gornall, C. J. Bardawill, and M. M. David, *J. Biol. Chem.* **177,** 751 (1949).
[8] E. C. Slater, *Biochem. J.* **45,** 1 (1949).
[9] T. E. King, *J. Biol. Chem.* **236,** 2342 (1961).
[10] D. Keilin, *Proc. Roy. Soc.* **B104,** 206 (1929).
[11] D. Keilin and E. F. Hartree, *Proc. Roy. Soc.* **B129,** 277 (1940); *Biochem. J.* **44,** 205 (1949).
[12] C. L. Tsou, *Biochem. J.* **49,** 362 (1951).

Step 2. Washing. The heart muscle is minced in a power-driven meat grinder. One and one-half kilograms of mince is washed with 30 liters of tap water in a plastic bucket by efficient stirring for 20 minutes. A mechanical stirrer under the trade name of "Laboratory Mixer, GT21" made by Gerald K. Heller Company, Las Vegas, Nevada, with a three-blade stainless steel propeller approximately 6 inches across, has been used satisfactorily. The mince is pressed and squeezed either by hand in cheesecloth or by a hydraulic press in heavy canvas. We have used Carver Laboratory Press, Model B, manufactured by Fred S. Carver, Inc., Summit, New Jersey, with success. This step is particularly important for the removal of noncytochrome hemoproteins; the water adhering immediately to the tissue which contains most soluble proteins can be removed only by hard squeezing or pressing. Usually the mince after six washings is light yellow and free of pink hue and the washing is colorless.

Step 3. Extraction with Phosphate. The mince is further mixed with 15 liters of 0.1 *M* phosphate buffer (the Sørensen type), pH 7.4, and stirred efficiently for 1 hour at 4°. It is then washed with 30 liters of deionized water as in step 2. Up to this stage, the hand-pressed mince weighs about 1200 g.

Step 4. Sand-Grinding. Into the unglazed bowl of a mechanical mortar are placed 1200 g of the mince, 650 ml of 0.02 *M* phosphate buffer, pH 7.4, and 600 g sand (No. S-25, Fisher Scientific Company). The mixture is ground for 60 minutes. Care must be taken that all the charge is actually ground, i.e., that no material sticks to the wall of the bowl or the surface of the pestle and thus escapes the mechanical action. At the end of grinding, the mass is diluted with additional 1350 ml 0.02 *M* phosphate buffer, pH 7.4, and mixed thoroughly.

A mechanical mortar manufactured by Pascall Engineering Company, Ltd., Gatewick Road, Crawley, Sussex, England, is employed satisfactorily. Keilin and Hartree[11] used an older model of the same type. The new model is more compact in dimensions, convenient in operation, and easy to clean. Three sizes are available: approximately 9×4, $14\text{-}\frac{1}{2} \times 5$, or $19 \times 5\text{-}\frac{1}{2}$ inches, inside measurements. The grinding is conducted at room temperature; external cooling is not needed at temperatures below 27°.

In small-scale preparations, a hand-operated porcelain mortar, approximately 10 inches in diameter with unglazed grinding surface, and pestle are used. The grinding time required is somewhat longer. But preparations with 100 g or less heart mince can be comfortably handled.

Step 5. Differential Centrifugations. The mixture from step 4 is centrifuged for 30 minutes at 2000 rpm in a refrigerated International centrifuge, size 3. The upper turbid layer is carefully collected without disturbing the pellet. The turbid suspension, about 1500 ml, is recentrifuged for 75 minutes at 21,000 rpm in a Spinco preparative centrifuge with rotor No. 21. The almost clear supernatant liquid is discarded, and the pellet is dispersed, with the aid of a Potter-Elvehjem homogenizer, in 300 ml borate–phosphate buffer[13] to give a final volume of 400 ml. The resulting preparation contains usually about 25 mg protein per milliliter. The buffer is prepared by mixing equal volumes of 0.15 M H_3BO_3 and 0.15 M Na_2HPO_4. Both centrifugations may be conducted either at room temperature or under refrigerated conditions. The minimum yield found is about 10 g. Higher yields are obtained when the separation of the turbid supernatant liquid in the first centrifugation is less sharp. Sacrifice of yield for higher quality is generally recommended.

One person can handle this amount of the preparation in less than 9 working hours. It is advisable to store the final preparation at 0–4° but not frozen.

Method 2

The phosphate washing (step 3) is omitted. Other steps are the same.

Method 3

Same as method 2 with the following exception. The turbid supernatant fluid after the first centrifugation is cooled to approximately 2° by addition of crushed ice with some ice remaining in the suspension and adjusted to pH 5.6 with 2 N acetic acid. The mixture is centrifuged at 2000 rpm for 15 minutes in a refrigerated International centrifuge, size 3, Model 194. The clear supernatant fraction is discarded, and the precipitate is washed with approximately 1.5 liters of 0.01 M KH_2PO_4. It is again centrifuged at 2000 rpm for 15 minutes. The residue is dispersed in borate–phosphate buffer as in Method 1.

When a refrigerated centrifuge is not available, addition of ice to the mixture serves the same purpose. The preparation is not stable at room temperature in media with pH lower than 6.8. In acidic mixture, the particles are very easily sedimented.

Method 4

The preparation obtained from method 1 is centrifuged for 60 minutes at 105,000 g. The pellet is suspended in 0.25 M sucrose to the original volume, and the mixture is again centrifuged. The residue collected is dispersed in 0.25 M sucrose to a desired concentration of protein.

[13] The use of borate buffer was first introduced by W. D. Bonner [*Biochem. J.* **56**, 274 (1954)]. Borate lengthens the useful life of the preparation.

Comments

Since our adaptation of the original method or variations as described, nearly 600 batches from ox, pig, and horse hearts have been made for various purposes in this laboratory. Preparations made from frozen heart do not show any noticeable difference from those started with the fresh tissue. However, it is not advised to store the tissue for more than 4 weeks, or to freeze them at dry-ice temperatures. Several samples of horse heart in 30-pound packages shipped from other cities in dry ice to this laboratory did not yield satisfactory preparations. On the other hand, fresh or frozen (at —25°) horse heart also gives normal preparations.

Properties[1]

The preparation has a complete assembly for the hydrogen (electron) transport from succinate and NADH through the respiratory carriers to molecular oxygen. During the electron transport, all the cytochromes (b, c_1, c, a, and a_3) undergo oxidation-reduction.

For the preparation according to method 1, the Q_{O_2} values with succinate as well as NADH in the presence of 30 μM cytochrome c are between 600 and 1200, whereas those with ascorbate for the cytochrome oxidase activity are 1000–2000 μl/mg per hour at 37°. The oxidation of NADH shows a high activation energy,[3] with a temperature coefficient of approximately 3; the corresponding value for succinate oxidation is only about 1.6. The oxidations of NADH and succinate by either cytochrome c or oxygen are nearly completely antimycin A sensitive.[14]

Succinate oxidase activity does not decrease within the "life" of the preparation, which is more than 1 week at 0–4°. On the other hand, NADH oxidase activity decreases with the storage time even at 0–4°.

The active components of the preparation have been quantitatively determined.[4,15] In addition to the cytochromes mentioned, the preparation also contains protein bound nonheme iron, copper, flavins, and coenzyme Q. Several active segments have been isolated from the preparation, such as succinate–cytochrome c reductase, the cytochrome b–c_1 particle (cf. this volume [40]), succinate dehydrogenase (cf. this volume [58]), and cytochrome oxidase. NADH dehydrogenases active only to artificial electron acceptors or cytochrome c (antimycin A insensitive) can be solubilized from the heart muscle preparation (cf. this volume [52]).

[14] M. Thorn, *Biochem. J.* 63, 420 (1956).

[15] T. E. King, K. S. Nickel, and D. R. Jensen, *J. Biol. Chem.* **239**, 1989 (1964).

In addition, "some energy-linked" reactions which may or may not be intermediates of the oxidative phosphorylation have also been demonstrated in the preparation.[16,17] Indeed, it is a good starting material for the isolation of soluble ATPase.[18] However, the heart muscle preparation does not have the capacity to synthesize ATP, and hence it is known as "nonphosphorylating" preparation.

The preparation does not catalyze the reactions of the Krebs cycle with the exception of succinate oxidation. Nor does it contain glycolytic enzymes.

[16] D. W. Haas, *Biochim. Biophys. Acta* **89**, 543 (1964).
[17] J. Kettman, *Biochem. Biophys. Res. Commun.* **19**, 237 (1965).
[18] J. Kettman and T. E. King, *Biochem. Biophys. Res. Commun.* **11**, 255 (1963).

[38] Preparation and Properties of Repeating Units of Mitochondrial Electron Transfer

By PAUL V. BLAIR

Principle

Most essential features of mitochondrial function such as electron transfer are retained by submitochondrial particulate preparations. Thus, the question arises as to the dimensions of the smallest functional units of the mitochondrion. How are these subunits oriented with respect to the total organization of phospholipids and proteins in the mitochondrion? The fragmentation of mitochondria by sonic irradiation into a particulate structured fraction with one set of enzymatic activities and a soluble fraction with another set of activities makes it unlikely that a single mitochondrial subunit encompasses all the mitochondrial functions; rather, it is likely that several fractions or subunits exist, each performing a specific function. The particular subunit carrying out the special function with which we are concerned is the one that contains the complete electron transfer chain—the apparatus for the transfer of electrons from succinate and DPNH to molecular oxygen.

It is assumed that oxidation-reduction components are arranged in an orderly sequence in the repeating electron transfer chains, each chain containing one molecule (or a multiple thereof) of each component. The four integrated complexes constituting the electron transfer chain is assumed to be bonded to the structural protein-phospholipid network in such a manner that it is vulnerable to dissociation with the appropriate reagents. The reagents found most suitable for fractionating the particulate material are those that tend to equilibrate the phospholipids among the various components and complexes (most likely through the modification of hydrophobic and ionic bonds) and thus to render the particles soluble. After the particles have been solubilized with detergents, the isolation of the units of electron transfer is carried out by salt precipitation and differential centrifugation.

Isolation of the Unit of Electron Transfer[1]

Beef heart mitochondria,[2] suspended in 0.25 *M* sucrose at an approximate protein concentration of 65 mg/ml, are diluted to a protein concentration of 45 mg/ml with 0.12 *M* KCl before they are frozen at —20°. The mitochondria may be maintained in this state for months without interfering with the isolation of the units of the electron transfer. The frozen suspension is thawed shortly before it is centrifuged for 20 minutes at 20,000 rpm (No. 30 rotor, Spinco Model L ultracentrifuge). The residue is suspended in 10 volumes of 0.12 *M* KCl; this mixture is homogenized before it is centrifuged at 20,000 rpm (Spinco, No. 30 rotor) for 20 minutes. The residue is collected in 0.66 *M* sucrose (final sucrose concentration about 0.30 *M*), and the protein concentration is adjusted to 30 mg/ml with 0.66 *M* sucrose. The suspension of KCl-washed mitochondria thus obtained are treated as follows: first, they are adjusted to pH 8.0 with Tris-HCl buffer (0.02 *M*); second, 0.3 mg of potassium cholate (as a 20% solution) per milligram of protein is added; third, 0.3 mg of potassium deoxycholate (as a 10% solution) per milligram of protein is introduced (both these detergents tend to equilibrate the phospholipid and to solubilize the particulate complexes of the mitochondrion); fourth, sufficient neutralized saturated ammonium sulfate (at 0–5°) is added slowly to bring the suspension to 33% saturation at this temperature; fifth, the turbid suspension is centrifuged at 30,000 rpm (Spinco, No. 30 rotor) for 15 minutes.

The grayish white residue (mainly structural protein) is discarded and the greenish red supernatant fluid is brought to 50% saturation by the slow addition of neutralized saturated ammonium sulfate (at 0–5°); a greenish red sticky precipitate commences to form at about 40% saturation. After this suspension has been centrifuged at 20,000 rpm (Spinco, No. 30 rotor) for 5 minutes, a floating greenish red fraction is collected from the tops of the tubes (after the almost colorless infranatant fluid is decanted). This residue, designated the *floating frac-*

[1] P. V. Blair, T. Oda, D. E. Green, and H. Fernández-Morán, *Biochemistry* **2,** 756 (1963).

[2] F. L. Crane, J. L. Glenn, and D. E. Green, *Biochim. Biophys. Acta* **22,** 475 (1956).

tion, has a composition (in terms of the components of the electron transfer chain) almost identical to that of the final preparation of units of electron transfer.

The *floating fraction* is diluted with 0.25 *M* sucrose to a concentration of 0.5 mg of protein per milliliter and is allowed to stand for 30 minutes. Almost immediately after dilution, the suspension becomes turbid as the reduced concentration of detergents decreases the solubility of the protein–phospholipid complexes and permits them to recombine. The turbid suspension is centrifuged at 30,000 rpm (Spinco, No. 30 rotor) for 4 hours. The relatively clear colorless supernatant liquid is decanted, and the greenish red residue is taken up in a buffered solution which is 0.05 *M* in Tris-HCl (pH 7.5) and 0.25 *M* in sucrose. This suspension of the particulate residue is dispersed by sonication; it is designated the *unit of electron transfer.* The unit of electron transfer has been, perhaps inappropriately, termed the "elementary particle" of the mitochondrion, a term now used to denote another repeating unit.[1,3] However, the original concept of the elementary particle is still valid provided this designation is understood to mean: (1) that the particle specified is capable of catalyzing the complete sequence of oxidative phosphorylation and is an integrated unit of minimal molecular weight; and (2) that it possesses all components essential both to electron transfer and to the coupling function. The particle described above may possess all the components necessary to carry out the sequence of reactions constituting oxidative phosphorylation, but demonstration of this functional property has not been attained; failure to do so may possibly be due to the disarrangement (or loss) of certain coupling components.

Assays

Several methods are employed for determining the concentration of the components of electron transfer in the isolated unit. Flavins are determined by a modification of the method described by Green and associates.[1,4] Estimations of the acid-extractable and acid-nonextractable flavin are believed to afford the best criteria of the amounts of flavin associated with the oxidations of DPNH and succinate, respectively. The cytochromes are determined by a spectrophotometric method in which ascorbate is used to reduce cytochromes c_1 plus c, and dithionite is added to reduce the remaining components.[5] Coenzyme Q is estimated,

[3] H. Fernández-Morán, T. Oda, P. V. Blair, and D. E. Green, *J. Cell Biol.* **22,** 63 (1964).

[4] D. E. Green, S. Mii, and P. Kohout, *J. Biol. Chem.* **217,** 551 (1955).

[5] W. S. Zaugg and J. S. Rieske, *Biochem. Biophys. Res. Commun.* **9,** 213 (1962).

after extraction with cyclohexane, by the difference in absorbancy at 275 mμ when reduction is accomplished with KBH_4.

Total iron is determined as the ferrous-bathophenathroline complex after the sample is wet-washed with concentrated H_2SO_4 and H_2O_2.[6] Copper is determined as the cuprous-neocuproine complex.[7] Phospholipids are extracted with chloroform–methanol, and elementary phosphorus content is used as an index of the amount of phospholipid.[8]

Probably the most useful and definitive assay for the unit of electron transfer is the functional one, i.e., a measure of the electron transfer activity expressed in terms of oxygen consumption. DPNH oxidase activity is measured polarographically[9] with the oxygen electrode (e.g., Gilson Medical Electronics Oxygraph). The oxygen content of 2.0 ml of suspension, containing all factors necessary for maximal activity, is calculated to be 0.83 microatoms of oxygen at 37 ±1°. The order of addition of materials to the jacketed reaction vessel is: (a) 1.80 ml of 0.02 *M* phosphate buffer (pH 7.4) containing 1×10^{-4} *M* EDTA; (b) 0.10 ml of enzyme (multienzyme complex) at a protein concentration of 0.5 mg per ml; (c) 0.08 ml of a 0.07 *M* solution of DPNH (Sigma); and (d) 0.02 ml of cytochrome *c* (Sigma, type III) at a concentration of 10 mg per ml.

The succinic oxidase activity of the unit of electron transfer is measured in a similar manner except that 0.08 ml of 0.10 *M* succinate is used as substrate, and coenzyme Q_2 in absolute ethanol is added. Both activities are expressed as microatoms of oxygen consumed per milligram of protein per minute.

Properties

The table summarizes the concentrations of the components concerned with electron transfer, which are found in the isolated unit of electron transfer. The concentration of acid-extractable flavin is 5.6% greater than that of the acid-nonextractable flavin. This suggests a contamination of the particle with primary dehydrogenase enzymes and enzyme complexes (e.g., α-ketoglutarate dehydrogenase). The percentage of phospholipid (45%) in the isolated multienzyme complex is much greater than it is in the mitochondria (about 30%). This accumulation of phospholipid is probably brought about by the procedure of fractionation; the phospholipids become equilibrated among the various components and com-

[6] G. F. Smith, W. H. McGurdy, and H. Diehl, *Analyst* **77**, 418 (1952).

[7] G. F. Smith and W. H. McGurdy, *Anal. Chem.* **24**, 371 (1952).

[8] S. Fleischer, H. Klouwen, and G. Brierley, *J. Biol. Chem.* **236**, 2936 (1961).

[9] For discussion see this volume [7].

Components of the Unit of Electron Transfer

Component	Concentration of component/mg protein
Total flavin	0.91 mμmoles
Acid-extractable	0.48 mμmoles
Acid-nonextractable	0.43 mμmoles
Total cytochromes	5.05 mμmoles
Cytochrome *a*	2.90 mμmoles
Cytochrome *b*	1.51 mμmoles
Cytochrome c_1	0.48 mμmoles
Cytochrome *c*	0.16 mμmoles
Total iron	12.20 mμmoles
Heme iron	5.05 mμmoles
Nonheme iron	7.5 mμmoles
Copper	3.32 mμmoles
Coenzyme Q	5.32 mμmoles
Phospholipid	0.83 mg

plexes and are thus solubilized. The lipid is then floated out with the electron transfer components because of its density and solubility.

The DPNH- and succinic oxidase activities of the isolated electron transfer unit are 8.4 and 3.6 microatoms of oxygen consumed per milligram of protein per minute, respectively, i.e., a 2.5-fold increase in each case over the specific activities in the sonicated solubilized mitochondria. The increase in specific activity parallels the purifications of the components of electron transfer. The preparation exhibits inhomogeneity with respect to particle size, but not with respect to chemical composition. The components of the electron transfer system and the oxidase activities remain constant irrespective of particle size. The variability in size is, therefore, likely due to the degree of aggregation of the particles.

The unit of electron transfer can be prepared either by isolating the four functional complexes individually, and permitting them to recombine,[10,11] or by isolating them collectively as described above. In both procedures the enzymatic activities and the components of electron transfer are preserved. In the mitochondrion electron transfer is coupled to the synthesis of ATP; however the isolated unit has no coupling ability or capacity to carry out other energy-linked functions normally diagnostic of intact mitochondria.

[10] L. R. Fowler and Y. Hatefi, *Biochem. Biophys. Res. Commun.* **5**, 203 (1961).
[11] L. R. Fowler and S. H. Richardson, *J. Biol. Chem.* **238**, 456 (1963).

[39] Preparation and Properties of Succinic–Cytochrome c Reductase (Complex II–III)

By HOWARD D. TISDALE

Succinate + 2 cytochrome *c* (Fe^{3+}) → fumarate + 2 cytochrome *c* (Fe^{++})

Assay Method

Principle. The enzyme-catalyzed reduction of cytochrome *c* by succinate proceeds at a linear rate under standard conditions over the first minute. The resulting increase in absorbancy at 550 mμ is followed during this linear phase.

This method of assay differs somewhat from that of Green and Burkhard[1] (who were the first to isolate this enzyme), since in practice it has been found to be more convenient and accurate.

Reagents

Potassium phosphate, 0.10 *M*, pH 7.4
NaN_3, 0.10 *M*
Disodium ethylenediamine tetraacetate (EDTA), 0.01 *M*
Bovine serum albumin (BSA), 10%
Potassium succinate, 0.10 *M*, pH 7.0
Ferricytochrome *c* (Sigma), 1%
M Sucrose, 0.88; 5 m*M* in potassium succinate

Procedure. Assays of enzymatic activity are carried out at 38° in a Beckman Model DU spectrophotometer equipped with a photomultiplier. The final reaction mixture contains 10 micromoles phosphate buffer (0.10 ml); 1 micromole NaN_3 (0.01 ml); 0.2 micromole EDTA (0.02 ml); 5 mg BSA (0.05 ml); 10 micromoles potassium succinate, pH 7.0 (0.10 ml); and water to a volume of 0.9 ml. The enzyme preparation to be assayed is diluted to a concentration of 100–200 μg protein per milliliter in a solution of 0.88 *M* sucrose 0.005 *M* in succinate. One microgram of enzyme is incubated with the assay mixture for 2 minutes at 38°. The reaction is initiated by adding 1 mg ferricytochrome *c* (0.1 ml) and the change in absorbancy is measured at 550 mμ against a control lacking only the enzyme. Absorbancy at 550 mμ is followed at 10-second intervals over the first 2 minutes. The extinction coefficient used for cytochrome *c* (reduced-oxidized) is $18.5 \times 10^6 M^{-1} cm^{-1}$. Specific activity is calculated as follows:

[1] D. E. Green and R. K. Burkhard, *Arch. Biochem. Biophys.* **92**, 2 (1961).

$$\frac{\Delta \text{ O.D./minute/mg protein}}{18.5} =$$

micromoles cytochrome *c* reduced/minute/mg protein

Method of Preparation

Beef heart mitochondria are isolated as described by Crane *et al.*[2] in 0.25 *M* sucrose and are frozen at —40° for a period of 1–7 days. Just prior to use the frozen mitochondrial suspension is thawed at room temperature and chilled to about 0°. The suspension (65–70 mg protein/ml) is diluted with 1/3 volume of 0.9% KCl, and cold *tert*-amyl alcohol is added with stirring to give a final concentration of 10%. The mixture is warmed at 20°, allowed to stand for 10 minutes, and then centrifuged for 10 minutes at 30,000 rpm in the Spinco Model L centrifuge using the No. 30 rotor. All subsequent centrifugations are also carried out in the Spinco No. 30 rotor.

The red supernatant fluid is discarded, and the fluffy red infranatant fraction is decanted and mixed with 3 volumes of cold 0.25 *M* sucrose containing 0.005 *M* succinate. The suspension is centrifged for 20 minutes at 30,000 rpm, and the well packed red residue is suspended in 4 volumes of the cold sucrose-succinate solution, and the mixture is homogenized in a Potter-Elvehjem homogenizer (glass-Teflon). The suspension is centrifuged for 30 minutes at 30,000 rpm; the well packed red residue is suspended in a few milliliters of the sucrose–succinate solution and stored overnight at 5°.

The stored red suspension is adjusted to a protein concentration of 20 mg/ml with the sucrose–succinate solution. Cold *tert*-amyl alcohol is added with stirring to give a final concentration of 10%, and the mixture is allowed to stand for 10 minutes at 20°. The mixture is centrifuged for 10 minutes at 30,000 rpm; the viscous red supernatant is decanted and centrifuged for 20 minutes at 30,000 rpm. The reddish brown sediment is washed with 15–20 volumes of sucrose–succinate solution to remove residual alcohol and is finally suspended in sucrose–succinate solution to a protein concentration of 15 mg/ml. Then 20% potassium cholate is added in an amount equivalent to 2 mg per milligram of protein. Sufficient solid potassium chloride is added to give a final concentration of 1.6 *M*, and the mixture is allowed to stand for 10 minutes at 0°. The optically clear solution is dialyzed under nitrogen for 5 hours against 12 volumes of sucrose–succinate solution buffered by 0.01 *M* potassium phosphate, pH 7.4, and is then centrifuged for 20 minutes at 30,000 rpm. The pink supernatant fluid is discarded, and the deep red residue is

[2] F. L. Crane, J. L. Glenn, and D. E. Green, *Biochim. Biophys. Acta* **22**, 475 (1956).

suspended in 0.88 *M* sucrose containing 0.005 *M* succinate. This suspension contains the purified succinic–cytochrome *c* reductase.

The yield and increase of activity during the purification of the succinic–cytochrome *c* reductase is illustrated in Table I. The purified preparation represents from a 12-fold up to a 25-fold increase in specific activity over that of mitochondria.

TABLE I
PURIFICATION OF SUCCINIC-CYTOCHROME *c* REDUCTASE

Fraction	Total protein (mg)	Specific activity[a]	Total units[b]
Mitochondria	10,000	1.0	10,000
First *tert*-amyl alcohol fraction	1,200	10.0	12,000
Second *tert*-amyl alcohol fraction	500	20	10,000
KCl extracted residue	200	25	5,000

[a] Micromoles cytochrome *c* reduced per minute per milligram of protein at 38°.
[b] Total units are represented as total protein multiplied by the specific activity.

Properties

Numerous preparations of this complex were found to contain F_S, cytochrome c_1, and cytochrome *b*, invariably in a stoichiometric ratio close to 1:1:3.[3] The DPNH flavoprotein (F_D) is either absent or present in a trace amount (Table II).

TABLE II
COMPOSITION OF SUCCINIC–CYTOCHROME *c* REDUCTASE

Component	Millimicromoles per milligram protein
Acid-extractable flavin (F_D)[a]	0–0.16
Non-acid-extractable flavin (F_S)[a]	1.0–1.43
Cytochrome c_1	1.0–1.27
Cytochrome *b*	3.0–4.20
Nonheme Fe	8.5–11.8

[a] F_D and F_S are the flavins that are specifically associated with the enzymatic complexes whose function it is to oxidize DPNH and succinate, respectively.

[3] H. D. Tisdale, D. C. Wharton, D. E. Green, *Arch. Biochem. Biophys.* **102,** 114 (1963).

[40] Preparations of Succinate–Cytochrome *c* Reductase and the Cytochrome *b*-c_1 Particle, and Reconstitution of Succinate–Cytochrome *c* Reductase

By TSOO E. KING

The principle of the preparations is based on the sequential fragmentation of the respiratory chain. The immediate aim is not to "purify" the fragmented particle (i.e., to increase its specific activity on the protein basis), but rather to cleave the chain systematically into segments as structurally complete as possible. Thus, functional manifestation may not be altered.

Succinate-Cytochrome *c* Reductase

Assay Method[1,2]

Principle. The enzymatic activity of succinate–cytochrome *c* reductase is determined according to overall reaction (1) or (2).

$$\text{Succinate} + 2 \text{ cytochrome } c^{3+} \rightarrow \text{fumarate} + \text{cytochrome } c^{2+} \quad (1)$$

$$\text{Succinate} + \text{DCIP} \rightarrow \text{fumarate} + \text{DCIPH}_2 \quad (2)$$

The reduction of both acceptors is followed spectrophotometrically.

Reagents

Phosphate buffer, 0.2 *M* (the Sørensen type), pH 7.4
EDTA, 0.003 *M* in water, pH 7.4
Succinic acid, 0.6 *M* in water, adjusted to pH 7.4 with NaOH
Cytochrome *c*, crystalline, 0.001 *M* of the oxidized form in water
2,6-Dichlorophenolindophenol (DCIP), 0.00053 *M* in water
Enzyme solution. It is appropriately diluted with 0.1 *M* phosphate buffer so that 0.05 ml gives an absorbance change between 0.05 and 0.3 per minute under the following assay conditions
Water, used in the assays and preparation, is redistilled in a glass apparatus from ordinary distilled and then deionized water.

Procedure. In a cuvette of 1-cm optical path are placed 1.5 ml phosphate buffer, 0.3 ml EDTA, 0.3 ml cytochrome *c* (or 0.3 ml DCIP), 0.1 ml succinate, and 0.75 ml water. The reaction is started by the addition of 0.05 ml enzyme. For cytochrome *c* assay the absorbance reading

[1] S. Takemori and T. E. King, *Biochim. Biophys. Acta* **64,** 192 (1962).
[2] S. Takemori and T. E. King, *J. Biol. Chem.* **239,** 3546 (1964).

of the experimental cuvette is adjusted to zero, whereas for DCIP assay it is set at about 0.6 against a blank which contains all components except the enzyme. Alternatively, water may be used as a blank for the DCIP assay. Spectrophotometric recording begins after about 5 seconds of stirring 0.05 ml enzyme into the mixture. The reaction is linear with time for at least 3 minutes. The activity for cytochrome *c* is followed by the increase of the absorbance at 550 mμ, and that for DCIP by the decrease of the absorbance at 600 mμ in a recording spectrophotometer. If only a manual spectrophotometer is available, readings may be taken at intervals of 30 seconds for 2 minutes.

The reaction mixture may be scaled down to a total volume of 1 ml with final concentrations of phosphate buffer, 0.1 *M*; EDTA, 0.3 m*M*; succinate, 20 m*M*; cytochrome *c*, 0.1 m*M* (or DCIP, 0.053 m*M*); and an appropriate amount of enzyme (usually 20 μl) to start the reaction.

For the starting material or the reductase contaminated with cytochrome oxidase, 1 m*M* cyanide should be present in the assay system to block the competitive reaction with oxygen. However, prolonged contact (more than a few minutes) of cyanide with cytochrome *c* must be avoided.

Calculation. The specific activity, i.e., micromoles of succinate oxidized per minute per milligram of protein, can be calculated from Eqs. (3) or (4).

$$\text{Specific activity (cytochrome } c \text{ as acceptor)} = \frac{\Delta A_{550}}{W \times 12.8} \quad (3)$$

$$\text{Specific activity (DCIP as acceptor)} = \frac{\Delta A_{600}}{W \times 7.00} \quad (4)$$

Here ΔA is the absorbance change per minute and W is the amount of enzyme in terms of milligrams of protein in the assay mixture of 3 ml. Specific activity expressed in this form is numerically equivalent to m*M* succinate oxidized per minute at 1 mg enzyme protein per milliliter, which is a proper unit for enzymatic velocity. The absorbancy index used for cytochrome *c* (reduced minus oxidized) is 19.2 mM^{-1} cm^{-1} at 550 mμ and that for DCIP is 21 mM^{-1} cm^{-1} at 600 mμ at room temperature.

Protein is determined by the biuret method.[3]

Preparation[1,2]

The heart muscle preparation is made from sand-grinding and differential centrifugation (see method 1 of [37] in this volume). This

[3] T. Yonetani, *J. Biol. Chem.* **236**, 1680 (1961).

preparation is suspended in phosphate–borate buffer, 0.1 M with respect to both components, and a protein concentration is adjusted to between 16 and 24 mg per milliliter. Subsequent operations are performed at 0–4° unless otherwise indicated.

A sodium cholate solution (10% weight per volume, adjusted to pH 7.4–7.8) is added to the heart muscle preparation to a final concentration of 1% cholate and solid ammonium sulfate to 0.25 saturation.[4] This mixture is adjusted[5] to pH 8.0 with 1 N NaOH. After the mixture has stood for 1 hour, more solid ammonium sulfate is added until it reaches 0.35 saturation. After 15 minutes, the suspension is centrifuged for 50 minutes at 23,000 g. The reddish-yellow supernatant fraction is collected and the greenish-brown residue is discarded. The residue contains very little of cytochromes b and c_1 and is a good source for the preparation of cytochrome oxidase.

The supernatant liquid is subsequently mixed with saturated ammonium sulfate solution to 0.51 saturation.[6] The saturated ammonium sulfate solution should be adjusted to pH 7.8 with NH_4OH before use. The mixture is allowed to stand for 15 minutes and then centrifuged for 30 minutes at 16,000 g. The precipitate is dissolved in 0.1 M phosphate buffer, pH 7.4, containing 0.5% cholate, with a volume approximately one-sixth of the original. It is then made to 0.3 saturation by the addition of saturated ammonium sulfate solution. After centrifugation at 16,000 g for 30 minutes, the supernatant fraction is made up to 0.48 saturation with saturated ammonium sulfate solution. Succinate–cytochrome c reductase is precipitated at ammonium sulfate concentrations between 0.30 and 0.48 saturation. The precipitate is dissolved in 0.1 M phosphate buffer, pH 7.4–7.8, to a protein concentration of approximately 20 mg/ml.[3]

[4] The term "saturation" refers to saturation at room temperature in spite of the fact that other temperatures are actually used in the operation. Thus, the figure of 767 g of ammonium sulfate per liter of solvent is the basis of the saturated solution in computation.

[5] This pH value refers to that at 0–4°.

[6] The volume of saturated ammonium sulfate solution required can be conveniently calculated from Eq. (5).

$$X = \frac{(S_f - S_i) \cdot 1000}{1 - S_f} \tag{5}$$

Here X is the volume in milliliters of saturated ammonium sulfate solution to be added to a 1-liter solution of the initial saturation of ammonium sulfate (S_i) to obtain the final saturation (S_f). For example, to bring 250 ml of 0.35 saturation to 0.51 saturation, 81.5 ml of saturated ammonium sulfate solution is required. Any volume change due to mixing is not considered significant for the computation.

Cholate and ammonium sulfate contaminated in the preparation can be largely removed by dialysis against 0.1 *M* phosphate buffer, pH 7.4–7.8, or phosphate-borate buffer in at least 500 times the volume of the reductase preparation for about 16 hours with efficient mixing. An equal effect is achieved by dialyzing the reductase solution against 10–15 times its volume of buffer on a rocking-type dialysis table with four or five changes, allowing 2–3 hours for each change.

Properties[1,2,7]

Enzyme Activities. Specific activities, at 23°, of the reductase prepared show 0.44 and 0.29 for cytochrome *c* and DCIP, respectively. In the presence of 42 μM coenzyme Q_6, the activities increase to 1.10 (cytochrome *c*) and 0.37 (DCIP) with turnover numbers of 2750 and 800, respectively, moles succinate oxidized per minute per mole of acid nonextractable flavin.[8] Phenazine methosulfate is also an effective acceptor.

Succinate–cytochrome *c* reductase does not show cytochrome oxidase activity or catalyze the oxidation of DPNH by cytochrome *c*, DCIP, or ferricyanide.

pH Optima. In the KH_2PO_4–$Na_2B_4O_7$ buffer system, the optimum pH of reductase is 7.7–7.9 for cytochrome *c* and 6.7 for DCIP.

Inhibitors. The enzymatic activity is competitively inhibited by malonate and fumarate. Antimycin A or 2-heptyl-4-hydroxyquinoline-*N*-oxide inhibits the cytochrome *c* activity but does not affect DCIP reaction. Thenoyltrifluoroacetone inhibits both activities. 2-(9-Cyclohexyl-*n*-nonyl)-3-hydroxy-1,4-naphthoquinone inhibits cytochrome *c* activity more than DCIP activity. The inhibition by these lipophilic compounds at low concentrations is competitive with the added coenzyme Q in the system. However, at high concentrations of inhibitors coenzyme Q causes little reversal effect.[7]

Kinetic Constants. Some kinetic constants at 23° are as follow: $K_m^{\text{cytochrome } c}$, 0.0042 m*M*; K_m^{DCIP}, 0.0075 m*M*; $K_m^{\text{succinate}}$, 0.23 m*M*; K_i^{malonate}, 0.005 m*M*; and K_i^{fumarate}, 0.3 m*M* (the latter three constants are determined with cytochrome *c* as the acceptor). These values are practically identical with those determined in the systems with the heart muscle preparation except K_m^{DCIP} which in the latter system is 0.019 m*M*. The reason for this difference has been discussed.[2]

Spectra. Absolute and difference spectra are shown in Figs. 1 and 2.

[7] S. Takemori and T. E. King, *Science* **144,** 852 (1964).

[8] For the determination of acid-nonextractable and acid-extractable flavins, see this volume [77].

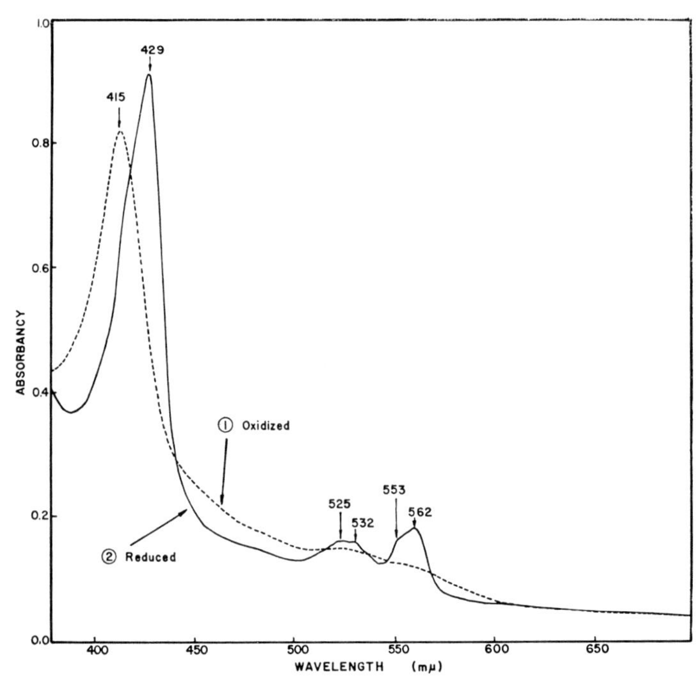

FIG. 1. Absolute spectra of succinate–cytochrome c reductase. Protein is at approximately 2 mg per milliliter of 0.1 M phosphate buffer, pH 7.4, containing 0.1% deoxycholate. Curve 1, untreated (oxidized) preparation; curve 2, reduced with sodium dithionite. Temperature, 23°.

Succinate can reduce at least 80% of the cytochromes b and c_1 in succinate–cytochrome c reductase.

Composition. Succinate–cytochrome c reductase contains, in terms of millimicromoles per milligram of protein, acid-nonextractable flavin, 0.40; cytochrome b, 1.97; cytochrome c_1, 1.21; nonheme iron, 9.75; and CoQ, 0.86. In addition it has 18.6% of lipid. No acid-extractable flavin[8] or heme a[9] has been detected in the preparation.

Stability. The reductase does not show an appreciable decrease in enzymatic activity after storage at 0° for 3 days.

Comments on Preparation. The specific activities described in a preceding section are averages of several preparations. Samples of higher activities are obtained by introducing additional cycles of ammonium sulfate fractionation. However, the turnover numbers based on either acid-nonextractable flavin or cytochromes remain practically constant.

[9] For the determination of heme a by pyridine hemochromogen or the direct spectrophotometric method, see this volume [76].

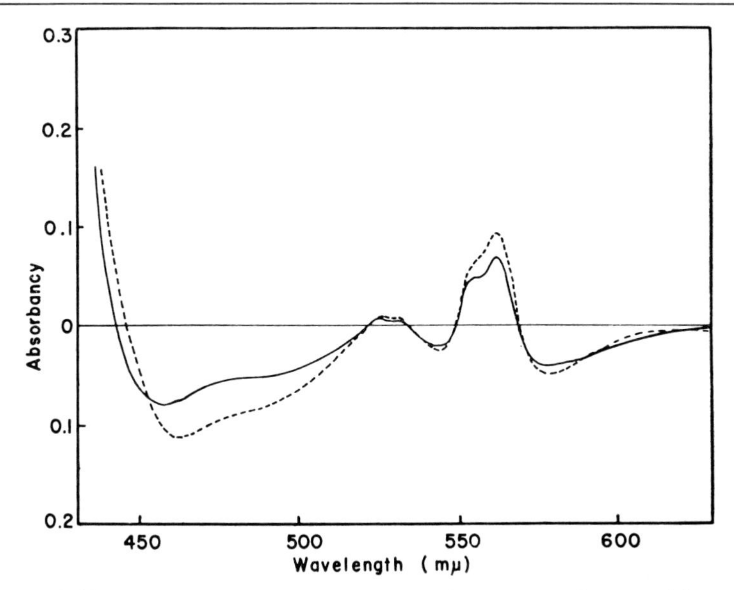

FIG. 2. Difference spectra of succinate–cytochrome c reductase. The system contains 70 mM phosphate buffer, pH 7.4; 0.3 mM EDTA; 0.23 M sucrose, and approximately 4.4 mg protein per milliliter. After the succinate (60 mM)-reduced spectrum is recorded, a small amount of sodium dithionite is added and the spectrum is again recorded. —, the succinate-reduced form minus the oxidized; ----, the enzyme further reduced by dithionite minus the oxidized. The horizontal solid line at 0 is the base line, i.e., the oxidized form minus the oxidized. Temperature, 23°.

The method described has been used in this laboratory for more than four years with no unusual difficulty. Contrary to several published reports, the ratio of cholate to protein is not important when the protein concentration is within the range specified. On the other hand, the absolute concentration of cholate is critical. High concentrations of cholate solubilize cytochrome oxidase. In these preparations, it is rather difficult to remove the contaminated cytochrome oxidase. Low concentrations of cholate decrease the yield of reductase. For the cytochrome oxidase preparation using the residue from 1% cholate extraction, a small amount of unextracted cytochromes b and c_1 may be removed without much effort. Indeed, some conventional methods for the preparation of cytochrome oxidase directly use 2% cholate extraction of heart mince.

Lipid content is always higher than the norm in those batches of the heart muscle preparation from the tissue of old or grain-fed ox. The hearts of these animals always seem to consist of solid, invisible (i.e., buried in the muscular tissue) lipid; thus, a minor modification is required. A prior treatment of the heart muscle preparation in a final concentration of 0.2% cholate (but without ammonium sulfate) followed by a washing with phosphate–borate buffer improves the yield of reductase. However, when the cholate-washed preparation is used, 0.9% (instead of 1%) of cholate concentration is recommended during the solubilization.

Preparation of the Cytochrome b-c_1 Particle and Reconstitution of Succinate–Cytochrome c Reductase

Assay Method[10,11]

Principle. Soluble succinate dehydrogenase or the cytochrome b-c_1 particle alone does not catalyze the oxidation of succinate by either cytochrome c or DCIP. The cytochrome b-c_1 particle in the presence of succinate dehydrogenase rapidly mediates the transport of hydrogen (electron) from succinate to cytochrome c or DCIP. The activity is then spectrophotometrically determined by following the reduction of cytochrome c or DCIP.

Reagents. In addition to the reagents described in the first part of the article, soluble succinate dehydrogenase, prepared through the stage of gel-eluates at about 1 mg/ml, is required. The dehydrogenase must be freshly made or freshly thawed from the frozen state stored at liquid nitrogen temperature. In either instance, it should not be more than 1 hour old. For the preparation of succinate dehydrogenase, see this volume [58].

Method of Assay. In a cuvette of 1-cm optical path are placed 1.5 ml phosphate buffer, 0.3 ml EDTA, an appropriate amount of the cytochrome b-c_1 particle and 0.3 ml cytochrome c (or 0.3 ml DCIP). Finally 0.1 ml succinate dehydrogenase together with water is added to 2.9 ml. The reaction is started by the addition of 0.1 ml succinate. Subsequent recording of the absorbance change and calculations of the specific activity are detailed under Assay Method in the first part of the article. Usually four cuvettes are made at the same time containing 0, 15, 30, and 45 μg of the cytochrome b-c_1 particle, respectively.

Preparation[1,2]

The succinate–cytochrome c reductase solution at about 10 mg protein per milliliter is adjusted to pH 9.5 at 0–4° with 1 N NaOH. The

[10] T. E. King and S. Takemori, *Biochim. Biophys. Acta* **64**, 194 (1962).
[11] T. E. King and S. Takemori, *J. Biol. Chem.* **239**, 3559 (1964).

mixture is incubated at 37° with gentle shaking for 60 minutes. It is then cooled to 0° and centrifuged at 78,000 g for 60 minutes. The precipitate is suspended in 0.1 M phosphate buffer, pH 7.4, with the aid of a Potter-Elvehjem homogenizer and again centrifuged. The pellet is dispersed in 0.1 M phosphate buffer, pH 7.4. The yield based on the cytochrome b content of the heart muscle preparation originally used is approximately 50%.

Reconstitution and Isolation of Reconstituted Reductase

The determination of the cytochrome b-c_1 particle activity by reconstitution as detailed under Assay Method uses excess succinate dehydrogenase. For the isolation of reconstituted succinate–cytochrome c reductase, the scale may be extended and the concentration increased. Usually a mixture is made containing phosphate buffer, 0.1 M, pH 7.4–7.8; succinate, 0.02 M; succinate dehydrogenase (specific activity of about 16 micromoles per minute per milligram), 0.5 mg/ml, and the cytochrome b-c_1 particle, 0.2 mg/ml. The excess succinate dehydrogenase in the system is removed by centrifugation and subsequent washing with 0.1 M phosphate buffer, pH 7.4. The reconstitution reaction takes place within the time of manipulation. The concentration of either succinate dehydrogenase or the cytochrome b-c_1 particle does not affect the time required for the reaction.

When the cytochrome b-c_1 particle is in excess and succinate dehydrogenase is limiting, the activity of the latter may be determined. The particle thus isolated by centrifugation is, however, deficient in succinate dehydrogenase.

Properties

The Cytochrome b-c_1 Particle.[1,2,10,11] The particle is free of enzymatic activity characteristic of succinate–cytochrome c reductase, assayed with cytochrome c or DCIP; succinate dehydrogenase, succinate–phenazine reductase, tested with ferricyanide or phenazine methosulfate; and DPNH dehydrogenase, determined by ferricyanide, cytochrome c, or DCIP. It is also devoid of cytochrome oxidase.

The cytochrome b-c_1 particle is insoluble but dispersible at low concentrations in aqueous solvents. High concentrations of sucrose facilitate the dispersion. The difference spectrum of the dithionite-reduced form minus the oxidized in 0.5 M sucrose is shown in Fig. 3. The particle contains cytochrome b, 2.4; cytochrome c_1, 1.4; nonheme iron, 6.3; and coenzyme Q, 0.8 in terms of millimicromoles per milligram of protein and also 19% lipid.

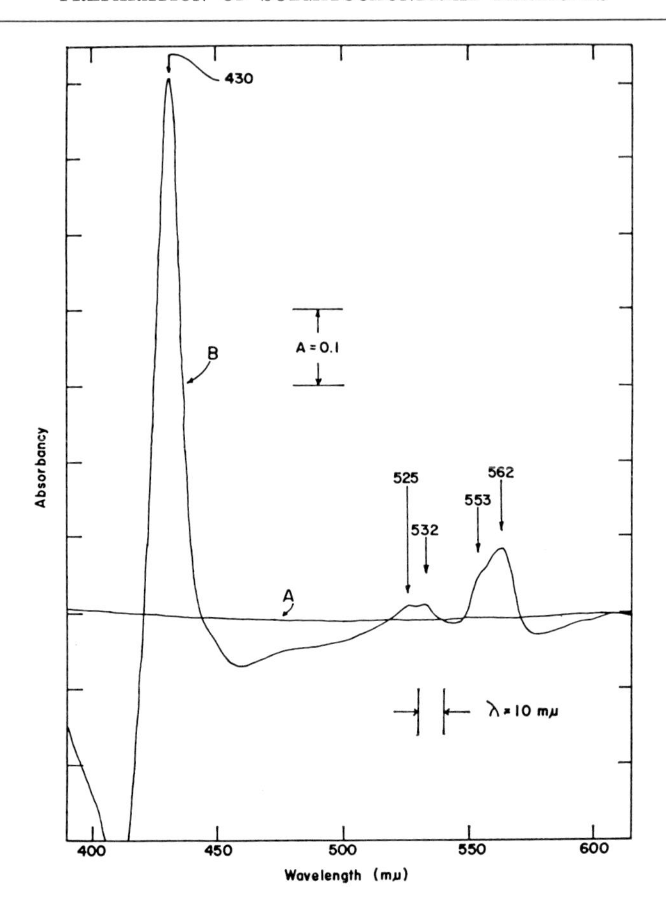

FIG. 3. Difference spectra of the cytochrome b-c_1 particle. The system contains 0.1 M phosphate buffer, pH 7.4, 0.5 M sucrose, and about 1.9 mg protein per milliliter. Temperature, 23°. Curve A, the oxidized form minus the oxidized; curve B, the dithionite-reduced form minus the oxidized.

Reconstituted Succinate–Cytochrome c Reductase.[10,11] Specific activities at 23° of the cytochrome b-c_1 particle prepared show 0.53 and 0.40 for cytochrome c and DCIP, respectively, in the reconstitution test. In the presence of 42 μM coenzyme Q_6, the activities increase to 1.1 (for cytochrome c) and 0.46 (for DCIP) with turnover numbers of 450 and 170, respectively, based on cytochrome b. On the basis[12] of cytochrome c_1, the corresponding turnover numbers are 780 for cytochrome c and 290 for DCIP. Turnover numbers based on incorporated acid-non-

[12] Although the turnover number for DCIP is cited, this fact should not be construed to mean that cytochrome c_1 is definitely a respiratory carrier for the DCIP reaction.

extractable flavin are 2500 (for cytochrome *c*) and 1180 (for DCIP) at 23°.

The enzymatic activity of reconstituted reductase is competitively inhibited by malonate and fumarate. Antimycin A or 2-heptyl-4-hydroxyquinoline-*N*-oxide inhibits the cytochrome *c* activity completely but does not affect DCIP activity. Thenoyltrifluoroacetone inhibits both activities. 2-(9-Cyclohexyl-*n*-nonyl)-3-hydroxy-1,4-naphthoquinone inhibits more cytochrome *c* activity than DCIP activity. Some kinetic constants at 23° are as follows: $K_m^{\text{cytochrome } c}$, 0.003 m$M$; K_m^{DCIP}, 0.0061; $K_m^{\text{succinate}}$ 0.3 mM; K_i^{malonate}, 0.0045 mM; and K_i^{fumarate}, 0.34 mM (the latter three constants are determined with cytochrome *c* as the acceptor).

The cytochrome b-c_1 particle reacts with succinate dehydrogenase, but not with succinate-phenazine reductase. Soluble succinate dehydrogenase is very unstable; once the dehydrogenase is incorporated into the cytochrome b-c_1 particle, it becomes as stable as intact succinate–cytochrome *c* reductase. The succinate dehydrogenase and the cytochrome b-c_1 in the reconstituted reductase cannot be separated by centrifugation nor dissociated by dilution.

[41] The Preparation and Properties of DPNH–Cytochrome c Reductase (Complex I–III of the Respiratory Chain)

By Y. HATEFI[1] and J. S. RIESKE[1]

$$\text{DPNH (NADH)} + 2\ \text{Ferricytochrome } c \rightarrow \text{DPN}^+ (\text{NAD}^+) + 2\ \text{Ferrocytochrome } c + 2\text{H}^+$$

Other DPNH–cytochrome *c* reductase preparations have been described previously.[1a,2] These preparations, however, either were not derived from the terminal respiratory chain or were not separated from other functional components of the respiratory chain. The respiratory DPNH–cytochrome *c* reductase is characterized by sensitivity to the respiratory inhibitors: Amytal, rotenone, antimycin A, and 2-*n*-heptyl-(or nonyl-) 4-hydroxyquinoline-*N*-oxide.

Assay Method

Principle. The reduction of ferricytochrome *c* by DPNH (NADH) is measured spectrophotometrically at 550 mμ. Alternatively, the oxidative

[1] During the preparation of this manuscript Y. Hatefi was at Pahlevi University, Shiraz, Iran; J. S. Rieske was at the Institute for Enzyme Research, The University of Wisconsin, Madison, Wisconsin.

[1a] H. R. Mahler, Vol. II [120]. See also A. F. Brodie, Vol. II [121].

[2] B. de Bernard, *Biochim. Biophys. Acta* **23**, 510 (1957).

of DPNH can be followed spectrophotometrically by a measurement of the decrease in absorbancy at 340 mμ.

Reagents

Potassium phosphate buffer, 1.0 *M*, pH 8
NaN_3 (neutralized) 0.1 *M*
DPNH (NADH), 0.01 *M*
Aqueous ferricytochrome *c* (Sigma, type III), 1%
Phospholipid mixture (15 mg/ml in 0.001 *M* EDTA)

Either mixed mitochondrial phospholipid[3] or asolectin,[4] a commercially available phospholipid, can be used. The lipid is dispersed into the solution of EDTA by the following procedure[3]: Asolectin (or mitochondrial phospholipid) is dissolved in *n*-butanol in the presence of 2 mg potassium deoxycholate per milligram of asolectin. The solution is dialyzed for 2 weeks against daily changes of 200 volumes of a solution of 0.01 *M* Tris-HCl, pH 8.0 and 0.001 *M* EDTA in water. The suspension of phospholipid is centrifuged for 10 minutes at 40,000 rpm (105,000 *g*). The supernatant suspension is exposed to sonic irradiation for 30 minutes. All these operations should be performed at near zero temperatures. The phospholipid dispersion can be stored for several months at near-zero temperatures.

Enzyme. A few minutes prior to assay, the enzyme is diluted to a concentration of 0.1–1.0 mg/ml in an aqueous medium that is 0.05 *M* in Tris-HCl, pH 8; 0.67 *M* in sucrose; and 0.001 *M* in histidine. For crude insoluble preparations that contain the enzyme it may be necessary to "solubilize" the protein with deoxycholate in order to achieve maximal activities.

Procedure. To each of two 1-ml quartz cuvettes are added 0.02 ml of 1.0 *M* phosphate buffer, pH 8; 0.02 ml of 0.1 *M* NaN_3; 0.06 ml of 1% ferricytochrome *c*; 0.01 ml of solution containing per milliliter 15 mg of lipid suspended in 0.001 *M* EDTA; 0.025 ml of 0.01 *M* DPNH; and water to a total volume of 1.0 ml. These solutions are allowed to reach temperature equilibrium in a jacketed, thermostatted (38°) chamber of the spectrophotometer. The reaction is started by an introduction of 0.01 or 0.02 ml of the diluted solution of enzyme into the sample cuvette. Absorbancy changes at 550 mμ are recorded at 15-second intervals for 90 seconds. Alternatively, the oxidation of DPNH may be followed by the decrease of absorbancy at 340 mμ. For this method the amount of

[3] S. Fleischer, G. Brierley, H. Klouwen, and D. B. Slautterback, *J. Biol. Chem.* **237**, 3264 (1962).

[4] Asolectin is a commercial preparation of mixed phospholipids obtained from Associated Concentrates, Inc., Woodside, New York.

DPNH added to the assay mixture is half of that used when the reduction of cytochrome c is estimated. All other conditions are unchanged.

Definition of Specific Activity. The specific activity of the enzyme can be expressed in terms of micromoles of cytochrome c reduced-min^{-1}-(mg enzyme protein)$^{-1}$. The specific activity is calculated by

$$\text{S.A.} = \frac{\Delta \text{ absorbancy (550 m}\mu\text{)}}{\Delta t} \times \frac{1}{\text{mg enzyme}} \times \frac{1}{18.5}$$

The first term usually is estimated as the slope of the linear portion of the curve obtained by a plot of the absorbancy change at 550 mμ vs. the time of reaction in minutes.

It must be noted that variations in specific activity may be observed if the preparation of the enzyme and its assay are not accurately carried out. Also differences in the degree of dispersion of the enzyme in different dilution mixtures may result in variations in activity measurements. Therefore it is advisable to assay the enzyme at the range of concentration given above so that comparison between one preparation and another would be possible. Because of these difficulties which are inherent in the assay of particulate enzymes, specific activities of crude preparations of the enzyme may yield only a rough approximation of the actual content of enzyme in the preparation. As a result a strict correlation between specific activity and the content of a prosthetic moiety in the enzyme molecule (i.e., FMN content in the case of the respiratory DPNH–cytochrome c reductase) may not be apparent at various stages of purification.

Although *added* phospholipid enhances the activity of DPNH–cytochrome c reductase, it is not absolutely required. However, if the enzyme is assayed without supplemented phospholipid, EDTA in the amount normally added with the phospholipid should be added.

Purification Procedure

The enzyme purification presented below has been described in most part by Hatefi *et al.*[5] All steps of the purification are carried out at 0–5°.

Source Material. Mitochondria from beef heart muscle[6] are suspended in 0.25 M sucrose and stored at −20° until used.

Step 1. Wash of the Mitochondria. The frozen suspension of mitochondria in 0.25 M sucrose is thawed and diluted with 0.25 M sucrose to a protein concentration[7] of about 30 mg/ml and then centrifuged at 20,000 rpm for 20–30 minutes. After centrifugation, the loosely packed

[5] Y. Hatefi, A. G. Haavik, and P. Jurtshuk, *Biochim. Biophys. Acta* **52,** 106 (1961).

[6] P. Blair, see this volume [12].

[7] All protein analyses were made by the biuret method of A. G. Gornall, C. J. Bardawill, and M. M. David, *J. Biol. Chem.* **177,** 751 (1949).

mitochondria are suspended in the Tris–sucrose–histidine medium (see Assay Method).

Step 2. Red-Green Separation of the Cytochromes. The suspension of mitochondria in the Tris–sucrose–histidine solution is diluted with the same solution to a protein concentration of 23 mg/ml. Potassium deoxycholate (10%, pH 9.0) is added to a level of 0.3 mg per milligram of protein followed by the addition of solid KCl to a level of 72 g per liter. When all the KCl is dissolved, the suspension is centrifuged for 30 minutes at 30,000 rpm in the No 30 rotor of the Beckman Model L ultracentrifuge. The green sediment contains crude cytochrome oxidase and may be saved for further purification of cytochrome oxidase.[8] The clear, red supernatant solution is diluted with 0.25 volumes of cold, distilled water and centrifuged for 30 minutes at 30,000 rpm. The small, brown-green pellet is discarded and the supernatant solution is collected.

Step 3. Dialysis. The supernatant solution from the previous step is poured into dialysis tubing of a bore diameter of about 2 cm and dialyzed for 3 hours against 8 volumes of 0.01 *M* Tris-HCl, pH 8.0. This dialysis step removes enough KCl and deoxycholate to cause the more insoluble proteins, which contain complexes I, II, and III, to precipitate. Most of the cytochrome *c* and other relatively soluble proteins remain in solution. The turbid dialyzate is centrifuged for 75 minutes at 30,000 rpm. The supernatant solution is discarded. The red pellet, hereafter referred to as S-1, is suspended in a small amount of the Tris–sucrose–histidine solution and stored overnight at —20°.

Step 4. Deoxycholate-Ammonium Acetate Fractionation. The thawed suspension of S-1 from step 3 is homogenized in a glass homogenizer fitted with a Teflon pestle, then diluted with the Tris–sucrose–histidine solution to a protein concentration of 10 mg/ml. At this point, potassium deoxycholate is added to a concentration of 0.5 mg per milligram of protein followed by the addition of 50% saturated ammonium acetate[9] to a concentration of 16.5 ml per 100 ml of protein solution. After being allowed to stand for 15 minutes at 0° the turbid solution is centrifuged for 30 minutes at 30,000 rpm. The residue should be a tightly packed, brownish white pellet; it is discarded. The supernatant solution is treated with 6.4 ml of the ammonium acetate solution per 100 ml of supernatant solution and again after 15 minutes is centrifuged for 30 minutes at 30,000 rpm. This time, the residue should consist of a loosely packed, light brown material overlaying a small, more dense, dark brown button.

[8] D. C. Wharton and A. Tzagoloff, this volume [45].

[9] The 50% saturated ammonium acetate is prepared by dissolving the contents of a fresh 1.0 pound (454 g) bottle of ammonium acetate (Reagent Grade) in 613 ml of water.

For this step, it may be necessary to standardize each fresh preparation of ammonium acetate solution with regard to the optimal amount to be added. The amount added may be varied from 6.0 to 6.5 ml per 100 ml of supernatant solution. An excessive amount of ammonium acetate will give rise to a large amount of dark colored sediment with a relatively small amount of light colored residue on top; too little ammonium acetate results only in a light brown, loosely packed residue. Other factors that appear to influence the concentration of ammonium acetate required for this step are the time involved in the precipitation and sedimentation of the fractions and the temperature of the solution during the fractionation. Centrifuge rotors that have warmed several degrees may induce precipitation of the enzyme at lower than expected concentrations of ammonium acetate.

The supernatant solution obtained from the above step is then treated with 3.2 ml of the ammonium acetate solution per milliliter of supernatant liquid. After 15 minutes at 0° the liquid is centrifuged as before. The reddish brown residue designated as R_4B is suspended in the Tris–sucrose–histidine buffer and may be stored at −20° with little loss of activity. The supernatant solution may be utilized for the preparation of complex III (reduced CoQ–cytochrome *c* reductase).[10]

Yields and activities of the enzyme at different stages of preparation are given in Table I.

TABLE I
YIELD AND ACTIVITIES OF PRINCIPAL FRACTIONS OBTAINED DURING PURIFICATION OF RESPIRATORY DPNH-CYTOCHROME *c* REDUCTASE

Fraction	Recovery of protein (%)	Purification of enzyme: FMN content[a]	Purification of enzyme: Specific activity[b]	Recovery of activity (%)
Mitochondria[c]	100	0.14	4	100
S-1	12–18	0.24	12–18	40–80
R_4B (DPNH-Cyt *c* reductase)	1.5–2.0	0.80	45–58	17–30

[a] Expressed as millimicromoles per milligram of protein. For details see: A. J. Merola, R. Coleman, and R. Hansen, *Biochim. Biophys. Acta* **73,** 638 (1963).

[b] Expressed as micromoles of cytochrome *c* reduced-min^{-1}-(mg protein)$^{-1}$. We have tested the preparation and assay of the enzyme in the hands of several individuals who had little or no previous familiarity with the procedures. The specific activities obtained by them ranged between 6 and 18 for S-1 and between 15 and 50 for R_4B.

[c] Assayed after treatment of the mitochondria with deoxycholate and KCl as in step 2 of the purification procedure.

[10] J. S. Rieske, this volume [44].

Properties

Composition. DPNH (NADH)–cytochrome *c* reductase as isolated by the above procedure consists of a mixture of two enzyme complexes that can be separated readily; complex I (DPNH–CoQ reductase)[11] and complex III (reduced CoQ–cytochrome *c* reductase). The enzyme as prepared appears to contain one molecule of complex I and one molecule of complex III, on the basis of the content of flavin and cytochrome $c + c_1$. The composition of the most active preparations is summarized in Table II.

TABLE II
COMPONENTS OF DPNH–CYTOCHROME *c* REDUCTASE[a]

Component	Concentration (units/mg protein)
Acid-extractable flavin[b]	0.83 (±0.07) mμmoles
Total flavin	0.97 (±0.05) mμmoles
Cytochrome *b*[c]	1.13 (±0.07) mμmoles
Cytochrome $c + c_1$[d]	0.81 (±0.04) mμmoles
Coenzyme Q	3.7 (±0.01) mμmoles
Nonheme Fe	14–16 mμmoles
Lipid	0.30–0.35 mg

[a] From Y. Hatefi, A. G. Haavik, and P. Jurtshuk, *Biochim. Biophys. Acta* **52,** 106 (1961).

[b] Essentially all FMN (cf. Table I).

[c] The content of cytochrome *b* is probably underestimated. Subsequent analyses of complex III have indicated that it contains very close to two molecules of cytochrome *b* to one molecule of cytochrome c_1; see W. S. Zaugg and J. S. Rieske, *Biochem. Biophys. Res. Commun.* **9,** 213 (1962).

[d] Although the analytical method used cannot distinguish between cytochromes *c* and c_1, practically all the cytochrome *c* is removed in prior steps in the purification.

Stability. A solution of the enzyme in the Tris–sucrose–histidine medium is quite stable when stored at —20°. Little loss of activity occurs over a period of several days.

Specificity. The enzyme appears to be specific for DPNH as the donor substrate. TPNH in the presence of the enzyme is oxidized at about 2% of the rate of oxidation of DPNH. Cytochrome *c* appears to be the only electron acceptor that is reduced in an antimycin A—sensitive pathway. Coenzyme Q_1, potassium ferricyanide, and methylene blue are efficient as electron acceptor substrates; however, the reduction of these compounds is not inhibited by antimycin A.

[11] Y. Hatefi and J. S. Rieske, this volume [43].

Contamination. The enzyme preparation is almost free of other oxidative activities catalyzed by mitochondria. Pyruvate + malate, α-ketoglutarate, β-hydroxybutyrate, and isocitrate dehydrogenase activities are virtually absent. The enzyme is also almost completely free of succinate–cytochrome *c* reductase activity and cytochrome *c* oxidase activity.

Spectral Properties. The difference spectrum (reduced minus oxidized) of the enzyme displays the α and β bands of cytochromes *b* (562 mμ, 532 mμ) and c_1 (554 mμ, 524 mμ) superimposed on a broad bleached area between 440 mμ and 520 mμ with the maximal bleaching at 460 mμ. Only about 50% of the bleaching at 460 mμ can be ascribed to the reduction of the flavoprotein in the enzyme.

Inhibitors. All the respiratory inhibitors that act between the site of oxidation of DPNH and the site of reduction of cytochrome *c* in the respiratory chain are inhibitors of DPNH–cytochrome *c* reductase. The two most potent inhibitors are rotenone and antimycin A. These inhibitors are stoichiometric inhibitors; both inhibit almost completely at concentrations approaching the concentration of flavoprotein or cytochrome c_1 in the enzyme. To attain this degree of inhibition, both inhibitors must be incubated a short time with a concentrated solution of enzyme before the enzyme is diluted for assay purposes.

The enzyme is inhibited to a lesser extent by sodium Amytal (50% inhibition at 0.4 mM), p-CMS (50% inhibition at 10 μM), 2-*n*-nonyl-4-hydroxyquinoline-*N*-oxide (>95% inhibition at 0.4 μM), and SN-5949[12] (50% inhibition at 0.25 μM).

[12] An antimalarial consisting of a 3-alkyl derivative of 2-hydroxy-1,4-naphthoquinone.

[42] Preparation and Properties of Succinate Dehydrogenase–Coenzyme Q Reductase (Complex II)[1]

By D. Ziegler and J. S. Rieske

Succinate + coenzyme Q → fumarate + (coenzyme Q)H_2

Assay Method

Principle. The rate of reduction of coenzyme Q_2 by succinate is determined indirectly by the measurement of the rate of reduction of 2,6-dichloroindophenol which is reduced rapidly by (coenzyme Q_{10})H_2 or

[1] D. M. Ziegler and K. A. Doeg, *Arch. Biochem. and Biophys.* **97,** 41 (1962).

(coenzyme Q_2)H_2. Under conditions of the assay the indophenol dye is not reduced in the absence of coenzyme Q.

Reagents

Potassium succinate, 1.0 *M*, pH 7
Potassium phosphate buffer, 1.0 *M*, pH 7.0
EDTA, 1.0 m*M*
Triton X-100, 10%
2,6-Dichloroindophenol, 0.1%
Coenzyme Q_2 (5 mg/ml in ethanol)
Enzyme. Immediately prior to assay, the enzyme is diluted with 0.25 *M* sucrose to a concentration of about 0.5 mg of protein per milliliter.

Procedure. The reaction mixture of 0.1 ml phosphate buffer, 0.015 ml of the indophenol solution, 0.02 ml of the succinate solution, 0.01 ml of the EDTA solution, 0.001 ml of Triton X-100, 0.002 ml of the coenzyme Q_2 solution, and enough water to bring the total volume to 1.0 ml is placed in a 1.0-ml cuvette of 1-cm light path. After temperature equilibration (38°) in a thermostatted spectrophotometer the reaction is started by the addition of 0.1–1.0 μg of enzyme protein (2–10 μg of the diluted enzyme). Absorbancy readings at 600 mμ are taken at 15-second intervals until at least 80% of the indophenol is reduced. In the absence of either enzyme or succinate the dye is not reduced.

Definition of Specific Activity. The specific activity is expressed in terms of micromoles succinate oxidized-min^{-1}-(mg of protein)$^{-1}$. In terms of the assay parameters the specific activity is expressed by:

$$\text{S.A.} = \frac{\Delta \text{ absorbancy } (600 \text{ m}\mu)}{\Delta t} \times \frac{1}{\text{mg enzyme}} \times \frac{1}{20.5}$$

The first term is usually estimated from the slope of the linear portion of the curve obtained by plotting the absorbancy change at 600 mμ versus the time of the reaction in minutes. The specific activity of the enzyme is independent of enzyme concentration up to absorbancy changes of 0.5–0.6 per minute.

Purification Procedure

Step 1. Preparation of the Mitochondrial Suspension. Mitochondria from beef heart are resuspended in an aliquot of the soluble protein fraction from the Sharples centrifuge effluent[2] (200 g wet weight of the mitochondrial residue per 200 ml of the soluble fraction). The suspension

[2] See P. V. Blair, this volume [12].

is treated with calcium chloride to a final concentration of 10^{-5} *M* and then stored at —20° until used.

Step 2. Incubation of the Mitochondrial Suspension. The mitochondrial suspension of step 1 is thawed (12–14 hours at 5°), treated with 1.0 *M* potassium phosphate, pH 7.4 (40 ml per 100 ml of suspension), and then homogenized in a glass-Teflon homogenizer. The protein concentration is adjusted to 70 mg/ml with 0.25 *M* sucrose. The suspension (about 600 ml) is transferred to a liter beaker and then placed in a thermostatted water bath, set at 38°. After 15 minutes, 20% (w/v) potassium cholate is added to a final concentration of 0.6 mg per milligram of protein. The incubation is continued for another 15 minutes, then the beaker is transferred to an ice bath.

Step 3. Ammonium Sulfate Fractionation. Immediately after being placed in the ice, solid ammonium sulfate is added to 0.30 saturation (16.4 g/100 ml). The mixture is stirred until the temperature drops to 5°. All subsequent operations are performed at 0–5° unless noted otherwise. The turbid solution is centrifuged for 15 minutes at 25,000 rpm in a Beckman Model L centrifuge; the residue is discarded. The supernatant solution is treated again with ammonium sulfate to 0.40 saturation (5.6 g/100 ml) and centrifuged as before. The residue is collected and dissolved in 0.25 *M* sucrose to give a concentration of 85–100 mg of protein per milliliter and then is dialyzed for 5 hours with internal and external stirring against 10 volumes of 0.25 *M* sucrose. The dialyzate is changed after 1 and 3 hours. After dialysis this fraction may be stored overnight at —20°.

Step 4. Solvent Extraction. The 0.30–0.40 ammonium-sulfate fraction is thawed, then treated with 10% potassium deoxycholate to a level of 0.75 mg per milligram of protein. The pH is adjusted to 7.3 with dilute HCl, and then cold (—10°) ethanol (46 ml of 95% ethanol/100 ml) is added rapidly. After the mixture is allowed to cool to 2° in the ice bath (addition of the alcohol causes a rise in temperature), it is centrifuged for 30 minutes at 40,000 rpm in the Beckman Model L centrifuge. The pale brown supernatant then is homogenized with an equal volume of cold (6°) cyclohexane and again is centrifuged for 30 minutes at 40,000 rpm. The resulting red pellet is resuspended in 0.25 *M* sucrose to a protein concentration of 30 mg/ml.

Step 5. Deoxycholate-Ammonium Sulfate Extraction. The suspension of protein from step 4 is treated with 10% (w/v) potassium deoxycholate to a level of 0.3 mg per milligram of protein and neutral, saturated ammonium sulfate to 0.10 saturation (0.11 ml per milliliter of suspension). The resulting mixture is centrifuged to remove insoluble, inactive protein and then is dialyzed briefly to remove ammonium sulfate. The resulting deep red solution contains the purified succinate dehydrogenase coenzyme Q reductase and may be stored in this solution at —20° for 1–2 weeks with little loss of activity. A summary of the purification procedure is given in Table I.

TABLE I
PURIFICATION OF THE SUCCINATE DEHYDROGENASE–COENZYME Q REDUCTASE (COMPLEX II)[a]

Fraction	Per cent protein	Per cent of total activity	Specific activity[b]
Mitochondria[c]	100	100	0.9
30–40% ammonium sulfate precipitate	18	80	3.8
Deoxycholate-ethanol extract	3.7	28	6.8
Cyclohexane precipitate	1.25	25	18.0
Deoxycholate-ammonium sulfate extract	0.37	23	56.0

[a] From D. M. Ziegler and K. A. Doeg, *Arch. Biochem. Biophys.* **97,** 41 (1962).
[b] Specific activity is expressed as micromoles of succinate oxidized-min^{-1}-(mg protein)$^{-1}$.
[c] Mitochondria were assayed after being dispersed with cholate and phosphate in proportions necessary for maximal activity.

Properties

Composition. Table II lists the component composition of succinate dehydrogenase–coenzyme Q reductase.

TABLE II
COMPOSITION OF SUCCINATE DEHYDROGENASE–COENZYME Q REDUCTASE (COMPLEX II)

Component	Concentration
Flavin[a]	4.6–4.8 μmoles/g protein
Heme (cytochrome *b*)[b]	4.8–5.0 μmoles/g protein
Iron (nonheme)	36–38 μmoles/g protein
Lipid	18–20% by weight protein

[a] Extractable by acid only after digestion of the protein with trypsin.
[b] Almost all the heme is extractable as protoheme. Spectrally, the hemoprotein is identified as principally cytochrome *b*.

Specificity. The enzyme catalyzes the direct reduction of coenzyme Q homologs and analogs. Coenzyme Q_2 appears to be the most efficient electron acceptor of the quinone compounds. Coenzyme Q_{10} in substrate quantities when supplemented with Triton X-100 and a lipid extract[3] is

[3] See Y. Hatefi and J. S. Rieske, this volume [41].

about 75% as efficient an electron acceptor as coenzyme Q_2; however, when used in smaller quantities to mediate in the reduction of ferricyanide, coenzyme Q_{10} is 86% as effective as coenzyme Q_2. As mediators for the reduction of ferricyanide, 2,3-dimethoxybenzoquinones with alkyl groups on the 3,6 or 5,6 positions are partially effective. Substituted naphthoquinones generally are ineffective as mediators. Phenazine methosulfate is effective as an electron acceptor but appears to function only in the region of the succinate dehydrogenase moiety of the complex.

Contamination. The purified enzyme does not contain detectable activities of DPNH, pyruvic, or β-hydroxybutyric dehydrogenase. The enzyme is free of cytochromes *a* and *c*, although some preparations may be contaminated with cytochrome c_1.

Spectral Properties. When reduced with dithionite the enzyme displays a typical spectrum of cytochrome *b* with α, β, and Soret absorption bands at 562, 530, and 426 mμ, respectively. When reduced with succinate, the difference spectrum (reduced minus oxidized) displays only a bleached area with a minimal absorption at 460 mμ. The flavin content of the enzyme can account only for 80% of this loss of absorption.

Inhibitors. Succinate dehydrogenase–coenzyme Q reductase is inhibited specifically by 2-thenoyltrifluoroacetone (97% inhibition at 10^{-4} *M*). However, the reduction of phenazine methosulfate by succinate as catalyzed by the enzyme is unaffected by this reagent. The enzyme is inhibited also by *p*-chloromercuriphenyl sulfonate (10^{-5} *M*) and by incubation with 1.0 m*M* cyanide.

[43] Preparation and Properties of DPNH-Coenzyme Q Reductase (Complex I of the Respiratory Chain)

By Y. Hatefi[1] and J. S. Rieske[1]

DPNH (NADH) + H^+ + coenzyme Q $\rightarrow$ DPN^+ (NAD^+) + (coenzyme Q)H_2

Assay Method[1a, 2]

Principle. The rate of oxidation of DPNH by coenzyme Q_1 is measured spectrophotometrically by following the decrease in absorbancy at 340 mμ as a function of time.

[1] See footnote 1, page 225.

[1a] From Y. Hatefi, A. G. Haavik, and P. Jurtshuk, *Biochem. Biophys. Res. Commun.* **3**, 281 (1960).

[2] Y. Hatefi, A. G. Haavik, and D. E. Griffiths, *J. Biol. Chem.* **237**, 1676 (1962).

Reagents

Potassium phosphate buffer, 1.0 *M*, pH 8
NaN_3 (neutralized), 0.1 *M*
DPNH, 0.01 *M*
Coenzyme Q_1 in water,[3] 1.0 m*M*
Asolectin suspension (15 mg/ml in 0.001 *M* EDTA)[4]
Enzyme. A few minutes prior to assay the enzyme is diluted to a concentration of 0.1–1.0 mg/ml in an aqueous medium that is 0.67 *M* in sucrose; 0.05 *M* in Tris-HCl, pH 8; and 0.001 *M* in histidine (hereafter referred to as Tris–sucrose–histidine buffer).

Procedure. To the sample cuvette (quartz, 1 ml) are added 0.02 ml 1 *M* potassium phosphate, pH 8; 0.02 ml of 0.1 *M* NaN_3; 0.012 ml of 10 m*M* DPNH; 0.05 ml of 1.0 m*M* coenzyme Q_1; 0.01 ml of the asolectin suspension; and water to a total volume of 1.0 ml. The reference cuvette contains the same mixture with the exception of DPNH, which is present in half the amount added to the sample cuvette. These solutions are allowed to reach temperature equilibrium in a jacketed, thermostatted (38°) chamber of the spectrophotometer. The reaction is started by the addition of the enzyme (0.01–0.02 ml) to the sample cuvette. Readings at 340 mμ are taken at 15-second intervals for 90 seconds.

Definition of Specific Activity. The specific activity of the enzyme can be expressed as micromoles of substrate (one electron) reduced-min^{-1}-(mg of enzyme protein).$^{-1}$ The specific activity is calculated by

$$\text{S.A.} = \frac{\Delta \text{ absorbancy (340 m}\mu)}{\Delta t} \times \frac{1}{\text{mg enzyme}} \times \frac{1}{3.11}$$

The first term may be estimated from the slope of the linear portion of the curve obtained by plotting the absorbancy change at 340 mμ versus the time of reaction in minutes.

Purification Procedure

The purification procedure as presented below has been described by Hatefi *et al.*[2] All steps of the purification are carried out at 0–2°.

Source Material. DPNH–cytochrome *c* reductase as described[4] is prepared from beef heart mitochondria and is stored at —20° as a solution in the Tris–sucrose–histidine buffer (see Assay Procedure).

Cholate-Ammonium Sulfate Fractionation. A suspension of DPNH-

[3] Coenzyme Q_1 (100 micromoles) is dissolved in 1.5 ml of absolute ethanol and the volume is made up to 100 ml with water.
[4] This volume [41].

cytochrome *c* reductase in the Tris–sucrose–histidine buffer is adjusted to a protein concentration (biuret) of 10 mg/ml. A solution containing 20% (w/v) of potassium cholate is added to a final concentration of 0.4 mg potassium cholate per milligram of protein. Then cold, saturated ammonium sulfate (neutralized) is added to a concentration of 0.39 saturation (0.65 ml per milligram of protein suspension). After it has stood in ice for 15 minutes, the turbid suspension is centrifuged for 15 minutes at 25,000 rpm in the No. 40 rotor of the Beckman Model L centrifuge. The supernatant fluid[5] is poured off and the residue is dissolved in the Tris–sucrose–histidine buffer. The protein concentration is adjusted to 10 mg/ml with the suspending medium and then saturated ammonium sulfate is added to a concentration of 0.36 saturation (0.56 ml per milliliter of protein solution). As before, the suspension is allowed to stand for 15 minutes at 0° and then it is centrifuged for 15 minutes at 30,000 rpm. The supernatant layer is discarded; the residue, which consists of DPNH–coenzyme Q reductase, is dissolved in the Tris–sucrose–histidine buffer. The solution of the enzyme is greenish yellow in color. It may be stored at —20° with little loss in activity.

For typical yield and activities of principal fractions see Table I.

TABLE I
YIELD, FLAVIN CONTENT, AND ACTIVITIES OF PRINCIPAL FRACTIONS OBTAINED DURING PURIFICATION OF DPNH-COENZYME REDUCTASE[a]

Fraction	Per cent protein	FMN content (mμmoles/ mg protein)	DPNH–Co Q reductase activity[b]
Mitochondria	100	0.14	2
S-1	12–18	0.24	7
DPNH–cytochrome *c* reductase	1.5–2.0	0.80	20
DPNH–coenzyme Q reductase	0.8–1.1	1.20	25

[a] From A. J. Merola, R. Coleman, and R. Hansen, *Biochem. Biophys. Acta* **73,** 638 (1963).

[b] Activity is expressed as micromoles DPNH oxidized-min^{-1}-(mg protein)$^{-1}$ at infinite coenzyme Q_1 concentration.

Properties

Composition. The composition of DPNH–coenzyme Q reductase is listed in Table II.

[5] (Reduced coenzyme Q)–cytochrome *c* reductase (complex III) may be isolated from this supernatant by a method described by Y. Hatefi, A. G. Haavik, and D. E. Griffiths [*J. Biol. Chem.* **237,** 1681 (1962)]. Also see this volume [44].

TABLE II
COMPOSITION OF DPNH-COENZYME Q REDUCTASE (COMPLEX I)[a]

Component	Concentration
Acid-extractable flavin	1.2–1.5 μmoles/g protein
FMN	1.2–1.5 μmoles/g protein
Coenzyme Q	4.2–4.5 μmoles/g protein
Nonheme iron	26 μmoles/g protein
Cytochromes $b + c_1$	0.1 μmoles/g protein
Lipid	0.22 mg/mg protein

[a] From Y. Hatefi, A. G. Haavik, and D. E. Griffiths, *J. Biol. Chem.* **237,** 1676 (1962).

Stability. When dispersed in the Tris–sucrose–histidine buffer (see Assay Method) the enzyme can be stored at —20° for a number of days with very little loss of activity. Treatment of the enzyme with organic solvents, high concentrations of detergents, and snake venoms (boiled *Naja naja, Crotalus*) destroys its activity.

Specificity. As with its parent particle, DPNH–cytochrome *c* reductase, DPNH–coenzyme Q reductase is absolutely specific for DPNH as the electron-donor substrate. Of the electron-acceptor substrates only the lower homologs of coenzyme Q, specifically coenzyme Q_1 and coenzyme Q_2, oxidize DPNH to any extent through an Amytal-sensitive pathway. In this respect, coenzyme Q_2 is only about one-fourth as efficient as coenzyme Q_1. Under the conditions of the assay, the oxidation of DPNH by coenzyme Q_{10} is negligible. It must be borne in mind that these specificities are applicable only to the specified conditions of the assay. Although coenzyme Q_{10} is the naturally occurring homolog of coenzyme Q in mammalian mitochondria, its insolubility in water limits its use in assay procedures that require substrate quantities of coenzyme Q. Although other quinones (i.e., 2,3-dimethoxy-5,6-dimenthylbenzoquinone; 2,3-dimethoxy-6-methylbenzoquinone; and menadione) are able to serve as oxidants of DPNH, the oxidation of DPNH by these compounds is not inhibited by Amytal. The enzyme also catalyzes the oxidation of DPNH by potassium ferricyanide although this reaction is not inhibited by Amytal.

Contamination. The usual preparations of the enzyme are practically devoid of succinate–coenzyme Q reductase, DPN–TPNH transhydrogenase, lipoic dehydrogenase, succinate–cytochrome *c* reductase, and cytochrome *c* oxidase activities. A residual DPNH–cytochrome *c* reductase activity is present because of a small contamination of the enzyme with cytochromes *b* and c_1.

Spectral Properties. The difference spectrum (reduced minus oxidized) of the enzyme displays a broad bleached region between 350 mμ and 600 mμ. Superimposed on this spectrum is an absorption peak at 430 mμ caused by the Soret bands of contaminating cytochromes. Only about 50% of the loss of absorption in the bleached region can be attributed to the reduction of the DPNH–dehydrogenase flavoprotein.

Electron paramagnetic resonance (EPR) spectra of the reduced enzyme taken at liquid nitrogen temperatures display a pronounced signal with microwave absorptions at $g_{\perp} = 1.94$ and $g_{\parallel} = 2.0$, a signal also a characteristic of DPNH-dehydrogenase.[6] DPNH–coenzyme Q reductase at intermediate states of oxidation also displays a complex EPR spectrum with multiple microwave absorptions between $g = 1.94$ and $g = 2.0$.[2]

Inhibitors. Similar to its parent particle, DPNH–cytochrome *c* reductase, DPNH–coenzyme Q reductase is inhibited by Amytal (>90% at 3 m*M*), *p*-chloromercuriphenyl sulfonate (>90% at 6 μ*M*), dicoumarol, and irreversibly by rotenone at concentrations approaching the concentration of flavin in the enzyme.

[6] H. Beinert, W. Heinen, and G. Palmer, *in* "Enzyme Models and Enzyme Structure," *Brookhaven Symp. Biol.* **15**, 229 (1962).

[44] Preparation and Properties of Reduced Coenzyme Q–Cytochrome c Reductase (Complex III of the Respiratory Chain)

By J. S. RIESKE[1]

$$(CoQ)H_2 + 2 \text{ ferricytochrome } c \rightarrow CoQ + 2 \text{ ferrocytochrome } c + 2H^+$$

Assay Method[1a]

Principle. The rate of reduction of cytochrome *c* by reduced coenzyme Q_2 is estimated from the amount of cytochrome *c* that is reduced (absorbancy change at 550 mμ) in a sample of the assay mixture that has been allowed to react for 10 seconds. The enzyme-catalyzed reaction is stopped by the addition of an appropriate stop reagent.

Reagents

Cytochrome *c* solution. A 1.5–2.0% solution of cytochrome *c* (Sigma type III) in 0.01 *M* phosphate, pH 7.4 is dialyzed against 0.01 *M*

[1] This manuscript was prepared while the author was at the Institute for Enzyme Research, The University of Wisconsin, Madison, Wisconsin.

[1a] Adapted from the procedure of D. E. Green and R. K. Burkhard, *Arch. Biochem. Biophys.* **92**, 312 (1961).

phosphate, pH 7.4 for 24 hours at 4°. After dialysis the cytochrome *c* is diluted to 10 mg/ml. This solution can be stored indefinitely at —20°.

Reduced coenzyme Q_2[2] [$(CoQ_2)H_2$]. Coenzyme Q_2 (10–50 mg in 5 ml ethanol) is added to 15 ml of a solution that is 0.1 *M* in phosphate, pH 7.4, and 0.25 *M* in sucrose. Two milliliters of cyclohexane and a pinch of solid sodium dithionite (sodium hydrosulfite) are added, after which the mixture is shaken vigorously in a stoppered graduated cylinder until the mixture is colorless. More than one addition of dithionite may be necessary. After the cyclohexane layer has formed completely on the surface of the aqueous phase, the cyclohexane layer is removed with a pipette and placed in a stoppered tube that is fitted for evacuation. The aqueous solution is extracted with 2-ml portions of cyclohexane two more times. The combined cyclohexane extracts are then evaporated *in vacuo* until only a light-yellow sirup remains in the bottom of the tube. This sirup is dissolved in absolute ethanol to a concentration of approximately 10 mg/ml. To retard autoxidation of the $(CoQ_2)H_2$ the alcoholic solution is acidified slightly with dilute HCl. This solution can be stored for several months at —20° with little autoxidation.

Bovine Plasma Albumin. A 20% solution (w/v in water) is dialyzed against distilled water for 48 hours at 4°; the solution then is diluted to final protein concentration of 10%. This solution can be stored indefinitely at —20°.

Assay Mixture. For routine assays the following mixture is convenient: 10 volumes of 0.1 *M* phosphate, pH 7.4; 5 volumes of 0.5 m*M* EDTA; 1 volume of 0.5 *M* NaN_3, pH 7.4; 0.22 volume of 10% bovine plasma albumin. This mixture can be prepared in liter quantities and stored at —20° until used. It can be stored a week or more at 0–4°.

Stop Reagent: 9 volumes of 2% Triton X-100[3]; 5 volumes of ethanol; 0.5 volume of 2.0 *M* Tris-HCl, pH 8; 0.5 volume of 0.1 *M* EDTA.

Enzyme. Immediately prior to assay a small sample (0.1–0.3 mg of protein) of the purified enzyme (well dispersed by deoxycholate in a Tris–sucrose–histidine buffer that is 0.05 *M* in Tris-HCl, pH 8, 0.67 *M* in sucrose, and 0.001 *M* in histidine) is diluted to

[2] 2,3-Dimethoxy-5-methyl-6-geranylbenzoquinol.

[3] Nonionic detergent manufactured by Rohm and Haas Co., Philadelphia, Pennsylvania.

3–5 ml with 0.05% potassium deoxycholate. For mitochondria or other nondispersed preparations it was found best to suspend the aliquot of enzyme in 0.25 *M* sucrose, add 10% potassium deoxycholate to a concentration of 1.0 mg/ml, and then clarify the turbid suspension by the addition of 1.0 *M* phosphate, pH 7.5 buffer or 1.0 *M* KCl to a concentration of 0.05–0.1 *M*.

Procedure. About 0.005 ml of the alcoholic solution of $(CoQ_2)H_2$ is added with shaking to a volume of the assay mixture which after the addition of the enzyme will give a final volume of 0.70 ml. After temperature equilibration in a thermostatted water bath at the temperature desired for the assay (normally 38°), 0.01–0.03 ml of the diluted enzyme is added. The reaction is started by the rapid addition with mixing of 0.3 ml of the ferricytochrome *c* solution. Ten seconds after the addition of the cytochrome *c*, 2.0 ml of the stop reagent is added with a syringe. The solution is transferred immediately to a 3-ml cuvette of 1.0-cm light path, and the absorbancy at 550 mμ is recorded. An identical sample of the assay system but with no enzyme present is used as a blank in the reference cuvette. Because some nonenzymatic reduction of cytochrome *c* in the blank occurs even after addition of the stop reagent, it is advisable to standardize a properly diluted solution of cytochrome *c* against a freshly prepared blank. This solution of cytochrome *c* is then used as a blank for zeroing the spectrophotometer in subsequent assays.

The amount of enzyme to be used in the assay should be that which will catalyze the reduction of between 0.01 and 0.05 micromole of cytochrome *c* in 10 seconds. In this assay, the proportionality between the amount of enzyme added and the amount of cytochrome *c* reduced was found to hold true only for amounts of cytochrome *c* reduced below 0.05 micromole. In the assay this would correspond to absorbancies below 0.30 after addition of the stop reagent.[3a]

Definition of Specific Activity. The specific activity of the enzyme is expressed as micromoles of cytochrome *c* reduced-min^{-1}-(mg enzyme $protein)^{-1}$. The specific activity is calculated from

$$\text{S.A.} = \frac{\Delta \text{ absorbancy (550 m}\mu)}{\text{mg of enzyme}} \times 0.973$$

The above equation is based on a 10-second reaction time and an

[3a] Since submission of this manuscript, it has been observed that the linearity of cytochrome *c* reduction with time can be extended to absorbancy values above 0.60 when the temperature of assay is lowered to 0°. Under these conditions the amount of enzyme used is increased 8-fold and the time of reaction is increased to 1 minute. In calculation of specific activities the factor "0.973" should be replaced by the number "0.162."

absorbancy index (reduced minus oxidized) of 18.5 mM^{-1} cm^{-1} for the 550 mμ band of cytochrome *c*.

Purification Procedure

Three procedures for the preparation of (reduced coenzyme Q) cytochrome *c* reductase have been described.[4-6] All three procedures utilize the S-1 fraction of the procedure described for the preparation of DPNH-cytochrome *c* reductase (R_4B as described in this volume [41]). The procedure described below is the method of choice because of its overall superiority with respect to purity, yield, and activity. This procedure utilizes the final supernatant solution obtained in the preparation of DPNH–cytochrome *c* reductase.

All operations are carried out at temperatures between 0 and 5°.

Step 1. Precipitation and Collection of Crude Enzyme from the Supernatant Solution of R_4B. Crystalline ammonium acetate (450 g) is dissolved in each liter of the supernatant solution as decanted from the R_4B residue. The turbid solution is allowed to stand for 10 minutes, then it is centrifuged for 40 minutes at 30,000 rpm in the No. 30 rotor of the Beckman Model L centrifuge. The supernatant solution is discarded; the packed residue is "dissolved" in a small volume of the Tris–sucrose–histidine buffer (see Assay Method). This solution of crude complex III can be stored at —20° until needed.

Step 2. Cholate-Ammonium Sulfate Fractionation. After being thawed, the crude enzyme is diluted to a protein concentration of 10 mg/ml with Tris–sucrose–histidine buffer. A 20% (w/v) solution of potassium cholate is added to a final concentration of 0.25 mg per milligram of protein. Neutral, saturated ammonium sulfate (0°) is added to 0.35 saturation (54 ml per 100 ml of solution). After 10 minutes the solution is centrifuged for 15 minutes at 30,000 rpm in the No. 30 rotor. The yellow-brown residue is discarded. If the residue is floating it is removed from the infranatant liquid by filtration of the solution through a pad of glass wool. The supernatant liquid is treated with saturated ammonium sulfate to 0.40 saturation (8.35 ml per 100 ml of solution) and is centrifuged as before. The floating oily material is removed by filtration through glass wool and discarded. Occasionally, most of the red-colored enzyme is precipitated at this point. If this occurs, the residue is dissolved in Tris–sucrose–histidine buffer and

[4] Y. Hatefi, A. G. Haavik, and D. E. Griffiths, *J. Biol. Chem.* **237**, 1681 (1962).
[5] J. S. Rieske, R. E. Hansen, and W. S. Zaugg, *J. Biol. Chem.* **239**, 3017 (1964).
[6] J. S. Rieske, W. S. Zaugg, and R. E. Hansen, *J. Biol. Chem.* **239**, 3023 (1964).

added back to the supernatant fluid. Additional potassium cholate is added (one-half the amount added to the crude enzyme) and the fractionation is continued. The clear red solution is brought to 0.42 saturation with saturated ammonium sulfate (3.45 ml per 100 ml of solution) and is centrifuged as before. This time the floating, oily material is brown-red; this material is removed from the infranatant solution and discarded. Finally, the desired protein is precipitated by the addition of saturated ammonium sulfate to 0.48 saturation (11.5 ml per 100 ml of solution) and is recovered as a solid, red pellet after centrifugation. Occasionally, the enzyme is found as a floating layer of protein; in this case it is recovered by filtration of the infranatant liquid through glass wool. For normal use the enzyme is dissolved in the Tris–sucrose–histidine and stored at —20°.

For typical yield and activities of principal fractions, see Table I.

TABLE I
YIELD AND ACTIVITIES OF PRINCIPAL FRACTIONS OBTAINED DURING PURIFICATION OF (REDUCED COENZYME Q)-CYTOCHROME *c* REDUCTASE (COMPLEX III)

Fraction	Per cent of protein recovered	Cyt. c_1 content (μmoles/g protein)	CoQ_2H_2-Cyt *c* reductase activity[a]
Mitochondria	100	0.2 ± 0.02	20–30
S-1	12–18	0.9 ± 0.05	70–100
Purified complex III	1.0–2.0	3.4 ± 0.2	300–600

[a] Expressed as micromoles of cytochrome *c* reduced-min^{-1}-(mg protein)$^{-1}$.

Properties

Composition. The composition of (reduced coenzyme Q)–cytochrome *c* reductase is summarized in Table II.

Stability. (Reduced coenzyme Q)–cytochrome *c* reductase can be stored as a solution in the Tris–sucrose–histidine buffer for weeks at —20° with little loss in activity. Exposure to pH 10 and heavy metals causes little loss of enzymatic activity. However, the enzyme is inactivated rapidly by concentrated organic solvents, high concentrations of detergents, and adsorption on surfaces (e.g., withdrawal of the enzyme into capillary pipette or passage of the enzyme through a chromatographic column).

Specificity. The reduced forms of coenzyme Q_1, coenzyme Q_2, and coenzyme Q_{10} all serve as electron-donor substrates with complex III.

TABLE II
COMPOSITION OF (REDUCED COENZYME Q)–CYTOCHROME *c* REDUCTASE

Component	Concentration[a] (per gram protein)
Cytochrome c_1	3.4 ± 0.2 μmoles
Cytochrome *b*	6.8 ± 0.4 μmoles
Nonheme iron	6.2 ± 0.5 μmoles
Total flavin	0.15–0.50 μmoles
Coenzyme Q[b]	1.0–4.0 μmoles
Phospholipid[b]	0.2–0.4 g

[a] These values are based on protein content as estimated by the biuret procedure of A. G. Gornall, C. J. Bardawill, and M. M. David [*J. Biol. Chem.* **177,** 751 (1949)] uncorrected for interference by heme. Slightly higher values may be obtained after correction for spectral interference by heme.

[b] The contents of flavin and lipid vary considerably according to the procedure used for the preparation of the enzyme. The lower values for the content of flavin coenzyme Q and phospholipid are near the values usually obtained for the preparation described in this article.

However, reduced coenzyme Q_{10} requires a supplementation of phospholipid in the assay medium in order to function with an efficiency even of 10% that of reduced coenzyme Q_2. Reduced coenzyme Q_1 is about 30% as efficient as reduced coenzyme Q_2.

Contamination. Preparations of the enzyme are essentially devoid of succinate–cytochrome *c* reductase, DPNH–cytochrome *c* reductase, and cytochrome *c* oxidase activities.

Spectral Properties. The enzyme displays a spectrum characteristic of both cytochromes *b* and c_1 with α peaks at 562 mμ and 554 mμ and Soret peaks at about 429 mμ and 418 mμ, respectively. Under aerobic conditions the cytochrome c_1 can be distinguished from cytochrome *b* by its selective reduction with ascorbate.

Electron paramagnetic resonance (EPR) spectra taken at −170° on the reduced enzyme shows a characteristic microwave-absorption maximum at $g_\perp = 1.90$ with less prominent absorptions at $g_\parallel = 2.0$ and at about $g = 1.84$[5] (see this volume [94]).

Activators and Inhibitors. (Reduced coenzyme Q)–cytochrome *c* reductase is inhibited almost completely by antimycin A at concentrations approaching the concentration of cytochrome c_1 in the solution of enzyme (1.5 μg per milligram of protein of complex III). For this stoichiometric inhibition, the enzyme must be at a relatively high concentration (>5 mg protein per milliliter) when exposed to the antimycin A. The enzyme is inhibited also by 2-*n*-alkyl-4-hydroxyquinoline-*N*-oxides (82% at 2 μg/ml for the nonyl derivative)[3] and 2-alkyl-3-hydroxy-1,4-naphthoquinones (100% at 30 μ*M* for SN 5949).[4]

The activity of the enzyme is stimulated by EDTA (0.1 m*M*) by a factor of 2–4 depending on the assay conditions.

Effect of pH. In the assay system described above, (reduced coenzyme Q)–cytochrome *c* reductase is active in the pH range from 6.0 to 8.5 with an optimal activity at pH 7.4.

[45] Cytochrome Oxidase from Beef Heart Mitochondria

By David C. Wharton and Alexander Tzagoloff

$$\text{Ferrocytochrome } c + 2\ H^+ + 1/2\ O_2 \rightarrow \text{ferricytochrome } c + H_2O$$

The hemoprotein nature of the terminal oxidase of cellular respiration was first recognized by Warburg[1] and by Keilin and Hartree.[2] Keilin and Hartree[2] named this enzyme cytochrome oxidase and identified it with cytochrome a_3. The isolation of cytochrome a_3 in an enzymatically active form has not been demonstrated thus far. A number of investigators, however, have succeeded in solubilizing and purifying the enzyme system which catalyzes the oxidation of ferrocytochrome *c* by oxygen.[3-9] The oxidation-reduction components of isolated cytochrome *c* oxidase are cytochrome *a*, cytochrome a_3, and copper.

Assay Method

Principle. Cytochrome *c* oxidase is most conveniently assayed by the spectrophotometric method of Smith.[10] The rate of oxidation of ferrocytochrome *c* is measured by following the decrease in the absorbancy of its α-band at 550 mμ.

Reagents

Potassium phosphate buffer, 0.1 *M*, pH 7.0.

[1] O. Warburg, *Biochem. Z.* **177,** 471 (1926).
[2] D. Keilin, and E. F. Hartree, *Proc. Roy. Soc.* **B125,** 171 (1938).
[3] K. Okunuki, I. Sekuzu, T. Yonetani, and S. Takemori, *J. Biochem.* (*Tokyo*), **45,** 847 (1958).
[4] L. Smith, and E. Stotz, *J. Biol. Chem.* **209,** 819 (1954).
[5] B. Eichel, W. W. Wainio, P. Person, and S. J. Cooperstein, *J. Biol. Chem.* **183,** 89 (1950).
[6] T. Yonetani, *J. Biol. Chem.* **235,** 845 (1960).
[7] S. Horie and M. Morrison, *J. Biol. Chem.* **238,** 1855 (1963).
[8] D. E. Griffiths and D. C. Wharton, *J. Biol. Chem.* **236,** 1850 (1961).
[9] L. R. Fowler, S. H. Richardson, and Y. Hatefi, *Biochim. Biophys. Acta* **96,** 103 (1962).
[10] L. Smith, *in* "Methods of Biochemical Analysis," (D. Glick, ed.), Vol. II, p. 427. Wiley (Interscience) New York, 1955.

Ferrocytochrome *c*, 1%. Cytochrome *c* (Sigma Chemical Co., type III) is dissolved in 0.01 *M* potassium phosphate buffer, pH 7.0. The solution is reduced with a few milligrams of potassium ascorbate. Excess ascorbate is removed by dialysis in size 8 Visking dialysis tubing against 0.01 *M* phosphate buffer, pH 7.0, for 18–24 hours with three changes of buffer. This cytochrome *c* remains 96–99% reduced up to several months or more.

Potassium ferricyanide, 0.1 *M*

Enzyme. Immediately before the assay, cytochrome *c* oxidase is diluted to a protein concentration of 0.3 mg/ml in buffered 0.25 *M* sucrose (0–4°)

Procedure. To each of two 1-ml cuvettes with a 10-mm light path add the following: 0.1 ml potassium phosphate buffer, 0.07 ml ferrocytochrome *c*, and 0.83 ml water. The blank cuvette is oxidized with 0.01 ml potassium ferricyanide. After temperature equilibration at 38°, the reaction is initiated by the addition of 10 μl of enzyme. The decrease in absorbancy is measured at 550 mμ every 15 seconds.

Definition of Activity. The activity of cytochrome *c* oxidase may be defined in terms of the first-order velocity constant.

$$k = 2.3 \log \frac{A\ (\text{time}_0)}{A\ (\text{time}_{0+1\ \text{min}})}\ \text{min}^{-1}$$

The specific activity is calculated from the known concentration of cytochrome *c* and enzyme in the assay mixture and the estimated first-order velocity constant.

$$\text{S.A.} = \frac{k\ (\text{concentration cytochrome } c)}{(\text{concentration enzyme})}$$

Alternative Methods of Assay. Cytochrome *c* oxidase can also be assayed by measuring oxygen uptake either manometrically or polarographically.[11]

Purification Procedure

Two procedures for purifying cytochrome *c* oxidase from beef heart mitochondria are described below. The enzyme obtained by procedure I is slightly less pure based on its activity and spectrum. This method, however, has the advantage of being less involved and is more easily adapted to small-scale preparations. All operations are carried out at 0–4° unless otherwise specified. Protein estimations are made by the biuret method of Gornall *et al.*[12]

[11] D. C. Wharton and D. E. Griffiths, *Arch. Biochem. Biophys.* **96**, 103 (1962).

[12] A. G. Gornall, C. J. Bardawill, and M. M. David, *J. Biol. Chem.* **177**, 751 (1949).

Procedure I[9,13]

Step 1. Washing of Mitochondria. Beef heart mitochondria prepared by the method of Crane *et al.*[14] or of Green and Ziegler[15] are suspended in 0.25 *M* sucrose at a protein concentration of 20 mg/ml and are then centrifuged at 78,000 *g* for 10 minutes. The supernatant is discarded and the mitochondrial pellet is suspended in a solution 0.66 *M* in sucrose, 0.05 *M* in Tris-chloride, pH 8.0, and 0.001 *M* in histidine (TSH) with the protein concentration adjusted to 23 mg/ml.

Step 2. Preparation of Crude Cytochrome Oxidase. To the mitochondrial suspension are added 10% (w/v) potassium deoxycholate, pH 7.5 (0.3 mg/mg protein) and solid KCl (7.2 g/100 ml). The clarified suspension is stirred for 10 minutes and centrifuged at 78,000 g for 20 minutes. The clear red supernatant is decanted and the green pellet is suspended in TSH, homogenized in a Potter-Elvehjem homogenizer (glass-Teflon), and recentrifuged at 78,000 *g* for 20 minutes. The washed pellet is resuspended in TSH with the protein concentration adjusted to 19 mg/ml. The crude cytochrome oxidase suspension may be stored frozen at —20°.

Step 3. Solubilization of Cytochrome Oxidase. Cytochrome oxidase is extracted from the crude material by adding 10% (w/v) potassium deoxycholate (0.5 mg/mg protein) and solid KCl (7.45 g/100 ml). The mixture is stirred for 10 minutes and then centrifuged at 78,000 *g* for 10 minutes. The clear green supernatant containing most of the cytochrome oxidase is dialyzed in size 20 Visking cellulose casing for 90 minutes against 30 volumes of 0.01 *M* Tris-HCl, pH 8.0.

Step 4. Fractionation with Ammonium Sulfate. Saturated ammonium sulfate, neutralized with NH_4OH (4°) is added to the turbid dialyzate to 16% of saturation. The suspension is centrifuged at 78,000 *g* for 10 minutes, and the resulting pellet is discarded. A second precipitate, containing cytochrome oxidase, is obtained by adding ammonium sulfate to 20% of saturation. This fraction is dissolved in 0.05 *M* Tris-HCl, pH 8.0, at a protein concentration of 20–30 mg/ml.

Step 5. Refractionation of Cytochrome Oxidase. A further purification of cytochrome oxidase is achieved by refractionation of the step 4 enzyme in the presence of potassium cholate. Saturated ammonium sulfate and 20% (w/v) potassium cholate, pH 7.5, are added to a final concentration of 27% and 3%, respectively. The mixture is incubated at

[13] A. Tzagoloff and D. H. MacLennan, *Biochim. Biophys. Acta* **99,** 476 (1965).
[14] F. L. Crane, J. L. Glenn, and D. E. Green, *Biochim. Biophys. Acta* **22,** 475 (1956).
[15] D. E. Green and D. M. Ziegler, see Vol. VI [58].

4° for 4 hours. During this incubation a brown colored precipitate containing residual cytochromes *b*, *c*, and c_1 as well as nonheme iron is formed. This material is discarded after centrifugation of the solution at 78,000 *g* for 10 minutes. The ammonium sulfate concentration of the supernatant is adjusted to 39% of saturation. The green oily precipitate obtained after centrifugation at 78,000 *g* for 10 minutes is dissolved in TSH.

Procedure II[8]

Step 1. Extraction of Mitochondria with tert-Amyl Alcohol. A suspension of beef heart mitochondria[14 or 15] in 0.25 *M* sucrose, having a protein concentration of 60–70 mg/ml is diluted with one-third volume of cold 0.9% (w/v) KCl. This mixture is warmed to 15° in a water bath, and *tert*-amyl alcohol is added at room temperature with continuous stirring to a final concentration of 10%. The temperature is maintained at 20° for 5 minutes and then cooled rapidly to 0–5° by immersing the container in a dry ice–acetone bath. During this operation the mixture must be stirred vigorously in order to avoid its freezing. The cooled mixture is centrifuged for 15 minutes at 78,000 *g*. The clear red supernatant and the fluffy red middle layer (containing the succinic dehydrogenase complex) are decanted, and the greenish brown residue is collected in the original volume of 0.9% KCl containing 1 m*M* potassium succinate. The suspended residue is homogenized with a Potter-Elvehjem homogenizer (glass-Teflon) and centrifuged for 5 minutes at 78,000 *g*. The resulting washed residue is suspended in 0.25 *M* sucrose containing 1 m*M* succinate, homogenized with a Potter-Elvehjem homogenizer, and the protein concentration is adjusted to 40–45 mg/ml by adding an appropriate volume of the 0.25 *M* sucrose solution.

Step 2. Extraction of Cytochrome Oxidase. Cytochrome oxidase is extracted from the greenish brown residue by adding 20% (w/v) potassium cholate, pH 7.2, to a concentration of 2 mg/mg protein and solid KCl to a concentration of 3 *M*. This mixture is stirred for 30 minutes and centrifuged for 5 minutes at 78,000 *g*. The turbid greenish supernatant is decanted, allowed to stand overnight in the refrigerator, and centrifuged for 10 minutes at 78,000 *g*. The greenish red supernatant is decanted and dialyzed in size 20 Visking cellulose casing for 3 hours against 0.01 *M* phosphate buffer, pH 7.4. The turbid dialyzate is centrifuged for 10 minutes at 78,000 *g*; the green residue is suspended in 0.25 *M* sucrose and homogenized with a Potter-Elvehjem homogenizer. After the protein concentration is adjusted to 10 mg/ml with 0.25 *M* sucrose, the cytochrome oxidase is solubilized by adding 10% (w/v) potassium deoxycholate, pH 7.6, to a concentration of 1 mg per milligram protein.

Step 3. Fractionation with Ammonium Sulfate. Cytochrome oxidase is purified by fractionation with ammonium sulfate. For the first step saturated ammonium sulfate (neutralized with NH_4OH at 4°) is added to give a concentration of 7% of saturation; the mixture is allowed to stand for 10 minutes and centrifuged for 10 minutes at 105,000 *g*. The clear greenish red supernatant is decanted and further purified by repeating the fractionation at ammonium sulfate concentrations of 15%, 24%, and 27% of saturation. The bulk of the cytochrome oxidase precipitates as a greenish red oil between 24 and 27% of saturation and is solubilized in 0.25 *M* sucrose, producing a clear reddish green solution.

Any contaminating cytochromes, such as cytochromes *b* and c_1, which may be present occasionally in the solution of cytochrome oxidase, may be removed by repeating the fractionation with ammonium sulfate.

The purification of cytochrome oxidase by this procedure is summarized in Table I.

TABLE I
YIELD AND ACTIVITIES OF CYTOCHROME OXIDASE PURIFIED FROM BEEF HEART MITOCHONDRIA (PROCEDURE II)

Fraction	Total heme *a* (μmoles)	Concentration heme *a* (mμmoles/mg protein)	*SA*[a]
1. Mitochondria	19.8	1.4	35.0
2. Washed amyl alcohol residue	18.0	1.8	12.1
3. Cholate-KCl supernatant	12.5	3.3	46.0
4. Cholate-KCl supernatant after standing overnight	10.5	3.4	35.0
5. Green residue	7.4	4.8	45.0
6. Ammonium sulfate fraction 24–27%	5.2	8.1	185.0
7. Refractionated ammonium sulfate fraction 24–27%	4.5	9.2	205.0

[a] Micromoles cytochrome *c* oxidized per minute per milligram of protein at infinite concentration of cytochrome *c* at 38°.

Properties

Storage and Stability. Preparations of cytochrome oxidase are stored best in 0.25 *M* sucrose (pH 7.0–7.5) at −15° at a protein concentration of 1 mg/ml. Preparations stored in this manner have retained maximal activity for 5–7 days. We have observed that cytochrome oxidase stored in phosphate buffer becomes turbid after a few days and that the enzyme appears to aggregate when stored for a time in Tris buffer.

Composition and Spectral Properties. The composition of cytochrome oxidase prepared by procedure II is presented in Table II. The heme to copper ratio is 1.1–1.2. The absorption spectrum of the oxidized enzyme has bands with maxima at 830 mμ, 599 mμ, and 423 mμ. The absorption spectrum of the reduced enzyme (dithionite) has bands with maxima at 605 mμ, 517 mμ, and 444 mμ.

TABLE II
COMPOSITION OF CYTOCHROME OXIDASE (PROCEDURE II)

Component	Content (conc. per mg protein)
Heme *a*	8.1–9.2 mμmoles
Iron	8.2–9.4 mμmoles
Copper	9.2–10.6 mμmoles
Lipid	0.20–0.28 mg
Deoxycholate	0.8–1.4 mg

Molecular Weight. Purified cytochrome oxidase exists as an aggregate of about 290,000 weight average molecular weight or about 230,000 weight average molecular weight when corrected for lipid. The aggregate can be dissociated to a weight of about 100,000 in the presence of 4–6 *M* urea.[16]

Reaction with Cytochrome c. Ferrocytochrome *c* donates electrons directly to cytochrome oxidase. The reaction as measured spectrophotometrically obeys first-order reaction kinetics. However, the first-order rate constant decreases as the total concentration of cytochrome *c* increases. The K_m value for cytochrome *c* is usually 1×10^{-5} *M* but occasionally with some preparations of cytochrome oxidase, has reached 8.5×10^{-6} *M*.[11]

Inhibitors. Cytochrome oxidase is inhibited by cyanide, azide, hydroxylamine, and sodium sulfide. Carbon monoxide also acts as an inhibitor in a reaction which is reversed by light. Most metal-chelating agents cause little or no inhibition.

[16] A. Tzagoloff, P. C. Chang, D. C. Wharton, and J. S. Rieske, *Biochim. Biophys. Acta* **96,** 1 (1965).

[46] Electron Transport Particle of Yeast

By BRUCE MACKLER

$$DPNH + \frac{1}{2}O_2 + H^+ \rightleftarrows DPN^+ + H_2O$$
$$\text{Succinate} + \frac{1}{2}O_2 \rightleftarrows \text{fumarate} + H_2O$$

Assay Method

Principle. DPNH oxidation by molecular oxygen is followed spectrophotometrically at 38° by measuring the rate of decrease in absorbancy at 340 mμ. The oxygen content of the solutions is sufficient to allow maximum rates of oxidation under the assay conditions. Succinoxidase activity is determined by use of an oxygen polarograph.[1]

Reagents

DPNH solution, 0.1%
Potassium phosphate buffer, 0.2 *M*, pH 7.5
Cytochrome *c* solution, 1%
Sodium succinate solution, 1 *M*
Enzyme, diluted in 8.5% sucrose solution as necessary

Procedure. The rate of oxidation of DPNH by oxygen at 38° is measured at 340 mμ spectrophotometrically or by means of an oxygen polarograph.[1] The complete system contains 0.1 ml of DPNH, 0.2 ml of 0.2 *M* phosphate buffer of pH 7.5, 0.02 ml of 1% cytochrome *c* solution, and water to a final volume of 1 ml. When the rate of DPNH oxidation is followed spectrophotometrically, the assay mixture is placed in 1 ml optical cuvettes of 10-mm light path. Reactions are started by addition of an appropriate amount of enzyme protein (usually about 4 μg). The oxidation of succinate by oxygen was measured by means of the oxygen polarograph as described above, with the exceptions that 0.05 ml of a 1 *M* solution of succinate was substituted for DPNH and approximately 1 mg of enzyme was added to start the reaction. Specific activity is expressed as micromoles of substrate oxidized per milligram of enzyme protein per minute. Protein is determined by the method of Lowry *et al.*[2]

Purification Procedure

Step 1. Bakers' yeast (2 kg) is washed twice with water by centrifugation and then diluted to a final volume of 2 liters by addition of a

[1] B. Chance and G. R. Williams, *J. Biol. Chem.* **217**, 383 (1955).
[2] O. H. Lowry, N. J. Rosebrough, A. L. Farr, and R. J. Randall, *J. Biol. Chem.* **193**, 265 (1951).

solution containing 0.01 *M* K_2HPO_4, 0.001 *M* EDTA (the disodium salt of ethylenediaminetetraacetic acid), and 8.5% sucrose. The suspension of yeast is added to approximately 4 liters of moist glass beads in a 6-liter stainless steel beaker immersed in an ice bath. All subsequent steps in the purification of the enzyme are performed at 0–5°. The suspension of yeast and glass beads[3] is homogenized for 30 minutes at top speed with a large overhead blender[4] equipped with a circular glass blade and then centrifuged at 1370 *g* in a refrigerated International centrifuge; the packed residue is discarded.

Step 2. The turbid supernatant fluid from step 1 is centrifuged at 3100 *g* for 30 minutes, and the residue is discarded. The supernatant fluid is then centrifuged at 30,000 rpm for 30 minutes in the No. 30 rotor of a Spinco ultracentrifuge, and the clear supernatant fraction is discarded. The residue is suspended in 250 ml of 8.5% sucrose solution and homogenized at top speed for 10 minutes in a VirTis high speed homogenizer.

Step 3. The material obtained after homogenization in the VirTis blender is diluted with 8.5% sucrose solution to a protein concentration of 20 mg/ml, and the pH is adjusted to 10 with 6 *N* KOH. The suspension is centrifuged in the No. 30 rotor of the Spinco ultracentrifuge at 30,000 rpm for 30 minutes. After centrifugation, three layers are observed; a clear supernatant layer which is discarded; a reddish brown middle layer which is saved; and the packed residue, which is discarded. The middle layer is diluted with 8.5% sucrose solution to a protein concentration of 15 mg/ml, and the pH is adjusted to 8.5 with 0.1 *N* HCl.

Step 4. The suspension obtained from step 3 is centrifuged in the No. 40 rotor of the Spinco ultracentrifuge at 15,000 rpm for 20 minutes, and the turbid supernatant suspension is recentrifuged at 40,000 rpm for 90 minutes. The final residue, which represents the purified ETP, is suspended in sufficient 8.5% sucrose solution to give a protein concentration of 20 mg/ml. Purification of the ETP is shown in the table.

Properties[5]

The enzyme is specific for DPNH and does not utilize TPNH as substrate. Preparations of yeast ETP contain negligible amounts of

[3] Superbrite Standard Glass Beads, Highway mix 0.0215–0.0066, from Minnesota Mining & Mfg. Co. Beads are washed in aqua regia and rinsed thoroughly and then dried by suction filtration before use.

[4] The blender is fitted with an overhead ½-horsepower motor and rotates at approximately 18,000 rpm without load.

[5] B. Mackler, P. J. Collipp, H. M. Duncan, N. A. Rao, and F. M. Huennekens, *J. Biol. Chem.* **237**, 2968 (1962).

PURIFICATION OF YEAST ETP

Preparation	Protein (mg)	DPNH oxidase: Specific activity	DPNH oxidase: Units of activity[a]	Succinoxidase: Specific activity	Succinoxidase: Units of activity[a]
Step 1	80,000	0.3	24,000	0.05	4000
Step 2	19,000	0.8	15,200	0.10	1900
Step 3	2,500	2.4	6,000	0.50	1250
Step 4	1,500	3.3	4,950	0.70	1050

[a] Units of activity: specific activity X total protein.

malic, isocitric, lactic, alcohol, lipoic, and butyryl-CoA dehydrogenases, TPNH–cytochrome *c* reductase, adenosine triphosphatase, and glucose 6-phosphatase. Preparations of ETP at pH 8.5 do not lose activity when stored at —20° for periods of up to 2 weeks. The enzyme contains the following components (millimicromoles per milligram of protein): flavin, 0.73; coenzyme Q, 2.5; nonheme iron, 0.80; copper, 1.3; cytochrome *b*, 1.2; cytochromes $a + a_3$, 0.90; cytochromes $c_1 + c$, 0.85; and lipid, 21% (determined as percentage of dry weight of the preparation).

[47] Preparation and Properties of the Heart Mitochondrial Electron Transporting Particles (Inner Membrane)

By KRYSTYNA C. KOPACZYK

General Considerations

Mitochondria, regardless of their origin, are composed of an inner and an outer membrane. It has been reported[1,1a] that the two membranes of beef heart mitochondria (BHM) differ in their functional properties and, on this basis, can be separated from each other quite effectively. The inner membrane contains the electron transfer chain and systems concerned with energy transduction, but lacks many of the ancillary enzymes of the mitochondrion such as those catalyzing oxidations of the citric acid cycle. The latter appear to be associated with the outer membrane.[1,1a]

[1] E. Bachmann, D. W. Allmann, and D. E. Green, *Arch. Biochem. Biophys.* **115**, 153 (1966).

[1a] D. W. Allmann, E. Bachmann, and D. E. Green, *Arch. Biochem. Biophys.* **115**, 165 (1966).

Terminology. The inner membrane ETP (electron transporting particles), described here, are designated EP_1 and EP_2 (elementary particles), in order to differentiate them from standard "ETP" preparations. The term "ETP," as commonly used, refers to a mitochondrial preparation which generally has the following properties: (1) contains the electron transfer chain; (2) may or may not phosphorylate (depending upon the isolation methods employed); and (3) is not free of outer membrane components. The EP_1 and EP_2 contain all the components of the electron transfer chain, but essentially no citric acid cycle activities. As visualized by electron microscopy, they are single membranes without tripartite repeating units.

Principle. Treatment of BHM with cholate and a subsequent short centrifugation lead to the isolation of EP_1 which can be fractionated further to yield EP_2. The latter differs from EP_1 in that it is essentially free of structural protein (SP).

$$BHM \rightarrow EP_1 \rightarrow EP_2 + SP$$

During the preparation of EP_2 the cytochromes are kept reduced with dithionite, since in this state they appear to be more stable; this is particularly true of cytochrome *b*.

Procedure

Reagents

0.25 *M* sucrose–0.01 *M* Tris, pH 7.8 (sucrose A)
0.25 *M* sucrose–0.01 *M* Tris, pH 7.5 (sucrose B)
Potassium cholate (recrystallized from 70% ethanol), 20% solution, pH 7.7–8.0
Potassium deoxycholate (recrystallized from 95% ethanol), 10% solution, pH 7.7–8.0 (DOC)
Saturated ammonium sulfate, pH 6.5–7.0
Disodium ethylenediaminetetraacetate (EDTA), 0.5 *M*, pH 7.7
Solid $Na_2S_2O_4$

Step 1. Washing of the Mitochondria. Beef heart mitochondria, prepared by the method of Crane *et al.*,[2] are homogenized with 4 volumes of sucrose A. The pH of this homogenate is usually 7.3–7.5 and is adjusted to 7.8 with 6 *N* KOH. The homogenate is then centrifuged at 30,000 rpm for 20 minutes (rotor No. 30), in a Spinco Model L ultracentrifuge. The resultant pellet consists of two layers (the dark, bottom layer containing

[2] F. L. Crane, J. L. Glenn, and D. E. Green, *Biochim. Biophys. Acta* **22**, 475 (1956).

heavy mitochondria and the top layer, light mitochondria). The entire pellet is homogenized in sucrose A and is recentrifuged as above. The pellet is then resuspended in sucrose A to a protein concentration of 27–30 mg/ml.

Step 2. Isolation of EP_1. To the suspension of washed mitochondria cholate (1.0 mg per mg of protein) is added, at 0° with stirring, and the mitochondrial suspension is immediately centrifuged at 40,000 rpm for 5 minutes (rotor No. 40). This centrifugation separates the suspension into three phases: (1) a dark-brown pellet, mostly unbroken mitochondria; (2) loosely sedimented material; and (3) an essentially clear supernatant. The supernatant together with the loosely packed material is poured off, and the mixture is recentrifuged at the same speed for 15 minutes. A red, soft pellet (EP_1) is thus obtained which constitutes about 17% of the starting mitochondrial protein.

Step 3. Washing of EP_1. The pellet is suspended in 10 volumes of sucrose A and is centrifuged at 40,000 rpm for 15 minutes. If the preparation is to be used for electron microscopy, this step is repeated twice. For the preparation of EP_2 one wash suffices. The supernatant from the first wash is pale yellow and contains cholate, as detected by precipitation with 6 N HCl. The supernatant from the second wash is colorless and contains no HCl-detectable cholate. On a protein basis, the washed EP_1 amounts to about 15% of the starting material (washed mitochondria). After cholate is removed, the pellet becomes firm and opaque.

Step 4. Removal of Structural Protein. Structural protein is removed by a modification of the method of Richardson *et al.*[3] The protein concentration of EP_1 is adjusted to 20 mg/ml with sucrose B, and the following additions are made in the order listed: (1) DOC, 1.5 mg per milligram of protein; (2) cholate, 0.75 mg per milligram of protein; (3) EDTA to 0.001 M; (4) $Na_2S_2O_4$, 1.2 mg/ml (EP_1 + DOC + cholate + EDTA); and (5) ammonium sulfate to 15% saturation. This mixture is then transferred into a graduated cylinder and is left at 0°, in the dark, for 15 hours. During this period, structural protein separates and forms a loose sediment occupying about one-half of the volume of the system. The sediment is removed by centrifugation at 10,000 rpm for 5–10 minutes (rotor No. 40). Structural protein thus obtained is an essentially white residue. The brown-green supernatant, which gelatinizes readily, contains the electron transfer chain.

Comments on Step 4. This procedure is carried out in a beaker at 0° with continuous, but slow, stirring. Dithionite, which must be com-

[3] S. H. Richardson, H. O. Hultin, and S. Fleischer, *Arch. Biochem. Biophys.* **105**, **254** (**1964**).

pletely dissolved before ammonium sulfate is added, lowers the pH of the system to about 7.3. When dithionite is first added, it causes surface precipitation of the bile salts. The stirring rate has to be slightly increased to dissolve this small amount of the precipitate.

Ammonium sulfate brings the pH of the system to 6.9–7.0. This pH is critical because, at higher values, the removal of structural protein from EP_1 is inefficient. The incubation period is also critical. (When the separation of structural protein from chain components has been attempted immediately after the addition of ammonium sulfate, about 75% of the original protein remains in the supernatant, whereas, after prolonged incubation, about 50% of the original protein remains in the supernatant.)

Step 5. Isolation of EP_2. Immediately after removal of the structural protein, saturated ammonium sulfate is added to the supernatant to bring it to 27–30% saturation. The mixture is then centrifuged at 40,000 rpm for 15 minutes (rotor No. 40). This treatment results in a clear, soft, dark-red pellet (EP_2) and a faintly pink supernatant. Since the pellet contains bile salts, it dissolves readily in sucrose solution. If desired, ammonium sulfate, most of the cholate, and some DOC can be removed by dialysis of the solution for 18 hours against sucrose A.

Analytical Methods and Activity Assays

Protein. The biuret method of Gornall *et al.*[4] is used for the determination of protein in all preparations.

Cytochromes. Cytochromes *a, b,* and c-c_1 (mainly c_1) are estimated from a direct spectrum taken with a Beckman DK-2 recording spectrophotometer. The extinction coefficient of 16.5 mM^{-1} cm^{-1} is used for calculating the content of cytochrome *a*.[5] Extinction coefficients of 21.4 and 19.6 mM^{-1} cm^{-1} are used for calculating cytochromes *b* and c-c_1, respectively. Appropriate corrections are made for the contribution of cytochromes c-c_1 to the aborbancy of cytochrome *b* and vice versa.

Copper and Iron. Copper is estimated as the cuprous-biquinoline complex.[6] Total iron is determined as the ferrous-bathophenanthroline complex,[7] after wet-ashing the samples with concentrated nitric and sulfuric acids. Heme iron is taken to be equal to the total heme content of all the cytochromes. Nonheme iron represents the difference between total iron and heme iron.

[4] A. G. Gornall, C. J. Bardawill, and M. M. David, *J. Biol. Chem.* **177**, 751 (1949).
[5] T. Yonetani, *J. Biochem.* (*Tokyo*) **46**, 917 (1959).
[6] D. C. Wharton and A. Tzagoloff, *J. Biol. Chem.* **239**, 2036 (1964).
[7] G. F. Smith, W. H. McGurdy, and H. Diehl, *Analyst* **77**, 418 (1952).

Flavin. The procedure of Blair *et al.*[8] is used, except that the ether extractions are omitted.

Phospholipids. Phospholipid is determined as phosphorus[9] in a chloroform-methanol extract, back extracted with 0.1 *M* KCl and Folch's "upper phase".[10] The phospholipid content of the samples is calculated by multiplying the observed number of micrograms of phosphorus by 25, on the assumption that the average molecular weight of the phospholipid is 775 and the average phosphorus content is 4%.

Enzyme Activity. Cytochrome oxidase is assayed by the procedure of Griffiths and Wharton.[11] Succinic oxidase and DPNH oxidase are determined essentially as reported by Blair *et al.*[8] All activities are measured polarographically (Gilson Oxygraph, Model KM). In order to make a valid comparison between EP_1 and EP_2, DOC (1.0 mg/mg protein) is added to EP_1 before assaying for activity.

Reconstitution of the Chain. The cytochrome bc_1 complex[12] is mixed with EP_2 (or EP_1) in the ratio of 1.5:1.0 on protein basis.[13] This system, after incubation at 0° for 15 minutes, is diluted to 1.0 mg of the enzyme protein (EP_1 or EP_2) per milliliter with 0.66 *M* sucrose, and is assayed for DPNH oxidase and succinic oxidase activities.

Properties

EP_2 accounts for about 55% of the EP_1 protein. With the exception of nonheme iron, the structural protein is essentially free from the constituents of the electron transfer chain. The properties of EP_1 and EP_2 are summarized in Tables I and II.

In order to determine whether the initially obtained DPNH oxidase and succinic oxidase activities are maximal in EP_1, the effect of the exogenous cytochrome bc_1 complex on these activities has been tested. The results of these experiments (Table II) suggest that the coenzyme QH_2-cytochrome *c* reductase (complex III) is partially damaged and that this lesion limits the manifestation of the full potential of the remainder of the chain in EP_1.

The high levels of bile salts and the long incubation, both necessary

[8] P. V. Blair, T. Oda, D. E. Green, and H. Fernandez-Moran, *Biochemistry* **2**, 756 (1963).

[9] P. S. Chen, T. Y. Toribara, and H. Warner, *Anal. Chem.* **28**, 1756 (1956).

[10] J. Folch, M. Lees, and G. H. Sloane-Stanely, *J. Biol. Chem.* **226**, 497 (1957).

[11] D. E. Griffiths and D. C. Wharton, *J. Biol. Chem.* **236**, 1850 (1961).

[12] Generously supplied by Dr. J. S. Rieske, Institute for Enzyme Research, University of Wisconsin, Madison, Wisconsin.

[13] D. E. Green and H. Tisdale, personal communication.

TABLE I
COMPOSITION OF EP_1 AND EP_2

Component	Concentration of component[a]	
	EP_1	EP_2
Total cytochromes	4.57	7.92
Cytochrome *a*	2.90	5.10
Cytochrome *b*	1.16	1.90
Cytochromes $c + c_1$[b]	0.51	0.92
Total iron	10.79	14.68
Heme iron	4.57	7.92
Nonheme iron	6.22	6.76
Copper	3.20	5.67
Total flavin	0.83	0.86
Phospholipids	0.30	0.58

[a] Millimicromoles per milligram of protein; phospholipids are expressed as milligrams per milligram of protein.
[b] Mainly cytochrome c_1.

TABLE II
ENZYMATIC ACTIVITIES

Activity measured	Specific activity[a]			
	EP_1	$EP_1 + bc_1$[b]	EP_2	$EP_2 + bc_1$[b]
Cytochrome oxidase	6.09	—	10.69	—
DPNH oxidase	4.23	9.13	None	None
Succinic oxidase	2.78	5.53	1.80	9.70

[a] Defined as microatoms of oxygen consumed per minute per milligram of enzyme protein.
[b] Reconstituted chain. These values are based on a limited number of experiments and they vary with the assay conditions. The optimal conditions have not yet been defined.

to remove most of the structural protein from EP_1, result in an EP_2 in which the DPNH-coenzyme Q reductase complement (complex I) and, probably, some of the complex III appear to be lost or irreversibly damaged. It has not been possible thus far to restore the DPNH oxidase activity of EP_2 with the exogenous cytochrome bc_1 complex, but not all the experimental alternatives have been exhausted. The succinic oxidase–cytochrome *c* reductase segment of the electron transfer chain is only partially inactivated and can be easily reconstituted. The remaining components of the chain appear to have been retained, as evidenced by the cytochrome *c* oxidase activity.

[48] Microbial Electron Transport Particle

By F. L. CRANE

An electron transport particle (ETP) from a microbial cell may be defined as any organized group of enzymes and cofactors which show a stoichiometric association and which carry out the complete oxidation of a metabolic substrate. The oxidation of succinate and DPNH by molecular oxygen is the major function found in electron transport particles from mammalian systems,[1,2] and a particle of this type which has been prepared by Bruemmer *et al.*[3] from *Azotobacter* is described here. It should be noted that similar particles may be prepared from other microbial sources. Presumably these particles will derive from the fragmented cell membrane or from extensions of the cell membrane into the interior of the cell.

The *Azotobacter* cells are harvested by centrifugation. They may be stored by freezing at —15° as a thick paste in plastic bottles. When ready for use the cells are thawed by shaking the bottle in a water bath.

Rupture of the cells is accomplished by sonication. Since a long exposure to sonic oscilation is required, it is best to provide some cooling arrangement during sonication or to sonicate in short burst of one-half minute interspersed with immersion in ice.

The thawed cell paste is suspended in an equal volume of 0.04 *M* potassium sulfate at 0–4°. With the Raytheon 10 kc sonic oscillator a total of 10 minutes' sonication is required for each batch of cells. The heating effects of sonication will require periods for cooling. With a Branson sonicator shorter periods may be sufficient and a sample batch of cells should be run to determine the best conditions.

Unbroken cells and larger cell fragments are separated by centrifugation at 25,000 *g* for 15 minutes (Spinco model L). The slightly turbid, tan supernatant is then centrifuged for 2 hours at 105,000 *g*. A tan residue should be obtained. These small particles are washed by suspension in 10 volumes of 0.05 *M* potassium sulfate and recentrifugation at 105,000 *g* for 2 hours. The washed small particles are then suspended in 4 volumes of 0.04 *M* potassium sulfate. These may be used directly for preparation of ETP or stored at —10°.

To prepare the ETP the suspension of small particles is further

[1] F. L. Crane, J. L. Glenn, and D. E. Green, *Biochim. Biophys. Acta* **22**, 475 (1956).
[2] D. N. Moury and F. L. Crane, *Biochem. Biophys. Res. Commun.* **15**, 442 (1964).
[3] J. H. Bruemmer, P. W. Wilson, J. L. Glenn, and F. L. Crane, *J. Bacteriol.* **73**, 113 (1957).

fractionated by differential centrifugation in the presence of ethanol, EDTA, and phosphate. All operations are carried out at 0–4°. After thawing, the suspension is homogenized in a glass homogenizer with Teflon pestle, 0.01 volume 0.1 *M* EDTA pH 7.4 is added. Then 0.11 volume of cold 95% ethanol is added gradually with constant mixing. The mixture is then stirred for 10 minutes. The phosphate ion concentration is then increased by adding 0.04 volumes of 0.5 *M* potassium phosphate buffer pH 7.4. The suspension is centrifuged for 5 minutes at 17,000 rpm in the No. 40 rotor of the Spinco L centrifuge (19,000 *g*). The cloudy tan supernatant contains the ETP. It is decanted carefully from the loosely packed tan residue. The cloudy supernatant is then diluted with an equal volume of 8.5% sucrose. Centrifugation at 105,000 *g* for 5 minutes (40,000 rpm in the 40 head, Spinco L) will separate the suspension into a tan packed residue, a loosely packed darker tan residue on top of the packed layer, and a slightly turbid supernatant liquid. The supernatant and loosely packed material are carefully decanted off and homogenized. This cloudy homogenized suspension is then centrifuged for 1 hour at 105,000 *g*. A red gel-like sediment should be obtained under a clear supernatant. The red gel (ETP) is homogenized in 5 volume of 8.5% sucrose. This suspension may be recentrifuged at 105,000 *g* for 30 minutes and then resuspended in 5 volumes of 8.5% sucrose as a washing procedure.

Succinoxidase and DPNH oxidase activity in these particles are determined by standard methods and have been described.[1] The ETP particles should show succinoxidase activity in the range of 1.5–2.4 micromoles succinate oxidized per minute per milligram of protein and 4.0–7.0 micromoles DPNH oxidized per minute per milligram of protein. Protein is determined by the biuret method.[4] Succinic–ferricyanide reductase and DPNH–ferricyanide reductase activities are present in the particles,[1] but the particles are inactive either in reduction or oxidation of mammalian cytochrome *c*. There is only slight activity with the mixed cytochrome *c* types from *Azotobacter*.

Other procedures for disruption of cells have not been thoroughly explored but will yield active particles. Thus Jones and Redfearn[5] have reported a preparation equivalent to the first stage small particles by using the French pressure cell for breaking the cells. Automated apparatus for this purpose[6] may make preparation of particles feasible on a large scale.

[4] A. G. Gornall, C. J. Bardawill, and M. M. David, *J. Biol. Chem.* **177,** 751 (1949).
[5] C. W. Jones and E. R. Redfearn, *Biochim. Biophys. Acta* **113,** 467 (1966).
[6] J. A. Duerre and E. Ribi, *Appl. Microbiol.* **11,** 467 (1963).

It should not be assumed that all microbes will yield particles of the type described here. The author and others[7] have found that sonication of *Escherichia coli* will yield heavy and light fragments which do not have identical electron transport activity or quinone content. Active particles have also been obtained by sonication from *Mycobacterium phlei*,[8] *Haemophilus parainfluenzae*,[9] *Bacillus stearothermophilus*,[10] and *Bacillus subtilis*.[11]

The oxidation of succinate or DPNH may not be the only criterion for defining an electron transport particle. For example, Asano *et al.*[12] have shown that the oxidation of malate is mediated by a flavoprotein which is associated with the electron transport particle from *M. phlei*.

[7] E. R. Kashket and A. F. Brodie, *J. Biol. Chem.* **238**, 2564 (1963).
[8] A. Asano and A. F. Brodie, *J. Biol. Chem.* **239**, 4280 (1964).
[9] D. C. White, *J. Biol. Chem.* **240**, 1387 (1965).
[10] R. J. Downey, *J. Bacteriol.* **84**, 953 (1962).
[11] R. J. Downey, *J. Bacteriol.* **88**, 904 (1964).
[12] A. Asano, T. Kaneshiro, and A. F. Brodie, *J. Biol. Chem.* **240**, 895 (1965).

[49] DPNH Oxidase of Heart Muscle

By BRUCE MACKLER

$$\text{DPNH} + \tfrac{1}{2}\text{O}_2 + \text{H}^+ \rightleftarrows \text{DPN}^+ + \text{H}_2\text{O}$$

Assay Method

Principle. DPNH oxidation by molecular oxygen is followed spectrophotometrically at 38° by measuring the rate of decrease in absorbancy at 340 mμ. The oxygen content of the solutions is sufficient to allow maximum rates of oxidation under the assay conditions.

Reagents

DPNH solution, 0.1%
Ethylenediaminetetraacetate solution, 0.005 *M*, pH 7.5
Potassium phosphate buffer, 0.2 *M*, pH 7.5
Enzyme. Dilute in 5% sucrose solution as necessary.

Procedure. Four-tenths milliliter of 0.2 *M* phosphate buffer, pH 7.5, 0.1 ml of 0.1% DPNH solution, 0.05 ml of 0.005 *M* ethylenediaminetetraacetate solution, and sufficient water to make a final volume of 1 ml are added to a small tube. The mixture is incubated at 38° for 2 minutes, then added to a 1-ml optical cuvette of 10-mm light path. The reaction is begun by addition of the enzyme preparation (20–40 μg of protein). The enzymatic reaction is followed at 38° by measuring the rate of decrease in absorbancy at 340 mμ in a spectrophotometer against a control cuvette containing water. Specific activity of the enzyme is defined as the micromoles of DPNH oxidized per minute per milligram of enzyme protein. Protein is determined by the biuret method[1] with addition of deoxycholate to a final concentration of 1%.

Purification Procedure

Preparation of Submitochondrial Particles. Fresh beef hearts are obtained at the slaughterhouse and packed in ice for transport to the laboratory. All subsequent steps for preparation of the enzyme are carried out at 0–5°. Hearts trimmed of fat and connective tissue are cut into small pieces and passed through an electric meat grinder. Then 500-g portions of ground tissue are placed in 4-liter stainless steel beakers, and 1200 ml of 8.5% sucrose solution containing 1.85 g of K_2HPO_4 per liter is added to each batch. The beaker is placed in ice and the suspension is homogenized with a high speed blender[2] at top speed for 3 minutes. The pH is maintained at 7–7.5 by the addition of 6 *N* KOH just as the blender starts. The amount of KOH added varies from about 2 to 5 ml with different batches of beef heart, and a test batch is run to determine the amount of KOH required.[3] The suspension is centrifuged for 30 minutes at 1200 *g*.[4] The turbid supernatant fluid is decanted through two layers of cheesecloth into a 20-liter plastic vat, and 4 liters of 0.9% KCl solution[5] is added for each 2 liters of supernatant solution. The suspension is passed through an air-driven refrigerated Sharples centrifuge[6] at an outflow rate of 50–60 ml per minute. More KCl as frozen cubes is added to the vat containing the inflowing mixture during the procedure in order to maintain a low temperature. The residue containing mitochondrial and submitochondrial fragments is removed from the bowl of the centrifuge, suspended in 5 volumes of 6% sucrose solution, and homogenized in a glass-Teflon motor-driven homogenizer. The final sus-

[1] B. Mackler and D. E. Green, *Biochim. Biophys. Acta* **21**, 1, 6 (1956).

[2] The blender is fitted with an overhead ½-horsepower motor and a large three-pointed steel blade which rotates at approximately 18,000 rpm without load. A Waring blendor is satisfactory for preparation of small amounts of enzyme.

[3] When the proper amount of KOH has been added, the addition of a drop of the homogenate to 2 ml of Bromthymol blue solution gives a green color.

[4] This step can be done conveniently in a Model PR-2 refrigerated International centrifuge with a 4-1 head.

[5] Part of this volume of KCl solution may be added as frozen cubes in order to maintain the enzyme preparation at low temperature.

[6] The Spinco preparatory ultracentrifuge may be used instead of the Sharples.

pension containing approximately 30 mg of protein per milliliter may be stored in a frozen state for further purification or used immediately.

Purification of DPNH Oxidase. The suspension is adjusted[7] to pH 10 after thawing by slow addition of 6 *N* KOH with stirring. The preparation is centrifuged in the No. 30 rotor of the Spinco preparative ultracentrifuge for 10 minutes at 10,000–12,000 rpm. The exact speed of centrifugation to be employed for a particular preparation must be determined by experiment. The residue should have a well defined upper border and the supernatant solution should be turbid after centrifugation. The residue is discarded, and the turbid supernatant solution is centrifuged in a No. 30 rotor for 30 minutes at 30,000 rpm. The supernatant solution is discarded, and the residue is suspended in about 100 ml of 5% sucrose to give a protein concentration of approximately 10 mg/ml. The pH is adjusted to 7 by slow addition of 0.2 *N* HCl with stirring, and the suspension is centrifuged in a No. 40 rotor of the Spinco ultracentrifuge for 10 minutes at 18,000 rpm. The packed residue is discarded, and the opalescent supernatant solution is centrifuged in a No. 40 rotor for 30 minutes at 40,000 rpm. The final residue containing the DPNH oxidase is suspended in 5% sucrose by homogenization and may be stored in the frozen state at —20°. About 1 g of enzyme protein is obtained from 350 ml of mitochondrial preparation; this represents a 3-fold increase in activity over the suspensions of mitochondria and a 300-fold increase in activity over crude homogenates of heart muscle.

Properties[1,8,9]

Preparations of DPNH oxidase contain traces, if any, of succinic, pyruvic, α-ketoglutaric, malic and isocitric dehydrogenases, fumarase, and the enzymes of fatty acid oxidation and glycolysis. The enzyme does not catalyze the oxidation of TPNH, and freshly prepared preparations react only very slowly to reduce added ferricytochrome *c* or oxidize added ferrocytochrome *c*. The preparations of enzyme suspended in sucrose solution and stored in the frozen state lose activity slowly, 50% of the activity being lost over a period of 3–4 weeks. Repeated freezing and thawing of the preparations increases the loss of enzymatic activity. Preparations of enzyme contain the following components (millimicromoles per milligram of protein): flavin, 0.4; DPN, 0.45; cytochrome *a*, 1.0; cytochrome *b*, 0.84; cytochromes $c_1 + c$, 0.76; cytochrome a_3, 0.63; coenzyme Q, 14; nonheme iron 6.2; copper, 2.4; and 35–40% lipid based on the dry weight of the enzyme.

[7] A pH meter adjusted for use at 5° is employed.

[8] R. Estabrook and B. Mackler, *J. Biol. Chem.* **224,** 637 (1957).

[9] B. Mackler and M. J. Latta, *Biochim. Biophys. Acta* **36,** 254 (1959).

[50] Acetone-Treated Particles (CoQ Deficient)

By ROBERT L. LESTER

One approach to the study of the function of lipids in respiratory particles has been to examine the consequences of direct removal of lipids with organic solvents. Acetone extraction of beef heart mitochondria has been extensively studied in this regard.[1-4] It was found that the final concentration of water in acetone during extraction determined the nature of the lipids extracted and consequently the nature of required additions to restore electron transport activity. Two standard preparations extracted with 4 and 10% water in acetone have been studied in most detail.

Standard Preparations

Preparation I. One volume of a suspension of "heavy" beef heart mitochondria[5] (25–50 mg protein per milliliter) in 0.25 *M* sucrose is added dropwise with stirring to 25 volumes of dry acetone. After it has stood for 5 minutes the suspension is centrifuged, and the residue is washed once with 25 volumes of dry acetone. The residue is suspended in 25 volumes of 0.88 *M* sucrose–0.01 *M* tris(hydroxymethyl)aminomethane (Tris) chloride (pH 7.5) and centrifuged. The pellet is washed once with 15 volumes of buffered 0.88 *M* sucrose (as above) and finally suspended in 1.5 volumes of the same sucrose solution and stored at −15°.

Preparation II. One volume of a 0.25 *M* sucrose suspension of "heavy" beef heart mitochondria[5] (25 mg protein per milliliter) is mixed rapidly with 24 volumes of a 15:1 (v/v) acetone:water solution yielding a final concentration of 10% water. After it has stood for 10 minutes with occasional stirring, the mixture is centrifuged and the supernatant solution is discarded. The residue is immediately shaken briefly but vigorously with 14 volumes of 0.88 *M* sucrose–0.01 Tris-chloride (pH 7.5), centrifuged and washed once, and stored as described for preparation I. The yield of protein should be about 50%. All volumes referred to are based on the volume of the original suspension of mitochondria. All operations are carried out at 0–5° with precooled solutions. Aside from

[1] R. L. Lester and S. Fleischer, *Biochim. Biophys. Acta* **47,** 358 (1961).
[2] R. L. Lester and A. L. Smith, *Biochim. Biophys. Acta* **47,** 475 (1961).
[3] S. F. Fleischer, G. Brierley, H. Klouwen, D. B. Slautterback, *J. Biol. Chem.* **237,** 3264 (1962).
[4] G. P. Brierley, A. J. Merola, and S. Fleischer, *Biochim. Biophys. Acta* **64,** 218 (1962).
[5] Y. Hatefi and R. L. Lester, *Biochim. Biophys. Acta* **27,** 83 (1958).

the H_2O-acetone ratio the critical variables, particularly for preparation II, are the concentration of the original suspension of mitochondria and the time of exposure to acetone.

Properties of Preparations I and II

Extraction with 96% acetone results in virtually complete loss of neutral lipids, notably coenzyme Q, α-tocopherol, and carotenoids.[1] Little, if any, phospholipid is removed. However, extraction with 90% acetone leads to an 80% loss of total phospholipids as well as complete loss of neutral lipids as in preparation I. This procedure removes about 90% of choline, ethanolamine and inositol phosphatides, but only 30% of the cardiolipin.[3]

Preparations I and II have no capacity to oxidize all substrates tested with oxygen as electron acceptor. However full, antimycin sensitive, succinoxidase activity can be restored in preparation I upon addition of cytochrome *c* and a coenzyme Q homolog. Preparation II requires the further addition of a lipid supplement to restore this activity.[1,3,4] A specific requirement for coenzyme Q (preparation I) and for coenzyme Q plus additional lipid (preparation II) are found for succinate–cytochrome *c* reductase activity.[1,3] Although DPNH–ferricyanide reductase activity is unimpaired by the acetone treatment, no additions thus far tested were able to restore DPNH–cytochrome *c* reductase or DPNH oxidase activity. In addition an activation effect of lipid has been demonstrated on preparation II for cytochrome *c* oxidase, succinate–CoQ reductase, and $CoQH_2$–cytochrome *c* reductase.[4]

Succinoxidase Assay of Preparations I and II. To the main compartment of a Warburg flask is added 100 micromoles K phosphate (pH 7.5), 150 micromoles K succinate, 0.2 mg cytochrome *c*, 0–1.5 mg particle protein; sucrose solution is added to a final volume of 3 ml and final molarity of 0.2. Coenzyme Q and other lipid supplements are added last. Coenzyme Q is added as an ethanolic solution; 0.03–05 ml volumes should be added rapidly and mixed well to achieve an efficient effect. After 9 minutes' equilibration at 30° (gas phase air) oxygen uptake readings are commenced and are found to be linear for at least 30 minutes. The amount of CoQ required to achieve maximal activity will depend on how good an emulsion has been formed and the time of preincubation with the particle. Generally, 0.1 micromole coenzyme Q_{10} should suffice.

Succinate–Cytochrome c Reductase Assay of Preparations I and II. To a final volume of 3.0 ml in a cuvette is added 180 micromoles K phosphate (pH 7.5), 1.5 mg cytochrome *c*, 15 μg enzyme protein, 10–20 μg coenzyme Q, and other lipids if necessary. After 10 minutes' preincubation at 30° the reaction is followed at 550 mμ after the addition of 0.1 ml of a solution containing 135 micromoles K succinate and 3 micromoles KCN.

The nature of the requirement for phospholipid supplements for Preparation II have been studied extensively.[3,4,6] It would appear that any well dispersed mitochondrial phospholipid can satisfy this requirement. Dispersion into a micellar state can be achieved either by treatment with butanol-cholate or by exposing a phospholipid suspension to ultrasonic irradiation.[2,6] Although variations are observed depending the lipid in question, such dispersed lipid preparations are required at levels of 1–5 mg lipid per milligram of particle protein to achieve maximal activity.

The requirements for coenzyme Q and other lipid have also been studied for these preparations in the reduction of various tetrazolium salts by succinate.[2]

[6] S. Fleischer and H. Klouwen, *Biochem. Biophys. Res. Commun.* **5**, 378 (1961).

[51] Resolved Mitochondrial Translocating Particles

By JOHN T. PENNISTON

Mitochondria are capable of actively translocating cations[1,2] and, in the presence of orthophosphate, of accumulating and retaining large amounts of alkaline earth phosphates.[3,4] The capability for active transport in a mitochondrial system is most easily monitored by observing the massive accumulation of radioactive calcium in the presence of phosphate. Calcium is the most rapidly taken up of the divalent cations, and the large amount of calcium-45 retained in a mitochondrial pellet isolated by centrifugation after incubation is easily measured. Accumulation of strontium,[5] manganese,[6] or magnesium[7] may also be observed, but

[1] F. D. Vasington and J. V. Murphy, *J. Biol. Chem.* **237**, 2670 (1962).
[2] B. C. Pressman, *Proc. Natl. Acad. Sci. U.S.* **53**, 1076 (1965).
[3] G. P. Brierley, E. Murer, and R. L. O'Brien, *Biochim. Biophys. Acta* **88**, 645 (1964).
[4] E. Carafoli, C. S. Rossi, and A. L. Lehninger, *J. Biol. Chem.* **240**, 2254 (1965).
[5] E. Carafoli, S. Weiland, and A. L. Lehninger, *Biochim. Biophys. Acta* **97**, 88 (1965).
[6] J. B. Chappell, M. Cohn, and G. D. Greville, *in* "Energy-Linked Functions of Mitochondria" (B. Chance, ed.), p. 219. Academic Press, New York, 1963.
[7] G. P. Brierley, E. Bachmann, and D. E. Green, *Proc. Natl. Acad. Sci. U.S.* **48**, 1928 (1962).

these ions are translocated more slowly and assays of their accumulations are more time consuming. Procedures for the preparation and assay of a resolved mitochondrial translocating particle are described below. For ion translocation this particle requires cytochrome *c* in addition to the assay system normally required for unresolved mitochondria. When a crude, nonparticulate mitochondrial extract is also added, the resolved particles translocate at rates comparable to those found with intact mitochondria.

Preparation of Translocating Particles

Prepare heavy beef heart mitochondria from slaughterhouse material on a large scale as described in this volume [12]. All materials used are kept on ice except during centrifugation, which is done with the rotor at 4°. Discard the supernatant liquid and the light mitochondria, and suspend the heavy mitochondrial pellet in a measured amount (about three pellet volumes) of cold distilled water. Homogenize briefly in a glass homogenizer with a Teflon pestle (this type of homogenizer is used throughout these procedures) and measure the volume of the smooth suspension. Note the increase in volume caused by the pellet and add enough more water so that the final volume of the suspension is nine times this volume increase. Age this suspension for 7 days at 0–2°. This aging can be done in a refrigerator, but if the temperature is too high (4° or above), bacterial growth may intervene. Such growth may be prevented by addition of penicillin G (0.2 mg/ml) and streptomycin (0.125 mg/ml).

After 7 days' aging, add 1/7 volume of 2 *M* sucrose (reagent grade) and mix, then freeze the suspension in a deep freeze compartment maintained at —20°. Keep the suspension frozen for about a month, thawing in warm water and immediately refreezing every week, for a total of at least four thawings. Then thaw and remove the particles from the suspension medium by centrifugation for 10 minutes at 30,000 rpm in the 30 rotor of a Spinco[8] Model L ultracentrifuge. Centrifugation times include the time required for acceleration of the rotor, but not the deceleration time. Pour off the clear supernatant fluid carefully, without pouring off the brown residue which is loosely packed at the bottom of the tube. Separate the brown residue from the small, tightly packed dark pellet beneath it by swirling the tube. Pour off and save the main part of the brown residue. Then separate the rest of the brown residue from the pellet with a glass rod, swirl with a few milliliters of added 0.25 *M* sucrose, and pool the portions of brown residue. Dilute this

[8] Spinco Division, Beckman Instrument Co., Palo Alto, California.

recovered material to the original volume of the suspension with 0.25 *M* sucrose and promptly homogenize for 1.5 minutes. Centrifuge the resulting suspension for 15 minutes at 10,000 rpm in the Spinco 30 rotor.

Carefully pour off the clear portion of the supernatant liquid remaining after this centrifugation, then swirl the tube to suspend the loose brown residue, which is poured off and saved. Discard the tightly packed brown pellet. The loose brown residue is put through this last-described centrifugal washing one more time and again diluted with 0.25 *M* sucrose and homogenized briefly.

Weigh enough reagent grade dibasic potassium hydrogen phosphate to make the suspension 0.5 *M* in this salt (126 mg salt per milliliter of suspension). Add the salt and homogenize until it is completely dissolved. At this stage the pH of the suspension should be 9.0 ± 0.2. Allow the particles to stand at 0° for 100 minutes in the presence of this high salt medium. After this extraction, centrifuge 15 minutes at 40,000 rpm in the Spinco 40 rotor to pack the particles tightly. Pour off the clear supernatant liquid and resuspend the entire pellet to the original volume in 0.25 *M* sucrose, 0.01 *M* in tris(hydroxymethyl)aminomethane, pH 8.0 (sucrose–Tris). Homogenize briefly and centrifuge again for 15 minutes at 40,000 rpm. Pour off the liquid and separate the lighter brown upper layer of the pellet with a glass rod, suspending it by swirling in a little sucrose–Tris. Dilute this suspension to the original volume with sucrose–Tris and adjust the pH to 7.7 ± 0.1 with hydrochloric acid. Repeat the last-described centrifugation and separation of the particles, finally suspending the resedimented upper layer in a minimum of sucrose–Tris and homogenizing briefly.

This suspension is the resolved mitochondrial translocating particle. It loses its activity within a few hours in suspension and should be stored frozen in many small tubes, to be used individually, thus avoiding repeated thawing of the preparation. Kept at —20°, it loses activity over a period of weeks, and is best stored in liquid nitrogen. The protein concentration of each preparation of particle should be about 20 mg/ml, as determined by the biuret method.

Preparation of Sonic Extract

Suspend the crude beef heart mitochondrial paste, described in this volume [12], in ten times its volume of sucrose–Tris, homogenize, and remove from suspension by passage through a Sharples super centrifuge. This preliminary washing may also be carried out by ordinary centrifugation. Suspend the washed paste in an equal volume of cold distilled water and sonicate 50-ml portions of the suspension for 5 minutes in a jacketed 100-ml beaker. Pass refrigerant at —4° through the jacket during sonication. A Branson[9] Sonifier, Model S-75, is used at maximum setting (about 7 amperes) with the horn tip within 3/16 inch of the bottom of the beaker, but not touching it. The temperature at the end of sonication should be 20–25°.

Centrifuge the sonicated suspension for 60 minutes at 30,000 rpm in the Spinco 30 rotor. Pour off the supernatant liquid and centrifuge it for 60 minutes at 50,000 rpm in the Spinco 50 rotor. Remove the clear portion of the supernatant liquid with a pipette. This crude sonic extract should be stored in liquid nitrogen, but is stable for a few weeks at —20°.

Assay of Ion Translocation

The assay system described here is easily adapted to any type of mitochondrial particle. It was developed from the system described by Brierley *et al.*,[10] and depends on the centrifugal separation of the particles from the assay medium following accumulation of calcium phosphate. The amount of calcium-45 accumulated is measured by determination of the radioactivity of the pellet. A blank without added oxidizable substrate and ATP allows correction for the nonspecific binding of the radioactive cation. In the measurement of calcium ion translocation, this blank is sufficiently low that the particles may be centrifuged directly from the assay suspension. With other ions it is preferable to wash the particles through a concentrated sucrose solution.

Reagents

Sucrose, 2 *M*, reagent grade

$MgCl_2$, 0.1 *M*, reagent grade

Imidazole buffer, 0.1 *M*, pH 7.0, adjusted to pH with HCl

Potassium phosphate, 0.1 *M*, pH 7.0: 61 ml of 1 *M* K_2HPO_4 and 39 ml of 1 *M* KH_2PO_4 diluted to 1 liter. Adjust the pH, if necessary by addition of 0.1 *M* K_2HPO_4 or 0.1 *M* KH_2PO_4.

Potassium succinate, 0.1 *M*, pH 7.0: succinic acid neutralized with KOH

Adenosine triphosphate, 0.1 *M*, pH 7.0: disodium ATP neutralized with KOH

Antimycin, 0.05 mg/ml plus oligomycin, 0.05 mg/ml in 95% ethanol (stop solution). These inhibitors may be obtained from the Wisconsin Alumni Research Foundation, Madison, Wisconsin.

[9] Branson Ultrasonic Corporation, Stamford, Connecticut.

[10] G. P. Brierley, E. Murer, and E. Bachmann, *Arch. Biochem. Biophys.* **105**, 89 (1964).

Cytochrome *c* 0.1 mg/ml, Sigma Chemical Co., St. Louis, Missouri 63118, type III, dissolved in water

$^{45}CaCl_2$ 3.3 mM, specific activity 0.02 curies/mole (about 20,000 cpm/micromole). Obtained from General Electric Co., Irradiation Products Operation, P. O. Box 846, Pleasanton, California 94566.

To 25-ml Erlenmeyer flasks kept on ice, add 0.30 ml of 2 M sucrose, 0.30 ml of 0.1 M $MgCl_2$, 0.10 ml of 0.1 M potassium phosphate, 0.10 ml of 0.1 M imidazole buffer, 0.10 ml of 0.1 M potassium succinate, 0.20 ml of 0.1 mg/ml cytochrome *c*, 0.10 ml of 0.1 M ATP, about 1 ml of sonic extract, and enough distilled water to make the final volume (including particle and $^{45}CaCl_2$, but not stop solution) 3.0 ml. Appropriate blanks should be carried out by omitting sonic extract, cytochrome *c*, ATP, and succinate. The optimal level of sonic extract should be determined for each extract preparation. Just before the incubation, thaw a portion of the translocating particle, add 2.5 mg of particle protein to each flask, and mix.

Using a stop watch, put the flasks on a shaking water bath at 37° at 10 second intervals. After each flask has been incubated for 3 minutes, start the ion translocation by adding 0.30 ml of 3.3 mM $^{45}CaCl_2$, using an automatic refilling syringe (Becton, Dickinson and Co., Rutherford, New Jersey— Cornwall continuous pipetting outfit used without two-way valve self filler). Stop ion translocation 2 minutes after the addition of $^{45}CaCl_2$ by adding from a syringe 0.15 ml of stop solution, and put the flask on ice.

Pour the reaction mixture into a thick-walled polypropylene centrifuge tube or other appropriate centrifuge tube and centrifuge for 10 minutes at 40,000 rpm in the Spinco 40 rotor. Pour off the supernatant liquid which contains most of the calcium-45 and dispose of it as radioactive waste. Wipe out the tube without disturbing the pellet and resuspend the pellet in 1.00 ml of water using a homogenizer. Pipette 0.20 ml of this suspension onto a 1-inch ground glass disk. Dry the sample on a hot plate, and determine the amount of calcium taken up by measuring the radioactivity with a thin window, gas flow Geiger counter. One micromole of $^{45}CaCl_2$ from the original solution is similarly prepared and counted as a standard. The blank without ATP, succinate, cytochrome *c*, and sonic extract is due almost entirely to mechanical trapping of calcium-45 and should be subtracted from the other values.

General

The most variable of the preparations described here is the sonic extract, which usually gives an increase of 1.5- to 2-fold over the ion accumulation rate with the otherwise complete system, including cytochrome *c*. Some preparations of extract have given no such increase and have inhibited when added in large quantities. The reason for this variability is unknown.

Less than 5% of the preparations of resolved translocating particle have proved inactive in calcium translocation, while the great majority show rates of 30–50 millimicromoles of calcium per milligram of protein per minute with ATP, succinate, and cytochrome *c* added. If the sonic extract is also added, the corresponding rate reaches 50–90. The rate for fresh heavy beef heart mitochondria under the same conditions is 220.

The blank with no ATP and no succinate should be in the range of 10–20 millimicromoles of calcium per milligram of protein regardless of the time of incubation. The control with ATP and succinate, but without cytochrome *c* and without sonic extract should be less than twice the blank. Occasional particle preparations have exceeded this figure, but these can be fully resolved with respect to the cytochrome *c* requirement by further extraction with dibasic potassium phosphate.

Section IV

Purified Respiratory Chain Oxidation-Reduction Components

Previously Published Articles from Methods in Enzymology Related to Section IV

Vol. I [121]. Succinic Dehydrogenase. Walter D. Bonner.
Vol. II [119]. Pyridine Nucleotide Transhydrogenase. Nathan O. Kaplan.
Vol. II [120]. DPNH Cytochrome *c* Reductase (Animal). Henry R. Mahler.
Vol. II [121]. DPNH Cytochrome *c* Reductase (Bacterial). Arnold F. Brodie.
Vol. II [122]. TPNH Cytochrome *c* Reductase from Yeast. Erwin Haas.
Vol. II [123]. TPNH Cytochrome *c* Reductase (Liver). B. L. Horecker.
Vol. II [124]. Diaphorases. Henry R. Mahler.
Vol. II [132]. Cytochrome *b* Group (Bacteria). A. M. Pappenheimer, Jr.
Vol. II [133]. Cytochrome *c* (Mammals). Karl-Gustav Paul.
Vol. II [134]. Cytochrome *c* (*Ustilago*). J. B. Neilands.
Vol. II [135]. Cytochrome *c* and Cytochrome *c* Peroxidase from *Pseudomonas fluorescens*. Howard M. Lenhoff and Nathan O. Kaplan.
Vol. II [136]. Assay of Catalases and Peroxidases. Britton Chance and A. C. Maehly.
Vol. V [59]. α-L-Glycerophosphate Dehydrogenase from Pig Brain. Robert L. Ringler and Thomas P. Singer.
Vol. V [76]. Choline Dehydrogenase from Rat Liver. Tokuji Kimura and Thomas P. Singer.
Vol. V [82]. Succinic Dehydrogenase. Paul Bernath and Thomas P. Singer.
Vol. VI [55]. Some Bacterial Cytochromes. Martin D. Kamen, Robert G. Bartsch, Takekazu Horio, and Henk de Klerk.
Vol. VI [59]. Electron-Transferring Flavoproteins from Pig Liver and Beef Heart. Helmut Beinert and William Lee.
Vol. IX [51]. Pyruvic (Cytochrome b_1) Dehydrogenase (Crystalline). F. Robert Williams and Lowell P. Hager.
Vol. IX [52]. Lipoyl Dehydrogenase from Pig Heart. V. Massey.
Vol. IX [56]. D(—)-Lactate Cytochrome *c* Reductase. Thomas P. Singer and Terenzio Cremona.
Vol. IX [57]. L-Lactate (Cytochrome) Dehydrogenase (Crystalline, Yeast). R. H. Symons and L. A. Burgoyne.

[52] Preparations and Properties of Soluble NADH Dehydrogenases from Cardiac Muscle

By Tsoo E. King and Robert L. Howard

At least seven NADH dehydrogenases have been solubilized from heart muscle. The solubilizing agents which have been used are ethanol, thiourea, and snake venom. The dehydrogenase solubilized by the digestion with *Naja naja* venom at 37° is different from that at 30° (for convenience they are referred to as the 37° enzyme and the 30° enzyme, respectively). Likewise, the enzyme prepared by Mackler *et al.* is different from the one reported by Mahler *et al.*[1] A recent paper[2] discusses the similarities and differences of these enzymes. In this article, methods for the preparation of the 37° enzyme, the 30° enzyme, the enzyme reported by Mackler *et al.*, and the thiourea enzymes are described.

Assay Methods

Principle

All known soluble NADH dehydrogenases can react only with artificial electron acceptors, cytochrome *c*, and lower homologs of coenzyme Q. The behavior of cytochrome *c* in the reaction with the soluble preparations is probably different[2,3] from that with NADH dehydrogenase in the particulate form, for example, ETP or the Keilin-Hartree preparation. Among the artificial acceptors, 2,6-dichlorophenolindophenol (DCIP), and ferricyanide are widely used. These reactions are conveniently followed spectrophotometrically. The activities are expressed as NADH oxidized either at a fixed concentration of the acceptor or extrapolated to its infinite concentration (V_{max}).

The 37° Enzyme[3]

Reagents

Glycylglycine-HCl buffer, 0.2 *M* in water, pH 8.5
NADH, 6 m*M* in 2 m*M* glycylglycine buffer, pH 8.5; use it within 4 hours

[1] H. R. Mahler, see Vol. II, p. 688.

[2] T. E. King, R. L. Howard, J. Kettman, B. M. Hegdekar, M. Kuboyama, K. S. Nickel, and E. A. Possehl, Comparison of soluble NADH dehydrogenases from the respiratory chain of cardiac mitochondria. *IUB Flavin Symp. Amsterdam, 1965,* p. 441. Elsevier, Amsterdam, 1966.

[3] T. E. King and R. L. Howard, *J. Biol. Chem.* **237,** 1686 (1962).

Cytochrome *c*, 1.05 m*M* in water
DCIP, 0.6 m*M* in water
Enzyme, diluted in 0.02 *M* bicarbonate so that 0.05 ml of the diluted sample causes an absorbance change of 0.05–0.2 per minute
Water used in all the assays and preparations described is distilled in an all-glass apparatus from ordinary distilled water, which is first deionized.

Procedure. To a cuvette of 1-cm optical path are added 0.1 ml cytochrome *c* (or 0.2 ml DCIP), 0.35 ml glycylglycine buffer, 0.10 ml NADH and 2.40 ml (or 2.30 ml with DCIP) water. The reaction is followed by recording the absorbance change at 550 mμ (or 600 mμ for DCIP), against a reagent blank, within 5 seconds after addition of 0.05 ml enzyme solution. The rate is determined from the linearized initial slope.

Units and Specific Activity. A unit of activity is defined as 1.0 micromole of NADH oxidized per minute at room temperature (23°). Specific activity is in units per milligram of protein. The absorbancy index used for cytochrome *c* (reduced minus oxidized) is 19.2 m$M^{-1} \times$ cm^{-1} at 550 mμ and for DCIP is 21 m$M^{-1} \times$ cm^{-1} at 600 mμ. Cytochrome *c* is considered as a one-electron equivalent oxidant, and DCIP as a two-electron equivalent. Thus $\Delta A_{550} \times 0.0262$ for cytochrome *c* or $\Delta A_{600} \times 0.0476$ for DCIP gives m*M* NADH oxidized. The fat-free dry weight[4] is used to measure total protein content of the heart muscle preparation. The method of Warburg and Christian[5] is used for the determination of the protein content of fractions obtained by chromatography, whereas the turbidimetric method[6] with trichloroacetic acid is employed for other fractions.

The 30° Enzyme[7,9]

Reagents

Phosphate buffer, 0.12 *M*, pH 7.4, at 30°, or,
Triethanolamine-HCl *buffer*,[7] 0.12 *M*, pH 7.8, at 30°

[4] E. C. Slater, *Biochem. J.* **45**, 1 (1949).
[5] O. Warburg and W. Christian, *Biochem. Z.* **310**, 384 (1941).
[6] T. E. King, *J. Biol. Chem.* **238**, 4037 (1963).
[7] In this paper, [S. Minakami, R. L. Ringler, and T. P. Singer, *J. Biol. Chem.* **237**, 569 (1962)], phosphate buffer is used for the assay but it was later found[8,9] that triethanolamine buffer is more satisfactory because of less steep slopes in the double reciprocal plots.
[8] T. Cremona and E. B. Kearney, *J. Biol. Chem.* **239**, 2328 (1964).
[9] R. L. Ringler, S. Minakami, and T. P. Singer, *J. Biol. Chem.* **238**, 801 (1963).

Potassium ferricyanide, 0.01 *M* in water

NADH, 4.5 m*M* in 2 m*M* phosphate buffer, pH 7.4; use it within 4 hours

Enzyme, appropriately diluted with 0.03 *M* phosphate buffer, pH 7.6 (at 0°) so that 0.05 ml gives a decrease of absorbancy of 0.1 to 0.2 per minute

Procedure. Pipette solutions into four cuvettes of 1-cm optical path according to the following scheme:

	Cuvette No.			
Solution	1	2	3	4
Ferricyanide (ml)	0.125	0.167	0.25	0.5
Buffer (ml)	1.0	1.0	1.0	1.0
H_2O (ml)	1.73	1.69	1.6	1.35

All cuvettes are placed in a water bath at 30° and brought to that temperature. No. 1 cuvette is then transferred to the thermostatted cell compartment of the spectrophotometer, and 0.1 ml NADH (at 30°) is added. A short recording of the nonenzymatic rate against a water blank is first obtained (not more than 0.015 absorbance change should be allowed during this period).[7] Then 0.05 ml enzyme is added, and the recording of the absorbance change at 420 mμ starts within 5 seconds. The reaction is followed for about 1 minute. The same procedure is then repeated with cuvettes Nos. 2, 3, and 4.

The rate of the reaction in the absence of enzyme is subtracted from the rate obtained in the presence of enzyme (i.e. the linearized initial slopes[10]). This corrected enzymatic rate is used for the calculation.

Units and Specific Activity. To convert ΔA_{420} to m*M* NADH oxidized, a factor of 0.5 is used.[11] The reciprocals of the activities are plotted against the reciprocals of initial ferricyanide concentrations according to Lineweaver and Burk.[12] The activity extrapolated to infinite concentration, i.e., $V_{max}^{ferricyanide}$, is reported. One unit of activity is defined as 1 μmole NADH oxidized per minute at 30°. Specific activity is units per milligram of protein. Protein is determined by the biuret method[13];

[10] The rate is nonlinear with time in the presence of the enzyme. Therefore, the first 10-second portion of the reaction is extrapolated to 1 minute for the calculation of the absorbance change.

[11] The factor used here is 0.5 and differs slightly from that listed in the other methods because an absorbance index[7] of 1.0 instead of 1.03 is used.

[12] H. Lineweaver and D. Burk, *J. Am. Chem. Soc.* **56,** 658 (1934).

[13] A. G. Gornall, C. J. Bardawill, and M. M. David, *J. Biol. Chem.* **177,** 751 (1949).

a coefficient[7] of 0.095 at 540 mμ for 1 mg protein in a 3-ml volume has been employed. However, this value should be checked before use because of variation of conditions.

The Enzyme by Mackler et al.[14,15]

Reagents

NADH, 1% in water; use it within 4 hours
Phosphate buffer, 1 *M*, pH 8.5; and also 0.2 *M*, pH 7.5
Potassium ferricyanide, 5 m*M* in water
Cytochrome *c*, 1% in 1 *M* phosphate buffer, pH 8.5
DCIP, 0.01% in 1 *M* phosphate buffer, pH 8.5
Enzyme, appropriately diluted so that 0.01 ml causes an absorbance change of 0.1–0.2 per minute

Procedure. In a cuvette of 1-cm optical path are placed 0.05 ml NADH, 0.1 ml of ferricyanide, 0.2 ml of 0.2 *M* phosphate buffer, pH 7.5, and water to 1.0 ml. The cuvette is placed in a water bath at 30°, brought to that temperature and then transferred to the thermostatted (30°) cell compartment of the spectrophotometer. The reaction is started by the addition of 0.01 ml enzyme. A spectrophotometric recording is made at 420 mμ against a water blank. Assays for DCIP and cytochrome *c* reductase activities are performed in a similar manner at 600 and 550 mμ, respectively, except that indophenol and cytochrome *c* solutions are substituted for ferricyanide solution, and 1 *M* phosphate buffer, pH 8.5, is substituted for the 0.2 *M* phosphate buffer and water.

Units and Specific Activity. A unit of activity is defined as 1.0 micromole NADH oxidized per minute. To convert ΔA_{420}, ΔA_{550}, or ΔA_{600} to m*M* NADH oxidized, a factor of 0.5, 0.0262, or 0.0476, respectively, is used. Specific activity is expressed in units per milligram of protein. Protein is determined by the biuret method.[13]

The Thiourea Preparations[2]

Reagents

Phosphate buffer, 0.2 *M* (the Sørensen type), pH 7.4
Tris-HCl buffer, 0.1 *M*, pH 8.5

[14] B. Mackler, *Biochim. Biophys. Acta* **50**, 141 (1961).
[15] We are greatly indebted to Dr. B. Mackler for reading the manuscript concerning the dehydrogenase isolated in his laboratory and for his valuable suggestions and information on unpublished observations.

NADH, 4 mM in 2 mM phosphate buffer, pH 7.4; use it within 4 hours

Potassium ferricyanide, 0.03 M in water; use within 4 hours

Cytochrome *c*, 0.84 mM in water

Enzyme, appropriately diluted so that 0.05 ml of the diluted enzyme gives absorbance change of 0.05–0.2 per minute in the assay. The diluents used are 0.02 M bicarbonate for the cytochrome *c* assay and 0.1 M PO_4, pH 7.4, for the ferricyanide assay.

Procedure. The system with cytochrome *c* as the acceptor contains 1.0 ml Tris buffer, 0.1 ml cytochrome *c*, and 0.1 ml NADH, and that with ferricyanide contains 1.0 ml phosphate buffer, 0.1 ml ferricyanide, and 0.1 ml NADH. Water is added to 2.95 ml. The reaction is started by the addition of 0.05 ml enzyme. Spectrophotometric recording of the absorbance change at 550 mμ for cytochrome *c* or 420 mμ for ferricyanide against a water blank commences within 5 seconds of the addition of the enzyme. The linearized initial slope, after subtraction of the non-enzymatic (reagent blank) rate, is used for calculation of the activity.

Units and Specific Activity. A unit of activity is defined as 1.0 micromole NADH oxidized per minute. To convert ΔA_{550} (cytochrome *c*) or ΔA_{420} (ferricyanide) to mM NADH oxidized, a factor of 0.0262 or 0.485, respectively, is used. The specific activity is expressed in units per milligram of protein. Protein is determined by the turbidimetric method.[6]

Preparation of the 37° Enzyme[3, 16]

The method to be described has been successfully used for several years in this laboratory. It involves digestion of the heart muscle preparation by snake venom at 37°, calcium phosphate gel adsorption, ammonium sulfate fractionation, and DEAE-cellulose chromatography.

All manipulations are conducted at 0–4° unless otherwise indicated.

Step 1. Solubilization of the Dehydrogenase. To 374 ml of the heart muscle preparation (20 mg protein per milliliter) (see method 1 in this volume [37] for the preparation) are added 289 ml water, and 25 ml *Naja naja* venom[17] and 5.1 ml of 0.1 M calcium chloride.[18] The mixture

[16] T. E. King and R. L. Howard, *Biochim. Biophys. Acta* **59,** 489 (1962)

[17] *Naja naja* venom (Ross Allen's Reptile Institute, Silver Springs, Florida), 241 mg dissolved in 25 ml 0.01 M phosphate buffer, pH 5.9, is heated in a boiling water bath for 10 minutes, then cooled to 5–10° before it is added to the heart muscle preparation.

[18] Calcium ion is actually not necessary for the solubilization.[3]

is adjusted to pH 7.4 with 2 N acetic acid and then placed in a water bath at 37° and gently stirred for 105 minutes. At the end of the incubation, the mixture is cooled to 5–10° and then centrifuged at 59,000 g for 75 minutes. The clear yellow supernatant solution obtained is carefully[19] collected.

Step 2. Calcium Phosphate Gel Absorption and Elution. The supernatant liquid obtained in step 1 (630 ml) is adjusted to pH 6.5 by the addition, with efficient mixing, of 2 N acetic acid and 2.1 ml calcium phosphate gel[20,21] (44 mg solid per milliliter) is then added. The mixture is stirred for 15 minutes and subsequently centrifuged for 10 minutes at 4000 g. The clear yellow supernatant fluid is collected, another 39 ml calcium phosphate gel is added, and the mixture is stirred and centrifuged as before. The supernatant liquid obtained is discarded. The gel is suspended in 90 ml of 0.2 M K_2HPO_4 by gentle homogenization, stirred for 15 minutes and then centrifuged for 10 minutes at 4000 g. The supernatant liquid is collected and the precipitate is again eluted with an additional 90 ml of 0.2 M K_2HPO_4 as above. The eluates are combined.

Step 3. Ammonium Sulfate Fractionation. To the pooled eluate fraction (175 ml), 26 g ammonium sulfate is added slowly with constant stirring. The mixture is allowed to stand with occasional stirring for 15 minutes and then centrifuged for 10 minutes at 12,000 g. The precipitate is discarded and the supernatant liquid is treated with 25 g ammonium sulfate in the same manner as the previous step. After centrifuging, the supernatant fraction is discarded and the pellet is suspended in 2.5 ml of 0.01 M phosphate buffer, pH 7.4. The suspension is then dialyzed against the same buffer for 4 hours.[22] During the dialysis, the suspension becomes clear; usually 5 ml of a dark amber solution is obtained.

Step 4. Column Chromatography on DEAE-Cellulose. A column, 1.2 × 40 cm, is packed with DEAE-cellulose[23] until a stable height of

[19] A white loosely packed precipitate is present as a thin layer over the firmly packed brown pellet. When the supernatant liquid is decanted, care is taken to avoid including this material in the supernatant fraction.

[20] See Vol. I, p. 98, for the preparation. It is usually aged for at least 3 weeks before use.

[21] Sufficient calcium phosphate gel is first added to remove about 10% of the activity present in the supernatant liquid, then additional gel is added until only 10% of the activity remains in the supernatant liquid.

[22] The buffer is changed each hour, 100 ml being used each time.

[23] DEAE-cellulose is obtained from the Brown Company or from the Eastman Kodak Company. The cellulose is first well washed with water and equilibrated with 0.01 M phosphate buffer, pH 7.4. The packed column is further washed with at least 200 ml of the same buffer before use.

16 cm is reached. Excess buffer on the top of the column is removed. The dialyzed enzyme, 4.5 ml from step 3, is carefully placed on the column and allowed to flow into the bed. The wall of the column is then rinsed with 5 ml 0.01 M phosphate buffer, pH 7.4. Gradient elution at constant pH (pH 7.4) is then employed by running 50 ml 0.1 M phosphate buffer followed by 0.2 M phosphate buffer into a mixing reservoir containing 100 ml of 0.01 M phosphate buffer. The flow rate is adjusted to approximately 15 ml per hour, and 50 fractions with 60 drops (about 3.8 ml) per fraction are collected.[24]

The individual fractions are assayed for enzymatic activity and protein content, and those with the highest specific activity are combined. The enzyme at this stage[24] shows a specific activity of 13.5–18.3 for cytochrome *c*. A summary of a representative preparation is given in Table I.

TABLE I
PURIFICATION OF THE 37° ENZYME

Fraction	Specific activity (unit/mg)		Recovery (%)		Activity ratio, cytochrome *c*: DCIP
	DCIP	Cytochrome *c*	DCIP	Protein	
Heart muscle preparation[a]	0.0866	0.117	100	100	1.35
Supernatant of venom digest	0.136	0.232	69	44	1.70
Gel eluate	0.841	1.44	44	4.5	1.71
Ammonium sulfate fraction (0.30–0.55 saturation)	1.69	2.74	24	1.2	1.62
First peak from chromatography	9.08	16.4	8	0.08	1.80

[a] Assay system for the heart muscle preparation uses 0.023 M phosphate buffer, pH 7.4, instead of glycylglycine buffer, and 1 mM KCN is also present.

Step 5. Rechromatography. Fractions obtained from step 4 may be further purified by rechromatography on DEAE-cellulose. For this purpose, the enzyme solution obtained in the first elution peak is dialyzed against 0.002 M phosphate, pH 7.4, for 4 hours.[22] The dialyzed enzyme is then placed on a DEAE-cellulose column similar to that used in step 4 except that it is equilibrated with 2 mM phosphate, pH 7.4. The

[24] Although 50 fractions are necessary to remove both active peaks from the column, the bulk of the activity is located in the first peak, (cf. Fig. 4 of King and Howard[3] for the elution pattern) for which the subsequent description is made.

enzyme is eluted at constant pH (7.4) by dropping 0.1 *M* phosphate buffer into a mixing reservoir containing 100 ml of 2 m*M* phosphate. The enzyme eluted in this manner usually has a specific activity approaching 20.

Properties of the 37° Enzyme[3,16]

The following properties refer to the first elution peak.

Physical Constants. In 0.1 *M* phosphate buffer, pH 7.4, a sample of the purified dehydrogenase (specific activity 19.8) shows an $S_{20,w}$ of 6.3 S. Free boundary electrophoresis of a sample of the purified enzyme (specific activity 18.3) at 4 mg protein per milliliter of glycylglycine-HCl buffer, pH 8.5, ionic strength 0.05, yields a single symmetrical peak after 97 minutes at 7.5 ma and 190 volts. The minimum molecular weight, based on FMN content, is estimated to be 1.2×10^5.

Appearance and Absorption Spectra. The purified dehydrogenase in concentrated solution is clear and dark amber. The absorption spectrum of the oxidized enzyme shows maxima at 450, 335, and 275 mμ with very small shoulders at 415 and 550 mμ. Samples with lower specific activity exhibit a distinct peak at 415 mμ.

Composition. The dehydrogenase contains nonheme iron, labile sulfide and FMN; the ratio of FMN:Fe is approximately 1:4. No lipid or FAD is detected.

Specificity. The purified enzyme catalyzes the oxidation of NADH by cytochrome *c*, DCIP, ferricyanide, and menadione, but coenzyme Q_6, coenzyme Q_{10}, lipoic acid, lipoamide, or oxygen shows no measurable activity. The relative rates ($V_{max}^{acceptor}$ in terms of NADH oxidized) of oxidation by these acceptors are: cytochrome *c*, 100; ferricyanide, 100; DCIP, 74; and menadione, 72. NADPH is not oxidized.

pH Optimum. The optimum pH for the reduction of cytochrome *c* is 8.8–9.0 in glycylglycine buffer. Under the same conditions, two ill-defined optima at pH 7.5 and at 8.5 occur in the DCIP assay.

Stability. The purified enzyme is highly unstable, losing in 20 hours storage in air at −15° approximately 25% of its activity toward either DCIP or cytochrome *c* and approximately 50% when stored at 0°. In the absence of oxygen it is relatively stable. No significant loss of activity is observed when the enzyme is stored *in vacuo* for 36 hours at 0°.

Fluorescence. The dehydrogenase possesses a fluorescence peak at 330 mμ at activation wavelength of 280 mμ. When activation at 370 mμ is used, a fluorescent peak occurs at 520 mμ. The fluorescence intensity of the latter is approximately 3% of an equivalent amount of free FMN.

Inhibitors and Activators. *p*-Chloromercuribenzoate, 10^{-6} *M*; *o*-phenanthroline, $10^{-3}\,M$; sodium arsenite, $10^{-2}\,M$; zinc chloride, $10^{-2}\,M$; dicoumarol, $10^{-3}\,M$ are all inhibitory to both cytochrome *c* and DCIP reactions. Cyanide, $10^{-2}\,M$; $CaCl_2$, $10^{-2}\,M$; $MgCl_2$, $10^{-2}\,M$; phosphate, $10^{-2}\,M$; or perchlorate, $10^{-2}\,M$ inhibits the cytochrome *c* activity but stimulates the DCIP reaction. Antimycin A (2 μg per milligram of protein) or Amytal, $5 \times 10^{-3}\,M$, does not affect the activity of the dehydrogenase.

Some Kinetic Constants. Some constants are summarized in Table II.

TABLE II
SOME KINETIC CONSTANTS OF THE 37° ENZYME[a]

Oxidant	$V_{max}^{acceptor}$	V_{max}^{NADH}	$K_m^{acceptor}$ (*M*)	K_m^{NADH} (*M*)	Turnover number ($\times 10^3$)
Cytochrome *c*	116	21[b]	7.8×10^{-5}	8.5×10^{-5}	7.1
2,6-Dichloro-phenolindophenol	43	14	13×10^{-5}	1.7×10^{-4}	5.3
Ferricyanide	114	20[b]	5.2×10^{-4}	7.3×10^{-5}	7.0
Menadione	41	25[b]	3.8×10^{-5}	4.6×10^{-4}	5.0

[a] V_{max}^{NADH} and $V_{max}^{acceptor}$ values are stated as micromoles of NADH and micromoles of acceptor reacted per minute per milligram of enzyme at about 23°. Turnover numbers are calculated from $V_{max}^{acceptor}$ values and expressed as moles of NADH oxidized per minute per mole of FMN in the enzyme.

[b] High concentrations of NADH are inhibitory. The values listed are estimated from the extrapolation of the linear part of the Lineweaver-Burk plot.

Preparation of the 30° Enzyme

The method to be described is that initially reported by Ringler *et al.*[9] and subsequently modified by Cremona and Kearney.[8] It involves the digestion of electron transport particle (ETP)[25] by *Naja naja* venom (or with partially purified phospholipase A) at 30° followed by ammonium sulfate fractionation, protamine sulfate precipitation, column chromatography on Sephadex G-200, and one or two sucrose gradient centrifugations.

Partial Purification of Phospholipase A[8]

An aqueous solution of *Naja naja* venom at 100 mg/ml is adjusted to pH 3.5–3.7 with 1 *N* H_2SO_4 at room temperature and then placed in a boiling water bath for 10 minutes. The mixture is cooled to room temperature and 1 *M* K_2HPO_4 is added to make the final phosphate concen-

[25] ETP is prepared by modification[9] of the published method [F. L. Crane, J. L. Glenn, and D. E. Green, *Biochim. Biophys. Acta* **22,** 475 (1956)].

tration 0.05 *M*. The suspension is adjusted to pH 7.6 with 3 *N* NH_4OH and then centrifuged for 15 minutes at 15,000 *g* at 0°. The protein concentration in the supernatant liquid ranges from 55 to 60 mg/ml. It is then chromatographed at 0° on a column of Sephadex G-75, previously equilibrated with 50 m*M* phosphate, pH 7.6. The void volume (V_0) of the column should be at least 30% greater than the sample volume. Elution is carried out with 50 m*M* phosphate, pH 7.6, and fractions equal to 25% of the V_0 are collected.

These fractions are assayed for protein by the biuret reaction[13] and for phospholipase A activity by the following method. In the main compartment of a Warburg flask are placed 0.28 ml of 0.3 *M* $NaHCO_3$, 0.1 ml of 8.5% NaCl, 2.0 ml 40% egg yolk suspension in 0.2% (v/v) Triton X-100, and 0.42 ml of water. To the side arm, are added 0.02 ml of 0.3 *M* $NaHCO_3$ and 0.18 ml diluted enzyme in 1% bovine serum albumin, and 50 m*M* phosphate buffer, pH 7.4. The evolution of CO_2 is followed at 38° under an atmosphere of 95% N_2–5% CO_2 for the 3 to 33 minute period after tipping in the enzyme. The enzyme concentration should be such that 20–60 μl of CO_2 is evolved during this period. Suitable reagent blanks and triplicate determination at two levels of the enzyme are recommended for accurate results. One unit of phospholipase A activity is defined as 1 μl of CO_2 liberated per 30 minutes in the assay system described. Specific activity is units per milligram of protein.

Specific activity of the enzyme thus prepared ranges from 1500 to 2600. Rechromatography can increase the activity to about 3600.

Purification of the Dehydrogenase

Ammonium sulfate used in the purification is first recrystallized from 1% EDTA and then from water. All manipulations are conducted at 0–4° except otherwise indicated.

Step 1. Preliminary Digestion. A suspension of ETP,[25] containing 15–17 mg protein per milliliter in 0.25 *M* sucrose–0.025 *M* phosphate buffer, pH 7.4, is rapidly brought to 30° by heating in a water bath. The mixture is treated at 30° with 0.67 to 1 μg of partially purified phospholipase A (specific activity 700–1000) per milligram of ETP protein and it is stirred for exactly 5 minutes, then transferred to a salt–ice water bath and cooled rapidly to between 5 and 10°. The suspension is centrifuged for 1 hour at 59,000 *g* in a No. 21 Spinco rotor. The supernatant fluid obtained is discarded, and the pellet and fluffy layer are resuspended, with the aid of a Potter-Elvehjem homogenizer, in sufficient 0.25 *M* sucrose–0.025 *M* phosphate buffer, pH 7.4, to yield the original volume of ETP.

Step 2. Solubilization of the Dehydrogenase. The suspension obtained in step 1 is rapidly brought to 30°, and 3.3 μg purified phospholipase A (specific activity, 2000) per milligram of ETP protein is added. Incubation at 30° with stirring is continued for 90 minutes, after this time the suspension is rapidly cooled to below 10° and centrifuged at 59,000 *g* as in step 1. The clear yellow supernatant liquid is collected, and the residue is discarded.

Step 3. Concentration and Dialysis. Solid ammonium sulfate is added to the supernatant liquid from step 2 to 0.4 saturation. The pH of the solution is maintained at pH 7.4 by 6 *N* NH_4OH during the addition of the ammonium sulfate. After 15 minutes of equilibration, the precipitate is collected by centrifuging the mixture for 30 minutes at 4000 *g* and is then dissolved in 0.03 *M* phosphate buffer, pH 7.6, to a protein concentration of 8–10 mg/ml. Mild homogenization facilitates the solution at this stage. The solution is dialyzed overnight against buffer. For 20–25 g ETP, the use of 20 liters of 0.03 *M* phosphate buffer, pH 7.6, for the first 3 hours of dialysis followed by another 20 liters for the remaining 6–9 hours is recommended.

Step 4. Treatment with Protamine Sulfate. The dialyzed enzyme (step 3) is centrifuged at 144,000 *g* for 10 minutes to remove any insoluble material. Protamine sulfate solution (1% weight per volume in 0.01 *M* phosphate, pH 7.4) is added to the clarified enzyme in the ratio of 0.04 mg protamine sulfate per milligram of protein.[26] After 10 minutes of equilibration, the precipitate formed is removed by centrifugation at 12,000 *g* for 10 minutes, and the supernatant liquid is collected.

Step 5. Ammonium Sulfate Fractionation. To the supernatant liquid from step 4 is added 20 g ammonium sulfate per liter.[27] After 15 minutes' equilibration, the precipitate is removed by centrifugation at 15,000 *g* for 10 minutes and the supernatant liquid obtained is mixed with an additional 160 g ammonium sulfate per liter. The precipitate is collected by centrifugation after 15 minutes' equilibration as before, and is suspended in 0.03 *M* phosphate buffer, pH 7.8, to a protein concentration of 13–15 mg/ml. This suspension is dialyzed overnight against the same buffer.[28]

Step 6. Column Chromatography on Sephadex G-200. The dialyzed enzyme from step 5 is clarified by brief centrifugation if necessary and then diluted to a protein concentration of 10–12 mg/ml with 0.03 *M*

[26] This is the optimal amount of protamine sulfate for the average preparation. A preliminary titration with protamine sulfate to remove the "maximal" impurities while leaving 90–95% of the activity in solution is recommended.

[27] During the addition of ammonium sulfate at this and subsequent stages, the pH of the solution is maintained at 8.0 by addition of 6 *N* NH_4OH.

[28] For 20 g ETP, a total of 6 liters of buffer is recommended. The buffer is changed twice during the dialysis.

phosphate buffer, pH 7.8. The enzyme solution is applied to a column[29] of Sephadex G-200 and eluted from the column with 0.03 *M* phosphate, pH 7.8 (at 0°). The excluded fraction is collected and used for subsequent purification.

The diluted enzyme solution is concentrated by adding 243 g ammonium sulfate per liter. After 15 minutes of equilibration, the precipitate is collected by centrifuging the mixture 10 minutes at 15,000 *g* and dissolved in 0.03 *M* phosphate, pH 7.8, to a protein concentration of about 30 mg/ml. The enzyme solution is then dialyzed against the same buffer overnight with two changes of 1 liter each.

Step 7. Gradient Centrifugation in Sucrose Solution. The dialyzed enzyme from step 6 is diluted[30] with 0.1 volume of 1 *M* glycine buffer, pH 10.0, at 0°. To each of three 5-ml tubes (of cellulose nitrate for Spinco centrifuge, Model L, rotor SW 39) containing a discontinuous sucrose gradient[31] from 13 to 50%, is added 0.5 ml of the diluted enzyme. The tubes are centrifuged at 40,000 rpm for 11 hours at 2°. The rotor is allowed to come to rest without braking and the tube contents removed immediately to prevent diffusion. Fractions, containing 10 drops each, are collected manually.[32] Usually 40 to 41 fractions are obtained from

[29] For 20 g of ETP, columns of 4.5–5 cm diameter are recommended. The Sephadex G-200 is equilibrated with 0.03 *M* phosphate buffer, pH 7.8, at 0°. The original method recommends that the void volume (V_0) of the column should be 20–25% greater than the volume of the sample to be chromatographed. According to the authors,[8] the activity is collected in the first V_0. However, there seems to be some disparities in defining void volume, V_0 [for example, the Sephadex manuals and J. R. Whitaker, *Anal. Chem.* **35,** 1950 (1963)]. The original procedure[8] does not precisely describe V_0 but refers to it only as "hold-back" or "void" volume. Actually, the excluded fraction contains the dehydrogenase. The term, excluded fraction, is conventionally used as the fraction which is not retained by the gel phase, i.e. K_D (the "distribution coefficient") = 0 (for K_D, see Sephadex Manual No. 2).

[30] Instead of diluting the sample as described, it may be passed through a Sephadex G-25 column equilibrated with 50 m*M* glycine buffer, pH 10.0, at 0°. It should be emphasized that when the enzyme is in glycine buffer, pH 10.0, the temperature must be kept between 0 and 2°. Higher temperatures cause denaturation of the enzyme.

[31] The sucrose gradient is prepared by placing 1 ml 50% (w/v) sucrose in the bottom of the centrifuge tube, then carefully layering 0.5-ml aliquots of 44%, 37%, 31%, 25%, 19%, and 13% sucrose solutions on top in the order given. All sucrose solutions are prepared in 50 m*M* glycine buffer, pH 10.0, at 0°. The tubes are then allowed to stand vertically 10–24 hours before addition of the enzyme solution.

[32] The collection tool is described by R. G. Martin and B. M. Ames [*J. Biol. Chem.* **236,** 1372 (1961)]. Approximately 15–20 minutes are required to collect the fractions from each tube. The temperature must be kept as close to 0° as possible during this process.

each tube. Corresponding fractions from each centrifuge tube are pooled and those containing colored material are assayed for activity.[33] Those fractions with the highest specific activity are combined, and sucrose is removed by passing the enzyme solution through a column of Sephadex G-25 equilibrated with 0.03 *M* phosphate buffer, pH 7.8. The enzyme solution thus obtained is concentrated at 0° to about 15–20 mg protein per milliliter in an ultrafiltration device manufactured by Carl Schleicher and Schull Company. The specific activity of the dehydrogenase at this stage is about 600.

Step 8. Second Sucrose Gradient Centrifugation. The purified enzyme prepared in step 7 is further purified by a second sucrose gradient centrifugation. To 5 ml centrifuge tubes, containing a discontinuous sucrose gradient[34] varied from 28 to 50%, are added 0.5-ml aliquots of the concentrated enzyme prepared in step 7. Centrifugation is the same as in step 7 except that the time is extended to 30 hours. The collection, assay, and concentration of the fractions are the same as described in step 7. The specific activity of the dehydrogenase at this stage is about 940. Table III is a summary of the preparation.

TABLE III
PURIFICATION OF THE 30° ENZYME[a,b]

Fraction	Total activity (units)	Specific activity (units/mg)
ETP	450,000	27.5
Venom extract	435,000	185
Fraction after ammonium sulfate	280,000	300
Fraction excluded from Sephadex G-200	220,000	360
After first sucrose gradient	176,000	660
After second sucrose gradient	90,000	940

[a] T. Cremona and E. B. Kearney, *J. Biol. Chem.* **239,** 2328 (1964).
[b] Activities are expressed as $V_{max}^{ferricyanide}$ at 30°.

Further Purification. Recently, Lusty *et al.*[35] have reported further modifications of the purification procedure presented above. Specific activities as high as 1500 can be obtained by these modifications. In step 4, the ratio of protamine sulfate is increased from 0.04 to 0.044 mg per milligram of protein. In step 6, the Sephadex G-200 is equilibrated with 0.05 *M* glycine buffer, pH 8.8, at 0°. In steps 7 and 8 the sucrose gradients

[33] These assays may be done at fixed ferricyanide concentration (1 m*M*).
[34] This sucrose gradient is prepared similarly to that described.[31] The sucrose solutions are, in ascending order, 50%, 44%, 41%, 37%, 34%, 31%, and 28%.
[35] C. J. Lusty, J. M. Machinist, and T. P. Singer, *J. Biol. Chem.* **240,** 1804 (1965).

are varied from 19% to 65%, and 28% to 65%, respectively. Centrifugation of the system at pH 10.0 is increased from the 11 hours used in step 7 to 17–20 hours. According to Lusty *et al.*[35] these changes yield samples of specific activity of about 1375 after the first sucrose gradient centrifugation and about 1500 after the second.

Our experience for the preparation of the 30° enzyme is discussed by King *et al.*[2]

Properties of the 30° Enzyme

Physical Constants. Sedimentation analysis of the purified enzyme[9] (specific activity 940–970) at pH 10.0 at 0° in the concentration range of 6–10 mg/ml yielded an $S_{20,w}$ of 14 S. The minimum molecular weight based on FMN content is 550,000. Less pure samples (specific activity of about 200), however, show practically one single band in electrophoresis.[9]

Appearance and Absorption Spectra. The dehydrogenase in concentrated solution is deep amber-brown. The absorption spectrum of the oxidized enzyme shows a maximum at 410 mμ. Reduction of the enzyme by NADH bleaches the color and decreases the absorbance in the 450–460 mμ and 410 mμ regions. Difference spectrum of the NADH-reduced enzyme minus oxidized shows a single broad band centering at 420–425 mμ.[9]

Electron Paramagnetic Resonance Spectra. EPR studies[36] with this enzyme indicate the presence of a NADH-inducible asymmetrical signal at $g = 1.94$ with a minor component at $g = 2.00$. NADPH at high concentrations also induces an EPR signal at $g = 1.94$.

Composition. The dehydrogenase (specific activity 940–970) contains FMN, 1.23 millimicromoles; and nonheme iron, 21 to 22 mμatoms per milligram of protein.[8]

Specificity. The partially purified enzyme (specific activity 200) has no appreciable rate with acceptors other than ferricyanide. Relative to ferricyanide (arbitrarily set at 100), the activity with cytochrome *c* is 0.01, and with lipoamide, 0.06. The ratio $V_{max}^{ferricyanide} : V_{max}^{cytochrome\ c}$ in purified preparations (specific activity 900) is about 6000.

Of the substrates tested, NADH is the most active followed by reduced nicotinamide-hypoxanthine dinucleotide (NHDH) at about 50% of the NADH rate. Other NADH analogs are only 1.5% or less of the NADH rate, and NADPH is completely inactive.[37]

[36] H. Beinert, G. Palmer, T. Cremona, and T. P. Singer, *J. Biol. Chem.* **240,** 475 (1965).

[37] S. Minakami, T. Cremona, R. L. Ringler, T. P. Singer, *J. Biol. Chem.* **238,** 1529 (1963).

Other Activities.[37] In addition to its ability to catalyze the oxidation of NADH by ferricyanide, the enzyme also possesses transhydrogenase activity when assayed with various NAD analogs.

pH Optimum. The optimum pH for the dehydrogenase at maximum velocity with respect to ferricyanide at 30° is 7.8 in triethanolamine buffer.[37] In phosphate buffer the double reciprocal plots approach 0 at pH values alkaline to about pH 7.4, thus an accurate estimation of the pH optimum in this medium is difficult. However, an alternate method utilizing the transhydrogenase assay yields a plateau in the region of pH 8.5–9.0 in phosphate buffer.[7]

Kinetics. K_m values[38] are for NADH, 1.08×10^{-4} *M*; AcPyADH,[39] 7.9×10^{-5} *M*; AcPyHDH, 4.8×10^{-5} *M*; NHDH, 4.0×10^{-5} *M*; PyAl-ADH, 6.4×10^{-5} *M*; TNADH, 2.4×10^{-5} *M*.

Inhibitors.[7,37,40] The purified dehydrogenase is inhibited 56% by 1.4 m*M* hypoxanthine-adenine dinucleotide, 100% by 3 m*M* FMN, 25% by 0.1 m*M* FAD, and 56% by AcPyAd when assayed at maximum velocity with respect to ferricyanide. In addition, NAD and NADH are inhibitors of the enzyme competitive with ferricyanide. Under certain conditions, *p*-chloromercuriphenylsulfonate or *p*-chloromercuribenzoate also inhibits the enzyme. Amytal, 3 m*M*; antimycin A, 10^{-6} *M*; cyanide, 1 m*M*; and azide, 60 m*M* have no effect.

Stability.[8] The purified enzyme (specific activity less than 500) is stable for many weeks when stored in concentrated form in the cold between pH 7 and 8. However, at specific activities greater than 500 the enzyme loses activity under these conditions within 24–48 hours.

Preparation of the Dehydrogenase of Mackler *et al.*[15]

NADH dehydrogenase is solubilized from a particulate NADH oxidase[41] by dilute ethanol at pH 4.8. The following procedure is sum-

[38] These values (K_m at 30°)[37] are calculated from infinite concentrations of both ferricyanide and the substrate by a Lineweaver-Burk plot. A straight line is obtained by plotting reciprocals of $V_{max}^{ferricyanide}$ values as a function of the reciprocals of substrate concentrations used. From this line, K_m values are computed in the conventional manner.

[39] The abbreviations are: AcPyAD, and AcPyADH, acetylpyridine-adenine dinucleotide and its reduced form; AcPyHDH, reduced acetylpyridine-hypoxanthine dinucleotide; NHDH, reduced nicotinamide-hypoxanthine dinucleotide; PyAlADH, reduced pyridinealdehyde-adenine dinucleotide; and TNADH, reduced thionicotinamide-adenine dinucleotide.

[40] T. Cremona and E. B. Kearney, *J. Biol. Chem.* **240,** 3645 (1965).

[41] For the preparation, see this volume [49].

marized from the papers by Mackler *et al.*[14,15,42,43] All procedures were performed at 0–5° except where otherwise indicated.

The particulate NADH oxidase preparation[41] in sucrose is centrifuged at 105,000 *g* for 60 minutes. The pellet is, with the aid of a Potter-Elvehjem homogenizer, suspended in water, recentrifuged, and finally dispersed in water, to a protein concentration of 30 mg/ml. This suspension is adjusted to pH 4.8 with dilute acetic acid, and absolute ethanol is added dropwise with constant stirring to a final concentration of 9%. The mixture is immediately placed in a water bath held at 44° and kept there for 15 minutes after it reaches 43°. The mixture is then rapidly cooled (in a salt–ice water bath) to 0° and readjusted to pH 7.0 with 1 *N* KOH. The suspension is centrifuged at 105,000 *g* for 15 minutes, and the clear yellow supernatant liquid is collected and lyophilized to dryness.

The lyophilized powder is dissolved in distilled water, and saturated ammonium sulfate solution (pH 7.0) is added in the ratio of 6 volumes ammonium sulfate to 4 volumes enzyme solution. After 5 minutes, of occasional stirring, the precipitate is collected by centrifuging for 15 minutes at 15,000 *g* and then dissolved in 0.02 *M* phosphate buffer, pH 7.5.

Properties of the Mackler Enzyme[14,15,42–44]

Physical Constants. The molecular weight[44] of the enzyme after further purification[15] is 70,000–90,000 as determined by the equilibrium sedimentation technique.

Appearance and Absorption Spectra. The enzyme in the oxidized form has maxima at 550 and 418 mμ and a shoulder at about 450 mμ. Reduction by NADH or hydrosulfite causes extensive bleaching of the color and decrease of absorbance in the 450 mμ and 418 mμ regions.

Composition.[15,44] The dehydrogenase contains 11 millimicromoles FMN and 20 mμatoms nonheme iron per milligram protein.

Specificity. The rates[14,44] of NADH oxidation are by ferricyanide 80–85; DCIP, 60–70; cytochrome *c*, 5.0; FAD, 5.0; FMN, 5.0; and O_2, 0.03 in the unit of micromoles NADH per minute per milligram of protein at 30°. K_m for NADH is 1.1×10^{-5} *M* with DCIP as acceptor.[14]

Inhibitors.[14] *p*-Chloromercuribenzoate at 10^{-4} *M* completely inhibits the enzymatic activity with any of the acceptors tested. Antimycin A, 1

[42] F. M. Huennekens, S. P. Felton, N. A. Rao, and B. Mackler, *J. Biol. Chem.* **236**, PC 57 (1961).

[43] N. A. Rao, S. P. Felton, F. M. Huennekens, and B. Mackler, *J. Biol. Chem.* **238**, 449 (1963).

[44] B. Mackler, *IUB Flavin Symp. Amsterdam, 1965*. Elsevier, Amsterdam, 1966.

μg/ml, inhibits 12% of the DCIP and FAD activities, but does not affect the reaction with cytochrome *c* or oxygen as acceptors. Amytal, 10^{-3} *M*, inhibits 30% of the DCIP and FAD activities but does not affect the reaction with cytochrome *c* or oxygen. Phosphate even at 1 *M* does not decrease the cytochrome *c* activity in great contrast to the Mahler enzyme.[1]

pH Optima. The optimum pH in 1 *M* phosphate buffer is approximately pH 8.5 for reactions with DCIP and cytochrome *c*. For reactions with ferricyanide in 0.02 *M* phosphate buffer the optimum is pH 7.5.

Stability. The enzyme is stable when stored at —70°. When stored at —20° there is a slow and progressive loss of activity after 1 week.[15]

After the manuscript had been submitted, we were informed by Dr. Mackler that this enzyme in the reduced form also shows EPR signal typical to nonheme iron at liquid helium temperatures.

Preparation of Thiourea Enzyme

Chapman and Jagannathan were the first to use thiourea as a solubilizing agent for the preparation of a NADH dehydrogenase from ETP. The procedure has not been published, except as a short preliminary note for private circulation.[45] We have used thiourea as a solubilizing agent and the Keilin-Hartree preparation as the starting material. Two active fractions are obtained. The following is based on the method worked out in this laboratory.[2]

All manipulations are conducted at 0–4° unless otherwise indicated. To 140 ml of the heart muscle preparation in 0.25 *M* sucrose (see method 4 in this volume [37]), containing approximately 60 mg protein per milliliter, is added a mixture of 140 ml 0.25 *M* sucrose; 140 ml 1.0 *M* K-phosphate buffer, pH 7.5; 21 ml 1.0 *M* Tris-HCl, pH 7.5; 4.2 ml 50 m*M* EDTA, pH 7.0; and 1.34 g thiourea. The final concentration of thiourea is 0.4 *M*. Immediately, the mixture is distributed to six 250-ml Erlenmeyer flasks which are placed in a salt–ice water bath. The flasks are gently swirled until the contents are nearly completely frozen. They are allowed to stand at —25° for 20 minutes, then thawed under tap water, and finally centrifuged at 105,000 *g* for 45 minutes.

The pellet is suspended in water containing an amount of thiourea to give a final concentration of 0.5 *M* after the mixture is diluted to 190 ml with water. The mixture is distributed into two 250-ml Erlenmeyer flasks, then rapidly frozen as before, and kept at —25° for 2 hours. The mixture is thawed, then centrifuged for 45 minutes at 105,000 *g*. The pellet is

[45] A. G. Chapman and V. Jagannathan, Scientific Memo 45, Information Exchange Group No. 1 (1963).

discarded. To the supernatant liquid (190 ml), are added 20 ml 0.1 *M* phosphate buffer, pH 7.4, and 62 g ammonium sulfate. The mixture is stirred 10 minutes, then centrifuged at 18,000 *g* for 10 minutes. The precipitate collected is dissolved in 37 ml 0.01 *M* phosphate buffer, pH 7.4. The clear, yellow solution is adjusted to pH 7.0 with *N* acetic acid and then 18.5 ml Alumina Cγ suspension[46] (27 mg solid per milliliter) is added. The mixture is gently stirred for 10 minutes and centrifuged at 10,000 *g* for 5 minutes. The gel precipitate is eluted twice, each time with 37 ml 0.1 *M* phosphate buffer, pH 7.4.

To the pooled eluates (74 ml) is added 21.8 g ammonium sulfate. After gentle stirring for 10 minutes, the precipitate is collected by centrifugation at 18,000 *g* for 10 minutes and dissolved in 6 ml 0.1 *M* phosphate buffer, pH 7.4. This solution is further fractionated by sequential additions of saturated ammonium sulfate to obtain the percentage saturation listed in Table IV. The precipitate from each fractionation is collected by centrifugation and dissolved in 1.5 ml phosphate buffer.

Table IV summarizes the various steps in the purification. Two frac-

TABLE IV
PURIFICATION OF NADH DEHYDROGENASE BY THE THIOUREA METHOD

Fraction	Specific activity (unit/mg): Ferricyanide	Specific activity (unit/mg): Cytochrome *c*	Recovery (%): Ferricyanide	Recovery (%): Cytochrome *c*	Recovery (%): Protein	Activity ratio, ferricyanide: cytochrome *c*
Thiourea supernatant	32.1	2.19	100	100	100	15
First ammonium sulfate fractionation	27.9	3.00	60.2	94.5	69	13
Alumina Cγ eluates	57.8	4.00	38.1	38.6	21	15
Second ammonium sulfate fractionation	56.5	5.00	27.5	35.0	15.7	11
Third ammonium sulfate fractionation						
0.0–0.30 saturation	27.3	0.92	1.36	0.67	1.59	30
0.30–0.35 saturation[a]	55.5	1.78	6.35	2.98	3.67	31
0.35–0.40 saturation[a]	74.7	2.99	2.87	1.70	1.23	25
0.40–0.45 saturation	73.2	3.90	2.97	2.32	1.29	19
0.45–0.50 saturation[b]	87.0	7.70	2.85	3.72	1.04	11
0.50–0.55 saturation[b]	91.5	10.9	4.16	7.25	1.45	8.4

[a] Combined and labeled as fraction A.
[b] Combined and labeled as fraction B.

[46] Vol. I, p. 97.

tions collected between 0.30 and 0.40 ammonium sulfate saturation differ from those between 0.45 and 0.55 saturation in the ratio of ferricyanide activity to cytochrome *c* activity. The pooling of fractions is made at 0.30 to 0.40 and 0.45 to 0.55 saturations and they are designated as fractions A and B, respectively.

Properties of Thiourea Enzyme

Sedimentation. Fraction B exhibits essentially one protein peak in a sedimentation experiment performed at 50,740 rpm at 4° for 145 minutes in a Spinco Model E ultracentrifuge. Two peaks are observed in fraction A; the slower peak travels at the same rate as the solitary peak in fraction B.

Appearance and Absorption Spectra. The concentrated enzyme solution is amber. The absorption spectrum of the oxidized form exhibits maxima at 450, 550, and about 420 mμ. Addition of NADH or dithionite bleaches the color and decreases the absorbance.

Composition. In millimicromoles or millimicroatoms per milligram of protein, fraction A contains FMN, 1.98, and nonheme iron, 27.7, whereas fraction B contains FMN, 4.83, and nonheme iron, 49.1. Both fractions also contain acid-labile sulfide.

EPR Signals. The dehydrogenase prepared by Chapman and Jagannathan[45] shows assymmetric EPR signals at $g = 1.94$ and 2.01. The enzymes prepared from the Keilin-Hartree preparation have not been tested for EPR behavior.

Specificity. Both fractions catalyze the oxidation of NADH by ferricyanide, cytochrome *c*, DCIP, menadione, vitamin K_1, FMN, the coenzymes Q_2, Q_3, Q_6, and Q_{10}, and molecular oxygen. NADPH is inactive. $V_{max}^{\text{ferricyanide}}$, $V_{max}^{\text{NADH (ferricyanide, 1 m}M)}$, and $V_{max}^{\text{NADH (cytochrome } c\text{, 0.028 m}M)}$ are 74, 83 and 2.5 respectively for fraction A; corresponding values for fraction B are 137, 200, and 15.

pH Optima. Both fractions show optimum pH at 8.5 for ferricyanide reaction. No pH optimum is found for cytochrome *c* reaction; the activity increases with increasing pH.

Inhibitors. For both fractions, the cytochrome *c* reaction is inhibited by phosphate, NADH, and *p*-chloromercuribenzoate (PCMS) whereas the ferricyanide reaction is depressed by NADH, PCMS, and FAD. These reactions are not inhibited by Amytal, 1,10-phenanthroline, 8-hydroxyquinoline, or atabrine.

Fluorescence. Both fractions show fluorescence with peaks at 330 mμ from 280 mμ activation, and at 520 mμ from 460 mμ activation. The latter fluorescence is presumably due to the FMN content of the enzyme but accounts for only approximately 3% of an equivalent amount of free FMN.

Stability. Both fractions are unstable in air at 0°. Storage *in vacuo* at —20° for 4 days decreases about 20% of the cytochrome *c* and 50% of the ferricyanide activities. The activities toward both acceptors remain practically the same after the samples have been in a liquid nitrogen refrigerator for 2 weeks.

General Remarks

In many aspects these soluble NADH dehydrogenases differ in varying degrees. It is inconceivable that all these exist *in situ* as separate entities. Evidence from a number of lines indicate that these dehydrogenases are derived from the same segment of the respiratory chain. The nature of the soluble enzyme isolated is dependent upon the method and conditions of solubilization employed.[2]

Acknowledgments

The experimental work was supported by grants from the National Science Foundation, the National Institutes of Health, the American Heart Association, and the Life Insurance Medical Research Fund.

[53] DPNH Dehydrogenase of Yeast

By BRUCE MACKLER

$$\text{DPNH} + 2\ \text{Fe(CN)}_6^{-3} \rightleftarrows \text{DPN}^+ + 2\ \text{Fe(CN)}_6^{-4} + \text{H}^+$$

Assay Method

Principle. Enzymatic activity was determined spectrophotometrically at 38° by measuring the rate of decrease of absorbancy at 420 mμ corresponding to the enzymatic reduction of ferricyanide.

Reagents

DPNH solution, 1%
Potassium ferricyanide solution, 0.005 *M*
Sodium azide solution, 0.1 *M*
Sodium acetate buffer, 0.2 *M*, pH 5.5
Enzyme. Dilute to required concentration with 0.02 *M* potassium phosphate buffer, pH 7.5.

Procedure. A mixture of 0.05 ml of 1% DPNH solution, 0.1 ml of 0.005 *M* ferricyanide solution, 0.1 ml of 0.1 *M* azide solution, 0.2 ml of

0.2 *M* acetate buffer, pH 5.5, and sufficient water to make a final volume of 1.0 ml is added to a 1-ml optical cuvette of 10-mm light path, and the mixture is brought to 38°. The optical density of the sample is read in a spectrophotometer every 15 seconds at 420 mμ against a cuvette containing only buffer and water, and the nonenzymatic rate of ferricyanide reduction is followed for a period of 1 minute. The enzymatic reaction is begun by addition of enzyme (usually 1–2 μg of protein) to the sample cuvette, and the optical density change is followed every 15 seconds at 38° for a period of 1–2 minutes. For calculation of activity, the nonenzymatic rate is subtracted from the enzymatic rate. Specific activity is expressed as micromoles of DPNH oxidized per milligram of enzyme protein per minute. Protein is determined by the method of Lowry *et al.*[1]

$$\text{Specific activity} = \frac{\Delta\text{OD per minute}}{1 \times 2 \times \text{mg protein in assay}}$$

Purification Procedure

ETP is prepared from Fleischmann's yeast as described previously.[2] All procedures are carried out at 0 to 5°. The ETP is washed with water by centrifugation and suspended in water (30 mg of protein per milliliter) and the pH is adjusted to 5.5 by slow addition of 1 *M* KH_2PO_4. Absolute ethyl alcohol is added to a final concentration of 9%; the suspension is incubated at 35° for 15 minutes and then cooled, and the pH is adjusted to 7.0 with 6 *N* KOH. The suspension is centrifuged in the No. 30 rotor of the Spinco Model L ultracentrifuge for 15 minutes at 30,000 rpm, and the clear supernatant is evaporated to dryness by lyophilization. The lyophilized residue is dissolved in a minimum amount of water and dialyzed overnight against 8 liters of 0.005 *M* phosphate buffer of pH 7.0. A column measuring 2.5 cm in diameter is packed to a height of 25 cm with DEAE-cellulose which has been repeatedly washed with water and equilibrated with the dialysis solution diluted with an equal volume of cold water. The dialyzed enzyme solution is placed on the column, the column is washed with 500 ml of diluted dialysis fluid, and gradient elution is applied. The mixing chamber contains 400 ml of the diluted dialysis fluid (0.0025 *M* phosphate buffer, pH 7.0), and the reservoir contains 200 ml of 0.05 *M* phosphate buffer of pH 7.0. The rate of flow of the column is such that 12 tubes of 9.0 ml each are obtained over a 1

[1] O. H. Lowry, N. J. Rosebrough, A. L. Farr, and R. J. Randall, *J. Biol. Chem.* **193**, 265 (1951).

[2] B. Mackler, P. J. Collipp, H. M. Duncan, N. A. Rao, and F. M. Huennekens, *J. Biol. Chem.* **237**, 2968 (1962).

hour period. Fractions of enzyme having the highest activities are found in tubes 40–60 and the fractions are combined and lyophilized. The lyophilized residue is dissolved in a minimum amount of water and dialyzed overnight against 8 liters of 0.01 M phosphate buffer of pH 7.0. A saturated solution of ammonium sulfate (pH 7.0) is added slowly with stirring to the enzyme solution to a final saturation of 70%. The suspension is centrifuged in the No. 40 rotor of the Spinco ultracentrifuge at 40,000 rpm for 15 minutes and the residue discarded. Sufficient solid ammonium sulfate is added to the supernatant solution to make it 100% saturated, and the suspension is allowed to stand with stirring at 5° for 10 minutes. The precipitate, or purified enzyme, is collected by centrifugation at 40,000 rpm for 15 minutes in a Spinco ultracentrifuge and dissolved in 0.02 M phosphate buffer of pH 7.5. Purification of the enzyme is shown in the table.

PURIFICATION OF YEAST DPNH DEHYDROGENASE

Preparation	Total Protein (mg)	Specific activity	Total activity
Yeast ETP	15,500	1.0	15,500
Crude extract after 1st lyophilization	1019	3.0	3057
DEAE-cellulose	11.4	52	593
After ammonium sulfate fractionation	3.0	182	546

Properties[3]

Preparations of enzyme catalyze the oxidation of DPNH but not TPNH and retain full activity on storage at −20° for several months. In addition to ferricyanide, the enzyme catalyzes the reduction of 2,6-dichlorophenolindophenol and cytochrome *c*. The enzyme may be assayed at pH 7.5 by using 0.2 ml of 0.2 M phosphate buffer of pH 7.5 in the assay mixture instead of acetate buffer. Since the enzyme loses some bound FAD during purification, maximal rates at pH 7.5 are obtained by addition of FAD (2 μg) to the assay cuvette. Preparations of enzyme contain 1.0 mole of FAD, 0.4 mole of nonheme iron, and between 1 and 2 sulfhydryl groups per mole of enzyme.

[3] H. M. Duncan and B. Mackler, *Biochemistry* **5**, 45 (1966).

[54] NADH–CoQ Reductase—Assay and Purification

By D. R. SANADI,[1] RICHARD L. PHARO,[1] and LOUIS A. SORDAHL[1]

The reaction catalyzed by the enzyme may be represented as follows:

$$NADH + H^+ + CoQ \rightarrow NAD^+ + CoQH_2$$

The identity of the electron carrier which is located between the NADH dehydrogenase flavoprotein and cytochrome *b* is still in dispute, although available evidence favors CoQ_{10} as the most likely compound. The methods for the assay of NADH-CoQ reductase activity in submitochondrial particles and soluble extracts using the higher isoprenologes of CoQ have been devised recently.[1a-3] It will become apparent during this description that the assay requires strict control of several variables since the substrate is essentially in a nonaqueous phase because of its low solubility. For routine measurement of activity during the preparation of the enzyme, menadione reductase activity, which is also described here, has been found to be useful.

Principle. The enzyme is assayed spectrophotometrically by recording the decrease in absorbancy at 340 mμ during oxidation of NADH by CoQ.

Reagents

Tris-sulfate, 0.5 *M*, pH 8.0. Tris(hydroxymethyl)aminomethane, 0.55 *M*, is adjusted to pH 8.0 with concentrated sulfuric acid and diluted to 0.5 *M*.

CoQ_6, 6 m*M*, or 4 m*M* CoQ_{10} in methanol. The latter is held at 30° to promote solubility.

Menadione (2-methyl-1,4-naphthaquinone), 10 m*M*, in methanol

KCN, 30 m*M*

NADH, 4 m*M*

Procedure. The following are added to a standard cuvette (1-cm light path) with shaking or stirring between each addition: enough water to

[1] This manuscript was prepared while the authors were at the Gerontology Branch, National Heart Institute, National Institutes of Health, Bethesda, Maryland, and The Baltimore City Hospitals, Baltimore, Maryland.

[1a] D. R. Sanadi, T. E. Andreoli, R. L. Pharo, and S. R. Vyas, *in* "Energy-Linked Functions of Mitochondria," (B. Chance, ed.), p. 26. Academic Press, New York, 1963.

[2] R. L. Pharo and D. R. Sanadi, *Biochim. Biophys. Acta* **85**, 346 (1964).

[3] R. L. Pharo, L. A. Sordhal, S. R. Vyas, and D. R. Sanadi, *J. Biol. Chem.* (in press).

give a final reaction volume of 3.0 ml, 0.3 ml Tris buffer, 0.05 ml quinone (CoQ or menadione), 0.1 ml KCN (when mitochondrial particles are used), and 0.1 ml NADH. The mixture is incubated for 2.5 minutes in a water bath at 30°. Enough enzyme is added to give an absorbance change with an initial rate of 0.20–0.35 per minute and quickly mixed by inverting the cuvette. The cuvette is placed in a Cary, Model 15, double-beam spectrophotometer with a 0.5 absorbance neutral-density screen filter in the reference path. The sample compartment of the spectrophotometer is maintained at 30°. The change in absorbance is recorded with a chart speed of 2 inches per minute. Reaction velocity is measured from the initial linear segment which generally lasts for 15–30 seconds with CoQ_6 and with menadione.

Unit and Specific Activity. A unit of reductase activity is defined as that amount of enzyme which is necessary to produce a decrease in NADH at the initial rate of 1 micromole/minute. Specific activity is given as units per milligram of protein, the latter being determined by the biuret reaction[4] in the case of mitochondrial particles and by the Folin method[5] for soluble preparations, with crystalline bovine serum albumin as the standard for both.

Precautions. (1) Since CoQ_6 and CoQ_{10} are "precipitated" immediately on addition to the aqueous medium, a turbidity develops which generally takes incubation for 2.5 minutes at 30° to become stabilized. Longer incubation should be avoided since an apparent decrease in activity results, presumably because the character of the precipitate changes. Since the absorbance due to the scattering suspension amounts to over 0.7, a spectrophotometer with a good monochrometer having low scattered-light contribution is necessary.

(2) The concentration of the enzyme and CoQ_6 are critical in the assay. Under the defined assay conditions, the total absorbance change varies from 0.2 to 0.4 before there is a large decrease in the reaction velocity. If the enzyme concentration is decreased below the indicated level, the reaction rate declines rapidly to almost zero (Fig. 1). However, the initial rate of the reaction is proportional to enzyme concentration. The difficulty has been found to be due to a nonspecific interaction between CoQ_6 and enzyme protein (as well as other protein) which results in loss of activity. For example, 50% inhibition is obtained in the presence of 5 μg bovine serum albumin, hemoglobin, or insulin in the reaction medium. The upper range of enzyme concentration (points *C* to *D* in Fig. 1) in the routine assay is actually at a concentration

[4] E. E. Jacobs, M. Jacob, D. R. Sanadi, and L. B. Bradley, *J. Biol. Chem.* **223**, 147 (1956).

[5] O. H. Lowry, N. J. Rosebrough, A. L. Farr, and R. J. Randall, *J. Biol. Chem.* **193**, 265 (1951).

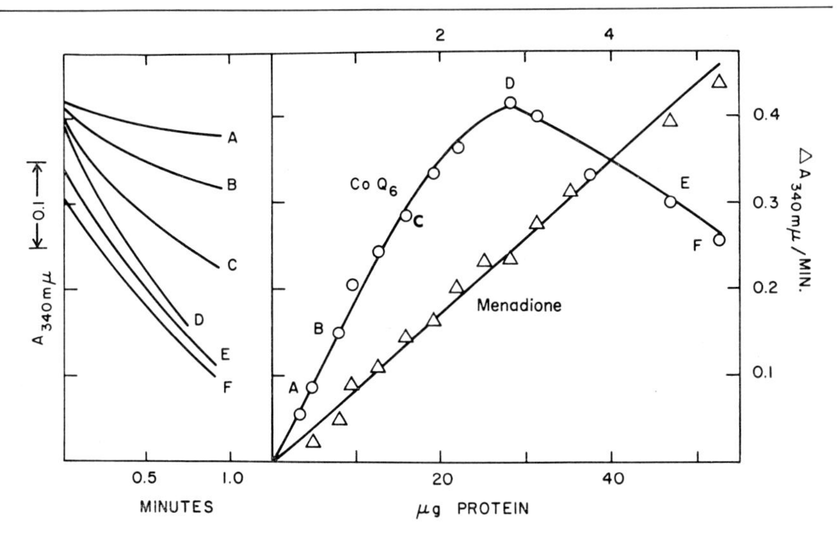

Fig. 1. Relationship between the concentration of the soluble NADH–CoQ reductase and activity. Assay conditions were as described in the text. The left block shows the course of the reaction at different enzyme levels. Curve *C* is in the range of routine assays.

where the activity is not entirely proportional to enzyme concentration, but close enough to it for meaningful data. At higher enzyme concentrations, the activity is again strongly inhibited.

(3) The incubation period of 2.5 minutes before addition of enzyme is adequate in most instances for attaining a steady level of turbidity. If there should be continued change, which can be determined by measuring the absorbance for a few seconds, the reaction mixture should be discarded.

The assay of menadione reductase activity is more simple and not subject to the above complications.

Preparation of Mitochondria and Submitochondrial Particles

Minced beef cardiac muscle is homogenized in 750-g lots with 2.25 liter of a medium containing 0.25 *M* sucrose, 13 m*M* Tris base, 13 m*M* K_2HPO_4, and 0.1 m*M* EDTA for 45 seconds at maximum speed in a 1-gallon Waring blendor. The homogenate is centrifuged at 1000 *g* for 15 minutes in a PR-2 International centrifuge. The supernatant fluid is filtered through several layers of cheesecloth into a chilled beaker and kept cold. The residue is rehomogenized with 1 liter of 0.25 *M* sucrose containing also 15 m*M* K_2HPO_4 for 30 seconds.[6] The homogenate is

[6] This procedure nearly doubles the yield of mitochondria. When heavy mitochondria are needed for routine assay of phosphorylation activity, the second extraction is generally omitted.

recentrifuged for 10 minutes at 1000 *g*, and the supernatant fluid is collected in the same beaker as before. The combined supernatant is passed through a Sharples centrifuge, spinning at 50,000 rpm (62,000 *g*) at a rate of 100 ml/minute. (If smaller amounts of tissue are used, the centrifugation may be carried out at 15,000 *g* for 15 minutes.) After 8 liters of the supernatant liquid have gone through, the mitochondrial sediment is scraped out of the bowl and suspended uniformly in 2 liters of 0.25 *M* sucrose by homogenization in a glass homogenizer with a Teflon pestle. The pH of the suspension is adjusted to 7.8 with 1 *M* Tris base, and the material is centrifuged at 25,000 *g* for 60 minutes in order to remove microsomal contamination. The washed mitochondria are suspended in 0.25 *M* sucrose to a protein concentration of 25 mg/ml and stored at —20°.

It may be noted that the heavy and light layers of mitochondria are not separated since no difference has been observed in the CoQ reductase preparations derived from them. If such a separation is desired, centrifugation at 15,000 *g* for 12–15 minutes is carried out in 0.25 *M* sucose.

The mitochondrial suspension is thawed, adjusted to pH 7.4, and disrupted by exposure to sonic oscillation in the "Branson Sonifier SW110" for 5 minutes at maximum power output. The vessel containing the mitochondria is kept in a bath at —10° during the disruption. The unbroken mitochondria and large particles are removed by sedimentation at 25,000 *g* for 10 minutes. The supernatant liquid is then centrifuged at 105,000 *g* for 45 minutes to yield particles from "sonicated" heavy and light mitochondria (abbreviated SP_{HL}) in the sediment. SP_{HL} is stored at —70° in 0.25 *M* sucrose.

The frozen SP_{HL} is thawed and diluted to a protein concentration of 8 mg/ml with 0.25 *M* sucrose. Phosphate buffer, 1 *M* at pH 6.8, is added to a concentration of 10 m*M* and the pH adjusted to 6.8 if it is not already at pH 6.8. After centrifugation at 39,000 *g* for 15 minutes, the pellet is suspended in water to 25 mg protein per milliliter. The pH is adjusted to 5.3 with 1 *M* acetic acid, and 95% redistilled ethanol is added in a volume equal to 11% of the volume of the particle suspension. The suspension is immersed in a bath at 45° with constant agitation and held at 43° for 15 minutes after the temperature of the contents reaches 43°. The extract is cooled to 10°, neutralized to pH 6.8 to 7.0 with 0.5 *M* NaOH, and centrifuged at 39,000 *g* for 30 minutes. The resulting alcoholic supernatant is clear, straw-colored, and contains the NADH-CoQ reductase. The extract may be lyophilized and stored at —70° for short periods.

The recovery of activity and the purification obtained by the procedure are shown in Table I. The large loss in CoQ_6 reduction activity in going from the particles to the extract may not be real since menadione reductase activity is recovered in high yield. The apparent loss may be explained on the basis that the lipid-rich mitochondrial particle can concentrate CoQ_6 and maintain a higher effective substrate concentration at the reductase active site than it is possible to achieve with the soluble enzyme.

TABLE I
EXTRACTION OF REDUCTASE FROM PARTICLES

	Menadione		Q_6	
Fraction	Specific activity	Total units	Specific activity	Total units
SP_{HL}	0.57	3530	0.41	2570
Extract	82.0	3540	16.0	680

[a] SP_{HL} stands for sonicated particles from heavy and light mitochondria.

The relative activities of the NADH-CoQ reductase with different electron acceptors is seen in Table II. The activity is clearly also a function of the solubility of the substituted CoQ derivatives in the aqueous assay medium. Thus, with decreasing solubility from CoQ_1 to CoQ_{10}, the activity also decreases.

TABLE II
ACTIVITY OF NADH-CoQ REDUCTASE WITH VARIOUS ELECTRON ACCEPTORS[a]

Fraction	Menadione	CoQ_1	CoQ_6	CoQ_{10}	Cytochrome *c*	$K_3Fe(CN)_6$
SP_{HL}	0.22[b]	0.31	0.32	0.27	1.9	8.8
Extract	77	40	15	8.7	26	5.5

[a] The assay conditions have been described for CoQ_6 and CoQ_{10} in the text. CoQ_1 was used as above in place of CoQ_6. Cytochrome *c* reduction was measured at 550 mμ in a medium containing 50 m*M* Tris-sulfate (pH 8.5), 0.1 m*M* cytochrome *c*, (1 m*M* KCN with SP_{HL}), and 0.13 m*M* NADH. The ferricyanide reduction activity was determined at 410 mμ in 50 m*M* Tris-sulfate (pH 7.5), 1.3 m*M* potassium ferricyanide (1 m*M* KCN with SP_{HL}), and 0.13 m*M* NADH. The initial reaction velocity was corrected for the nonenzymatic reaction rate measured under similar conditions.

[b] Values are stated as micromoles per minute per milligram of protein.

Comments on the Preparation. It is important to use in the preparation submitochondrial particles which have not been exposed to drastic treatment. If the particles are exposed to higher pH or *Naja naja* venom, the CoQ_6 reduction activity decreases rapidly with little or no change in menadione reduction activity. Extracts made from such particles, as well as the NADH dehydrogenases prepared by previously described procedures,[7-10] had 0–5% of the CoQ_6 reduction activity compared to the preparation described here.

The CoQ_6 and menadione reduction activities of the soluble reductase are not too stable. Often, CoQ_6 reductase activity is lost faster than the menadione reductase activity. Even as a lyophilized powder at −70°, 50% of the activity is lost in about a week.

[7] H. R. Mahler, N. K. Sarkar, L. P. Vernon, and R. A. Alberty, *J. Biol. Chem.* **199,** 585 (1952).
[8] T. E. King and R. L. Howard, *J. Biol. Chem.* **237,** 1686 (1962).
[9] R. L. Ringler, S. Minakami, and T. P. Singer, *J. Biol. Chem.* **238,** 601 (1963).
[10] B. Mackler, *Biochim. Biophys. Acta* **50,** 141 (1961).

[55] Electron Transfer Flavoprotein

By JOHN R. CRONIN, WILHELM R. FRISELL, and COSMO G. MACKENZIE

Sarcosine + sarcosine dehydrogenase (flavin) → glycine + formaldehyde + dehydrogenase (flavin·2 H)

Dehydrogenase (flavin·2 H) + ETF[1] (FAD) → dehydrogenase (flavin) + ETF (FAD·2 H)

Assay Method

Principle. The preparative and assay methods for the sarcosine dehydrogenase have been described previously.[2-4] The quantitative determination of ETF in the presence of excess sarcosine and sarcosine dehydrogenase is based on the measurement of the rate of reduction of DCPIP.[1]

Reagents

Potassium phosphate, 7.5 m*M*, pH 7.5
Potassium cyanide, 0.02 *M*, adjusted to pH 8.0 with HCl
DCPIP, 0.8 m*M*
Sarcosine, 0.1 *M*
FAD, 1.0 m*M*

[1] Abbreviations: ETF, electron transfer flavoprotein; DCPIP, 2,6-dichlorophenolindophenol; DEAE-cellulose, diethylaminoethyl-cellulose.
[2] See Vol. V [100b].
[3] D. D. Hoskins and C. G. Mackenzie, *J. Biol. Chem.* **236,** 177 (1961).
[4] W. R. Frisell and C. G. Mackenzie, *J. Biol. Chem.* **237,** 94 (1962).

Procedure. The following solutions are added to two cuvettes: 0.05–0.2 ml of solutions of sarcosine dehydrogenase (in excess) and ETF, 0.2 ml potassium cyanide, 0.1 ml FAD, 0.2 ml DCPIP, and potassium phosphate buffer to bring the volume to 2.4 ml. Following incubation at 37° for 5 minutes, 0.2 ml of buffer is added to the "endogenous" reaction mixture, and the zero time reading is recorded. During the next minute interval, 0.2 ml of the sarcosine solution is added to the second cuvette and its initial reading is taken. Changes in optical density are recorded thereafter at intervals of 1.5 minutes and are corrected for the changes observed in the cuvette containing no sarcosine. The photometer is set at zero, at both the zero-time and succeeding readings, against a cuvette containing all reagents except sarcosine and DCPIP.

Definition of Units and Specific Activity. Based on the following relationship between absorbance and dye concentration at 600 mμ in a 3.0 ml volume

$$\text{millimicromoles of dye reduced} = \frac{\Delta A_{600\ \text{m}\mu}}{0.00645}$$

one unit is defined as the amount of enzyme which will cause the reduction of 1 millimicromole of DCPIP in 1 hour. Specific activity is expressed as units per milligram of protein.

Purification Procedures

Presented in this section are several chromatographic procedures for purifying the ETF of rat liver mitochondria. In addition to the methods published previously, in which the ETF is released from the mitochondria by sonic irradiation,[5] newer procedures are also described for chromatographic purification of ETF obtained by osmotic shock. The two preparative methods most commonly employed in our recent studies on ETF are those beginning with the osmotic supernatant, followed by chromatography on either equilibrated or unequilibrated DEAE-cellulose. The data for the latter preparations are summarized in the table.

Isolation of Rat Liver Mitochondria, Sonic Disintegration, and Removal of Electron Transfer Particles. The procedures for the isolation of the "soluble compartment" of the mitochondria have been described previously.[2-4,6] The supernatant fraction obtained from centrifugation of the sonically irradiated mitochondria will be referred to below as the "sonic supernatant." All the ETF in the mitochondria is located in this fraction.[3,5,6]

[5] W. R. Frisell, J. R. Cronin, and C. G. Mackenzie, *J. Biol. Chem.* **237,** 2975 (1962).
[6] W. R. Frisell, M. V. Patwardhan, and C. G. Mackenzie, *J. Biol. Chem.* **240,** 1829 (1965).

Osmotic Shock of Mitochondria. The mitochondrial fraction is suspended in 7.5 mM potassium phosphate (pH 7.5) to give a final volume, in milliliters, equal to 3 times the original weight of the liver (v/w) and is centrifuged at 78,000 g for 30 minutes. The supernatant is replaced by an equal volume of cold distilled water, and the mitochondrial pellet is evenly suspended by gentle homogenization in a Potter-Elvehjem apparatus. The suspension is stirred for 15 minutes in a water bath at 37°, chilled to 0°, and centrifuged at 100,000 g for 1 hour. The resulting supernatant fraction constitutes the "osmotic supernatant."

Ammonium Sulfate Fractionation. The supernatant fraction from sonic irradiation of the mitochondria is fractionated with ammonium sulfate to yield the 0–40%, 40–60%, and 60–80% fractions as described previously.[3] The ETF is concentrated in the fraction obtained between 60 and 80% of saturation.

The osmotic supernatant is fractionated by the addition of solid ammonium sulfate, with stirring in an ice-jacketed beaker. The salt concentration is brought to 60% of saturation by the addition of 3.54 g per 10 ml of solution. The suspension is stirred for 15 minutes, then centrifuged at 15,000 g for 10 minutes. The supernatant is decanted and is made 80% saturated in ammonium sulfate by the addition of 1.32 g of the salt per original 10 ml of the supernatant solution. After it has been stirred for 15 minutes, the precipitate is sedimented by centrifugation as described above.

The ETF in the 60–80% ammonium sulfate fractions is stable for about 1 week when these fractions are stored as wet precipitates at −18°.

Preparation of Chromatographic Absorbents. (a) DEAE-cellulose: DEAE-cellulose (Brown Co., Berlin, New Hampshire) is suspended in 0.5 N NaOH–0.5 N NaCl to give 30 g per 300 ml of solution. The suspension is stirred for 1 hour, and the cellulose is collected on Whatman No. 4 paper on a Büchner funnel. This material is then resuspended on the funnel and filtered with 125-ml portions of 0.5 N HCl–0.5 N NaCl, 0.5 N NaOH–0.5 N NaCl, and 0.5 N HCl–0.5 N NaCl. The DEAE-cellulose is finally suspended in 125 ml of 0.5 N NaOH–0.5 N NaCl and is stirred for 3 hours. After filtration, the cake is resuspended in 125 ml of 0.5 N NaOH–0.5 N NaCl, diluted with water to 1.5 liters, and allowed to stand overnight. The supernatant is decanted, and the cellulose is collected by filtration and washed on the funnel with five 200-ml portions of 50%, 75%, and absolute ethanol, and finally with distilled water until the filtrate gives a negative test for chloride. The resulting cake is suspended in 500 ml of water and stored at 4°.

When required, the DEAE-cellulose is equilibrated with potassium phosphate by passing 0.375 M potassium phosphate solution through the column until the pH of the effluent has returned to the pH of the buffer being applied. The column is finally washed exhaustively with the potassium phosphate buffer in which the enzyme preparation is to be dissolved.

(b) Hydroxylapatite gel: The gel is prepared according to the procedure of Tiselius *et al.*[7] and is stored at room temperature under 1 m*M* potassium phosphate, pH 6.7.

(c) Calcium phosphate gel is made by the method of Keilin and Hartree.[8]

(d) Cellulose powder: The powder (Whatman Standard Grade) is repeatedly suspended in distilled water, and decanted until the material is reasonably free of the finer particles. The cellulose is stored under distilled water at room temperature.

Packing of Chromatographic Columns. The DEAE-cellulose columns are packed and washed under full vacuum with the water pump. The protein and eluent solutions are applied under sufficient vacuum at the water pump to give a flow rate of 1–2 ml per minute. The calcium phosphate–gel mixture is poured into the column and allowed to pack by gravity. Flow rates through these columns by gravity can be adjusted by varying the height of the eluent vessel above the column.

Chromatographic Purification of ETF from Sonic Supernatant. (a) CALCIUM PHOSPHATE GEL-CELLULOSE. 100 mg of the 60–80% ammonium sulfate fraction of the sonic supernatant is dissolved in 15 ml of 7.5 m*M* potassium phosphate, pH 7.6, and dialyzed 2 hours against the same buffer, with one change at 1 hour. Of this solution, 16 ml (80 mg of protein) is applied to the column of calcium phosphate gel-cellulose (1.8×10 cm; gel:cellulose, 1:9 by weight, mixed by overhead stirring). The eluates are collected by gravity in aliquots of 5.0 ml, with the following order of addition of elution buffers to the column: water, 100 ml; 0.04 *M* phosphate, pH 7.6, 150 ml; 0.05 *M* phosphate, pH 7.6, 150 ml; 0.1 *M* phosphate, pH 7.6, 125 ml; 0.1 *M* phosphate, 5% ammonium sulfate, pH 7.6, 75 ml. The ETF is eluted with the 0.05 *M* phosphate and is generally concentrated in tubes 60–75. Compared with the 60–80% ammonium sulfate fraction, the purification by this procedure is 10-fold.

(b) HYDROXYLAPATITE. The 60–80% ammonium sulfate fraction is dissolved in 10 ml of 1 m*M* potassium phosphate, pH 7.2, and dialyzed against a 50-fold volume of the same buffer for 1 hour, with one change. The dialyzed solution is centrifuged at 2300 *g* for 10 minutes. Ten milliliters of the supernatant solution, containing 50 mg of protein, is mixed

[7] A. Tiselius, S. Hjertén, and Ö. Levin, *Arch. Biochim. Biophys.* **65**, 132 (1956).
[8] D. Keilin and E. F. Hartree, *Proc. Roy. Soc.* **B124**, 397 (1938).

with 2.7 g of hydroxylapatite, stirred magnetically at 4° for 15 minutes, and then centrifuged 5 minutes at 2300 g. The ETF is eluted from the gel by resuspending the pellet successively in two 19-ml portions of phosphate buffers, pH 7.2, each of the following molarities: 0.09, 0.15, 0.20, and 0.25. The ETF of highest activity is eluted in the second resuspension of the pellet in the 0.15 M phosphate and represents about a 3.5-fold purification compared to the activity in the 60–80% ammonium sulfate fraction.

(c) EQUILIBRATED DEAE-CELLULOSE. The 60–80% ammonium sulfate fraction is dissolved in 2.5 mM potassium phosphate, pH 7.5, to give 6 mg of protein per milliliter in a total volume of 5 ml, and is then dialyzed against a 100-fold volume of the same buffer for 2 hours, with one change. For a preliminary purification by negative absorption, the solution may be passed first through a carboxymethyl-cellulose column (0.9×12 cm) and washed down with 8 ml of 2.5 mM phosphate, pH 7.5. The eluate is dialyzed against the same buffer for 1 hour and is then applied to a DEAE-cellulose column (0.9×12 cm), which has been equilibrated with 2.5 mM phosphate, pH 7.5. The column is washed with 25 ml of 2.5 mM phosphate, pH 7.5, and 25 ml of 0.1 M phosphate, pH 6.6. The ETF of highest activity appears in the 5–10 ml effluent of the 0.1 M phosphate buffer. Compared to the activity in the 60–80% ammonium sulfate fraction, the purification by this procedure is about 4.5-fold.

Chromatographic Purification of ETF from the Osmotic Supernatant. (a) UNEQUILIBRATED DEAE-CELLULOSE. The 60–80% fraction obtained from 80 ml of osmotic supernatant is dissolved in 6 ml of distilled water and dialyzed against a 650-ml volume of water for 3 hours with changes at ¾, 1½, and 2¼ hours. After dialysis, the volume of the solution is adjusted to 10 ml with water and the pH is brought to 7.5. This solution, adjusted in volume to contain no more than 3 mg of protein per milliliter, is then applied to a DEAE-cellulose column, which previously has been washed extensively with distilled water. Elution is carried out with water, 25 ml; 0.01 M NaCl, 20 ml; 0.04 M NaCl, 35 ml; and 1.0 M NaCl, 25 ml. The ETF is usually contained in the tenth through twentieth milliliter eluted after applying the 0.04 M NaCl.

(b) EQUILIBRATED DEAE-CELLULOSE. The 60–80% ammonium sulfate fraction from 130 ml of osmotic supernatant is dissolved in 7 ml of 1 mM potassium phosphate, pH 7.3, and is dialyzed against this buffer for 3.5 hours with changes at 40 and 90 minutes. The dialyzed preparation is diluted to 12 ml with 1 mM potassium phosphate, pH 7.3, and the pH is adjusted to 7.5. A 9.0-ml aliquot of this solution, containing a total of about 50 mg of protein, is then applied to the column (5.5×1.7 cm) of

DEAE-cellulose which has been equilibrated previously with 1 mM potassium phosphate, pH 7.3. Elution is carried out first with 1 mM potassium phosphate, pH 7.3. Under these conditions the ETF is not adsorbed and appears in the combined eluates, between volumes 5.5 and 22 ml. This eluate is then applied to another column, identical with the first, and elution is carried out with the following buffers, all of pH 7.3: 1 mM potassium phosphate, 60 ml; 1 mM potassium phosphate–0.1 M KCl, 50 ml; and 1 mM potassium phosphate–0.5 M KCl, 45 ml. The ETF activity appears in the early stages of elution with the 1 mM phosphate–0.1 M KCl, generally in tubes 15–17 (5.6 ml per tube).

(c) Equilibrated DEAE-Cellulose with Gradient Elution. The 60–80% ammonium sulfate fraction from approximately 80 ml of osmotic supernatant is dissolved in 6.0 ml of 1 mM potassium phosphate, pH 7.3, and dialyzed against the same buffer for 3 hours, with changes at ¾, 1¼, and 2¼ hours. The preparation is diluted to 12 ml with the same buffer and the pH is adjusted to 7.3. Ten milliliters (containing about 30 mg of protein) is applied to a DEAE-cellulose column (5.5 × 1.7 cm), previously equilibrated with 1 mM phosphate, pH 7.3. The column is eluted with 25 ml of 1 mM potassium phosphate, pH 7.3, and linear gradient elution is then begun with 1 mM potassium phosphate, pH 7.3 (150 ml initially in mixing chamber), with increasing salt concentration to 1 mM potassium phosphate–0.05 M KCl, pH 7.3, (150 ml in other compartment). The ETF activity appears in the eluate volume from 125–250 ml, with the peak activity at about 200 ml.

Summary of Purification Procedures. Typical data for the purification of ETF, beginning with the osmotic supernatant of the mitochondria and employing either equilibrated or unequilibrated DEAE-cellulose, are presented in the table.

Purification of ETF in Osmotic Supernatant

Preparation	Specific activity	Protein (mg/ml)	Total activity	Per cent of original activity
Osmotic supernatant	708	5.8	4106	100
60–80% $(NH_4)_2SO_4$ fraction	1650	1.2	1980	48
Unequilibrated DEAE-cellulose column, 0.04 M NaCl eluate	2526	0.25	618	15
Equilibrated DEAE-cellulose column				
Initial eluate (column No. 1)	1980	0.36	714	17
1 mM phosphate–0.1 M KCl eluate (column No. 2)	13200	0.036	474	12

Properties

Distribution. ETF is located in liver mitochondria and is absent from other fractions of the liver cell.[6] The richest sources of the enzyme found to date are rat,[5] pig,[9] and monkey liver.[10] An ETF has also been found in bacterial systems.[11]

Prosthetic Group. The visible absorption spectrum of rat liver ETF shows a maximum at 410 mμ and a broad shoulder at 450 mμ.[5] The spectrum thus resembles that of several known metalloflavoenzymes.[12]

The activity of the purified ETF, like that of the original 60–80% ammonium sulfate fraction, can be stimulated by the addition of FAD.[3,5] Moreover, the flavin of ETF preparations can be released and isolated quantitatively as FAD by treatment with 8% trichloroacetic acid.[5,10]

Specificity. The only substrates found for ETF to date are the substrate-reduced sarcosine and dimethylglycine dehydrogenases[5] and the acyl-coenzyme A dehydrogenases.[13] Moreover, the ETF preparations from rat liver mitochondria and of pig liver have been found to be interchangeable in their function toward the reduced dehydrogenases.[9] Thus far, all the preparations of ETF have also exhibited DPNH–DCPIP diaphorase activity and a somewhat lower DPNH–cytochrome *c* reductase activity.[5] However, since ETF can be distinguished chromatographically from various DPNH and TPNH dehydrogenases of the soluble compartment of the mitochondria[12] as well as those in the membranous fraction[6] and cannot be replaced by these enzymes, it is concluded that ETF is not "diaphorase" or "cytochrome *c* reductase" as conventionally defined.

In addition to DCPIP, ETF can also reduce soluble cytochrome *c*.[5] These reactions are not inhibited by Amytal or antimycin A. The sarcosine oxidase system, which is inhibited by antimycin A, but not by Amytal, can be reconstituted using the purified dehydrogenase, ETF, and a preparation of the mitochondrial respiratory chain free of these enzymes.[12]

Physical Constants. When the concentration of either sarcosine dehydrogenase or ETF is held constant and the concentration of the other is varied, the double reciprocal plots of the observed rates of reactions as functions of concentrations are consistent with Michaelis-Menten

[9] H. Beinert and W. R. Frisell, *J. Biol. Chem.* **237,** 2988 (1962).
[10] D. D. Hoskins and R. A. Bjur, *J. Biol. Chem.* **240,** 2201 (1965).
[11] A. Gelbard and D. S. Goldman, *Arch. Biochem. Biophys.* **94,** 228 (1961).
[12] W. R. Frisell, J. R. Cronin, and C. G. Mackenzie, *in* "Flavins and Flavoproteins" (E. C. Slater, ed.), p. 367. Elsevier. Amsterdam, 1966.
[13] F. L. Crane and H. Beinert, *J. Biol. Chem.* **218,** 717 (1956).

kinetics, indicating the formation of enzyme–substrate complexes. Assuming that, in the initial steady state, all the dehydrogenase is in the reduced form and that its molecular weight is 300,000,[4] these data give a K_s of $1\text{–}2 \times 10^{-6} M$ for the reduced sarcosine dehydrogenase as the substrate for the ETF.[12]

Stability. The purified ETF is stable in solution, at 4° or frozen, for only a few hours. Although it withstands lyophilization, the enzyme preparations in the dried form cannot be stored at −18° for more than a week without appreciable loss of activity.

[56] DT Diaphorase

By LARS ERNSTER

In 1958, Ernster and Navazio[1] briefly reported the occurrence of a highly active diaphorase in the soluble fraction of rat liver homogenates which catalyzes the oxidation of NADH and NADPH at equal rates. Purification and some basic properties of the enzyme were described in 1960 by Ernster *et al.*,[2] who named the enzyme DT diaphorase from its reactivity with both NADH and NADPH (at that time DPNH and TPNH).[3] The same authors subsequently published a detailed report[4] on DT diaphorase, including method of purification, assay conditions, data regarding kinetics, electron acceptors, activators, inhibitors, as well as a comparison of the enzyme with various diaphorases and quinone reductases earlier described in the literature. Metabolic aspects of DT diaphorase are treated in a series of papers[5–10] and summarized in a recent review.[11]

[1] L. Ernster and F. Navazio, *Acta Chem. Scand.* **12**, 595 (1958). See also L. Ernster, *Federation Proc.* **17**, 216 (1958).

[2] L. Ernster, M. Ljunggren, and L. Danielson, *Biochem. Biophys. Res. Commun.* **2**, 88 (1960).

[3] The Enzyme Nomenclature of 1964 of the International Union of Biochemistry recommends the systematic name reduced-NAD(P):(acceptor) oxidoreductase (EC 1.6.99.2) for DT diaphorase.

[4] L. Ernster, L. Danielson, and M. Ljunggren, *Biochim. Biophys. Acta* **58**, 171 (1962).

[5] T. E. Conover and L. Ernster, *Acta Chem. Scand.* **14**, 1840 (1960); *Biochim. Biophys. Acta* **58**, 189 (1962).

[6] T. E. Conover and L. Ernster, *Biochem. Biophys. Res. Commun.* **2**, 26 (1960); *Biochim. Biophys. Acta* **67**, 268 (1963).

[7] L. Ernster, *in* "Biological Structure and Function," (T. W. Goodwin and O. Lindberg, eds.), Vol. 2, p. 139. Academic Press, New York, 1961.

Distribution

DT diaphorase is widely distributed in the animal kingdom. An interesting exception is the pigeon which seems to lack DT diaphorase.[10] The enzyme occurs most abundantly in the liver. Relatively high activities are also found in kidney and brain, whereas heart and skeletal muscle are relatively poor in DT diaphorase activity. An appreciable DT diaphorase activity is also found in Ehrlich ascites tumor cells.[12]

Upon tissue fractionation, the bulk of the DT diaphorase activity, about 95%, is found in the cell-sap.[1,2,4,13] Mitochondria, at least from liver, contain a small but consistent portion of the DT diaphorase, corresponding to 2–3% of the total cellular activity. This fraction of the enzyme is truly mitochondrial as shown by the fact that it reacts only with intramitochondrial reduced pyridine nucleotides and is inaccessable to those added from the outside.[6] In addition, there seems to be a difference in electrophoretic mobility between the enzymes purified from mitochondria and from the cell-sap.[14] A small portion of DT diaphorase, 2–3%, is also found in the microsomal fraction from rat liver.[13] However, this portion of the DT diaphorase is not associated with the vesicles derived from the endoplasmic reticulum but seems to belong to a special vesicle fraction whose cytologic origin is not yet clarified.[15]

Purification

DT diaphorase has been purified from rat liver[2,4] which is the richest source of the enzyme found so far. As a starting material, the supernatant fraction is collected from a 0.25 *M* sucrose homogenate of rat liver (containing about 0.3 g of liver, wet weight, per milliliter of homogenate) which is obtained after centrifugation of the homogenate in a Spinco

---

[8] T. E. Conover, L. Danielson, and L. Ernster, *Biochim. Biophys. Acta* **67,** 254 (1963).

[9] B. Kadenbach, T. E. Conover, and L. Ernster, *Acta Chem. Scand.* **14,** 1850 (1960). See also T. E. Conover, *in* "Biological Structure and Function" (T. W. Goodwin and O. Lindberg, eds.), Vol. 2, p. 169. Academic Press, New York, 1961.

[10] L. Danielson and L. Ernster, *Nature* **194,** 155 (1962).

[11] L. Ernster and C. P. Lee, *Ann. Rev. Biochem.* **33,** 729 (1964).

[12] G. Dallner and L. Ernster, *Exptl. Cell Res.* **27,** 368 (1962). See also E. E. Gordon, L. Ernster and G. Dallner, in preparation.

[13] L. Danielson, L. Ernster, and M. Ljunggren, *Acta Chem. Scand.* **14,** 1837 (1960).

[14] F. Märki and C. Martius, *Biochem. Z.* **334,** 293 (1961).

[15] G. Dallner, S. Orrenius, and A. Bergstrand, *J. Cell Biol.* **16,** 426 (1963). See also G. Dallner, *Acta Pathol. Microbiol. Scand.,* Suppl. 166, 1963.

centrifuge, rotor 40, at 100,000 *g* for 60 minutes. The purification procedure involves, (a) precipitation with acetone, (b) fractionation with ammonium sulfate, and (c) chromatography on a DEAE-cellulose column. The detailed procedure (cf. Ernster *et al.*[4]) is given below.

The clear supernatant fraction obtained from the homogenate of 25–100 g of rat liver is poured with vigorous stirring into 9 volumes of —12° cold acetone in a glass beaker. A sticky, reddish precipitate is immediately formed and can be separated by simple decantation. The precipitate, which contains about 50% of the protein and 60% of the DT diaphorase activity of the initial supernatant, is dried in a desiccator *in vacuo* over silica gel at 4° overnight. The dry residue is scraped out of the beaker, weighed, and suspended in 25 ml 0.01 *M* phosphate buffer, pH 6.4, per gram. The suspension is transferred into a dialysis bag and dialyzed against running tap water (4°) overnight in order to remove the bulk of sucrose present in the residue. The dialyzed material is cleared by centrifugation. During this whole procedure practically no enzyme activity is lost.

The clear solution is fractionated with neutralized saturated ammonium sulfate in three stages: at 40, 55, and 65% saturation. DT diaphorase activity is concentrated in the precipitate obtained between 55 and 65% saturation which contains about 50% of the enzyme activity present in the clear solution before fractionation and has a specific activity on a protein basis of about sixfold that of the original supernatant. This precipitate is transferred, with the aid of a minimal volume of water, to a dialysis bag and dialyzed against running tap water until practically all ammonium sulfate is removed.

A chromatography column 1 cm in diameter and 20 cm in height is prepared using a suspension of approximately 2.5 g DEAE-cellulose (DEAE-20, Sigma) in 0.5 *M* NaOH. The column is washed with approximately 250 ml 0.5 *M* NaOH, followed by 250 ml 0.3 *M* sodium phosphate buffer, pH 6.4, and finally with 250 ml 0.01 *M* sodium phosphate buffer, pH 6.4. The dialyzed enzyme solution is transferred to the top of the column. When the bulk of the solution has run into the column, 0.01 *M* sodium phosphate buffer, pH 6.4, is added on the top and allowed to run through for about 2 hours (flow rate approximately 8 ml/hour). The effluent is collected and assayed for enzyme activity; no DT diaphorase should be found in the effluent. Continuous gradient elution is then started, with 150 ml 0.01 *M* sodium phosphate buffer, pH 6.4, in the mixing flask, and 250 ml 0.3 *M* of the same buffer in the reservoir; the flow rate again is 8 ml/hour. Fractions are collected at 15-minute intervals and assayed for DT diaphorase activity. The protein content of the samples is estimated spectrophotometrically at 280 mμ. The bulk of the DT diaphorase usually appears in fractions 14–18, which can be recognized also by their faint yellow color. The most active fraction contains DT diaphorase at a specific activity which is 300–450 times higher than that of the initial liver supernatant before precipitation with acetone. The enzyme solution can be kept in the frozen state for several years without appreciable loss of activity.

Properties

DT diaphorase is an FAD-containing flavoprotein, which catalyzes the oxidation of NADH and NADPH by various dyes and quinones with a maximal velocity of the order of 10^7 moles per mole of flavin per minute. Bovine serum albumin (0.07%), polyvinylpyrrolidone (5%), and certain nonionic detergents such as Tween-20 or -60 (0.2%) are activators of DT diaphorase; the concentrations in parentheses are those required for maximal activation. The activation is reversible and implies both an elevation of the V_{max} of the enzyme and an increase of its affinity for NADH and NADPH. The need for an activator increases during the purification and storage of the enzyme, probably because of the removal or destruction of a naturally occurring activator. The V_{max} of the 300- to 450-fold purified, freshly prepared enzyme is increased usually 3- to 10-fold by the addition of one of the aforementioned activators, and its K_m for NADH and NADPH is simultaneously decreased 2- to 3-fold. With the fully activated enzyme, the K_m is 8.3×10^{-5} M for NADH, and 4.4×10^{-5} M for NADPH. The maximal velocities with NADH and NADPH are equal. The enzyme activity is also influenced by the ionic strength; for example, 0.25 M Tris or phosphate buffer, pH 7.5, in the assay system yields an activity about one-third of that observed in 0.01 M buffer. Enzyme activity is maximal in the pH range 7.2–9.1.

Various electron acceptors for DT diaphorase are listed in Table I. 2,6-Dichlorophenolindophenol (DCPIP) and certain benzo- and naphthoquinones are the best electron acceptors, whereas methylene blue and ferricyanide are relatively inefficient, and cytochromes c and b_5 are practically inactive as electron acceptors. Among the quinones, those without a side chain in the 3-position are most active, and the activity decreases with increasing length of the side chain. Substitution of a methyl group in the 2-position, or, in the case of benzoquinones, by methoxy groups in the 5- and 6-positions (as in coenzyme Q) has little influence on the efficiency as electron acceptors. On the other hand, substitution of a hydroxy group in the 2- or 3-position renders the quinones inactive as electron acceptors for DT diaphorase. Both DCPIP and the various quinones inhibit DT diaphorase activity when used above a certain concentration. The inhibition is less marked in the presence than

TABLE I
ELECTRON ACCEPTORS FOR DT DIAPHORASE[a]

Electron acceptor	Concentration, required for maximal activity (μM)	Relative maximal activity[b]
2,6-Dichlorophenolindophenol	40	100
Methylene blue	180	46
1,4-Naphthoquinone	17	184
1,2-Naphthoquinone	33	109
2-Methyl-1,4-naphthoquinone (vitamin K_3, menadione)	17	233
Vitamin K_1	133	0.012
p-Benzoquinone	67	336
2-Methyl-1,4-benzoquinone	67	260
2,6-Dimethyl-1,4-benzoquinone	33	265
5,6-Dimethoxy-2-methyl-1,4-benzoquinone (coenzyme Q_0, ubiquinone-0)	50	239
Coenzyme Q_2 (ubiquinone-10)	267	52
Ferricyanide	667	75
Cytochrome *c*	40	0.01

[a] From L. Ernster, L. Danielson, and M. Ljunggren, *Biochim. Biophys. Acta* **58,** 171 (1962).

[b] Taking the activity with 2,6-dichlorophenolindophenol as 100. All values refer to measurements made in the presence of 0.07% bovine serum albumin as activator.

in the absence of added activator. Certain quinones are able to mediate electron transfer from DT diaphorase to cytochrome *c* or coenzyme Q_{10}. In both cases, vitamin K_3 and 1,4-naphthoquinone are the best mediators.

Among known inhibitors of DT diaphorase (Table II), dicoumarol is most potent. The inhibition is competitive with respect to NADH and NADPH, with a K_i value of the order of 10^{-8} M. This value varies somewhat with the nature of the electron acceptor and of the activator used; the dicoumarol inhibition is strongest with benzoquinones as electron acceptors and with Tween-20 or -60 as the activator. On the other hand, the inhibition is independent of the concentration of the electron acceptor. Other inhibitors of DT diaphorase are of three categories: certain sulfhydryl reagents (*p*-chloromercuribenzoate, *o*-iodosobenzoate); thyronine derivatives (thyroxine, deaminothyroxine, 3,3′,5-triiodothyronine); and flavins and flavin antagonists (FAD, FMN, atabrine, chlorpromazine). Amytal, rotenone, and various iron-chelating agents are without effect on the activity of DT diaphorase.

TABLE II
INHIBITORS OF DT DIAPHORASE[a]

Inhibitor	Concentration required for half-inhibition (μM)
Dicoumarol	0.001–0.1[b]
p-Chloromercuribenzoate	100
o-Iodosobenzoate	700
Thyroxine	60
3,3′,5-Triiodothyronine	60
Deaminothyroxine	22
FAD	500
FMN	7000[c]
Atabrine	2000
Chlorpromazine	500[c]

[a] From L. Ernster, L. Danielson, M. Ljunggren, *Biochim. Biophys. Acta* **58,** 171 (1962).
[b] Inhibition competitive with respect to NADH and NADPH. Extent of inhibition also varies with the nature of the electron acceptor and the activator used.
[c] Approximately 40% inhibition.

The reaction catalyzed by DT diaphorase involves the 4*A* hydrogen atom of both NADH and NADPH.[16] The enzyme does not catalyze an exchange of hydrogen atoms between the reduced pyridine nucleotides and water.

Assay

DT diaphorase is assayed spectrophotometrically in a reaction mixture containing 0.01–0.05 *M* Tris or phosphate buffer, pH 7.5–8.0, 0.2–0.3 m*M* NADH or NADPH, a suitable concentration of an electron acceptor (0.04 m*M* DCPIP, *or* 0.05–0.1 m*M* vitamin K_3 or 1.4-naphthoquinone or *p*-benzoquinone or coenzyme Q_0; *or* 0.01 m*M* vitamin K_3 or 1.4-naphthoquinone plus 33 μM cytochrome *c*), and an activator (0.07% bovine serum albumin; *or* 0.2% Tween-20 or -60; *or* 5% polyvinylpyrrolidone), in a final volume of 3 ml. The reaction is started by the addition of the enzyme, and is followed by recording either the reduction of the acceptor at a suitable wavelength (DCPIP, 600 mμ; cytochrome *c*, 550 mμ), or, in the case of the quinones, the oxidation of the reduced pyridine nucleotides at 340 mμ. The following extinction coefficients (liter/mole/cm) are used: 6.22×10^3 at 340 mμ for NADH and NADPH;

[16] C. P. Lee, N. Simard-Duquesne, L. Ernster, and H. D. Hoberman, *Biochim. Biophys. Acta* **105,** 397 (1965).

2.1×10^4 at 600 mμ for DCPIP; and 1.85×10^4 at 550 mμ for reduced minus oxidized cytochrome *c*.

In nonpurified preparations, such as cell fractions, where other diaphorases as well may be present, DT diaphorase can suitably be assayed with NADPH as the substrate, DCPIP as the electron acceptor, both added in concentrations as indicated above, and with or without 10^{-6} *M* dicoumarol present in the assay system. The dicoumarol-sensitive part of the activity is a fairly selective measure of the DT diaphorase activity of the preparation.

Relation to Other Diaphorases and Related Enzymes

DT diaphorase closely resembles the "phylloquinone reductase" or "vitamin K reductase" described by Martius and associates.[17] Initially it seemed to differ from the latter enzyme with regards to its very poor reactivity with vitamin K_1 and its sensitivity to *p*-chloromercuribenzoate. However, Martius and Märki[14,18] later revised the properties of vitamin K reductase in both these respects. It was therefore concluded[4,7] that the two enzymes most probably are identical. DT diaphorase also shows some resemblence to the "quinone reductase"[19] and "menadione reductase"[20] described by Wosilait and associates, but these enzymes react less rapidly with NADPH than with NADH, and their sensitivity to dicoumarol is far below that of DT diaphorase. A "brain diaphorase" with properties very similar to those of DT diaphorase has been described by Giuditta *et al.*[21]

Purified DT diaphorase is devoid of pyridine nucleotide transhydrogenase activity.[4] Its stereospecificity[16] clearly distinguishes it from the pyridine nucleotide transhydrogenase first described by Kaplan *et al.*,[22] which is 4*A* specific with respect to NADH and 4*B* specific with respect

[17] C. Martius and R. Strufe, *Biochem. Z.* **326,** 24 (1954). See also C. Martius, *Biochem. Z.* **326,** 26 (1954); *Angew. Chem.* **67,** 161 (1955); *Deut. Med. Wochschr.* **83,** 1701 (1958); *Ciba Found. Symp. Regulation of Cell Metabolism,* p. 194, Little Brown, Boston, Massachusetts, 1959; *Ciba Found. Symp. Quinones Electron Transport,* p. 312, Little Brown, Boston, Massachusetts, 1961; *Proc. 5th Intern. Congr. Biochem., Moscow, 1961,* Vol. V, p. 225, Pergamon, Oxford, 1963. See also F. Märki, Thesis No. 2908, Eidg. Techn. Hochschule, Zürich, 1959.

[18] C. Martius and F. Märki, *Biochem. Z.* **333,** 111 (1960).

[19] W. D. Wosilait and A. Nason, *J. Biol. Chem.* **206,** 255 (1954). See also W. D. Wosilait, A. Nason, and A. J. Terrell, *J. Biol. Chem.* **206,** 271 (1954).

[20] W. D. Wosilait and A. Nason, *J. Biol. Chem.* **208,** 785 (1954). See also W. D. Wosilait, *J. Biol. Chem.* **235,** 1196 (1960).

[21] A. Giuditta and H. J. Strecker, *Biochem. Biophys. Res. Commun.* **2,** 159 (1960); W. Levine, A. Giuditta, S. Englard, and H. J. Strecker, *J. Neurochem.* **6,** 28 (1960). A. Giuditta and H. J. Strecker, *Biochim. Biophys. Acta* **48,** 10 (1961).

[22] N. O. Kaplan, S. P. Colowick, and E. F. Neufeld, *J. Biol. Chem.* **205,** 1 (1953).

to NADPH.[16] On the same basis it has been possible to eliminate[16] DT diaphorase as a participant (cf. McGuire *et al.*[23]) in the energy-linked and non-energy-linked pyridine nucleotide transhydrogenase reactions catalyzed by submitochondrial particles.

Metabolic Aspects

Based on the early finding[24] that dicoumarol uncouples mitochondrial oxidative phosphorylation, the hypothesis has been advanced by Martius[17,18] that vitamin K reductase is involved in electron transport and oxidative phosphorylation in the NADH-cytochrome *b* region of the respiratory chain. The identification of vitamin K reductase with DT diaphorase,[4,7] and subsequent studies of the relationship of DT diaphorase to the respiratory chain,[6-8,10] have failed to support Martius' hypothesis. These studies have demonstrated that, on the contrary, DT diaphorase is not a member of the respiratory chain, and can even constitute a bypass of the normal, phosphorylating electron-transfer pathway between NADH and cytochrome *b*, provided that an artificial electron mediator, vitamin K_3, is present. Such a bypass can be established if the respiration of liver mitochondria, supported by a pyridine nucleotide-linked substrate, is blocked by amytal or rotenone, and a catalytic amount of vitamin K_3 is added.[6] The resulting respiration is highly sensitive to dicoumarol and its phosphorylating efficiency is 1 P:O unit lower than that of the original system; i.e., it involves a specific bypass of coupling site I of the respiratory chain via DT diaphorase and vitamin K_3. Interestingly, this "shunt" seems to require vitamin K_3 specifically, despite the fact that DT diaphorase as such can react with a great variety of both naphtho- and benzoquinones as electron acceptors. Various aspects of the DT diaphorase- and vitamin K_3-mediated oxidation of reduced pyridine nucleotides by the respiratory chain have been the subject of extensive studies with both mitochondria[5,6] and submitochondrial particles[7,8] as well as with whole cells.[9,12]

The physiological function of DT diaphorase is not yet known. It has been suggested[4] that DT diaphorase may participate in the biosynthesis of the natural forms of vitamin K and coenzyme Q in animal tissues. This process has been shown[25] to involve a condensation between reduced vitamin K_3 or coenzyme Q_0, respectively, and the pyrophosphate

[23] J. McGuire, L. Pesch, and H. Fanning, *Nature* **200,** 71 (1963).

[24] C. Martius and D. Nitz-Litzow, *Biochim. Biophys. Acta* **12,** 134 (1953); **13,** 289 (1954).

[25] W. Stoffel and C. Martius, *Biochem. Z.* **333,** 440 (1960). See also H.-G. Schiefer and C. Martius, *Biochem. Z.* **333,** 454 (1960).

ester of a polyisoprenoid alcohol. The role of DT diaphorase would be to bring about the reduction of vitamin K_3 and coenzyme Q_0. There are also indications[26] of a possible role of DT diaphorase in the biological detoxication of quinones, which must undergo reduction prior to conjugation with sulfate or glucuronate.

[26] P. Hochstein and L. Ernster, unpublished observations, 1963.

[57] Beef Heart TPNH-DPN Pyridine Nucleotide Transhydrogenases

By NATHAN O. KAPLAN

This enzyme, which also has been termed the TD transhydrogenase, is a mitochondrial enzyme and can catalyze the following reversible reaction[1-3]:

$$\text{TPNH} + \text{DPN} \rightleftharpoons \text{TPN} + \text{DPNH}$$

Although the reaction is reversible, the rate of reaction in the forward direction is roughly ten times as great as the backward rate. Hence, the reaction usually is measured in the forward direction.

Method of Assay

Since both DPNH and TPNH have the same absorption maxima, the enzyme cannot be assayed directly unless a coenzyme analog is used. The most suitable method appears to be the oxidation of TPNH by AcPyDPN according to the following equation:

$$\text{TPNH} + \text{AcPyDPN} \rightleftharpoons \text{TPN} + \text{AcPyDPNH}$$

The reduced acetylpyridine analog has a maximum at 363 mμ in contrast to TPNH and DPNH, whose maximum is at 340 mμ.[2,3] By following the increase at 375 mμ, where there is practically no contribution of TPNH, the reduction of AcPyDPN can easily be determined. The extinction coefficient of reduced AcPyDPN at 375 mμ is taken as 5.1.

Reagents

Potassium phosphate buffer, pH 6.5
Reduced TPN

[1] N. O. Kaplan, S. P. Colowick, and E. F. Neufeld, *J. Biol. Chem.* **205**, 1 (1953).
[2] A. M. Stein, N. O. Kaplan, and M. M. Ciotti, *J. Biol. Chem.* **234**, 979 (1959).
[3] B. Kaufman and N. O. Kaplan, *J. Biol. Chem.* **236**, 2133 (1961).

AcPyDPN
KCN

Method. The reaction mixture of a 1-ml final volume was composed of 100 micromoles of the phosphate buffer, 0.5 micromoles of TPNH, 1 micromole of acetylpyridine DPN and 1 micromole of KCN. The assay was carried out at room temperature in 1-ml cells of 1-cm light path. The amount of enzyme used was sufficient to give an increase in optical density of between 0.03 and 0.1. A unit of enzyme activity is defined as that amount of coenzyme reduced per minute per milligram of protein.

Enzyme Purification[4]

A highly purified TD enzyme from beef heart has been obtained by sucrose gradient centrifugation with a zonal Spinco ultracentrifuge. All steps in the purification procedure (outlined in the table) were carried out at 0–4°.

One beef heart obtained from freshly killed animal was freed of fat, quartered, and passed through an electric meat grinder. The ground heart muscle (1250 g) was homogenized for 1 minute at top speed in a large Waring blendor with 4000 ml of cold 0.25 *M* sucrose containing 0.01 *M* K_2HPO_4. The homogenate was adjusted to pH 8.5 during homogenization with 6 *N* KOH and centrifuged for 15 minutes at 1000 *g* in the large head of a Servall centrifuge. The supernatant was carefully decanted through cheesecloth. The mitochondria were sedimented by centrifuging for 15 minutes at 8000 *g* in the same centrifuge. The mitochondrial paste was resuspended by brief homogenization with a Teflon pestle in 5 volumes of 0.25 *M* sucrose containing 0.005 *M* Tris-HCl, pH 7.5, and 0.001 *M* EDTA. The mitochondrial suspension was resedimented by centrifugation at maximal speed in a Servall small head and resuspended in 5 volumes of 0.25 *M* sucrose containing 0.005 *M* Tris-HCl, pH 7.5, and 0.001 *M* EDTA and were stored frozen. A chilled solution of digitonin in 0.005 *M* Tris-HCl, pH 7.5, prepared by heating in a water bath, was added to the mitochondrial suspension. The final concentration of digitonin was 1 g per gram of mitochondrial protein. After incubation in an ice bath for 2 hours with occasional stirring, the suspension was centrifuged for 30 minutes at the maximal speed in a Servall small head. The supernatant suspension was centrifuged for 1 hour at 30,000 rpm in a No. 30 head of the Spinco centrifuge, Model L.

The high speed supernatant (1 g in 50–80 ml) was applied to a sucrose density gradient from 20% to 40% in a Beckman-Spinco zonal ultra-

[4] K. Satoh, T. Kawasaki, and N. O. Kaplan, in preparation.

centrifuge, ZU-II, equipped with a B-IV rotor, it was then centrifuged for 12 hours at 40,000 rpm. After the centrifugation, 20-ml fractions were collected. A typical profile for protein and various enzyme activities is shown in Fig. 1. The TD-transhydrogenase peak was pooled and freed of sucrose by passage through a Sephadex G-25 column (4 × 75 cm) saturated with 0.005 *M* Tris-HCl, pH 7.5, and 0.001 *M* EDTA, and eluted with the same buffer. The eluate was concentrated by Carbowax to give appropriate volume for the next sucrose gradient centrifugation. The digitonin solution, prepared as before, was added to the concentrated preparation to make the cloudy solution clear. The final concentration of digitonin was 1% (w/v) of the enzyme solution.

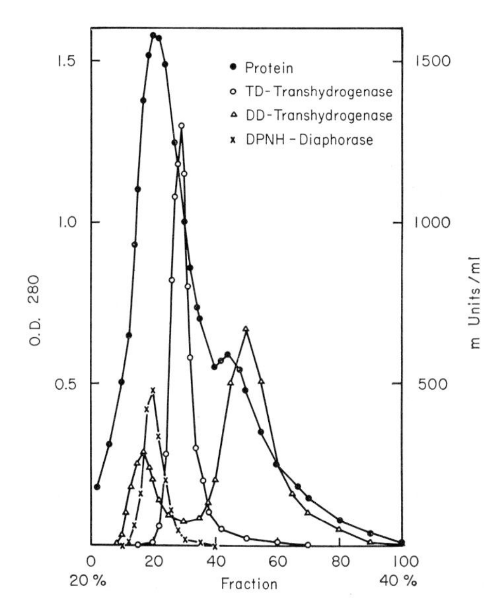

FIG. 1. Distribution of enzyme in first sucrose gradient.

The enzyme preparation was again subjected to the sucrose density centrifugation as described above. A typical pattern for the second sucrose density gradient is shown in Fig. 2. The TD-transhydrogenase peak was pooled and freed of sucrose by the Sephadex G-25 column as before. The eluate was concentrated by Carbowax. A digitonin solution of a final concentration of 1% (w/v) was added to the concentrated solution for clarification.

The preparation obtained was applied to a third sucrose density gradient, 15–35%, and centrifuged for 8 hours in the zonal ultracentrifuge. A typical pattern for the third sucrose density gradient is shown

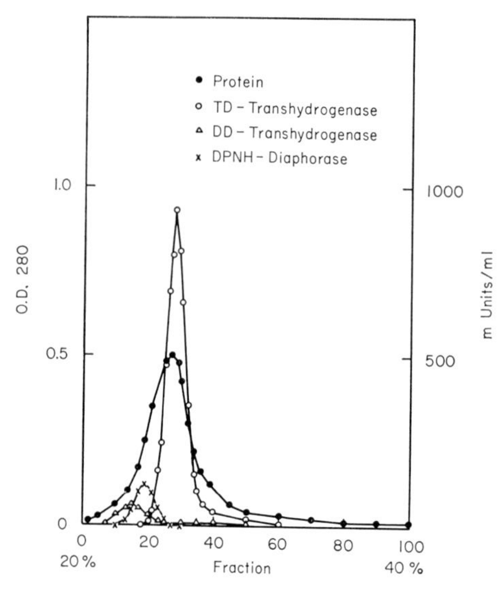

FIG. 2. Distribution of enzyme in second sucrose gradient.

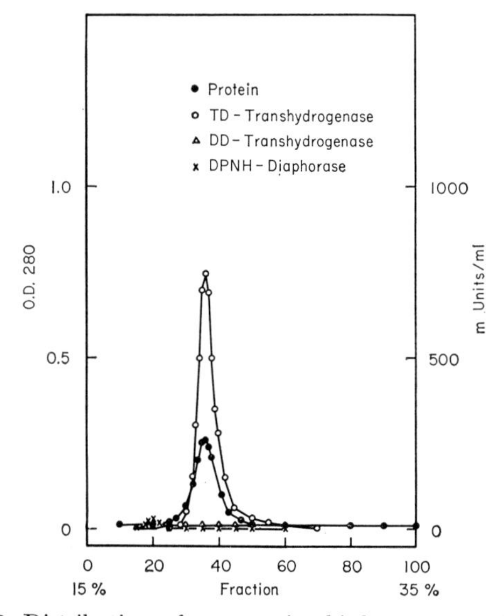

FIG. 3. Distribution of enzyme in third sucrose gradient.

in Fig. 3. The TD-transhydrogenase peak was pooled and the concentration of sucrose was adjusted to 0.25 M with 0.005 M Tris-HCl, pH 7.5, and 0.001 M EDTA. Unless otherwise indicated, this preparation was used for experiments described below. A summary of the purification procedure is shown in the table. A fourth sucrose gradient centrifugation

PURIFICATION OF TD TRANSHYDROGENASE FROM BEEF HEART MITOCHONDRIA

	TD Transhydrogenase				
Purification step	Total units (μmole/min)	Specific activity (unit/min/mg)	Recovery (%)	Total DD transhydrogenase	Units DPNH diaphorase
Digitonized mitochondria	3050	0.14	100	7620	4300
Supernatant, 100,000 *g*, 1 hour	2075	0.23	68	5650	3600
Sucrose gradient step					
1st	900	1.11	30	108	135
2nd	600	2.1	20	24	36
3rd	400	3.5	13	<4	<4

gave only a slightly higher specific activity. Attempts to purify the TD-transhydrogenase with organic solvents, snake venom extracts, and pancreatic lipase to remove possible lipid moieties of the enzyme have to date failed.

Properties of the Purified Enzyme

The purified enzyme has negligible activities with respect to the DPNH-DPN (DD) transhydrogenase, DPNH diaphorase, and DPNH–cytochrome *c* reductase. The purified TD enzyme also does not promote a TPNH-TPN reaction (TT transhydrogenase). The enzyme catalyzes a stereospecific reversible hydride transfer from the B side of TPNH to the A side of DPN, according to the following equation[5]:

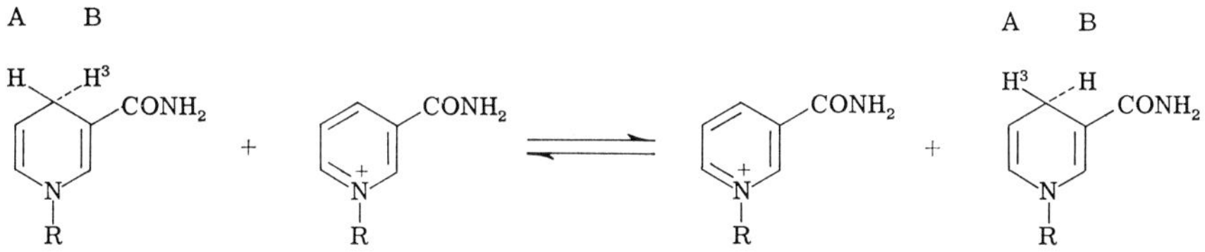

A B / H, H^3, $CONH_2$, N–R + $CONH_2$, $\overset{+}{N}$–R ⇌ $CONH_2$, $\overset{+}{N}$–R + A B / H^3, H, $CONH_2$, N–R

$$TPNH + DPN \rightleftharpoons TPN + DPNH$$

The enzyme has a pH optimum of about 6.0. There appears to be no metal requirement for the TD enzyme, and the enzyme does not contain any flavin. The enzyme is very sensitive to sulfhydryl reagents. A preliminary estimation of the molecular weight indicates that the enzyme is a large molecule with a weight between 250,000 and 300,000.

[5] T. Kawasaki, K. Satoh, and N. O. Kaplan, *Biochem. Biophys. Res. Commun.* **17,** 648 (1964).

The purified enzyme, although soluble in the presence of digitonin, sediments out after removal of the digitonin by dialysis.

Preparation of Antibody[5]

An antibody has been prepared against the enzyme in rabbits. It was obtained by an intramuscular injection of the transhydrogenases which was followed by two intravenous injections. After an initial antibody was obtained, a booster course of injections (three times in 1 week) was given after a rest period of 3 weeks. This resulted in a fairly potent antibody where only one band was detectable on Ouchterlony plates. This antibody will react with the TD transhydrogenase in submitochondrial particles, but not with other enzymes.[5]

[58] Preparation of Succinate Dehydrogenase and Reconstitution of Succinate Oxidase

By TSOO E. KING

Succinate dehydrogenase is solubilized, by butanol, from the Keilin-Hartree heart muscle preparation preincubated with a moderate concentration of succinate. The soluble dehydrogenase is reactive with the cytochrome system. This method is also applicable to other starting material, such as heart and liver mitochondria, but less effectively. Another method for preparing succinate dehydrogenase active to the cytochrome system is by means of the alkaline cleavage of the heart muscle preparation pretreated with succinate.[1,2]

Assay[2]

Principle. The activity of soluble succinate dehydrogenase is assayed by the technique of reconstitution. In the presence of the "cytochrome" particle, the dehydrogenase is incorporated into the particle to form an active succinate oxidase. Reconstituted succinate oxidase catalyzes the oxidation of succinate by oxygen to fumarate and water just as intact oxidase does. Conventional methods, such as manometry, are then used for the determination.

Succinate dehydrogenase can also catalyze the oxidation of succinate by a number of artificial electron acceptors. This property has been used for the determination of the artificial activity of the enzyme. Most

[1] Readers are referred to the bibliographies cited in references listed in footnotes 2, 5, 7, and 10 for other more complete references.

[2] T. E. King, *J. Biol. Chem.* **238,** 4037 (1963).

of the acceptors are first used for succinate-phenazine reductase. They are: (A) manometric method with a constant level of phenazine methosulfate (PMS) by following oxygen consumption; (B) manometric method with several levels of PMS for V_{max}; (C) spectrophotometric method with several levels of PMS coupled with 2,6-dichlorophenol-indophenol (DCIP) for V_{max} by following the reduction of DCIP at 600 mμ; (D) spectrophotometric method with constant level of ferricyanide by recording the decrease of $Fe(CN)_6^{3+}$ absorption at 420 mμ; (E) spectrophotometric method with several levels of ferricyanide for V_{max}; and (F) a recent spectrophotometric method with Wurster's blue (WB), the semiquinone of *N,N,N′,N′*-tetramethyl-*p*-phenylenediamine (TMPD) by following the decrease at 611 mμ.

The overall reactions for methods A and B are shown in reactions (1) and (1a), for C in (2) and (2a), for D and E in (3), and for F in (4).

$$\text{Succinate} + \text{PMS} \rightarrow \text{fumarate} + \text{PMSH}_2 \tag{1}$$

$$\text{PMSH}_2 + \text{O}_2 \rightarrow \text{PMS} + \text{H}_2\text{O}_2 \tag{1a}$$

$$\text{Succinate} + \text{PMS} \rightarrow \text{fumarate} + \text{PMSH}_2 \tag{2}$$

$$\text{PMSH}_2 + \text{DCIP} \rightarrow \text{PMS} + \text{DCIPH}_2 \tag{2a}$$

$$\text{Succinate} + 2\ \text{Fe(CN)}_6^{3+} \rightarrow \text{fumarate} + 2\ \text{Fe(CN)}_6^{2+} + 2\ \text{H}^+ \tag{3}$$

$$\text{Succinate} + 2\ \text{WB} \rightarrow \text{fumarate} + 2\ \text{TMPD} \tag{4}$$

Here, only methods C and E will be described in detail. It must be emphasized that the activity measured with artificial electron acceptors does not parallel that by the method of reconstitution.

Reagents

Phosphate buffer, 0.2 *M* (the Sørensen type), pH 7.8

Crystalline bovine serum albumin, 1% in water

Succinic acid, 0.6 *M* in water, adjusted to pH 7.8 with NaOH

DCIP, 0.0015 *M* in water

Phenazine methosulfate, 0.009 *M* in water, stored in dark and frozen; avoid undue exposure to light

Potassium ferricyanide, 0.03 *M* in water, freshly prepared

Potassium cyanide, 0.045 *M* in water, freshly prepared

Cytochrome *c*, 0.0009 *M* in water

"Cytochrome" particle. The cytochrome particle used in reconstitution contains all components of succinate oxidase except succinate dehydrogenase. It is sometimes referred to as "the alkali-treated heart muscle preparation." The heart muscle preparation (see method 1 of this volume [37]) containing approximately 10 mg of protein per milliliter of Na_2HPO_4–H_3BO_3 buffer (50 m*M* each), is adjusted to pH 9.3 with 1 *N* NaOH at room temperature. Care must be taken during the addition of NaOH; mixing must be efficient. After incubation at 37° for 60 minutes with gentle shaking, the mixture is then cooled to room temperature and readjusted to pH 7.8 with 2 *N* acetic acid. The preparation can be used as such, or the precipitate obtained from centrifuging can be resuspended in 0.1 *M* phosphate buffer, pH 7.8. In either preparation, 0.3 ml should give the oxygen consumption of less than 10 μl in 60 minutes in the absence of succinate dehydrogenase.

Succinate dehydrogenase, properly (depending upon the assay) diluted with oxygen-free[3] phosphate buffer containing 0.1% albumin, is kept under inert atmosphere at 0° (cf. the table).

Water used in the assay and in the preparation is redistilled from a glass apparatus from ordinary distilled and then deionized water.

Methods. RECONSTITUTION. In the main compartment of a Warburg flask are placed 0.1 ml cytochrome *c*, 1.5 ml phosphate buffer, and 0.3 ml of the alkali-treated heart muscle preparation. In the side arm 0.2 ml succinate is pipetted. An amount of succinate dehydrogenase to give 10–80 μl of oxygen uptake per 10 minutes is then added to the main compartment of the flask, which is immediately swirled for a few seconds. Water is added to the main compartment up to 2.8 ml. Four or five levels of the dehydrogenase are tested; a blank containing no succinate dehydrogenase and another blank containing the highest level of the dehydrogenase but no alkali-treated heart muscle preparation are also made.

After temperature equilibration at 37° for 8 minutes, the succinate is tipped into the main compartment. Readings are taken for 1 hour at 10-minute intervals. The oxygen uptake is linear with time; occasionally a very slight deviation from linearity occurs after the first 30 minutes. Net values of oxygen uptake are plotted against the amount of succinate dehydrogenase present in each flask. The linear portion of the curve is used for the calculation of the specific activity.

PHENAZINE METHOSULFATE. In a 1-cm optical path cuvette are placed 0.75 ml phosphate buffer, 0.1 ml cyanide, 0.2 ml succinate, 0.1 ml DCIP, 0.3 ml albumin, and an appropriate amount of phenazine methosulfate. Water is then added to make 2.95 ml. The reaction is started by the addition of 0.05 ml succinate dehydrogenase which is diluted to give an absorbancy change, in the assay, between 0.05 and 0.20 unit per minute. The spectrophotometric recording at 600 mμ against a water blank begins at 5 seconds after the addition of the enzyme. The linearized initial slope is used for computation. The levels of PMS used are 0.0,

[3] The so-called "oxygen-free" solutions or gel suspension used in this article refer to those of which the dissolved oxygen is displaced by inert gas.

0.06, 0.1, 0.2, 0.4, and 0.6 ml per cuvette. The activity is dependent on PMS concentration. Consequently, the concentration of the enzyme added to each cuvette may be varied, if necessary. In addition, a significant amount of the nonenzymatic reaction must be corrected by separate experiments at each level of PMS in the absence of the dehydrogenase. The nonenzymatic reaction also increases with the increase of PMS concentration.

FERRICYANIDE. In a cuvette of 1-cm optical path are placed 1.5 ml phosphate buffer, 0.2 ml succinate, 0.3 ml albumin, and an appropriate amount of ferricyanide. The content is then diluted with water to 2.95 ml. The reaction is started by the addition of 0.05 ml diluted succinate dehydrogenase to give an absorbancy change between 0.03 and 0.12 unit per minute. The spectrophotometric recording at 420 mμ against a water blank commences at 5 seconds after the addition of the enzyme. The linearized initial slope is used for computation. The levels of ferricyanide used are 0.0, 0.05, 0.1, 0.15, 0.20, and 0.4 ml per cuvette. The activity is strongly influenced by ferricyanide concentration. Consequently the concentration of the enzyme added to each cuvette should be varied, if necessary. For the most accurate work, an almost negligible nonenzymatic reaction may be corrected by separate experiments without the enzyme.

Calculation. To convert observed change to the activity in the unit of mM succinate oxidized, the following are used: 0.0298 $\times$ μl of oxygen uptake for the reconstitution, 0.485 $\times$ $\Delta A_{420\,m\mu}$ for the ferricyanide method, and 0.0476 $\times$ $\Delta A_{600\,m\mu}$ for the PMS method. The specific activity is expressed as micromoles of succinate oxidized per minute per milligram of enzyme protein. V_{max} values for artificial electron acceptors are expressed in the same units. They are computed from extrapolation to the infinite concentration of the acceptor by a conventional plot of the reciprocal of the activities (velocities) against the reciprocal of PMS or ferricyanide concentrations used according to Lineweaver and Burk.[4]

Comments.[2,5,6] When the situation requires simultaneous assays, each operator should handle one method of determination. For reconstitution, manometric method is preferred because several levels of succinate dehydrogenase may be tested at the same time. For "true" activity, the dehydrogenase should be used immediately after the preparation or immediately after thawing from the frozen state in which it is stored at liquid nitrogen temperature. However, when the dehydrogenase is incorporated into the cytochrome particle, it is stable.

[4] H. Lineweaver and D. Burk, *J. Am. Chem. Soc.* **56**, 658 (1934).
[5] T. E. King, *J. Biol. Chem.* **238**, 4032 (1963).
[6] Unpublished observations from this laboratory.

Since the methods for V_{max} using artificial electron acceptors require some time to finish, especially when several concentrations of the enzyme are tested, it is advisable to store the enzyme under inert atmosphere at 0°. In the phenazine methosulfate method, although the activity, either obtained at fixed concentrations of PMS or extrapolated for V_{max} values, is fairly linear to the enzyme concentration, these straight lines do not go through the origin even if the data are corrected for the nonenzymatic reaction.[3] Whether this is due to the complication of the competitive reaction of reduced PMS with oxygen is not known. According to reports, the competitive reaction is negligible under the conditions used. It may also be noted that PMS is used as an intermediate electron carrier, but its concentration is many times higher than that of DCIP, the final acceptor.

The plots of enzymatic activity tested at ferricyanide concentrations between 0.5 and 5 mM or expressed in V_{max} values are linear with the amount of the enzyme present. These straight lines pass through the origin in contrast to the phenazine method. However, concentrations of ferricyanide higher than 5 mM inhibit the reaction as shown in the upward curve in the double reciprocal plot.

The precision of these spectrophotometric methods is approximately 15% in our hands. The low precision of the ferricyanide method may be caused partly by high $K_m^{ferricyanide}$ value and inhibitory behavior of ferricyanide at high concentrations. The difficulties in PMS assay are due mainly to the rapid deviation of the activity from linearity with time and the nonenzymatic reaction between PMS and DCIP.

Preparation[2,6,7]

Four hundred sixty-five milliliters of the Keilin-Hartree preparation from beef heart (see method 1 of this volume [37] for the preparation) containing about 10 mg protein per milliliter of 50 mM H_3BO_3-50 mM Na_2HPO_4 buffer is mixed with 35 ml 0.6 M succinate. The mixture is allowed to stand 1–2 hours at 0–4°. Subsequent operations are conducted at 0–4° unless otherwise specified. The mixture is adjusted with 1 N NaOH to pH 9 and immediately transferred to a 3-neck flask fitted with a mechanical stirrer, a thermometer, an addition-funnel, and an inert gas inlet. The opening at the stirrer serves as the outlet. The flask is placed in a salt–ice water bath. After the mixture is adjusted to pH 9, it should be operated as quickly as possible under inert atmosphere (helium or argon).

To the mixture is added with stirring 100 ml n-butanol (prechilled to −20°) in such a way that the temperature of the mixture can be

[7] D. Keilin and T. E. King, *Proc. Roy. Soc.* **B152**, 163 (1960).

maintained between -1° and $+1^\circ$. Efficient, but gentle, stirring is essential, and freezing of the mixture near the wall must be avoided. At the end of the addition, the stirring is continued for an additional 15–30 minutes, and the temperature is maintained at -1° to $+1^\circ$. The mixture is then centrifuged in an International Centrifuge, size 2, at 2000 rpm, for 30–45 minutes. The centrifuge is welded with an inlet about 2.5 cm in diameter for helium or argon and has an outlet in the cover. The air in the centrifuge is first displaced by a rapid stream of inert gas.

The clear aqueous middle layer thus separated by centrifuging is siphoned out. It is immediately adjusted with oxygen-free[3] 2 N acetic acid to pH 6.0. "Oxygen-free" calcium phosphate gel[8] is added at 4 mg (dry weight of the gel) per milliliter of the aqueous layer. The mixture is gently stirred for 10 minutes under an inert atmosphere and then centrifuged for 10 minutes at 2000 rpm.

The gel precipitate is suspended, with the aid of a Potter-Elvehjem homogenizer, in oxygen-free phosphate buffer, 0.1 M, pH 7.8, at the proportion of about three-fourths of the volume of the initial aqueous layer. The suspension is stirred gently for 15 minutes under an inert atmosphere and then centrifuged 10 minutes at 2000 rpm. The yield up to this stage, based on acid nonextractable flavin, is about 70%.

The gel eluate is fractionated with solid ammonium sulfate, and the fraction between 0.3 and 0.55 saturation is collected. A second fractionation with ammonium sulfate is performed and the fraction between 0.32 and 0.48 saturation is collected. The precipitate is dissolved in oxygen-free phosphate buffer, 0.1 M, pH 7.8.

The enzyme prepared either through the gel-eluate step or further fractionated with ammonium sulfate for reconstitution study should be used immediately or stored in small vials in a liquid nitrogen refrigerator. The table serves as a general guide to allowable storage time in air for various purposes.

Comments. The method described has been satisfactorily used in this laboratory for many years at the scale ranging from 25 ml to 3 liters of the heart muscle preparation. Some laboratories also use EDTA in all steps. The protein concentration in the heart muscle preparation and the time for standing with succinate are not critical.

Other types of centrifuges with higher speeds requiring less centrifugation time may be employed in all steps. Capped centrifuge bottles may be used instead of centrifugation under inert atmosphere. The need for sharp separation of the aqueous solution after butanol solubilization cannot be overemphasized. Sacrifice of the yield rather than the risk

[8] We usually use the gel which is aged for at least 3 weeks; see Vol. I, p. 98, for the preparation.

COMPARISON OF THE STABILITY OF SUCCINATE DEHYDROGENASE WITH RESPECT TO ITS RECONSTITUTIVE AND ARTIFICIAL ACTIVITIES[a]

Time of storage in air at 0° (hours)	Activity in reconstitution (37°)		Phenazine methosulfate (36°)			Ferricyanide (27°)		
			V_{max}^{PMS}			$V_{max}^{ferricyanide}$		
	Micromole/min/mg	%	Micromole/min/mg	%	K_m^{PMS} (mM)	Micromole/min/mg	%	$K_m^{ferricyanide}$ (mM)
0	16	100	14.2	100	0.48	4.1	100	11
3	3.2	20	13.8	97	0.51	4.3	105	10
6	0.2	1	10.4	73	0.36	3.5	86	5.6
24	0.0	0	5.0	35	0.34	1.7	41	3.0

[a] Soluble succinate dehydrogenase at about 1 mg/ml is prepared through the gel-eluate stage. Some variation in stability has been observed from batch to batch.

of contamination with particles or excessive butanol to the aqueous layer is advised. Once the particle gets into the solution, satisfactory preparations can rarely be made.

The enzyme preparation through the gel-elution stage is suitable for many kinds of tests and can be prepared within 2.5 hours after contact with butanol. One objection of the gel eluate is low concentration of the enzyme, usually about 1 mg/ml. However, the protein concentration can be significantly increased by decreasing the volume of phosphate buffer used for elution. Additional steps of ammonium sulfate fractionation can enable the making of much more concentrated enzyme solutions. At the same time, this method of fractionation increases the flavin, iron, and "labile sulfide" contents, but increase in the reconstitutive activity is not directly proportional.

On very rare occasions, the precipitate from ammonium sulfate fractionation floats on the surface of the solution, possibly the result of unsatisfactory separation of the aqueous layer after the first centrifugation or rapid and uneven addition of ammonium sulfate. These batches should be discarded.

Reconstitution of Succinate Oxidase[2,7]

In the system described in the Assay section, succinate dehydrogenase is added in limiting amounts. For isolation of reconstituted succinate oxidase, it is more convenient to use an excess amount of the dehydrogenase.

The alkali-treated heart muscle preparation at 10 mg protein per milliliter is mixed with two volumes of freshly prepared succinate dehydrogenase (at the gel-eluate stage with about 1 mg protein per milliliter and specific activity of 16). The mixture can be allowed to stand or is subjected to centrifugation immediately at 30,000 rpm for 1 hour in a Spinco centrifuge. The supernatant liquid containing excess succinate dehydrogenase is discarded and the precipitate is suspended in the original volume of phosphate buffer, 0.1 *M*, pH 7.8, with aid of a Potter-Elvehjem homogenizer. The suspension is again centrifuged and this washing process is repeated two more times. The reconstituted oxidase is stable.

Properties[2,6,7,9,10]

These properties are for succinate dehydrogenase from bovine heart.

Specificity, Activities and Stability. Succinate dehydrogenase, in contrast to succinate phenazine reductase, can react with the "cytochrome particle" to form succinate oxidase. It also catalyzes succinate oxidation by ferricyanide, phenazine alkyl sulfate, or Wurster's blue, but not significantly by molecular oxygen, methylene blue, cytochrome *c*, DCIP, triphenyltetrazolium chloride, coenzyme Q_6 or coenzyme Q_{10}. The comparison of the activities and stability of the enzyme at the gel-eluate stage is shown in the table. The specific activity in reconstitution at 37° is about 16 micromoles succinate oxidized per minute per milligram of protein. This value corresponds to a turnover number of 7500 moles of succinate oxidized per minute per mole of the acid nonextractable flavin incorporated into the "cytochrome particle." The table also shows that the reconstitutive activity is not parallel to the artificial activities.

Spectra. Concentrated succinate dehydrogenase is a dark amber, clear solution. Its spectra are shown in Fig. 1 (see pp. 330 and 331).

Composition. Succinate dehydrogenase after two cycles of fractionation with ammonium sulfate contain, on the average, acid nonextractable flavin,[11] 3.6; nonheme iron, 30; and labile sulfide,[12] 28 millimicromoles per milligram of protein. The ratio of flavin:Fe:sulfide is approximately 1:8:8. Veeger and co-workers have found the ratio of 1:8:4 for their preparations from pig heart (see King[10] for the discussion of the disparity). Succinate dehydrogenase is free of lipid, hemes, coenzyme Q, and acid-extractable flavin.

[9] T. E. King, *Biochem. Biophys. Res. Commun.* **16**, 511 (1964).

[10] T. E. King, *Advan. Enzymol.* **28**, 155 (1966).

[11] Acid-extractable flavin refers to the flavin attached to the protein which can be cleaved by and thus extracted to the trichloroacetic acid or perchloric acid. Acid-nonextractable flavin does not show this behavior. However, the acid nonextractable, i.e., "covalently" bound, with protein, may be rendered soluble after proteolytic digestion. For their determinations, see this volume [77].

[12] Operationally defined; see this volume [98] for its determination.

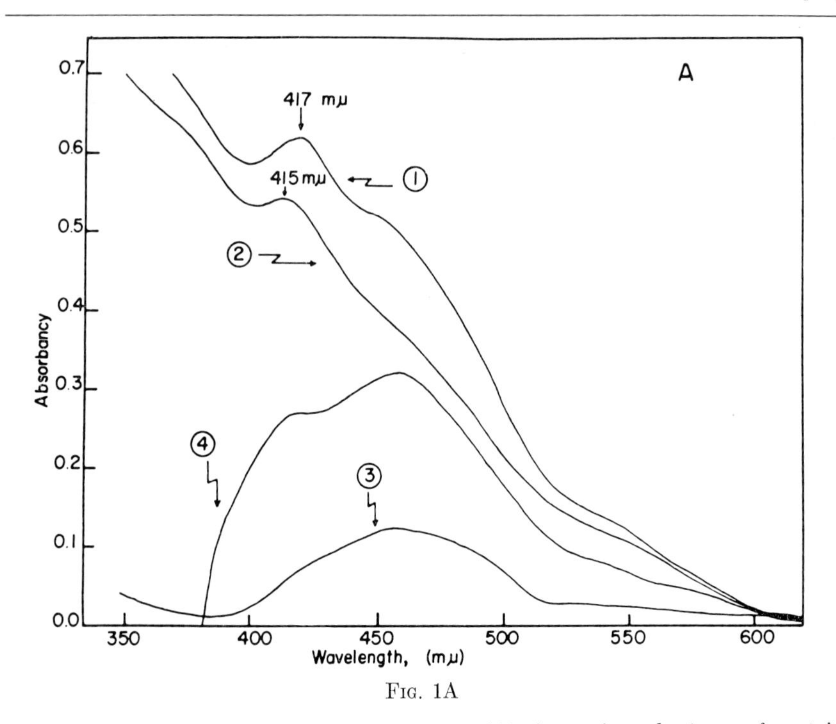

FIG. 1A

FIG. 1. Spectra of succinate dehydrogenase. (A) Approximately 4 mg of protein per milliliter of 50 mM phosphate buffer, pH 7.8. (B) Approximately 1 mg/ml. Curve *1*, oxidized (as prepared); *2*, reduced with 20 mM succinate; *3*, difference spectrum (oxidized minus succinate-reduced); and *4*, difference spectrum (oxidized minus dithionite-reduced). The absolute spectrum of the dithionite-reduced sample is not shown.

Inhibitors. The artificial activities of the dehydrogenase are also inhibited by malonate, fumarate, and pyrophosphate. At 20–23°, K_i^{malonate} and $K_i^{\text{pyrophosphate}}$ are 0.045 mM and 0.23 mM, respectively, using ferricyanide as an acceptor, whereas $K_m^{\text{succinate}}$ is 1.3 mM. The artificial reactions are not inhibited by antimycin A. Cyanide treatment of the soluble dehydrogenase under Tsou's condition destroys its reconstitutive activity but does not affect the artificial reactions. Incubation of succinate dehydrogenase with oxidized glutathione under the conditions according to Hopkins *et al.* abolishes both activities.

Reconstitution. When succinate dehydrogenase is reincorporated into the "cytochrome particle," it becomes an integral part of the respiratory chain and reacquires all the properties of its "endogenous" form, such as insolubility, stability, and reactivity toward the cytochrome system. Reconstituted succinate oxidase shows normal oxidation and reduction

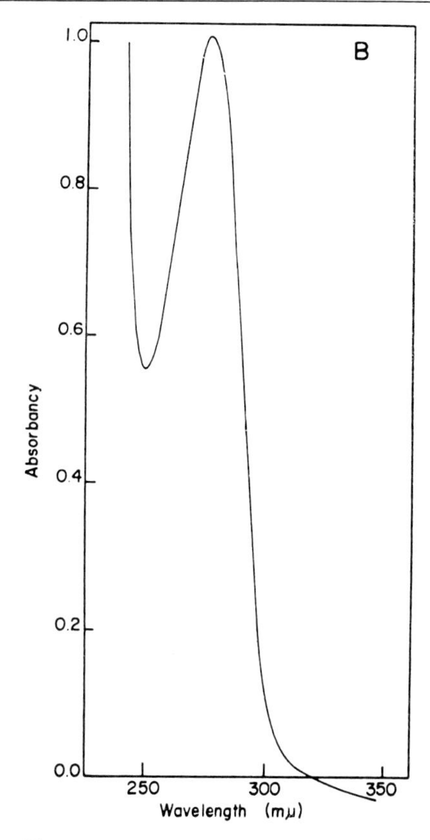

FIG. 1B. For legend see opposite page.

of its cytochrome components and is susceptible to all inhibitors, such as antimycin A and thenoyltrifluoroacetone, in the same way as is the succinate oxidase system of the untreated heart muscle preparation (see King[10] for other information).

Acknowledgment

The experimental work was supported by grants from the National Science Foundation, the National Institutes of Health, the American Heart Association, and the Life Insurance Medical Research Fund.

[59] Cytochrome Oxidase: Beef Heart[1]

By TAKASHI YONETANI

O_2 + 2 ferrocytochrome *c* + 2 H^+—2 ferricytochrome *c* + 2 H_2O

Assay Method

Principle. For cytochrome oxidase, a hemoprotein,[2] the purity index, which is defined as the ratio of absorbance at 422 mμ to that at 280 mμ ($A_{422}:A_{280}$), is a convenient measure to estimate an approximate purity of the enzyme during purification. The enzyme activity is assayed by measuring the initial rate of the aerobic oxidation of ferrocytochrome *c* catalyzed by this enzyme.

Reagents

Potassium phosphate buffer, 0.2 *M*, pH 6.0 containing 1 m*M* EDTA
Ferrocytochrome *c*, 1.0 m*M*, in 0.01 *M* potassium phosphate buffer, pH 7.0 containing 1 m*M* EDTA and saturated with nitrogen.[3]
Cytochrome oxidase, 0.5 μ*M*, in 0.01 *M* potassium phosphate buffer, pH 7.4, containing 0.1% Tween-80

Activity Assay. A set of six reaction mixtures of the following compositions are made in cuvettes of 1.0-cm light path: 1.00 ml of the buffer, pH 6.0, varying amounts of ferrocytochrome *c* (for example, 5, 10, 20, 40, 60, and 100 μl to give final concentrations of 2.5, 5, 10, 20, 30, and 50 μ*M*, respectively) and distilled water to a final volume of 2.00 ml. The measurement is carried out at 25° in a recording spectrophotometer at 550 mμ. The reaction is initiated by mixing in 5 μl of 0.5 μ*M* cytochrome oxidase. The initial rate of the absorbance decrease of 550 mμ is recorded, from which the initial turnover rate of the enzyme (moles of ferrocytochrome *c* oxidized per mole of enzyme hematin per second) was calculated by the use of the following extinction coefficients:

$$\Delta\epsilon_{(\text{reduced-oxidized})} \text{ at } 550 \text{ m}\mu = 19.6 \text{ m}M^{-1} \text{ cm}^{-1} \text{ for heart cytochrome } c^4$$
$$\epsilon_{(\text{oxidized})} \text{ at } 422 \text{ m}\mu = 79 \text{ m}M^{-1} \text{ cm}^{-1} \text{ for cytochrome oxidase}^5$$

Reciprocal turnover rates are plotted against reciprocal concentrations of

[1] Supported by a research grant (GM12202-02) from the United States Public Health Service.
[2] D. Keilin and E. F. Hartree, *Proc. Roy. Soc.* **B127,** 167 (1939).
[3] T. Yonetani and G. S. Ray, *J. Biol. Chem.* **240,** 3392 (1965).
[4] T. Yonetani, *J. Biol. Chem.* **240,** 4509 (1965).
[5] T. Yonetani, *J. Biol. Chem.* **235,** 845 (1960); **236,** 1680 (1961).

ferrocytochrome *c* according to Lineweaver-Burk procedure,[6] to determine a maximal turnover rate of the enzyme at infinite ferrocytochrome *c* concentration.

Procedure

Since cytochrome oxidase contains a dichroic heme, called heme *a*, as its prosthetic group, solutions of cytochrome oxidase exhibit distinctive colors: light brown in dilute solutions and deep brownish red in concentrated solutions. The enzyme is, therefore, readily traced during purification by following these colors.

A suspension of mitochondrial fragments, so-called particulate preparation, is prepared from fresh beef heart according to the method described by Smith.[7]

Step 1. Extraction of Cytochrome Oxidase. To the particulate preparation are added 10% neutralized cholic acid and solid ammonium sulfate (the amounts to be added are 100 ml and 88 g per 400 ml, respectively). The latter is added slowly with mechanical stirring over a period of 30 minutes. The pH of the mixture is kept at 7.4 with 3 *M* NH_4OH. After it has stood for 1 hour, the mixture is centrifuged at 20,000 *g* for 10 minutes. The brown supernatant is brought to 50% saturation by further addition of ammonium sulfate (the amount to be added is 127 g per liter of the supernatant). The mixture is immediately centrifuged at 10,000 *g* for 10 minutes. The precipitate is taken up to a total volume of 150 ml in 0.1 *M* potassium buffer, pH 7.4 containing 2% cholate. The concentration of cytochrome oxidase in this solution should be 30–50 μM.

Step 2. Ammonium Sulfate Fractionation. To the solution (150 ml) are slowly added 50 ml of saturated ammonium sulfate to bring the mixture to 25% saturation (the amount to be added is 33.3 ml per 100 ml). During the addition the pH of the mixture is maintained at 7.4 with 3 *M* NH_4OH. The mixture is allowed to stand at 0° for 10 hours. During the standing, most of cytochrome *b* and c_1 become precipitated. After the centrifugation at 10,000 *g* for 10 minutes, the deeply brownish red supernatant is made up to 35% saturation with saturated ammonium sulfate (the amount to be added is 15.4 ml per 100 ml). The brown precipitate is taken up to a total volume of 75 ml with the phosphate buffer containing 2% cholate. The solution, which should contain 50–70 μM cytochrome oxidase, is refractionated several times with ammonium sulfate in the same manner to collect the fractions precipitated between

[6] H. Lineweaver and D. Burk, *J. Am. Chem. Soc.* **56**, 658 (1934).
[7] L. Smith, see Vol. II [130].

26 and 33% saturation, until the following purity test is passed. The precipitate at 33% saturation obtained hereafter forms usually deeply brown pellets with an appearance of an oily paste.

Step 3. Examination of Purity. A small portion of each precipitate obtained at 33 or 35% saturation is taken up in 0.1 *M* potassium phosphate buffer, pH 7.4 containing 1% Tween-80 and examined for its purity spectrophotometrically, before proceeding to further refractionations. Cytochrome oxidase is reduced very slowly even with dithionite, so that the absorption spectrum of the dithionite-reduced preparation should be measured at least 10 minutes after the addition of the reducing agent. The purity criteria are: (a) complete removal of cytochrome *b* and c_1, which have absorption bands at 560 mμ and 552 mμ, respectively, in the reduced state; (b) complete removal of modified cytochrome oxidase, which is indicated by a shoulder at around 422 mμ in the absorption spectrum of the dithionite-reduced preparation; (c) the ratio of the absorbancy at 445 mμ (reduced) to that at 422 mμ (oxidized) should be 1.25 or more; and (d) the ratio of the absorbancy at 280 mμ (oxidized) to that at 445 mμ (reduced) should be 2.5 or less.

Step 4. Reactivation of Cytochrome Oxidase. Because of the inhibitory effect of cholate, the specific activity of the purified enzyme in the cholate solution is extremely low. Cholate has to be replaced by noninhibitory detergents such as Emasol-4130 or Tween-80. If the above-mentioned criteria are satisfied, the precipitate obtained at 33% saturated ammonium sulfate is dissolved in 10 ml of 0.1 *M* potassium phosphate buffer, pH 7.4 containing 1% Tween-80. The solution is dialyzed against 1 liter of 0.01 *M* potassium phosphate buffer, pH 7.4 containing 0.1% Tween-80 at 0° for 10 hours. The dialyzed solution should be clear and transparent. The concentration of cytochrome at this stage should be accurately determined; it is in the range of 100–200 μ*M*. The enzyme solution can be stored either at 0° for 2 weeks or at −15° for several months. In the latter case, the solution should be divided into appropriate vials to avoid repeated freezing and thawing with each use.

Properties

Beef heart cytochrome oxidase prepared by this procedure has a minimal molecular weight of 100,000 per heme *a* and contains approximately 10% unidentified lipids. Heavy metal constituents of this enzyme are copper and heme *a* iron: the copper to iron ratio is 1.0 to 2.0 moles per mole. The enzyme preparation is homogeneous on ultracentrifugation and is apparently in a polymeric state.[8] The maximal turnover rate of the

[8] S. Takemori, I. Sekuzu, and K. Okunuki, *Biochim. Biophys. Acta* **51,** 464 (1961).

enzyme in the overall reaction is in the range of 60–200 per second at 25° and pH 6.0. Absorption spectra of the enzyme and its dithionite-reduced form are shown in Fig. 1. This enzyme is reversibly inhibited by cyanide, azide, and carbon monoxide.[2] Although only one kind of heme, heme *a*, is extracted from the enzyme by acid-acetone treatments, the heme absorption bands at 605 mμ and 445 mμ of this enzyme behave differently toward oxygen and these inhibitors.[2,9] The molar CO-binding capacity of the enzyme is reported to be 0.5 mole or less per mole of heme *a*.[10]

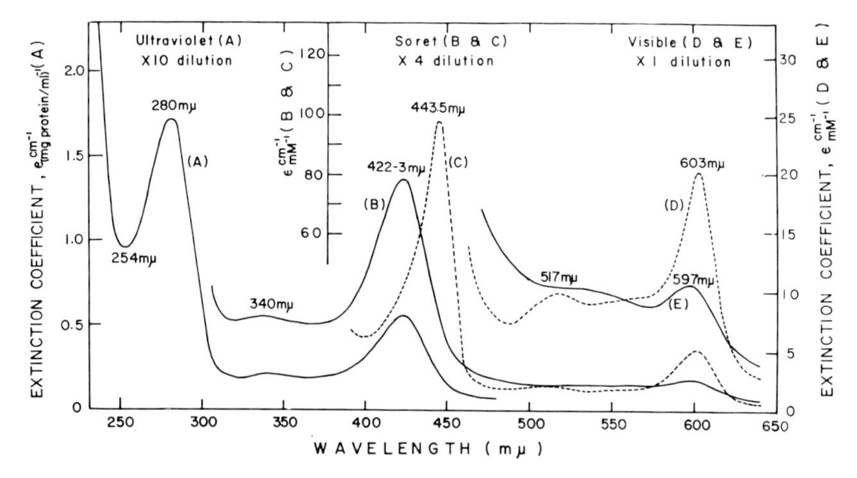

FIG. 1. Absorption spectra of cytochrome oxidase. Solid lines: oxidized preparation; dotted lines: dithionite-reduced preparation. From T. Yonetani, *J. Biol. Chem.* **236**, 1680 (1965).

[9] B. Chance, *Discussions Faraday Soc.* **20**, 205 (1955).
[10] D. H. Gibson and C. Greenwood, *Biochem. J.* **86**, 541 (1963).

[60] Cytochrome *c* Peroxidase (Bakers' Yeast)[1]

By TAKASHI YONETANI

$$H_2O_2 + 2 \text{ ferrocytochrome } c + 2 H^+ \rightarrow 2 \text{ ferricytochrome } c + 2 H_2O$$

Assay Method

Principle. For cytochrome *c* peroxidase, a protohemoprotein,[2,3] the purity index, which is defined as the ratio of absorbance at 408 mμ to that at 280 mμ (A_{480}:A_{280}), is a convenient measure to estimate an approximate purity of the enzyme during purification. The enzyme activity was assayed by measuring the initial rate of the peroxidatic oxidation of ferrocytochrome *c* catalyzed by this enzyme.

Reagents

Sodium acetate buffer, pH 6.0, ionic strength of 0.1
H_2O_2, 18 m*M*, freshly prepared by a 500-fold dilution of 30% H_2O_2
Ferrocytochrome *c*, approximately 1 m*M*, more than 90% reduced
Cytochrome *c* peroxidase, 0.1–1 μ*M*, which corresponds to absorbance at 408 mμ of the enzyme of 0.01–0.1 cm^{-1}

Procedure. The procedure of assay of the cytochrome *c* peroxidase is essentially identical to that described for cytochrome oxidase[4] except for the use of a different buffer and an extra addition of 20 μl of 18 m*M* H_2O_2 to the reaction mixture to give a final concentration of 180 μ*M* H_2O_2. The initial turnover rate of the enzyme (moles of ferrocytochrome *c* oxidized per mole of enzyme hematin per second) was calculated by the use of the following extinction coefficients:

$$\Delta\epsilon_{(\text{reduced-oxidized})} \text{ at } 550 \text{ m}\mu = 19.6 \text{ m}M^{-1} \text{ cm}^{-1} \text{ for heart cytochrome } c^5$$
$$\epsilon_{(\text{oxidized})} \text{ at } 408 \text{ m}\mu = 93.0 \text{ m}M^{-1} \text{ cm}^{-1} \text{ for cytochrome } c \text{ peroxidase}^3$$

Reciprocal turnover rates are plotted against reciprocal concentrations of ferrocytochrome *c* according to Lineweaver-Burk procedure,[6] to determine a maximal turnover rate at infinite cytochrome *c* concentration and 180 μ*M* H_2O_2.

[1] Supported by a research grant (GM12202-02) from the United States Public Health Service.
[2] A. M. Altschul, R. Abrams, and T. R. Hogness, *J. Biol. Chem.* **136,** 777 (1940). See also R. Abrams, A. M. Altschul, and T. R. Hogness, *J. Biol. Chem.* **142,** 303 (1942).
[3] T. Yonetani and G. S. Ray, *J. Biol. Chem.* **240,** 4503 (1965).
[4] This volume [59].
[5] T. Yonetani, *J. Biol. Chem.* **240,** 4509 (1965).
[6] H. Lineweaver and D. Burk, *J. Am. Chem. Soc.* **56,** 658 (1934).

Procedure[3]

A purified preparation of cytochrome *c* peroxidase is made from commercial bakers' yeast (National Yeast Corporation, New Jersey) according to the method described here.

Step 1. Autolysis in the Presence of Ethyl Acetate. Five kilograms of the press yeast are dried at 20° for 10–20 hours until 30–35% of the original weight is lost. The half-dried yeast is well mixed with 500 ml of ethyl acetate. The resulting thick paste is allowed to stand overnight at 4°.

Step 2. Extraction. The paste is homogeneously suspended in 5 liters of distilled water. The suspension is stirred at 20° for 3 hours and centrifuged at 5000 *g* for 15 minutes. The supernatant (about 5 liters) is saved. Solid ammonium sulfate (760 g per liter) is added to the supernatant to saturate it 100%. The mixture is stirred at 20° for 2 hours and centrifuged at 13,000 *g* for 25 minutes. The following procedures are carried out at 4°. The loosely packed solid material, which is floating on the top of the liquid phase, is carefully separated from the latter and homogenized by gentle stirring with a glass rod. The thick homogenate (approximately 500 ml) is transferred into dialysis tubes by the aid of washing with a minimal amount of 0.1 *M* phosphate buffer, pH 6, and dialyzed against several changes of distilled water and finally against 5 m*M* phosphate buffer, pH 6.

Step 3. Concentration by DEAE-Cellulose Absorption. The dialyzed preparation is centrifuged at 5000 *g* for 10 minutes to remove insoluble materials. The supernatant (approximately 1 liter) is passed through a column (5×10 cm) of the cellulose equilibrated with 5 m*M* phosphate buffer, pH 6. The enzyme is absorbed at the top of the column as a conspicuous brown band. The column is washed with 500 ml of 5 m*M* phosphate buffer, pH 6. The brown band is eluted with 0.5 *M* phosphate buffer, pH 6. The eluate (approximately 150 ml) is dialyzed against several changes of 5 m*M* phosphate buffer, pH 6.

Step 4. Chromatography on DEAE-Cellulose Column. The dialyzed solution is placed on a column (2×50 cm) of DEAE-cellulose equilibrated with 5 m*M* phosphate buffer, pH 6. The column, at the top of which the enzyme is absorbed, is washed successively with 1 liter of 0.05 *M* phosphate buffer, pH 6, and with an appropriate volume of 0.1 *M* phosphate buffer, pH 6, until the brown layer spreads down and reaches the bottom of the column. Then the brown layer is eluted with 0.5 *M* phosphate buffer, pH 6. The colored eluates are combined and dialyzed against 5 m*M* phosphate buffer, pH 6. The dialyzed solution is kept either at 0° or at −20° for a prolonged storage period. If too dilute, the enzyme can be readily concentrated by absorption on and elution from a small amount of the cellulose. The enzyme appears to be stable at 0° for at least 2 months. The enzyme can be stored at −15° for a considerably longer period of time without inactivation. The yield is usually 3–4 micromoles of the enzyme from 5 kg of the press yeast. These values correspond to an overall recovery of 25–30%.

Properties[3,5]

Purified cytochrome *c* peroxidase has a minimum molecular weight of 49,000 per mole of enzyme hematin. This enzyme forms dissociable enzyme–inhibitor complexes with cyanide, azide, and fluoride. Only one type of stable complex is formed between this enzyme and H_2O_2 (cf. Fig. 1) with a molar stoichiometry of 1:1. The oxidative titration of the

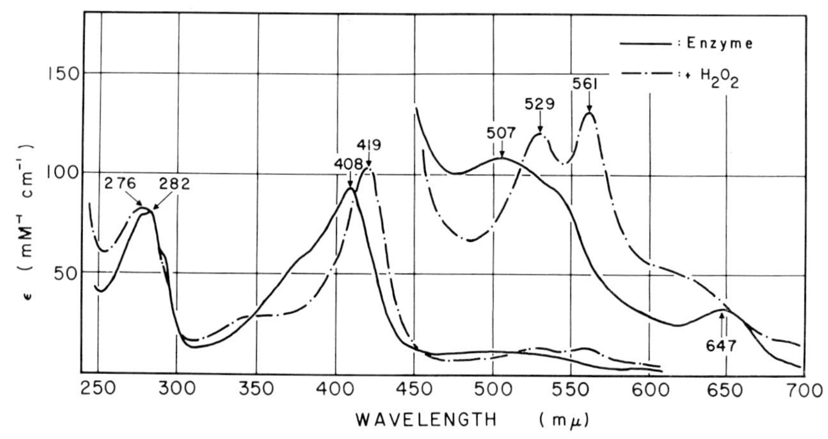

FIG. 1. Absorption spectra of cytochrome *c* peroxidase and its H_2O_2 complex (= complex II). From T. Yonetani, *J. Biol. Chem.* **240,** 4509 (1965).

DONOR SPECIFICITY OF CYTOCHROME *c* PEROXIDASE[a]

Donor	Maximum turnover rate $(V_m/e)^b$	$K_m{}^b$
Horse heart ferrocytochrome *c*	2000 sec^{-1}	5 μM
Bakers' yeast ferrocytochrome *c*	1500 sec^{-1}	10 μM
Ascorbate	3 sec^{-1}	11 mM
Guaiacol	4.4 sec^{-1}	10 mM
Pyrogallol	8.0 sec^{-1}	5 mM

[a] From T. Yonetani and G. S. Ray, *J. Biol. Chem.* **240,** 4503 (1965).

[b] Apparent values at 180 μM H_2O_2 and an infinite concentration of donors in sodium acetate buffer, pH 6.0, with an ionic strength of 0.05 at 23°.

enzyme–peroxide complex with ferrocytochrome *c* indicates that the complex has two oxidizing equivalents with respect to free peroxidase.[5] As shown in the table, this enzyme has a highly selective donor specificity toward ferrocytochrome *c*.[2,3,7,8]

[7] A. C. Maehly, Vol. II [143].

[8] B. Chance, *Advan. Enzymol.* **12**, 153 (1951).

[61] Cytochrome *c* from Vertebrate and Invertebrate Sources

By E. MARGOLIASH and O. F. WALASEK

Preparation Procedure

Principle. Cytochrome *c* is extracted from ground and homogenized tissue with a dilute solution of aluminum sulfate at pH 4.5. At slightly alkaline pH, excess aluminum ions are precipitated as the hydroxide, and exchanged for three monovalent ammonium ions. This permits the direct collection of the protein from the extract on a cation exchange resin. The cytochrome *c* is purified by $(NH_4)_2SO_4$ fractionation, cation exchange chromatography, and crystallization. Other methods, involving extraction with trichloroacetic[1-4] or sulfuric acids,[5-7] extraction at neutral pH following the decomposition of the tissue in acetic acid[8-12] or by treat-

[1] D. Keilin and E. F. Hartree, *Proc. Roy. Soc.* **B122**, 298 (1937).

[2] D. Keilin and E. F. Hartree, *Biochem. J.* **39**, 289 (1945).

[3] D. Keilin and E. F. Hartree, *in* "Biochemical Preparations" (E. G. Ball, ed.), Vol. 2, p. 1. Wiley, New York, 1952.

[4] E. Margoliash, *in* "Biochemical Preparations" (D. Shemin, ed.), Vol. 5, p. 33. Wiley, New York, 1957.

[5] H. Theorell, *Biochem. Z.* **279**, 463 (1935).

[6] H. Theorell, *Biochem. Z.* **285**, 207 (1936).

[7] S. Paléus, *Acta Chem. Scand.* **14**, 1743 (1960).

[8] B. Hagihara, I. Morikawa, I. Sekuzu, and K. Okunuki, *J. Biochem.* (*Tokyo*) **45**, 551 (1958).

[9] B. Hagihara, M. Yoneda, K. Tagawa, I. Morikawa, M. Shin, and K. Okunuki, *J. Biochem.* (*Tokyo*) **45**, 565 (1958).

[10] B. Hagihara, K. Tagawa, I. Morikawa, M. Shin, and K. Okunuki, *J. Biochem.* (*Tokyo*) **45**, 725 (1958).

[11] K. Okunuki, *in* "A Laboratory Manual of Analytical Methods in Protein Chemistry" (P. Alexander and R. J. Block, eds.), Vol. I, p. 32. Pergamon Press, London, 1960.

[12] B. Hagihara, I. Morikawa, K. Tagawa, and K. Okunuki, *in* "Biochemical Preparations" (C. S. Vestling, ed.), Vol. 5, p. 1. Wiley, New York, 1958.

ment with a sucrose-saponin solution and acetone,[13] have been described. The method of extraction described below as well as those just cited are suited to vertebrate and invertebrate tissues but are generally ineffective with plant materials, such as wheat germ, molds, and bakers' yeast. In these cases special autolysis and extraction procedures are required[14-22] even though the properties of the cytochromes *c* are much the same as those of the vertebrate and invertebrate proteins and similar purification schemes can be used.

Preparation of Resin. Amberlite Type II (Rohm and Haas Co., Philadelphia) 200–400 mesh is cycled through the Na^+ and H^+ forms, using 2×8 volumes of 2 *N* NaOH, followed by five washings with deionized water and 2×8 volumes of 2 *N* HCl, followed by five washings with deionized water. The supernatant solution is decanted at each washing after the bulk of the resin has settled, repeatedly removing the finer particles. The cycle is repeated once and in the H^+ form the resin is washed with 5×8 volumes of acetone. The acetone is removed *completely* by 10–15 water washings; the resin is cycled twice more and brought to the final Na^+ form. After five water washings the resin is equilibrated at pH 8.0–8.2 using 5 *N* H_3PO_4 and an automatic titrator (Model TTTlc, Radiometer, Copenhagen). Equilibration is allowed to proceed for 5–6 hours, under vigorous stirring. After settling, the resin is washed twice with water and suspended in two volumes of 0.02 *M* Na/Na phosphate buffer pH 8.0. If after several hours the pH of the suspension is more alkaline than 8.2, the equilibration procedure is repeated. The resin can be stored indefinately in the dilute phosphate buffer at 4°.

[13] M. Morrison, T. Hollocher, R. Murray, G. Marinetti, and E. Stotz, *Biochim. Biophys. Acta* **41**, 334 (1960).

[14] D. Keilin, *Proc. Roy. Soc.* **B106**, 418 (1930).

[15] D. R. Goddard, *Am. J. Botany* **31**, 270 (1944).

[16] B. Hagihara, T. Horio, J. Yamashita, M. Nozaki, and K. Okunuki, *Nature* **178**, 629 (1956).

[17] B. Hagihara, K. Tagawa, I. Morikawa, M. Shin, and K. Okunuki, *J. Biochem. (Tokyo)* **46**, 321 (1959).

[18] M. Nozaki, T. Yamanaka, T. Horio, and K. Okunuki, *J. Biochem. (Tokyo)* **44**, 453 (1957).

[19] R. Nunnikhoven, *Biochim. Biophys. Acta* **28**, 108 (1958).

[20] A. R. Wasserman, Hemoprotein 550 mμ and cytochrome *c* of wheat germ. Dissertation, University of Wisconsin, 1960.

[21] B. A. Hardesty, The biochemistry of cytochrome *c* in relation to maternally inherited phenotypes of *Neurospora*. Dissertation, California Institute of Technology, 1961.

[22] J. Lustgarten, Studies on cytochrome *c*. Dissertation, McGill University, 1962.

Preparation of Cytochrome *c*

Step 1. Coarsely ground tissue is homogenized in a Waring blendor, usually for 15–45 seconds, with 3 volumes of 0.3% $Al_2(SO_4)_3 \cdot 17\ H_2O$. The extent of homogenization must be adjusted to the ease with which the tissue is dispersed. If too fine a suspension is produced the subsequent filtration will be impeded, and if the suspension is too coarse extraction will be incomplete. With a tissue as fragile as liver, passage through a meat grinder is sufficient, whereas at the other extreme, skeletal muscles from older animals may require as much as 1–2 minutes of homogenization. Extremely tough tissues are best ground in the semifrozen state.

The mixture is adjusted at pH 4.5 by adding, if necessary, more of the $Al_2(SO_4)_3$ solution. With some tissues it may be necessary to bring the pH up to 4.5, using 5 *N* NaOH. With tissues that have a particularly high buffering capacity, such as insect thoraces,[23] 5 *N* H_2SO_4 can be used to adjust the pH. Extraction is allowed to proceed with occasional stirring, at room temperature, for 1½–2 hours. *All further operations are carried out in the cold room at 4°.* The extract is filtered through fluted filter paper. Of several varieties tested, No. 588, coarse fluted filter paper, 25-inch diameter (C. Schleicher and Shuell Co., Keene, New Hampshire) has been found to give the most satisfactory filtration rate.

Step 2. The filtrate is adjusted to pH 8.2–8.5 with concentrated ammonia solution, filtered clean through fluted filter paper with the aid of 10 g Hyflo-Super-Cel (Johns Manville Co., New York) per liter of extract, and rapidly passed, under suction or with positive pressure, through a wide column of Amberlite IRC-50 prepared as given above. Columns 15 × 10 cm are suitable for 20–30 liters of extract. The cytochrome *c* collects in the reduced, salmon pink, form. The column is washed with 10 volumes of 0.02 *M* Na/Na phosphate buffer pH 8.0, and either the protein is eluted with the same buffer containing 0.5 *M* NaCl or the colored resin is scraped off, poured into a smaller column, the column washed with the dilute buffer, and the cytochrome *c* eluted with the NaCl solution. If carefully carried out, the latter procedure is likely to yield a more manageable smaller eluate volume.

Step 3. The cytochrome *c* is oxidized with the minimal amount of $K_3Fe(CN)_6$, and the solution is brought to the maximal concentration of $(NH_4)_2SO_4$ that will not precipitate the cytochrome *c*, usually 0.7–0.8 saturation. This saturation will depend on the particular cytochrome *c* being purified (for example, the tuna protein usually precipitates above

[23] S. K. Chan and E. Margoliash, *J. Biol. Chem.* **241**, 335 (1966).

0.7 saturation, while the horse, beef, and pig proteins remain in solution up to 0.85 saturation), the exact pH of the solution and the concentration of cytochrome *c*. Since the latter two factors are like to vary somewhat from preparation to preparation, it is best to test a small portion before precipitating the bulk of the extract. The precipitate is filtered off and the filtrate is dialyzed against deionized water brought to pH 8.5 with ammonia solution, adding 200–300 mg solid Na_2HPO_4 per liter of solution, inside the dialysis bags. Dialysis is continued for 2 days with 6 changes of outer solution.

Step 4. The cytochrome *c* is collected and chromatographed on a column of Amberlite IRC-50 at pH 8.0, under a linear gradient of Na^+ concentration, from 0.02 *M* Na/Na phosphate buffer, pH 8.0, to the same buffer containing 0.5 *M* NaCl. Columns 3×60 cm are suitable for 1–3 g of protein and 2–3 liters of the dilute, and concentrated solutions can be used on each side of the gradient mixing device. Before the gradient is started, a few drops of 1 *M* $K_3Fe(CN)_6$ are added to the solution at the top of the column to convert all the protein to the ferric form. The cytochrome *c* should come off in a single peak, *showing 0% carbon monoxide combination throughout* (see note 3, below) and is best collected in a fraction collector. If the material obtained is not pure, steps 3 and 4 are repeated. If it is not intended to crystallize the protein, it can at this point be lyophilized following thorough dialysis (5 days, twice daily changes of outer solution) against distilled water, adjusted to pH 8.5 with ammonia solution.

Step 5. The column fractions are pooled and the cytochrome *c* may often be crystallized directly from them. However, if the solution contains less than 1% cytochrome *c*, as judged by the absorbance of the ferrous form at 550 mμ (see section on Properties, below), it is best to dialyze it overnight, concentrate the protein on a small Amberlite IRC-50 column, and elute it with the 0.02 *M* phosphate buffer containing 1 *M* NaCl. In either case the solution is brought to a concentration of 1% protein in the dilute pH 8.0 buffer, NaCl is added to a final concentration of 1.0–1.5 *M*, and finely powdered $(NH_4)_2SO_4$ is carefully added until a faint haze can just be detected, foaming being avoided as much as possible. The protein will usually crystallize within 30 minutes to 2 days, and the crystals continue to grow for several weeks. Leaving the suspension at room temperature overnight before returning it to 4°, several times in succession, often hastens this process. After the initial crystallization, the solution can gradually be fully saturated with $(NH_4)_2SO_4$. Even under such conditions from 5% to 30% of the cytochrome *c* remains in the supernatant, depending on the solubility of the particular protein used. Crystalline suspensions keep indefinitely at 4°.

Notes

1. *Starting Materials and Yields.* Cytochrome *c* has been prepared by the $Al_2(SO_4)_3$ extraction method from hearts and skeletal muscles of numerous vertebrates, from pig kidney, liver, and brain, from whole lampreys, and from several insect materials (whole screw-worm flies, *Drosophila,* and moth thoraces). Storage of the tissues in plastic bags for 8 months at —15° did not lead to deterioration of the protein or to decreased yields. The yield from hearts varies from about 4 micromoles/kg for frog hearts to 25 micromoles/kg for horse hearts, most mammalian hearts yielding from 16–20 micromoles/kg. Mammalian skeletal muscles yield about 4 micromoles/kg; the yield from bird flight muscles and insect thoraces is comparable to that from vertebrate hearts.

2. *Preparation Procedure.* To avoid preparation artifacts (see note 3) it is important to proceed as fast as possible to the point (step 3) at which the solution is brought to 0.7–0.8 saturation with $(NH_4)_2SO_4$. The first three steps should be carried out the same day. Keeping at a minimum the time the extract is left at pH 4.5 decreases the probability of contamination with poorly separable deamidated forms, while proteolytic enzymes which might be extracted with the cytochrome *c* and attack the polypeptide chain are separated at the $(NH_4)_2SO_4$ precipitation step. If filtration of the original extract (step 1) is too slow, it is preferable to continue the preparation with the first portion of the filtrate, discarding the remaining parts.

It is not possible to use dialysis tubing indiscriminately as the majority of rolls are too permeable to cytochrome *c*. The authors have regularly tested samples from individual rolls, by dialysis of cytochrome *c* (ca. 10^{-4} *M*) against water for 3 days. The protein passes through the cellophane particularly rapidly in very low ionic strength solutions.[23a]

For those cytochromes *c* which are relatively less soluble than others in $(NH_4)_2SO_4$ solutions, the concentration of $(NH_4)_2SO_4$ attainable in step 3 is sometimes too low to ensure purification. In such cases the gel filtration procedure of Flatmark[24] is usually satisfactory. The authors have used it following step 4, as follows: 1–2 g of the protein in 10–20 ml 0.05 *M* NH_4HCO_3 is applied to a 3 × 150 cm column of Sephadex G-75, bead form (Pharmacia, Uppsala), equilibrated with the same solvent. The column is developed with the solvent at flow rates of 10–15 ml/hour. A faintly colored impurity commonly separates in front of the main

[23a] It has also been found that boiling dialysis tubing in 0.1 *M* NH_4HCO_3 for 30 minutes considerably decreases the permeability to cytochrome *c*. The inside of the dialysis tubing should be well wetted before boiling.

[24] T. Flatmark, *Acta Chem. Scand.* **18,** 1517 (1964).

cytochrome *c* band. The protein solution can then either be dialyzed and lyophilized or crystallized following, if necessary, concentration and elution from a small Amberlite column.

3. *Tests of Purity and Homogeneity—Artifactual Forms.* The indices of purity that are usually employed are the iron content (0.45% for pure cytochrome *c*[25,26]) and the ratio of the absorbance of the ferrous protein reduced with dithionite, at 550 mμ, to that of the ferric protein at 280 mμ. Among the numerous methods which have been described for the determination of iron in heme proteins the simple procedure recently given by Cameron[27] has been found by the authors to be well suited to cytochrome *c*, giving reproducible and accurate results.

The absorbance ratio is determined on a solution of the protein in water, not containing buffer salts. Readings are taken successively at 280 mμ and 550 mμ. A few crystals of Na_2HPO_4 are then dissolved in the solution in the spectrophotometer cell, the protein is reduced with minimal dithionite, and the reading at 550 mμ is taken again, making sure the wavelength is actually at the sharp 550 mμ maximum. From the values of the extinction coefficients and the readings at 550 mμ the percentage of the original sample in the ferrous form can be calculated. From this value and the extinction coefficients of the reduced and oxidized protein at 280 mμ, the reading at 280 mμ can be corrected to that which would have been given by the fully oxidized protein. See the table for the extinction coefficients. For most cytochromes *c* which contain a single residue of tryptophan in the polypeptide chain, the absorbance ratio (550 mμ, reduced/280 mμ, oxidized) is 1.25, while for cytochromes *c* which carry two residues of tryptophan, such as the tuna protein, this ratio is 1.04–1.06.[28] With fresh preparations absorbance ratios as high as 1.4 have occasionally been observed,[29,30] but on standing for several days in solution the values return to 1.25. Reasons for this are unknown.

It cannot be too strongly emphasized that these criteria of purity are *not* criteria of homogeneity and do not distinguish between native cytochrome *c* and deamidated or polymeric artifactual forms[31,32] of the protein. Various forms of electrophoresis[33,34] and chromatography on

[25] E. Margoliash, *J. Biol. Chem.* **237**, 2161 (1962).
[26] S. K. Chan and E. Margoliash, *J. Biol. Chem.* **241**, 335 (1966).
[27] B. F. Cameron, *Anal. Biochem.* **11**, 164 (1965).
[28] G. Kreil, *Z. Physiol. Chem.* **334**, 154 (1963).
[29] E. Margoliash, unpublished observations, 1964.
[30] W. B. Elliott, personal communication, 1965.
[31] E. Margoliash, *Brookhaven Symp. Biol.* **15**, 266 (1962).
[32] E. Margoliash and J. Lustgarten, *J. Biol. Chem.* **237**, 3397 (1962).
[33] S. Paléus and K. G. Paul, *in* "The Enzymes" (P. D. Boyer, H. Lardy, and K. Myrbäck, eds.), 2nd ed., Vol. 8, p. 97. Academic Press, New York, 1963.
[34] E. Margoliash and A. Schejter, *Advan. Protein Chem.* **21**, 113 (1966).

SOME PHYSICOCHEMICAL PROPERTIES OF CYTOCHROME *c*

λ	Absorption spectrum[a] Fe⁺⁺ (ϵmM)	Absorption spectrum[a] Fe⁺⁺⁺ (ϵmM)	Electrophoretic mobilities at pH 6.9 and 0°[c] (cm²/sec/volt × 10⁵)	Isoionic point at 20°[d] (pH)	Sedimentation coefficient,[e] $S_{20,w}$ × 10¹³ (sec⁻¹)	Diffusion coefficient,[e] $D_{20,w}$ × 10⁷ (cm²/sec)
550 mμ	29.5[b]	9.0	Human, 9.52	10.04 ± 0.04	1.83	13.0
416 mμ	129.1	88.8	Tuna, 8.75	—	—	—
410 mμ	106.1	106.1	Pigeon, 7.90	—	—	—
280 mμ	31.0	23.2	—	—	—	—

[a] Except for the extinction value of the ferrous protein at 550 mμ, the values are taken from E. Margoliash and N. Frohwirt [*Biochem. J.* **71,** 570 (1959)] and are for the horse heart protein.

[b] According to B. F. VanGelder and E. C. Slater, *Biochim. Biophys. Acta* **58,** 593 (1962).

[c] G. H. Barlow and E. Margoliash, *J. Biol. Chem.* **241,** 1473 (1966). After removal of bound ions by electrodialysis the electrophoretic mobilities were measured in Tris-cacodylate buffer, μ = 0.1.

[d] G. H. Barlow and E. Margoliash, *J. Biol. Chem.* **241,** 1473 (1966). The value listed represents the isoionic points for cytochromes *c* from ten different species.

[e] E. Margoliash and J. Lustgarten, *J. Biol. Chem.* **237,** 3397 (1962). The values are for horse heart cytochrome *c*.

cation exchange resins have been extensively used as indices of homogeneity. Electrophoresis is complicated by the appearance of multiple anomalous boundaries due to anion binding,[35] so that electrophoretic mobilities are influenced by the particular cytochrome *c* tested, its concentration, the nature of the buffer anion used, and its concentration.[35] Nevertheless, electrophoresis will distinguish deamidated and polymeric forms[31,32] from the native material, both types of artifactual species having lower cathodic mobilities than the native protein. In the hands of the authors column chromatography on Amberlite IRC-50 under a linear Na^+ concentration gradient, as described above (Preparation Procedure, step 4), has proved to be the most valuable criterion of homogeneity. Deamidated forms tend to separate at the leading edge of the native chromatographic peak, while the polymers appear as well separated bands behind it.[31,32] Since both types of artifacts are autoxidizable and combine to varying extents with carbon monoxide they can be readily detected by following the carbon monoxide combination throughout the elution pattern,[31,32] by the method of Tsou[36] (see Fig. 1).

[35] G. H. Barlow and E. Margoliash, *J. Biol. Chem.* **241,** 1473 (1966).

[36] C. L. Tsou, *Biochem. J.* **49,** 362 (1951).

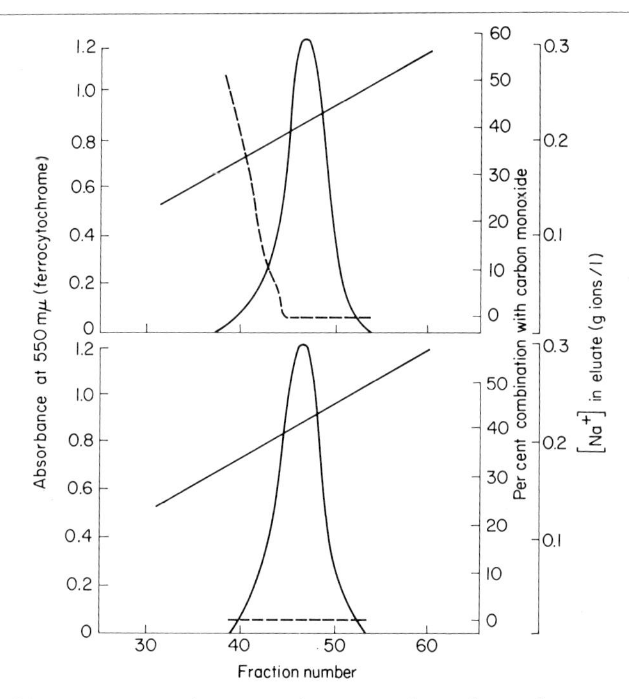

FIG. 1. Chromatograms of monomeric preparations of cow heart cytochrome *c*. *Top panel:* A preparation containing deamidated material. *Bottom panel:* A completely native preparation. The straight lines represent the Na^+ concentration in the eluate and the dashed curves the percentage combination with carbon monoxide. From E. Margoliash, *Brookhaven Symp. Biol.* **15,** 266 (1962).

A third type of artifact, occasionally encountered, consists of native cytochrome *c* apparently complexed with a basic protein occurring in the original extract. This complex chromatographs as a separate fraction behind the pure native protein.[34] The purity indices will serve to detect the contamination with a non-cytochrome *c* protein, and repetition of the $(NH_4)_2SO_4$ precipitation step or gel filtration usually provide satisfactory purifications.

Properties of Cytochrome c

Cytochrome *c* is a basic protein with a molecular weight near 12,300. It consists of a single polypeptide chain 103–108 amino acid residues long, to which a single heme prosthetic group is covalently attached by thioether bonds[37,38] formed by the addition of the sulfhydryl groups of two cysteinyl residues in positions 14 and 17 across the vinyl side chains

[37] H. Theorell, *Enzymologia* **4,** 192 (1937).
[38] H. Theorell, *Biochem. Z.* **298,** 242 (1938).

of the porphyrin ring. These thioether bonds can be broken by treatment with heavy-metal salts at acid pH.[39] It has recently been found possible to react the apoprotein thus obtained with protoporphyrinogen, re-forming the thioether bonds, let the porphyrinogen autoxidize to the prophyrin, and reintroduce the iron to obtain, in 8% yield, the native, enzymatically active protein.[40] In the vertebrate cytochromes *c* the amino-terminal residue is acetylglycine while the proteins from non-vertebrate sources carry several extra residues in place of the acetyl. Complete amino acid sequences have been determined for the cytochromes *c* from horse,[25,41] man,[42] pig,[43,44] chicken,[45] bakers' yeast,[46] cow,[47]

Acetyl-Gly-Asp-Val-Glu-Lys-Gly-Lys-Lys-Ile-Phe-Val-Gln-Lys- (10)

CyS-Ala-Gln-CyS-His-Thr-Val-Glu-Lys-Gly-Gly-Lys-His-Lys-Thr- (20)
(CyS ... CyS linked by HEME)

Gly-Pro-Asn-Leu-His-Gly-Leu-Phe-Gly-Arg-Lys-Thr-Gly-Gln- (30 ... 40)

Ala-Pro-Gly-Phe-Thr-Tyr-Thr-Asp-Ala-Asn-Lys-Asn-Lys-Gly- (50)

Ile-Thr-Trp-Lys-Glu-Glu-Thr-Leu-Met-Glu-Tyr-Leu-Glu-Asn- (60 ... 70)

Pro-Lys-Lys-Tyr-Ile-Pro-Gly-Thr-Lys-Met-Ile-Phe-Ala-Gly-Ile- (80)

Lys-Lys-Lys-Thr-Glu-Arg-Glu-Asp-Leu-Ile-Ala-Tyr-Leu-Lys-Lys- (90 ... 100)

Ala-Thr-Asn-GluCOOH (104)

FIG. 2. Amino acid sequence of horse heart cytochrome *c*. Based on E. Margoliash, E. L. Smith, G. Kreil, and H. Tuppy, *Nature* **192**, 1125 (1961); E. Margoliash, *J. Biol. Chem.* **237**, 2161 (1962).

[39] K. G. Paul, *Acta Chem. Scand.* **4**, 239 (1950).

[40] S. Sano and K. Tanaka, *J. Biol. Chem.* **239**, PC3109 (1964).

[41] E. Margoliash, E. L. Smith, G. Kreil, and H. Tuppy, *Nature* **192**, 1125 (1961).

[42] H. Matsubara and E. L. Smith, *J. Biol. Chem.* **238**, 2732 (1963).

[43] E. Margoliash, S. B. Needleman, and J. W. Stewart, *Acta Chem. Scand.* **17**, S250 (1963).

[44] J. W. Stewart and E. Margoliash, *Can. J. Biochem.* **43**, 1187 (1965).

[45] S. K. Chan and E. Margoliash, *J. Biol. Chem.* **241**, 507 (1966).

[46] K. Narita, K. Titani, Y. Yaoi, and H. Murakami, *Biochim. Biophys. Acta* **77**, 688 (1963).

[47] K. T. Yasunobu, T. Nakashima, H. Higa, H. Matsubara, and A. Benson, *Biochim. Biophys. Acta* **78**, 791 (1963).

tuna fish,[28,48] rattlesnake,[49] rabbit,[50] a saturniid moth, *Samita cynthia*,[26] a rhesus monkey,[51] dog,[52] the great gray kangaroo,[53] snapping turtle,[54] several birds,[55] the screw-worm fly.[55] The proteins from the kidney, liver, brain, and skeletal muscles of the pig have been found to be identical to that from pig heart.[44] The amino acid sequence of a typical cytochrome *c*, that from horse heart,[25,41] is given in Fig. 2. An extensive discussion of structural, evolutionary, and functional aspects of cytochrome *c* has recently been given by Margoliash and Schejter,[34] some useful physico-chemical data are listed in the table, and a drawing of the absorption spectrum of the protein in the visible and ultraviolet regions is shown in Fig. 3.[56]

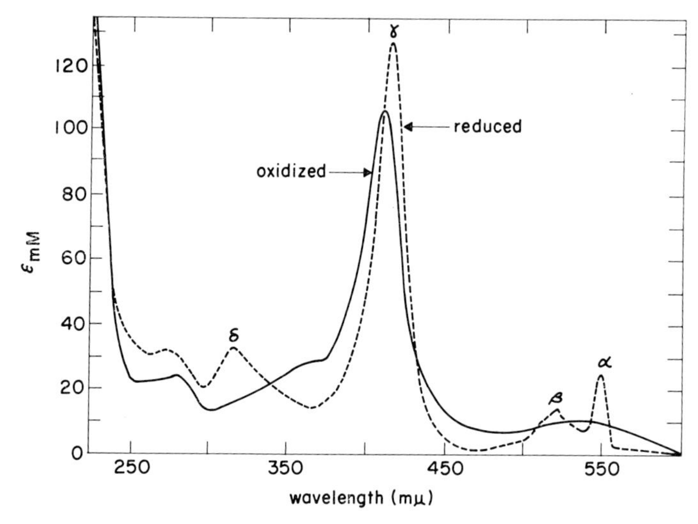

FIG. 3. Absorption spectrum of horse heart cytochrome *c*, calculated from the data of E. Margoliash and N. Frohwirt, *Biochem. J.* **71**, 570 (1959).

[48] G. Kreil, *Z. Physiol. Chem.* **340**, 86 (1965).
[49] O. P. Bahl and E. L. Smith, *J. Biol. Chem.* **240**, 3585 (1965).
[50] S. B. Needleman and E. Margoliash, *J. Biol. Chem.* **241**, 853 (1966).
[51] J. A. Rothfus and E. L. Smith, *J. Biol. Chem.* **240**, 4277 (1965).
[52] M. A. McDowall and E. L. Smith, *J. Biol. Chem.* **240**, 4635 (1965).
[53] C. Nolan and E. Margoliash, *J. Biol. Chem.* **241**, 1049 (1966).
[54] S. K. Chan, I. Tulloss, and E. Margoliash, *Biochemistry* (in press).
[55] S. K. Chan, I. Tulloss, and E. Margoliash, unpublished experiments, 1966.
[56] E. Margoliash and N. Frohwirt, *Biochem. J.* **71**, 570 (1959).

[62] Isolation and Properties of Cytochrome c_1

By J. S. RIESKE[1] and H. D. TISDALE

Several procedures for the isolation of cytochrome c_1 from other mitochondrial hemoproteins have been reported.[1a-4] The procedure described below[5] is at present the method of choice because of the yield and purity of the cytochrome c_1 obtained.

Assay

Principle. The concentration of cytochrome c_1 (in solution) is estimated from absorbancy measurements of its α band at 554 mμ. Cytochrome *c* must be extracted from the preparation prior to the assay.

Procedure. Details of the analysis of the mitochondrial cytochromes are given in this volume [76]. It should be pointed out, however, that cytochrome c_1 in the presence of sodium dodecyl sulfate may not be reduced completely by ascorbate. Therefore, for direct spectral measurements on solutions containing sodium dodecyl sulfate, complete reduction by dithionite should be employed. Contamination with cytochromes *a* and *b* may be estimated by the differential extraction and analysis of hemes.

Purification Procedure

Unless otherwise noted, all operations are carried out at 0–5°.

Step 1. Preparation of KCl-Mitochondrial Suspension. Beef heart mitochondria are prepared by the method of Crane *et al.*[6] (see this volume [12]) except for the following modification. The supernatant liquid from ground beef heart is diluted in 0.9% KCl instead of 0.25 *M* sucrose. The sediment (mitochondrial paste) from the Sharples centrifuge is suspended in an equal volume of 0.9% KCl at a protein concentration of 60–65 mg/ml. This suspension may be stored at −20° until used.

Step 2. Deoxycholate Extraction. Potassium deoxycholate (as a 10%, w/v, solution) is added to the suspension of mitochondria in an amount sufficient to give a final concentration of 0.4 mg per milligram of protein.

[1] See footnote 1, page 239.

[1a] E. Yakushiji and K. Okunuki, *Proc. Imp. Acad.* (*Tokyo*) **17**, 38 (1941).

[2] D. E. Green, J. Jarnefelt, and H. Tisdale, *Biochim. Biophys. Acta* **31**, 34 (1959).

[3] R. P. Glaze and M. Morrison, *Federation Proc.* **19**, 34 (1960).

[4] I. Sekuzu, Y. Orii, and K. Okunuki, *J. Biochem.* (*Tokyo*) **48**, 214 (1960).

[5] R. Bomstein, R. Goldberger, and H. Tisdale, *Biochim. Biophys. Acta* **50**, 527 (1961).

[6] F. L. Crane, J. L. Glenn, and D. E. Green, *Biochim. Biophys. Acta* **22**, 475 (1956).

The suspension is centrifuged for 1 hour at 30,000 rpm in the No. 30 rotor of a Beckman Model L centrifuge (100,000 *g*); the residue, which contains almost all the cytochrome *a*, is discarded.

Step 3. Detergent-Ammonium Sulfate Extraction. The supernatant solution of step 2 is treated with potassium cholate (added as a 20%, w/v, solution) to a concentration of 1.0 mg per milligram of protein; Duponol C[7] (added as a 10%, w/v, solution) to a concentration of 20 mg/ml; and solid ammonium sulfate to a concentration of 170 mg/ml in the order listed. The suspension is centrifuged for 20 minutes at 30,000 rpm; the residue is discarded. The supernatant liquid is treated with solid ammonium sulfate (80 mg/ml) to a final concentration of 0.45 saturation. The suspension is centrifuged for 20 minutes at 30,000 rpm (100,000 *g*); the supernatant liquid, which contains practically all the cytochrome *c*, is discarded.

Step 4. Precipitation of Extraneous Protein. The sedimented protein remaining from step 3 is suspended in 0.1 *M* potassium phosphate, pH 7.4, to a protein concentration of 20 mg/ml. Neutralized, saturated ammonium sulfate is added to give a concentration of 0.05 saturation. The solution is warmed to 40° and maintained at this temperature for 5 minutes. The copious precipitate is removed by centrifugation (100,000 *g* for 10 minutes) and discarded.

Step 5. Precipitation of Cytochrome b. The supernatant solution from step 4 is treated with potassium cholate (added as a 20% solution) to give a final concentration of 3.0 mg per milligram of protein. Neutral, saturated ammonium sulfate is added to a final concentration of 0.21 saturation. The slightly turbid solution is warmed to 30° and is maintained at this temperature for 5 minutes. The precipitate, which contains most of the cytochrome *b*, is removed by centrifugation (10 minutes at 100,000 *g*). If desired, this crude cytochrome *b* may be used for the preparation of purified cytochrome *b* (see this volume [63]). The supernatant solution still contains small amounts of cytochrome *b*. This residual cytochrome *b* is removed by successive application of the heating (30° for 5 minutes) and centrifugation procedure at ammonium sulfate levels of 0.30, and 0.33 saturation, respectively. Each of the residues from the heating steps is discarded; the supernatant solution contains crude, although spectroscopically pure, cytochrome c_1.

Step 6. Extraction of a Cholate-Cytochrome c_1 Complex. The solution of crude cytochrome c_1 of step 5 is treated with potassium cholate (20%) to give a final concentration of 20 mg/ml; the pH is adjusted to 7.1, and the solution is stored overnight at −20°. After being thawed, the

[7] Sodium dodecyl (lauryl) sulfate, obtained from E. I. duPont de Nemours.

mixture is treated with sodium dithionite (10 mg/ml) and then is allowed to stand at 0° for 2 hours; the precipitated cytochrome c_1 is recovered by sedimentation at low speed (2000–5000 *g*). The orange-pink sediment is washed three times with ten volumes of cold, 2 *M* KCl, each time being sedimented at low speed. The washed sediment then is extracted repeatedly with small volumes of cold, distilled water until very little pink color remains in the sedimentable residue. The extracts, containing the cytochrome c_1; are pooled and then the pooled solution is adjusted to pH 6.0–6.5 with 6 *N* KOH.

Step 7. Removal of Bile Salts with Calcium Phosphate Gel. The protein of the pooled extracts is adsorbed onto calcium phosphate gel.[8] The protein–gel suspension then is washed with cold, distilled water (by alternate homogenizations and sedimentations) until the wash solution contains no bile salts (indicated by an absence of precipitated bile acids upon acidification). The cytochrome c_1 is eluted from the gel with 1.0 *M* potassium phosphate, pH 7.4 in a manner analogous to the washing procedure. The purified cytochrome c_1 may be stored at −20° as a buffered solution or, alternatively, in the dry state after lyophilization. A preparation of higher purity (25–27 mμmoles per milligram of protein) but lower yield may be obtained by a modification of the above procedure.[5] A summary of the purification is given in Table I.

TABLE I
PURIFICATION PROCEDURE

Step	Fraction	Recovery of protein (%)	Specific content Cyt c_1	Recovery of Cyt c_1 (%)
1	Mitochondria	100	0.20–0.25	100
4	Cytochrome $b + c_1$	15	1.50	77
5	Crude cytochrome c_1	6	3.0–4.0	75
7	$Ca_3(PO_4)_2$ eluate	0.3–0.4	13.0–19.0	13–23

Properties[2,5]

Composition. Table II lists the composition of highly purified cytochrome c_1. The absence of heme extractable by acidified acetone indicates that the preparation is not contaminated with cytochromes *a* and *b*.

Stability. Purified cytochrome c_1 suffers no alteration in solubility or spectral properties during prolonged storage at −20°, or after repeated freezing and thawing. Lyophilization or dialysis against distilled water or unbuffered salt solution may result in a lowered ratio of heme to

[8] T. Singer, *Arch. Biochem. Biophys.* **29,** 190 (1950).

TABLE II
COMPOSITION OF THE MOST HIGHLY PURIFIED CYTOCHROME c_1

Analysis	Amount
Total iron	27 μg atoms/g protein
Heme	27 μmoles/g protein
Heme extractable with acid-acetone	Undetectable
Total flavin	Undetectable
Phospholipid	<3.6%

protein. EDTA (0.001 M final concentration) protects the cytochrome against loss of heme during dialysis. The reduced cytochrome is not autoxidizable; also, no spectral change is effected by treatment of the cytochrome with carbon monoxide.

Spectral Properties. Isolated and purified cytochrome c_1 in the reduced form displays an absorption spectrum with maxima at 554 mμ, 524 mμ, and 418 mμ for the α, β, and Soret bands, respectively. A small, broad band is located at 472 mμ. Absorbancy indices for 554 mμ minus 540 mμ (reduced) and 554 mμ (reduced minus oxidized) are 17.5 mM^{-1}-cm^{-1} and 17.1 mM^{-1}-cm^{-1}, respectively. Unlike cytochrome b (see this volume [63]), cytochrome c_1 undergoes no significant alteration, either qualitative or quantitative, in spectral properties during separation from its parent particle.[9]

Physicochemical Properties. Purified cytochrome c_1 is a soluble protein under moderate conditions of pH, temperature, and ionic strength. Although the hemoprotein is homogeneous by both electrophoretic and sedimentation criteria, usually it exists in solution as a hexamer; however, in the presence of 0.0005 M sodium dodecyl sulfate the protein is dissociated to the monomeric form. The monomer has a measured molecular weight (calculated from sedimentation data) of 53,000 which compares well with a minimal molecular weight (calculated from the heme content) of 52,000.

The oxidation potential, E'_0, of purified cytochrome c_1 is about +0.22 volt.[2]

Enzymatic Properties. Cytochrome c_1 does not replace cytochrome c in enzymatic systems in which cytochrome c functions as a substrate; however, oxidation-reduction equilibrium is established rapidly between oxidized or reduced cytochrome c and reduced or oxidized cytochrome c_1, respectively.

[9] W. S. Zaugg and J. S. Rieske, *Biochem. Biophys. Res. Commun.* **9**, 213 (1962).

[63] Purification and Properties of Cytochrome *b*

By J. S. RIESKE[1] and H. D. TISDALE

Cytochrome *b* has been separated from other mitochondrial hemoproteins by several procedures[1a-3]; however, the purification procedure described below[4] has yielded preparations of cytochrome *b* of highest purity as yet reported.[4a] Also, a crude cytochrome *b* (12–18 millimicromoles of heme per milligram of protein) but free from other hemoproteins can be prepared by cleavage of complex III (see this volume [44]).

Assay

Principle. Cytochrome *b* is assayed spectrophotometrically either from absorbancy measurements of its α band at 563 mμ or from absorbancy measurements of the pyridine hemochrome derived from the heme group of the cytochrome.

Procedure. See this volume [76].

Purification Procedure

Step 1. Preparation of a Suspension of KCl-Mitochondria. See this volume [62]. The KCl-mitochondria are diluted to 30 mg of protein per milliliter with 0.9% KCl and stored at —20° until used.

Step 2. Extraction with Deoxycholate. Unless noted otherwise, all operations are carried out in the cold (0–5°) and all centrifugations are made in a Beckman Model L centrifuge at 30,000 rpm using a No. 30 rotor.

The thawed, homogenized mitochondria are treated with potassium deoxycholate[5] to give a concentration of 0.45 mg per milligram of protein. The suspension is centrifuged for 1 hour and the brownish-green sediment is discarded.

[1] See footnote 1, page 239.

[1a] G. Hübscher, M. Kiese, and R. Nicholas, *Biochem. Z.* **325,** 223 (1954).

[2] I. Sekuzu and K. Okunuki, *J. Biochem.* (*Tokyo*) **43,** 107 (1956).

[3] D. Feldman and W. W. Wainio, *J. Biol. Chem.* **235,** 3635 (1960).

[4] R. Goldberger, A. L. Smith, H. Tisdale, and R. Bomstein, *J. Biol. Chem.* **236,** 2788 (1961).

[4a] *Note Added in Proof:* Since the submission of the manuscript, a preparation of cytochrome *b* from beef heart muscle with a higher content of heme has been described [K. Ohnishi, *J. Biochem.* (*Tokyo*) **59,** 1 (1966)]. This procedure employs a digestion in the presence of bacterial proteinase to solubilize the crude cytochrome *b* prior to further purification by ammonium sulfate fractionation and chromatography.

[5] A 10% (w/v) aqueous solution prepared from deoxycholic acid that has been recrystallized from a 50% ethanol-water solution.

Step 3. Fractional Precipitation with Ammonium Sulfate. The supernatant liquid of step 2 is treated with potassium cholate[6] to a concentration of 1.0 mg per milligram of protein, sodium dodecyl sulfate[7] to a concentration of 2% (w/v), and solid ammonium sulfate to 0.30 saturation (21 g/100 ml of solution) in the order listed. The suspension is centrifuged for 15 minutes, and the sediment is discarded. Solid ammonium sulfate is added to the supernatant liquid to 0.50 saturation (35 g/100 ml) and centrifuged for 20 minutes. The supernatant liquid is discarded. The brownish-red pellet is dissolved in 0.1 M potassium phosphate buffer, pH 7.4 to a protein concentration of 15 mg/ml.

Step 4. Precipitation of Extraneous Protein. The final solution of step 3 is heated to 40° and is kept at this temperature for 5 minutes; the precipitate that is formed is removed by a 15-minute centrifugation and discarded. Solid ammonium sulfate is added to give a slightly turbid suspension (10.6 g/100 ml of suspension) after which the suspension is heated for 5 minutes at 35°. This step yields a very copious precipitate that is removed by a 15-minute centrifugation.

Step 5. Precipitation of Crude Cytochrome b. The deep red supernatant solution of step 4 contains both cytochrome *b* and c_1. This solution is treated with potassium cholate (3.0 mg per milligram of protein) and sufficient neutral, saturated ammonium sulfate to give a slightly turbid suspension (usually 15 ml per 100 ml of suspension is sufficient). The suspension is heated to 40° and maintained at this temperature for 5 minutes. A copious precipitate of crude cytochrome *b* is formed during this treatment and is subsequently recovered as a deep red pellet by centrifugation (a centrifugal force of a few thousand *g* is sufficient to sediment this precipitate). The supernatant solution contains impure cytochrome c_1 that may be utilized in the purification of this cytochrome (see this volume [62]). The pellet of crude cytochrome *b* is homogenized in 0.25 *M* sucrose, pH 7.5, and is diluted to a protein concentration of 10 mg/ml with the same sucrose solution.

Step 6. Sodium Dodecyl Sulfate-Ammonium Sulfate Fractionation. The suspension of crude cytochrome *b* is treated with sodium dodecyl sulfate[7] to a final concentration of 0.3% (w/v). Solid ammonium sulfate (2.6 g per 100 ml of suspension) is added; the suspension is centrifuged as before, and the supernatant liquid is discarded. The residue is suspended in 0.25 *M* sucrose, pH 7.5, and is adjusted to a protein

[6] A 20% (w/v) aqueous solution prepared from cholic acid that has been recrystallized from a 50% ethanol-water solution.

[7] Purchased under the trade name Duponol C from E. I. duPont de Nemours. The detergent is used as a 10% (w/v) aqueous solution.

concentration of 10 mg/ml as before. Again sodium dodecyl sulfate is added to a concentration of 0.3% (w/v), then the clarified suspension is frozen at —20° for no less than 8 hours. After being thawed the suspension is centrifuged for 15 minutes to remove the faint opalescence. The clear supernatant solution is treated with solid ammonium sulfate (2.6 g per 100 ml of solution) and is centrifuged. The orange supernatant solution is discarded; the residue is suspended in 0.25 *M* sucrose, pH 7, and diluted to 10 mg of protein per milliliter. As before, the suspension of protein is clarified with sodium dodecyl sulfate 0.3% (w/v) and is then treated with solid ammonium sulfate (2.6 g per 100 ml of solution). The suspension is centrifuged for 15 minutes; the residue, which consists of the purified cytochrome *b* is suspended in 0.1 *M* potassium phosphate, pH 7.4 and stored as such at —20°. A summary of the purification procedure is given in the table.

PURIFICATION PROCEDURE

Step	Protein fraction	Cytochrome *b* content (μmoles/g protein)	Recovery (%)
1	Mitochondria	0.5–0.6	100
4	Cytochrome *b*-c_1 complex	3.5–4.0	50–60
5	Crude cytochrome *b*	10.0–12.0	35–40
6	Purified cytochrome *b*	30.0–36.0	25–30

Properties[4]

Stability. Cytochrome *b* as isolated by the procedure above is very stable under the usual conditions of storage. Several months of storage at —20° with repeated freezing and thawing had no observable effect on the spectral properties or the solubility characteristics of the cytochrome. A 10-minute exposure to temperatures approaching 70° is required in order to cause extensive denaturation of the cytochrome (as judged by the spectral response of the reduced homoprotein to carbon monoxide). However, dialysis of the cytochrome against distilled water or unbuffered salt solutions may lead to a reduction of the heme to protein ratio.

Although the purified cytochrome *b* appears to be very stable with respect to further change, a comparison of the properties of the purified cytochrome with the properties of the particle-bound cytochrome indicates that extensive modifications in certain properties may occur during isolation. The absorbancy index (reduced minus oxidized) at 562–563 mμ changes[8] from 23.4 mM^{-1} cm^{-1} in the intact particle to 13.2

[8] W. S. Zaugg and J. S. Rieske, *Biochem. Biophys. Res. Commun.* **9**, 213 (1962).

mM^{-1} cm^{-1} (see footnote 4) in the purified cytochrome.[9] Also, a precipitous drop in the oxidation potential of the cytochrome occurs during isolation (see below).

Purity. Cytochrome *b* of highest purity contains 35 μgatoms of iron per gram of protein, which is in agreement with a measured heme content of 36 μgatoms per gram of protein. Contamination with flavin, lipid, and diaphorase activity is negligible or not measureable. Contamination with cytochromes *a* or c_1 is not detectable spectroscopically. The preparation of cytochrome *b* when dispersed in the cationic detergent, cetyl dimethylethylammonium bromide (see below) displays a single boundary during sedimentation in the ultracentrifuge.

Spectral Properties. The oxidized form of the cytochrome displays absorption maxima at 538 mμ and 418 mμ. When reduced with dithionite the cytochrome displays α, β, and γ absorption bands at 562, 532, and 429 mμ, respectively. Isosbestic points are located at 572, 550, 540, and 521 mμ.

Physicochemical Properties. Cytochrome *b* is insoluble in ordinary aqueous media, and is dispersed only sparingly in solutions of bile salts. However, the cytochrome is soluble at high pH's (pH $>$ 12) or in the presence of strong detergents (e.g. sodium dodecyl sulfate or cetyl-dimethylethylammonium bromide). Only the cationic detergents disperse the cytochrome in a monomolecular state. The oxidation potential, E'_0, of the purified cytochrome is estimated to be —0.34 volt.[10] This is contrasted to a potential of about +0.06 volt for particle-bound cytochrome *b*.[11] When complexed with mitochondrial structural protein (see this volume [71]), the potential of purified cytochrome *b* is raised as indicated by its ready reducibility by reduced coenzyme Q[10] which has an estimated potential in the region of the potential of particle-bound cytochrome *b*.[12]

[9] These measurements were made with a solution of the cytochrome in 1 m*M* sodium dodecyl sulfate and 10 m*M* potassium phosphate, pH 7.4.

[10] R. Goldberger, A. Pumphrey, and A. Smith, *Biochim. Biophys. Acta* **58,** 307 (1962).

[11] J. P. Straub and J. P. Colpa-Boonstra, *Biochim. Biophys. Acta* **60,** 650 (1962).

[12] E. C. Slater, *"Ciba Found. Symp. Quinones Electron Transport,"* pp. 415 and 416. Little, Brown, Boston, Massachusetts, 1961.

[64] Preparation and Properties of a Respiratory Chain Iron-Protein

By J. S. RIESKE[1]

Two properties of the $(CoQ)H_2$–cytochrome *c* reductase complex (complex III) of the respiratory chain[1a] are its content of nonheme iron and its display of an electron paramagnetic resonance (EPR) signal at $g = 1.90$. Both the nonheme iron and the EPR signal have been identified with an iron-protein that appears to be an oxidation–reduction component of the complex.[2] This iron-protein will be designated as the "$g = 1.90$" iron-protein.

Assay

Principle. The "$g = 1.90$" iron-protein is identified by its content of nonheme iron, a characteristic absorption spectrum, and a characteristic EPR spectrum. These properties also can be utilized for the assay of this iron-protein; however they have shown no consistent, quantitative interrelationship from preparation to preparation. These discrepancies may be ascribed to the labile nature of the iron-containing prosthetic group, especially in circumstances where the protein is separated from the complex. Therefore, only estimates of iron-protein of a comparative nature are possible at present.

Reagents

Potassium phosphate, 1.0 *M*, pH 7
Sodium bathophenanthroline sulfonate, 0.2% in water[3]
Solution, 0.5 m*M*, of standard iron in dilute HCl. A weighed piece (ca. 55 mg) of pure iron wire is dissolved in about 5 ml of hot, concentrated HCl. The final dilution is made with distilled water.

Analysis of Iron Content. To a 1-ml cuvette of 1-cm light path are added 0.50 ml of 1 *M* phosphate, pH 7; a solution of iron-protein containing 0.2–0.4 mg of protein[4]; and enough water to bring the total volume to 0.8 ml. At this point a few crystals of sodium dithionite (hydrosulfite)

[1] See footnote 1, page 239.
[1a] See this volume [44].
[2] J. S. Rieske, R. E. Hansen, and W. S. Zaugg, *J. Biol. Chem.* **239,** 3017 (1964).
[3] Obtained from the G. Frederick Smith Chemical Co., Columbus, Ohio.
[4] Protein values were determined by the method of O. H. Lowry, N. J. Rosenbrough, A. L. Farr, and R. J. Randall, *J. Biol. Chem.* **193,** 265 (1951). For the iron-protein, these values are 30% lower than those obtained by the biuret procedure of A. G. Gornall, C. J. Bardawill, and M. M. David, *J. Biol. Chem.* **177,** 751 (1949).

are added. After a few moments, 0.20 ml of a 0.2% solution of sodium bathophenanthroline sulfonate is mixed with the contents of the cuvette. A reference solution is prepared with all ingredients except the iron-protein. The absorbancy at 535 mμ of the sample vs. the reference solution is recorded immediately. The sample and reference cuvettes together with two cuvettes containing the same reagents but with about 10 and 20 mμgatoms of standard iron, respectively, in place of the iron-protein are heated a few moments in boiling water. After the cuvettes have cooled to room temperature, the absorbancies at 535 mμ of the solutions in the sample and standard cuvettes vs. the reference solution, respectively, are recorded. The specific concentration of protein-bound iron in the sample of iron-protein may be calculated by the following expression:

$$\frac{\mu\text{g atoms Fe}}{\text{g protein}} = \frac{A_{535}\text{ (sample after heating)} - A_{535}\text{ (sample before heating)}}{A_{535}\text{ (standard Fe)}} \times \frac{\mu\text{g atoms Fe (standard)}}{\text{g protein sample}}$$

The specific iron content of purified, succinylated iron-protein is as high as 85 μgatoms per gram of protein.[4]

Specific Absorbancy. Purity of the iron-protein also may be estimated by the specific absorbancy of the reduced protein. Either 400 mμ or 500 mμ can be used as the absorbing wavelength. At 400 mμ the ratio of the absorbancy (1-cm light path) to the protein concentration in milligrams per milliliter of the purified iron-protein has given values of about 0.15. With fresh preparations containing a relatively high content of iron (70–85 mμgatoms per milligram of protein), values as high as 0.25 have been obtained. Considerable contamination by heme groups, as indicated by a large absorption peak at 410 mμ, will interfere with this determination; in this case, 500 mμ is the more satisfactory wavelength to use. At 500 mμ, the specific absorbancies will be about 60% of values given for 400 mμ.

Specific Intensity of EPR Signal. EPR spectroscopy[5] is the only method to date that is sufficiently specific and sensitive to identify the "$g = 1.90$" iron-protein in mitochondria and submitochondrial fractions and to measure its relative concentration in these preparations. The samples to be assayed (about 0.2 ml containing 10–20 mg of protein) are transferred (by use of a calibrated plastic tube attached to a syringe) into the sample tubes. These tubes are matched tubes of quartz, sealed at one end, with a bore diameter of 3 mm and a wall thickness of 0.5 mm. The samples after suitable treatment (see below) are frozen in

[5] See this volume [94].

liquid nitrogen and then are examined at −176° in the EPR spectrometer. Spectra are recorded in the form of the first derivative of the microwave-absorption curve. For each sample, three consecutive spectra are recorded; the sample after oxidation with a droplet of 0.05 *M* ferricyanide, the sample reduced with a few crystals of solid potassium ascorbate, and the sample further reduced with a few crystals of sodium dithionite. Between treatments the quartz tube containing the sample is thawed by a quick immersion in ice water in order to avoid breakage of the tube. The relative concentration of the "$g = 1.90$" iron-protein in the sample is estimated from the displacement of the signal at $g = 1.90$, from the signal baseline (where $dE/dt = 0$) in the ascorbate-reduced sample.[6] Treatment with dithionite causes the appearance of additional signals at $g = 1.94$ originating from contamination with other species of iron-protein associated with the DPNH-dehydrogenase and the succinate dehydrogenase flavoproteins. A useful reference for the EPR signal of "$g = 1.90$" iron-protein is the purified (CoQ)H_2–cytochrome *c* reductase complex since the compound giving rise to the EPR signal at $g = 1.90$ is quite stable in the intact complex. The complex contains 2 gatoms of nonheme iron per unit molecular weight or approximately one molecule of iron-protein per molecule of complex.[6]

Purification Procedure[7]

Source Material. Purified (CoQ)H_2–cytochrome *c* reductase (complex III).[1a]

Step 1. Separation of Crude Iron-Protein from Complex III. A solution of complex III (10–20 mg of protein per milliliter) in a buffer that is 0.67 *M* in sucrose, 1.0 m*M* in histidine, and 0.05 *M* in Tris-HCl, pH 8, is treated with antimycin A (4–6 micromoles per gram of protein). This solution is then mixed with 0.5 volume of 10% sodium taurocholate (or 20% potassium cholate) and 0.5 volume of saturated ammonium sulfate (final concentration of 0.25 saturation) and is incubated for 30–40 minutes at 20°. The precipitated, brown protein is sedimented by centrifugation for 10 minutes in a low-speed centrifuge (1000–5000 *g*). The residue is suspended in 0.01 *M* phosphate, pH 7.5 and again centrifuged. The pellet of protein is suspended in a few milliliters of 0.01 *M* phosphate, pH 7.5. All subsequent steps are carried out at 0–5°.

[6] Derived on the basis that complex III contains one molecule of cytochrome c_1 per unit complex. See J. S. Rieske, W. S. Zaugg, and R. E. Hansen, *J. Biol. Chem.* **239**, 3023 (1964).

[7] J. S. Rieske, D. H. MacLennan, and R. Coleman, *Biochem. Biophys. Res. Commun.* **15**, 338 (1964).

Step 2. Succinylation of the Crude Iron-Protein. The suspension of crude iron-protein is placed in a small, ice-cooled beaker and stirred with a small magnetic bar. The pH is monitored by use of a pH meter. The suspension of protein then is treated with successive increments of 5–10 mg of succinic anhydride, and the pH is maintained between pH 7 and 8 by the dropwise addition of 1 *N* KOH. Each portion of the anhydride is allowed to react completely (as indicated by no further acid production) before the next addition. After maximal clarification of the suspension is obtained a small insoluble residue is removed by centrifugation.

Step 3. Chromatography of the Succinylated, Crude Iron-Protein on Sephadex G-100 Polydextran Gel. The solution of succinylated iron-protein is placed on a column (2.5 × 120 cm) of Sephadex G-100[8] gel which had been equilibrated with 0.05 *M* phosphate, pH 7.5. The protein is eluted with this same buffer at a flow rate of 30–40 ml per hour. The eluted solution is collected in ca. 2.5-ml fractions with an automatic collector. The protein content of the fractions is monitored by absorption measurements at 278 mμ. The protein is eluted as two bands, a fast-moving major band followed by a minor band consisting of the purified iron-protein. The fractions containing the second band of protein are pooled and treated with ammonium sulfate (to 0.70 saturation) to precipitate all the protein; the protein is sedimented by centrifugation at 120,000 *g* for 10 minutes. The brown pellet of succinylated iron-protein is dissolved in a buffer that is 0.05 *M* in phosphate, pH 7.5, and 0.25 *M* in sucrose and stored at −20°. These conditions of storage, however, may lead to aggregation of the protein into polymeric forms (see below). A summary of the yields of protein, iron, and EPR signal at $g = 1.90$ in protein fractions obtained at different stages of purification are given in the table.

Properties

Stability. The EPR signal at $g = 1.90$ of the isolated iron-protein diminishes severalfold in strength after 2–3 days at 0°. Although a general loss of color and of protein-bound iron[9] also occurs, these losses are not nearly as pronounced as the loss in EPR signal. Storage at −20° preserves the intensity of the EPR signal for an indefinite time; however, the protein undergoes aggregation upon freezing and thawing.

[8] Sephadex G-100 is obtained from Pharmacia Fine Chemicals Inc., Piscataway, New Market, New Jersey.

[9] Iron that has separated from its protein-iron binding is estimated as the iron which is available for immediate chelation with iron chelators.

PURIFICATION PROCEDURE[a]

Fraction	Per cent recovery		EPR signal at $g = 1.90$	Fe:protein (μg atoms:g protein)
	Protein[b]	Iron		
Complex III	100	100	100	8
Crude iron-protein	7.5	44	12	40
Succinylated crude iron-protein	6.5	39	10.5	37
Sephadex G-100, 2nd band	1.4	16	5.4	63

[a] The values in this table were obtained in a single preparation.
[b] Protein values are based on analysis by the method of Lowry *et al.* (see text footnote 4).

Composition. The "$g = 1.90$" iron-protein contains 60–80 μgatoms of iron per gram of protein.[4] Upon acidification or heating 0.7–0.8 mole of sulfide per gram atomic weight of iron is released.[10] The succinyl content of the succinylated protein is about 430 micromoles per gram.

Contamination. The isolated iron-protein usually contains a slight contamination with heme which shows up as a small absorption peak at 410 mμ. Contamination with copper is evidenced by a Cu(II) signal in the EPR spectrum of the ferricyanide-oxidized protein. The copper content is about 1 μgatom per gram of protein. No flavin was detectable in the isolated iron-protein. The protein is homogeneous on the basis of sedimentation studies with the analytical ultracentrifuge.

Physicochemical Properties. The molecular weight of the succinylated iron-protein is estimated to be about 30,000. On the basis of this molecular weight, the protein contains two atoms of iron per molecule. The oxidation-reduction potential E'_0 (pH 7), of the iron-protein is about 0.22 volt, and only one electron per molecule appears to be involved in its oxidation or reduction.

Spectral Properties. The absorption spectrum of the "$g = 1.90$" iron-protein displays absorption maxima at 575 mμ and 460 mμ, and a shoulder at 315 mμ. Reduction of the protein with mild reducing agents (e.g., ascorbate) reduces the absorbancy throughout the visible range and yields a spectrum with small maxima at 380 mμ, 420 mμ, and 515 mμ; also the shoulder at 315 mμ disappears.

The EPR spectrum (first derivative of the microwave absorption curve) of the protein after treatment with reducing agents displays inflections that correspond to prominent absorption maxima at $g_{\perp} = 1.90$, $g_{\parallel} = 2.0$, and at about $g = 1.85$. The EPR signal of this iron-protein

[10] R. Coleman, J. S. Rieske, and D. C. Wharton, *Biochem. Biophys. Res. Commun.* **15**, 345 (1964).

is distinguished from the signals of the $g = 1.94$ type (given by iron-proteins associated with certain dehydrogenase flavoproteins; e.g., succinic and DPNH dehydrogenase flavoproteins of the respiratory chain) by a much greater broadness of the signal and the presence of the signal component at about $g = 1.85$.

[65] Isolation of Adrenal Cortex Nonheme Iron Protein

By T. OMURA,[1] E. SANDERS, D. Y. COOPER, and R. W. ESTABROOK

Nonheme iron protein (NHI) is present in the mitochondrial fraction of adrenal cortex.[1a-3] In mitochondria, it is associated with a flavoprotein (Fp) which has TPNH-dehydrogenase activity, and these two components constitute a reductase system which transfers electrons from TPNH to cytochrome P-450. Both the NHI and Fp can be readily solubilized from mitochondrial particles by sonication.

Assay Method

The absorption spectrum of oxidized NHI has three broad peaks at 320, 415, and 455 mμ, respectively. If the sample is free of other colored materials, the concentration of NHI may be calculated from the absorption at these peaks using the extinction coefficients listed in the table.

POSITION OF PEAKS AND EXTINCTION COEFFICIENTS (ϵ) PER MILLIMOLE OF IRON IN THE ABSORPTION SPECTRUM OF OXIDIZED NHI

Wavelength (mμ)	ϵ_{Fe} (cm^{-1} mM^{-1})
320	7.0
415	5.6
455	4.8

Pure NHI protein contains about 110 millimicromoles of iron per milligram of protein, and the ratio of optical density at 415 mμ to that at 280 mμ is about 0.8.[1]

Purified NHI and Fp constitute a reductase system which reduces

[1] This manuscript was prepared while the author was at the Johnson Research Foundation, University of Pennsylvania, Philadelphia, Pennsylvania.

[1a] T. Omura, E. Sanders, D. Y. Cooper, O. Rosenthal, and R. W. Estabrook, *in* "Non-Heme Iron Proteins: Role in Energy Conversion" (A. San Pietro, ed.), p. 401, Kettering Symp. Antioch Press, Yellow Springs, Ohio, 1965.

[2] T. Omura, R. Sato, D. Y. Cooper, O. Rosenthal, and R. W. Estabrook, *Federation Proc.* **24,** 1181 (1965).

[3] K. Suzuki and T. Kimura, *Biochem. Biophys. Res. Commun.* **19,** 340 (1965).

cytochrome P-450 as well as cytochrome *c* by TPNH. Without the addition of NHI, purified Fp does not reduce cytochrome *c* but does reduce the dye, dichlorophenolindophenol. NHI is specifically required by Fp, as demonstrated by the complete ineffectiveness of FAD, FMN, and plant ferredoxin, in the catalysis of the reduction of cytochrome *c* by Fp and TPNH. Therefore, the reduction of cytochrome *c* by Fp affords a highly sensitive method for the quantitative detection of NHI. However, quantitative assay of NHI by this method is difficult because the reaction rate is dependent on the concentration of Fp as well as that of NHI, though the rate is almost directly proportional to the concentration of either NHI or Fp.[2]

Purification Procedure

The Fp and NHI may be purified by the treatment of adrenal cortex mitochondria as described in Fig. 1. An alternative method to purify NHI, starting from acetone-dried mitochondria of pig adrenals, has been described by Suzuki and Kimura.[3]

All steps in this procedure must be carried out at about 4°.

Medullas are separated from the cortex of fresh beef adrenal glands, and the cortex scraped with a scalpel blade from the outside capsule. The minced cortex is then washed with cold 0.25 *M* sucrose to remove blood, and homogenized with 4 volumes of 0.25 *M* sucrose, using a Potter glass homogenizer equipped with a Teflon pestle. The homogenate is centrifuged at 3000 rpm for 10 minutes, the precipitate is discarded, and the supernatant is further centrifuged at 8000 rpm for 10 minutes. The sedimented mitochondrial fraction is resuspended by gentle homogenization in one-half the original volume of 0.25 *M* sucrose. This suspension is centrifuged at 8000 rpm for 10 minutes. Precipitated mitochondria are suspended in distilled water and diluted to a final volume of about 200 ml per kilogram of original minced adrenal cortex. The concentration of the final mitochondrial suspension is 25–30 mg of protein/ml.

The water suspension of mitochondria is sonicated (20 kc, 3 amp) for 10 minutes, using a Branson Instruments Inc. "Sonifier," in portions of 20 ml. The sonicated material is then centrifuged at 40,000 rpm for 30 minutes. The dark red pellet is discarded, and the slightly turbid red supernatant (S-1), which contains most of the NHI and Fp from the original mitochondria, is centrifuged at 50,000 rpm for 100 minutes. The precipitate, (P-2) which contains mainly cytochrome P-450, is suspended in 0.01 *M*, potassium phosphate buffer, pH 7.5, and stored frozen. The supernatant fluid (S-2) is retained for use in purification of both NHI and Fp. This supernatant fluid may be frozen and stored for more than a month without loss of NHI or Fp.

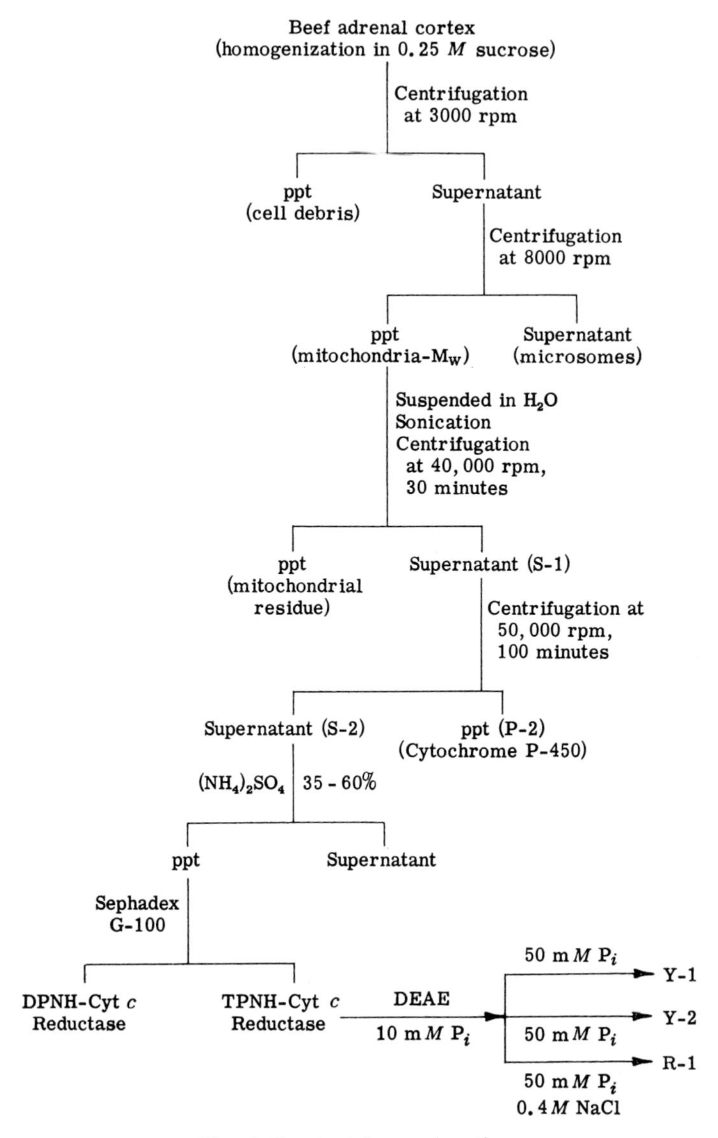

FIG. 1. See text for explanation.

Fractionation of the supernatant fluid by ammonium sulfate is carried out by the addition of a saturated solution of the salt which has been adjusted to pH 7.5 with 1.0 *M* tris(hydroxymethyl)aminomethane (Tris). Ammonium sulfate is added dropwise. The pH is maintained at about 7.5 by the addition of 1.0 *M* Tris. When a saturation of 35% is reached, the solution is centrifuged at 10,000 rpm for 10 minutes, and the precipitate is discarded. The ammonium sulfate saturation is then increased to 60%. After centrifugation, the pellet is suspended in 0.01 *M* phosphate buffer, pH 7.5, to a total volume of 30–50 ml. The supernatant can be saved for further fractionation and recovery of more NHI protein.

The solution obtained by dissolving the 35–60% ammonium sulfate precipitate is applied to a column of Sephadex G-100[4] (5 × 70 cm) which has been equilibrated with 0.01 *M* phosphate buffer, pH 7.4. Elution is carried out using the same buffer, collecting 20 ml fractions at a flow rate of 1 ml per minute. A yellow band with strong DPNH–cytochrome *c* reductase activity is eluted initially, followed by a brown-colored band containing all the Fp as well as most of the NHI. The measurement of TPNH–cytochrome *c* reductase activity is a convenient method for locating the active fractions. However, since the reduction of cytochrome *c* by TPNH, catalyzed by the eluate, involves two protein components, NHI and Fp, the reduction rate is not proportional to the amount of eluate added. The fractions with high TPNH–cytochrome *c* reductase activity are combined and put onto a column of DEAE-cellulose[5] (2 × 60 cm, packed under moderate pressure) which has been equilibrated with 0.01 *M* phosphate buffer. The column is eluted with a non-linear concentration gradient of phosphate buffer prepared by connecting a flask containing 100 ml of 0.01 *M* buffer with another flask containing 100 ml of 0.05 *M* buffer. Fractions of 10 ml are collected at a flow rate of about 0.5 ml per minute. The yellow band moves down separating into two components (Y-1 and Y-2), the major one moving more slowly than the minor. Using TPNH-diaphorase activity measurement, dichlorophenol-indophenol (DCPIP) as an electron acceptor, the amount of enzyme present in the eluate is determined. After combining the most active fractions, the Fp is concentrated by the addition of ammonium sulfate to 60% saturation. Precipitated Fp is dissolved in a small volume of 0.01 *M* phosphate buffer and dialyzed overnight against the same buffer. When kept frozen, purified Fp is highly stable and may be stored for more than one month without loss of activity. The minor yellow component (Y-1) has almost identical enzymatic activities as the major fraction (Y-2).

After the elution of Fp, the column is washed with 50 ml of 0.05 *M* phosphate buffer pH 7.5 containing 0.15 *M* NaCl. The brown band of NHI begins to move slowly down the column. The concentration of

[4] Sephadex G-100 was obtained from the Sigma Chemical Co.

[5] DEAE-cellulose, anion exchanger, 0.9 meq/g of medium mesh, was obtained from the Sigma Chemical Co., St. Louis, Missouri.

NaCl in the eluent is then increased gradually, using 100 ml of 0.05 *M* phosphate buffer containing 0.15 *M* NaCl and 100 ml of 0.05 *M* phosphate buffer containing 0.50 *M* NaCl. The concentration of NHI in the eluted fractions may be determined by observing the optical density at 455 mμ. The most concentrated fractions are combined, diluted with 5 volumes of 0.05 *M* Tris buffer, pH 7.5, and put onto a small DEAE-cellulose column (2 $\times$ 10 cm, packed with moderate pressure) previously equilibrated with the same Tris buffer. A solution of 0.05 *M* Tris buffer, pH 7.5, containing 0.5 *M* NaCl is used to elute the NHI protein from the column. Concentrated NHI is collected and dialyzed overnight against 0.05 *M* Tris buffer, pH 7.5. The yield of purified NHI is 7–15 mg of protein from 1 kg of beef adrenal cortex.

The dialyzed solution of NHI is fairly stable. It may be kept in a refrigerator over 10 days without any change in the absorption spectra. The presence of a high concentration of phosphate is harmful to the stability of NHI, the color being gradually bleached. When frozen in phosphate buffer, the protein is decolorized.

In order to obtain more NHI, the concentration of ammonium sulfate in the supernatant from the 60% precipitate in the original fractionation is increased to 80%. The precipitate is dissolved in about 100 ml 0.01 *M* phosphate buffer, pH 7.5, and dialyzed overnight against the same buffer. After dilution of the dialyzate with 2 or 3 volumes of 0.01 *M* phosphate buffer, pH 7.5, solution is absorbed onto a DEAE-cellulose column (5 $\times$ 30 cm) and washed with 50 ml 0.01 *M* phosphate buffer. Direct elution with 50 ml 0.05 *M* phosphate buffer, pH 7.5, containing 0.15 *M* NaCl is carried out, followed by 50 ml of 0.05 *M* phosphate buffer containing 0.5 *M* NaCl. As described above, the NHI protein collected is diluted fivefold, concentrated by passing through a small DEAE-cellulose column, dialyzed against Tris buffer overnight, and stored in the refrigerator.

Properties

Purified NHI is homogeneous on analysis with the analytical ultracentrifuge ($S_{20,w} = 1.9$ S). The molecular weight determined by a Sephadex-filtration method is 19,000, indicating the presence of 2 atoms of iron in one molecule of NHI.

The absorption spectrum of oxidized NHI is very similar to that of plant ferredoxin. However, NHI cannot replace plant ferredoxin in the photophosphorylation reactions of chloroplast preparations. Plant ferredoxin is inactive in the catalysis of the transfer of reducing equivalents from adrenal Fp to cytochrome *c* or cytochrome P-450. The absorption peaks in the visible region (see table) disappear upon reduction with

Fp and TPNH, or with dithionite. NHI is readily autoxidizable. When reduced by Fp and TPNH, reduced NHI is re-oxidized rapidly by molecular oxygen, and the original spectrum of oxidized pigment is restored. A rather high oxidation-reduction potential (E'_0) of +0.15 volt at pH 7.4 has been reported,[3] though NHI cannot be reduced by ascorbate at neutral pH's.

In addition to nonheme iron, the NHI protein contains acid-labile sulfur. By adding strong acid to the solution of NHI, hydrogen sulfide is evolved. The content of labile sulfur is about 0.8 mole per mole of iron.

When examined by electron paramagnetic resonance spectroscopy, the reduced NHI shows a strong signal at $g = 1.94$. Oxidized NHI shows no EPR signal. Since the detection of NHI by spectroscopic observation is difficult in the presence of other colored materials, this electron paramagnetic resonance signal may be useful to detect and determine NHI in crude samples.

The purified TPNH-nonheme iron protein reductase has a molecular weight of about 60,000 as determined by a Sephadex-filtration method. The flavin prosthetic group is FAD with one mole of FAD per 60,000 gm of protein.

[66] The Preparation of Bacterial Cytochrome b_1

By Lowell P. Hager and Samir S. Deeb

Cytochrome b_1 was discovered and studied by Keilin and Harpley[1,2] as a principal cytochrome component of *Escherichia coli* and related organisms. It was early recognized that cytochrome b_1 in these organisms was associated with an insoluble cellular component having a high lipid content. This component, which is generally referred to as the particulate fraction, is obtained by high speed centrifugation of ruptured cell preparations. It is generally conceded that this particulate fraction represents cell membrane fragments. Cytochrome b_1 is firmly bound to this fragment; however, it has been possible to liberate considerable quantities of the bound cytochrome by either brief exposure of the particulate preparation to trypsin[3] or by prolonged sonic oscillation. Soluble cytochrome

[1] D. Keilin, *Nature* **133**, 290 (1934); *Compt. Rend. Soc. Biol.* **97**, (appendix), 39 (1927).

[2] D. Keilin and C. H. Harpley, *Biochem. J.* **35**, 688 (1941).

[3] F. R. Williams and L. P. Hager, *Biochim. Biophys. Acta* **38**, 566 (1960).

b_1 preparations thus derived from the particulate fraction may be purified by conventional techniques.[4,5]

Assay Method

Principle. The assay is based on the increase in absorbance at 427.5 mμ when cytochrome b_1 is converted to the reduced form. An excess of sodium dithionite serves as the reductant in this reaction. On the basis of the difference spectra between the oxidized and the reduced forms of crystalline cytochrome b_1, a molar extinction coefficient of 6×10^4 was calculated for the difference spectra between the oxidized and reduced forms at the 427.5 mμ absorption peak.[5] The determination of cytochrome b_1 in crude extracts by this method is relatively inaccurate. The particulate nature of the preparation gives rise to light-scattering problems which affect the different spectral measurements. In addition, the presence of flavoproteins in the crude preparations also gives rise to ambiguous results since they undergo spectral changes upon oxidation and reduction in this region of the spectrum.

Preparation of the Enzyme

Growth Medium. The cells are grown on a medium containing 0.2% NH_4Cl; 0.4% glucose; 0.25% sodium glutamate; 0.0005% yeast extract; 0.15% KH_2PO_4; 1.35% Na_2HPO_4; 0.02% $MgSO_4 \cdot 7\ H_2O$; 0.001% $CaCl_2$, and 0.00005% $FeSO_4 \cdot 7\ H_2O$. The phosphate, glutamate, and glucose are sterilized in separate containers and added to the sterile salt solution. We have routinely used an acetate-requiring mutant of *E. coli* strain W (mutant 191-6) for the isolation of cytochrome b_1 since large amounts of the particulate fraction obtained from this mutant were available from other experiments.[6] For the growth of the acetate-requiring mutant, the growth medium is further supplemented with 20 micromoles of potassium acetate per millimeter. The cells are grown under forced aeration in 100-liter batches in a 50-gallon fermentor.

Inoculum. Escherichia coli mutant 191-6,[7] derived from strain W, is grown on a complex medium containing 1% tryptone, 1% yeast extract, 0.5% K_2HPO_4, and 0.3% glucose. Cells from the complex media are used to inoculate a 10-liter batch of regular growth medium. The 10-liter culture is grown to stationary phase under conditions of forced

[4] S. S. Deeb and L. P. Hager, *Federation Proc.* **21,** 49 (1962).

[5] S. S. Deeb and L. P. Hager, *J. Biol. Chem.* **239,** 1024 (1964).

[6] A. D. Gounaris and L. P. Hager, *J. Biol. Chem.* **236,** 1013 (1961).

[7] This was obtained from Professor B. Davis, Department of Bacteriology and Immunology, Harvard Medical School, Boston, Massachusetts.

aeration (in a New Brunswick fermentor), and this 10-liter culture serves as inoculum for the 100 liter culture.

Growth. Cells are grown to the stationary phase (5–8 hours) and are harvested on a Sharples centrifuge. The yield is approximately 10 g of cell paste per liter of culture medium.

Purification Procedure

Step 1. Preparation of Cell Extract. Six hundred grams of cell paste is suspended in 500 ml of 0.02 *M* potassium phosphate buffer, pH 7 (referred to hereafter as standard phosphate buffer) and mixed thoroughly in a Servall Omnimixer. Six hundred milliliters of washed glass beads (100 μ in diameter) is added to the cell suspension.[8] The cell-glass bead mixture is then ground in an Eppenbach colloid mill (rotor-stator setting of 0.030) for 30 minutes at 15–20° with continuous recycling. The resulting slurry is centrifuged at 15,000 *g* for 10 minutes to remove the glass beads.[9] The supernatant fluid (supernatant 1) is saved. The precipitate (together with the glass beads) is washed by resuspension in 1000 ml of standard phosphate buffer and is centrifuged at 15,000 *g* for 10 minutes. The supernatant fluid (supernatant 2) is again saved. The precipitate (consisting mostly of glass beads) is again resuspended in 1000 ml of standard phosphate buffer and allowed to sediment for 1 hour. The supernatant fluid (supernatant 3) is decanted and saved. Supernatants 1 and 2 are combined and centrifuged at 15,000 *g* for 1 hour. Three layers are formed during the centrifugation: a clear liquid phase; a dense, viscous particulate phase; and a precipitate of cell debris.

Step 2. First Ammonium Sulfate Precipitation. The crude cell-free extract is brought to 0.25 ammonium sulfate saturation by the addition of 14.4 g of ammonium sulfate for 100 ml of crude extract. The suspension is stirred for 4 hours. The precipitate is removed by centrifugation at 15,000 *g* for 75 minutes. The supernatant fluid is removed and saved for isolation of soluble proteins. The precipitate consists primarily of membrane and cell-wall fragments and will be referred to as the particulate fraction.

Step 3. Release of Cytochrome b_1 from Particulate Fraction. The particulate fraction resulting from 700 g of cells is suspended in 700 ml of standard phosphate buffer and mixed in a Servall Omnimixer. The

[8] The glass beads were obtained from Minneapolis Mining and Manufacturing Co., Minneapolis, Minnesota.

[9] All operations involving the use of enzyme preparations were carried out at 3–4° unless otherwise specified.

suspension is centrifuged at 100,000 g for 90 minutes in the Spinco Model L centrifuge. The supernatant fraction is saved, and the reddish-brown gelatinous precipitate is suspended in 700 ml of standard phosphate buffer again using a Servall Omnimixer. This material is then subjected to sonic oscillation in a Raytheon 10-kc sonic oscillator in 40-ml batches for 20-minute periods. After sonic oscillation, the suspension is centrifuged at 100,000 g for 90 minutes. The supernatant fraction containing soluble cytochrome b_1 is saved, and the precipitate is again subjected to sonic oscillation to release more cytochrome b_1 from the particulate fraction. In all, each batch of particulate fraction is subjected to three successive sonic oscillations, followed each time by a 100,000 g centrifugation. Resulting supernatants are combined to yield the soluble cytochrome b_1 fraction.

Step 4. Chromatography of Soluble Cytochrome b_1 Fractions on Calcium Phosphate Gel-Cellulose Columns. Calcium phosphate gel columns are prepared by mixing 300 ml of a 30 mg per milliliter calcium phosphate gel suspension, prepared in the manner described by Swingle and Tiselius[10] with 500 ml of a 10% (w/v) suspension of cellulose powder in phosphate buffer. A 4 $\times$ 12 inch column is packed with the calcium phosphate gel-cellulose mixture to a column height of approximately 7 inches. The column is first washed with 500 ml of standard phosphate buffer, followed by the application of approximately 1 liter of the soluble cytochrome b_1 fraction containing approximately 6 mg of protein per milliliter. The column is eluted successively with 800-ml portions of 0.02 M, 0.1 M, and 0.3 M potassium phosphate buffer, pH 7. Eighteen milliliter fractions are collected. Cytochrome b_1 is eluted by the 0.3 M phosphate buffer. The tubes containing cytochrome b_1 were pooled and the contents dialyzed against 50 volumes of standard phosphate buffer (0.02 M) for 15 hours.

Step 5. Second Chromatography on Calcium Phosphate Gel. Further purification of cytochrome b_1 was achieved by a second chromatography step on a calcium phosphate gel-cellulose column. A 1 $\times$ 24 inch column was packed to a height of 15 inches with the calcium phosphate gel-cellulose mixture described previously. After equilibration of the gel column with 500 ml of 0.02 M standard phosphate buffer, the dialyzed cytochrome b_1 fractions from the first column are applied to the second column. The column is successively washed with 100 ml of 0.02 M and 100 ml of 0.1 M standard phosphate buffer. Cytochrome b_1 is eluted by a linear ionic strength gradient prepared by placing 350 ml of 0.1 M standard phosphate buffer in the mixing chamber and 300 ml of 0.3 M

[10] S. M. Swingle and A. Tiselius, *Biochem. J.* **48**, 171 (1951).

standard phosphate buffer in the reservoir. Fifty fractions, each containing 12 ml, are collected. Cytochrome b_1 usually appears in fraction 20 to 30.

Step 6. Crystallization of Cytochrome b_1. The cytochrome b_1 containing fractions from the second calcium phosphate gel-cellulose column having a specific activity of 0.4 and higher are pooled, and solid ammonium sulfate is added to 0.5 saturation (31.2 g per 100 ml of column eluate). The mixture is stirred for 8 hours to assure complete precipitation of cytochrome b_1. The precipitated cytochrome b_1 is collected by centrifugation at 34,000 *g* for 30 minutes and resuspended in standard phosphate buffer at a protein concentration of 20 mg/ml. This fraction is dialyzed for 3 hours against 500 ml of a 10% ammonium sulfate solution. Following the 3-hour dialysis step, the dialysis is continued for 12 more hours against a slowly increasing ammonium sulfate concentration accompanied by the dropwise addition ($\sim$ 1 ml/min) of 1 liter of a saturated solution of ammonium sulfate to the dialyzate. Recrystallizations are performed by dissolving the cytochrome b_1 in standard phosphate buffer at a concentration of 20 mg protein per milliliter and repeating the ammonium sulfate dialysis. Occasional batches of cytochrome b_1 yield an amorphous precipitate rather than crystals during the first dialysis against ammonium sulfate. In these cases, crystallization is subsequently achieved by repeating the ammonium sulfate dialysis

TABLE I
PURIFICATION SUMMARY FOR CYTOCHROME b_1

Fraction	Total volume (ml)	Total protein (mg)	Cytochrome b_1 (mg)	Specific activity[a]	Yield (%)
Crude extract[b]	2625	81,375	—	—	—
Particulate fraction[c]	700	18,200	137.5	0.007	100
Soluble b_1 fraction	2800	16,100	115.7	0.007	84
Pooled fractions from column 2	215	152.6	62.2	0.400	45.2
Ammonium sulfate fraction, 0–0.5 saturated	4.5	86.4	49.0	0.567	35.6
First crystals	3.0	48.3	46.5	0.962	33.8
Second crystals	3.75	40.5	40.7	1.000	29.6
Third crystals	6.75	37.1	37.7	1.015	27.2

[a] Specific activity is defined as milligrams of cytochrome b_1 per milligram of protein.

[b] Prepared from 700 g of *Escherichia coli* cell paste.

[c] Cytochrome b_1 determinations on the crude extract are relatively inaccurate. The determination of overall yield is based on the assumption that the particulate fraction contains 100% of the cytochrome b_1 of the original cell extract.

TABLE II
SPECTRAL PROPERTIES OF CRYSTALLINE CYTOCHROME b_1

Form	Absorption peaks		
	Maxima (mμ)	Minima (mμ)	$\epsilon \times 10^{-3}$ (M^{-1} cm^{-1})
Reduced	557.5	—	22.1
	527.5	—	15.2
	425	—	127
Oxidized	564	—	8.3
	532	—	11
	418	—	107
	365	—	42.5
Difference (reduced minus oxidized)	557.5	—	16
	527.5	—	6
	427.5	—	60
	—	409	−25
	—	455	−5.3
			−1.5

steps. Recrystallizations are invariably successful with the dialysis technique. A summary of the purification of cytochrome b_1 is given in Table I. The spectral properties are presented in Table II.

Properties

Crystalline cytochrome b_1 has a molecular weight at neutral pH of approximately 500,000 as determined by hydrodynamic methods. Heme determinations indicate that cytochrome b_1 contains 8 moles of iron protoporphyrin IX per 500,000 molecular weight unit. Thus, on the basis of heme determinations, the minimum molecular weight of the crystalline cytochrome b_1 would be approximately 60,000, and the 500,000 molecular weight species would then represent an octomer. This conclusion is also supported by dissociation of the cytochrome b_1 at high pH values. Under these conditions the octomeric species dissociates into a monomeric species having a molecular weight of approximately 60,000. There is a drastic change in the oxidation-reduction potential of cytochrome b_1 associated with the octomer-monomer transition. Octomeric b_1 has an oxidation-reduction potential (E'_0) of —0.34 volt whereas the monomeric species has the 0.0 volt potential which is also characteristic of crude cytochrome b_1 preparations.[4] During purification of cytochrome b_1, the b_1 separates from a colorless protein component which has been termed, potential modifying protein.[4] The readdition of potential modifying protein to crystalline b_1 results in a rise to E'_0 values to more positive values.[4]

[67] The Preparation and Properties of Cytochrome b_{562} from *Escherichia coli*

By LOWELL P. HAGER and EIJI ITAGAKI

Cytochrome b_{562} has been recognized as a major hemoprotein component of anaerobically grown *Escherichia coli* cells,[1,2] and more recently it has been shown that significant amounts of b_{562} are also present in aerobically grown *E. coli* cells.[2] In contrast to cytochromes b_1, a_1, a_2 and o of *E. coli*, cytochrome b_{562} is located among the soluble cell constituents rather than in the membrane fraction. The presence of a soluble *b*-type cytochrome having the identical or very similar spectral properties of *E. coli* cytochrome b_{562} has been reported for *Bacterium anitratum*,[3] *Serratia marcescens*,[1] and a green alga species.[4] Preliminary studies indicate that cytochrome b_{562} can serve as an electron acceptor for various flavoprotein-linked oxidations[5]; however, a precise biochemical role for b_{562} remains to be established.

Assay Method

Principle. Cytochrome b_{562} is determined spectrophotometrically from the absorption spectra of dithionite-reduced samples using the millimolar extinction coefficients listed in Table I of this paper. It is not possible to determine directly the cytochrome b_{562} content of crude extracts by

TABLE I
SPECTRAL PROPERTIES OF CYTOCHROME b_{562}

Absorption peak	Reduced form: Wavelength (mμ)	Reduced form: Millimolar extinction coefficient	Oxidized form: Wavelength (mμ)	Oxidized form: Millimolar extinction coefficient
α	562	31.6	564	9.7
β	531.5	17.4	530	10.6
Soret	427	180.1	418	117.4
δ	324	—	365	—
Reduced form − oxidized form at 562 mμ				24.6

[1] T. Fujita and R. Sato, *Biochim. Biophys. Acta* **77,** 690 (1963).
[2] E. Itagaki and L. P. Hager, *Federation Proc.* **24,** 545 (1965).
[3] J. G. Hauge, *Arch. Biochem. Biophys.* **94,** 308 (1961).
[4] S. Katoh and A. San Pietro, *Biochem. Biophys. Res. Commun.* **20,** 406 (1965).
[5] E. Itagaki and L. P. Hager, unpublished results, 1966.

the spectrophotometric method since crude extracts invariably contain some cytochrome b_1 in addition to b_{562}. In these cases, an aliquot (2–5 ml) of the crude extract is diluted with 2 volumes of water and is charged on a small calcium phosphate–cellulose column (0.6 cm $\times$ 3.5 cm) which has been equilibrated with 0.02 M potassium phosphate buffer, pH 7 (hereafter referred to as standard phosphate buffer). Cytochrome b_{562} is eluted from the column with 0.3 M potassium phosphate buffer, pH 7, and the cytochrome b_{562} eluate is then assayed spectrophotometrically.

Preparation of the Enzyme

Growth Medium. Escherichia coli B cell paste grown aerobically on a complex nutrient medium can be purchased in 1–75 pound batches from the Grain Processing Corp., Muscatine, Iowa (GPC cells). Cell paste is then stored at −12° in a deep freeze until used. The GPC cells are grown in a medium containing 4% amber ECH or NZ-amine (enzymatically hydrolyzed casein),[6] 1.5% Basamine Busch (autolyzed yeast extract),[7] 0.3% K_2HPO_4, 0.07% KH_2PO_4, 0.8% dextrose, and 1.0% salt stock to a level of 40 g of cell paste per liter of culture medium.

Preparation of the Crude Extract

Step 1. Extraction of Cytochrome b_{562}.[8] For extraction of cytochrome b_{562} on a small scale, the frozen cells (1 kg) are thawed, suspended in 3 liters of cold standard phosphate buffer, and treated with a Branson Sonic-Oscillator (Model S-110) for 30 minutes in batches of 300 ml each. The sonicated cell suspension is centrifuged for 1 hour at 13,000 g to obtain the crude extract. For the large scale preparation of cytochrome b_{562}, the frozen cell paste ($\sim$ 5 kg) is thawed in the cold room and then blended into a heavy cell suspension using a minimal amount of water. The heavy cell suspension is mixed with cold acetone ($\sim$ 20 liters at −20°) under conditions of vigorous stirring. The cells are collected by filtration on a Büchner funnel, washed with cold ether, and dried over silica gel in large vacuum desiccators. The acetone-dried cells are suspended in 20 liters of standard phosphate buffer and stirred in a cold room for 12 hours. After extraction, the cell suspension is centrifuged for 30 minutes at 13,000 g. The pellets are discarded and the supernatant fraction is designated as the crude cell extract.

Step 2. Absorption of Cytochrome b_{562} on Calcium Phosphate Gel-Cellulose Columns. The following purification procedure is based on a

[6] This was obtained from Sheffield Chemical Company.

[7] This was obtained from Anheuser-Busch, Inc., St. Louis, Missouri.

[8] All manipulations are carried out in a 4° cold room unless otherwise noted.

crude extract obtained from 1 kg of *E. coli* cell paste. The crude extract is dialyzed overnight against 50 liters of deionized water and then is divided into three equal portions. Each portion is adsorbed on a large calcium phosphate gel-cellulose column (5 cm $\times$ 6 cm) which had been equilibrated with standard phosphate buffer. Cytochrome b_{562} is tightly and sharply bound to the upper part of these columns. Each column is first washed with 1 liter of standard phosphate buffer, and then the cytochrome b_{562} is eluted from each column with 800 ml of 0.3 *M* potassium phosphate buffer, pH 7. This first column fractionation purifies the cytochrome b_{562} approximately 17-fold over the crude extract and completely separates the b_{562} from the small cell fragments which contain cytochrome b_1.

Step 3. Second Calcium Phosphate Gel-Cellulose Column. The eluates from the first calcium phosphate gel column are combined and dialyzed against 50 liters of deionized water. Cytochrome b_{562} is then adsorbed on a second calcium phosphate gel-cellulose column (2.6 cm $\times$ 25 cm) under the same conditions as described above. The second column is washed with 2 liters of standard phosphate buffer and 2 liters of 0.1 *M* potassium phosphate buffer, pH 7, to remove flavoproteins. The cytochrome b_{562} is again removed batchwise by elution with 400 ml of 0.3 *M* potassium phosphate buffer, pH 7. A concentrated cytochrome b_{562} preparation is obtained by this procedure ($\sim$ 6 millimicromoles of cytochrome b_{562} per milliliter).

Step 4. Chromatography on DEAE-Cellulose Column. The eluate from the second calcium phosphate gel-cellulose column is dialyzed overnight against 20 liters of deionized water. Cytochrome b_{562} is then adsorbed onto a DEAE-cellulose column (4 cm $\times$ 14 cm) which has been previously equilibrated with 0.002 *M* potassium phosphate buffer, pH 7. First the DEAE-cellulose column is washed with 5 liters of 0.002 *M* potassium phosphate buffer, pH 7; then cytochrome b_{562} is eluted with 1 liter of standard phosphate buffer and 1 liter of 0.03 *M* potassium phosphate buffer, pH 7. Under these elution conditions, cytochrome b_{562} moves slowly through the column and is eluted at a low protein concentration. All fractions containing b_{562} are pooled and dialyzed overnight against deionized water ($\sim$ 50 liters for the combined fractions).

Step 5. Second Column Chromatography on DEAE-Cellulose Column. The dialyzed fraction from step 4 is sharply adsorbed on a small DEAE-cellulose column (2.5 cm $\times$ 25 cm) in the same manner as described in step 4, and the blood red colored DEAE-cellulose is removed from the top of the column and packed into a new small column (0.9 cm $\times$ 27 cm). The small column is eluted with 3.5 ml of 1.0 *M* potassium phosphate buffer, pH 8, to yield a highly concentrated cytochrome fraction.

Step 6. Crystallization of the Oxidized Form. Finely powdered ammonium sulfate is added to the concentrated cytochrome preparation obtained in step 5 to a final concentration of 60% ammonium sulfate saturation. The cytochrome solution is constantly stirred during the addition of ammonium sulfate and then is put aside to stand for 20 minutes in an ice bath. The solution is centrifuged for 20 minutes at 1400 *g* to remove a white precipitate; this precipitate is then discarded. The concentration of ammonium sulfate in the supernatant fraction is increased to 80% saturation, and the cytochrome preparation is allowed to stand for 1–2 days in the cold room for crystallization. At this time fine needle-type crystals of cytochrome b_{562} appear in the bottom of the test tube. To stimulate further crystallization, more ammonium sulfate is added to near 100% saturation. After all the cytochrome has precipitated as crystals, the crystals are collected by brief centrifugation, and the supernatant fraction is discarded. Recrystallization of cytochrome b_{562} is accomplished by dissolving the crystals in a small amount (2–3 ml) of 0.1 *M* potassium phosphate, pH 7, and centrifuging briefly to remove any insoluble residues. The cytochrome is again crystallized by the addition of finely powdered ammonium sulfate to about 80% saturation, and the preparation is allowed to stand in the cold room.

The overall yield of cytochrome b_{562} is approximately 36% at the stage of the second crystallization when 14–15 mg of the crystalline cytochrome b_{562} is obtained from 1 kg of frozen *E. coli* cell paste.

Reduced crystals of cytochrome b_{562} may be readily obtained by adding 2 to 3 drops of a fresh solution of sodium dithionite ($\sim$ 5 mg/ml) to a solution of cytochrome b_{562} in 80% saturated ammonium sulfate. A summary of the purification and crystallization of cytochrome b_{562} is shown in Table II.

TABLE II
PURIFICATION OF CYTOCHROME b_{562} FROM *E. coli B*

Fraction	Volume (ml)	Protein (mg)	Cyt b_{562} (μmoles)	Purification (-fold)	Yield (%)
Crude extract	3760	94,000	3.06	1	100
First calcium phosphate gel column	2400	5,520	3.03	17	99
Second calcium phosphate gel column	420	1,260	2.48	62	81
DEAE-cellulose column	1600	53	1.78	1050	58
Concentration on DEAE-cellulose	3.5	47	1.64	1090	54
First crystallization	1.6	41	1.44	1103	47
Second crystallization	1.8	34	1.12	1000	36

Properties

Spectrophotometric Properties of Cytochrome b_{562}. The crystalline cytochrome preparation shows absorption bands at 562 mμ (α band), 531 mμ (β band), 427 mμ (Soret), and 324 mμ (δ band) in the reduced form and at 564 mμ (α band), 531 mμ (β band), 418 mμ (Soret), and 363 mμ (δ band) in the oxidized form. A summary of the extinction coefficients for the hemoprotein at the various absorption peaks is shown in Table I.

Physical Properties of Cytochrome b_{562}. The sedimentation constant ($S_{20,w}$) of cytochrome b_{562} based on an extrapolated value to zero protein concentration is 1.64 S. A diffusion constant ($D_{20,w}$) of 11.9 D has been calculated graphically from ultracentrifuge schlieren peak height to peak area ratios. Based on the above data and assuming a partial specific volume of 0.75, the molecular weight of cytochrome b_{562} in solution at neutral pH is 12.7×10^3. The molecular weight of b_{562} determined by the Archibald modification[9] of the sedimentation equilibrium method gives a value of 11.7×10^3.

Chemical Properties. The heme prosthetic group of cytochrome b_{562}

TABLE III
AMINO ACID COMPOSITION OF CYTOCHROME b_{562}

Amino acid residue	Nearest integer
Lys	11
His	2
Arg	4
Asp	16
Thr	6
Ser	3
Glu	14
Pro	5
Gly	4
Ala	15
Val	5
Met	2
Ile	3
Leu	9
Tyr	2
Phe	2
	103

[9] See H. K. Schachman, Vol. IV, p. 32.

is iron protoporphyrin IX. Apo-b_{562}, prepared by treatment with acid-acetone at low temperature,[10] can be readily converted to a holoenzyme by the addition of hemin chloride.

The oxidation-reduction potential (E'_0) of cytochrome b_{562}, as determined by coupling with the tolylene blue–leuco tolylene blue system at pH 7.0 is 0.113 volt.

Cytochrome b_{562} is readily reduced by chemical reductants such as sodium borohydride, sodium dithionite, and ascorbic acid. It is very slowly autoxidizable in the presence of air. The half-life of a reduced solution of b_{562} in air at room temperature is measured in terms of hours.

The amino acid composition of cytochrome b_{562} is shown in Table III. There are several unusual features of the amino acid content of cytochrome b_{562}. For example, the absence of cystine residues indicates that there are no disulfide bridges in this molecule. There are four amino acids (histidine, methionine, tyrosine, and phenylalanine) which are present at a level of 2 residues per molecule. The low molecular weight of cytochrome b_{562} and the favorable amino acid composition of this hemoprotein indicate that it should present an ideal molecule for detailed studies in the chemistry of b-type cytochromes.

[10] S. C. Harrison and E. R. Blout, *J. Biol. Chem.* **240**, 299 (1965).

Section V

Isolation and Determination of Other Mitochondrial Constituents

Previously Published Articles from Methods in Enzymology Related to Section V

Vol. III [114]. General Procedure for Isolating and Analyzing Tissue Organic Phosphates. Carlos E. Cardini and Luis F. Leloir.

Vol. III [120]. Chromatographic Separation of ATP, ADP, and AMP. Waldo E. Cohn.

Vol. III [122]. Assay of Adenosine Triphosphate. Bernard L. Strehler and W. D. McElroy.

Vol. III [124]. Isolation of Diphosphopyridine Nucleotide and Triphosphopyridine Nucleotide. Arthur Kornberg (parts I and II) and B. L. Horecker (part III).

Vol. III [127]. Enzymatic Preparation of DPNH and TPNH. Gale W. Rafter and Sidney P. Colowick.

Vol. III [128]. Procedures for Determination of Pyridine Nucleotides. Margaret M. Ciotti and Nathan O. Kaplan.

Vol. III [132]. Assay of Coenzyme A. G. David Novelli.

Vol. III [140]. Preparation and Enzymatic Assay of FAD and FMN. F. M. Huennekens and S. P. Felton.

Vol. III [141]. Fluorometric Assay of FAD, FMN, and Riboflavin. Helen B. Burch.

Vol. IV [34]. Preparation and Analysis of Labeled Coenzymes. S. P. Colowick and N. O. Kaplan.

Vol. VI [36]. Isolation and Photoinactivation of Quinone Coenzymes. Arnold F. Brodie.

Vol. VI [111]. Measurement of Pyridine Nucleotides by Enzymatic Cycling. Oliver H. Lowry and Janet V. Passonneau.

[68] Isolation and Determination of Ubiquinone

By E. R. REDFEARN

Introduction

The ubiquinones or coenzymes Q are a family of lipid-soluble benzoquinones that are widely distributed in living organisms. They are 2,3-dimethoxy-5-methyl-1,4-benzoquinones with isoprenoid side chains of various lengths (30–50 carbon atoms) in the 6-position. They are designated by the number of isoprenoid units in the side chain as Q-10, Q-9, and so on to Q-6. Ubiquinone is found in relatively high concentrations in animal and plant mitochondria, where it appears to function as an electron carrier in the respiratory chain. It is also present in certain bacteria, either alone or together with naphthoquinones of the vitamin K_2 group (menaquinones).

Methods for the large-scale isolation and purification of ubiquinone from animal tissues and bacteria have already been described in a previous volume of this series.[1] The method to be described here is a semimicro assay which enables the concentration and oxidation–reduction state of ubiquinone to be determined in mitochondria and other particulate preparations.[2] It is based on the facts that ubiquinone is readily soluble in organic solvents and can thus be separated fairly selectively from other tissue components and that it has characteristic absorption spectra in the oxidized and reduced states. The method consists essentially of the following steps: (a) the simultaneous termination of any enzymatic reaction and denaturation of enzyme proteins without destroying the ubiquinone or altering its oxidation-reduction state; (b) extraction of the lipid with light petroleum; (c) removal of interfering lipids (e.g., phospholipids), by partitioning the extract between light petroleum and methanol; and (d) spectrophotometric determination of the ubiquinone. The advantages of the method are that it is a simple and rapid assay, that only small samples of the tissue preparations are required, and that the oxidation–reduction state of the quinone is preserved during the isolation procedure. It is thus very convenient not only for the determination of the concentration of ubiquinone, but also for kinetic studies where it is necessary to assay its redox state in a large number of samples. Its main disadvantage is that it cannot be used when the tissue preparation contains relatively large amounts of material, not

[1] A. F. Brodie, Vol. VI [36].

[2] A. M. Pumphrey and E. R. Redfearn, *Biochem. J.* **76**, 61 (1960).

removed by the partitioning procedure, which absorbs in the same spectral region as ubiquinone and undergoes an absorption change on chemical reduction. This problem is not encountered with most animal and plant mitochondria, but it does become serious in the case of those bacterial preparations which contain menaquinones in addition to ubiquinone.

Assay Method

Reagents

Spectroscopically pure ethanol. Zinc dust (20 g) and potassium hydroxide (40 g) are added to ethanol (1 liter), and the mixture is refluxed for at least 1 hour before distillation. The fraction distilling over at 78° is collected.

Light petroleum (b.p. 40–60°), analytical grade, free from aromatic hydrocarbons. If necessary, small amounts of aromatic hydrocarbons can be removed by passing the light petroleum through a column (20 × 1.5 cm) of silica gel (200-mesh) and then distilling.

Methanol, analytical grade redistilled

Pyrogallol, analytical grade

Sodium borohydride, analytical grade

Chloroauric acid ($HAuCl_4 \cdot 3\ H_2O$), 2% (w/v) aqueous solution

Procedure. A portion (1 ml) of the fresh tissue preparation (e.g., mitochondria, heart muscle preparation) containing 10–20 mg of protein is placed in a 15-ml glass-stoppered test tube or centrifuge tube. The preparation is denatured by the rapid addition from a hypodermic or spring-loaded pipetting syringe of cold (—20°) methanol (4 ml). When the method is used for the determination of the redox state of ubiquinone, as in kinetic studies, it is recommended that pyrogallol be added to the methanol (1 mg/ml) used for the initial denaturation. This protects any ubiquinol present from oxidation during the extraction procedure. Ubiquinol is fairly easily autoxidizable, and oxidation is particularly liable to occur in the denatured preparation before extraction into light petroleum. It is important also that the methanol–pyrogallol mixture should be at —20° and added rapidly to ensure that enzymatic reactions are arrested immediately. A general precaution which should be mentioned here is that care must be taken to avoid contact of the methanol and other solvents with rubber or plastic stoppers, tubing, and valves. Otherwise extraneous ultraviolet absorbing substances may be picked up. For this reason apparatus must be all glass and metal.

Immediately after the methanol addition to the preparation, light petroleum (5 ml) is added, and the stopper is placed in the tube. The tube is then agitated vigorously by hand or on a Vortex Junior mixer for 1 minute. The tube is then given a short spin (2–3 minutes) in the angle-head of a bench centrifuge to separate the layers. The upper light petroleum layer is then transferred by means of a Pasteur pipette to another 15-ml glass-stoppered test tube. The denatured residue is then extracted again with a further portion of light petroleum (3 ml), the separation of layers and transfer of the upper layer being carried out as before. Virtually complete extraction of the ubiquinone is obtained with two light petroleum extractions of the denatured preparation provided the original protein concentration does not exceed 20 mg/ml. For aliquots containing higher concentrations of protein, the number of petrol extractions should be increased; e.g., for concentrations from 20 to 40 mg/ml, at least three extractions are necessary. The combined light petroleum extracts are then treated with 2 ml of 95% (v/v) methanol and the mixture shaken gently for 30 seconds. After separation of the layers, the light petroleum layer is transferred to a 10 ml beaker which is then placed in a vacuum desiccator where the solvent is removed under reduced pressure. Alternative methods may be used for evaporating the solvent, e.g., blowing-off under a stream of nitrogen or use of a rotary evaporator.

After evaporation of the solvent, the residual lipid is dissolved in spectroscopically pure ethanol (3 ml). Complete solution is ensured by placing the beaker on a water bath at 60° and stirring with a glass-rod for about 30 seconds. The spectrum of the ethanolic solution is then determined in a 1-cm light-path cuvette in a manual or recording spectrophotometer in the range 230–320 mμ. The presence of ubiquinone is indicated by selective absorption with a maximum at 275 mμ. The curve should be characteristic of the spectrum of ubiquinone, but some end absorption will increase the extinction value of the minimum at 236 mμ and will tend to shift it to a slightly longer wavelength. Nevertheless, the persistence is usually good; ratios of the maximum to the minimum as high as 3 have been obtained with heart muscle preparations. The ubiquinone is reduced to ubiquinol by the addition of a few crystals (approximately 0.2 mg) of sodium borohydride. This is conveniently done by adding the sodium borohydride from the end of a small glass rod and stirring rapidly. The mixture is then left for 2 minutes before the spectrum is redetermined. The reduction of ubiquinone by sodium borohydride is relatively slow, and some unidentified intermediates are formed. The reduction to the quinol is however usually complete in 2 minutes. Also, after this time evolution of bubbles of hydrogen, which can give rise to false readings, has generally ceased. After the spectrum has been redetermined in the same spectral region as before, the difference in absorbance at 275 mμ (ΔE_{275}) is found. The concentration of ubiquinone is then calculated using the molecular extinction coefficient for the difference in absorption of the oxidized and reduced forms of ubiquinone $(\epsilon_{ox} - \epsilon_{red})_{275} = 12{,}250$. Ubiquinone concentrations are usually expressed as micromoles per gram of protein and may be calculated from the equation:

$$\text{micromoles UQ per gram protein} = \frac{3 \times \Delta E_{275} \times 10^3}{12.25 \times \text{mg protein/ml}}$$

The $\Delta\epsilon_{275}$ value of 12,250 was obtained from data on ubiquinone-50 (Q — 10),[3] but it has been found that this also represents the mean value of $\Delta\epsilon_{275}$ calculated from the $\Delta E_{1\text{cm}}^{1\%}$ values given for Q-10, Q-9, Q-8, and Q-7.[4] Thus the same figure may be used for all the naturally occurring homologs.

In studies on the kinetics of ubiquinone reactions and in the determination of concentration where the quinone initially is not in the completely oxidized state, it is necessary to oxidize the quinone chemically after extraction. This is done by addition to the light petroleum extract 0.1 ml of 2% (w/v) chloroauric acid followed by intermittent shaking for 2 minutes. The extract is then washed four times with 3 ml of water and treated as previously described.

It has been claimed previously[2] that treatment of mitochondrial preparations with alkali (final concentration 0.33 N NaOH) gave higher values of ubiquinone concentration than untreated preparations, and this was attributed to the release of a more tightly bound form of the quinone. Recent experiments have suggested that another interpretation of this result is that alkali facilitates the extraction of ubiquinone by light petroleum, but that if an efficient agitation or shaking procedure is used all the ubiquinone is extracted without the need for the addition of alkali.

The method described has been used successfully for the determination of the concentration and oxidation–reduction state of ubiquinone in mitochondria derived from a number of animal and plant tissues[2] and in particulate preparations isolated from bacteria.[5] The method has also been modified to determine ubiquinone in mitochondrial suspensions containing as little as 2 mg of protein.[6]

[3] F. W. Hemming, Ph.D. Thesis, University of Liverpool (1958).

[4] R. L. Lester, Y. Hatefi, C. Widmer, and F. L. Crane, *Biochim. Biophys. Acta* **33**, 184 (1959).

[5] C. W. Jones and E. R. Redfearn, *Biochim. Biophys. Acta* **113**, 467 (1966).

[6] L. Szarkowska and M. Klingenberg, *Biochem. Z.* **338**, 674 (1963).

[69] Isolation, Characterization, and Determination of Polar Lipids of Mitochondria

By GEORGE ROUSER and SIDNEY FLEISCHER

Introduction

The purpose of this chapter is to present the combined methodology for the isolation and determination of polar lipids from mitochondria and submitochondrial preparations. Procedures are presented both for micro scale and somewhat larger scale work. The sequence of presentation is largely that of the actual operations in the laboratory. Included are extraction, general laboratory techniques (working under nitrogen, evaporation of solutions, weighing, and storage of lipids), removal of nonlipid contaminants from lipid extracts, isolation by diethylaminoethyl (DEAE) cellulose column chromatography, quantitative analysis by thin layer chromatography, and procedures for characterization of lipids.

The procedures have been used for the most part with mitochondria isolated from animal organs, particularly those from heart, kidney, liver, and brain. More extensive work is required to demonstrate conclusively that these mitochondrial preparations are indeed completely representative of those from other animal organs. The methods will probably prove to be adequate for mitochondria from other animal organs since the procedures have been used successfully for examination of whole organ extracts from most organs.

Mitochondria from animal organs have been found to contain three principal polar lipid classes: phosphatidylcholine (lecithin), phosphatidylethanolamine, and diphosphatidylglycerol (cardiolipin).[1] The structures of these three lipids are shown below in schematic form.

```
    O
    ||
R—C—O—CH2
          |
    O     |
    ||    |
R—C—O—CH
          |
          |       O
          |       ||
       H2C—O—P—O—CH2—CH2
                 |           |
               O(-)     (+)N(CH3)3
```

Phosphatidyl choline

```
    O
    ||
R—C—O—CH2
          |
    O     |
    ||    |
R—C—O—CH
          |
          |       O
          |       ||
       H2C—O—P—O—CH2—CH2
                 |           |
               O(-)       (+)NH3
```

Phosphatidyl ethanolamine

[1] S. Fleischer and G. Rouser, *J. Am. Oil Chemists Soc.* **42**, 588 (1965).

```
       O                           OH
       ||                          |
R—C—O—CH2            H2C—O—P—O—CH2
       O       |             |     ||      |        O
       ||      |             |     O       |        ||
R—C—O—CH            HC—OH         HC—O—C—R
               |      O      |               |        O
               |      ||     |               |        ||
            H2C—O—P—O—CH2         H2C—O—C—R
                      |
                      OH
```

Diphosphatidyl glycerol

A fourth phospholipid, phosphatidylinositol, in which inositol is present in place of choline or ethanolamine, is a minor component of mitochondria. Trace components including free fatty acid and uncharacterized acidic phospholipids also occur (see Figs. 1 and 2).

The composition of highly purified mitochondria is readily visualized by two-dimensional thin layer chromatography (TLC) as shown in Figs. 1 and 2, where two different two-dimensional TLC systems serve to separate the major components as well as trace components of preparations. The several less polar or "neutral" lipid components migrate to the solvent front with these TLC systems for polar lipids. Whole organs and other subcellular particles may contain other common lipid classes includ-

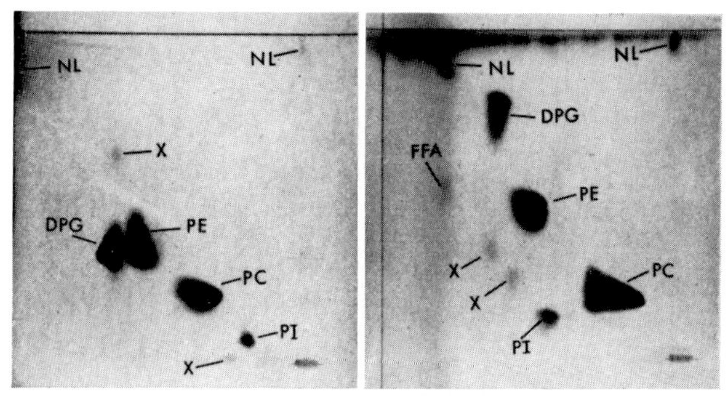

FIGS. 1 and 2. Two-dimensional thin layer chromatograms of heavy beef heart mitochondria lipid. The lipid extracts (600 μg) were applied in the lower right corner. The chromatogram in Fig. 1 was developed first with chloroform/methanol/28% aqueous ammonia 65:35:5 (vertical direction), air-dried for 10 minutes, and then developed with chloroform/acetone/methanol/acetic acid/water 5:2:1:1:0.5 (horizontal direction). Figure 2 illustrates results obtained by developing first with chloroform/methanol/water 65:25:4 (vertical direction) followed by 1-butanol/acetic acid/water 60:20:20 (horizontal direction). Spots were developed with the sulfuric acid–dichromate char reagent and heat (see text). Abbreviations: *NL*, neutral (less polar) lipids; *X*, uncharacterized; *PE*, phosphatidylethanolamine; *DPG*, diphosphatidylglycerol; *PC*, phosphatidylcholine; *PI*, phosphatidylinositol; FFA, free fatty acids.

ing sphingomyelin, phosphatidylserine, cerebroside, and sulfatide. The positions of these lipid classes are shown in Figs. 3 and 4. These photographs demonstrate the positions of the other lipid classes that may be encountered in impure mitochondrial preparations from some organs. The presence of cerebroside and sulfatide in mitochondrial preparations from nervous tissue can arise by contamination with myelin. Other subcellular structures in the nervous system are rich in sphingomyelin and phosphatidylserine, and contamination with these structures will give rise to the presence of sphingomyelin and phosphatidylserine. Such contamination is readily disclosed by two-dimensional TLC.

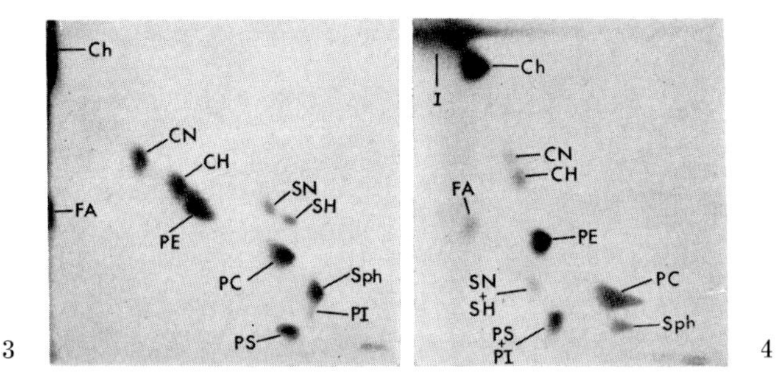

FIGS. 3 and 4. Two-dimensional chromatograms of human brain lipids to illustrate positions of lipids not seen in pure mitochondria (cerebrosides, sulfatides, phosphatidylserine, sphingomyelin). Figure 3 is from 300 μg of adult brain lipid developed as described for Fig. 1, and Fig. 4 is developed as described for Fig. 2, with 300 μg of lipid of immature brain (age 5½ months). Abbreviations: *Ch,* cholesterol; *CN* and *CH,* cerebroside with normal and hydroxy fatty acids, respectively; *SN* and *SH,* sulfatide with normal and hydroxy fatty acids, respectively; *PS,* phosphatidylserine; *Sph,* sphingomyelin; *FA,* fatty acids; *I,* impurities; others as for Figs. 1 and 2.

Mitochondrial polar lipids are very susceptible to alteration. The vinyl ether linkage of the plasmalogen forms of phospholipids is hydrolyzed by weak acid. In plasmalogens, the hydrocarbon chain of a fatty aldehyde joined by an α,β-unsaturated ether linkage replaces the fatty acid ester linkage on carbon 1. Plasmalogen forms of phosphatidylethanolamine and phosphatidylcholine occur in some organs. In weak base, hydrolysis or transesterification of the ester linkage obtains. Mitochondrial lipids generally contain a great deal of polyunsaturated fatty acid (fatty acids containing two to six double bonds) and are quite susceptible to air oxidation. These particular labilities of the mitochondrial lipids make it especially important to use very mild procedures. Strong acids and strong bases must be avoided, column fractions containing even such a weak acid as acetic acid should be worked up as soon as they are obtained, and air should be excluded by working in an atmosphere of pure nitrogen as much as possible. Careful attention to avoidance of unnecessary contamination with heavy metals is also necessary since metals catalyze peroxidation of lipids. In actual practice the procedures for working under an atmosphere of pure nitrogen in most of the steps in handling mitochondrial lipids do not present great difficulties nor does the avoidance of exposure of the lipid preparations to acids, bases, or heavy metals.

I. Preparation of Lipid Extracts[2-5]

A. Reagents

Reagent grade chloroform, methanol, and glacial acetic acid redistilled from glass. Chloroform is distilled into enough methanol to make a final concentration of 0.2% for stabilization.

Aqueous ammonia is best prepared in the laboratory by bubbling gaseous ammonia from an ammonia cylinder into cold distilled water to the desired weight (about 28% ammonia). The ammonia should be prepared fresh every 1–3 weeks and stored in a tightly capped plastic bottle.

Solutions for washing of lipid extracts include 0.017% and 0.034% magnesium chloride. Equilibrated "upper phase" and "lower phase" are prepared by mixing in a separatory funnel (equipped with a Teflon stopcock) chloroform, methanol, and 0.034% magnesium chloride in the ratio of 8:4:3. The methanol-water rich upper phase and the chloroform-rich lower phase are respectively designated equilibrated "upper" and "lower" phases.

Sephadex G-25 coarse, beaded (Pharmacia Fine Chemicals, Inc., New York)

Pure nitrogen (at least 99.998%)

B. General Comments

The most popular single mixture for extraction of lipids is chloroform-methanol 2:1. We have found this to be satisfactory for mitochondrial lipids. Extraction is done at room temperature and all operations are performed as much as possible under an atmosphere of pure nitrogen to minimize air oxidation.

[2] J. Folch, M. Lees, and G. H. Sloane-Stanley, *J. Biol. Chem.* **226**, 497 (1957).
[3] S. Fleischer, H. Klouwen, and G. Brierley, *J. Biol. Chem.* **236**, 2936 (1961).
[4] G. Rouser, G. Kritchevsky, D. Heller, and E. Lieber, *J. Am. Oil Chemists Soc.* **40**, 425 (1963).
[5] A. N. Siakotos and G. Rouser, *J. Am. Oil Chemists Soc.* **42**, 913 (1965).

C. Extraction Procedures[2-4]

1. *Small Samples.* Two milliliters or less of the mitochondrial or enzyme suspension, usually in 0.25 *M* sucrose, are homogenized with 20 volumes of chloroform/methanol 2:1 in a Potter-Elvehjem type homogenizer equipped with a Teflon pestle. The extract is filtered through a sintered-glass filter of coarse porosity with vacuum from a water aspirator. The residue is scraped from the filter and reextracted twice more with ten volumes of solvent each time. The total chloroform/methanol filtrate contains lipid as well as nonlipid materials and is then treated to remove nonlipid materials (Section I,E). Homogenization in a nitrogen atmosphere is readily accomplished with the combined use of a polyethylene bag and a polyethylene "housing" prepared from a 1-liter polyethylene bottle with the base piece removed. The neck of the housing is supported close to the chuck of the motor. The polyethylene bag, attached to the housing by means of a rubber band, encloses the homogenizer fitted onto the pestle. During homogenization the system is flushed with nitrogen supplied through polyethylene tubes passing into the polyethylene "housing."

2. *Larger Samples.* Samples up to 125 ml (such as mitochondrial paste obtained from a Sharples centrifuge bowl) are extracted with a Waring blendor. Glass containers of different volumes are available. The enzyme preparation is added to 20 volumes of chloroform/methanol 2:1 in a Waring blendor. Blending is carried out for 5–15 seconds at a time for five times over a period of 5 minutes. The extract is maintained near room temperature by immersing the homogenizer in ice between blending periods. Insoluble solids are removed by filtration through a sintered-glass filter of coarse porosity under suction from a water aspirator. The residue is extracted twice more in the same fashion with 10 volumes of chloroform/methanol 2:1 each time and finally with 10 volumes of chloroform/methanol 7:1 saturated with 28% aqueous ammonia. Use of the latter solvent mixture ensures complete extraction of diphosphatidylglycerol. The number of volumes of solvent added in each case is based on the original volume of sample; i.e., for each gram wet weight or milliliter of packed mitochondrial mass either 20 or 10 ml of solvent is used.

D. Concentration, Weighing, and Storage of Samples

1. *Concentration and Weighing with a Semimicro Balance.* Lipid extracts and column fractions are concentrated under reduced pressure and in the cold with a rotary or flash evaporator. For large volumes a

rotary type evaporator (Universal model, Buchler Instruments, 1327-16th St., Fort Lee, New Jersey) is particularly useful. For evaporating smaller amounts of material the Calab Model C flash evaporator (1165-67th St., Oakland, California) is much simpler to control. The system is first thoroughly flushed with nitrogen, evacuation and gassing being controlled with a 3-way stopcock. Evaporation is begun by allowing the solution to cool rapidly as evaporation occurs. Enough heat is supplied from the water bath to prevent the sample flask from accumulating a large amount of ice. Evaporation of 500 ml of most solvents can be accomplished in 30 minutes or less at low temperature and under nitrogen. Reduced pressure is provided by a vacuum pump of good capacity with a cold trap (Methyl Cellosolve–Dry Ice). Evaporation is more efficient with two cold traps in series.

The final evaporation and weighing can be accomplished quite simply in 50- and 100-ml round-bottom flasks (14/35 standard taper). The sample is evaporated to visual dryness. The flask is then placed in a vacuum desiccator over potassium hydroxide pellets and evacuated for 15 minutes. Nitrogen is added, and the desiccator is allowed to temperature equilibrate next to the balance for 15 minutes or more before the sample is weighed on a semimicro or other analytical balance. The sample is then immediately dissolved in chloroform/methanol 2:1 (5–10 mg/ml) and stored under nitrogen in a freezer (—20° or below).

2. *Reextraction and Storage of Lipid in the Dry State.* The lipid extract (Section I,C) is evaporated to dryness in a rotary (flash) evaporator (Section I,D,1). The dry solids are then reextracted with a small volume of chloroform/methanol 2:1. Complete extraction of lipid is ensured by intimate mixing in a Potter-Elvehjem homogenizer equipped with a Teflon pestle. The insoluble nonlipid residue is removed by filtration (sintered-glass filter) or by low speed centrifugation in a clinical centrifuge. The insoluble residue is washed several times with chloroform/methanol 2:1 for quantitative removal of all lipid. Completeness of extraction can be checked by thin layer chromatography (Section III). The residue is then discarded. The lipid concentrate in a mortar is then transferred to a bell jar or vacuum desiccator equipped with a nitrogen inlet and a vacuum attachment and the remaining solvent is removed under reduced pressure and a gentle stream of nitrogen. The lipid is thoroughly dried over potassium hydroxide pellets and mixed thoroughly by repeatedly scraping it from the sides of the mortar with a spatula followed by pressing the lipid with a pestle against the sides of the mortar. These steps can be conducted in a plastic bag flushed with nitrogen. The dried lipid can then be stored under nitrogen either in a

sealed tube or in a vacuum desiccator in a freezer (—20° or lower). This procedure has been particularly useful for lipid extracts which contain large amounts of inorganic materials (sugars and salts) such as those obtained from mitochondrial preparations sedimented from 0.25 M sucrose. These crude total lipid extracts are more stable than highly purified phospholipids. Total mitochondrial lipids thus prepared and stored have been found functional for lipid reactivation studies several years after their initial preparation. For these studies, samples were placed in a vacuum desiccator that was flushed with nitrogen several times by evacuation with a vacuum pump and filling with pure nitrogen. The desiccator was then filled with nitrogen to atmospheric pressure and left in a freezer. Lipid extracts can also be stored in solution (Section I,D,1).

3. *Concentration without Evaporation to Dryness and Weighing with a Microbalance.* Exposure of lipids to solid surfaces in the dry state may cause alteration or decomposition of lipids. These changes occur particularly when samples are dried prior to weighing. These undesirable alterations in samples can be eliminated by weighing an aliquot only with a microbalance. The aliquot is discarded after weighing. The bulk portion of sample by this procedure is never dried completely. This procedure is recommended for samples to be analyzed quantitatively.

The combined lipid extracts are carefully concentrated to a small volume in the cold under nitrogen using a rotary evaporator. Water is displaced by repeated addition of methanol and then chloroform with concentration to a small volume after each addition and with care to avoid complete evaporation to dryness. Finally, the moist solids are treated with enough chloroform/methanol 2:1 (or other desired solvent) to give a solution of approximately 10 mg/ml. Water (1–5%) can be added to ensure solution of all or most of the solids. The final concentration and the total weight of the extract is determined by weighing the solids of a 50–200 μl aliquot of the solution. An aliquot is drawn into a 50–100 μl Hamilton syringe and transferred to a 5–6 mg aluminum cup. The tare weight of the cup is first determined by heating it for about 3 minutes at 60–80°, cooling it in a desiccator over potassium hydroxide pellets for about 3 minutes, and then weighing it on a Cahn microbalance after equilibration inside the balance for 1 minute. The sample is then transferred to the cup, which is again put through the same cycle of heating, cooling, and weighing. This procedure is satisfactory for weighing of lipid extracts in general and for most preparations of polar lipids. Shorter-chain fatty acids and fatty acid methyl esters may be lost by evaporation. Some free fatty acids and longer-chain methyl esters can be weighed by evaporation of solvent under a stream of nitrogen without heat.

Samples are preferably stored for prolonged periods of time in solution in sealed tubes. The tubes are flushed with pure nitrogen, then sealed and kept at —20° or lower in the dark.

E. Removal of Nonlipid Contaminants

1. *Folch Partition.*[2] The procedure is useful for smaller volumes of extracts (Section I,C,1). Five milliliters of extract is mixed with 1 ml of 0.034% $MgCl_2$ in a glass-stoppered centrifuge tube, nitrogen is bubbled through the solvents, and the stoppered tube is inverted about 100 times. Two phases are obtained, the upper phase representing about 40% of the total volume. Separation of phases is aided by centrifugation (International clinical centrifuge speeds are sufficient). The upper layer should be clear before it is removed and discarded. Care should be taken not to remove any interface material, if present. Equilibrated "upper phase" (see Reagents) is added (approximately 1 ml per 3 ml lower phase) and the partition is repeated 3 times. Finally, enough methanol is added to form one phase with residual upper phase, and any insoluble material is removed by filtration through a sintered-glass filter of medium porosity.

2. *Sephadex Column Chromatography.*[5] The procedure is used for larger samples and lipid extracts prepared as described in Section I,C,2.

a. ELUTION MIXTURES

(1) chloroform/methanol 19:1 saturated with water (about 5 ml/liter)
(2) chloroform/methanol 19:1, 850 ml; glacial acetic acid, 170 ml; water, 25 ml
(3) chloroform/methanol 9:1, 850 ml; glacial acetic acid, 170 ml; water, 42 ml
(4) methanol/water 1:1

Solvents for (1), (2), and (3) are mixed several times in a separatory funnel. The clear lower phase is used. Mixture (4) must be mixed and allowed to stand for at least 1 hour before use to ensure temperature equilibration.

b. SOLVENT EQUILIBRATION AND DEGASSING OF GEL. Sephadex G-25 (about 50 g is required for a column 2.5 cm × 30 cm) is placed in 200 ml of methanol/water 1:1 in a flask covered with aluminum foil and is allowed to stand overnight for equilibration. Dissolved gases are removed from the solvent-equilibrated gel under suction from a vacuum pump with swirling (to prevent bumping) for 30–60 seconds.

c. COLUMN PREPARATION. Columns of various diameters and heights are useful for various quantities of lipid and different purposes. Columns 1 cm in diameter and 10–30 cm in height are useful for 250 mg or less of sample, whereas columns 2.5 cm i.d. and 10–30 cm in height are useful for larger samples. Samples weighing several grams have been successfully separated on these columns. Directions are given for 2.5 × 30 cm columns as most generally useful for total extracts. Directions for preparation and elution of 1 cm × 10 cm columns for removal of salts from DEAE column fractions are given in Section I,E,2,f.

A chromatography tube approximately 2.5 cm (i.d.) and 40 cm long fitted with a Teflon stopcock and a 1-liter reservoir for solvent is used with a glass-wool plug moistened with methanol/water 1:1 to retain the gel. The gel slurry in methanol/water 1:1 is poured into the tube, the walls of the tube are washed with methanol/water 1:1, and the gel bed is washed with 500 ml of the same solvent at a gravity flow rate of 5–6 ml/min. To avoid bubble formation, the wash solvent should be mixed and allowed to come to room temperature before use. The final column height is adjusted to 30 cm by aspiration of excess gel. Dry, moderately fine grained, reagent grade sand (previously washed with methanol/water 1:1) is poured to a height of about 1½ inches when 2–3 inches of solvent still remain over the bed. The sand is stirred with a glass rod to ensure that the top of the bed is even, care being taken not to stir into the gel bed. The layer of sand prevents floating of the gel when chloroform-containing solvent mixtures are applied and prevents disturbance of the bed during solvent additions.

Careful washing of the column before sample application is necessary to prevent the appearance of "fines" in fractions 3 and 4. The methanol/water 1:1 solvent used for packing the column is allowed to pass completely to the top of the sand layer. The column is then washed through two cycles with each of the four eluting mixtures using volumes of 500 ml, 1000 ml, 500 ml, and 1000 ml for mixtures (1), (2), (3), and (4), respectively. Wash solvents can be passed through the bed at 5–6 ml per minute during working hours, the rate being adjusted downward at the end of the day to prevent the column from running dry overnight. The appearance of irregular opaque areas in the column can be ignored since column performance is not affected to any appreciable extent. This appearance can largely be avoided by prolonged equilibration of the column in methanol/water 1:1 prior to replacement with elution mixture (1). Each solvent change results in the appearance for a time of a two-phase system in the effluent (at the solvent front). The solvent mixtures properly saturated with water do not produce significant changes in column height or solvent flow rates.

d. Sample application. Chloroform/methanol extracted lipid (2 g or less) containing water-soluble contaminants is applied in a small volume of chloroform/methanol 19:1 saturated with water (5 ml water per liter) and lipid and any suspended solids are rinsed in with the same solvent. Samples containing large amounts of suspended solids can be transferred conveniently to the column with a 5 or 10 ml blowout-type pipette with the opening of the tip enlarged to prevent blockage. The transfer pipette is rinsed inside and out with solvent delivered from a 1-ml pipette after transfer of sample is complete. A glass wool plug (washed with chloroform/methanol/water) is added after sample application to further protect the column surface and prevent sand from moving up into the solvent reservoir.

e. Elution of column. After sample application, bulk fractions are collected at a flow rate of 3 ml/min. The four elution mixtures are used sequentially collecting 500 ml, 1000 ml, 500 ml, 1000 ml, respectively.

Columns are preferably allowed to stand in methanol/water 1:1 for about 48 hours before reuse, 500 ml chloroform/methanol 19:1 saturated with water being passed through the bed just prior to application of another sample.

Mitochondrial lipid is eluted in fraction 1. Traces of decomposed or altered lipids may appear in fractions 2 and 3. Water-soluble nonlipids (salts, sucrose, etc.) are eluted in fraction 4. With brain lipid as sample, fractions 2 and 3 contain gangliosides.

f. Removal of salt introduced into lipid fractions during DEAE column chromatography. Salts introduced during DEAE-cellulose chromatography (Section II) are removed by Sephadex column chromatography using a short column and only two eluting mixtures. Lipid is eluted with chloroform/methanol 19:1 saturated with water, and methanol/water 1:1 is then used to clear the column of salt prior to reuse. The general procedure described in sections a to d above is used. A slurry of Sephadex is poured to a height of 10 cm in a tube 1 cm in diameter. The bed is washed with 50 ml of elution mixture (1), and then 50 ml of elution mixture (4) (two cycles), and then with 50 ml of elution mixture (1). The sample is applied in this latter mixture and elution is carried out with 50 ml of elution mixtures (1) and (4). Elution mixture (4) clears the column of salts. Columns can be reused repeatedly without repacking provided they are thoroughly cleared of salt by washing with elution mixture (4) and not allowed to run dry. Loads up to approximately 250 mg of lipid plus salt can be applied to columns 1 cm in diameter; for larger loads, columns of greater diameter should be used.

II. Diethylaminoethyl (DEAE) Cellulose Column Chromatography[4,6]

A. Reagents

Reagent grade chloroform, methanol, glacial acetic acid, and ammonia as described in Section I,A

Selectacel DEAE, regular grade (Brown Co., Berlin, New Hampshire)

Ammonium acetate prepared by mixing just prior to use the calculated amounts of freshly prepared aqueous ammonia (Section I,A) and redistilled glacial acetic acid into the chromatography solvents.

B. General Comments

Lipid extracts can be separated into groups based upon net charge, relative polarity, and types and numbers of functional groups using DEAE column chromatography. Lipid mixtures can be separated rapidly into three classes: (1) neutral lipids (elution with chloroform); (2) zwitterion type (no net charge, elution with methanol); and (3) acidic phospholipids (elution with chloroform/methanol/ammonium acetate). A more extensive elution scheme is used to separate the individual polar lipid classes. The elution scheme shown in the table is suitable for mitochondrial lipid extracts. It is to be stressed that other elution schemes may be preferable for other mixtures of lipids.

C. Selection of Proper DEAE-Cellulose for Lipid Chromatography

DEAE preparations vary in coarseness of grade as well as total ion exchange capacity. The coarser grades are the most generally useful for lipid separations. We have used the regular grade Selectacel DEAE. Finer grades such as types 20 and 40 Selectacel are satisfactory for some purposes and are readily packed into a chromatography tube without the manual pressure required for the regular grade. These finer grades are generally less satisfactory, however, since fines from the adsorbent may appear in the column effluent with certain solvents and because the finer grades tend to pack more solidly upon repeated use until flow rates are seriously reduced. It is desirable to select preparations that have the highest rated number of equivalents per gram although the texture of the material is generally more important for successful chromatography than the relatively small differences in number of milliequivalents per gram of the various preparations.

[6] G. Rouser, G. Kritchevsky, C. Galli, and D. Heller, *J. Am. Oil Chemists Soc.* **42**, 215 (1965).

ELUTION OF MITOCHONDRIAL LIPIDS FROM DEAE COLUMNS[a]

Solvent	Volume (column volumes)	Components
(1) $CHCl_3$	5	Neutral (nonionic) lipids[b]
(2) $CHCl_3/CH_3OH$, 9:1	8	Phosphatidylcholine (lecithin)
(3) $CHCl_3/CH_3OH$, 7:3	9	Phosphatidylethanolamine
(4) CH_3OH	9	Nonlipid (primarily sucrose)
(5) $CHCl_3$/HAc[c]	10	"Altered" PE[d] and free fatty acid
(6) HAc	10	Phosphatidylserine, if present
(7) CH_3OH	3	Little or no lipid, used to remove HAc
(8) $CHCl_3/CH_3OH/NH_3$[e]	10	Acidic lipids related to diphosphatidylinositol,[f] phosphatidylinositol
(9) $CHCl_3/CH_3OH/NH_3$[e] + 0.01 *M* NH_4Ac[g]	10	Diphosphatidylglycerol
(10) CH_3OH	5	Diphosphatidylglycerol may appear here[d]

[a] Columns 20 cm high, column volumes for 2.5 and 4.5 cm (i.d.) columns, respectively, 75 and 230 ml with corresponding flow rates of 3 and 10 ml/min.

[b] Does not include some acidic materials such as free fatty acids eluted with $CHCl_3$ from silicic acid columns.

[c] HAc, glacial acetic acid.

[d] See text for discussion.

[e] Chloroform/methanol 4:1 containing 20 ml/liter of 28% by weight aqueous ammonia prepared by bubbling gaseous ammonia into distilled water at 0–4°C.

[f] These acidic lipids are possibly formed by decomposition of diphosphatidyl glycerol since rechromatography of pure diphosphatidyl glycerol gives rise to the same components.

[g] Ammonium acetate prepared by mixing the proper amounts of redistilled acetic acid and freshly prepared 28% ammonia as in (e).

D. Washing of DEAE-Cellulose

DEAE preparations contain various impurities that must be removed prior to use for column chromatography. The adsorbent is washed with 1 *N* aqueous HCl followed by water to neutrality and then 0.1 *N* KOH followed by washing with water to neutrality. Three cycles of acid and base are used. Exposure to acid and base should be as brief as possible, and usually no more than three bed volumes of acid or base are required per cycle. Washing may be carried out on a Büchner or a sintered-glass filter and is most rapid when both filter paper and cheesecloth or gauze (4 layers) are used to retain the DEAE to prevent clogging of pores during filtration.

The DEAE is then converted to the acetate form by washing with 3 bed volumes of redistilled glacial acetic acid, and excess acid is removed by washing with 3 bed volumes of redistilled methanol. The adsorbent is removed from the filter, spread over a clean glass surface, air-dried in an area free of fumes and dust, and finally dried to constant weight in a vacuum desiccator over potassium hydroxide pellets. It is important to weigh carefully the amount of DEAE used if reproducible results are to be obtained. Ion exchange celluloses can hold surprisingly large amounts of water or other solvent and appear to be dry, and therefore care must be exerted to obtain a constant weight upon drying.

E. Packing of DEAE Columns

The thoroughly dried DEAE preparation is weighed and then treated with glacial acetic acid overnight. This treatment aids in breaking up aggregates and facilitates packing in a uniform manner into the chromatography tube. Any visible aggregates of DEAE still remaining after overnight treatment with glacial acetic acid are gently broken up with a glass rod prior to transfer of the slurry to the chromatography tube.

A chromatography tube of double thickness Pyrex glass 30 cm in length fitted with a Teflon stopcock and a 1-liter solvent reservoir is used. We have used columns 1, 2.5, and 4.5 cm internal diameter. Columns 2.5 cm. (i.d.) are the most generally useful. Columns 1 cm in diameter are used for relatively small-scale chromatography, and columns 4.5 cm in diameter are especially useful for isolation of relatively large amounts of lipids. The amount of DEAE required for packing columns of various sizes is readily determined by comparing the ratios of surface areas. We use 2.4, 15, and 50 g of the regular grade Selectacel DEAE for 1, 2.5, and 4.5 cm (i.d.) columns, respectively.

A clean glass wool plug is placed in the chromatography tube. The plug is held in place by a glass rod and a portion of the slurry of DEAE in glacial acetic acid is passed into the chromatography tube. After the first addition of DEAE, the rod is withdrawn and packing is continued. Approximately 5 equal portions can be packed to give a satisfactory column. After each addition of DEAE, the excess acid is forced out under nitrogen pressure, the DEAE bed is pressed lightly and uniformly with a large-bore glass rod, the uppermost portion of the bed after application of pressure is then gently stirred to free the bed of very tightly packed adsorbent, and the next addition of slurry is made. The procedure is repeated until all the slurry has been transferred quantitatively to the chromatography tube. At this stage the bed height may be 18–22 cm. The bed should not be allowed to run dry at any stage of column preparation. The glacial acetic acid is allowed to run just to the top of the column, and excess acid is removed by washing the column carefully with 3–5 bed volumes of methanol. Methanol is then replaced stepwise by washing through the column 3 bed volumes of chloroform/methanol 1:1 followed by 3–5 bed volumes of chloroform. The column thus prepared is ready for testing to determine evenness of packing.

F. Testing of DEAE Columns

Each newly packed DEAE column should be tested to determine that it will give satisfactory performance. If a column is packed in an uneven manner, some portions of the bed will not be in equilibrium with other portions, and overlap of fractions will result. Improper solvent changes or allowing a column to run dry may create numerous small channels (not visible to the eye) through which lipid may pass very rapidly causing extensive fraction overlap. These defects can be disclosed by a simple test with a lipid such as cholesterol. Cholesterol should be freshly recrystallized from ethanol to avoid the presence of oxidation products that may be retained by the column.

Testing of a 2.5 × 20 cm DEAE column is accomplished as follows. A solution of 10–30 mg of cholesterol in 5–10 ml of chloroform is applied to the column. Fractions of 10 ml volume are collected beginning at the point of application of the sample. Each fraction is tested for cholesterol either by evaporation of a 1-ml aliquot and observation of solids or by a color test such as the Liebermann-Burchard reaction. Columns are judged to be completely satisfactory when cholesterol appears in tube 8. If the DEAE has been packed more tightly, e.g., to 18 rather than 20 cm, cholesterol may appear in tube 7 whereas if the packing has been somewhat looser, e.g., to a height of 22 cm, the first appearance of cholesterol in a fully satisfactory column may be in tube 9. If the appearance of cholesterol is earlier than tube 6, the column should be repacked. This simple test should be carried out just prior to application of the sample. If the test shows the column to be satisfactory, cholesterol is eluted quantitatively and quickly with 3 bed volumes of chloroform, after which the sample can be applied. If the column must be repacked, the DEAE is extruded and repacked as a slurry in chloroform. Both testing of the column and elution of remaining cholesterol should be carried out with a 2.5 cm column at a flow rate of 2.5–3.0 ml/min. The flow rate for the 4.5-cm column is 10 ml/min. One column volume is equal to 70 ml and 230 ml for columns 2.5 and 4.5 cm in diameter, respectively.

G. Sample Sizes for DEAE Columns

The recommended loads for mitochondrial lipids are 200 mg for 2.5 × 20 cm columns and 500 mg for 4.5 × 20 cm columns. It is preferable to first free the lipid extract of protein and water-soluble nonlipid by Sephadex column chromatography (Section I,E,2), although we have chromatographed total mitochondrial lipids that contained sizable amounts of sucrose (30% by weight) directly by DEAE column chromatography and obtained sucrose quantitatively by elution with methanol. Care must be taken not to overload DEAE columns. The optimum load for each column is determined empirically for each mixture of lipids. Once the characteristics of a particular column for one type of sample have been determined, very reproducible results are obtained and it is necessary only to collect bulk fractions.

H. Elution Sequence

The recommended eluting solvents for mitochondrial lipids from DEAE columns are shown in the table. The volumes of all eluting solvents except chloroform/methanol 9:1 can be increased without fraction overlap since elution is generally an all-or-none phenomenon; i.e., regardless of the volume of solvent used, the next fraction is not eluted. The exception is the elution of phosphatidylethanolamine with chloroform/methanol 9:1. The point at which phosphatidylethanolamine appears in the column effluent with this solvent is determined by the load of total lipid and more particularly by the amount of phosphatidylethanolamine. The greater the amount of phosphatidylethanolamine, the more rapidly this lipid appears in the column effluent. At very high loads, phosphatidylethanolamine may not be separated from phosphatidylcholine, whereas at suitably low loads separation is wide and easily obtained. If phosphatidylethanolamine is found in the chloroform/methanol 9:1 fraction, the load of lipid should be reduced and the exact eluting volume for a particular column determined first by collecting small (e.g., 10 ml) fractions in order to determine the correct bulk volume for the chloroform/methanol 9:1 fraction. The appearance of phosphatidylethanolamine is detected by reaction with ninhydrin. For the test 0.1 ml of column effluent is mixed with 0.1 ml of 0.1% ninhydrin in 1-butanol and 0.1 ml of lutidine or pyridine and heated to boiling. The appearance of a purple color is a positive ninhydrin test and demonstrates the presence of phosphatidylethanolamine in the eluate. The elution mixture is then changed to chloroform/methanol 7:3 for more efficient elution of phosphatidylethanolamine. Salts and sucrose, if not previously removed, are eluted with methanol.

The lipids remaining on the column after elution with methanol are largely acidic lipids, although chloroform/acetic acid 3:1 elutes a form of phosphatidylethanolamine. This form is indistinguishable from phosphatidylethanolamine eluted with chloroform/methanol 7:3 except by DEAE column chromatography. Free fatty acids if present are also eluted in chloroform/acetic acid 3:1. If very large amounts of fatty acids, i.e., 3% or more, are present in a lipid extract (not the case with mitochondrial lipids), free fatty acid will appear in the chloroform/methanol 9:1 eluate.

Variability of elution and fraction overlap are avoided in DEAE column chromatography by careful adherence to the procedure described. Of great importance are selection of the proper grade of ion exchange cellulose, proper washing, proper sample load, and use of pure reagents. Impurities in commercially available ammonia and ammonium acetate can seriously impair column performance and cause variability of elution of diphosphatidylglycerol. Extraneous substances that have a similar effect develop with time in ammonia prepared from gaseous ammonia (Section I,A). It is essential therefore that only fresh reagents be used. Depending upon the presence of other substances and their amounts, diphosphatidylglycerol may be eluted in fractions 8, 9, or 10 of the table. Column performance is determined by thin layer chromatography (Section III).

III. Separation and Quantitative Analysis by Thin Layer[6-9] Chromatography

A. Reagents

Silicic acid (Silica Gel Plain, Warner-Chilcott Laboratories, 200 S. Garrard Blvd. Richmond, California)

Magnesium silicate, synthetic (Allegheny Industrial Chemical Co., P. O. Box 786, Butler, New Jersey)

Chloroform, methanol, glacial acetic acid, and aqueous (28%) ammonia prepared as described in Section I,A

Char spray, potassium dichromate (0.6%) in sulfuric acid (55% by weight)

Perchloric acid, reagent grade, 72%

Ammonium molybdate, reagent grade, 2.5%

Ascorbic acid, 10%

Na_2HPO_4, reagent grade

Rhodamine 6G, stock solution (500 mg dye/500 ml distilled water), stored in a refrigerator

Phospholipid detection spray[10]: Solution I: To 1 liter of 25 *N* H_2SO_4, 40.1 g of MoO_3 is added and the mixture is boiled gently

[7] G. Rouser, C. Galli, E. Lieber, M. L. Blank, and O. S. Privett, *J. Am. Oil Chemists Soc.* **41**, 836 (1964).

[8] G. Rouser, A. N. Siakotos, and S. Fleischer, *Lipids* **1**, 85 (1966).

[9] A. N. Siakotos and G. Rouser, *Anal. Biochem.* **14**, 162 (1966).

[10] J. Dittmer and R. Lester, *J. Lipid Res.* **5**, 126 (1964).

until the MoO_3 is dissolved. Solution II: To 500 ml of Solution I, 1.78 g of powdered molybdenum is added and the mixture is boiled gently for 15 minutes. The solution is cooled and decanted from any residue that may be present. Equal volumes of solutions I and II are mixed and the combined solution is mixed with 2 volumes of water. The final solution is greenish yellow in color. If too little water is used it will be blue, and if too much, yellow. The spray is stable for months.

Ninhydrin reagent, 0.1% in 1-butanol

2,4-Lutidine, redistilled

α-Naphthol (recrystallized from hexane/chloroform) 0.5% solution in methanol/water 1:1

B. General Comments

Thin layer chromatography (TLC) provides a convenient and rapid means for the separation and precise quantitative analysis of phospholipids of mitochondrial preparations. Since most of the phosphorus of very pure mitochondrial preparations is derived from four phospholipids (phosphatidylcholine, phosphatidylethanolamine, diphosphatidylglycerol, and phosphatidylinositol), one-dimensional TLC is adequate, although two-dimensional TLC is required for recognition of trace components (see Figs. 1 and 2) and for impure preparations containing other lipid classes arising from other subcellular particles.

Two procedures, charring and transmission densitometry[6] or aspiration and phosphorus analysis of spots,[8] are available for quantitative analysis of mitochondrial phospholipids after separation by TLC. The phosphorus analysis method is relatively simple and accurate and is recommended as the preferred procedure.

Several adsorbents are commercially available for TLC. Commercially prepared Silica Gel G containing calcium sulfate is not recommended for mitochondrial lipids since diphosphatidylglycerol does not give a compact spot with the adsorbent. Less spreading is obtained with a modified Silica Gel G prepared by mixing Silica Gel plain (without binder) with 10% by weight reagent grade calcium sulfate. A better adsorbent giving very compact spots of acidic lipids is prepared from Silica Gel plain and magnesium silicate.[7] Adsorbosil 3 (Applied Science Laboratories, State College, Pennsylvania) is a useful commercially available mixture of silicic acid and magnesium silicate.

C. Preparation of Adsorbent

Silicic acid (180 g) is mixed with magnesium silicate (20 g). Porcelain grinding balls ½ inch in diameter (approximately 1 pound) are added and the mixture is heated at 150° for 6 hours. The mixture is allowed to cool in the absence of air and then is rotated on a ball mill for 2 hours.

D. Spreading of Plates

A slurry of 20 g of adsorbent in 60 to 70 ml of water is used to spread five 8 × 8-inch glass plates with a Desaga fixed distance (0.25 mm) spreader. Plates are air dried, heat activated at 120° for 30 minutes, and cooled for 20 minutes just prior to application of samples.

E. Sample Application

Mitochondrial lipids dissolved in chloroform/methanol 2:1 (5–10 mg/ml) are applied over a 1-cm line with a 50 μl Hamilton syringe. For two-dimensional chromatography the sample is applied to the lower right portion of the plate about 1 inch from each edge. A line is scraped with a spatula about ½ inch from the top of the plate as a limiting height for each solvent front. For one-dimensional TLC, spots of 100–200 μg are satisfactory. For two-dimensional TLC, 200–600 μg of lipid is spotted.

F. Development of Chromatograms

Immediately after spotting, plates are placed in chromatography chambers 10¾ × 2¾ × 10½ inches ($l \times w \times h$) lined on all sides with solvent-saturated Whatman 3 MM filter paper. About 200 ml of the desired solvent mixture is placed in each chamber, and the paper liners are saturated about 30 minutes before insertion of the plates by tilting the chamber first to one side and then the other.

G. Solvent Systems for Thin Layer Chromatography

1. *One-Dimensional Systems*
   (a) chloroform/methanol/water 65:25:4
   (b) chloroform/acetone/methanol/acetic acid/water 5:2:1:1:0.5
2. *Two-Dimensional Systems*
   System A:
   (1) chloroform/methanol/water 65:25:4
   (2) 1-butanol/acetic acid/water 60:20:20
   System B:
   (1) chloroform/methanol/28% aqueous ammonia 65:35:5
   (2) chloroform/acetone/methanol/acetic acid/water 5:2:1:1:0.5

All solvent ratios are volume:volume. Acidic solvent mixtures are always used for the second dimension in two-dimensional chromatography since some decomposition of lipids usually takes place during drying after development with acidic solvents. Chromatograms are air-dried for 10 minutes between development in the first and second dimensions when the humidity is below 35%, or for 10 minutes in chromatography chambers flushed with dry nitrogen at higher humidity values.

H. Detection Reagents for Lipids after Thin Layer Chromatography

1. *General Detection Reagents.* a. SULFURIC ACID-POTASSIUM DICHROMATE CHARRING. The plates are sprayed lightly with reagent, then heated for 20–30 minutes at 180° in a forced-draft oven. The lipids appear as brown to black spots on a white background.

b. IODINE VAPOR. Iodine crystals are placed in a glass jar, such as a chromatography chamber of the type used for TLC. The TLC plate is placed in the chamber saturated with iodine vapor and after several minutes brown spots appear on a white background. The spots fade rapidly but can be maintained by placing a clean glass plate over the stained chromatogram.

c. RHODAMINE 6 G. For TLC plates, 1 ml of stock solution is mixed with 49 ml of 2 *N* KOH just before use and sprayed evenly to wet the adsorbent layer completely. The chromatogram is observed while wet under shortwave ultraviolet light (Mineralite Model SL 2537, Ultraviolet Products Inc., 5114 Walnut Grove, San Gabriel, California).

2. *Specific Spray Reagents.* a. PHOSPHORUS DETECTION SPRAY FOR PHOSPHOLIPIDS.[8] Plates are sprayed lightly until the adsorbent is uniformly wet. Phosphorus-containing lipids show up immediately as blue spots on a white or light blue background. The intensity of the color may increase on standing. With time the background darkens to a deep blue and the spots are obscured.

b. NINHYDRIN SPRAY FOR AMINO GROUPS. Three parts of ninhydrin solution in butanol are mixed with one part of lutidine or pyridine immediately before use. The chromatogram is sprayed until uniformly wet and heated at 100–120° for several minutes. The progress of color development should be observed at intervals to avoid overheating and loss of color. Purple spots on a white background are obtained. Best results are obtained when chromatograms are developed with acidic solvents and sprayed before all the acid has evaporated from the adsorbent.

c. α-NAPHTHOL SPRAY FOR SUGAR-CONTAINING LIPIDS (GLYCOLIPIDS).[5] This reagent is sprayed as a fine mist until the chromatogram is damp. The plate is air dried and sprayed very lightly with a fine spray of 99% (by weight) sulfuric acid. Color development is achieved by either of two procedures. Heating the chromatogram at 120° until maximum color development takes place results in production of bluish purple spots for glycolipids and yellow spots for other polar lipid classes (cholesterol gives a gray-red spot). If the chromatogram is covered with a glass plate to prevent drying and is heated at 120°, the glycolipids give somewhat stronger bluish purple spots and other lipids give little or no color. Sucrose reacts and will give a spot if not removed prior to TLC.

I. Determination of Phospholipids by Phosphorus Analysis[8]

Spots are detected by spraying with the potassium dichromate–sulfuric acid reagent, followed by heating for 30 minutes at 180° in a forced-draft oven. After development, spots are circled and lettered for identification and several blank areas corresponding in size to the sample spots are marked off. The chromatogram is photographed (Polaroid camera), and the spots are recovered by aspiration. Typical two-dimensional chromatograms are shown in Figs. 1 and 2.

Aspiration of the spots directly into 30-ml Kjeldahl digestion flasks is accomplished by fitting a rubber stopper with two plastic tubes (polyethylene) such as those used in plastic wash bottles. One tube with a pointed end serves as the intake and the other tube is attached to a water pump for suction. Adsorbent is prevented from passing out of the digestion flask during aspiration by adding 0.9 ml of 72% perchloric acid (used subsequently for digestion) to the flask to act as a liquid trap by moistening the lower bulb portion of the flask and by insertion of a 1-cm square of Kimwipe or similar light weight paper into the end of the suction tube to serve as a filter. After aspiration, the plastic tubes are tapped or flushed with nitrogen to remove any dry powder and the paper filter is pushed into the flask with a wire plunger.

Digestion of the flask contents is carried out on an electrically heated Kjeldahl rack with water aspirator suction to remove any escaping fumes. The heaters are adjusted to give gentle refluxing (heavy white vapors appear in the flask) so that digestion is complete in about 20–30 minutes.

After digestion, the sides of the flask are rinsed with 5 ml of distilled water; 1 ml of 2.5% ammonium molybdate solution is added, the flask is swirled for mixing, 1 ml of 10% ascorbic acid solution is added, and finally 2 ml of distilled water is added. The solution is transferred to a centrifuge tube, heated in a boiling water bath for 5 minutes, and cooled; suspended adsorbent is removed by centrifugation for 5–10 minutes. Samples and blanks are transferred to cuvettes and the optical density is determined at 820 mμ after zero adjustment with water. Sensitivity can be increased by using a 10-ml digestion flask and one-half of the specified amounts of reagents. Glassware should be acid cleaned. The reagent blank should be maintained below O.D. 0.010 for best quantitation.

Corrected optical densities are determined by subtraction of the reading obtained from a blank area corresponding in size to that of the sample. The values are then converted to μg of phosphorus using a factor derived from a standard curve prepared with Na_2HPO_4. The factor in our laboratories is 11.0 for standard amounts and 5.50 for half amounts of reagents. Molar ratios of phospholipids are obtained by expression of results as percentage of the total phosphorus. Determination of total phosphorus is conveniently accomplished by spotting 50–100 μg of sample in a blank area (upper right corner) after development with both solvents. The total sample is then charred, etc., along with the spots on the chromatogram.

The phosphorus analysis procedure must be used cautiously with lipid extracts from organs that have not been investigated by other procedures since some spots may represent more than one lipid class and new lipids may be encountered. Two chromatograms of brain lipids (Figs. 3 and 4) developed with solvent systems A and B are included for the purpose of indicating the positions of lipids not found in pure mitochondria. Large errors may be introduced if, prior to spotting, lipid samples are not handled with care to prevent oxidation and/or hydrolysis producing artifacts that may migrate to different positions.

J. Isolation of Phospholipids by TLC and Sephadex Column Chromatography

Isolation of small amounts of phospholipids using TLC for separation is accomplished by applying 8 spots of 100 μg each of mitochondrial lipid to one plate followed by development with chloroform/methanol/water 65:25:4. The sample can be applied to the origin as a continuous streak using the Radin-Pelick sample streaker (Applied Science Laboratories, State College, Pennsylvania). Immediately after the developing solvent has evaporated from a plate (2–3 minutes), the plate is sprayed with water as a detection reagent (white spots on a semitranslucent background), the areas are marked off and scraped (while wet) from the plate with a razor blade into glass-stoppered flasks. Lipids are eluted from the moist adsorbent from each area of one plate by shaking vigorously with 10 ml of chloroform/methanol 2:1 saturated with water followed by filtration (sintered-glass filter, medium porosity). After filtration, the adsorbent is again treated with 10 ml of the same solvent mixture. The combined filtrates are evaporated in the cold by means of a rotary evaporator (Section I,D,1) flushed initially with pure nitrogen. The lipid thus obtained may be contaminated with adsorbent and salts. These are removed by passing through a 1 cm (i.d.) $\times$ 10 cm Sephadex column as described in Section I,E,2 for removal of salts from lipid fractions.

IV. Identification of Lipids

Elution from DEAE columns (Section II) in the characteristic manner, migration to the proper positions by two-dimensional thin layer chromatography (Section III), and reactivity to specific spray reagents provided the basis for characterization of mitochondrial phospholipids. Chromatographic identifications can be confirmed with lipids isolated by DEAE column or thin layer chromatography by pressing with potassium bromide into a pellet for infrared examination[4] or by hydrolysis with acid or base followed by paper and thin layer chromatography for separation of hydrolysis products.

[70] Removal and Binding of Polar Lipids in Mitochondria and Other Membrane Systems

By SIDNEY FLEISCHER and BECCA FLEISCHER

Procedures for the removal of phospholipids from mitochondria[1-4] were developed as a result of a desire to understand the contribution of such lipids to structure and function. Methods had to be developed both for the gentle removal of phospholipids from particles and for reinsertion of the lipid. In this way the following two criteria for lipid involvement could be satisfied: (1) removal of lipid leads to loss of enzymatic function; and (2) rebinding of lipid can be correlated with restoration of enzymatic activity. The methods to be described have been worked out

[1] Mitochondria from beef heart contain 25% lipid by weight of which greater than 90% is phospholipid.[2,3] The phospholipids are diphosphatidylglycerol (cardiolipin), phosphatidylcholine, and phosphatidylethanolamine in the approximate molar ratio of 1:4:4. In addition, phosphatidylinositol, a minor component, accounts for about 3% of the phospholipid.[2,3] The formulas of these phospholipids as well as their isolation and characterization are given elsewhere in this volume.[4] The term phosphatid*yl* is not used as differentiated from phosphatid*al* but is meant to include phospholipids containing vinyl ethers as well as diacyl fatty acids.

The lipid content of beef heart, kidney and liver mitochondria varies, paralleling approximately the concentration of the components of the electron transfer chain (0.32, 0.24, and 0.18 mg lipid per milligram of protein, respectively). The composition of the lipid from these mitochondria is the same in that most of the lipid (>90%) is phospholipid; the phospholipid composition is both qualitatively and quantitatively very similar.[3]

[2] S. Fleischer, H. Klouwen, and G. Brierley, *J. Biol. Chem.* **236**, 2936 (1961).

[3] S. Fleischer, G. Rouser, B. Fleischer, A. Casu, and G. Kritchevsky, *J. Lipid Res.* in press.

[4] G. Rouser and S. Fleischer, this volume [69].

specifically for beef heart mitochondria. A number of mitochondrial enzymes related to the electron transport process have been shown to have a requirement for lipid for enzymatic activity by the above criteria (cf. Table I). It would thus appear that for these enzyme systems the functional enzyme is a complex of lipid and protein, i.e., a lipoprotein.[5,6]

The principles of the methods to be described are applicable to other membrane systems though each system may require some change of procedure. Details may have to be varied to obtain optimal removal of lipid. In addition, conditions for optimal stability and other properties of the enzymes have to be taken into account in the methodology for removal and reinsertion of the lipid.

The methodology is divided into four categories: (I) methods for removal of lipid; (II) methods for preparing microdispersions of polar lipids; (III) assay of enzymatic activity in the presence of added phospholipid; and (IV) methodology for rebinding phospholipid.

I. Removal of Lipid from Mitochondria

To evaluate whether an enzyme requires lipid for function, the preparation is treated to reduce its lipid content. The preparation which is deficient in lipid thus serves as the test system to which phospholipid can be added and reactivation of enzyme activity is measured. Sufficiently mild conditions have to be used for removal of the lipid so as to allow restoration of enzymatic activity with the readdition and rebinding of the lipid. Test systems for demonstrating a lipid requirement in a number of enzymatic activities of beef heart mitochondria are summarized in Table I.

The first successful test system for demonstrating a requirement of phospholipid[7] for succinate–cytochrome *c* reductase activity was obtained by extracting mitochondria with 10% water in acetone (procedure I,A).[8,9] This treatment extracts both neutral lipid as well as phospholipid (better than 80% of the lipid). Thus, coenzyme Q_{10}, the neutral lipid oxidation–reduction component which mediates electron transport be-

[5] S. Fleischer, *Abstracts, 6th Intern. Congr. Biochem. New York, 1964,* VIII-S2.

[6] D. E. Green and S. Fleischer, *in* "Horizons in Biochemistry" (M. Kasha and B. Pullman, eds.). Academic Press, New York, 1962.

[7] S. Fleischer, G. Brierley, H. Klouwen, and D. B. Slautterback, *J. Biol. Chem.* **237,** 3264 (1962).

[8] R. L. Lester and S. Fleischer, *Biochim. Biophys. Acta* **47,** 358 (1961).

[9] This procedure has also been used to demonstrate a dependency of phospholipid for the three segments of the electron transfer chain which together constitute succinate oxidation: i.e., succinate → coenzyme Q; reduced coenzyme Q → cytochrome *c*; and reduced cytochrome *c* → oxygen. See G. P. Brierley, A. Merola, and S. Fleischer, *Biochim. Biophys. Acta* **64,** 218 (1962).

TABLE I

TEST SYSTEMS FOR DEMONSTRATING A LIPID REQUIREMENT FOR ENZYMATIC ACTIVITY OF BEEF HEART MITOCHONDRIA

Section no.	Type of preparation (extraction procedure)	Lipid extracted: Neutral lipid	Lipid extracted: Phospholipid	Requirement	Enzyme system	References
I	Neutral lipid depleted[a] (4% water in acetone)	+	−	CoQ	Succinate → cyt *c* reductase	d,e
I, A	Lipid deficient (10% water in acetone)	+	+ (>80%)	Phospholipid[b]	Succinate → cyt *c* reductase	d,f
I, B	Lipid depleted (10% water + ammonia in acetone)	+	+ (>95%)	Phospholipid[b]	Succinate → cyt *c* reductase	g
I, C	Lipid dependent (phospholipase A + washing with serum albumin)	−	+ (to >80%)	Phospholipid	a) DPNH → cyt *c* reductase	h,i
				Phospholipid	b) Succinate → cyt *c* reductase	h,i
				Phospholipid	c) Mg^{++} stimulated ATPase	j,k,l
				Lecithin	d) β-OH butyric dehydrogenase[c]	m

[a] Described by Lester (this volume [50]).
[b] Also requires CoQ for enzymatic activity.
[c] The activity which is released into the supernatant is devoid of electron transfer activity.
[d] R. L. Lester and S. Fleischer, *Biochim. Biophys. Acta* **47,** 358 (1961).
[e] R. L. Lester and S. Fleischer, *Arch. Biochem. Biophys.* **80,** 470 (1959).
[f] S. Fleischer, G. Brierley, H. Klouwen, and D. B. Slautterback, *J. Biol. Chem.* **237,** 3264 (1962).
[g] S. Fleischer and D. B. Slautterback, unpublished observations.
[h] S. Fleischer, A. Casu, and B. Fleischer, *Federation Proc.* **23,** 2305 (1964).
[i] S. Fleischer, A. Casu, and B. Fleischer, unpublished observations.
[j] A. Casu, B. Fleischer, and S. Fleischer, *Federation Proc.* **25,** 413 (1966).
[k] A. Casu, B. Fleischer, and S. Fleischer, unpublished observations.
[l] The reader is referred to the paper by Y. Kagawa and E. Racker [*J. Biol. Chem.* **241,** 2467 (1966)], who have reported a phospholipid stimulation of ATPase and are of the opinion that phospholipid serves to tie up an ATPase protein inhibitor in CF_0, rather than being directly required for ATPase activity.
[m] B. Fleischer, A. Casu, and S. Fleischer, *Biochem. Biophys. Res. Commun.* **24,** 189 (1966).

tween succinate and cytochrome *c*, must also be added to the test system before activity can be restored.[10,11]

Mitochondria can be rendered lipid free (>95% of the phospholipid extracted) by treatment with 10% water in acetone supplemented with ammonia. The ammonia is effective in extracting the residual phospholipid cardiolipin (I,B).[12,13]

Some enzymes do not survive extraction with solvents. An alternate procedure involves treatment of mitochondria with phospholipase A, followed by washing with bovine serum albumin (I,C).[14–17] Phospholipase A degrades the mitochondrial bound phospholipid:

$$(\text{phospholipid})_{\text{mitochondria}} \xrightarrow[\text{Ca}^{++}]{\text{phospholipase A}} (\text{lysophosphatide} + \text{fatty acid})_{\text{mitochondria}} \quad (1)$$

The by-products of the reaction, i.e. the lysophosphatides and the fatty acids, are bound to the mitochondrion and may then be removed by washing the preparation with serum albumin:

$$(\text{lysophosphatide} + \text{fatty acid})_{\text{mitochondria}} \xrightarrow[\text{albumin}]{\text{serum}} (\text{lysophosphatide} + \text{fatty acid})_{\text{serum albumin}} + (\text{lipid deficient})_{\text{mitochondria}} \quad (2)$$

It is with the use of this procedure that a requirement for phospholipid in DPNH oxidation,[14,15] and Mg^{++} stimulated ATPase activity[16,17] could be demonstrated, and the requirement of phospholipid for succinate oxidation could be confirmed.[14,15] By varying the time of incubation with phospholipase A, the amount of phospholipid which is to be removed can be controlled. In this manner a phospholipid requirement is first manifest for DPNH oxidation, then for Mg^{++}-stimulated ATPase activity, and finally for succinate oxidation; the latter requirement begins to appear after about two-thirds of the phospholipid has been extracted.

Treatment of mitochondria with phospholipase A results in the release of some enzymes into the supernatant.[18,19] In this manner the

[10] R. L. Lester, this volume [50].
[11] R. L. Lester and S. Fleischer, *Arch. Biochem. Biophys.* **80**, 470 (1959).
[12] S. Fleischer and D. B. Slautterback (submitted to *Lipids*).
[13] S. Fleischer, B. Fleischer, and W. Stoeckenius, *J. Cell Biol.* **32**, 193 (1967).
[14] S. Fleischer, A. Casu, and B. Fleischer, *Federation Proc.* **23**, 2305 (1964).
[15] S. Fleischer, A. Casu, and B. Fleischer, unpublished observations.
[16] A. Casu, B. Fleischer, and S. Fleischer, *Federation Proc.* **25**, 413 (1966).
[17] A. Casu, B. Fleischer, and S. Fleischer, unpublished observations.
[18] R. L. Ringler, S. Minakami, and T. P. Singer, *J. Biol. Chem.* **238**, 801 (1963).
[19] B. Fleischer, A. Casu, and S. Fleischer, *Biochem. Biophys. Res. Commun.* **24**, 189 (1966).

enzyme β-hydroxybutyric dehydrogenase, which has a specific lecithin requirement for enzymatic activity[20] can be separated free from the readily sedimentable electron transfer activity.[19] The released enzyme has a specific requirement for lecithin for enzymatic activity.[19]

Oxidative phosphorylation has thus far not been restored to any of the lipid-deficient preparations by the addition of phospholipid.

Reagents and Preparations

Beef heart mitochondria isolated and purified according to Hatefi and Lester[21]

0.88 *M* sucrose–0.01 *M* Tris-chloride pH 7.5

Acetone, reagent grade

Glycyl glycine buffer, 0.05 *M*, pH 7.4

Beef serum albumin, 1%, in 0.05 *M* glycyl glycine buffer, pH 7.4

Calcium chloride, 0.4 *M*

Albumin wash solution: 1% crystalline bovine plasma albumin (Armour Pharmaceutical Co.), in 0.25 *M* sucrose, 1 m*M* EDTA, and 0.05 *M* glycyl glycine, pH 7.4

Sucrose, 0.25 *M*, neutralized to pH 7.4 with a solution of potassium hydroxide

Phospholipase A, prepared from lyophilized *Naja naja* venom, which can be obtained from the Miami Serpentarium Laboratories, Miami, Florida.[22,23] Fifty milligrams of snake venom is dissolved in 10 ml of 0.02 *M* acetate buffer pH 5.0. The mixture is heated 5 minutes in a boiling water-bath, chilled, and centrifuged 10 minutes at 30,000 rpm in a Spinco No. 30 rotor. The supernatant is carefully decanted and diluted with 4 volumes of cold 1% serum albumin in glycyl glycine buffer, pH 7.4. One ml of the "phospholipase A preparation" is therefore equivalent to one mg of venom. The enzyme preparation may be stored in small aliquots in the frozen state and is stable for months.

A. Extraction with 10% Water in Acetone ("Lipid-Deficient" Preparation)[8]

All operations are carried out at 0–4°. Three milliliters of purified beef heart mitochondria (25 mg protein per milliliter in 0.25 *M* sucrose) is added to 72 ml containing 67.5 ml of dry acetone + 4.5 ml of distilled water and mixed in an Erlenmeyer flask. The final concentration is thus

[20] I. Sekuzu, P. Jurtshuk, Jr., and D. E. Green, *J. Biol. Chem.* **238,** 975 (1963).

[21] Y. Hatefi and R. L. Lester, *Biochim. Biophys. Acta* **27,** 83 (1958).

[22] S. W. Edwards and E. G. Ball, *J. Biol. Chem.* **209,** 619 (1954).

[23] C. Klibansky and A. De Vries, *Biochim. Biophys. Acta* **70,** 176 (1963).

10% water in acetone. The mixture is allowed to stand 10–12 minutes with occasional swirling and is then centrifuged at low force (1–2 minutes at top speed in an International clinical centrifuge). The supernatant is decanted, and the pellet is rapidly mixed with 75 ml of 0.88 *M* sucrose—0.01 *M* Tris-chloride, pH 7.5. Usually about 18–20 minutes have elapsed from the beginning of extraction to the addition of the sucrose buffer to the residue. The mixture is then homogenized in a glass homogenizer using a Teflon pestle and centrifuged 5 minutes at 18,000 rpm in a Spinco No. 30 rotor. The wash is repeated once more, and the pellet is finally resuspended in 3.0 ml of the same buffer. The yield of protein is about 50% and the bound phosphorus of the preparation is usually around 3.7 μg per milligram of protein[24] (cf. Table II). This is equivalent to about 15% of the original phospholipid remaining. The residual phospholipid consists mainly of cardiolipin.[7]

TABLE II
PROPERTIES OF SOLVENT-EXTRACTED MITOCHONDRIA

Type	Bound phosphorus (μg P/mg protein)	Residual[b] phospholipid (%)	Succinate → cyt *c* reductase[a]: With CoQ_{10}	Succinate → cyt *c* reductase[a]: With CoQ_{10} + MPL	Succinate → cyt *c* reductase[a]: Reactivation ratio[c]	References
I Neutral lipid depleted	16	~100	1.1	1.2	1.1	[e]
I,A Lipid-deficient (Prep. 1,A)	3.8	16	0.18	1.2	6.7	[f,g]
I,B Lipid-depleted (Prep. 1,B)	2.5	6.9	0.02	0.58	29	[h]
I,B Lipid-depleted[d] (Prep 1,B)	1.9	1.6	0.01	0.22	22	[h]

[a] Micromoles of cytochrome *c* reduced per minute per milligram of protein at 30°C.
[b] The total bound phosphorus of beef heart mitochondria is 18 μg per milligram of protein, of which 13 μg P/mg protein is lipid phosphorus, 3.5 μg P/mg protein is acid extractable, 1.7 μg P/mg protein is extracted neither by acid nor organic solvents.
[c] The reactivation ratio is the ratio of the rates obtained when assayed in the presence and absence of phospholipid, both rates being measured in the presence of added coenzyme Q_{10}.
[d] 0.03 ml of concentrated ammonia per 75 ml of extraction mixture.
[e] R. L. Lester and S. Fleischer, *Arch. Biochem. Biophys.* **80,** 470 (1959).
[f] R. L. Lester and S. Fleischer, *Biochim. Biophys. Acta* **47,** 358 (1961).
[g] S. Fleischer, G. Brierley, H. Klouwen, and D. B. Slautterback, *J. Biol. Chem.* **237,** 3264 (1962).
[h] S. Fleischer and D. B. Slautterback, unpublished observations.

[24] Beef heart mitochondria contain about 18 μg per milligram of protein of which 13 is phospholipid phosphorus, 3.5 is acid extractable, and 1.7 is neither acid nor solvent extractable.[7]

B. Extraction of Mitochondria with 10% Water in Acetone Containing Ammonia ("Lipid-Depleted" Preparation)[12]

Mitochondria may be rendered depleted of lipid, i.e., depleted of greater than 95% of their lipid, by a variation of the acetone extraction procedure outlined in I,A. The procedure is identical to that described in I,A except that 12 μl of concentrated ammonia (28% w/v) is added per 100 ml of water-acetone extraction mixture.

Comments on Procedures A and B

The preparations obtained by procedures I,A and I,B are designed to deplete mitochondria of their lipid content so as to produce a phospholipid requirement for succinate–cytochrome *c* reductase activity (cf. Table II). "Lipid-deficient" preparations have a greater residual phospholipid content as well as a greater residual enzymatic activity (measured in the presence of CoQ_{10}) than does the "lipid-depleted" preparation. On the other hand, the more thorough extraction is at the expense of the maximal activity which can be restored with phospholipid.

It should be stressed that even with the precise conditions described in procedures I,A and I,B, the characteristics of the preparations which are obtained depend in part on the pretreatment of the original mitochondria. To obtain optimal results, i.e., minimal activity of succinate–cytochrome *c* reductase when assayed in the presence of CoQ_{10} alone, and maximal activity when assayed in the presence of phospholipids and CoQ_{10}, it is necessary to conduct pilot runs with each batch of mitochondria with regard to the time of extraction (I,A and I,B) as well as the amount of ammonia to be used in the extraction mixture (I,B). Extraction times of 10–25 minutes have been used. The amount of ammonia to be used depends in part on the freshness of the ammonia (fresh concentrated ammonia is 28% w/v). Additional variables which affect the efficiency of extraction are the concentration of the mitochondria to be extracted, the temperature at which extraction is carried out, as well as the amount of water in the extraction mixture. The efficiency of extraction is increased by lowering the concentration of mitochondria to be extracted and is reduced by lowering the temperature at which the extraction occurs. As the concentration of water in the extraction mixture approaches 4%, the amount of phospholipid which is extracted becomes vanishingly small. The extraction efficiency differs for mitochondria from sources other than beef heart mitochondria. The water content of the acetone as well as the time of extraction should be varied for optimal results with new preparations. The intactness of the preparation also effects its extractability; e.g., some phospholipid is also removed when beef heart mitochondrial vesicles are extracted with 4% water in acetone.[13]

C. Removal of Phospholipid by Digestion with Phospholipase A Combined with a Serum Albumin Wash Procedure ("Phospholipid-Dependent" Preparations)[14, 15]

Beef heart mitochondria, 10 mg protein in 0.4 ml of 0.25 *M* sucrose, are diluted sequentially to a final volume of 1 ml with the following reagents: (a) 0.40 ml of 0.05 *M* glycyl glycine buffer, pH 7.4; (b) 0.01 ml 0.40 *M* $CaCl_2$ and (c) 1.0% serum albumin in 0.05 *M* glycyl glycine and (d) phospholipase A. Additions c and d should add to a combined volume of 0.10 ml. A control sample containing no phospholipase A is always run simultaneously. The reaction mixture is preincubated 1–2

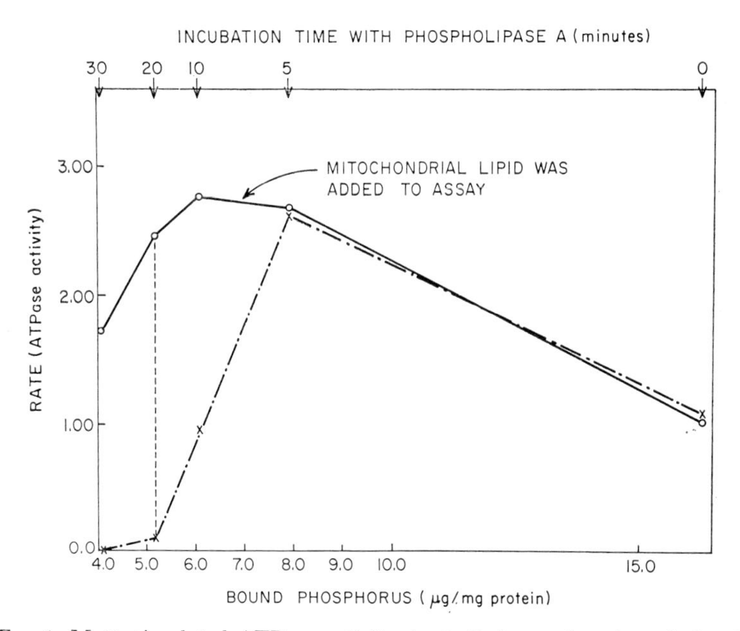

FIG. 1. Mg^{++}-stimulated ATPase activity is studied as a function of the time of incubation with phospholipase A. This type of study enables the selection of an incubation time with optimal phospholipid dependency. A compromise time (vertical dashed line) is selected at which there is close to maximal activation and little residual activity (assayed in the absence of added phospholipid). The conditions are as described in the text (cf. I,C); the preparations were washed with serum albumin four times to remove fatty acids and lysophosphatides which are formed during treatment. The lipid dependency begins when the bound phosphorus is reduced to approximately 8 μg P/mg protein. The solid line and the dot-dashed line refer to activity measured in the presence and absence of added phospholipid. [A. Casu, B. Fleischer, and S. Fleischer, *Federation Proc.* **25,** 413 (1966); also work in preparation].

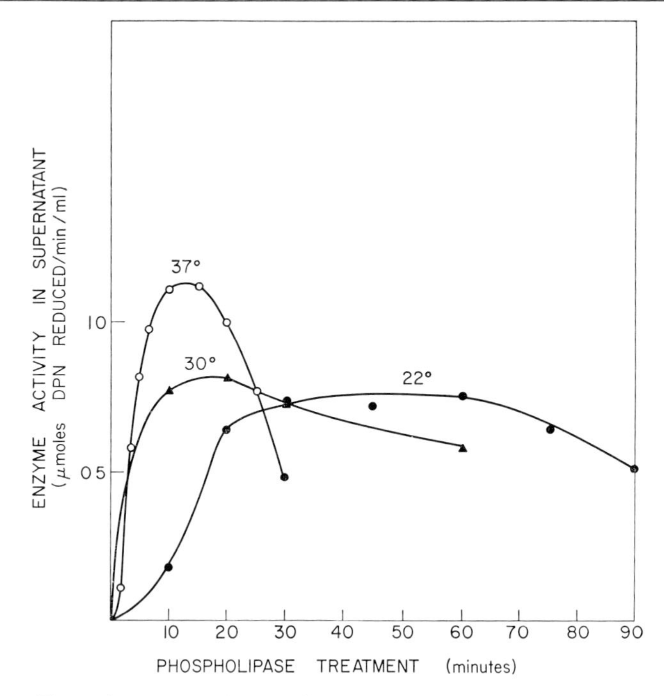

FIG. 2. The release of β-hydroxybutyric apodehydrogenase by phospholipase digestion of mitochondria. Mitochondria, to a final concentration of 10 mg protein per milliliter were incubated with phospholipase A as described in the text. At the end of the incubation time, samples were removed from the mixture, adjusted to 0.009 M with respect to EDTA, and sedimented at 59,000 g for 15 minutes (cf. I,C,d). The supernatants were assayed in the presence of 0.2 mg of total mitochondrial lipid, microdispersed as described for procedure II,A [B. Fleischer, A. Casu, and S. Fleischer, *Biochem. Biophys. Res. Commun.* **24**, 189 (1966)].

minutes at the desired temperature, and the reaction is begun with the addition of the phospholipase A. The time and temperature of the incubation are variables and are determined by the enzymatic activity for which a phospholipid requirement is to be induced (cf. Figs 1 and 2). The reaction is terminated by dilution of the samples with 9 volumes of cold serum albumin wash solution. The mixture is centrifuged at 40,000 rpm for 15 minutes in a No. 40 Spinco rotor. This preparation is referred to as "one time washed." The residue is resuspended in 10 volumes of the same wash solution using a Potter-Elvehjem type glass homogenizer equipped with a Teflon pestle, and the mixture is recentrifuged as before. The wash procedure is repeated a third and a fourth time in order to remove last traces of fatty acids and lysophosphatides which result from the treatment. A final wash with 0.25 M sucrose is desirable to remove serum albumin so that an accurate protein determination for the particle can be obtained. The preparation is resuspended in 0.25 M sucrose. All operations subsequent to the incubation with phospholipase A are carried out at 0–4°.

Variations in the activity of each batch of phospholipase A which is prepared as well as slight differences in the pretreatment of the mitochondria make it necessary to calibrate each enzyme preparation against each preparation of mitochondria. Four micrograms of protein of phospholipase A equivalent to the original venom per milligram of protein of the particle is a good starting ratio. Once calibrated, however, identical conditions effect quite reproducible results. Treatment of mitochondria with phospholipase A for 30 minutes at 37° followed by four serum albumin washes yield a bound phosphorus of approximately 3.8 μg per milligram of protein. This corresponds to 84% depletion of phospholipids.

Procedure for Obtaining Particles with an Optimal Phospholipid Requirement. The response of an enzymatic activity to phospholipase digestion can arbitrarily be divided into three stages. The extent of treatment with phospholipase A to obtain these stages may differ for each enzyme system (cf. Fig. 1):

(1) Short-term stage. Digestion with phospholipase A induces an "opening phenomenon," i.e., the enzymatic activity increases over that observable with the fresh or even the frozen preparation. The magnitude of increase in the latent activity depends in part on the pretreatment of the mitochondria. Is it a fresh preparation? How many times has it been frozen? The increase in activity seems to be due, at least in part, to the elimination of permeability barriers and may be different for each enzymatic activity. A detectable increase in the reactivation ratio (R.R.) may or may not be observed depending upon the particular enzyme in question. R.R. is defined as the ratio of enzymatic activity in the presence and absence of phospholipid in the assay; CoQ should be added to the assays when measuring substrate–cytochrome c reductase activity.

Even small amounts of fatty acids can be quite inhibitory to some enzymes (e.g., DPNH–cytochrome c reductase).[15] It is desirable, therefore, to wash with serum albumin at least one time before assaying the preparation at this stage.

(2) Optimal stage. The enzymatic activity falls sharply with increased phospholipase A digestion, but enzymatic activity is largely reactivable with added phospholipid.

(3) Prolonged stage. When the optimal stage is exceeded, the maximal activity which can be reactivated with added phospholipid is diminished.

In principle the optimal time for treatment of mitochondria with phospholipase A is that point at which there is no enzymatic activity in the treated preparation in the absence of added phospholipid and at which a maximal enzymatic rate ("opened rate") is obtained when assayed in the presence of phospholipid. This type of preparation is more an ideal than a reality. A curve varying the time of phospholipase A digestion as a function of enzyme activity in the presence and absence of added phospholipid is shown in Fig. 1. A compromise time is selected at which there is close to maximal activation and little residual activity (assayed in the absence of added phospholipid). The temperature of incubation as well as the amount of phospholipase A which is used are additional parameters which may be varied.

Comments on Specific Lipid-Requiring Enzymes. In general, to assay enzymatic activity, one wash with the albumin wash solution is necessary. This serves to remove fatty acids which inhibit many enzymatic activities. Multiple washes are required to remove the lysophosphatides quantitatively. The removal of lysophosphatides thus allows a correlation of enzymatic activity with bound phosphorus. Moreover, in the case of succinate–cytochrome *c* reductase, the lysophosphatides can support electron-transport activity and therefore the removal from the preparation is necessary to show a lipid requirement for this activity.

a. DPNH–CYTOCHROME *c* REDUCTASE.[14,15] Fatty acids produced by the phospholipase digestion are strongly inhibitory to this activity. One wash is generally sufficient to remove most of the fatty acids. Multiple washes are required to remove the lysophosphatides. Not all the lysophosphatides need be washed out to obtain an optimal lipid response. It would thus appear that the lysophosphatides do not support this enzymatic activity, nor are they potent inhibitors (cf. Table III). Temperatures of phospholipase treatment of 30° and 37° have been used. Incubation at the lower temperature is perhaps a bit easier to control.

b. SUCCINATE–CYTOCHROME *c* REDUCTASE.[14,15] More thorough digestion conditions (in the range of 30 minutes at 37°) are required to induce a phospholipid requirement for succinate–cytochrome *c* reductase as compared with DPNH–cytochrome *c* reductase. Moreover this activity is not lost until the bound lysophosphatides are removed by multiple washes with serum albumin. Lysophosphatides apparently can support succinate–cytochrome *c* reductase activity. As much as two-thirds of the phospholipid can be removed before a strong reactivation with phospholipid is observable (cf. Table III). This observation could be interpreted in one of two ways: (1) the amount of lipid which is required for this enzymatic activity is very small; and/or (2) the phospholipid associated with this activity is attacked much later than is most of the other phospholipid of the mitochondria.

TABLE III

DPNH → Cytochrome c and Succinate → Cytochrome c Reductase Activity in Lipid-Dependent Preparations[a]

No.	Experiment	Condition	Number of washes	Bound phosphorus[c]	DPNH → cyt c^b		Succinate → cyt c^b	
					CoQ	MPL + CoQ	CoQ	MPL + CoQ
1	Control	10 min at 37°	4	16.5	1.24	1.43	1.26	1.23
1	Phospholipase A	10 min at 37°	1	11.4	0.28	1.43	1.38	1.25
			4	5.7	0.25	1.38	1.13	1.29
2	Phospholipase A	20 min at 30°	4	7.4	0.20	1.73	—	—
3	Phospholipase A	30 min at 37°	1	11.5	0.04	0.35	0.71	1.51
			4	4.2	0.03	0.21	0.31	1.30

[a] Prepared by incubating mitochondria with phospholipase A and washing with serum albumin to remove fatty acids and lysophosphatides (cf. Section I,C) [S. Fleischer, A. Casu, and B. Fleischer, *Federation Proc.* **23,** 2305 (1964); A. Casu, B. Fleischer, and S. Fleischer, *Federation Proc.* **25,** 413 (1966); S. Fleischer, A. Casu, and B. Fleischer, in preparation; A. Casu, B. Fleischer, and S. Fleischer, in preparation].

[b] Micromoles of cytochrome c reduced per minute per milligram of protein at 30°.

[c] Micrograms of phosphorus per milligram of protein.

c. Mg^{++} STIMULATED ATPASE ACTIVITY.[16,17] The study of this activity is complicated because fatty acids stimulate this activity both in the lipid-deficient preparation as well as in untreated mitochondria. However, the reactivation of enzymatic activity with phospholipid, in contrast to that obtained with fatty acids, is sensitive to oligomycin and is not reversed by the addition of serum albumin to the assay. The optimal condition for phospholipase digestion is around 20 minutes at 32° (cf. Fig. 1).

d. β-HYDROXYBUTYRIC DEHYDROGENASE.[19] This enzymatic activity is released into the supernatant upon digestion with phospholipase A. At the end of the incubation time (approximately 12 minutes at 37°), the reaction can be stopped by adjusting the concentration of EDTA to 9 mM. The mixture is sedimented at 59,000 g for 15 minutes. The super-

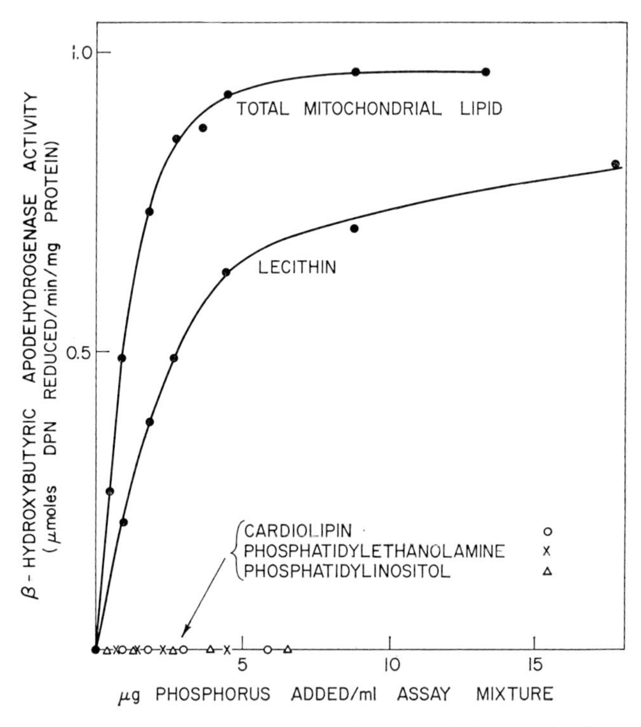

FIG. 3. Activation of β-hydroxybutyric apodehydrogenase by phospholipid micelles. The enzyme was prepared by the treatment of mitochondria with phospholipase A at 37° for 12.5 minutes (as in Fig. 2). The enzyme was partially purified by precipitation with ammonium sulfate and reprecipitated twice in the presence of excess serum albumin as described for procedure I,C,d [B. Fleischer, A. Casu, and S. Fleischer, *Biochem. Biophys. Res. Commun.* **24**, 189 (1966)].

natant contains the enzyme devoid of electron transfer activity. The release of β-hydroxybutyric dehydrogenase activity as a function of time and temperature is shown in Fig. 2. At temperatures lower than 37°, the release of the enzyme is not as efficient. The decrease of activity in the supernatant is probably due to inactivation of the enzyme in the supernatant. Only 50% of the enzyme is released by this treatment, the remainder is not released from the sediment. Both the particle bound and the released enzymes show a complete dependence upon added phospholipid for activity and indeed show the same specificity for lecithin (Fig. 3).

The released enzyme can be partially purified from the supernatant by precipitation with 50% saturated ammonium sulfate. The enzyme in this purified form, as well as in the original supernatant, loses most of its activity when stored in the frozen state for several days. It is considerably stabilized by reprecipitation with 50% saturated ammonium sulfate in the presence of a 1% crystalline bovine serum albumin solution.[19] When reprecipitated twice in this manner, the preparation is devoid of cytochromes, exhibits no malic or isocitric dehydrogenase activity, and no DPNH or succinate–cytochrome *c* reductase activity, and has a specific activity of 1.2 (μmoles DPNH formed/min/mg). The α-ketoglutaric and pyruvic dehydrogenase activities; however, are enriched in such a purified β-hydroxybutyric dehydrogenase preparation.

II. Microdispersions of Phospholipids in Water[25-28]

General Principles. Phospholipids have a negligible solubility in water, yet they can be oriented into macromolecular arrays so that they are practically clear or slightly opalescent in appearance. Aqueous microdispersions of phospholipids may be handled as though they were soluble. The difficulties of working with lipids in aqueous systems are thus overcome.[25]

The ability to form such microdispersions is referable to the amphipathic[29] nature of the phospholipid molecule itself, i.e., there is both a polar and a nonpolar end to the molecule. A thousand or more phospholipid molecules can be arranged into an array such that the polar ends

[25] S. Fleischer and H. Klouwen, *Biochem. Biophys. Res. Commun.* **5**, 378 (1961).
[26] H. Thierfelder and E. Klenk, *in* "Die Chemie der Cerebroside und Phosphatide." Springer, Berlin, 1930.
[27] H. G. Bungenberg de Jong and R. F. Westerkamp, *Biochem. Z.* **234**, 347 (1931).
[28] J. H. Law, H. Zalkin, and T. Kaneshiro, *Biochim. Biophys. Acta* **70**, 143 (1963).
[29] G. S. Hartley, *in* "Progress in the Chemistry of Fats and Other Lipids" (R. T. Holman, W. O. Lindberg, and T. Maldin, eds.), Vol. 3, p. 19. Pergamon, London, 1955.

of the molecules face the water while the hydrophobic fatty acid groups are buried within the interior.[30] Generally phospholipids which form such microdispersions contain unsaturated fatty acids.[25] Some bacterial phospholipids whose fatty acids contain cyclopropane groups instead of double bonds can also readily be microdispersed.[28] It would appear that *cis* double bonds or cyclopropane groups serve to put kinks into the fatty acid moieties of the phospholipid thereby favoring a fluid rather than paracrystalline arrangement. Synthetic phospholipids containing short chain fatty acids are soluble and therefore do not form such arrays.

The microdispersions of phospholipids allow a hydrophobic milieu to exist in an aqueous environment. This serves to explain in part the "solubilization" of neutral lipids in water in the presence of phospholipids.[31] For example, when total mitochondrial lipids[32] are microdispersed in water, the neutral lipids are "solubilized" in the presence of the phospholipids. The addition of neutral lipids to aqueous test systems or the observation of spectral properties of compounds such as CoQ_{10} in water can readily be accomplished after their "solubilization" with phospholipids. The amount of neutral lipid which can be "solubilized" in the presence of phospholipid depends on the particular neutral lipid. Approximately 3% of CoQ_{10} and 20% of cholesterol per weight of phospholipid are "solubilized" in water.[31]

The lipids of the mitochondrion readily form microdispersions in water both in mixtures as well as individually.[25] All but phosphatidylethanolamine form clear or slightly opalescent microdispersions. Phosphatidylethanolamine forms only a highly turbid preparation. Microdispersions have been prepared from the lipids of other membranes as well, from bacterial, plant, and animal sources. We have also formed clear microdispersions of purified polar lipids from nonmitochondrial sources including phosphatidic acid, phosphatidyl glycerol, and mono- and digalactosyl diglyceride.[33]

The microdispersions are prepared in dilute buffer at pH 8 and in the presence of EDTA. A pH of 8 was chosen because ester bonds (of phospholipids) become more labilized at higher pH; at pH's lower than 8.0 the microdispersion becomes more unstable as reflected by increases

[30] S. Fleischer, A. Wasserman, P. Yang, R. M. Bock, and J. Anderegg, unpublished studies.

[31] S. Fleischer and G. Brierley, *Biochem. Biophys. Res. Commun.* **5**, 367 (1961).

[32] Total mitochondrial lipids refers to the lipid extract from mitochondria after the lipid has been freed of nonlipid material. This mixture contains mostly phospholipids but some neutral lipids as well. Mitochondrial phospholipid refers to the phospholipid fraction when the neutral lipids have been separated from the total mitochondrial lipid.

[33] S. Fleischer, unpublished studies.

in light scattering. EDTA, which chelates heavy metals, is used to minimize peroxidation. Microdispersions prepared as described are stable for many months.[25]

The authors favor procedure II,A as the method which yields the most reproducible microdispensions. The other procedures (II,B and II,C) have the advantage that no bile acids have been introduced into the phospholipid and any ambiguity which may be referable to bile acids can thus be eliminated. Moreover the latter procedures are more rapid.

Procedures. Microdispersions of phospholipids in water are spontaneously formed by bringing the phospholipids into intimate contact with the water. This can be achieved by: (1) dissolving them in a solvent which is miscible with water and replacing the solvent by dialysis versus water; and (2) by mechanical means such as by sonic irradiation or passage through a French pressure cell.

Reagents

Potassium cholate, 20%, pH 7.5
1-Butanol
Butanol-cholate mixture. This mixture is prepared by mixing 1-butanol and 20% potassium cholate (pH 7.5) in a volume ratio of 87:13.
Tris-EDTA buffer, 0.02 *M* Tris-acetate, 1 m*M* ethylenediamine tetraacetate (EDTA), pH 8.0
Ethanol, reagent grade
Acetone, reagent grade
Chloroform, reagent grade. Chloroform is redistilled and methanol is added to a final concentration of 0.2% (v/v).
Methanol, reagent grade, redistilled
Purified nitrogen

A. Microdispersion of Phospholipid Dissolved in a Butanol–Cholate Mixture or Organic Solvents by Dialysis versus Aqueous Buffer[25]

The phospholipid dissolved in organic solvent (usually chloroform or chloroform–methanol, 2:1) is evaporated to dryness at reduced pressure with the aid of a flash evaporator.[34] The temperature of the water bath should be no higher than room temperature.

[34] Soybean phosphatides (Asolectin) from Associated Concentrates Inc., Woodside, New York are an inexpensive source of lipids which readily form microdispersions. Asolectin is supplied in the form of granules and therefore does not have to be dried. The Tris-EDTA can be added directly to the Asolectin and the mixture homogenized.

The dried lipid is weighed and dissolved to a final concentration of 25 mg/ml or less in a mixture of 1-butanol and 20% potassium cholate, pH 7.5, mixed in a ratio of 87:13 (v/v).[35] When the lipids are completely dissolved, the mixture is dialyzed at 4° in a rocker dialyzer (Fig. 4)

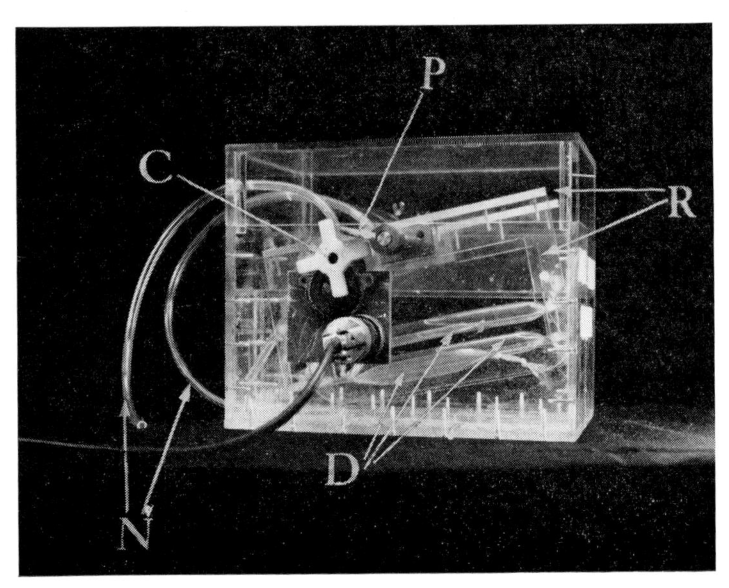

FIG. 4. Rocker dialyzer used in the preparation of aqueous phospholipid microdispersions [S. Fleischer and H. Klouwen, *Biochem. Biophys. Res. Commun.* **5**, 378 (1961) (cf. II,A). The unit shown contains two separate compartments; each contains its own separate "D"-shaped Lucite rack (*R*) with pegs onto which dialysis bags (*D*) are mounted. Enough liquid is added to each compartment to cover the dialysis bags. The system is covered and prepurified nitrogen (*N*) is bubbled through the liquid during dialysis to approximate anaerobic conditions.

Both racks are supported at the top center portion by a single shaft to which they are affixed. Perpendicular to one end of the shaft is a paddle (*P*). The rocking motion of the rack arises from the counterclockwise rotation of a four-toothed cam (*C*) which activates the paddle. The weighted left side of the rack ordinarily tilts downward; however, when the four-toothed cam pushes the paddle upward, the rack tilts the other way. One complete rotation of the four-toothed cam, therefore, causes the rack to rock, back and forth, four times. The cam rotates 75 revolutions per hour. The rocking motion serves to mix the dialyzate as well as the contents within the dialysis bag. The latter requires that an air bubble be trapped within the bag which moves back and forth with the rocking motion.

[35] Solvents that are miscible with water, such as ethanol and acetone, have been used instead of the butanol-cholate mixture. It has been our experience that the butanol-cholate mixture is superior for some lipids. The use of acetone is further limited because phospholipids containing saturated fatty acids are not readily soluble in this solvent.

against at least 100 volumes of Tris-EDTA buffer. The dialysis fluid is changed once each 24 hours and the dialysis continued for a total of 7 days with prepurified nitrogen bubbling through the dialyzate.

The last step is to clarify the microdispersion by centrifugation. The force used varies with the preparation. Sedimentation at 35,000 rpm in the No. 40 Spinco rotor for one-half hour is frequently used; this is equivalent to a force of 80,000 *g* in the middle of the tube. Some preparations, e.g., diphosphatidylglycerol and phosphatidylinositol are practically clear initially and clarification is not necessary. Such preparations can be centrifuged at even higher forces with negligible formation of a sediment. Phosphatidylethanolamine, on the other hand, is highly turbid and cannot be sedimented at this force without loss of most of the lipid. In the study of the rebinding of phosphotidylethanolamine, to particles requiring phospholipid for enzyme activity, the centrifugal forces used to clarify the phospholipid microdispensions are just slightly in excess of that needed to sediment the particles (cf. Procedure IV).

B. Microdispersion by Ultrasonic Irradiation

The lipid in organic solvent is dried in a Potter-Elvehjem type homogenizer by blowing prepurified nitrogen at reduced pressure over the solvent containing the lipid. This is conveniently accomplished in a vacuum desiccator equipped with an opening in the lid to take a rubber stopper. The rubber stopper is fitted with an inlet for nitrogen and an outlet which leads to a vacuum pump. Tris-EDTA buffer is added to obtain a final concentration of 10–25 mg lipid per milliliter. The sample is homogenized in a Potter-Elvehjem type homogenizer equipped with a Teflon pestle. The "lipid homogenate" is then sonic irradiated. We have used both a Raytheon sonic oscillator (10 kc-250 watts, Raytheon Manufacturing Co., Waltham, Massachusetts), and a Branson Probe Sonifier (Branson Instruments, Inc., Stamford, Connecticut).

The Raytheon unit is convenient for sonic irradiation of 30 ml, as maximum cavitation is obtained with this volume. The lipid homogenate is placed directly into the chamber. The chamber is equipped for thermostatting and the temperature is maintained at about 2–4°. Smaller volumes may be sonic irradiated by placing the sample in a stoppered cellulose nitrate tube and bringing the total volume in the chamber to 30 ml. The level of sample in the tube and the liquid in the chamber should be equal. The machine is tuned to a "frying sound." Sonication is continued until maximum clarity is obtained. Five to ten minutes is usually adequate.

In using the probe type sonic irradiator, the probe is inserted directly into the lipid homogenate. The container is placed into an ice bath so that the heat generated during sonic irradiation can be dissipated. The rosette cooling cells which are made available through Branson Instruments offer an efficient means of heat exchange and are particularly useful in maintaining low temperature. Cells of several sizes are available which accommodate volumes ranging from several milliliters to several hundred milliliters. A beaker or plastic tube can serve to contain the sample as well, but care must be exercised in keeping the sample from heating up rapidly. Small samples in a plastic tube are given several short bursts of 10–15 seconds each with additional intermittent cooling.

Several types of probe are available. The "step horn" type allows maximum agitation and may be used with low viscosity materials such as aqueous suspensions of phospholipids. A microtip attachment makes possible the treatment of only 2 ml of sample. The power input is varied to the maximum agitation which does not result in foaming or splashing.

The final step is to clarify the preparation by means of centrifugation (cf. II,A).

C. Microdispersion with the French Pressure Cell[36]

Microdispersions of phospholipids can be simply and rapidly prepared with the use of a French pressure cell. The American Instrument Co., Silver Spring, Maryland, supplies both the French pressure cell as well as a motor-driven laboratory press which is used together with the cell to obtain pressures of up to 20,000 psi. The lipid homogenate is poured into the cell (5–40 ml can be accommodated). The contents of the cell are compressed with a force of 20,000 psi with the use of the motor-driven press. The liquid, under pressure, is then allowed to flow dropwise out of the cell via a small opening. One pass through the cell is sufficient to give a fine microdispersion. Two passes are used as insurance that all liquid passed through dropwise and uniformly. It is a worthwhile precaution to first pass Tris-EDTA through the steel cell as a means of removing metal impurities.

The final step is to clarify the preparation by means of centrifugation (cf. II,A).

D. "Solubilization" of Neutral Lipids with Phospholipids As Mixed Microdispersions[31]

The neutral lipid is mixed together with phospholipid in chloroform-methanol (2:1, v/v). The mixture is evaporated to dryness under nitrogen using reduced pressure. A microdispersion is then prepared as described in either II,A, B, or C. The weight ratio of neutral lipid to

[36] B. Fleischer and S. Fleischer, unpublished studies.

phospholipid which can be solubilized varies with the particular neutral lipid.

III. Assay of Enzymatic Activity in the Presence of Added Lipid

The assay procedures for measuring the effect of lipid by lipid-requiring enzyme preparations (cf. Section I) are described below. Enzymatic activity is assayed in the presence and absence of added lipid.

Phospholipids can be added without difficulty in the form of aqueous microdispersions (cf. Section II).[7,25] It is necessary to incubate the assay mixture for 10 or 15 minutes, depending upon the assay, to allow the lipid to fully interact with the enzyme preparation.

Coenzyme Q_{10} or water-insoluble inhibitors such as rotenone and antimycin are dissolved in ethanol and added to the assay mixture containing the enzyme preparation, i.e., the sequence of addition should be enzyme first and then the sparingly soluble material in ethanol. When both phospholipid and CoQ_{10} are to be added, the sequence of addition should be; the enzyme preparation, then the phospholipid, and lastly, CoQ_{10} "and/or other water insoluble substances." This sequence of addition allows for more efficient adsorption of the water-insoluble material by the enzyme and yields more reproducible results. Small amounts of material in ethanolic solution can be accurately added with the use of a Hamilton syringe. The ethanolic solution should be squirted into the assay mixture under the the surface of the liquid and the mixture should immediately be shaken vigorously. The final concentration of ethanol in the assay mixture can be kept to 0.5–1.0% by volume.

In contrast with the acetone extracted preparations which are devoid of CoQ, preparations which are made deficient in phospholipid with the phospholipase A-serum albumin wash procedure should not require addition of CoQ for electron transport from substrate to cytochrome *c*. In actuality, there is some stimulation of activity with the addition of CoQ (cf. Fig. 5). It is desirable, therefore, to assay enzymatic activity under each of the following conditions (a) no lipid additions; (b) added CoQ; (c) added phospholipid; and (d) added CoQ and phospholipid.

The amount of phospholipid and/or CoQ which is recommended to be added to the assay is more than enough to obtain maximal activity. In the characterization of a new lipid-requiring enzyme preparation, it is desirable to study the enzymatic activity as a function of the amount of phospholipid which is added. Typical plots of this kind are shown in Figs. 3 and 5. There is a rapid initial rise in enzymatic activity with added lipid; when maximal activation is reached, the activation curve forms a plateau.

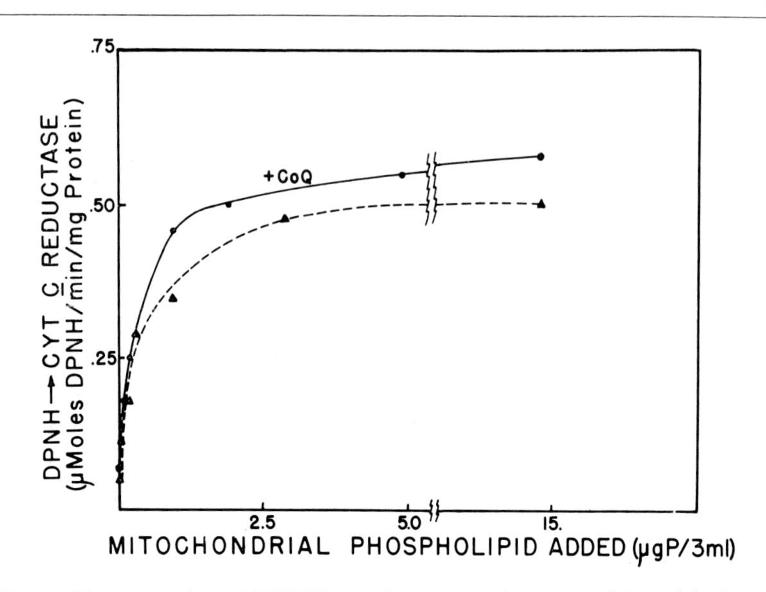

FIG. 5. The activation of DPNH–cytochrome *c* reductase activity with the addition of lipid, in a lipid dependent preparation. The preparation was made deficient in lipid by treatment with phospholipase A followed by a series of serum albumin washes (cf. I,C,a). There is an initial rapid rise in enzymatic activity with increased amounts of added phospholipid. The curve forms a plateau once maximal activity is reached. A similar curve is obtained when both CoQ and phospholipid are added, but the maximal activity is somewhat greater [S. Fleischer, A. Casu, and B. Fleischer, *Federation Proc.* **23**, 2305 (1964); also work in preparation].

Preparations which are made lipid-deficient by solvent extraction exhibit a requirement both for CoQ and phospholipid. The amounts of CoQ and phospholipid which are required for optimal activity are related to one another. In the presence of an excess of phospholipid, less CoQ is required for maximal activity and vice versa (cf. Fig. 4 of Brierley *et al.*[9]).

It is desirable, wherever applicable, to demonstrate that the activity which is restored by the addition of phospholipid is inhibited by the same selective inhibitors which inhibit the unaltered preparations. Indeed, the inhibition curves of the unaltered enzyme and of the lipid restored activity are practically superimposable; that is, succinate–cytochrome *c* reductase is inhibited by antimycin A,[7] DPNH–cytochrome *c* reductase by rotenone, amytal, and antimycin A,[14,15] and Mg^{++}-stimulated ATPase activity by oligomycin,[16,17] at the same levels which inhibit the original native enzymes.

Procedures A and B. Assay of DPNH or Succinate-Cytochrome c Reductase Activity[37,38]

$$\text{DPNH} + 2 \text{ ferricytochrome } c \leftrightarrow \text{DPN}^+ + \text{H}^+ + 2 \text{ ferrocytochrome } c \quad (3)$$

$$\text{Succinate} + 2 \text{ ferricytochrome } c \leftrightarrow \text{fumarate} + 2 \text{ ferrocytochrome } c \quad (4)$$

DPNH or succinate-cytochrome *c* reductase activity is measured spectrophotometrically at 30° by the increase in optical density at 550.5 mμ, in the presence of excess DPNH or succinate. The rate of increase is associated with the rate of reduction of the alpha band of cytochrome *c*.

Reagents

Potassium phosphate buffer, 1.0 *M*, pH 7.4

Ferricytochrome *c*, 1%. Cytochrome *c* (type III) from horse heart is obtained commercially from Sigma Chemical Co., St. Louis, Missouri. It is mostly in the oxidized form. When a lot arrives which is partially reduced, it can readily be converted to the oxidized form by adding a slight excess of potassium ferricyanide. The latter is then separated from the cytochrome *c* by dialysis or by passing the mixture through a G-25 Sephadex column.

Enzyme preparation (cf. procedures I,A, B, and C). The preparation is generally diluted immediately before use to approximately 0.40 mg/ml with a solution of the same composition as that in which it was originally stored.

Phospholipid, aqueous microdispersion, approximately 250–300 μg per milliliter in 0.02 *M* Tris-acetate, 1 m*M* EDTA, pH 8.0 at 4° (cf. Section II).

0.02 *M* Tris-acetate-1 m*M* EDTA, pH 8.0 at 4°

Coenzyme Q_{10}, 0.2%, in ethanol

Succinate-cyanide mixture. An equal volume of 1 *M* potassium succinate, pH 7.4, is mixed with 0.1 *M* potassium cyanide. The potassium cyanide solution is prepared fresh daily.

Sodium azide, 0.10 *M* is prepared fresh daily

DPNH (w/v), 0.2%, neutralized to pH 7.4

Procedure. In a typical protocol the reagents are added sequentially as follows: (1) distilled water to a final volume of ml; (2) 0.06 ml of potassium phosphate buffer; (3) 0.05 ml of ferricytochrome *c*; (4) 0.02

[37] H. Edelhoch, O. Hayaishi, and L. J. Teply, *J. Biol. Chem.* **197,** 97 (1952).

[38] D. E. Green, S. Mii, and P. M. Kohout, *J. Biol. Chem.* **217,** 551 (1955).

ml of enzyme preparation; (5) 10 μl of phospholipid or Tris-EDTA buffer; (6) 5 μl of CoQ or ethanol.

The mixture is incubated for 10 minutes at 30° and the reaction is then initiated with the addition of either: (procedure A) 0.02 ml of succinate-cyanide mixture (for the assay of succinate-cytochrome *c* reductase activity); or (procedure B) 0.02 ml of sodium azide followed by 0.10 ml of DPNH at 30° (for the assay of DPNH-cytochrome *c* reductase activity). The increase in optical density at 550.5 mμ is measured in a recording spectrophotometer. Specific activity is expressed as micromoles of cytochrome *c* reduced per minute per milligram of enzyme preparation. The extinction coefficient for cytochrome *c* is taken as 18.5 mM^{-1} cm^{-1}.

C. Assay of Mg^{++}-Stimulated ATPase Activity[39]

$$ATP + H_2O \xleftrightarrow{Mg^{++}} ADP + H_3PO_4 \tag{5}$$

The hydrolysis of ATP is measured at 30°, in the presence of excess added ATP. The reaction is terminated by the addition of acid and the released inorganic phosphate is determined after it has been partitioned free from the organic phosphate.[40]

Reagents

Tris-acetate buffer, 1.0 *M*, pH 7.4
Magnesium chloride, 0.05 *M*
Phospholipid aqueous microdispersion, approximately 250–300 μg per milliliter in 0.02 *M* Tris-acetate–1 m*M* EDTA, pH 8.0 at 4°
0.02 *M* Tris-acetate-1 m*M* EDTA, pH 8.0 at 4°
Enzyme, approximately 10 mg protein per milliliter (cf. I,C,c)
Bovine plasma albumin, 1%, crystalline Armour Pharmaceutical Co., Kankakee, Illinois
Oligomycin, 0.08% (w/v) in ethanol; obtained from the Wisconsin Alumni Research Foundation, Madison, Wisconsin
ATP, 0.1 *M* pH 7.4
Perchloric acid, 1.5 *M*

Procedure. In a typical protocol the reagents are added sequentially as follows: (1) distilled water to a final volume of 1.5 ml; (2) 0.05 ml of Tris-acetate buffer; (3) 0.05 ml of magnesium chloride; (4) 0.03 ml of phospholipid or Tris-acetate buffer; (5) 0.42 mg protein of the enzyme

[39] H. A. Lardy and H. Wellman, *J. Biol. Chem.* **201**, 357 (1953).

[40] D. Lindberg and L. Ernster, *in* "Methods of Biochemical Analysis" (D. Glick, ed.), Vol. 7, p. 1. Wiley (Interscience), New York.

preparation. The mixture is incubated for 10 minutes at 30°. The reaction is initiated with the addition of 0.05 ml of ATP. The reaction has been standardized for 4 minutes at 30° using 0.42 mg protein of the enzyme preparation. The reaction is terminated with the addition of 0.5 ml perchloric acid. The tubes are centrifuged in the clinical centrifuge for 5 minutes. One milliliter of the supernatant is analyzed for the release of inorganic phosphate.[40] A zero time control is prepared in the same way as that described for the enzyme assay but with the enzyme omitted until after the addition of the perchloric acid. Any inorganic phosphate in the constituents of the assay are subtracted using the zero time control value. Specific activity is expressed as micromoles phosphates released per minute per milligram of enzyme protein. Fatty acids also stimulate Mg^{++}-stimulated ATPase activity.[41,42] However, such activation is reversed by the addition of bovine serum albumin (0.1 ml) and is insensitive to oligomycin[43] (5 μl) inhibition, in contrast with the activation which is restored with phospholipids.

D. Assay of β-Hydroxybutyric Dehydrogenase Activity[44]

$$\beta\text{-Hydroxybutyrate} + \text{DPN}^+ \leftrightarrow \text{acetoacetate} + \text{DPNH} + \text{H}^+ \quad (6)$$

β-Hydroxybutyrate dehydrogenase activity is measured spectrophotometrically at 37° by following the rate of increase in optical density at 340 mμ associated with the reduction of DPN^+ to DPNH in the presence of excess β-hydroxybutyrate.

Reagents

Enzyme cocktail contains; 1.0 ml of 1.0 *M* Tris-HCl, pH 8.0; 0.4 ml of 2% bovine serum albumin; 1.0 ml of 0.01 *M* EDTA, pH 7.4; 2.0 ml of 0.02 *M* DPN (neutralized to pH 7.4); 0.6 ml of 95% ethanol and 1.0 ml 1.0 *M* cysteine·HCl, pH 7–8. The cysteine·HCl is neutralized immediately before beginning the assays and is not used beyond 4 hours. When particles containing electron transfer activity are assayed, 0.6 ml of antimycin (0.7 mg/ml ethanol) is substituted for the 0.6 ml ethanol.

Phospholipid aqueous microdispersion, approximately 250–300 μg per milliliter in 0.02 *M* Tris-acetate–1 m*M* EDTA pH 8.0 at 4°

[41] B. C. Pressman and H. A. Lardy, *J. Biol. Chem.* **197**, 547 (1952).
[42] P. Borst, J. A. Loos, E. J. Christ, and E. C. Slater, *Biochim. Biophys. Acta* **62**, 509 (1962).
[43] H. A. Lardy, D. Johnson, and W. C. McMurray, *Arch. Biochem. Biophys.* **78**, 587 (1958).
[44] P. Jurtshuk, Jr., I. Sekuzu, and D. E. Green, *J. Biol. Chem.* **238**, 3595 (1963).

0.02 M Tris-acetate–1 mM EDTA pH 8.0 at 4°
Sodium β-hydroxybutyrate, 0.2 M, pH 8.0 (DL-β-hydroxybutyrate sodium salt is obtained from Calbiochem, Los Angeles, California)
The enzyme (cf. I,C,d under procedure)

Procedure. In a typical protocol the reagents are added sequentially as follows: (1) 0.30 ml of the enzyme cocktail; (2) distilled water to a final volume of 1.0 ml; (3) 0.02 ml of phospholipid or Tris-EDTA buffer and (4) Enzyme approximately 0.10 mg mitochondria or the singly washed sediment from phospholipase A digestion or 0.025 mg of the purified enzyme (cf. I,C,d).

The mixture is incubated 15 minutes at 37°. The reaction is initiated with the addition of 0.10 ml of β-hydroxybutyrate (at 37°). The increase in optical density at 340 mμ is measured with a recording spectrophotometer. Specific activity is expressed as μmoles DPNH formed per minute per milligram of enzyme protein. The extinction coefficient of DPNH is taken as 6.22 mM^{-1} cm.$^{-1}$

IV. Rebinding of Lipid to Lipid-Deficient Enzyme Preparations[7]

In demonstrating a requirement for lipid in a lipid-deficient preparation, one is plagued with the possibility that loss of activity is not actually due to loss of phospholipid and that reactivation may be unrelated to the readdition of the phospholipid. The early pioneers who used phospholipases to study phospholipid requirements were aware of this problem.[22,45-47] Fatty acids which are released by the action of phospholipase A are potent inhibitors of some enzymes. Loss of activity could be related to the production of such inhibitors or to the loss of structural integrity which attends such treatment. Addition of phospholipid could thus result in the reversal of inhibition. Indeed, microdispersions of phospholipids can be used in the same way as the albumin wash solution (cf. I,C) to remove fatty acids which are released by digestion with phospholipase A.[33]

In developing procedures for the removal of lipid to study lipid requirements, we have attempted to develop procedures which remove lipid without producing breakdown products, or to remove such products if they are formed.

The case for a genuine requirement for phospholipid can be made

[45] A. P. Nygaard and J. B. Sumner, *J. Biol. Chem.* **200,** 723 (1953).
[46] H. L. Tookey and A. K. Balls, *J. Biol. Chem.* **220,** 15 (1956).
[47] E. Petrushka, J. H. Quastel, and P. G. Sholefield, *Can. J. Biochem. Physiol.* **37,** 975 (1959).

stronger with the demonstration that the restoration of enzymatic activity is dependent on *rebinding* of the lipid. Such a correlation has been made for succinate[7] and DPNH–cytochrome *c* reductase,[14,15] and Mg^{++}-stimulated ATPase activity.[16,17]

The study of the rebinding of phospholipid is made possible with the use of aqueous phospholipid microdispersions. The lipid-deficient particle is mixed with phospholipid in aqueous medium. After a brief period of incubation, the particle is sedimented free from the unbound phospholipid; the phospholipid microdispersion behaves in this regard as a water-

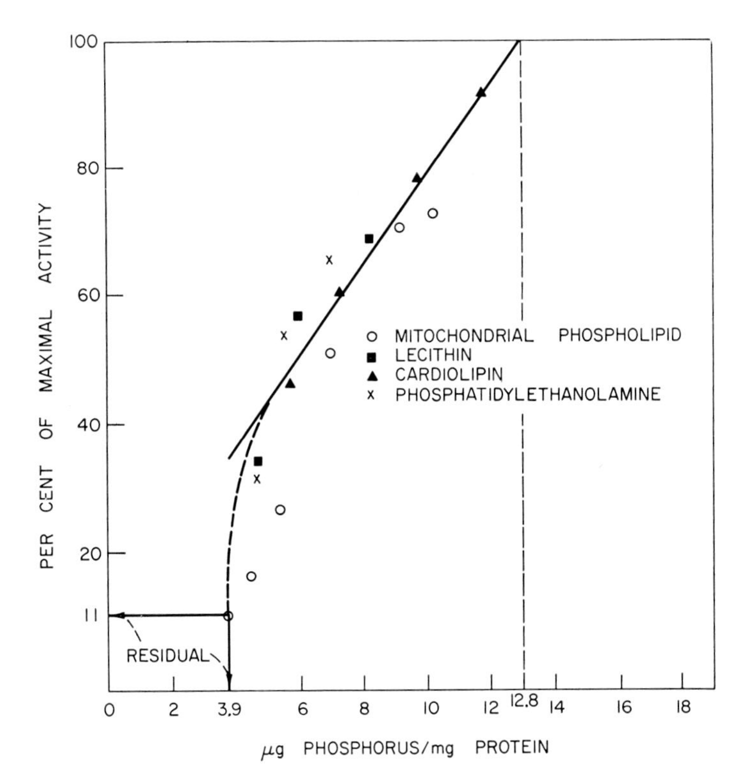

FIG. 6. The reactivation of phospholipid-deficient mitochondria as a function of rebinding of phospholipids (cf. IV) [S. Fleischer, G. Brierley, H. Klouwen, and D. B. Slautterback, *J. Biol. Chem.* **237**, 3264 (1962)]. The preparation was made lipid-deficient by extraction with aqueous acetone (cf. I,A) and lipid was rebound as described (cf. IV). The percentage of maximal activity is plotted as a function of bound phosphorus per milligram of protein. The preparations with phospholipid rebound are assayed in the presence of added CoQ. Maximal activity is defined as the rate of succinate-cytochrome *c* reductase which is obtained in the presence of an excess of added phospholipid.

soluble material. The sediment is washed, and assayed for bound phosphorus and enzymatic activity. The extent of rebinding is concentration dependent. Therefore, by increasing the phospholipid concentration, a series of preparations can be obtained with increasing amounts of rebound phospholipid. A typical plot showing reactivation of enzymatic activity as a function of rebinding of phospholipid is shown in Fig. 6. There is a small residual amount of phospholipid in the solvent-extracted preparations as well as residual succinate–cytochrome *c* reductase activity. Reactivation of enzymatic activity is correlated with the rebinding of phospholipid. Complete reactivation is obtained when the amount of phospholipid which is bound equals approximately that which was present originally. With the phospholipase A–serum albumin wash procedure, two-thirds of the phospholipid can be removed before an appreciable phospholipid requirement is observable (cf. I,C,b).[15] The rebinding process, however, is not so selective. Phospholipid is rebound by all the protein which had lost lipid, both enzymatic protein and structural protein.[48] When the full quota of the phospholipid for the entire preparation is reinserted, the activity is fully restored, i.e., the activity of the preparation is maximal when assayed in the presence or absence of added phospholipid.

Reagents and Preparations

Phospholipid-deficient preparations (cf. I,A; I,B; I,C), 10 mg protein per milliliter
0.88 *M* sucrose–0.01 *M* Tris·HCl buffer, pH 7.5
Sucrose, 0.25 *M*, is neutralized to pH 7.4 with a solution of potassium hydroxide
Phospholipid aqueous microdispersion, approximately 250–300 μg per milliliter in 0.02 *M* Tris-acetate–1 m*M* EDTA, pH 8.0 at 4°
Tris-EDTA buffer; 0.02 *M* Tris-acetate–m*M* EDTA, pH 8.0 at 4°

Procedure for the Rebinding of Phospholipid.[7] One milliliter of enzyme preparation (I,A or I,B), in 0.88 *M* sucrose–0.01 *M* Tris·HCl, pH 7.5, is mixed with 2.0 ml containing 0.88 sucrose–0.01 *M* Tris·HCl, pH 7.5, in a Potter-Elvehjem type glass homogenizer equipped with a Teflon pestle. Three milliliters of Tris-EDTA buffer containing varying amounts of phospholipid are added. The mixture is incubated at 31° for 15 minutes with periodic homogenization by hand. Fourteen milliliters of cold sucrose-Tris buffer are then added and the homogenized mixture is sedimented for 10 minutes at 20,000 rpm [No. 40 Spinco rotor (average

[48] S. Richardson, H. O. Hultin, and S. Fleischer, *Arch. Biochem. Biophys.* **105**, 254 (1964).

force 26,000 g)]. The pellet is homogenized in 10 ml sucrose-Tris buffer, resedimented and resuspended in 1.5 ml of the sucrose-Tris buffer solution.

Preparations described in I,C are in 0.25 M sucrose; the procedure is thus modified slightly. One milliliter of enzyme preparation is mixed with 2.0 ml, containing 0.63 ml of 2 M sucrose and 1.37 ml of water. The remaining steps are the same except that 0.25 M sucrose is used for the washes instead of the sucrose-Tris buffer. Sedimentation for 20 minutes at 20,000 rpm (No. 40 Spinco rotor) is sometimes preferable.

The extent of phospholipid is dependent on the phospholipid to protein ratio in the mixture. For maximal rebinding of phospholipid a large excess of phospholipid, i.e., ratios of the order of 100 μg lipid phosphorus per milligram of protein are used. The amount of protein is not critical, as little as 3.0 mg protein has been used in the mixture. The indication is, however, that recombination is more efficient with higher concentrations of lipid and protein.

[71] Preparation of Mitochondrial Structural Protein

By David W. Allmann, Albert Lauwers,[1] and Giorgio Lenaz

The concept of a structural protein in the mitochondrion arose from the observation that the oxidation-reduction components of the electron transfer chain could account for only about 25% of the total protein.[1a,2] It was also known that during the purification of the electron transport enzymes, the major contaminant was an essentially colorless, insoluble protein,[1a-3] which appeared to be intimately associated with the electron transport components. In 1961 Green and his colleagues[1a,2] isolated from mitochondria a colorless protein which they characterized and named structural protein. This protein represented about 35–45% of the mitochondrial protein.

This structural protein(s) has been prepared by two methods: one method is that described by Criddle *et al.*,[2] and the other method is that

[1] This manuscript was prepared while the author was at the Institute for Enzyme Research, University of Wisconsin, Madison, Wisconsin.

[1a] D. E. Green, H. D. Tisdale, R. S. Criddle, and R. M. Bock, *Biochem. Biophys. Res. Commun.* **5**, 109 (1961).

[2] R. S. Criddle, R. M. Bock, D. E. Green, and H. D. Tisdale, *Biochemistry* **1**, 827 (1962).

[3] P. V. Blair, T. Oda, D. E. Green, and H. Fernández-Morán, *Biochemistry* **2**, 756 (1963).

described by Richardson *et al.*[4] These two preparations, while yielding a white insoluble protein, do not yield proteins of identical properties or purity. The Richardson protein, for example, has the capacity to bind phospholipid at neutral pH whereas the Criddle protein has to be solubilized at high pH before phospholipids are bound.[4] This property of phospholipid binding at neutral pH was thought to be due to the "nativeness" of the Richardson protein. This discrepancy of the phospholipid binding has yet to be resolved. These two preparations of structural protein are not of the same purity as judged by polyacrylamide electrophoresis. The Criddle protein is at least 75–80% pure while the Richardson protein is more heterogeneous.

This protein or class of proteins has now been demonstrated to be present in numerous other subcellular membranes, such as microsomes,[5] erythrocyte stroma,[6] chloroplasts,[5,6] *Neurospora* mitochondria,[7] yeast mitochondria,[7] the inner membrane of beef heart mitochondria,[3,8] and the outer membrane of beef heart mitochondria.[9]

Preparation of Structural Protein

Reagents

Beef heart mitochondria[10]
Potassium cholate, pH 7.8 (twice recrystallized from ethanol), 20%
Potassium deoxycholate, pH 7.8 (twice recrystallized from ethanol), 10%
Ammonium sulfate
KCl, 0.9%
0.25 *M* sucrose–0.01 *M* Tris-acetate, pH 7.8 (sucrose-Tris)
Sodium dodecyl sulfate, 10%
Potassium hydroxide
Sodium dithionite
Butanol
Methanol
Urea, 8 *M* (recrystallized)
Trichloroacetic acid, 0.4% in methanol

[4] S. H. Richardson, H. O. Hultin, and S. Fleischer, *Arch. Biochem. Biophys.* **105,** 254 (1964).
[5] S. H. Richardson, H. O. Hultin, and D. E. Green, *Proc. Natl. Acad. Sci. U.S.* **50,** 821 (1963).
[6] R. S. Criddle and L. Park, *Biochem. Biophys. Res. Commun.* **17,** 74 (1964).
[7] D. O. Woodward and K. D. Munkres, *Proc. Natl. Acad. Sci. U.S.* **55,** 872 (1966).
[8] K. Kopaczyk, J. F. Perdue, and D. E. Green, *Arch. Biochem. Biophys.* **115,** 215 (1966).
[9] A. Lauwers, D. W. Allmann, G. Lenaz, and D. E. Green, unpublished observations.
[10] F. L. Crane, J. L. Glenn, and D. E. Green, *Biochim. Biophys. Acta* **22,** 475 (1956).

Preparation of Structural Protein As Described by Richardson et al.[4] A modification of the method of Richardson *et al.*[4] is described. Beef heart mitochondria prepared by the method of Crane *et al.*[10] (cf. this volume [12]) are suspended in 0.25 *M* sucrose and mixed with one-third volume of 0.9% KCl. The suspension is homogenized and centrifuged at 79,000 *g* for 20 minutes. The red supernatant and the dark pellet at the bottom of the tube are discarded. The mitochondria (freed of soluble protein) are suspended in sucrose-Tris and the protein concentration adjusted to 20 mg/ml. The mitochondrial suspension is frozen for at least 15 hours.

The mitochondria are solubilized by the slow addition of cholate (with stirring) to a concentration of 0.75 mg per milligram of protein, and deoxycholate to a concentration of 1.5 mg per milligram of protein. The solution is centrifuged at 79,000 *g* for 10 minutes, and the clear supernatant containing the cytochromes and structural protein is separated from the dark pellet. Sodium dithionite is added to the clear supernatant to a concentration of 1.2 mg/ml and ammonium sulfate to 0.12 saturation. The sodium dithionite is added to keep the cytochromes in the reduced state, which ensures a more complete separation of structural protein from the cytochromes. The pH is maintained at 7.5 by addition of Tris-chloride buffer. The solution is incubated at room temperature for 90 minutes or at 0–4° for at least 15 hours. The suspension is centrifuged for 20 minutes at 79,000 *g*. The whitish precipitate is homogenized and washed twice with sucrose-Tris.

To remove the remaining bile salts and phospholipid, the structural protein suspension at a protein concentration of 20 mg/ml in sucrose-Tris is extracted with 20 volumes of 90% acetone (final concentration) at 0–4°. After the suspension has stood for 3 minutes, the major portion of the supernatant is decanted and the settled protein is homogenized in a Potter-Elvehjem homogenizer. The decanted supernatant is added back, and the protein is sedimented at low speed centrifugation. The extracted structural protein is washed twice with sucrose-Tris and suspended in sucrose-Tris.

Preparation of Structural Protein As Described by Criddle et al.[2] A modification of the method of Criddle *et al.*[2] is described. Beef heart mitochondria or heavy beef heart mitochondria (cf. this volume [12]) are suspended in 0.25 *M* sucrose at a protein concentration of 20 mg/ml. The mitochondria are solubilized by the slow addition of deoxycholate (with stirring) to a concentration of 2.0 mg per milligram of protein, then of cholate to a concentration of 1.0 mg per milligram of protein, and finally of sodium dodecyl sulfate to a concentration of 0.75 mg per milligram of protein. The solution is centrifuged at 79,000 *g* for

10 minutes. The dark pellet is discarded. The pH is adjusted to pH 9.0 with KOH; then sodium dithionite is added, 1.2 mg/ml, and ammonium sulfate to 0.12 saturation, the pH being maintained at 9.0, to ensure a more complete separation of the structural protein. The copious, very light green protein is sedimented at 79,000 *g* for 10 minutes. The sedimented protein is suspended in 0.25 *M* sucrose. The yield of the insoluble protein is usually 35–45% of the mitochondrial protein.

The bile salts and phospholipid are removed by adding deoxycholate (1 mg per milligram of protein), ammonium sulfate (20% v/v), and butanol (20% v/v) to the structural protein in sucrose-Tris (20 mg/ml). The suspension is stirred and filtered in a Büchner funnel. The protein "cake" is washed with 0.25 *M* sucrose (1–2 times the original volume) in order to remove the ammonium sulfate and butanol. The protein is then washed with about 10 volumes of 75% methanol at 50° and the protein is dried by suction. The dried protein is suspended in water and dialyzed against a large volume of water for about 20 hours at 0–4°.

Purification of the Structural Protein. The structural protein at this stage is about 75–80% pure as judged by polyacrylamide gel electrophoresis following solubilization in phenol-acetic acid-urea as described by Takayama *et al.*[11] (cf. this volume [103]). The impurities can be removed by extraction with urea as follows. The lyophilized structural protein is homogenized in 8 *M* urea pH 5.5 (in order to reduce cyanate formation) at a protein concentration of 5–10 mg/ml. The suspension is stirred for 30 minutes and centrifuged at 100,000 *g* for 15 minutes. The yellow-green, gelatinous pellet is twice reextracted in 8 *M* urea. The urea-soluble fraction contains most of the impurities together with some of the main component. The extracted residue is washed several times with water. The remaining colored contaminants (mostly heme as determined by the spectrum of the pyridine) can be removed with three washes with 0.4% trichloroacetic acid in methanol. The structural protein is then washed with neutral methanol and water. The protein suspension is dialyzed against water and lyophilized. The yield of structural protein at this stage is 60–75% of the material subjected to the purification procedure. The same purification procedure has successfully been applied to the Richardson protein.

Properties of the Structural Protein

The structural protein as prepared by the second procedure outlined above is essentially free of bile salts and lipid phosphorus. It is also free

[11] K. Takayama, D. H. MacLennan, A. Tzagoloff, and C. D. Stoner, *Arch. Biochem. Biophys.* **114,** 223 (1966).

of such oxidation-reduction components as flavin and cytochromes. The structural protein is virtually insoluble at neutral pH, but it can be solubilized in 67% acetic acid, in 0.1% sodium dodecyl sulfate at pH 11.5, or in 8 *M* urea in 0.1 *N* NaOH; it is only slightly soluble at neutral pH in 8 *M* urea, but most of the contaminating proteins can be solubilized. Structural protein can also be rendered water soluble by succinylation with succinic anhydride.[12]

Although the purified structural protein has not been found to have any known enzymatic activity, it has the capacity to combine with water-soluble proteins and certain small molecules. Mitochondrial proteins such as cytochrome c_1, b, and a,[2] cytochrome c,[13] and malic dehydrogenase[7] can form water-soluble complexes with structural protein. The only nonmitochondrial protein that is able to form a water-soluble complex with structural protein is myoglobin.[2,14] Structural protein also binds phospholipid[4] and small molecules, such as adenine nucleotides.[5,7]

The purity of the structural protein has been judged by polyacrylamide disc electrophoresis. The crude Criddle protein as prepared by the modifications described here is about 75–80% pure as judged by electrophoresis whereas the crude Richardson protein is more heterogeneous (about 40–50% pure). Both these mixtures of proteins can, however, be purified as described above, and structural protein is thus obtained in a form that is better than 95% pure. The yield of the Richardson structural protein is, however, much less.

Woodward and Munkres[7] have determined the amino acid composition of beef heart, *Neurospora*, and yeast mitochondrial structural protein. They found very few differences in the composition of these three proteins. There is no cystine in the protein, but the protein contains 4–10 half-cystine residues depending on the source of the mitochondria.

The structural protein is homogeneous in the model E analytical ultracentrifuge and homogeneous according to immunochemical criteria.[7] The protein has a molecular weight of 22,000–23,000 as determined by several independent methods.[2,7] Criddle *et al.* have determined that the N-terminal amino acid is *N*-acetyl serine and the C-terminal amino acid is leucine.[2,15]

[12] D. H. MacLennan, A. Tzagoloff, and J. S. Rieske, *Arch. Biochem. Biophys.* **109**, 383 (1965).

[13] D. L. Edwards and R. S. Criddle, *Biochemistry* **5**, 583 (1966).

[14] D. L. Edwards and R. S. Criddle, *Biochemistry* **5**, 588 (1966).

[15] R. S. Criddle, D. L. Edwards, and T. G. Petersen, *Biochemistry* **5**, 578 (1966).

[72a] Isolation of a Mitochondrial Membrane Fraction Containing the Citric Acid Cycle and Ancillary Enzymes

By David W. Allmann and Elisabeth Bachmann[1]

In the preparation of submitochondrial particles from beef heart mitochondria, interest has usually been centered on the isolation and purification of particles concerned with the electron transfer process. The classical preparation of the electron transfer particles (ETP) is composed of membranous vesicles comprised of the characteristic tripartite repeating units described by Fernández-Morán.[1a] This membrane fraction is more concentrated than the mitochondrion in respect to the oxidation–reduction components of the electron transfer chain, but significantly less concentrated in respect to the enzymes of the citric acid cycle. Another membrane fraction also can be isolated which contains the citric acid cycle enzymes and no electron transfer components. This latter membrane fraction, when examined by electron microscopy with phosphotungstic acid as a negative stain, shows vesicular structures but lacks the tripartite structure of the ETP. The preparation is rich not only in the enzymes of the citric acid cycle, but also the enzyme systems responsible for the oxidation and elongation of fatty acids and for substrate phosphorylation. The fractionation procedure can be monitored by observing the decrease in electron transfer components or the increase in the enzymes of the citric acid cycle.

Reagents for Fractionation

Sucrose–0.01 *M* Tris-Cl, 0.25 *M*, pH 7.8 (sucrose-Tris)
Vitamin E (α-*d*-tocopherol), 100 mg/ml in 95% ethanol
EDTA, 0.2 *M* (ethylenediamine tetraacetic acid, neutralized)
Potassium cholate, 20% (recrystallized twice), pH 7.8

Isolation Procedure[2,3,4]

Heavy beef heart mitochondria were prepared as described previously[5,6] and suspended in sucrose-Tris. The protein concentration was

[1] This manuscript was prepared while the author was at the Institute for Enzyme Research, University of Wisconsin, Madison, Wisconsin.
[1a] H. Fernández-Morán, *Circulation* **26**, 1039 (1963).
[2] E. Bachmann, D. W. Allmann, and D. E. Green, *Arch. Biochem. Biophys.* **115**, 153 (1966).
[3] D. W. Allmann, E. Bachmann, and D. E. Green, *Arch. Biochem. Biophys.* **115**, 165 (1966).
[4] D. E. Green, E. Bachmann, D. W. Allmann, and J. F. Perdue, *Arch. Biochem. Biophys.* **115**, 172 (1966).
[5] F. L. Crane, J. Glenn, and D. E. Green, *Biochim. Biophys. Acta* **22**, 475 (1956).
[6] Y. Hatefi and R. L. Lester, *Biochim. Biophys. Acta* **27**, 83 (1958).

adjusted to 30 mg/ml and the mitochondrial suspension was aged overnight (approximately 18 hours) at 0–5° and then supplemented with vitamin E (1 mg per 100 mg protein) and EDTA (2 mM) to prevent lipid peroxidation. The mitochondrial suspension was adjusted to pH 8.0 with KOH, and cholate was added to a final concentration of 0.1 mg per milligram of protein (the pH remaining at 8.0). The suspension (20-ml portions in a 30-ml beaker) was exposed to sonic irradiation for 15–30 seconds in the Branson sonifier at the maximal power setting. The entire procedure was carried out at 0–5°.

The irradiated suspension was centrifuged at 6500 g for 10 minutes in a Spinco Model L centrifuge No. 40 rotor. The residue (R_1) was composed of unbroken mitochondria. The supernatant and the loosely layered residue were combined (S_1) and centrifuged for 30 minutes at 104,000 g. The supernatant and the loosely layered residue (light yellow) was removed by swirling with a few milliliters of the supernatant. The suspension was centrifuged for 60 minutes at 150,000 g. The clear supernatant (S_3) was carefully removed by pipette. The lightly colored, loosely packed residue (F_3) was then separated from the dark, reddish, well packed residue (R_3) by swirling with a few milliliters of sucrose-Tris. The F_3 suspension was diluted with about five volumes of sucrose-Tris and recentrifuged for 60 minutes at 150,000 g. The clear supernatant (S_4) was removed by pipette, and the light yellowish residue (F_4), which was loosely packed (fluffy), and the dark red residue (R_4) were then removed and suspended in sucrose-Tris.

An alternate procedure for the fragmentation of the mitochondria involved sonic irradiation for 45–60 seconds of a mitochondrial suspension, which had been aged, supplemented with EDTA, and Vitamin E as above, but to which no cholate was added. The resulting suspension was fractionated according to the same procedure as described. After the first 150,000 g centrifugation, the clear supernatant (S_3) was removed by pipette to within about 1–1.5 ml of the bottom of the centrifuge tube. The fraction designated as F_3 was found as a loose fluffy layer on the top of the well packed red residue (R_3). The F_3 fractions from each of the centrifuge tubes were pooled, diluted with 1–2 volumes of sucrose-Tris, and recentrifuged for 60 minutes at 150,000 g. The resulting clear supernatant (S_4) was removed by pipette leaving about 1–2 ml of supernatant which contained the fluffy layer (F_4). The F_4 fraction and the R_4 fraction (the tightly packed red residue) were removed separately and suspended in sucrose-Tris.

Characteristics of the F_4 Fraction[2,3,4]

The F_4 fraction obtained by the cholate procedure accounted for 15–20% of the starting protein in the S_1 fraction, while the F_4 fraction obtained by sonication in the absence of cholate accounted for 3–6% of the S_1 fraction. The S_1 fraction is taken as 100% since the first residue (R_1) is composed mainly of unbroken mitochondria. A measure of the separation of the electron transfer chain from the F_4 fraction is the reduction in cytochrome *a* content in the latter fraction and a corresponding increase in the cytochrome *a* content of the R_2 and R_3 fractions. A comparison of distribution of protein and cytochrome *a* among the various fractions obtained by each of the two preparative procedures is presented in Table I. The other cytochromes and coenzyme Q were correspondingly reduced in the F_4 fraction.

TABLE I
DISTRIBUTION OF PROTEIN AND CYTOCHROME *a* IN THE SUBMITOCHONDRIAL FRACTIONS[a]

	Method of disrupting the mitochondria					
	Cholate + sonication			Sonication		
		Cytochrome *a*			Cytochrome *a*	
Fraction	Protein recovery (%)	Millimicromoles/mg protein	Recovery (%)	Protein recovery (%)	Millimicromoles/mg protein	Recovery (%)
HBHM[b]	160	1.20	—	300	1.20	—
S_1	100	1.20	100	100	1.30	100
R_2	35	2.10	57	50	1.70	65
R_3	20	1.60	23	20	1.95	30
R_4	5	1.40	5	5	1.40	5
$S_3 + S_4$	22	0.10	1.0	20	0.05	0.01
F_4	18	0.50	5.0	5	0.01–0.2	<1

[a] Protein was determined by the biuret method [A. G. Gornall, C. J. Bardawill, and M. M. David, *J. Biol. Chem.* **177,** 751 (1949)] and cytochrome *a* by the method of T. Yonetani [*J. Biochem.* (*Tokyo*) **46,** 917 (1959)].
[b] HBHM = heavy beef heart mitochondria.

The complete sequence of enzymes catalyzing the citric acid cycle (with the exception of succinic dehydrogenase) was found to be associated with the F_4 fraction (Table II). At least 30–50% of each enzyme activity was recovered in the F_4 fraction obtained by the cholate procedure (sometimes the recovery of the α-ketoglutaric dehydrogenase activity approached 70%). These enzyme activities appear to be intimately associated with the F_4 fraction since they were not removed by further sucrose washing; for example, 65% of the isocitric acid dehydrogenase activity in the F_3 fraction (obtained after the first 150,000 *g* centrifuga-

TABLE II
SPECIFIC ACTIVITIES OF SOME CITRIC ACID CYCLE ENZYMES IN SUBMITOCHONDRIAL FRACTIONS[a]

	Method of disruption of the mitochondria					
	Cholate + sonication			Sonication		
Fraction	I-DH	M-DH	α-K-DH	I-DH	M-DH	α-K-DH
S_1	1205[b]	1610	139	550	830	120
R_2	323	563	47	114	310	20
R_3	265	344	175	146	275	228
$S_3 + S_4$	5900	6275	111	4900	7120	120
F_4	1050	1531	311	2010	2250	1030

[a] Abbreviations: I-DH = isocitric dehydrogenase; M-DH = malic dehydrogenase; α-K-DH = α-ketoglutaric dehydrogenase.

[b] Values are stated as millimicromoles of DPN or TPN reduced per minute per milligram of protein.

tion) was retained after two washings (5–10 volumes of sucrose-Tris). The cholate procedure resulted in a higher yield of the enzymes of the citric acid cycle than did sonication alone. On the other hand, the F_4 prepared by the latter procedure was less contaminated by the electron transfer components but contained all the citric acid cycle enzymes (10–15% of each enzyme activity). The citric acid cycle enzymes were essentially absent in the R_2 and R_3 fractions which were enriched in the electron transfer components. Other enzyme systems such as those catalyzing substrate level phosphorylation, fatty acid oxidation and fatty acid elongation, were found to be associated with the F_4 fraction. The F_4 fraction (obtained by sonication alone) was also more concentrated than heavy beef heart mitochondria in respect to cholesterol, acid-extractable flavin, RNA, and adenine nucleotides.

Examination of the R_2, R_3, and F_4 fractions in the electron microscope using phosphotungstic acid as a negative stain revealed that each fraction is vesicular in nature. The essential morphological difference between these two membrane fractions is the lack of tripartite structures on the F_4 membrane while the R_2 and R_3 vesicles (are enriched in the electron transfer chain) are composed of membranous vesicles comprised of a tripartite subunit structure described by Fernández-Morán.[1a] All the membrane fractions (R_2, R_3, and F_4) contain 25–30% phospholipid on a dry weight basis.

Addendum

A third method, developed to separate the two mitochondrial membrane systems, involved digestion of heavy beef heart or liver mitochondria with *Crotalus atrox* phospholipase.[6a]

Heavy beef heart mitochondria[5,6] were aged overnight at 0–5° at a protein concentration of 30 mg/ml in sucrose-Tris. To 600 mg of protein 25 micromoles of $CaCl_2$, 2.5 micromoles of potassium EDTA pH 7.5, and 5 ml of phospholipase solution were added. The pH of the mixture was adjusted to 8.5 with *N* KOH. The incubation was carried out at 38° in a Dubnoff metabolic shaker for the desired length of time (the standard length of incubation was 60 minutes). The mixture was then cooled and centrifuged for 30 minutes at 35,000 *g*. The supernatant S_1 was removed and the residue resuspended in the original volume of sucrose-Tris. The homogenized suspension of the residue was centrifuged for 30 minutes at 78,000 *g* and the residue (R_2) resuspended in sucrose-Tris. The second supernatant fluid S_2 was pooled with S_1, and the combined supernatants centrifuged for 20 minutes at 78,000 *g* to eliminate contamination with R_2 material.

The collected supernatant was then centrifuged for 60 minutes at 150,000 *g* and separated into a sedimentable, membranous fraction K, and a nonsedimentable, nonmembranous fraction S. The R_4,[3,7] derived from R_2 by repeated washings, accounts for about 72% of the starting protein and is an essentially pure fraction of inner mitochondrial membrane, containing all the enzymatic functions of electron transfer (95% of DPNH-oxidase and succinic dehydrogenase, 98% of the Mg^{++} stimulated ATPase and 95% of the cytochrome *a* complement). The starting material HBHM was always taken to be 100%. The membranous K-fraction, accounts for about 8% of the starting protein, is free of inner membrane components, but contains the multi-enzyme complexes of the citric acid cycle system, as α-ketoglutaric dehydrogenase (62%), pyruvic dehydrogenase (49%), and β-hydroxybutyric dehydrogenase (50%). Between 5 and 15% of the small molecular enzymes of the citric acid cycle enzymes remain associated with the K-membrane. About 75–85% of these activities (isocitric dehydrogenase, malic dehydrogenase, fumarase, aconitase, and condensing enzyme) are formed as soluble proteins in the S-fraction. This fraction accounts for 15% of the protein in the starting material (Table III).

[6a] Lyophilized venom of *Crotalus atrox* from the Ross Allen Reptile Institute was suspended in water at a concentration of 10 mg/ml. The pH of the suspension was adjusted to 5.0 with *M* acetate buffer, pH 4.8. The suspension was heated for 5 minutes in a boiling water bath, cooled, and centrifuged for 10 minutes at 105,000 *g*. The clear supernatant fluid was collected and adjusted to pH 7.0 with *M* Tris-chloride. The final protein concentration was about 1 mg/ml.

[7] E. Bachmann, G. Lenaz, and D. E. Green, in preparation.

TABLE III
DISTRIBUTION OF SOME INNER AND OUTER MEMBRANE ACTIVITIES AFTER PHOSPHOLIPASE DIGESTION OF MITOCHONDRIA[a]

		Activities				
		Outer membrane		Inner membrane		
Fraction	Protein	Isocitric dehydrogenase[b]	α-Ketoglutaric dehydrogenase[b]	Succinic dehydrogenase[b]	DPNH Oxidase[b]	Mg^{++} ATPase[b]
HBHM	100	1051	95	210	2.77	537
R_4	72	11	12	300	5.4	650
S_3	15	10,470	23	0	0	5
K	8	660	542	0	0.47	43
HBLM	100	111	51	88	—	160
R_4	36	0	2	219	—	415
S_3	46	218	42	2	—	0
K	12	59	285	10	—	65

[a] See text footnotes 3, 7, and 8.
[b] Specific activities of the enzymes are expressed as mμmoles or μatoms of substrate reduced per minute per mg protein.

The same separation can be applied to beef liver mitochondria (HBLM) resulting in a similar fractionation and distribution of enzymic activities between the two mitochondrial membranes.[8]

[8] E. Bachmann, J. F. Perdue, and D. E. Green, in preparation.

[72b] Isolation and Purification of the Outer Membrane and Inner Membrane of Liver Mitochondria

By D. F. PARSONS[1] and G. R. WILLIAMS

The following method[1a, 2] developed out of electron microscope observations on thin sections of swollen mitochondria. Large amplitude swelling of liver, heart, brain, kidney, and plant mitochondria produced rupture and partial detachment of the outer membrane. The inner mem-

[1] This manuscript was prepared while the author was at the Department of Medical Biophysics, University of Toronto, Toronto, Ontario, Canada.
[1a] D. F. Parsons, G. R. Williams, and B. Chance, *Ann. N.Y. Acad. Sci.* **137**, 643 (1966).
[2] D. F. Parsons, G. R. Williams, W. Thompson, D. F. Wilson, and B. Chance, *Proc. Round Table Disc. Mitochondrial Structures and Compartmentation, Bari, Italy, May 23–26, 1966.* Adriatica Ed., Bari, Italy, in press.

brane remained intact as a swollen "ghost." It proved possible to obtain a crude separation of the outer and inner membranes by differential centrifugation. The inner membrane ghosts sedimented at low speeds and the outer membrane fragments at high speeds. Further purification was obtained by sucrose density gradient centrifugation of the crude preparations. The bouyant density of the outer membrane (density = 1.13–1.14) proved sufficiently different from that of the inner membrane (density = 1.21) to allow clean separations of the two membranes.

The method requires some attention to detail to minimize break-up of the inner membrane during swelling and subsequent procedures. Extensive disintegration of the inner membrane at any stage causes significant contamination of the outer membrane with small pieces of inner membrane. The yields of membrane obtained are small (about 5 mg of outer membrane and 5–15 of inner membrane per 100 g of guinea pig liver), but sufficient for most enzymatic studies. The procedure worked best with mouse, rat, or guinea pig liver. The highest yields and cleanest preparations of membranes were obtained from guinea pig liver.

Procedure

Ten guinea pigs are guillotined after starving 18 hours (water ad libitum), and the livers (about 100 g) are homogenized in 0.28 *M* sucrose containing 1.0 m*M* tris(hydroxymethyl)aminomethane hydrochloride and 0.1 m*M* EDTA, at pH 7.2. Gall bladders are removed before homogenization. The livers are chopped to pieces 4–6 mm in size and washed with medium until free from blood. (All operations are carried out in the cold room, at 4°, and all solutions are kept at 0° or 4°.) The chopped and washed liver pieces are mixed with 10 times their volume of medium and homogenized with a Teflon pestle and glass vessel. The pestle is rotated at 150 rpm, a heavy-duty stirring motor being used. Only two passes are made. In each pass, the vessel is pushed firmly up against the pestle until the pestle touches the bottom of the glass vessel. No attempt is made to homogenize all the liver fragments. The homogenate has a pH of 7.0–7.2 and does not require neutralization. However, because of possible breakdown of the mitochondria by lysosomal enzymes, traces of bile salts and endogenous phosphatases, the preparation is manipulated as quickly as possible.

The homogenate (about 1 liter) is spun in the GSA Servall head, in 250 ml polycarbonate bottles, at a speed of 2000 rpm (650 *g*) for 10 minutes. The supernatant is removed and spun at 9500 rpm (14,600 *g*) for 12 minutes. The "fluffy layer" is washed off the pellets. The pellets are resuspended in 320 ml of medium and spun in the Servall SS 34 rotor at 2000 rpm (500 *g*) for 10 minutes. The supernatant is removed with a syringe and spun at 4000 rpm (1940 *g*) for 8 minutes and then accelerated to 9000 rpm (9750 *g*) for another 2 minutes. The fluffy layer is washed off the pellets, which are then suspended in 320 ml of medium and the 4000 rpm/9000 rpm high speed spin repeated. The two-speed centrifugation step minimizes microsomal contamination. A yield of 2.2% of mitochondria (wet weight of mitochondria as a percentage of wet weight of liver used) is obtained.

The mitochondria from about 100 g of liver are suspended slowly in 320 ml of 20 m*M* phosphate buffer, pH 7.2, containing 0.02% bovine serum albumin (the latter serves to reduce breakage of the inner membrane during swelling). After it has swelled for 20 minutes, the freed outer membrane is concentrated by differential centrifugation. The suspension is first spun for 20 minutes at 17,000 rpm (35,000 *g*), and the pellets are suspended in 320 ml of the same swelling medium. Most of the swollen inner membrane "ghosts" are removed by spinning at 4000 rpm (1900 *g*) for 15 minutes. The pellet (LSP) is saved as a crude inner membrane fraction. The supernatant is removed with a syringe (avoiding suspended material), and the outer membranes are sedimented by spinning at 17,000 rpm (35,000 *g*) for 20 minutes. The light brown pellets are suspended in 40 ml of 20 m*M* phosphate buffer (pH 7.2) without bovine serum albumin, to form the input to the sucrose gradient.

The discontinuous gradient is formed in Spinco SW 25.2 (60 ml) bucket rotor tubes, by successively layering 12 ml of 51.3% (w/v) sucrose (density = 1.192), 20 ml of 37.7% (w/v) sucrose (density = 1.142), 12 ml of 23.2% (w/v) sucrose (density = 1.094) and 12 ml of the input. All sucrose solutions are made up in 20 m*M* phosphate buffer, pH 7.2. The rotor is run at 23,500 rpm (65,000 *g*) for 90 minutes. Tubes are balanced by weighing, the Spinco antiprecession device is used, and the rotor is decelerated with the brake on.

Fractions are removed from the gradient with a horizontally directed pipette (bent Pasteur pipette). All the fractions are mixed well with three times their volume of water, and the membranes are sedimented by spinning at 17,000 rpm (35,000 *g*) for 45 minutes. The pellets are resuspended in 0.25 *M* sucrose (no phosphate buffer). The input is also suspended in 0.25 *M* sucrose, so that all fractions (including the original mitochondria) can be compared in the same medium. For most assays the membranes are resuspended in 1.3 ml of 0.25 *M* sucrose.

Assay

The preparations can be satisfactorily checked for cross contamination by two methods. First, the color of the outer membrane pellet is examined by transmitted light. If free of inner membrane they are clear and light brown in color. In contrast, contaminated outer membrane preparations resemble the inner membrane in being yellow and opales-

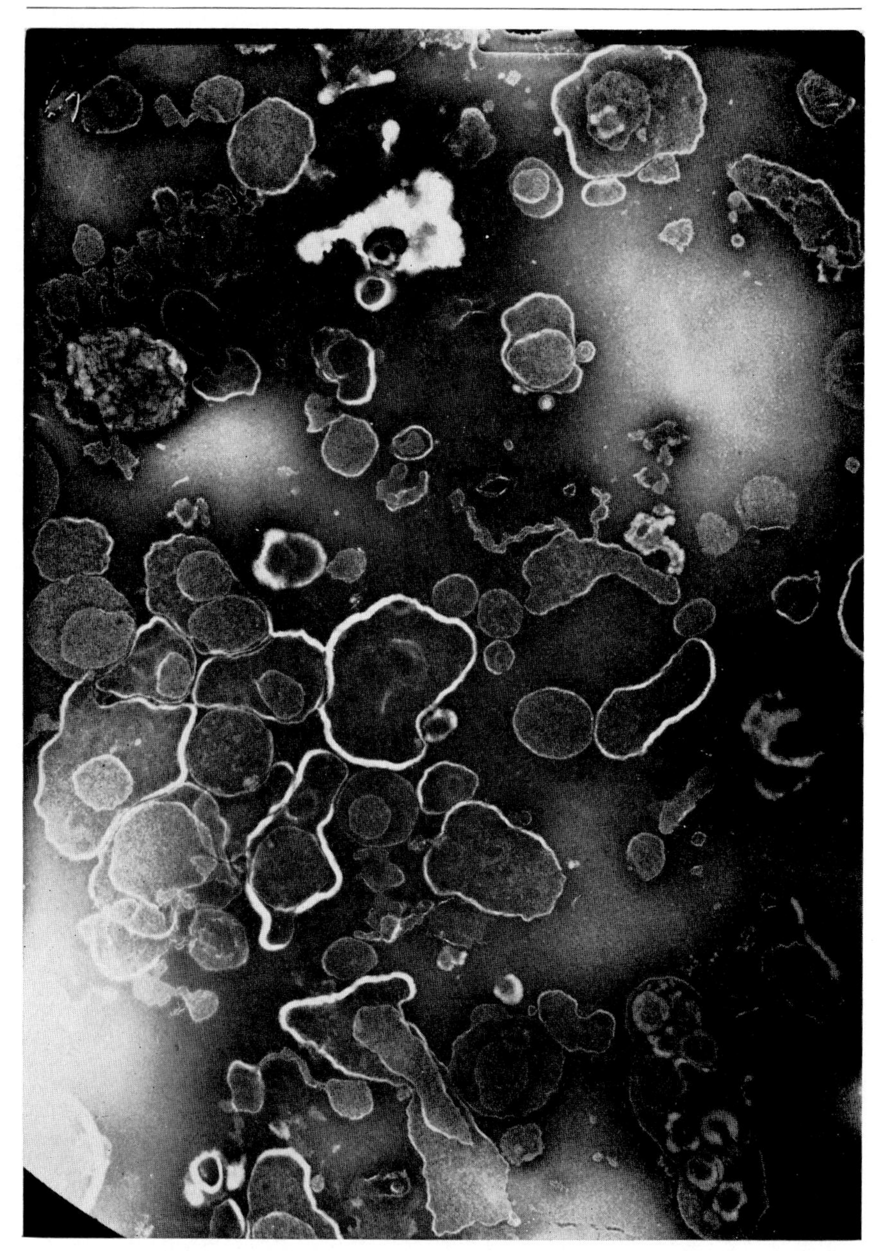

FIG. 1. Low magnification view of negatively stained (nonfixed) purified outer membrane from guinea pig liver mitochondria. The outer membrane has a slightly granular surface but no 90 Å subunits. Little or no contamination by inner membrane with 90 Å subunits is present. Magnification: $\times$29,000.

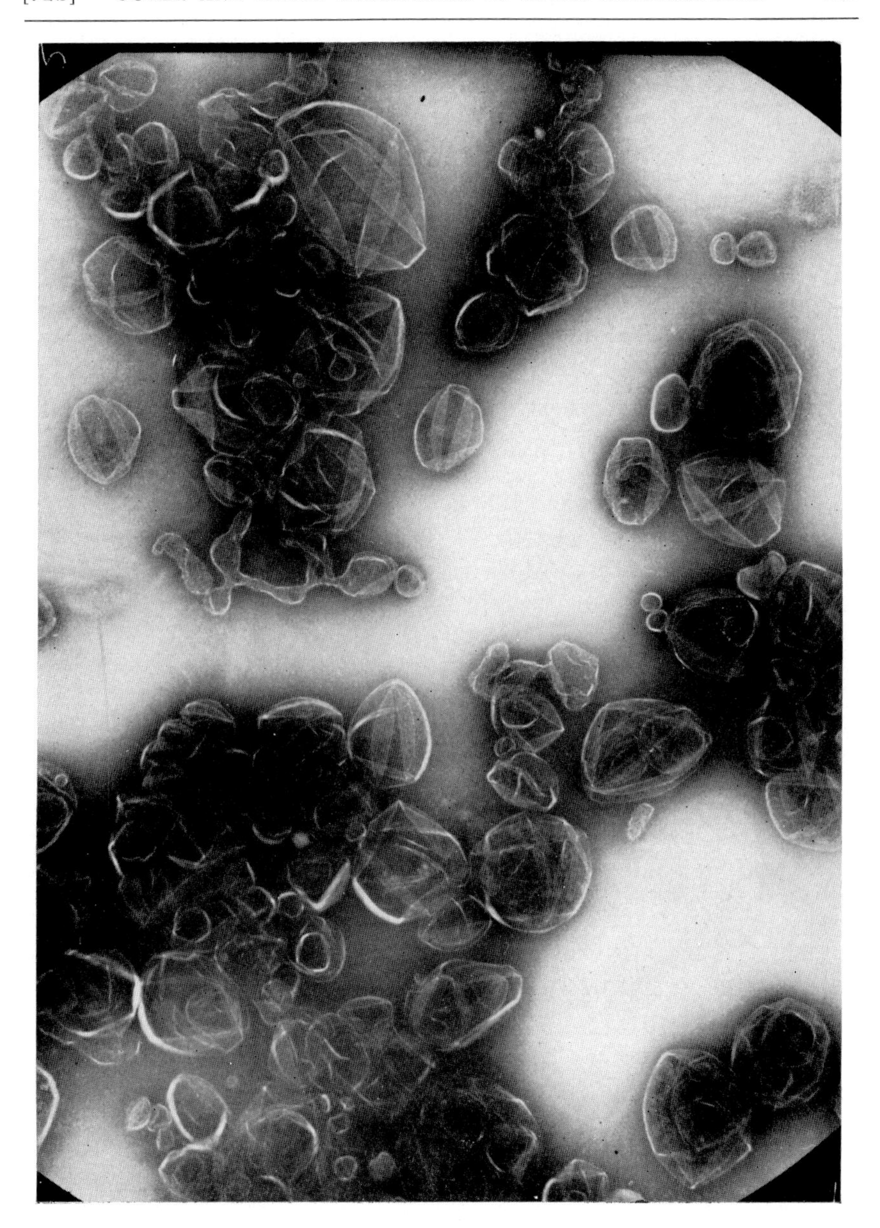

FIG. 2. A preparation similar to that of Fig. 1, but the outer membrane was fixed with osmium tetroxide before negative staining. The outer membrane has a characteristic "folded bag" appearance not obtained with microsomal membranes. Thin sections show that the outer membranes re-form closed vesicles after detachment from the mitochondria. Magnification: ×29,000.

cent. Secondly, the outer membrane fraction can be negatively stained by surface spreading in 3% potassium phosphotungstate (pH 6.8) as described in this volume [101]. The nonfixed outer membrane appears as in Fig. 1. When prefixed with osmium tetroxide (the 0.25 *M* sucrose membrane suspension is mixed with an equal volume of Palade's osmium tetroxide[3] solution and fixed at 0° for 30 minutes), the outer membrane has the different appearance shown in Fig. 2. The outer membrane assumes a "folded bag" form due to collapse of the spherical (re-fused) outer membrane fragments. Endoplasmic reticulum does not assume this appearance. After this period of fixation folded outer membrane can readily be distinguished from nonfolded endoplasmic reticulum, and also from fragments of inner membrane which have 90 Å subunits attached.

The inner membrane contains all the oxidative phosphorylation linked cytochromes. Hence, outer membrane preparations can be checked for inner membrane contamination by spectrophotometric or enzymatic assay of cytochrome oxidase.[2] The outer membrane is the site of a $NADH_2$ cytochrome b_5 reductase.[2,4] Hence endoplasmic reticulum contamination is conveniently estimated by assay of $NADPH_2$-cytochrome *c* reductase.

In future work it may well turn out that other methods can be used to assay separately for the outer and inner membranes. The lipid composition of the two membranes is already known to be different.[2] To date, only a few comparisons have been made between enzymes of the smooth endoplasmic reticulum and the outer membrane.

[3] G. E. Palade, *J. Exptl. Med.* **95**, 285 (1952).

[4] G. L. Sottocasa and L. Ernster, *Proc. Round Table Disc. Mitochondrial Structures and Compartmentation, Bari, Italy, May 23–26, 1966*. Adriatica Ed., Bari, Italy, in press.

[72c] Separation and Some Enzymatic Properties of the Inner and Outer Membranes of Rat Liver Mitochondria

By GIAN LUIGI SOTTOCASA, BO KUYLENSTIERNA, LARS ERNSTER, and ANDERS BERGSTRAND

The present procedure for the separation of the inner and outer membranes of rat liver mitochondria has been developed[1,2] in the course of

[1] G. L. Sottocasa and L. Ernster, *Abstr. 2nd Meeting Federation European Biochem. Soc., Vienna, 1965* p. 112.

See also G. L. Sottocasa and B. Kuylenstierna, *Abstr. 3rd Meeting Federation European Biochem. Soc., Warsaw, 1966* p. 118.

[2] G. L. Sottocasa, B. Kuylenstirena, L. Ernster, and A. Bergstrand, *Proc. Round*

an attempt to characterize cytochemically the enzyme system catalyzing the antimycin A-, amytal-, and rotenone-insensitive, nonphosphorylating oxidation of exogenous NADH by cytochrome *c*, known to occur in isolated liver mitochondria.[3-5] It was found[2] that brief exposure of isolated rat liver mitochondria to sonic oscillation, followed by centrifugation on a sucrose gradient, resulted in the separation of a particulate "light" subfraction from the bulk of the mitochondria which exhibited a high rotenone-insensitive NADH-cytochrome *c* reductase activity but was devoid of rotenone-sensitive NADH-cytochrome *c* reductase and other respiratory chain-linked enzyme activities. A similar subfraction was obtained when the mitochondria were subjected to swelling and contraction—rather than sonication—prior to density gradient centrifugation. The latter treatment had been shown by Parsons[6] to lead to a selective dilation and disruption of the outer mitchondrial membrane, and was subsequently used by him and his colleagues[7,8] to prepare purified mitochondrial outer and inner membrane fractions (cf. also this volume [72b]). Since both the sonication and the swelling-contraction procedure yielded only partial separation of the rotenone-insensitive NADH-cytochrome *c* reductase, we elaborated[2] a combined procedure, which is described below. It results in morphologically well defined outer and inner membrane fractions with a quantitative recovery that makes it suitable for the study of intramitochondrial distribution of enzymes and other chemical constituents.

Separation Procedure

Rat liver mitochondria are prepared by differential centrifugation from a 10% (w/v) homogenate in 0.25 *M* sucrose. After sedimentation

Table Disc. Mitochondrial Structures and Compartmentation, Bari, Italy, May 23–26, 1966. Adriatica Ed., Bari, Italy, in press; *J. Cell Biol.* **32**, 415 (1967).

[3] A. L. Lehninger, *J. Biol. Chem.* **190**, 345 (1951); *Phosphorus Metabolism*, **1**, 344 (1951); *Harvey Lectures Ser.* **49**, 176 (1955).

[4] C. de Duve, B. C. Pressman, R. Gianetto, R. Wattiaux, and F. Appelmans, *Biochem. J.* **60**, 604 (1955).

[5] L. Ernster, O. Jalling, H. Löw, and O. Lindberg, *Exptl. Cell Res.*, Suppl. 3, p. 124 (1955); L. Ernster, *Exptl. Cell Res.* **10**, 721 (1956); L. Ernster and O. Lindberg, *Ann. Rev. Physiol.* **20**, 13 (1958); L. Ernster, *Biochem. Soc. Symp.* **16**, 54 (1959); L. Ernster, G. Dallner, and G. F. Azzone, *J. Biol. Chem.* **238**, 1124 (1963).

[6] D. F. Parsons, *Intern. Rev. Exptl. Pathol.* **4**, 1 (1965).

[7] D. F. Parsons, G. R. Williams, and B. Chance, *Ann. N.Y. Acad. Sci.* **137**, 643 (1966).

[8] D. F. Parsons, G. R. Williams, W. Thompson, D. F. Wilson, and B. Chance, *Proc. Round Table Disc. Mitochondrial Structures and Compartmentation, Bari, Italy, May 23–26, 1966*. Adriatica, Ed., Bari, Italy, in press.

of the nuclear fraction at 600 g for 15 minutes, mitochondria are sedimented from the supernatant by centrifugation at 6500 g for 20 minutes. The fluffy layer is carefully discarded, and the pellet is washed twice, with one-half and one-fourth of the initial volume of sucrose.

The mitochondria prepared by the above procedure should be practically free from microsomal contamination. The extent of the latter can be estimated by determining the specific activity (on the protein basis) of the preparation with respect to glucose 6-phosphatase or some other exclusively microsomal enzyme,[2] and by comparing this value with that found for microsomes prepared from the same homogenate. Microsomes are obtained by centrifuging the 6500 g supernatant, obtained above, first at 15,000 g for 15 minutes (in order to remove residual mitochondria) and then at 105,000 g for 60 minutes; the surface of the microsomal pellets is rinsed with 0.25 M sucrose to remove the bulk of adhering cell sap. The specific activities of glucose 6-phosphatase and other exclusively microsomal enzymes in the mitochondrial fraction should be less than 5% of those in the microsomes; in contrast, the rotenone-insensitive NADH cytochrome c reductase (which occurs in both the mitochondria and microsomes) exhibits a specific activity in the mitochondrial fraction that is 25–30% of that of the microsomes.

Mitochondria containing ca. 50 mg protein are suspended in 7.5 ml of 10 mM Tris-phosphate buffer, pH 7.5, by means of a Teflon pestle fitted into the centrifuge tube. After standing at 0° for 5 minutes, during which time the mitochondria undergo swelling, 2.5 ml of a solution containing 1.8 M sucrose, 2 mM ATP, and 2 mM $MgSO_4$ is added to the suspension. A visible increase in turbidity immediately appears, due to contraction of the mitochondria. After another 5 minutes at 0°, the suspension is subjected, in aliquots of 3.5 ml, to sonic oscillation at 3 amperes with a Branson Sonifier for 15 seconds at 0°. The total volume of the sonicated suspension (10.5 ml; final concentration of sucrose, 0.45 M) is layered over 15 ml 1.18 M sucrose in a 34-ml Spinco centrifuge tube and centrifuged in rotor SW-25 at 24,000 rpm for 3 hours.

After centrifugation, 3 subfractions can be distinguished: a tightly packed, dark-brown pellet at the bottom of the tube ("heavy" subfraction); a pinkish-yellow band at the interface of the two sucrose layers ("light" subfraction); and a clear yellow supernatant in the 0.45 M sucrose layer of the gradient ("soluble" subfraction). The 1.18 M sucrose layer of the gradient is water clear and free of protein. The bulk of the supernatant is sucked off by means of a capillary (U-shaped, in order to avoid turbulence), followed by the interface band. The remaining lower sucrose layer is then discarded. The surface of the bottom pellet is rinsed with several portions of 0.25 M sucrose, and the pellet is finally suspended in 0.25 M sucrose with the aid of a Teflon pestle.

The interface band separated as above contains, by necessity, a portion of the supernatant fraction. The volume of this portion can be calculated by subtracting the volume of the separated supernatant from that of the originally added upper sucrose layer. In computing the contents and activities of various constituents in the "light" and "soluble" mitochondrial subfractions, correction must be made for the amount of supernatant present in the separated interface band.

Alternatively to the 2-layer gradient system described above, the 3 subfractions may be separated on a 3-layer gradient, consisting of 10 ml 0.45 *M* sucrose (containing the swollen-contracted and sonicated mitochondria), 5 ml 0.76 *M* sucrose, and 10 ml 1.32 *M* sucrose. After centrifugation as in the foregoing case, the "heavy" subfraction again appears as a tightly packed pellet at the bottom of the tube, whereas the "light" subfraction forms an interface between the 0.76 and 1.32 *M* sucrose layers. The "soluble" subfraction is confined to the 0.45 *M* sucrose layer. The "light" and "soluble" subfractions can thus be separated completely, and no correction in computing their contents and activties of various constituents is necessary.

Properties

Electron microscopic examination of the "heavy" and "light submitochondrial fractions reveals the following (Figs. 1–5): Osmium-fixed specimens of the "heavy" subfraction (Fig. 1) consist of relatively large vesicles (average diameter 0.8 μ) bordered by a single membrane. In many instances the vesicles contain smaller, round or elongated profiles bordered by a single membrane of the same thickness as the surrounding vesicle, and probably representing sections of cristae. Negatively stained specimens of the same subfraction (Fig. 3) show mitochondrial images in the stage of bursting, with protrusions of unfolding cristae. At higher magnifications (Fig. 5a) a coating of mushroom-like repeating units, similar to those first described by Fernández-Morán,[9] can be discerned on the surface of the cristal membranes. The osmium-fixed "light" subfraction (Fig. 2) consists of relatively small vesicles (average diameter 0.2 μ) bordered by a single membrane and devoid of inner structures. As observed with the negative straining technique (Fig. 4), the "light" subfraction consists of more or less flattened vesicles. The outer surface of the vesicles is slightly irregular, but higher magnification reveals no mushroom-like repeating units (Fig. 5b). These pictures, which are similar to those obtained by Parsons *et al.*,[6-8] are consistent with the interpretation that the "heavy" subfraction represents the inner membrane system of the mitochondria, with some of the matrix contents re-

[9] H. Fernández-Morán, *Circulation* **26**, 1039 (1962).

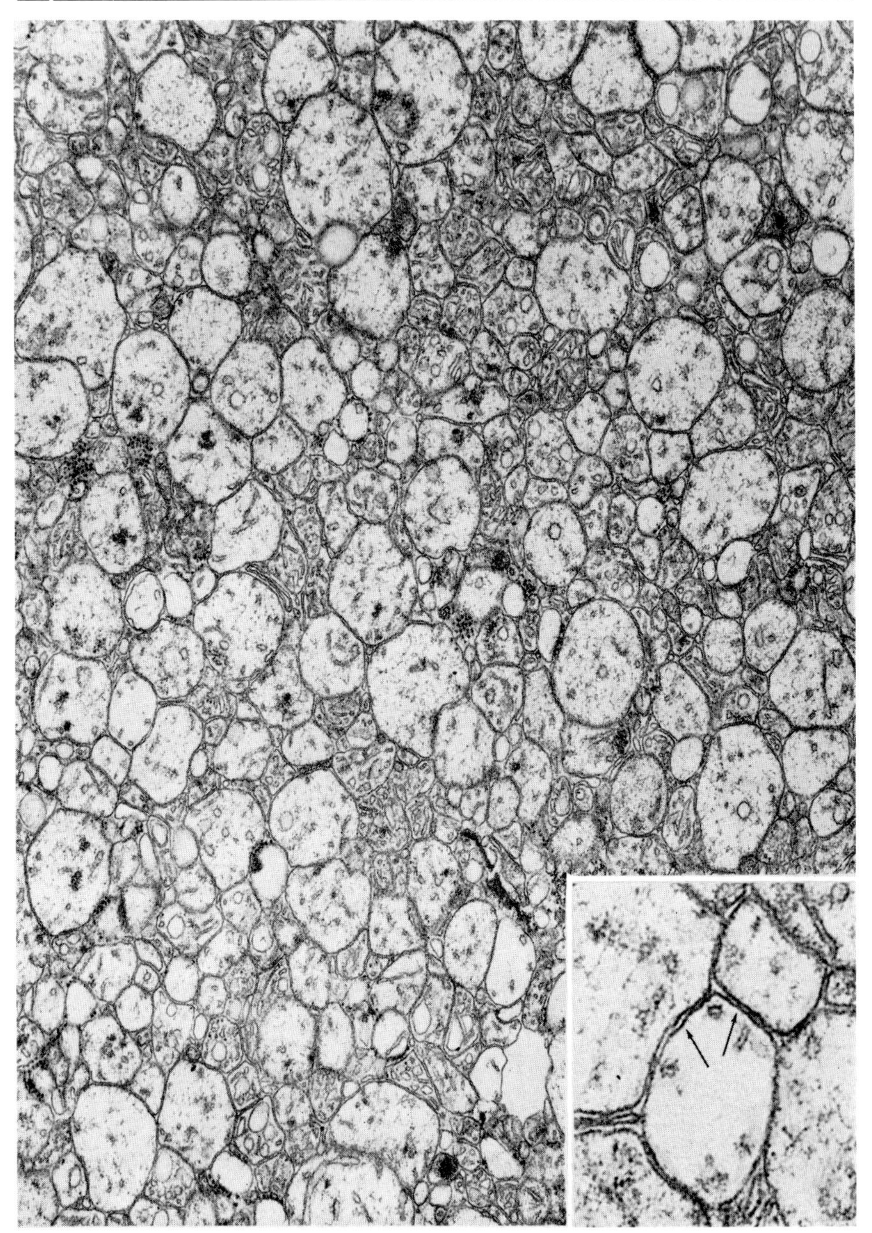

FIG. 1. Electron micrograph of osmium-fixed specimen of the "heavy" mitochondrial subfraction. ×30,000. *Inset:* Higher magnification of same, showing that apparent "double" membranes (arrows) are derived from different adjacent vesicles which are bordered by single membranes. ×80,000. From G. L. Sottocasa, B. Kuylenstierna, L. Ernster, and A. Bergstrand, *J. Cell Biol.* **32,** 415 (1967). Reproduced with the permission of *The Journal of Cell Biology.*

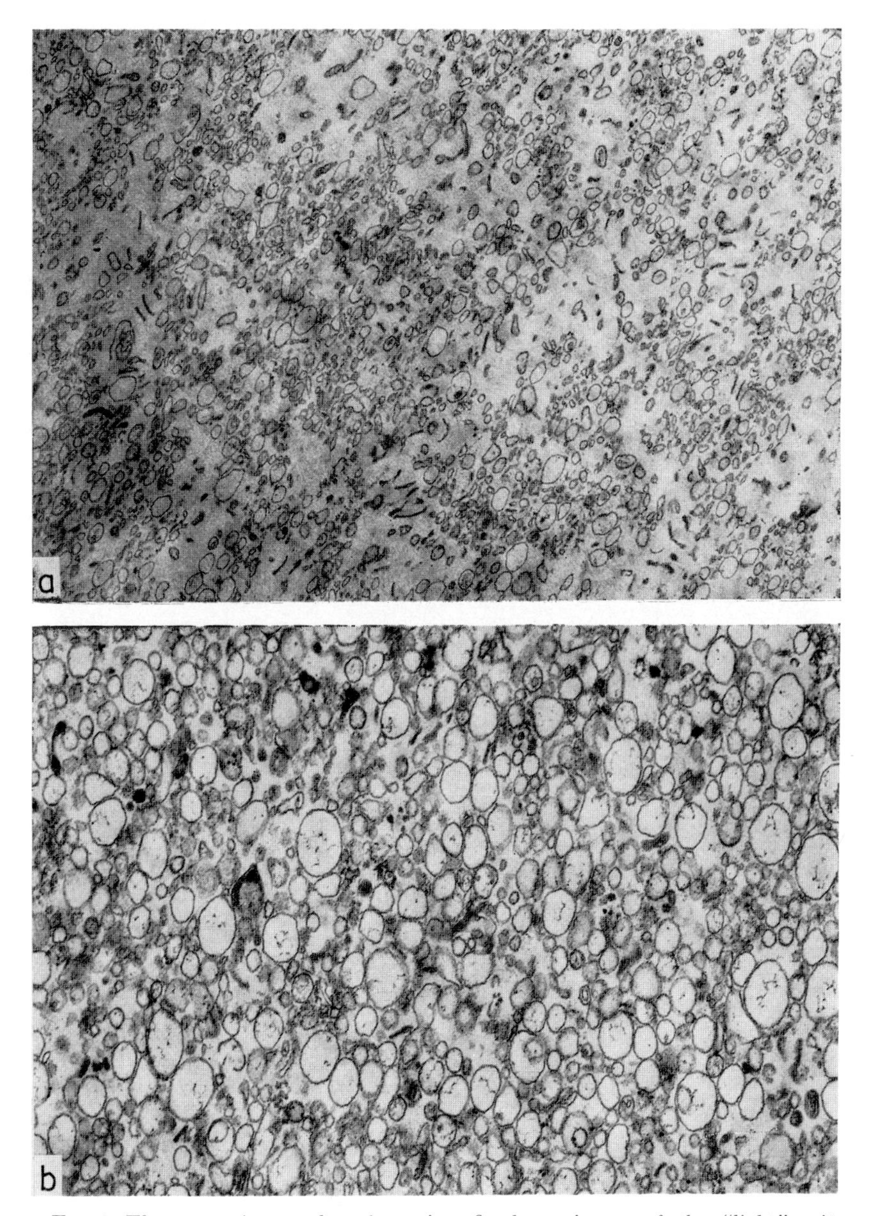

FIG. 2. Electron micrographs of osmium-fixed specimens of the "light" mitochondrial subfraction. (a) upper and (b) lower part of the recentrifuged interface band. ×30,000. From G. L. Sottocasa, B. Kuylenstierna, L. Ernster, and A. Bergstrand, *J. Cell Biol.* **32**, 415 (1967). Reproduced with the permission of *The Journal of Cell Biology.*

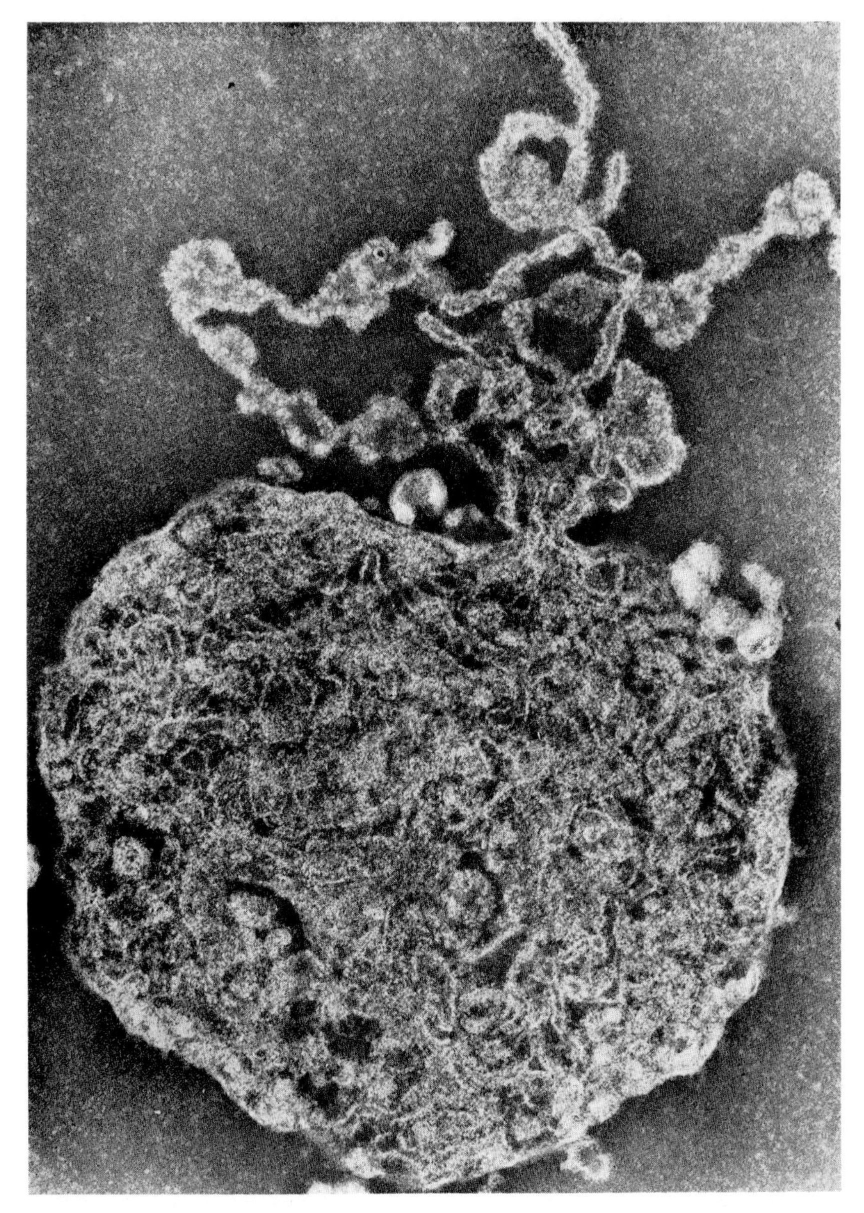

FIG. 3. Negatively stained specimen of the "heavy" mitochondrial subfraction. $\times$86,000. From G. L. Sottocasa, B. Kuylenstierna, L. Ernster, and A. Bergstrand, *J. Cell Biol.* **32**, 415 (1967). Reproduced with the permission of *The Journal of Cell Biology*.

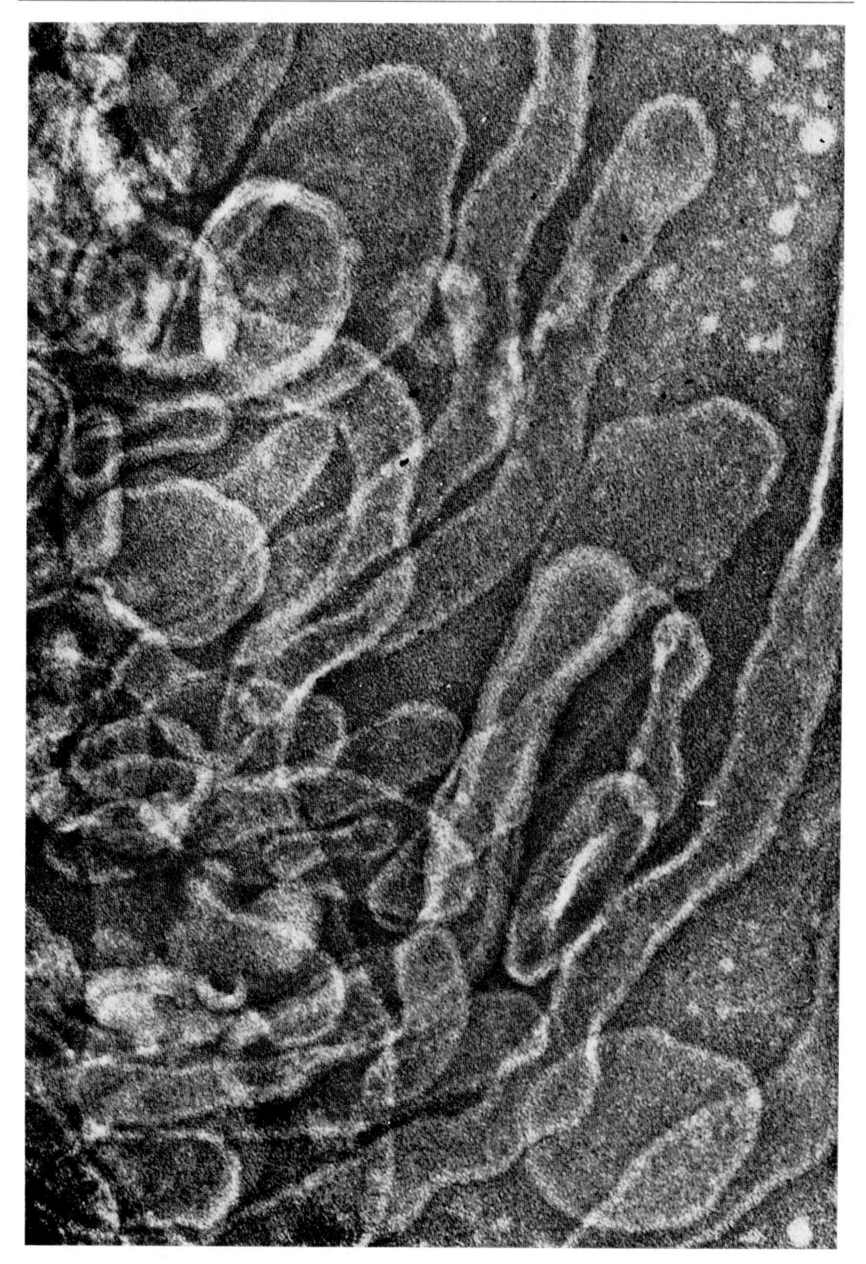

FIG. 4. Negatively stained specimen of the "light" mitochondrial subfraction. ×165,000. From G. L. Sottocasa, B. Kuylenstierna, L. Ernster, and A. Bergstrand, *J. Cell Biol.* **32,** 415 (1967). Reproduced with the permission of *The Journal of Cell Biology*.

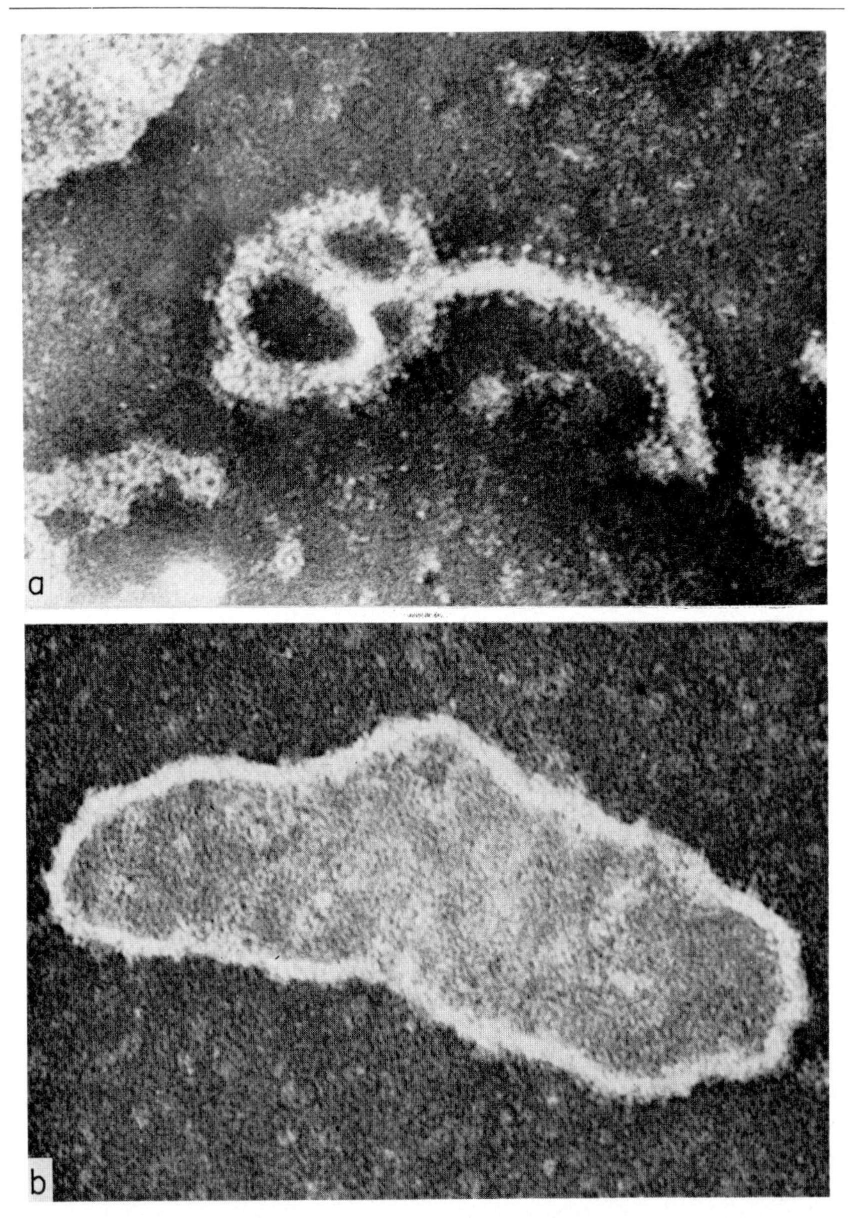

FIG. 5. Negatively stained specimens of the "heavy" and "light" mitochondrial subfractions. (a) "Heavy" subfraction. ×240,000. (b) "Light" subfraction. ×278,000. From G. L. Sottocasa, B. Kuylenstierna, L. Ernster, and A. Bergstrand, *J. Cell Biol.* **32,** 415 (1967). Reproduced with the permission of *The Journal of Cell Biology.*

tained, whereas the "light" subfraction consists of vesiculated derivatives of the outer membrane. The "soluble" subfraction probably contains part of the matrix contents, together with any material originating from the space between the outer and inner membranes or released from these in the course of the fractionation procedure.

The distribution of some enzyme activities in the 3 subfractions is

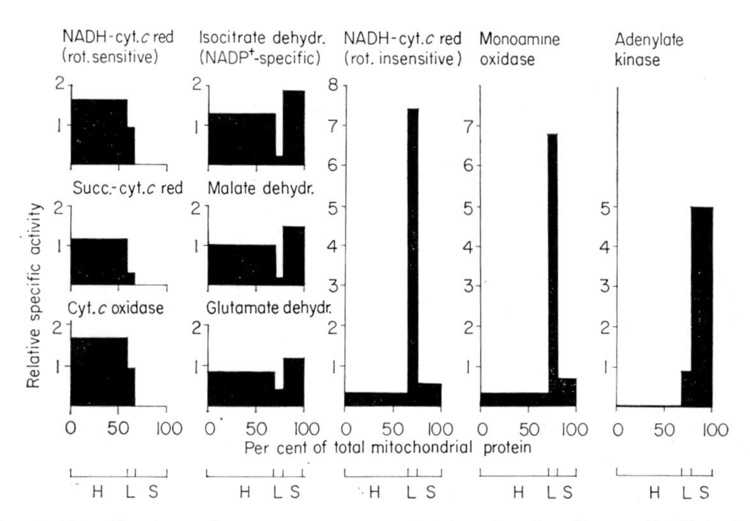

FIG. 6. Distribution of some enzyme activities in the "heavy," "light," and "soluble" subfractions. The ordinates represent relative specific activities on the protein basis, taking the specific activity of the swollen-contracted and sonicated mitochondria before separation of the subfractions as 1. The abscissas indicate the percentages of the total mitochondrial protein in the "heavy" (*H*), "light" (*L*), and "soluble" (*S*) subfractions. Protein contents were estimated with the method of O. H. Lowry, N. J. Rosebrough, A. L. Farr, and R. J. Randall [*J. Biol. Chem.* **193,** 265 (1951)]. Cytochrome *c* oxidase, succinate-cytochrome *c* reductase, and the rotenone-sensitive and rotenone-insensitive NADH-cytochrome *c* reductase activities were assayed as described by G. L. Sottocasa, B. Kuylenstierna, L. Ernster, and A. Bergstrand [*J. Cell Biol.* **32,** 415 (1967)]. Nicotinamide nucleotide-linked dehydrogenase activities were assayed spectrophotometrically at 340 mμ, in a medium consisting of 50 m*M* Tris-chloride buffer, pH 7.5, and the following additions: 10 m*M* DL-isocitrate, 0.1 m*M* $MnCl_2$, and 0.2 m*M* $NADP^+$ in the case of isocitrate dehydrogenase; 5 m*M* oxaloacetate and 0.1 m*M* NADH in the case of malate dehydrogenase; and 5 m*M* α-ketoglutarate, 5 m*M* NH_4Cl, and 0.1 m*M* NADH in the case of glutamate dehydrogenase. Monoamine oxidase was assayed by the method of C. W. Tabor, H. Tabor, and S. M. Rosenthal [*J. Biol. Chem.* **208,** 645 (1954)] as modified by C. Schnaitman, V. G. Erwin, and J. W. Greenawalt [*J. Cell Biol.* (in press)]. Adenylate kinase was assayed spectrophotometrically at 340 mμ in a reaction mixture containing 50 m*M* Tris-chloride buffer, pH 7.5, 5 m*M* ADP, 10 m*M* glucose, 10 units hexokinase, 10 units glucose 6-phosphate dehydrogenase, 5 m*M* $MgSO_4$, and 0.2 m*M* $NADP^+$. All enzyme assays were carried out at 30°.

summarized in Fig. 6. The data are presented in a diagrammatic fashion adopted from de Duve *et al.*[4] by plotting the relative specific activities (taking the activities found with the swollen-contracted and sonicated mitochondria before the density gradient centrifugation as 1) versus the percentage of the total mitochondrial protein recovered within each of the "heavy" (*H*), "light" (*L*), and "soluble" (*S*) subfractions. The "heavy" subfraction represents an average of 67%, the "light" 9%, and the "soluble" subfraction 24% of the total mitochondrial protein. The total area of each diagram is equal to the percentage recovery of a given enzyme activity.

The respiratory chain-linked enzyme activities, including cytochrome *c* oxidase, succinate-cytochrome *c* reductase, and the rotenone-sensitive NADH-cytochrome *c* reductase, are concentrated in the "heavy" subfraction, only a small portion of these activities being found in the "light" subfraction. The "soluble" subfraction is devoid of respiratory chain-linked enzyme activities. This pattern of distribution is consistent with the generally held view that the enzymes of the respiratory chain are associated with the mitochondrial inner membrane.

A number of nicotinamide nucleotide-linked dehydrogenase activities, including the malate, glutamate, and $NADP^+$-specific isocitrate dehydrogenases, are recovered to a large extent in both the "heavy" and the "soluble" subfractions, again only negligible activities being found in the "light" subfraction. The specific activities in the "soluble" subfraction are consistently higher than those in the "heavy" subfraction. This distribution pattern suggests that these enzymes are localized within the mitochondrial matrix contents, part of which has been released in the course of the fractionation. The present findings are in agreement with those of Parsons *et al.*,[8] who also find no appreciable citric acid cycle enzyme dehydrogenase activities associated with their outer membrane preparation from liver mitochondria. Furthermore, Norum *et al.*,[10] using the present fractionation procedure, have reported a distribution pattern of glutamate dehydrogenase very similar to that shown in Fig. 6. They have also found that another nicotinamide nucleotide-linked enzyme, β-hydroxybutyrate dehydrogenase, is firmly associated with the inner membrane fraction.

The rotenone-insensitive NADH-cytochrome *c* reductase activity is concentrated in the "light" subfraction—i.e., in the outer membrane—with a specific activity 7–8 times that of the unfractionated mitochondria. These results have been confirmed by Parsons *et al.*[8] As already

[10] K. R. Norum, M. Farstad, and J. Bremer, *Biochem. Biophys. Res. Commun.* **24**, 797 (1966).

indicated by earlier studies of Raw and associates,[11] this enzyme system is similar to the one found in microsomes,[12] consisting of the flavoprotein, NADH-cytochrome b_5 reductase,[13] and cytochrome b_5.[14] Indeed, when the "light" subfraction obtained with the present procedure is supplemented with NADH, it reveals a difference spectrum versus the unsupplemented "light" subfraction that is characteristic of cytochrome b_5, exhibiting absorption maxima at 556, 526, and 424 mμ; addition of $Na_2S_2O_4$ does not increase these absorption bands. These findings indicate that the outer mitochondrial membrane contains both a cytochrome b_5-like hemoprotein and an enzyme catalyzing the reduction of the latter by NADH. The reductase is similar to the microsomal NADH-cytochrome b_5 reductase[15] in that it involves the 4A hydrogen atom of NADH[1,2] (in contrast to the respiratory chain-linked NADH dehydrogenase, which is 4B-specific[16]). On the other hand, in accordance with earlier observations of Raw *et al.*,[11] the mitochondrial reductase differs from the microsomal one in regard to a comparatively greater sensitivity to dicoumarol. Likewise the mitochondrial hemoprotein differs from the microsomal cytochrome b_5 in not being reducible by cysteine.[11] Moreover, as reported by Parsons *et al.*,[8] the low-temperature spectrum of the mitochondrial outer-membrane hemoprotein shows α-peaks at 551 and 558 mμ, as compared with those at 552 and 557 mμ found with the microsomal cytochrome b_5.

Data in the table illustrate the distribution of the cytochrome b_5-like hemoprotein in the 3 subfractions. The values have been deduced from difference spectra, assuming an extinction coefficient equal to that of cytochrome b_5. The total amount of "cytochrome b_5" recovered is 163 millimicromoles per gram of total mitochondrial protein; this value is approximately one-fourth of that found with liver microsomes.[2] Over

[11] I. Raw, R. Molinari, D. Ferreira do Amaral, and H. R. Mahler, *J. Biol. Chem.* **233,** 225 (1958); H. R. Mahler, I. Raw, R. Molinari, and D. Ferreira do Amaral, *J. Biol. Chem.* **233,** 230 (1958); I. Raw and H. R. Mahler, *J. Biol. Chem.* **234,** 1867 (1959); I. Raw, N. Petragnani, and O. Camargo-Nogueira, *J. Biol. Chem.* **235,** 1517 (1960).

[12] G. H. Hogeboom, *J. Biol. Chem.* **177,** 847 (1949); G. H. Hogeboom and W. C. Schneider, *J. Natl. Cancer Inst.* **10,** 983 (1950).

[13] C. F. Strittmatter, *in* "Hematin Enzymes" (J. E. Falk, R. Lemberg, and R. K. Morton, eds.), p. 461. Pergamon Press, London, 1961.

[14] P. Strittmatter, *in* "The Enzymes" (P. D. Boyer, H. A. Lardy, and K. Myrbäck, eds.) Vol. VIII, p. 113. Academic Press, New York, 1963.

[15] G. R. Drysdale, M. J. Spiegel, and P. Strittmatter, *J. Biol. Chem.* **236,** 2323 (1961).

[16] C. P. Lee, N. Simard-Duquesne, L. Ernster, and H. D. Hoberman, *Biochim. Biophys. Acta* **105,** 397 (1965); L. Ernster, H. D. Hoberman, R. L. Howard, T. E. King, C. P. Lee, B. Mackler, and G. L. Sottocasa, *Nature* **207,** 940 (1965).

DISTRIBUTION OF "CYTOCHROME b_5" UPON SUBFRACTIONATION OF RAT LIVER MITOCHONDRIA BY VARIOUS PROCEDURES

Procedure:	Swelling-contraction, sonication			Sonication			Swelling-contraction		
		"Cytochrome b_5" (mμmoles/g)			"Cytochrome b_5" (mμmoles/g)			"Cytochrome b_5" (mμmoles/g)	
Subfraction	Protein (% of total)	Subfr. protein	Total protein	Protein (% of total)	Subfr. protein	Total protein	Protein (% of total)	Subfr. protein	Total protein
"Heavy"	58	0	0	84	—	—	87	—	—
"Light"	9	152	13	8	572	47	3	716	20
"Soluble"	33	478	150	8	212	17	10	850	87
"Light" + "soluble"	42	388	163	16	392	64	13	820	107

90% of the cytochrome is found in the "soluble" subfraction, and the remainder in the "light" subfraction, corresponding to a ca. 3-fold concentration of the cytochrome in the former as compared with the latter. The "heavy" subfraction is devoid of "cytochrome b_5." The relatively high concentration of "cytochrome b_5" in the "soluble" subfraction is probably the result of a release of the cytochrome from the outer membrane during the swelling-contraction procedure, as indicated by data obtained with subfractions prepared by the partial procedures, i.e., only sonication or only swelling-contraction (for details concerning the partial procedures, see references cited in footnote 2). With these procedures, only part of the mitochondrial "cytochrome b_5" is recovered in the "light" and "soluble" subfractions, 64 and 107 millimicromoles per gram of total mitochondrial protein with the sonication and the swelling-contraction procedure, respectively. However, with the sonication procedure, more than 75% of the "cytochrome b_5" recovered is found in the "light" subfraction, and its concentration is over 2.5 times higher than in the "soluble" subfraction. Conversely, with the swelling-contraction procedure, ca. 80% of the cytochrome is recovered in the "soluble" subfraction and its concentration is 1.2 times higher in this than in the "light" subfraction.

The "light" subfraction also contains in a 6–7-fold concentrated form the enzyme monoamine oxidase, which has previously been shown to be present in mitochondria.[17] Schnaitman *et al.*[18] have recently found a concentration of monoamine oxidase in a subfraction obtained from rat liver mitochondria after controlled osmotic lysis which, on the basis of electron microscopic evidence, was identified as a derivative of the outer mitochondrial membrane. They pointed out that monoamine oxidase may serve as a suitable marker for this membrane. The present results are in full accordance with these findings and add strong support to the conclusion that the "light" subfraction obtained by the present procedure indeed originates to a large extent from the outer mitochondrial membrane.

The liver-mitochondrial adenylate kinase activity is concentrated in the "soluble" subfraction, with a specific activity 5-fold over that of the unfractionated mitochondria. Very little or no adenylate kinase activity is found in the "heavy" subfraction, and only a low (though not insignifi-

[17] G. Cotzias, and V. Dole, *Proc. Exptl. Biol. Med.* **78,** 157 (1951); G. Rodriguez de Lores Arnaiz and E. de Robertis, *J. Neurochem.* **9,** 503 (1962); P. Baudhuin, H. Beaufay, Y. Rahman-Li, O. Sellinger, R. Wattiaux, P. Jacques, and C. de Duve, *Biochem. J.* **92,** 179 (1963); E. Oswald and C. Strittmatter, *Proc. Exptl. Biol. Med.* **114,** 668 (1963); V. Gorkin, *Pharmacol. Rev.* **18,** 115 (1966).

[18] C. Schnaitman, V. G. Erwin, and J. W. Greenawalt, *J. Cell Biol.* in press.

cant) activity in the "light" subfraction. This pattern of distribution suggests that adenylate kinase either is located in the space between the inner and outer mitochondrial membranes, or is associated with the outer membrane and released therefrom during the fractionation procedure; if it were located in the matrix inside the inner membrane, one would expect to recover a considerable portion of it in the "heavy" subfraction, in a manner similar to the pyridine nucleotide-linked dehydrogenases. The location of adenylate kinase near the mitochondrial surface substantiates an early postulate of Siekevitz and Potter,[19] and is consistent with the more recent observation of Chappell and Crofts[20] that the adenylate kinase reaction, unlike most mitochondrial enzyme reactions involving extramitochondrial ADP or ATP as substrate, is unaffected by atractyloside, i.e., takes place outside of the atractyloside-sensitive "adenylate translocase" system (cf. Klingenberg and Pfaff[21]). Of interest in this connection are also the recent findings of Norum *et al.*[10] that the ATP-dependent fatty acyl-CoA synthetase is associated with the outer membrane, and the palmityl-CoA carnitine transferase with the inner membrane of rat liver mitochondria. These findings, again, are consistent with the earlier conclusion of Chappell and Crofts[20] that the fatty acid activation by the ATP-dependent enzyme takes place outside the "carnitine barrier" of the mitochondria, and that the β-oxidation of fatty acids takes place inside this barrier; activated fatty acids are transported into the mitochondria as acyl-carnitines.[22] The location of the ATP-dependent acyl-CoA synthetase in the outer mitochondrial membrane[10] is interesting in view of the fact that atractyloside inhibits fatty acid activation by externally added ATP in intact mitochondria.[20] Earlier work of Klingenberg and Pfaff[21] has indicated that the "atractyloside barrier" is located in the inner mitochondrial membrane. Studies of the effects of atractyloside on mitochondrial preparations devoid of outer membrane may be helpful in clarifying this problem.

It is noteworthy that a number of enzymes so far found to be associated with the outer membrane of liver mitochondria, namely NADH-cytochrome b_5 reductase, cytochrome b_5, monoamine oxidase, and the ATP-dependent fatty acyl-CoA synthetase, are also present in liver microsomes. Other liver-microsomal enzymes, such as NADPH-cytochrome *c* reductase, cytochrome P-450, or glucose 6-phosphatase, seem to

[19] P. Siekevitz and V. R. Potter, *J. Biol. Chem.* **215**, 237 (1955).

[20] J. B. Chappell and A. R. Crofts, *Biochem. J.* **95**, 707 (1965).

[21] M. Klingenberg and E. Pfaff, *in* "Regulation of Metabolic Processes in Mitochondria," Vol. 7, p. 180. BBA Library, 1966.

[22] J. Bremer, *J. Biol. Chem.* **237**, 3628 (1962); I. B. Fritz, *Advan. Lipid Res.* **1**, 285 (1963).

be absent from the mitochondrial outer membrane.[2] It may be speculated that the outer mitochondrial membrane arises from endoplasmic membranes, which, after undergoing this specialization, lose some of their enzymatic functions; in fact, the three last-mentioned enzymes are well documented to be subject to vigorous control by substrates and hormones. Alternatively, the two types of membrane may arise from a common origin, after which each of them may acquire special additional enzymatic complements according to its physiological function.

Some of the enzyme distribution patterns described above are strikingly different from those recently reported by Green and associates[23] for beef heart mitochondria (cf. also this volume [72a]). These authors have concluded that all enzymes involved in the citric acid cycle and in fatty acid metabolism are associated with the outer mitochondrial membrane. It appears unlikely that there would exist such fundamental differences in intramitochondrial enzyme topography between two animal species or organs. More probably, these discrepancies reflect differences in the separation methods employed and/or in the criteria chosen for the morphological identification of the resulting subfractions.

[23] E. Bachmann, D. W. Allmann, and D. E. Green, *Arch. Biochem. Biophys.* **115,** 153 (1966); D. W. Allmann, E. Bachmann, and D. E. Green, *Arch. Biochem. Biophys.* **115,** 165 (1966); D. E. Green, E. Bachmann, D. W. Allmann, and J. F. Perdue, *Arch. Biochem. Biophys.* **115,** 172 (1966).

[73] Determination of Nonheme Iron, Total Iron, and Copper

By PHILIP E. BRUMBY and VINCENT MASSEY

Introduction

An extensive variety of methods has been described for the determination of iron and copper in biological materials,[1-3] some of which have been covered in a previous volume in this series.[4] However, most of them require amounts of these elements prohibitively large for the

[1] E. B. Sandell, "Colorimetric Determination of Traces of Metals," 3rd ed. Wiley (Interscience), New York, 1959.

[2] L. M. Melnick, *in* "Treatise on Analytical Chemistry" (I. M. Kolthoff and P. J. Elving, eds.), Vol. 2, Part II, Section A, p. 247. Wiley (Interscience), New York, 1962.

[3] W. C. Cooper, *in* "Treatise on Analytical Chemistry" (I. M. Kolthoff and P. J. Elving, eds.), Vol. 3, Part II, Section A, p. 1. Wiley (Interscience), New York, 1961.

[4] R. Ballentine and D. D. Burford, Vol. III, p. 1002.

analysis of a number of oxidizing enzymes or other samples available in very small quantities. For this reason, methods of analysis more appropriate for use with such small samples have been developed, and some of them are presented below. In addition to methods chosen for their sensitivity, others have been selected for use in determinations of the valency state of the elements in the sample.

As a good account of the general techniques and precautions used in colorimetric methods for metal analysis is already available,[1] this topic will not be considered here. It must be emphasized however, that a test for interference should always be made by adding copper or iron to the sample and evaluating its recovery, before the validity of any determination can be accepted. Also, the metal content should always be determined by reference to a calibration curve prepared under the same conditions. Values obtained by extraction techniques should be compared with those by a total digestion method for each type of sample when analyses are first undertaken.

Iron Determinations

Stable, strongly colored complexes of the hexacovalent type $Fe^{++} L_3$ are formed in a weakly acidic, neutral, or weakly alkaline medium between iron and 1,10-phenanthroline and various derivatives of this compound. The high absorbance of these complexes makes the phenanthrolines useful in the determination of trace amounts of iron. A number of elements, especially divalent metals, also form complexes and cause interference. Fortunately most of these elements, e.g., cadmium, silver, bismuth, are not likely to be present in biological samples. Interference is also caused by groups which form complexes with iron, e.g. phosphate and pyrophosphate. By careful control of pH and by adopting certain other precautions, many interferences may be minimized or eliminated[1,2,5,6] (see the table). Solutions of phenanthroline are stable for several weeks at 0°. They should be discarded if at all colored.

Nonheme Iron

Nonheme iron may be extracted quantitatively from most materials with trichloroacetic acid, e.g. NADH dehydrogenase,[7] succinic dehydrogenase,[8] and the liberated iron determined as described in method A. For materials from which all the nonheme iron is not extracted by tri-

[5] W. B. Fortune and M. G. Mellon, *Ind. Eng. Chem., Anal. Ed.* **10,** 60 (1938).
[6] G. F. Smith, W. H. McCurdy, and H. Diehl, *Analyst* **77,** 418 (1952).
[7] H. R. Mahler and D. G. Elowe, *J. Biol. Chem.* **210,** 165 (1954).
[8] V. Massey, *J. Biol. Chem.* **229,** 763 (1957).

INTERFERENCES IN THE DETERMINATION OF IRON BY 1,10-PHENANTHROLINE[a-d]

Element or ion	Tolerance limit for 2 ppm iron (ppm)	pH	Notes
Cd	50	—	Form slightly soluble complexes with 1,10-phenanthroline and reduce the intensity of the color. Interference is diminished by using a larger excess of reagent
Zn	10	—	
Hg (II)	1	—	
Hg (I)	10	3–9	
Be	50	3.0–5.5	Below pH 3.0 a stable complex is formed. Above pH 5.5 the hydroxide is precipitated
Mo (VI)	100	5.5–9.0	Produces turbidity below pH 5.5
W	5	—	Decreases color intensity
Cu	10	2.5–4.0	Interference reduced by using 4,7-diphenyl-1,10-phenanthroline
Ni	2	—	Produces change in color and increase in absorbance below 540 mμ
Co	10	3.0–5.0	Produces yellow color
Sn (II)	20	2.0–3.0	—
Sn (IV)	50	2.5	—
Zr	10	—	—
Cr	25	—	—
Mn (II)	200	—	—
Ag	—	—	Precipitates formed
Bi	—	—	—
Oxalate	500	6.0–9.0	—
Tartrate	500	3.0–9.0	—
Fluoride	500	4.0–9.0	—
Perchlorate	—	—	Forms precipitate if present in more than small amounts. This may be overcome by the addition of pyridine which forms a complex with perchlorate[e]
Pyrophosphate	50	6.0–9.0	Interference reduced by standing 1 hour or more
Phosphate	20	2.0–9.0	In the presence of aluminum with phosphate, iron is carried down with an aluminum phosphate precipitate. Addition of citrate after addition of 1,10-phenanthroline and reducing agent, and before pH adjustment overcomes this difficulty

[a] E. B. Sandell, "Colorimetric Determination of Traces of Metals," 3rd ed. Wiley (Interscience), New York, 1959.

[b] L. M. Melnick, *in* "Treatise on Analytical Chemistry" (I. M. Kolthoff and P. J. Elving, eds.), Vol. 2, Part II, Section A, p. 247. Wiley (Interscience), New York, 1962.

[c] W. B. Fortune and M. G. Mellon, *Ind. Eng. Chem., Anal. Ed.* **10,** 60 (1938).

[d] G. F. Smith, W. H. McCurdy, and H. Diehl, *Analyst* **77,** 418 (1952).

[e] B. F. Cameron, A Comparative Study of Hemoglobin M, Ph.D. Thesis, University of Pennsylvania, Philadelphia, 1962.

chloroacetic acid, ethanol after dithionite may be used (method B). By carrying out the color reaction in the presence and the absence of a reducing agent, method A may be used to determine the valency state of the iron in the trichloroacetic acid extract. It should be noted, however, that liberation of protein groups during the extraction process may cause changes in the oxidation state of the iron, so that the valency in the extract may be quite different from that in the sample. For example, hydrogen sulfide produced upon acidification of nonheme iron proteins containing inorganic or labile sulfide[8,9] will reduce iron to the ferrous state. This may be overcome by treatment of the sample with trichloroacetic acid containing a mercurial. The concentration of mercurial required may vary from sample to sample and should be ascertained by using a range of conditions. A concentration of 0.05 *M* *p*-chloromercuriphenylsulfonic acid in the trichloroacetic acid precipitant is often satisfactory.

Method A (20–300 millimicromoles)

The method described is that presented by Massey.[8]

Reagents

Trichloroacetic acid, 20% (w/v) ± 0.05 *M* *p*-chloromercuriphenylsulfonic acid
1,10-Phenanthroline, 0.1% (w/v)
Ascorbic acid, 0.06 *N*
Acetic acid, 0.06 *N*
Ammonium acetate, saturated solution
Standard iron solution, 1.78×10^{-2} *M*: 1.0 g of electrolytic Fe or Fe wire is dissolved in 50 ml 1:3 HNO_3, boiled to expel oxides of nitrogen, and diluted to 1 liter. This solution is diluted to 10^{-3} to 10^{-5} *M* with 5% (w/v) trichloroacetic acid.

Procedure. Into a centrifuge tube, pipette 1.5 ml of sample containing 20–300 millimicromoles of iron and 0.5 ml of 20% trichloroacetic acid. Mix and allow to stand 10 minutes at room temperature, then centrifuge at 5000 *g* for 10 minutes. Withdraw the supernatant and pipette 0.4-ml aliquots of this extract or standard iron solution into 1-ml cuvettes containing 0.36 ml of water and 0.15 ml of 1,10-phenanthroline, or 0.51 ml of water (sample blank). For estimation of ferrous iron only, add 0.05 ml of acetic acid or for ferrous + ferric iron, 0.05 ml of ascorbic acid. Finally add 0.04 ml of saturated ammonium acetate, mix, and read the absorbance at 510 mμ against distilled water.

[9] R. W. Miller and V. Massey, *J. Biol. Chem.* **240,** 1453 (1965).

A modification of this method which gives better iron recoveries under some conditions has been described.[9] The reagents are added directly to the sample in trichloroacetic acid, the colored complex is extracted into *n*-amyl alcohol, and the absorbance is read at 510 mμ against *n*-amyl alcohol.

Method B (2–30 millimicromoles)

This method was developed by Doeg and Ziegler[10] for estimation of iron in mitochondrial preparations.

Reagents

Sodium dithionite, 0.2% (w/v) prepared immediately before use by dissolving 20 mg of sodium dithionite in 10 ml of water through which pure N_2 is bubbled vigorously before and during solution

Ethanol, 95%

4,7-Diphenyl-1,10-phenanthroline, 0.2% (w/v) in 95% ethanol

Sodium acetate, 1 *M*, pH 4.6, prepared by adjusting 1 *M* sodium acetate to pH 4.6 with *N* acetic acid and made iron free by repeated extraction with 4,7-diphenyl-1,10-phenanthroline (0.083%, w/v, in isoamyl alcohol) in the presence of thioglycolic acid (1.0 ml per liter) and washing afterward with several portions of isoamyl alcohol

Standard iron solution. The solution prepared as above may be used, diluting to 10^{-3} to 10^{-5} *M* with water.

Procedure. Pipette 0.1-ml aliquots of sample or standard containing 2–30 millimicromoles of nonheme iron into stoppered test tubes. Add 0.1 ml dithionite solution and 0.7 ml 95% ethanol. Stopper the tubes and mix vigorously on a mechanical vortex shaker. Add 0.05 ml of diphenylphenanthroline or 0.05 ml of 95% ethanol (sample blank) and 0.05 ml of sodium acetate, pH 4.6. Stopper the tubes and mix vigorously once again, then place in a 38° water bath for 5 minutes to permit maximum color development. Centrifuge the tubes, and read the absorbance of the supernatant at 535 mμ against a reagent blank.

Total Iron

For the determination of total iron in biological samples, the material for analysis is most commonly subjected to an acid digestion prior to carrying out the color reaction. A number of different methods for di-

[10] K. A. Doeg and D. M. Ziegler, *Arch. Biochem. Biophys.* **97**, 37 (1962).

gesting the sample and developing the color are described (methods A, B, and C). A method is also described for the determination of total iron using an extraction procedure in place of the acid digestion (method D).

Method A (5–50 millimicromoles)

This method, developed by Beinert,[11] employs wet oxidation of the sample by nitric acid followed by evaporation to dryness. The absence of chloride in samples to be ashed by this procedure is desirable since dry heating in the presence of chloride results in loss of iron by volatilization. The dissolved digest is reduced with ascorbic acid and the color is developed with 1,10-phenanthroline.

Reagents

Nitric acid, concentrated or redistilled
Hydrochloric acid, 2 *N*
Ammonium acetate, saturated solution
Ammonia, 12.6% (w/v), concentrated ammonia diluted with equal volume of water
Ascorbic acid, 1% (w/v), prepared freshly for each determination
1,10-Phenanthroline, 0.1% (w/v)
Standard iron solution, prepared as above, diluting to 10^{-3} to 10^{-5} *M* with water

Procedure. Pipette sample or standard containing 5–50 millimicromoles of iron into a 150 × 17 mm Pyrex test tube (previously cleaned with boiling HCl and glass-distilled water), add 0.2 ml concentrated HNO_3, and evaporate the contents to dryness taking care that none of the sample is lost by spattering. As soon as drying is complete, cool the tube, and add another 0.2 ml portion of nitric acid. Repeat the digestion until the dried residue is completely colorless. Cool the tube and add 0.2 ml of 2 *N* HCl. Dissolve the residue very thoroughly by gently warming and swirling the tube. Low recoveries result if this operation is inadequate, but care must be taken to prevent evaporation of water. Add 0.65 ml of water, warm, and swirl again, then add 0.15 ml of 1,10-phenanthroline, 0.05 ml of ascorbic acid, 0.02 ml of saturated ammonium acetate, and 0.03 ml of ammonia. Mix the contents well and read the absorbance at 510 mμ against a reagent blank. In the presence of certain ions such as phosphate and pyrophosphate, reading of the absorbance may be delayed several hours until the net difference between the blank and the sample no longer increases.

[11] H. Beinert, personal communications, 1965.

Method B (50–500 millimicromoles)

The following method is based on that of Cameron[12] with modification by Smith.[13] The sample is ashed with perchloric acid and hydrogen peroxide at 100°, excess perchlorate is complexed with pyridine and the iron is reduced with sodium dithionite for the color reaction with 1,10-phenanthroline.

Reagents

Perchloric acid, 60% (w/w)
Hydrogen peroxide, 30%
Pyridine
Sodium dithionite. The solution prepared as described above may be used.
Standard iron solution, prepared as above diluting to 10^{-3} to 10^{-5} M with water
1,10-Phenanthroline, 0.2% (w/v)

Procedure. Weigh out or pipette the sample containing 50–500 millimicromoles of iron into a 10-ml Pyrex volumetric flask. Concentrate if necessary to a volume of about 0.1 ml. Add 0.2 ml of 60% perchloric acid and immerse in a boiling water bath, adding hydrogen peroxide dropwise as necessary until the digest is completely colorless. When this occurs, heat for 30 minutes more to effect complete breakdown of hydrogen peroxide, then cool to 40–50° (15–20 minutes at room temperature), add 0.5 ml of pyridine, 1.0 ml of sodium dithionite, and a further 0.5 ml of pyridine, mixing after each addition. Add 1.0 ml of 0.2% phenanthroline and make up to 10 ml with water. Read the absorbance at 510 mμ against a reagent blank.

Method C (2–30 millimicromoles)

The following method is a modification of that of Peterson.[14] The sample is ashed with sulfuric, nitric, and perchloric acids and the iron is reduced with thioglycolic acid, complexed with 4,7-diphenyl-1,10-phenanthroline, and extracted into isoamyl alcohol.

Reagents

Sulfuric acid, concentrated
Nitric acid, concentrated or redistilled

[12] B. F. Cameron, A Comparative Study of Hemoglobin M, Ph.D. Thesis, University of Pennsylvania, Philadelphia, 1962.
[13] M. H. Smith, personal communication, 1965.
[14] R. E. Peterson, *Anal. Chem.* **25**, 1337 (1953).

Perchloric acid, 60% (w/w)
Thioglycolic acid, 1% (v/v)
Sodium acetate, saturated solution adjusted to pH 6.0 with glacial acetic acid, and made iron-free as above
4,7-Diphenyl-1,10-phenanthroline, 0.083% (w/v) in isoamyl alcohol
Standard iron solution, prepared as above diluting to 10^{-3} to 10^{-5} M with water

Procedure. Pipette samples or standards containing 2–30 millimicromoles of iron into 150 × 17 mm stoppered Pyrex test tubes (previously cleaned by mock ashing when first used, or by boiling in HCl for reuse, finally flushing with water). Add 0.1 ml of concentrated H_2SO_4 and 0.1 ml of concentrated HNO_3 and heat to fumes of sulfuric acid. Cool, add another 0.1-ml portion of HNO_3, and heat as above. Cool again and add 0.05 ml of 60% perchloric acid and heat (without boiling) for 3–5 minutes. Cool and add 0.5 ml of water, 0.25 ml of 1% thioglycolic acid, 1.5 ml of saturated sodium acetate, and 1.0 ml of 4,7-diphenyl-1,10-phenanthroline. Stopper the tubes and shake on a mechanical vortex mixer for 3 minutes, then centrifuge to separate the layers. Transfer the upper layers to cuvettes and read the absorbances at 535 $m\mu$ against a reagent blank. Between readings cuvettes may be rinsed with 100% ethanol, then air dried.

Method D (2–30 millimicromoles)

This method was developed by Doeg and Ziegler[10] for estimating iron in mitochondria. Iron is released into solution by treatment of the sample with thioglycolic and acetic acids and extracted as the complex with 4,7-diphenyl-1,10-phenanthroline into isoamyl alcohol.

Reagents

Thioglycolic acid, 5% (v/v)
Acetic acid, glacial
Sodium acetate, saturated pH 6.0, prepared and made iron free as above
4,7-Diphenyl-1,10-phenanthroline, 0.083% (w/v) in isoamyl alcohol
Standard iron solution, prepared as above diluting to 10^{-3} to 10^{-5} M with water

Procedure. Pipette 0.1-ml portions of the sample or standard containing 2–30 millimicromoles of iron into stoppered test tubes, and add

0.1 ml of 5% thioglycolic acid and 0.2 ml of glacial acetic acid. Stopper the tubes and agitate vigorously on a mechanical vortex mixer for several minutes. Recovery is low if mixing is inadequate. Add 0.28 ml saturated sodium acetate, 0.32 ml water and 1.0 ml diphenylphenanthroline (or in the case of the sample blank, 1.0 ml isoamyl alcohol). Stopper the tubes, agitate them vigorously as above, centrifuge to separate the phases, transfer the upper layers to cuvettes, and read the absorbances at 535 mμ against isoamyl alcohol.

Copper Determinations

There are a number of reagents which form more or less specific colored complexes with copper, so that interference can usually be overcome by selection from the methods of determination available. This is especially so for biological samples which are generally free from interfering elements, and preliminary separations are seldom required. If a separation is unavoidable, Sandell[1] and Cooper[3] may be consulted for information about useful techniques.

As for iron determinations, the color reaction with copper is usually preceded by digestion of the sample (methods A and B). An extraction method has also been described and may be more convenient in some cases. It has also been used to determine the valency state of the copper in the sample.[15,16]

Method A (5–50 millimicromoles)

This method is a micro adaptation of that of Martens and Githens[17] and uses zinc dibenzyldithiocarbamate for complexing with the copper. This reagent is superior to the more commonly used sodium diethyldithiocarbamate since the copper complex is more light stable and the extraction may be made from a digest of higher acidity. This feature reduces inteference by ferric iron, nickel, and cobalt. Antimony, bismuth, mercury, and silver combine with the reagent and inhibit copper extraction.

Reagents

Sulfuric acid, concentrated
Nitric acid, concentrated or redistilled
Hydrogen peroxide, 30%
Zinc dibenzyldithiocarbamate, 0.01% (w/v) in reagent grade carbon tetrachloride

[15] G. Felsenfeld, *Arch. Biochem. Biophys.* **87**, 247 (1960).
[16] D. E. Griffiths and D. C. Wharton, *J. Biol. Chem.* **236**, 1850 (1961).
[17] R. I. Martens and R. E. Githens, Sr., *Anal. Chem.* **24**, 991 (1952).

Copper standard, 10^{-3} M. Dissolve 0.2497 g of clear uneffloresced crystals of $CuSO_4 \cdot 5\ H_2O$ in water, add 100 ml of N HCl, and make up to 1 liter with water. When required dilute to 10^{-4} to 10^{-6} M with 0.1 N HCl.

Procedure. Weigh out or pipette sample and standards containing 5–50 millimicromoles of copper into stoppered 150 × 17 mm Pyrex tubes (previously cleaned by mock ashing, washing with dibenzyldithiocarbamate solution, and rinsing with CCl_4). Add 0.1 ml of concentrated sulfuric acid and heat carefully until charring begins. Cool, add 0.1 ml of concentrated nitric acid and continue heating, making further additions of nitric acid as charring recurs, until upon heating to fumes of sulfuric acid no charring takes place. Cool and add 1 drop of hydrogen peroxide and heat again to fumes of sulfuric acid, repeating this process until the digest is colorless. Cool again, add 2 ml of water and boil to fumes of sulfuric acid. Cool and dilute with 3.5 ml of water. Add 1.0 ml of dibenzyldithiocarbamate, stopper the tube, and agitate vigorously on a vortex shaker for 3 minutes. Centrifuge to separate the layers, withdraw the lower layer with a fine-tipped pipette into a 1-ml cuvette, and read the absorbance at 435 mμ against a reagent blank.

A modification of this method in which copper is extracted by boiling the sample in N sulfuric acid instead of ashing has also been described.[18]

Method B (1–10 millimicromoles)

In spite of its high color and poor stability, dithizone is frequently used in copper determinations because of its extreme sensitivity. It also has the advantage of complex formation in dilute mineral acid. Silver, gold, palladium, bismuth, mercury, and iron interfere, but nickel, cobalt, lead, zinc, and cadmium are without effect unless present in high concentrations. If interference is encountered, preliminary extraction of the digest with dithizone in carbon tetrachloride must be made. The organic layer containing the copper is then washed twice for 2–3 minutes with an equal volume of 0.1 N HCl with bromide added for removal of mercury, or iodide for silver and bismuth. The organic solvent is then evaporated, and the residue is redigested.[1] The procedure described below was developed by Beinert.[11]

Reagents

Sulfuric acid, concentrated
Hydrogen peroxide, 30%
Dithizone 0.001–0.0012% (w/v) in carbon tetrachloride. The latter

[18] I. Stone, R. Ettinger, and C. Gantz, *Anal. Chem.* **25**, 893 (1953).

must be reagent grade and have passed the dithizone test. The 0.001% solution is advantageously prepared immediately before use by diluting a 0.01% (w/v) solution. The concentrated reagent is quite stable if kept cold and dark.

Sulfuric acid, 0.05 N

Standard copper solution, 10^{-3} M, prepared and diluted as above

Procedure. Use pipettes and tubes cleaned in hot aqua regia, washed with water followed by dithizone, then rinsed thoroughly with carbon tetrachloride, and finally air dried. Pipette samples containing about 1–10 millimicromoles of copper into 150 × 15 mm stoppered Pyrex tubes. Add 0.1 ml of concentrated H_2SO_4 and evaporate to fumes of sulfuric acid, taking care to avoid excessive foaming. Cool, add 1 drop of hydrogen peroxide, heat to fumes of sulfuric acid again and reflux until the walls of the tubes are free from particles. Repeat this process until, after refluxing, the digest is clear and colorless. Heat more strongly and drive off all the sulfuric acid. As soon as this is complete, cool the tubes and add 5 ml 0.05 N H_2SO_4. Because the dithizone complex is unstable, it is essential to exclude light as much as possible and process each sample quickly (if necessary, independently). Add 1.0 ml of 0.001% (w/v) dithizone solution, stopper the tube, and agitate vigorously on a vortex mixer for 3 minutes. Centrifuge to separate the layers, withdraw the lower layer with a fine-tipped pipette, and place in a 1-ml cuvette. Read the absorbance at 510 mμ against carbon tetrachloride.

Method C (10–50 millimicromoles)

A method has been devised for extraction and estimation of copper with glacial acetic acid and biquinoline. Since the biquinoline forms the colored complex only with cuprous copper, the method may be useful for estimating the valency state of the copper in the native material, provided certain precautions are taken, by carrying out the determination in the presence and absence of a reducing agent.[15,16] The determination is remarkably free from interference; only cyanide, thiocyanate, and oxalate interfere appreciably and must be absent. The complex is stable for several days. The method presented is taken from those of Felsenfeld,[15] Griffiths and Wharton,[16] and Fowler *et al.*[19]

Reagents

2,2′-Biquinoline, 0.1% (w/v) in glacial acetic acid

Hydroxylamine hydrochloride, 10% (w/v)

[19] L. R. Fowler, S. H. Richardson, and Y. Hatefi, *Biochim. Biophys. Acta* **64,** 170 (1962).

1-Hexanol

Standard copper solution, 10^{-3} *M*. Prepared as previously described, but diluting to 10^{-4} to 10^{-6} *M* with the medium in which the biological sample is dissolved.

Procedure. If the valency of the copper in the sample is to be determined, cupric ions must be protected from reduction by sulfhydryl groups in the sample by the additions of ethylenediaminetetraacetic acid, disodium salt, to 0.02 *M*. Alternatively, the sulfhydryl groups may be blocked by titration with *p*-chloromercuribenzoate before the analysis is made.[15] Pipette 1 ml of the sample or standard containing 10–50 millimicromoles of copper into stoppered Pyrex test tubes and add 1.0 ml of biquinoline reagent, or 1.0 ml of glacial acetic acid (for sample blank). For total copper add also 0.05 ml of hydroxylamine hydrochloride, or for cuprous copper only, 0.05 ml of water. Mix, allow to stand 5 minutes, then add 2 ml of 1-hexanol. Stopper the tubes and agitate vigorously on a vortex shaker for 1 minute. Separate the phases by low speed centrifugation, then transfer a portion of the organic phase to a cuvette and read the absorbance at 540 mμ against 1-hexanol.

A modification of the method not involving extraction of the complex into an organic solvent has also been used.[16] Ethanol, 0.95 ml, is added in place of the hexanol and the absorbance is read at 535 mμ.

[74] The Fluorometric Determination of Mitochondrial Adenine and Pyridine Nucleotides

By Ronald W. Estabrook, John R. Williamson, Rene Frenkel, and Pabitra K. Maitra[1]

Spectrophotometric methods have been developed[1a] to measure the extent of reduction of the cytochromes or pyridine nucleotides of mitochondria during reactions of oxidative phosphorylation. Such methods, however, cannot distinguish between DPNH and TPNH, nor can they evaluate modifications in the intramitochondrial balance of ATP, ADP, and AMP. Determination of the concentrations of oxidized and reduced diphospho- and triphosphopyridine nucleotides, as well as adenine nucleotides, can only be accomplished by rapidly terminating mitochondrial reactions with acid or alkali followed by extraction and assay of the various forms of the nucleotides.

[1] This manuscript was prepared while the author was at the Johnson Research Foundation, University of Pennsylvania, Philadelphia, Pennsylvania.

[1a] B. Chance, see Vol. IV, p. 273.

The intramitochondrial concentration of either adenine or pyridine nucleotide is generally between 1 and 10 millimicromoles per milligram of protein. Evaluation of changes in concentrations of these nucleotides under conditions of experiments designed to measure other mitochondrial activities, such as oxygen utilization or cytochrome oxidation and reduction, requires sufficiently sensitive methods to quantitatively determine 5×10^{-11} mole of nucleotide. A variety of methods have been employed for the quantitative microanalysis of adenine and pyridine nucleotides. Ciotti and Kaplan[2] measured the concentration of oxidized pyridine nucleotides by conversion of these compounds to a highly fluorescent derivative with either methyl ethyl ketone or concentrated alkali. Lowry *et al.*[3] have recently introduced a technique whereby the nucleotide serves as a rate-limiting component in the coupled recycling of two enzyme reactions. This method permits the measurement of relatively high levels of an accumulated product effectively amplifying the concentration of the nucleotide. Although the concentrations of mitochondrial adenine and pyridine nucleotides are generally much lower than that assayable using conventional spectrophotometric methods, Klingenberg and Slenczka[4] have extended the "optical test" introduced by Warburg[5] for use with the highly sensitive dual wavelength spectrophotometer. Since reduced pyridine nucleotide is readily detected by its fluorescence, coupled enzymatic methods, based on the methods of Bücher,[6] were developed[7] to measure changes in concentrations of mitochondrial adenine and pyridine nucleotides. These methods have been extended[8] for the measurement of the concentrations of a number of other intermediates of metabolism.

Apparatus

Any one of a number of commercial fluorometers may be conveniently used for the measurement of reduced pyridine nucleotide fluorescence. In general, a sample is irradiated by an intense source of monochromatic light, for example the 365 mμ emission line from a mercury lamp. The emitted fluorescent light is detected by a photomultiplier after trans-

[2] M. M. Ciotti and N. O. Kaplan, see Vol. III, p. 890.

[3] O. H. Lowry, J. V. Passonneau, D. W. Schulz, and M. K. Rock, *J. Biol. Chem.* **236,** 2746 (1961).

[4] M. Klingenberg and W. Slenczka, *Biochem. Z.* **331,** 486 (1943).

[5] O. Warburg and W. Christian, *Biochem. Z.* **314,** 399 (1943).

[6] T. Bücher, unpublished procedures. See "Sammlung von Laboratoriumvorschriften, Biochemica Boehringer," C. F. Boehringer und Soehne, GmbH, Mannheim. Also see "Methods of Enzymatic Analysis" (H. U. Bergmeyer, ed.). Academic Press, New York, 1963.

[7] R. W. Estabrook and P. K. Maitra, *Anal. Biochem.* **3,** 369 (1962).

[8] P. K. Maitra and R. W. Estabrook, *Anal. Biochem.* **7,** 472 (1964).

mitting through a secondary filter such as a Wratten 2C.[9] The instrument employed should fulfill the following criteria or contain the following characteristics: (1) A stable, intense source of monochromatic light of wavelength 340 mμ to 365 mμ. (2) External outlets for conveniently connecting an amplifier and recorder. (3) A scale expander unit together with a means of electrical offset for compensation of nonspecific background fluorescence.[7] The scale expander unit should be sufficient to permit recording full scale a fluorescence change equivalent to 0.1 μM DPNH. (4) Sufficient stability in the electronic circuitry to permit detection of 5×10^{-11} mole of reduced pyridine nucleotide per milliliter of the assay mixture at a noise level of less than 10 to 1. (5) A means of easily adding reactants to the reaction cuvette.[10] (6) Temperature control of the reaction cuvette. (7) A convenient means of cleaning the instrument of dust and lint. (8) If turbid solutions are ever employed, the instrument must have favorable optical geometry of the actinic and emitted light,[11] i.e., the photomultiplier employed for detecting the fluorescent light is placed so that the emitted light originates from the surface of the cuvette exposed to the activating light beam.

The details of the electronic circuitry associated with the instruments in the authors' laboratory have been described elsewhere.[7]

Termination of the Reaction

The validity of extrapolating the results obtained upon analysis of extracts of mitochondria, to the conditions prevailing during removal of the sample, depends upon the success of rapidly terminating all enzymatic reactions. Oxidized pyridine nucleotides as well as adenine nucleotides are determined from acid extracts of mitochondria. With mitochondria, perchloric acid is favored, since potassium perchlorate can be largely removed in the subsequent neutralization of the extract and the remaining potassium perchlorate does not significantly interfere with the enzymatic assays. In contrast, neutralized trichloroacetic acid may inhibit some enzymes used in the subsequent fluorometric assays, such as glucose 6-phosphate dehydrogenase.

A 2-ml aliquot of dilute mitochondrial suspension (usually 1–2 mg of protein per milliliter) is added to 1 ml of a 15% solution of perchloric acid in a heavy glass-walled centrifuge tube. The suspension is mixed

[9] Wratten Gelatin Filter obtained from the Eastman Kodak Company, Rochester, New York.

[10] Reactants, such as small volumes of enzyme, are easily added from the tip of a flattened stirring rod.

[11] An Eppendorf photometer with fluorometer attachment (Netheler and Hinz, Hamburg, Germany) has the desired optical geometry for measuring fluorescence of turbid suspensions.

vigorously for a few seconds and then stored in ice until centrifuged. After a sufficient number of samples have been obtained in this manner, denatured protein is removed by centrifugation at about 12,000 *g* for 10 minutes, using a Sorvall RC-2 refrigerated centrifuge. Aliquots (2 ml) of the supernatant fluid are carefully pipetted from the protein pellet and transferred to a second heavy-walled glass centrifuge tube. The supernatant is neutralized either by the slow addition of 1 ml of a solution 0.4 *M* with respect to triethanolamine (TRA) pH 7.4 and 1.8 *N* with respect to KOH, or, alternatively, by titration with a solution 0.5 *M* with respect to TRA and 6 *N* with respect to K_2CO_3. The neutralizing solutions must be added slowly and with constant mixing in order to avoid localized regions of alkaline pH. It is imperative that the pH of the neutralized sample be between 6.5 and 7.0. This is ascertained with a pH meter or by removing a drop of the supernatant fluid with a glass rod and testing with suitable pH paper. After remaining in ice for about 5 minutes, a heavy white precipitate of potassium perchlorate should be visible. The exact composition of the neutralizing solution may require some adjustment, as determined by first carrying out preliminary tests with a 2-ml aliquot of a mixture prepared by adding 1 ml of the 15% perchloric acid solution to 2 ml of the buffer mixture used in the reaction employing mitochondria.

The neutralized sample is centrifuged for 5 minutes at 8000 *g* to remove the precipitated potassium perchlorate. The clear supernatant is carefully decanted into test tubes and stored in the cold until assayed for oxidized pyridine nucleotides or adenine nucleotides.

Since reduced pyridine nucleotides decompose in acid solution, it is necessary to terminate the mitochondrial reactions in alkali. A 2-ml aliquot of a mitochondrial suspension (1–2 mg protein per milliliter) is added to a heavy-walled glass centrifuge tube containing 0.5 ml of a freshly prepared 2 *N* KOH solution in ethanol (prepared by dissolving 11.2 g of KOH pellets in about 10 ml of water and diluting to 100 ml with 99% ethanol). The alkaline-treated samples are heated at 55° for 60 seconds, cooled, and then titrated to pH 8.4 with a 1 *M* solution of unneutralized triethanolamine HCl (pH ca. 5.5) which is added slowly and with vigorous stirring. The neutralized samples are stored in ice and, after a sufficient number of samples have been obtained, they are centrifuged at 25,000 *g* for 10 minutes in a refrigerated Servall RC-2 centrifuge. The supernatant fluid is carefully decanted from the pellet of protein to test tubes, and the neutralized alkaline extracts stored in the cold.

Assay for ATP, ADP, and AMP

The content of ATP in acid extracts of mitochondria may be determined by two alternate methods. Both methods utilize coupled enzymatic systems ultimately involving the reduction or oxidation of pyridine nucleotide. The first method involves the quantitative phosphorylation of glucose by ATP with hexokinase and the simultaneous measurement of TPN reduction in the presence of glucose 6-phosphate dehydrogenase. An aliquot of the neutralized perchloric acid extract of mitochondria (usually 0.1 ml or 0.2 ml) is diluted to 2.0 ml with a reaction mixture containing 50 m*M* triethanolamine buffer of pH 7.4, 10 m*M* KCl, 10 m*M* $MgCl_2$, 5 m*M* EDTA of pH 7.4, 10 m*M* glucose, and 0.10 m*M* TPN. The sample is placed in a cuvette in the fluorometer and the background fluorescence is compensated by the associated electrical offset circuit. An aliquot of glucose 6-phosphate dehydrogenase (about 2 μg) is next added to ensure that glucose 6-phosphate, possibly present in the sample, is suitably determined. After establishment of a stable baseline of fluorescence, an aliquot of hexokinase (about 1 μg) is added to the reaction mixture and the increase in fluorescence associated with TPN reduction determined. The concentration of enzymes used should be sufficient to complete the reaction in about 3 minutes. As with all the assays described here, a second sample should be used to which an aliquot of a standard solution of ATP (previously standardized spectrophotometrically) has been added. The extent of fluorescence change can be related to that obtained with a standard solution of ATP and the concentration of ATP directly determined.

The second method for determination of ATP involves the coupled enzyme system of phosphoglycerokinase and triosephosphate dehydrogenase as described by Bücher.[6] An aliquot of the neutralized perchloric acid extract of mitochondria (usually 0.1 ml or 0.2 ml) is diluted to 2.0 ml in a buffer mixture containing 110 m*M* triethanolamine buffer of pH 7.4, 10 m*M* KCl, 1 m*M* $MgCl_2$, 5 m*M* cysteine or 5 m*M* mercaptoethanol, 6 m*M* of 3-phosphoglyceric acid (tricyclohexylammonium salt), and about 5 μ*M* DPNH. The mixture is placed in a cuvette in a fluorometer and the fluorescence of DPNH compensated by the associated offset circuit. The decrease in fluorescence occurring on addition of glyceraldehyde phosphate dehydrogenase (about 40 μg) and phosphoglycerate kinase (about 10 μg) is then recorded. After completion of the reaction, an aliquot of a standardized solution of ATP can then be added to calibrate the change in fluorescence. This method, however, is less specific than the above method with glucose 6-phosphate dehydrogenase.

The ADP content of mitochondria can be determined by the coupled enzymatic reactions involving pyruvate kinase and lactic dehydrogenase. An aliquot of the neutralized perchloric acid extract (0.1–0.5 ml) is diluted to 2.0 ml with a buffer mixture containing 50 m*M* potassium phosphate buffer of pH 7.0, 10 m*M* KCl, 5 m*M* $MgCl_2$, 10 μM DPNH, and 0.8 phosphoenolpyruvate (tricyclohexylammonium salt).[12] It is most important not to prepare this reaction mixture and store it, as the phosphoenolpyruvate decomposes to pyruvate in the presence of magnesium ions. As described above, the sample mixture is placed in a cuvette in the fluorometer and electrically compensated for the fluorescence of DPNH. An aliquot of lactic dehydrogenase (about 2 μg) is added and the decrease due to pyruvate present in the sample is determined. An aliquot of pyruvate kinase (about 10 μg) is then added and the further decrease in fluorescence is a measure of the ADP present in the sample. A second aliquot of enzymes should always be added at the end of the reaction to determine the extent of the fluorescence due to enzyme addition alone.

The concentration of AMP can be determined on the same sample used for ADP analysis by adding ATP (about 5 μM) and myokinase (about 20 μg). The extent of decrease in fluorescence associated with ADP formation by the myokinase reaction is equal to two times the concentration of AMP in the sample. The extent of fluorescent change is calibrated by carrying out the same assay procedure with a sample fortified with an aliquot of a standardized solution of ADP and AMP.[13]

Assay for DPN and TPN

Using the principles developed above, the content of DPN and TPN in the neutralized perchloric acid extracts can be determined by measuring the increase in fluorescence associated with reduction of the pyridine nucleotides.

For the determination of DPN, an aliquot (0.1–0.5 ml) of the neu-

[12] The assay can also be carried out with 50 m*M* TRA buffer, pH 7.4, instead of phosphate, but it has been consistently observed that the reactions in phosphate buffer are faster and show a much sharper cutoff point. It is important to use the minimum possible amounts of enzymes when phosphate is employed in order to avoid the precipitation of $Mg(NH_4)PO_4$ in the cuvette. If necessary, the phosphate concentration can be reduced to 30 m*M*.

[13] Commercially available samples of DPNH may contain 5–15% AMP as an impurity. Recently M. Höfer and W. Hempfling (unpublished method) have developed a method whereby 10 mg of DPNH is dissolved in 3 ml of 0.1 *M* triethanolamine buffer, pH 8.2, and 100 μg of bacterial alkaline phosphatase, Type III (Sigma Chemical Co., St. Louis, Missouri) is added. The reaction mixture is incubated at 37° for 5 minutes, and the reaction is terminated by the addition of 0.4 ml of 1 *N* KOH. The alkaline phosphatase is destroyed by a further 5-minute incubation at 37°, and the pH is adjusted to 8.2 by the careful addition of HCl. The volume is adjusted to 5 ml with 0.1 triethanolamine buffer, pH 8.2, to give a final concentration of approximately 2.5 m*M* DPNH.

tralized acid extract is diluted to 2.0 ml with a buffer mixture containing 0.2 *M* glycine, 0.4 *M* hydrazine hydrate, and 0.15 *M* ethanol adjusted to pH 9.0 with KOH. After compensation for the background fluorescence of the sample, the increase in fluorescence occurring on addition of alcohol dehydrogenase[14] (about 15 μg) is determined. After completion of the reaction, an aliquot of a standardized solution of DPN is added to the reaction mixture to directly calibrate the extent of fluorescence change.

The TPN content of mitochondria is determined by diluting an aliquot of the neutralized perchloric acid extract to 2.0 ml in a buffer mixture containing 0.1 *M* triethanolamine buffer, pH 7.4, and 0.01 ml of 0.1 *M* glucose 6-phosphate. The increase in fluorescence occurring on addition of glucose 6-phosphate dehydrogenase (about 5 μg) is determined as described above. As with the determination of DPN, it is necessary to make an addition of a standard solution of TPN to the reaction mixture as a means of calibrating the fluorescent change.

Determination of Reduced Pyridine Nucleotides

The concentrations of DPNH and TPNH in neutralized alkaline extracts of mitochondria can be determined fluorometrically in the sample by measuring the decrease in fluorescence associated with the enzymatic oxidation of the reduced pyridine nucleotides. An aliquot of the neutralized alcoholic-KOH extract of mitochondria is diluted to 2.5 ml with a mixture containing 0.1 *M* phosphate buffer of pH 7.4, 33 m*M* acetaldehyde, 4 m*M* neutralized α-ketoglutarate, and 4 m*M* NH_4Cl. The DPNH content of the sample is determined by adding alcohol dehydrogenase[14] (about 2.5 μg). After establishing the extent of decrease in fluorescence associated with DPNH oxidation, an aliquot of glutamic dehydrogenase (about 1 μg) is added and the further decrease in fluorescence is a measure of the TPNH content of the sample. Rather than using alcohol dehydrogenase and acetaldehyde, comparable concentrations of lactic dehydrogenase and pyruvate may be used, although this latter enzyme system will react slowly with TPNH.

The neutralized alkaline extract of mitochondria usually contains a relatively high concentration of fluorescent materials not attributable to DPNH or TPNH. This is compensated for by use of the electric offset circuit described previously.[7]

In order to standardize the assays, it is necessary to add a known amount of reduced pyridine nucleotide to a second sample. After the

[14] A fresh solution of alcohol dehydrogenase must be used for this assay. The preferable source of alcohol dehydrogenase is the lyophilized enzyme available from the Sigma Chemical Co., St. Louis, Missouri.

sample has been placed in the cuvette with the appropriate buffer and substrate and a steady baseline obtained, a known amount of DPNH or TPNH (usually 10 μl of a 0.1 mM solution, equivalent to 1 millimicromole) is added and the increase in fluorescence is recorded. This fluorescence is subsequently subtracted from the total change after the enzyme oxidation of the reduced nucleotide is completed. This corrected value usually agrees within 5% with the value obtained with the sample without standard added.

The reduced pyridine nucleotides in the extract are not stable for long periods of time, and should be assayed as soon as possible, i.e., storage of the sample at 4° overnight or freezing of the sample results in a lowered recovery of reduced pyridine nucleotides.

General Comments

The fluorometric method described above is a very sensitive method of measuring the content of mitochondrial adenine and pyridine nucleotides in acid and alkaline extracts. Because the method is very sensitive, it is necessary to determine the range of linearity of each assay condition and to suitably select concentrations of extracts which do not contain nucleotide levels exceeding the linear range. Most extracts of mitochondria contain fluorescent substances contributing to the background fluorescence of the samples. In particular, dust and lint from laboratory tissues are the principal culprits in contributing to the instability of fluorescence during an assay. Rigid precautions must be taken to reduce the content of debris in the assay mixtures. Each assay should be measured with an internal standard of added concentrations of previously standardized adenine or pyridine nucleotides. In addition, any fluorescence contributed by the enzyme solutions added to initiate the reactions must be accounted for by adding an equal amount of the enzyme at the end of the assay.

It is important to equilibrate the reaction cuvettes to the temperature of the fluorometer sample compartment before starting each assay; otherwise a noticeable drift is observed for 3–5 minutes. The usual practice is to prepare a cuvette containing all necessary components and place it inside the fluorometer while another sample is being analyzed. This avoids the unnecessary delay mentioned above.

The most critical criteria for the success of a series of assays is the constancy of the sum of adenine nucleotides (ATP + ADP + AMP = constant) or of pyridine nucleotides (DPN + DPNH = constant and TPN + TPNH = constant) when mitochondria are subjected to changes in their respiratory activity, i.e., during uncoupling, ion uptake, etc.

The concentrations of enzymes required in the assays are dependent upon the purity and activity of the enzyme preparations employed. The ranges used in the author's laboratory are indicated above. However, the concentrations employed are generally adjusted each day by carrying out a few preliminary assays to determine that the proper range is used.

[75] Assay of Nucleotides and Other Phosphate-Containing Compounds by Ultramicro Scale Ion-Exchange Chromatography

By HANS W. HELDT and MARTIN KLINGENBERG

General Remarks

The method of ultramicro scale ion-exchange chromatography described here is based on the work of Schnitger *et al.*[1] It has been modified extensively, and automatic measurement of ultraviolet absorbancy, ^{32}P activity, and total phosphate content has been added.[2] The method as described has given reliable results for the assay of mitochondrial components. It has been particularly useful for the separation and detection of isotopically labeled components.[3] It must be stressed that in principle the chromatographic procedure might be adapted to a larger scale with commercially available chromatographic equipment.

The method enables the quantitative measurement of nucleoside mono-, di-, and triphosphates, both reduced and oxidized pyridine nucleotides, and other phosphate-containing compounds, e.g., sugar phosphates. The assay of reduced pyridine nucleotides is based on the measurement of their corresponding acid degradation products (adenosine diphosphate ribose and adenosine 2′-phosphate 5′-diphosphate ribose, respectively), which are formed quantitatively under the conditions of extraction and column chromatography employed.[4] By the use of ^{32}P measurement in suitable cases, minute amounts of compounds can be detected which would otherwise be too small for detection by UV absorbancy or total phosphate assay.[3]

[1] H. Schnitger, K. Papenberg, E. Ganse, R. Czok, T. Bücher, and H. Adam, *Biochem. Z.* **332**, 167 (1959).

[2] H. W. Heldt and M. Klingenberg, *Biochem. Z.* **343**, 433 (1965).

[3] H. W. Heldt, *in* "Regulation of Metabolic Processes in Mitochondria" (E. Quagliarello, E. C. Slater, S. Papa, and J. M. Tager, eds.), p. 51 (BBA Library, Vol. 7). Elsevier, Amsterdam, 1966.

[4] H. W. Heldt, M. Klingenberg, and K. Papenberg, *Biochem. Z.* **342**, 508 (1965).

Procedure

The whole procedure of ion-exchange chromatography as described below is summarized in Fig. 1.

Preparation of Extract. Rat liver mitochondria (3–8 mg of protein per milliliter of suspension) incubated in sucrose medium in the presence of 20 mM triethanolamine buffer, pH 7.2, 2 mM EDTA, 0.5–1 mM ^{32}P-phosphate (specific activity 20–30 $\times$ 10^6 cpm/micromole), and substrate are deproteinized by addition of 0.2 volume 3 M $HClO_4$. After being

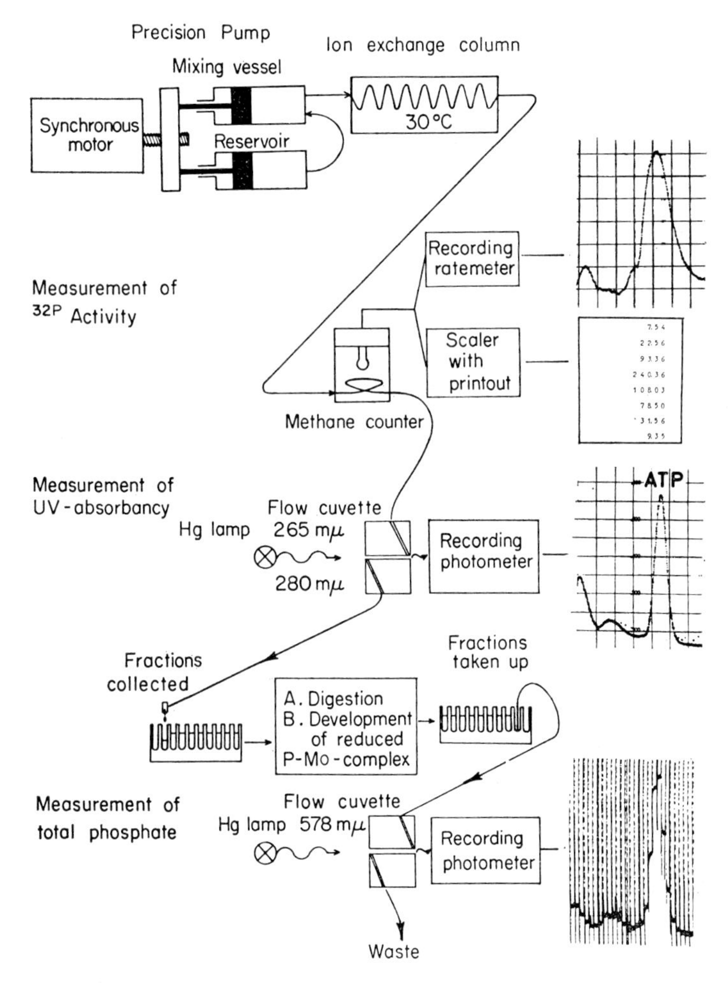

FIG. 1. Summary of ion-exchange chromatography procedure.

stirred for 10 minutes at 0° the protein precipitate is centrifuged off and the acid-soluble extract is neutralized with 3 *M* KOH to pH 6–7.

Preparation of Ion Exchange Resin. Fifty grams of Dowex 1 × 8, <400 mesh, is purified from very fine and very large particles by successive decanting. The sample is then washed on a column (2 cm in diameter) with 1000 ml 1 *N* HCl, 1000 ml 1 *N* NaOH, and 1000 ml 3 *M* sodium formate-formic acid, pH 4, and finally with distilled water. The ion exchanger thus prepared is kept in a refrigerator for stock.

Filling the Columns. A polyethylene tube (inner diameter 1.1 mm, 200 cm long), is narrowed at the end by a clamp with a small wad of quartz wool placed before it. The tubing is wound in a spiral around an aluminum block. The ion exchanger, suspended in 5 parts of water, is pressed into the tubing. Before use, the packed columns are washed with 20 ml 9 *M* formic acid + 0.9 *M* ammonium formate and afterward with distilled water until they are neutral. During chromatography the column is kept in a water bath at 30°.

Application of the Extract to the Column and Elution of the Column. The extract (equivalent to 3–8 mg of protein) is pumped through the prepared column by means of a peristaltic pump with a flow of 0.6 ml per hour, followed by 1.5 ml of distilled water. The column is then attached to a pumping system designed by Schnitger (unpublished), consisting of two syringe type piston pumps. The first piston, containing the reservoir, has one outlet, and the second piston, representing the mixing vessel, has one inlet and one outlet. The reservoir is connected by a polyethylene tube to the mixing vessel, and the column is connected to the outlet of the mixing vessel. Both pistons have the same diameter (3.5 cm) and are driven forward at the same speed by a synchronous motor. Thus at the same time 1 part of the content of the reservoir is transferred to the mixing vessel and 2 parts of the contents of the mixing vessel are pumped through the column, and so a linear concentration gradient is obtained. Mixing in the vessel is achieved as follows. Built into the cap of the vessel is a tube containing a cylinder made of a Teflon-coated magnet. Both ends of the tube are connected with the vessel. By means of an outside rotating magnet, the magnetic cylinder is brought into a backward and forward motion. Thus a small amount of the content of the vessel is sucked in at one end of the tube, and the same amount is ejected into the vessel at the other end, and vice versa, causing thorough mixing.

For the chromatographic assay shown in Fig. 2 the reservoir contained 9 *M* formic acid + 0.9 *M* ammonium formate and the mixing vessel contained 22 ml distilled water at the beginning. The flow rate through the column was 0.578 ml per hour. In other experiments columns of smaller diameter (θ 0.8 mm) have been used. Under this condition, the sensitivity of the recordings is increased twofold, at half the flow rate and content of the mixing vessel. Furthermore it is possible to shorten the time required for a chromatographic run by shortening the column, increasing the flow rate, and decreasing the volume of the mixing vessel. This concurs, however, with a decrease in the resolution of the chromatogram.

Recording of ^{32}P Activity. To measure ^{32}P activity continuously, a thin polyethylene tube of 0.5 mm diameter is inserted into the counting chamber of a methane counter. This flow cell contains a fluid volume of about 20 μl. As ^{32}P is a strong β emitter (E_{max} 1.68 Mev), the loss of ^{32}P activity by absorption of the walls of the tube (0.17 mm thickness) is less than 25%. Radioactivity is measured by a scaler and a digital recorder, which types the counts stored at 10-minute intervals. For quick survey of the chromatogram, ^{32}P activity is simultaneously recorded by a strip chart recorder attached to a logarithmic rate meter. A logarithmic chart recording is shown in Fig. 2. For quantitative evaluation of a radioactive peak the recorded digits are added up, the background being subtracted, and the sum is corrected for radioactivity decay.

Ultraviolet Recording. In the effluent of the column, ultraviolet absorbancy is recorded continuously with an Eppendorf photometer provided with a monochromator. The cuvette (obtained from Fa. Netheler & Hinz, Hamburg, Germany) has a light path of 10-mm length and 1-mm diameter and contains about 8 μl of fluid. It is automatically measured clockwise at two different wavelengths (265 and 280 mμ), the absorbancy being recorded with a multichannel chart recorder. For quantitative evaluation the area of the peaks is measured by planimetry. Molar absorbancy coefficients of nucleotides are given in a Pabst circular.[5]

Measurement of Total Phosphate

Reagents

Digestion mixture: 200 ml conc. H_2SO_4 and 100 ml 60% $HClO_4$ are added to 200 ml H_2O.

Molybdate reagent: 20.8 g sodium acetate·3 H_2O together with 3.75 g ammonium molybdate are dissolved in 1000 ml H_2O. Before use, 9 parts of the solution are added to 1 part of 10% ascorbic acid in water.

[5] Pabst Laboratories, Milwaukee, Wisconsin: Ultraviolet absorption spectra of 5′-ribonucleotides, *Circ. OR.* **10** (1956).

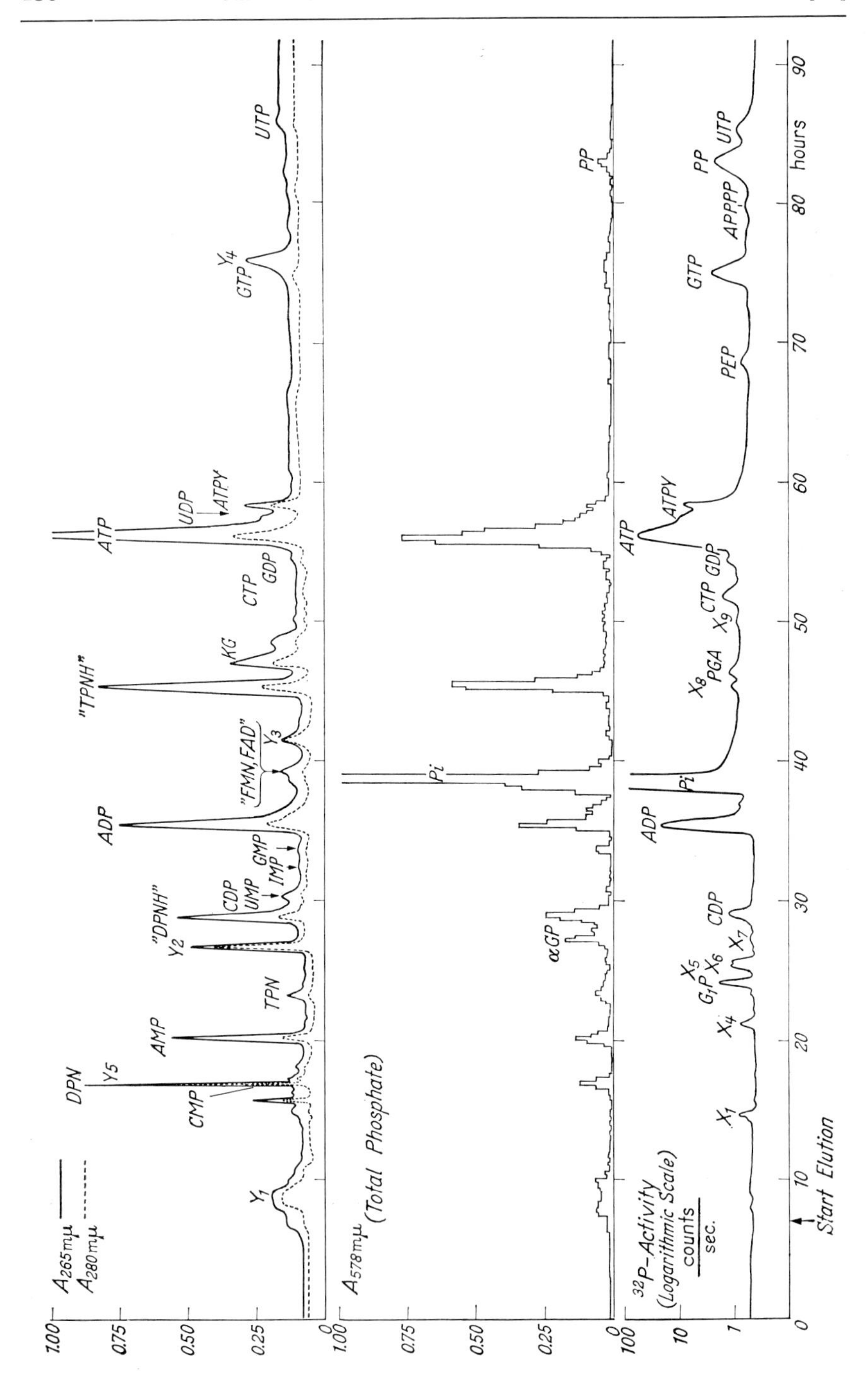
A265mμ
A280mμ
1.00
0.75
0.50
0.25
0
Y1
CMP
DPN
Y5
AMP
TPN
Y2
"DPNH"
CDP
UMP
IMP
GMP
ADP
"FMN, FAD"
Y3
"TPNH"
KG
CTP
GDP
ATP
UDP
ATPY
GTP
Y4
UTP
A578mμ (Total Phosphate)
αGP
Pi
PP
32P-Activity
(Logarithmic Scale)
counts
sec.
100
10
1
X1
X4
G1P
X5
X6
X7
CDP
ADP
Pi
X8
PGA
X9
CTP
GDP
ATP
ATPY
PEP
GTP
APPPP
PP
UTP
0
10
20
30
40
50
60
70
80
90
hours
Start Elution

The effluent is collected in Teflon racks where each rack contains 10 separate fractions. The fraction collector has been described in detail by Schnitger *et al.*[1] Nine racks and one rack with blank and standards are put on a heating block, the fractions being evaporated at 100° within 1 hour; 0.05 ml of digestion mixture is added, and the racks are heated for 2 hours at 165°. While the racks are cooling in an ice bath, 0.5 ml molybdate reagent is added to each fraction. The reduced blue phosphomolybdate complex is developed by heating the racks for 2 hours at 38° in a water bath. The racks are then placed in a modified fraction collector, the colored fractions being automatically taken up and pumped through a 1-cm flow cuvette. Absorbancy is measured at 578 mμ by an Eppendorf photometer and recorded by a chart recorder. For quantitative evaluation of the peaks, the phosphate recordings are added and corrected for the base line. The phosphate content is calculated by the calibration obtained from the standard being assayed at the same time.

Comments

Stability of DPN and TPN during Chromatography. A small decay of oxidized pyridine nucleotides (about 5%), yielding the same products as the reduced pyridine nucleotides (ADPR and ADPRP), might occur.[3] This has to be taken into account if the ratios of DPN:DPNH or TPN:TPNH are very large. This disadvantage, however is compensated for by the advantage of having a reliable method of measuring the sum of both reduced and oxidized nucleotides in one extract.

Sensitivity of the Method. The technique of ultramicro scale ion-exchange chromatography allows the detection of as little as 10^{-10} mole of nucleotide by recording the UV absorbancy. After labeling the mitochondria with ^{32}P-phosphate, continuous detection of ^{32}P in the chromatographic eluate may further increase the sensitivity of detection by more than 10.[2]

FIG. 2. Ion exchange chromatogram from rat liver mitochondria equivalent to 7 mg of protein [H. W. Heldt and M. Klingenberg, *Biochem. Z.* **343**, 433 (1965)]. The mitochondria had been incubated aerobically for 5 minutes at 18° in the presence of 5 m*M* ketoglutarate and 0.5 m*M* (^{32}P) phosphate (specific activity 22.4×10^6 cpm/micromoles). Besides the common abbreviations for nucleotides, the following were used: *KG,* ketoglutarate; *αGP,* α-glycerophosphate; *G_1P,* glucose 1-phosphate; *PGA,* phosphoglyceric acid; *PEP,* phosphoenolpyruvate; *APPPP,* adenosine tetraphosphate; *Pi,* inorganic phosphate; *PP,* inorganic pyrophosphate; *ATPY,* a different form of ATP, formed artificially during chromatography. Peaks marked with abbreviations enclosed in quotation marks are due to acid breakdown products of the corresponding substances. Peaks marked with *X* and *Y* have not been identified yet.

[76] The Quantitative Determination of Mitochondrial Hemoproteins

By J. S. RIESKE[1]

Techniques for the assay of respiratory enzymes, including the cytochrome components of these enzymes, have been described in a previous volume.[1a] This article will be limited to the quantitative determinations of hemoproteins as found in mammalian mitochondria and in preparations derived from these mitochondria. In such cases, the hemoproteins may be found mixed in varying proportions. Also, the methods described in this article are confined to procedures that may be used with instruments that are available in most chemical laboratories. Another method for the quantitative analysis of mixed cytochromes in mitochondria is that reported by Williams.[2]

Principle. Mitochondria or mitochondrial preparations are treated with bile salts to yield optically clear solutions. The concentrations of cytochromes $c + c_1$ and cytochrome *a* are estimated spectrophotometrically after preferential reduction with ascorbate. The content of cytochrome *b* is in turn estimated after reduction of the cytochrome with dithionite. The content of cytochrome *c* alone is estimated spectrophotometrically after extraction into a water–ethanol–salt mixture.

Reagents

Sodium taurocholate,[3] 10% (w/v)
Potassium deoxycholate, 10% (w/v) in deoxycholic acid[4]
Potassium cholate, 20% (w/v) in cholic acid[4]
Potassium ferricyanide, 0.05 *M*
Potassium ascorbate, crystalline[5]
Sodium dithionite, hydrosulfite
Antimycin A, 5 mg/ml, in absolute ethanol
Potassium cyanide, 0.1 *M*, neutralized

Procedure

Solubilization of the Mitochondrial Protein. A clarified solution of the mitochondrial proteins is prepared by the following procedure.[6] The

[1] See footnote 1, page 239.
[1a] See Vol. IV [12].
[2] J. N. Williams, *Arch. Biochem. Biophys.* **107,** 535 (1964).
[3] Synthesized by a modification of the method of A. Norman, *Arkiv Kemi* **8,** 331 (1955).
[4] Recrystallized from 50% ethanol.
[5] J. S. Rieske, R. E. Hansen, and W. S. Zaugg, *J. Biol. Chem.* **239,** 3017 (1964).
[6] P. V. Blair, T. Oda, and D. E. Green, *Biochemistry* **2,** 756 (1963).

mitochondria are treated with 1.0 μg of antimycin A per milligram of protein and then suspended in Tris-HCl (0.1 M, pH 8.0) to a protein concentration of 5–10 mg/ml. Potassium deoxycholate (10%) and potassium cholate (20%) are added in amounts equal to 0.5 mg per milligram of protein and 1.0 mg per milligram of protein, respectively. If slight turbidity remains, the solution is centrifuged briefly at about 100,000 g to sediment the insoluble material. Turbidity may also result from the precipitation of some structural protein.[7] This precipitation occurs to a significant extent at elevated temperatures (20–40°) in the presence of a high ionic strength (e.g., 10–20% saturation of ammonium sulfate). This method of solubilization works well for mitochondria that have been isolated from beef heart; modifications of this procedure may be required for mitochondria that have been isolated from other tissues.

Determination of Cytochromes $c + c_1$ *and* a. One-milliliter aliquots of the dispersion of mitochondria, prepared by the above procedure, are placed in two 1.0-ml cuvettes and each is then treated with 0.01 ml of 0.05 M potassium ferricyanide and 0.01 ml of 0.1 M potassium cyanide. The cuvettes are placed in a spectrophotometer equipped with a cooled chamber (0–5°). The solution in the sample cuvette then is treated with a few milligrams of solid potassium ascorbate, and the change in absorbancy at 552 mμ, with the untreated solution as reference, is recorded immediately. The concentration of cytochromes $c + c_1$ in the sample is estimated from the equation

$$\text{Cyt } c + c_1 \text{ (m}\mu\text{moles/ml)} = \frac{\Delta A_{552} \text{ (ascorbate-reduced minus oxidized)} \times 1000}{14.5 \text{ (footnote 8)}} \quad (1)$$

The concentration of cytochrome a may be estimated in the same sample by a measurement of the change in absorbancy at 605 mμ. If the preparation is deficient in cytochrome c, it may be necessary to allow several minutes for the cytochrome a to be reduced completely. Also, a small degree of bleaching occurs at 605 mμ due to the reduction of cytochrome c, cytochrome c_1, and iron-protein; however, the bleaching at 605 mμ contributed by the iron-protein is considered to be negligible. In order to correct the absorbancy change at 605 mμ for the interference by cytochromes c and c_1 about 8% of the total absorbancy change at 552 mμ is added to the absorbancy change (ascorbate-reduced minus oxidized) at 605 mμ. The content of total cytochrome a is estimated with Eq. (2).

[7] D. E. Green, H. D. Tisdale, R. S. Criddle, P. Y. Chen, and R. M. Bock, *Biochem. Biophys. Res. Commun.* **5**, 109 (1963).

[8] This absorbancy index is the mean value of the indices of 15.5 mM^{-1} cm^{-1} and 13.6 mM^{-1} cm^{-1} at 552 mμ for cytochrome c and cytochrome c_1, respectively (cf. footnotes 14 and 15).

$$\text{Cyt } a \text{ (m}\mu\text{moles/ml)} = \frac{1000}{10.4 \text{ (footnote 9)}} [\Delta A_{605} \text{ (ascorbate-reduced minus oxidized)} + 0.08\, \Delta A_{552} \text{ (ascorbate-reduced minus oxidized)}] \quad (2)$$

In solutions that are optically clear the concentration of cytochrome *a* may be estimated from the difference in absorbancy of the reduced (dithionite) preparation between 605 mμ and 630 mμ. Turbidity will exaggerate the absorbancy reading at 605 mμ in comparison to that at 630 mμ because of the increased scattering of light at shorter wavelengths. The absorbancy index for A_{605} (reduced) minus A_{630} (reduced) is about 16.5 mM^{-1} cm^{-1}.[9]

Determination of Cytochrome b.[10] An estimation of cytochrome *b* in mitochondrial preparations may be made as an extension of the determination of cytochromes $c + c_1$. This procedure is based on the phenomenon that in an aerobic system at 0° the rate of reduction of cytochrome *b* by ascorbate is balanced by its rate of autoxidation. Two cuvettes, sample and reference, containing protein samples that are solubilized in the manner required for the analysis of cytochromes $c + c_1$ are treated with solid potassium ascorbate; next, the sample cuvette is treated with a few grains of dithionite. After maximal reduction of cytochrome *b* is obtained, the absorbancy changes at 563 mμ and 577 mμ are recorded. In cases where difference spectra cannot be taken, the reference and sample cuvettes must be reversed for the reading at 577 mμ since a bleaching occurs at this wavelength. It is essential that the cuvettes be kept cold during this operation, otherwise some reduction of cytochrome *b* may occur in the reference cuvette resulting in an underestimation of the cytochrome. It also is essential that the cytochrome *a* be reduced completely in both samples, otherwise the trough at 577 mμ will not be developed fully. The index for the absorbancy difference between 563 mμ and 577 mμ in the difference spectrum is 28.5 mM^{-1} cm^{-1}. This analysis may be summarized with Eq. (3).

$$\text{Cyt } b \text{ (m}\mu\text{moles/ml)} = \frac{1000}{28.5} \times [\Delta A_{563} \text{ (dithionite minus ascorbate)} + \Delta A_{577} \text{ (ascorbate minus dithionite)}] \quad (3)$$

The above method is applicable only to mitochondrial preparations in which complex III[11] retains its structural integrity. The cytochrome *b* of damaged complex III may undergo alterations that result in a lowered absorbancy index for the cytochrome. Treatment with antimycin A protects complex III and its cytochrome *b* against the bile salts used for

[9] D. E. Griffiths and D. C. Wharton, *J. Biol. Chem.* **236,** 1850 (1961).
[10] W. S. Zaugg and J. S. Rieske, *Biochem. Biophys. Res. Commun.* **9,** 213 (1962).
[11] See this volume [44].

clarification, but not against strong detergents, such as the alkyl sulfates, or against the effects of organic solvents.

In cases where degradation of complex III has occurred (detectable by a shift of the α peak of cytochrome *b* to shorter wavelengths), it may be necessary to separate the cytochrome *b* from the other cytochromes and to determine its heme content after conversion of the heme group to the pyridine hemochrome. The submitochondrial fraction (final concentration of protein, 7–30 mg/ml) is treated with sodium taurocholate (2–3 mg per milligram of protein), and sodium acetate–acetic acid buffer, pH 5.8 (final concentration, 0.14 *M*). Saturated ammonium sulfate is added, with constant stirring, to a final concentration of 25–28% by volume. The solution should be only slightly turbid immediately after addition of the ammonium sulfate. If marked precipitation occurs, the concentrations of protein and of ammonium sulfate may require lowering. The mixture (in a 12-ml conical centrifuge tube) is incubated at 30° for 40 minutes, after which the precipitated cytochrome *b* is sedimented in a clinical centrifuge and washed with water. The washed precipitate is suspended in water (to 5–10 mg protein per milliliter) and dissolved by making the suspension 0.25 *M* in KOH and 3% in potassium cholate. A suitable aliquot of the solution is pipetted into a 1.0-ml cuvette of 1-cm light path to which is added water, pyridine, and 1 *M* KOH in the proportions required to give 1.0 ml of solution that is 0.1 *M* in KOH and 30% in pyridine. A few grains of dithionite are added and the solution is covered with mineral oil. After a few minutes the absorbancy of the pyridine hemochrome is recorded at 556 mμ. The absorbancy index at 556 mμ of pyridine ferroprotoporphyrin IX is 34.7 mM^{-1} cm^{-1}.[12]

The amount of residual cytochrome *b* in the supernatant liquid from the incubation mixture can be estimated by the usual ascorbate-dithionite method as described above.

Determination of Cytochrome c. Because of the similarity in the spectra of cytochrome c_1 and cytochrome *c*, it is necessary to separate them physically in order to obtain individual determinations. Cytochrome *c* is extracted from mitochondria and mitochondrial subfractions by the following procedure.[6]

The mitochondrial preparation is suspended in 0.05 *M* Tris-HCl, pH 8.0, and then treated with bile salts (0.60 mg of cholate and 0.70 mg of deoxycholate per milligram of protein) and ammonium sulfate to 10% saturation (0°). The final protein concentration is about 60 mg/ml. Two milliliters of absolute ethanol is added to 2 ml of the suspension,

[12] K. G. Paul, H. Theorell, and A. Akeson, *Acta Chem. Scand.* 7, 1284 (1953).

and the mixture is homogenized. Immediately, the mixture is warmed at 38° for 5 minutes, after which it is cooled to —20°. After 10 minutes at —20°, the mixture is centrifuged at top speed in a clinical centrifuge. An aliquot of the supernatant solution is buffered at 0.1 *M* with Tris-HCl, pH 7.5 (1.8 ml extract and 0.2 ml of 1.0 *M* Tris-HCl). The concentration of cytochrome *c* in this alcohol extract is determined spectrophotometrically from the difference in absorbancy at 550 mμ between the reduced (dithionite) and oxidized (ferricyanide) cytochrome. The absorbancy index used is 18.5 mM^{-1} cm^{-1}.[13]

Direct Determination of Cytochrome c_1. The content of cytochrome c_1 may be estimated directly by the "ascorbate" method (cf. Eq. 1) in mitochondrial preparations from which the cytochrome *c* has been extracted. The absorbancy index for cytochrome c_1 (reduced minus oxidized) at 554 mμ is 17.1 mM^{-1} cm^{-1}.[14] Submitochondrial particles that have undergone two or more precipitations in the presence of bile salts or alcohols during preparation usually are substantially free of cytochrome *c*. Also, practically all the cytochrome *c* may be extracted from whole mitochondria by two or three extractions of the hypotonically swelled mitochondria with 0.15 *M* potassium chloride.[15] The cytochrome *c*-depleted mitochondria then may be solubilized with bile salts and assayed for cytochromes *a*, *b*, and c_1 by the usual method.

Differential Extraction of Hemes.[16] After certain treatments, some mitochondrial fractions are refractory to solubilization by mild detergents such as bile salts. In these cases, in order to estimate the concentrations of the cytochromes it is necessary to perform a differential extraction of the hemes with acidified acetone. The hemes are then estimated in the form of pyridine hemochromes.

The sample of protein (about 20 mg) is washed or dialyzed to remove excess sucrose and bile salts; it is then homogenized in turn with 10-ml portions of cold (0°) acetone; a cold 2:1 mixture of chloroform and methanol; and finally cold acetone again. The hemes of cytochromes *a* and *b* are then extracted from the protein by three successive homogenizations with a mixture of 5 ml cold acetone and 0.05 ml of 2.4 *N* HCl. The pooled acetone extracts are evaporated to near dryness *in vacuo*, and immediately the residue is dissolved in about 1 ml of pyridine. All the operations are performed in dim light and in the cold (0°) in order to minimize decomposition of the hemes. The pyridine solution of

[13] E. Margoliash and N. Frohwirt, *Biochem. J.* **71,** 570 (1959).

[14] D. E. Green, J. Järnefelt, and H. D. Tisdale, *Biochim. Biophys. Acta* **31,** 34 (1959).

[15] See this volume [78].

[16] Based on the procedure of R. E. Basford, H. D. Tisdale, J. L. Glenn, and D. E. Green, *Biochim. Biophys. Acta* **24,** 107 (1957).

the hemes is added to an equal volume of 0.2 N KOH and mixed. If any turbidity is present, the solution is centrifuged briefly in a clinical centrifuge and the clear supernatant is recovered. The pyridine solution is treated with a small drop of dilute potassium ferricyanide (0.05 M), after which the solution is transferred to a 1.0-ml cuvette and the absorbancy at 556 mμ is recorded (water as a blank). Next, the solution is treated with a few grains of dithionite and covered with mineral oil; after a few minutes the absorbancies at 556 mμ and 587 mμ are recorded. The concentration of the hemes of cytochromes a and b in the cuvette are calculated with the following equations.

$$\text{Cyt } a \text{ heme (m}\mu\text{moles/ml)} = \frac{A_{587} \text{ (reduced)} \times 1000}{24 \text{ (footnote 17)}} \quad (4a)$$

$$\text{Cyt } b \text{ heme (m}\mu\text{moles/ml)} = \frac{\Delta A_{556} \text{ (reduced minus oxidized)} \times 1000}{30 \text{ (footnote 18)}} \quad (4b)$$

The protein residue, after extraction with HCl acetone, retains cytochromes $c + c_1$. The total content of heme of these cytochromes is estimated after direct conversion of the protein-bound hemes to their pyridine hemochromes. The protein residue is homogenized in a mixture containing equal volumes of pyridine and 0.2 N KOH. The concentration of heme from cytochromes $c + c_1$ is estimated from the difference in absorbancy of the reduced and the oxidized hemochrome at 550 mμ. Because of the turbidity usually encountered in this procedure, it is recommended that a difference spectrum (dithionite-reduced vs. ferricyanide-oxidized) be obtained with a recording spectrophotometer. The index for ΔA_{550} (reduced minus oxidized) is 19.1 mM^{-1} cm^{-1}.[19]

The validity of cytochrome analysis by this method is dependent upon the complete extraction and preservation of the hemes of cytochromes a and b. With purified complex IV (cytochrome oxidase)[20] and complex III ($CoQH_2$–cytochrome c reductase)[21] of beef heart, this procedure has yielded values for the content of cytochromes a and b, respectively, somewhat lower than those determined by direct spectrophotometry.[22–24] A modification of the above procedure has been described recently[25] in which the recovery of extracted heme a from purified cytochrome oxidase is in good agreement with the content of heme a as determined directly.

[17] W. A. Rawlinson and J. H. Hale, *Biochem. J.* **45**, 247 (1949).
[18] W. S. Zaugg and J. S. Rieske, unpublished observation, 1962.
[19] H. Theorell, *Biochem. Z.* **285**, 207 (1936).
[20] See this volume [45].
[21] See this volume [44].
[22] D. E. Green and D. C. Wharton, *Biochem. Z.* **338**, 335 (1963).
[23] J. S. Rieske and W. S. Zaugg, unpublished observations, 1962.
[24] Y. Hatefi, A. G. Haavik, and D. E. Griffiths, *J. Biol. Chem.* **237**, 1681 (1962).
[25] M. Morrison and S. Horie, *Anal. Biochem.* **12**, 77 (1965).

[77] Quantitative Determination of Mitochondrial Flavins

By N. APPAJI RAO, S. P. FELTON, and F. M. HUENNEKENS

General Considerations

Three different forms of flavin are present in mitochondria: FAD, FMN, and a covalently bound form of FAD[1] that is associated with succinate and sarcosine dehydrogenases.[2] There is little, if any, free riboflavin in mitochondria. Prior to assay, the flavins must be separated from the protein and lipid of the mitochondria under conditions that minimize destruction or interconversion (e.g., FAD → FMN). This is accomplished by treating a sample of the mitochondria (or any of the particulate preparations derived from mitochondria) with trypsin,[3,4] followed by heat and acid-denaturation. The total flavin obtained under these conditions is estimated by the decrease in absorbancy at 450 $m\mu$ upon reduction with hydrosulfite.[5] A second sample of the mitochondria is treated the same manner except that the trypsin step is omitted; the difference in total flavin between these two values is taken as the amount of peptide-bound flavin. The extracts are analyzed for FAD and FMN using the D-amino acid apo-oxidase and the TPNH-cytochrome *c* apo-reductase, respectively. The bound flavin released by tryptic digestion does not respond in either enzymatic assay system.

Preparation of Extracts from Mitochondria or Mitochondrial Subparticles and Measurement of Total Flavin

Reagents

Trypsin, crystalline (Worthington)
Perchloric acid, 50%
KOH, 6 *N*
Sodium hydrosulfite

In a 15-ml conical glass centrifuge tube, 6 ml of a suspension of

[1] E. B. Kearney, *J. Biol. Chem.* **235**, 865 (1960).

[2] Succinate dehydrogenase: T. P. Singer and E. B. Kearney, *in* "The Enzymes" (P. D. Boyer, H. Lardy, and K. Myrbäck, eds.), 2nd ed., Vol. 7, p. 383. Academic Press, New York, 1963. Sarcosine dehydrogenase: W. R. Frisell and C. G. Mackenzie, *J. Biol. Chem.* **237**, 94 (1962).

[3] D. E. Green, S. Mii, and P. M. Kohout, *J. Biol. Chem.* **217**, 551 (1955).

[4] A mixture of trypsin and chymotrypsin has also been used to release the peptide-bound flavin (see also footnote 1).

[5] Total flavin in these extracts can also be measured fluorimetrically. see D. F. Wilson and T. E. King, *J. Biol. Chem.* **239**, 2683 (1964).

mitochondria containing about 30 mg per milliliter of protein[6] (sample A) is treated with 30 mg. of trypsin.[7] The tube is agitated vigorously for 1 minute (e.g., with a Vortex Jr. mixer) to dissolve the trypsin. A second aliquot of the mitochondria (sample B) is treated in the same manner except that the trypsin is omitted. The tubes are incubated at 38° for 90 minutes, then heated at 100° for 6 minutes and cooled in an ice bath. Cold perchloric acid, 1.5 ml, is added with stirring to each sample. After centrifugation of the tubes in a clinical centrifuge for 5 minutes, the supernatant fluids are recovered by decantation. The precipitates are resuspended with vigorous mixing in 2 ml of water, and after centrifugation the second supernatant solutions are added to the first solutions. The clear yellow solutions are adjusted to pH 7.0 with KOH, and the insoluble potassium perchlorate is removed by centrifugation. The supernatant solutions are then lyophilized to dryness and the residues are redissolved in 2.0 ml of water.[8]

One milliliter each of samples A and B are placed in cuvettes having a light path of 1.0 cm. The absorbancy of each solution is measured in a Beckman spectrophotometer, Model DU, at 450 mμ against a blank containing water. After addition of 1 mg of sodium hydrosulfite to each of the three cuvettes, the absorbancy of each solution is again measured. The concentration of total flavin in each extract is calculated from the decrease in absorbancy at 450 mμ using a differential extinction coefficient[9] (oxidized minus reduced) of $9.8 \times 10^3 M^{-1}$ cm^{-1}.

Enzymatic Assay for FAD

D-Amino acid oxidase is isolated as the apoenzyme from pig kidney.[10] FAD in the above extracts is quantitated by its ability to restore oxidase

[6] Protein is measured by the biuret method [A. G. Gornall, C. J. Bardawill, and M. M. David, *J. Biol. Chem.* **177**, 751 (1949)], with cholate added to solubilize the particulate material.

[7] Mitochondria are ordinarily prepared as suspensions in sucrose, KCl, or phosphate which are usually maintained near neutrality. If not, the pH of the suspension should be adjusted to 7 before the trypsin digestion step.

[8] The above procedure may be scaled down, or the lyophilization procedure omitted, depending upon the amount of flavin in the original particulate preparation and the sensitivity of the instrument for measuring absorbancy. The suspensions may be deproteinized, alternatively, with trichloroacetic acid at a final concentration of 10%. Trichloroacetic acid is then removed by repeated extraction with ether or by passage of the acid supernatant solution through a 1 × 10 cm column of Florisil. In the latter method, the adsorbed flavins at the top of the column are washed with 25 ml of water and then eluted with a minimum volume of 10% pyridine. The eluate is extracted several times with chloroform, lyophilized to dryness, and reconstituted to the desired volume.

activity in comparison with known amounts of FAD. Reaction (1),

$$\text{D-Alanine} + H_2O + O_2 \rightarrow \text{pyruvate} + NH_3 + H_2O_2 \quad (1)$$

catalyzed by the holoenzyme, is followed either by: (a) measuring oxygen consumption; or (b) measuring DPNH disappearance using a coupled system[11] in which the pyruvate formed in reaction (1) is converted to lactate (reaction 2) via lactate dehydrogenase.

$$\text{Pyruvate} + \text{DPNH} + H^+ \rightleftharpoons \text{lactate} + \text{DPN}^+ \quad (2)$$

Reagents

D-Amino acid apooxidase. Dissolve 250 mg of the partially purified protein[10] in 10 ml of 0.05 *M* pyrophosphate buffer, pH 8.3.
Pyrophosphate buffer, 0.05 *M*, pH 8.3
DL-Alanine, 1.0 *M*, pH 8.3
FAD standard solution, 1.0×10^{-6} *M* (concentration determined spectrophotometrically at pH 7.0 using an extinction coefficient[10] at 450 mμ of 11.3×10^3 M^{-1} cm^{-1})
DPNH, 4×10^{-3} *M*
Phosphate buffer, 1.0 *M*, pH 7.5
Crystalline muscle lactate dehydrogenase (Sigma), 1 mg/ml
Crystalline catalase (Worthington), 1500 units/ml

Manometric Assay. The following reagents are added, in order, to a Warburg manometer cup: 0.5 ml of pyrophosphate buffer, 0.1 ml of DL-alanine, 0.1–1.0 ml of the extract to be tested for FAD, 0.2 ml of 6.0 *N* sodium hydroxide (center well), 0.5 ml of enzyme, and water to make 3.0 ml. Several cups are prepared which contain graded amounts of FAD in place of the extract. Two additional "blank" cups are prepared in which the FAD and the DL-alanine, respectively, are omitted. Manometry is carried out at 30°. Oxygen uptake values are averaged for successive 10-minute intervals over the period of 1 hour. The values for any cup containing the complete system should be corrected for the two blanks. Under these conditions, FAD at a final concentration of 3.3×10^{-7} *M* will produce an oxygen uptake of about 40 μl per 10 minutes.[12] The activity is linear with FAD concentrations up to about 2.0×10^{-7} *M*,

[9] V. Massey and B. E. P. Swoboda, Information Exchange Group No. 1, *Sci. Memo* No. 52 (1963).

[10] F. M. Huennekens and S. P. Felton, Vol. III, p. 950.

[11] C. DeLuca, M. M. Weber, and N. O. Kaplan, *J. Biol. Chem.* **223**, 559 (1956).

[12] The reaction can also be followed using an oxygen microelectrode. Depending upon the size of reaction vessel, the same reaction mixture (except for 6 *N* NaOH) can be used, or the amounts can be scaled down appropriately. For each run, a blank rate of oxygen uptake is determined with all components present except FAD. Then either known amounts of FAD, or the extract to be tested, are added and the rate of oxygen uptake is measured from the linear portion of the trace.

and several concentrations should be used to establish a standard curve.

Spectrophotometric Assay.[11,13] The following components are added to a cuvette having a light path of 1 cm: 0.1 ml of DPNH, 0.1 ml of phosphate buffer, pH 7.5, 0.1 ml of lactate dehydrogenase, 0.1 ml of catalase, 0.1 ml of D-amino acid apo-oxidase, 0.1–1.5 millimicromoles of FAD (or the extract to be tested) and water to make 3.0 ml. The optical blank is identical except for the omission of DPNH. DL-Alanine, 0.1 ml, is added to each cuvette in order to initiate the reaction, and the absorbancy change at 340 mμ is recorded over a 5-minute period. The response due to FAD present in the extract is compared to that produced by known amounts of FAD.

Enzymatic Assay for FMN

TPNH cytochrome *c* reductase[14] is purified from brewers' yeast and then treated to dissociate the bound FMN.[10]

Reagents

TPNH cytochrome *c* aporeductase[10]
Phosphate buffer, 0.1 *M*, pH 7.3
Cytochrome *c*, 10 mg/ml
TPNH, 4.8×10^{-4} *M*
FMN standard solution, 2.0×10^{-6} *M* (concentration determined spectrophotometrically at pH 7.0 using an extinction coefficient[15] at 450 mμ of $12.2 \times 10^{3}\ M^{-1}\ cm^{-1}$)

Spectrophotometric Assay. The following components are added, in order, to a cuvette having a light path of 1 cm: 0.2 ml of phosphate buffer, 0.1 ml of cytochrome *c*, 0.1 ml of enzyme, 0.1 ml of the extract to be tested for FMN, and water to make 3.0 ml. After the solution has been allowed to stand for 5 minutes at room temperature, 0.1 ml of TPNH is added, and the optical density at 550 mμ is measured at 1-minute intervals against a blank cell without cytochrome *c*. An additional blank should be run with FMN omitted, and the value for the experimental cell corrected accordingly. The net change in absorbancy is linear over a 5-minute period. The activity is linear with FMN concentration up to 1.0×10^{-7} *M*, and several concentrations[16] should be used to construct the standard curve.

[13] N. A. Rao, S. P. Felton, F. M. Huennekens, and B. Mackler, *J. Biol. Chem.* **238,** 449 (1963).

[14] E. Haas, B. L. Horecker, and T. Hogness, *J. Biol. Chem.* **136,** 747 (1940).

[15] L. Whitby, *Biochem. J.* **54,** 437 (1953).

[16] For maximum accuracy of results, standards of FMN in this assay, or of FAD in the previous assay, should contain the same amount of perchlorate (or trichloroacetase) as the extract being assayed (see Massey and Swoboda,[9] for a discussion of this point).

FLAVIN CONTENT OF MITOCHONDRIA AND RELATED PARTICULATE PREPARATIONS[a]

Preparation	Flavin content[b]			Reference
	FAD	FMN	Peptide flavin	
Mitochondria	0.27	0.08	0.11	*c*
	0.10	0.03	0.18	*d*
	0.31	0.15	0.20	*e*
Mitochondria (P)	0.33	0.07	—	*f*
Mitochondria (Y)	0.27	0.01	0.32	*g*
Nagarse mitochondria	0.31	0.24	—	*d*
	—	0.17	—	*h*
Heavy mitochondria	—	0.16	—	*h*
Keilin-Hartree preparation	0.15	0.09	0.12	*c*
	—	—	0.14	*i*
Keilin-Hartree preparation (P)	0.29	0.04	—	*f*
ETP	0.13	0.09	0.27	*d*
	0.15	0.23	0.28	*e*
ETP (P)	0.24	0.10	—	*f*
DPNH–cytochrome *c* reductase	0.06	0.80	0	*d*
	—	0.77	—	*j*
DPNH–Q reductase	0.04	1.30	0	*d*
	—	1.20	—	*j*

[a] All preparations are from beef heart except those marked (P) and (Y) which are from pig heart and yeast, respectively.

[b] Values are given as millimicrograms per milligram protein.

[c] T. E. King, R. L. Howard, D. F. Wilson, and J. C. R. Li, *J. Biol. Chem.* **237,** 2941 (1962).

[d] N. A. Rao, S. P. Felton, F. M. Huennekens, and B. Mackler, *J. Biol. Chem.* **238,** 449 (1963).

[e] D. E. Green and D. C. Wharton, *Biochem. Z.* **338,** 335 (1963).

[f] V. Massey and B. E. P. Swoboda, Information Exchange Group No. 1, *Sci. Memo* No. 52 (1963).

[g] B. Mackler, P. J. Collipp, H. M. Duncan, N. A. Rao, and F. M. Huennekens, *J. Biol. Chem.* **237,** 2968 (1962).

[h] A. J. Merola and R. Coleman, Information Exchange Group No. 1, *Sci. Memo* No. 48 (1963).

[i] T. P. Singer, J. Hauber, and O. Arrigoni, *Biochem. Biophys. Res. Commun.* **9,** 150 (1962).

[j] A. J. Merola, R. Coleman, and R. Hansen, Information Exchange Group No. 1, *Sci. Memo* No. 58 (1963).

Amount of Various Flavins in Heart Mitochondria and Related Preparations

Using methods similar to those outlined above, the values shown in the table have been reported for the flavin content in beef and pig heart mitochondria, in derived particulate preparations, and in yeast mitochondria. Some of the values for peptide flavin may require slight revision since, as Wilson and King[5] have pointed out, heme peptides are also released by proteolytic digestion. Likewise, the small amounts of FAD found[13] in the particulate DPNH–cytochrome *c* and DPNH–Q reductases may have been due to contaminating enzymes in the preparations assayed.

Paper Chromatography of Mitochondrial Flavins

In the solutions described above under preparations of extracts, flavins can also be distinguished from each other and semiquantitated by chromatography on DEAE-cellulose[13] or by paper chromatography.[17] Variations in the R_f value of flavin nucleotides, depending on the manner in which the sample was deproteinized, have been discussed elsewhere.[13]

[17] G. L. Kilgour, S. P. Felton, and F. M. Huennekens, *J. Am. Chem. Soc.* **79,** 2254 (1957).

[78] Extraction and Estimation of Cytochrome c from Mitochondria and Submitochondrial Particles

By GIORGIO LENAZ and DAVID H. MACLENNAN

Cytochrome *c* has been described by Green[1] as a "mobile" component of the electron transfer chain. The removal of cytochrome *c* from mitochondria, however, requires rigorous extraction procedures. Cytochrome *c* forms an *in vitro* complex with phospholipid,[2,3] and this bond may exist *in vivo*.[3,4] Cytochrome *c* may also interact with structural protein.[5]

In order to estimate cytochrome *c* in mitochondria quantitatively, it is necessary to remove it, free of the other cytochromes whose absorption spectrum coincides closely with that of cytochrome *c*. The quantitative extraction of cytochrome *c* from mitochondria is accomplished with

[1] D. E. Green, *Comp. Biochem. Physiol.* **4,** 81 (1962).

[2] D. E. Green and S. Fleischer, *Biochim. Biophys. Acta* **70,** 554 (1963).

[3] M. L. Das and F. L. Crane, *Biochemistry* **3,** 696 (1964).

[4] K. S. Ambe and F. L. Crane, *Science* **129,** 98 (1959).

[5] D. L. Edwards and R. S. Criddle, *Biochemistry* **5,** 583 (1966).

organic solvents or detergents.[6-8] This method, however, yields a mitochondrial residue which is largely denatured. A milder procedure for the removal of most of the mitochondrial cytochrome *c* involves salt extraction.[9] This treatment leads to a loss of substrate oxidation and coupled phosphorylation, but the reintroduction of cytochrome *c* into salt-extracted mitochondria reestablishes oxidative and phosphorylative activities.

Extraction and Estimation of Cytochrome c with Bile Acids, Salt, and Ethanol[7]

Reagents

Tris-HCl, 0.02 *M*, pH 8.0
Tris-HCl, 1.0 *M*, pH 7.5
Potassium cholate, 20% (w/v) in cholic acid, pH 8.0
Potassium deoxycholate, 10% (w/v) in deoxycholic acid, pH 8.0
Ethanol, absolute
Ammonium sulfate, saturated, pH 8.0 (0°)

Procedure. Mitochondria, or any enzyme preparation derived from mitochondria, are sedimented by centrifugation. The supernatant fluid is decanted and the pellet is suspended as a thick slurry at a protein concentration of 200 mg/ml in a solution 0.02 *M* in Tris-HCl pH 8.0 at 0°. To each 0.3 ml of the suspension are added 0.18 ml of 20% potassium cholate (0.6 mg of potassium cholate per milligram of protein) and 0.42 ml of 10% potassium deoxycholate (0.70 mg of potassium deoxycholate per milligram of protein); after mixing, 0.1 ml of saturated ammonium sulfate is added to make the suspending medium 10% saturated with respect to ammonium sulfate. The protein concentration should be 60 mg/ml at this stage. The protein concentration can be determined by the biuret method of Gornall *et al.*[10] An equal volume of absolute ethanol is added to the suspension, and thorough mixing is accomplished with a mechanical test tube mixer. Immediately after mixing, the preparation is incubated in a 38° water bath for 5 minutes and subsequently is incubated at —20° for 10 minutes. A clear supernatant fluid containing cytochrome *c* is obtained by centrifugation at maximal speed in a clinical centrifuge. One-tenth milliliter of a solution 1 *M* in Tris-HCl pH 7.5 is

[6] S. H. Richardson and L. R. Fowler, *Arch. Biochem. Biophys.* **103**, 567 (1963).
[7] P. V. Blair, T. Oda, D. E. Green, and H. Fernández-Morán, *Biochemistry* **2**, 756 (1963).
[8] E. G. Ball and O. Cooper, *J. Biol. Chem.* **226**, 755 (1957).
[9] E. E. Jacobs and D. R. Sanadi, *J. Biol. Chem.* **235**, 531 (1960).
[10] A. G. Gornall, C. J. Bardawill, and M. M. David, *J. Biol. Chem.* **177**, 751 (1949).

added to each 0.9 ml of the supernatant fluid. The concentration of cytochrome *c* in the solution is determined spectrally by the difference in absorbancy at 550 mμ between the oxidized and reduced forms of cytochrome *c*. An extinction coefficient of 18.5 mM^{-1} cm^{-1} is used.[11] This extraction and measurement of cytochrome *c* should be performed rapidly since the cytochrome *c* is denatured by prolonged contact with the extraction medium.

Extraction of Cytochrome c from Mitochondria with Salt

The extraction procedure described below is similar to that reported by Jacobs and Sanadi[9] for rat liver mitochondria. We have found this procedure to be effective in removing most of the cytochrome *c* from heavy beef heart mitochondria (HBHM).

Principle. In the extraction procedure mitochondria are first swollen in a hypotonic solution of KCl, and cytochrome *c* is subsequently extracted with isotonic KCl.

Reagents

0.25 *M* Sucrose, 0.01 *M* in Tris-acetate, pH 7.5 (4°)
Preserving mixture for HBHM: 0.25 *M* sucrose, 0.01 *M* in Tris-acetate pH 7.5, 1 m*M* in ATP, 1 m*M* in succinate, and 1 m*M* in $MgCl_2$ (4°)
KCl (4°), 0.015 *M*
KCl (4°), 0.15 *M*

Procedure. All the steps in this procedure are carried out at 4°. Beef heart mitochondria are prepared by the large-scale method of Crane *et al.*[12]; the heavy mitochondria (HBHM) are isolated by the method of Hatefi and Lester[13] with the exception that the pH is adjusted to 7.8 prior to centrifugation. HBHM are kept frozen at —20° overnight, or for a longer period of time, at a protein concentration of 40 mg/ml, in the preserving mixture. Upon thawing, the pH of the suspension is adjusted to 7.8 with *N* KOH and the suspension is centrifuged at 26,000 *g* for 10 minutes in the No. 30 rotor of the Spinco Model L ultracentrifuge. The supernatant and the light infranatant layer (light mitochondria) are removed. The heavy residue is suspended in a solution 0.015 *M* in KCl, homogenized with a glass-Teflon homogenizer, and the protein concentration is adjusted to 20 mg/ml. The mitochondria are permitted to swell in the hypotonic medium for 10 minutes, and are then centrifuged

[11] E. Margoliash, *Biochem. J.* **56**, 535 (1954).
[12] F. L. Crane, J. Glenn, and D. E. Green, *Biochim. Biophys. Acta* **22**, 675 (1956).
[13] Y. Hatefi and R. L. Lester, *Biochim. Biophys. Acta* **29**, 630 (1958).

at 105,000 g for 15 minutes in the No. 40 rotor of the Spinco ultracentrifuge. The colorless supernatant and a small dark pellet of unswollen mitochondria at the bottom of the tube are discarded. The swollen mitochondria are resuspended in a solution of 0.15 M in KCl, homogenized as above, and the protein concentration is adjusted to 20 mg/ml. After 10 minutes the suspension is centrifuged at 105,000 g for 15 minutes in the No. 40 rotor of the Spinco. The red supernatant containing cytochrome c is decanted, and the mitochondrial pellet is reextracted by the same procedure with 0.15 M KCl two more times. After the final centrifugation, the supernatant is removed and a small fluffy layer is separated from the densely packed mitochondria and discarded. The extracted mitochondria are suspended at a protein concentration of 20 mg/ml in a solution 0.25 M in sucrose and 0.01 M in Tris-acetate pH 7.5. Cytochrome c can be estimated in the 0.15 M KCl washings by measuring the spectrum, using a millimolar extinction coefficient of 18.5 cm^{-1} at 550 mμ for the reduced minus oxidized forms (see also Jacobs and Sanadi[9] for purification of the extract). The amount of unextracted cytochrome c can be estimated by reextracting the mitochondria with n-butanol and salt according to the procedure of Richardson and Fowler[6] or according to the procedure of Blair *et al.*[7] described above. In general, however, the amount of cytochrome c left in the extracted mitochondria is too small to be accurately measured. The recovery of cytochrome c and of heavy mitochondria and the oxidative properties of the cytochrome c deficient preparations are described in Table I.

Properties of Cytochrome c-Deficient Mitochondria. In the KCl-extracted HBHM the rate of substrate oxidation is 15% or less of the

TABLE I
EXTRACTION OF CYTOCHROME *c* FROM HEAVY BEEF HEART MITOCHONDRIA

Extraction medium	Mitochondrial protein treated (mg)	Millimicromoles of cytochrome c recovered	Millimicromoles of cytochrome c extracted/mg protein
0.015 M KCl	1280	4.46	0.003
0.15 M KCl	1140	387.77	0.340
0.15 M KCl	1060	41.82	0.039
0.15 M KCl	960	3.13	0.003
			0.385
Oxidation rate of extracted particles (microatoms of oxygen consumed/min/mg protein at 30° with succinate as substrate).			0.023
Mitochondrial protein recovered after removal of light layer (mg)			800

original rate and the measured P:O ratios are low. When cytochrome *c* is added at a concentration of 3–5 μg per milligram of mitochondrial protein, oxidation and phosphorylation rates are both increased to values close to those of unextracted HBHM. The increase in the phosphate esterification rate resulting from the addition of cytochrome *c* to the extracted HBHM is greater than the increase in the oxygen consumption rate, so that there is an actual increase of the P:O ratios.[14]

Preparation of Cytochrome *c*-Deficient Submitochondrial Particle

Extraction with KCl by the procedure described above does not remove cytochrome *c* from submitochondrial particles.[15] It is, however, possible to prepare cytochrome *c*-deficient ETP_H by sonication of KCl-extracted mitochondria. To retain the highest phosphorylation rates in the resultant submitochondrial particles the conditions of preparation described by Hansen and Smith[16] for ETP_H (Mg, Mn) are used. The addition of cytochrome *c* to these deficient particles increases the oxidation rates only slightly and has no effect on phosphate esterification. ETP_H with normal oxidative and phosphorylative capacity can, however, be prepared by sonicating KCl-extracted mitochondria which have been reconstituted with cytochrome *c*.

Reagents

Sonication medium: 0.25 *M* sucrose, 0.01 *M* in Tris-acetate pH 7.5, 1 m*M* in ATP, 1 m*M* in succinate, 5 m*M* in $MgCl_2$, and 10 m*M* in $MnCl_2$ (4°)

Cytochrome *c*, 2% (w/v) solution in 0.01 *M* phosphate buffer pH 7.0; the solution is kept frozen

Preserving mixture for ETP_H (0.25 *M* sucrose, 0.01 *M* in Tris-acetate pH 7.5, 1 m*M* in ATP, 1 m*M* in succinate, 4 m*M* in $MgCl_2$, and 2 m*M* in GSH (4°)

Procedure. KCl-extracted HBHM prepared as described above are suspended at a protein concentration of 20 mg/ml in the medium for sonication. The pH is adjusted to 7.6 with *N* KOH. Samples of 25 ml in a 50-ml beaker surrounded by ice are sonicated for a period of 30–60 seconds at the maximal output of a Branson Probe Sonifier. The pH is readjusted to 7.6 with *N* KOH, and the suspension is centrifuged at 26,000 *g* for 10 minutes in the No. 40 rotor of the Spinco Model L ultracentrifuge. The residue from this centrifugation can be resonicated and

[14] D. H. MacLennan, G. Lenaz, and L. Szarkowska, *J. Biol. Chem.* **241,** 5251 (1966).
[15] G. Lenaz and D. H. MacLennan, *J. Biol. Chem.* **241,** 5260 (1966).
[16] M. Hansen and A. L. Smith, *Biochim. Biophys. Acta* **81,** 218 (1964).

then centrifuged as above. The combined supernatants are centrifuged at 105,000 g for 40 minutes in the No. 40 rotor, and the pellet is resuspended in the preserving mixture for ETP_H.[16]

Preparation of Cytochrome c-Reconstituted ETP_H. KCl-extracted HBHM are suspended in the same medium described above, with the exception that cytochrome c is added at a concentration of 10 μg per milligram of protein (10.0 μl of the 2% solution per milliliter of suspension). The same procedure as for the preparation of the cytochrome c-deficient ETP_H is then followed.

Cytochrome c has to be added *before* the KCl-extracted HBHM are disrupted by sonication. If it is added after sonication only a slight increase in oxidation rate is observed.

Properties. The oxidation rate of the cytochrome c-deficient ETP_H ranges from 40 to 50 millimicroatoms of oxygen consumed per minute per milligram of protein at 30° when the substrate is succinate. The phosphorylation rate is near 30 millimicromoles per minute per milligram of protein so that P:O ratios are near 0.6. These rates cannot be significantly increased with exogenously added cytochrome c. In the ETP_H prepared from KCl-extracted mitochondria reconstituted with cytochrome c the oxidation rate is about 270 millimicroatoms of oxygen consumed per minute per milligram of protein. The phosphorylation rate is about 300 millimicromoles per minute per milligram of protein and P:O ratios are greater than 1.0. A summary of the recoveries of cytochrome c from mitochondria and submitochondrial particles using the two extraction methods is presented in Table II.

TABLE II

EXTRACTION OF CYTOCHROME c FROM MITOCHONDRIA AND SUBMITOCHONDRIAL PARTICLES WITH THE BILE ACID–SALT–ETHANOL MIXTURE AND WITH KCl

Enzyme system extracted	Millimicromoles of cytochrome c extracted from each milligram of protein with the use of: Bile acid, salt, ethanol	KCl
Mitochondria	0.45[a]	0.385[b]
ETP_H	0.36[a]	Not measurable[c]

[a] P. V. Blair, T. Oda, D. E. Green, and H. Fernández-Morán, *Biochemistry* **2,** 756 (1963).

[b] D. H. MacLennan, G. Lenaz, and L. Szarkowska, *J. Biol. Chem.* in press (1966).

[c] G. Lenaz and D. H. MacLennan, *J. Biol. Chem.* in press (1966).

[79] Preparation and Properties of a Factor Conferring Oligomycin Sensitivity (F_0) and of Oligomycin-Sensitive ATPase ($CF_0 \cdot F_1$)

By YASUO KAGAWA[1]

Assay Method

Principle. The assay of F_0 activity is based on the inhibition of ATPase activity of coupling factor 1 (F_1) by oligomycin in the presence of F_0.[1] The ATPase activity of purified coupling factor 1 is not inhibited by oligomycin. After adsorption to F_0, the ATPase activity becomes sensitive to oligomycin and to certain other energy-transfer inhibitors.

Reagents

ATP, 0.2 *M*, pH 7.4

$MgSO_4$, 0.1 *M*

Potassium phosphoenolpyruvate, 0.1 *M*, pH 7.4[2]

Sucrose, 0.5 *M*

EDTA, 0.5 *M*, pH 7.4

Tris-sulfate, 1.0 *M*, pH 7.4

Oligomycin A [mol. wt. = 425 (footnote 3)], 0.1 m*M* or rutamycin [mol. wt. = 439 (footnote 4)], 0.1 m*M*. Dilute 10 μl of a 10 m*M* solution prepared in methanol to 1.0 ml with H_2O.

Pyruvate kinase, 3.2 mg/ml in 0.01 *M* Tris-sulfate, pH 7.4[5]

Soybean phospholipid (azolectin) 1%. A 10% solution prepared as described[6] was diluted 1:10 with water.

Solution A: 1.0 ml ATP, 2.0 ml of $MgSO_4$, 10.0 ml of potassium phosphoenolpyruvate, 10.0 ml of Tris-sulfate, 2.0 ml of pyruvate kinase, and 15.0 ml of water.

Solution B: 50 ml of sucrose, 1 ml of Tris-sulfate, 0.2 ml of EDTA, 2 ml of ATP, and 46.8 ml of water

Coupling factor 1 solution: Coupling factor 1 suspended in am-

[1] This manuscript was prepared while the author was at the Department of Biochemistry, The Public Health Research Institute of the City of New York, Inc., New York.

[1a] Y. Kagawa and E. Racker, *J. Biol. Chem.* **241**, 2461 (1966).

[2] Prepared from the monocyclohexylammonium salt as described by V. M. Clark and A. J. Kirby, *Biochem. Prep.* **11**, 103 (1966).

[3] H. A. Lardy, P. Witonsky, and D. Johnson, *Biochemistry* **4**, 552 (1965).

[4] R. Q. Thompson, M. M. Hoehn, and C. E. Higgens, *in* "Antimicrobial Agents and Chemotherapy—1961" (M. Finland and G. M. Savage, eds.), p. 474 (referred to as A272 in this reference). Am. Soc. Microbiol., Detroit, Michigan, 1961.

[5] Purchased from C. F. Boehringer and Sons, New York.

[6] J. M. Fessenden and E. Racker, this volume [35].

monium sulfate[7] is centrifuged and the resulting pellet is dissolved in Solution B to a final protein concentration of 100 μg/ml.

Procedure for F_0. F_0 is incubated in the presence of F_1 (1–2 μg) and 5 millimicromoles of oligomycin in a final volume of 0.8 ml, containing 0.025 *M* Tris-sulfate, pH 7.4, at 30° for 5 minutes. The amount of F_1 used is capable of catalyzing the release of about 1 micromole of inorganic phosphate from ATP in 10 minutes. The mixture is assayed for ATPase activity by the addition of 0.2 ml of Solution A. After 10 minutes at 30°, 0.1 ml of 50% trichloracetic acid is added and, after centrifugation, inorganic phosphate is determined on a 0.5–ml aliquot of the supernatant solution. Determination of inorganic phosphate is carried out by the procedure of Lohmann and Jendrassik[8] scaled down to 5 ml. F_1 is completely insensitive to oligomycin in the absence of F_0. The inhibition of F_1 in the presence of oligomycin increases with the amount of F_0 added but levels off after 50% inhibition is reached.[1a] In order to obtain accurate values, titration of F_0 activity close to the 50% inhibition point are required with each preparation. ATPase activity in the absence of oligomycin serves as the control (100% activity).

Procedure for CF_0 and $CF_0 \cdot F_1$. The assay of these preparations is similar to that of F_0 except that F_1 was incubated with 100 μg of CF_0 protein for 3 minutes prior to the addition of 0.01 ml of the 1% phospholipid suspension. Addition of phospholipids after F_1 gives rise to a consistently greater sensitivity to oligomycin than does the reverse order of addition. In the case of $CF_0 \cdot F_1$, 100 μg of the preparation was used in the presence of 100 μg of phospholipid.

Definition of Unit and Specific Activity. A unit of ATPase activity is defined as that amount of enzyme which catalyzes the turnover of 1 micromole of substrate per minute under the special assay condition described above. Specific activity is expressed as units per milliliter of protein. A unit of F_0 activity is defined as the amount of protein which results in a 50% inhibition of 0.1 unit of ATPase activity of F_1 in the presence of 5 millimicromoles of oligomycin. Specific F_0 activity is expressed as units per milligram of F_0 protein.

Preparation

Factor Conferring Oligomycin Sensitivity (F_0)

Step 1. Preparation of Submitochondrial Particles. Beef heart mitochondria (light layer) are diluted with water and 0.1 *M* pyrophosphate

[7] M. E. Pullman and H. S. Penefsky, Vol. VI [34].

[8] K. Lohmann and L. Jendrassik, *Biochem. Z.* **178,** 419 (1926).

buffer, pH 7.4, to a final protein concentration of 20 mg/ml and a final sodium pyrophosphate concentration of 0.01 *M*. Then 30-ml batches of these suspensions are exposed for 2 minutes to sonic oscillation in a 10-kilocycle Raytheon sonic oscillator. After removal of the heavy pellet by centrifugation at 26,000 *g* for 15 minutes, the resulting supernatant solution is again centrifuged at 105,000 *g* for 45 minutes. The pellet, which represents 32–47% of the mitochondrial protein, is washed and resuspended in 0.25 *M* sucrose to yield a suspension containing about 50 mg protein per milliliter.

Step 2. Trypsin Treatment. The following mixture is incubated at 30° for 45 minutes: 70 ml of the submitochondrial particles (48 mg/ml); 244.2 ml of water; 16.8 ml of 1 *M* Tris-sulfate, pH 8.0; 1.1 ml of 0.5 *M* EDTA, pH 8.0; 3.9 ml of 0.2% trypsin in 0.001 *N* H_2SO_4; and 0.05 *M* Tris-sulfate, pH 8.0, to yield a protein concentration of 10 mg/ml. The digestion is stopped by the addition of 16.8 ml of 0.2% trypsin inhibitor dissolved in 0.01 *M* Tris-sulfate, pH 8.0. Usually step 2 is followed directly by step 3, but when the trypsin particles are isolated by centrifugation at 105,000 *g* for 1 hour, the protein yield from the submitochondrial particles is about 80%.

Step 3. Urea Treatment. The above mixture is chilled to 0°, and an equal volume (352.8 ml) of 4 *M* urea is added. After 45 minutes at 0°, the mixture is centrifuged at 30,000 rpm in the No. 30 rotor of the Spinco Model L centrifuge for 45 minutes. Each pellet is carefully rinsed with 2 ml of 0.25 *M* sucrose, and suspended by homogenization in 0.25 *M* sucrose to a final volume of 300 ml. The total period of exposure of particles to 2 *M* urea is about 2 hours, including the time required for deceleration of the rotors and the manipulation of 24 centrifuge tubes of two rotors. The homogenate is centrifuged again as described above. The pellets are washed and combined, followed by homogenization in 0.25 *M* sucrose to yield a suspension of 15.2 mg of protein per milliliter. The protein yield from the submitochondrial particles at this stage is usually about 52–65%.

Step 4. Sonic Oscillation and Centrifugation. The homogenate is exposed to sonic oscillation in 30-ml batches in a 10-kilocycle Raytheon sonic oscillator for 2 minutes. The mixture is centrifuged at 40,000 rpm in the Spinco No. 40 rotor for 1 hour. The supernatant fluid is collected (F_0). The yield of F_0 is usually 30% of the protein of the trypsin-urea-treated particles and the specific activity is between 8 and 10 units per milligram. The precipitate also contains F_0 activity, but because of the particulate nature was not used as extensively as the supernatant solution. The summary of the preparation is shown in Table I.

TABLE I
PREPARATION OF OLIGOMYCIN SENSITIVITY-CONFERRING FACTOR

Step	Volume (ml)	F_0 activity: Units mg	F_0 activity: Total units	ATPase activity (units/mg)	Protein (mg/ml)	Total P content (μmoles P/mg)
1. Submitochondrial particles	70	11	—	1.97	48.0	0.780
2. Trypsin-treated particles	336	—	—	5.21	10.0	0.780
3. Trypsin-urea-treated particles	120	15.5	28,200	0.560	15.2	0.924
4. Exposure to sonic oscillation						
a. Whole sonicate	120	12.5	22,800	0.39	15.2	0.942
b. Supernatant solution (F_0)	105	8.9	7,850	0.37	8.7	1.137
c. Precipitate	27	12.6	7,550	0.31	21.6	—

Oligomycin Sensitivity-Conferring Factor Prepared in the Presence of Cholate (CF_0)

Extraction of Trypsin-Urea-Treated Particles with Cholate and Fractionation with Ammonium Sulfate. Trypsin-urea-treated particles are obtained as described above. To 120.4 ml of trypsin-urea-treated particles (21.0 mg/ml), 13.7 ml of 20% potassium cholate, pH 8.0, and 3.4 ml of saturated ammonium sulfate solution, pH 7.4, are added with gentle mixing at 0°. The final concentrations of cholate is 2% and of ammonium sulfate 0.1 *M*. After incubation for 1 hour at 0°, the mixture is centrifuged in a Spinco 30 rotor at 30,000 rpm for 1 hour. To the clear yellow supernatant fluid, one-half volume of saturated ammonium sulfate solution, pH 7.4, is added. After 30 minutes at 0°, the mixture is centrifuged as described above. Each pellet is rinsed with 2 ml of 0.25 *M* sucrose and suspended in 0.25 *M* sucrose. The suspension is divided into small aliquots to avoid undue inactivation by freezing and thawing and is stored at −70°.

Oligomycin Sensitive ATPase ($CF_0 \cdot F_1$)

Extraction of trypsin-treated particles with cholate and fractionation. Trypsin-treated particles are obtained essentially as described above (step 2), with trypsin at a concentration of 30 μg per 10 mg of protein per milliliter. After the reaction has been stopped by addition of trypsin inhibitor (150 μg/ml), the particles are centrifuged at 105,000 *g* for 1 hour. Extraction of the particles with 2% cholate and fractionation with

TABLE II
PREPARATION OF CF_0 AND $CF_0 \cdot F_1$

Sample	Volume (ml)	Activity (units/mg) F_0	Total F_0 (units)	Protein (mg/ml)	Total P[c] (μmoles/mg)
TU particles[d]	135.7	8.7 (3.9)[a]	22,700	19.0	0.825
First precipitate	50.0	2.5[b]	4,750	38.0	0.388
CF_0	12.0	11.2[b]	3,840	28.6	0.0455
		ATPase			
T particles	25.0	3.24		38.4	1.09
$CF_0 \cdot F_1$	1.37	1.49[b]		27.3	0.148

[a] After addition of cholate.
[b] Measured in the presence of equal weight of soybean phospholipids.
[c] In TU-particles about 95% of the total P was phospholipid-P. In a previous preparation of CF_0 and $CF_0 \cdot F_1$ which had been analyzed the phospholipid-P represented only 50% and 65%, respectively, of the total P content.
[d] TU = trypsin-urea treated; T = trypsin treated.

ammonium sulfate is as described above for preparation of CF_0. A summary of the preparation of CF_0 and $CF_0 \cdot F_1$ is shown in Table II.

Properties

Respiratory Enzymes and Phospholipid Content. F_0 preparations contain the entire respiratory chain and have a phospholipid content which is slightly higher than that of submitochondrial particles. On the other hand, preparations of CF_0 and $CF_0 \cdot F_1$ have no DPNH or succinate oxidase activity and contain little phospholipid, but they are capable of binding about 1 micromole of phospholipid phosphorus per milligram of protein.

Binding of F_1. The binding capacity is about 15 μg of F_1 per milligram of F_0 protein. CF_0 can bind as much as 100 μg of F_1 per milligram of protein, resulting in a preparation of a latent ATPase which is activated by the addition of phospholipid. The activated ATPase activity is sensitive to oligomycin.

Effect of F_0 and CF_0 on F_1.[9] Like F_1, the complexes of $F_0 + F_1$, $CF_0 + F_1$, and $CF_0 \cdot F_1$ hydrolyze ATP, ITP, GTP, and UTP. Although there are some quantitative differences between the soluble enzyme and the insoluble complexes in the relative rates of hydrolysis of the various nucleotides, the qualitative properties are similar.

Mn^{++}, Co^{++}, Fe^{++}, and Ca^{++} can substitute for Mg^{++}. However, again quantitative differences compared with the reaction catalyzed by soluble

[9] Y. Kagawa and E. Racker, *J. Biol. Chem.* **241**, 2467 (1966).

F_1 exist (e.g., Ca^{++} is less effective as an activator with the complexes). Oligomycin, rutamycin, tri-*n*-butyltin chloride, and Dio-9 inhibit the ATPase activity of $F_0 + F_1$, $CF_0 \cdot F_1$, and $CF_0 + F_1$ but have little or no effect on free F_1. On the other hand, the mitochondrial inhibitor of ATPase[10] inhibits both free and bound F_1.

Stability. Combination of the cold-labile F_1 with F_0 stabilizes F_1 at 0° for several hours. Preparations of F_0 have been kept at —70° for several months with little loss of activity; however, CF_0 preparations deteriorate in a few days. Addition of phospholipid stabilizes CF_0 somewhat.

Electron Microscopy. In electron micrographs CF_0 appears as an amorphous preparation which becomes vesicular on addition of phospholipid. When F_1 is added to such preparations, characteristic inner membrane spheres appear along the membranous vesicles.

[10] G. C. Monroy and M. E. Pullman, this volume [80].

[80] Preparation and Properties of a Naturally Occurring Inhibitor of Mitochondrial ATPase

By Gladys C. Monroy and Maynard E. Pullman

General Principle

Treatment of beef heart mitochondria with dilute alkali releases a protein of low molecular weight which is a specific inhibitor of mitochondrial ATPase. The inhibitor suppresses the ATPase activity of coupling factor 1 (F_1) without affecting the ability of the latter to recouple phosphorylation to oxidation in F_1-deficient submitochondrial particles.[1,2]

Assay for Inhibitor

The assay for the inhibitor is based upon the inhibition of the soluble ATPase. However, it should be mentioned that the effectiveness of a given amount of inhibitor is the same whether it is measured with the soluble enzyme or with the ATPase of submitochondrial particles.[1] The later is considerably easier to prepare and can be used instead of the

[1] See Vol. VI [34].

[2] The coupling factor will be referred to as ATPase or F_1 depending on whether hydrolytic activity or coupling in oxidative phosphorylation is being considered.

soluble ATPase in the assay system described below. The inhibitor does not inhibit ATPase in intact mitochondria.

The assay is carried out in two steps—an incubation of the ATPase with the inhibitor followed by an assay for the remaining ATPase activity.

Reagents

Tris-SO_4, 1.0 *M*, pH 7.4
Sodium ATP, 0.2 *M*, pH 7.4
$MgSO_4$, 0.1 *M*
Potassium phosphoenolpyruvate,[3] 0.1 *M*, pH 7.4
Pyruvate kinase,[4] 0.8 mg/ml in 0.01 *M* Tris, pH 7.4
Sucrose, 0.25 *M*, containing 0.01 *M* imidazole, pH 7.0

A mixed assay medium is freshly prepared each day and contains for *each* tube 0.05 ml of Tris-SO_4, 0.005 ml of ATP, 0.01 ml of $MgSO_4$, 0.05 ml of phosphoenolpyruvate, 0.04 ml of pyruvate kinase, and 0.645 ml of H_2O.

Procedure. A solution of the ATPase is prepared as described previously,[5] except that a solution containing 250 m*M* sucrose, 10 m*M* imidazole, pH 7.0, and 2 m*M* ATP is used to dissolve the ammonium sulfate precipitate. The volume of the enzyme solution is adjusted to contain 30–40 units[6] of ATPase per milliliter (0.3–0.5 mg of protein[7] per milliliter). Under these conditions, the ATPase is stable for over 8 hours at room temperature. Immediately before assay of the inhibitor, the enzyme is diluted 1:5 with 250 m*M* sucrose containing 10 m*M* imidazole, pH 7.0, so that a 0.03-ml aliquot contains approximately 0.2 unit of ATPase and 0.3 μmole of imidazole. This amount of enzyme is incubated for 10 minutes at 30° in a final volume of 0.2 ml, with an amount of inhibitor (diluted with water) which gives between 30% and 70% inhibition. At the end of the incubation period, 0.8 ml of the mixed

[3] The monocyclohexylammonium salt of phosphoenolypyruvate was synthesized and converted to the potassium salt as described by V. M. Clark and A. J. Kirby, *Biochem. Prep.* **11,** (1966).
[4] Purchased from C. F. Boehringer and Soehne, Mannheim, Germany.
[5] Vol. VI [34], bottom of page 282.
[6] A unit of ATPase activity is defined as the amount of enzyme which catalyzes the hydrolysis of 1 micromole of ATP per minute. A unit of inhibitor activity is defined as the amount of protein which results in a 50% inhibition of 0.2 unit of ATPase.
[7] Determined according to O. H. Lowry, N. J. Rosebrough, A. L. Farr, and R. J. Randall, *J. Biol. Chem.* **193,** 265 (1951).

assay medium, warmed to 30°, is added and the incubation period is continued for an additional 5 minutes. The reaction is stopped by the addition of 2 ml of 5 *N* H_2SO_4, containing 2.5% ammonium molybdate, and P_i is determined[8] on the whole reaction mixture.

Inhibitor activity is most accurately measured using amounts which give no more than 70% inhibition of the enzyme.[9]

Preparation and Purification of the Inhibitor

Light layer mitochondria prepared from beef heart according to the method of Green *et al.*[10] are stored at —20° for at least 2 weeks in 0.25 *M* sucrose, 0.01 Tris, pH 7.4. After thawing, 48 g of mitochondrial protein[11] is diluted to a final volume of 2.4 liters with cold distilled water. To each of five 480-ml aliquots, 24 ml of 1 *N* KOH are added while the suspension is mixed rapidly with a magnetic stirrer. One minute after all the alkali has been added, the pH of the suspension is adjusted to 7.4 with approximately 4.5 ml of 5 *N* H_2SO_4. The suspension is placed on ice until all five aliquots are similarly treated. All subsequent operations are conducted at 4°. The five aliquots are pooled and centrifuged for 30 minutes at 20,000 *g*. To the somewhat turbid, yellow supernatant solution (approximately 1.6 liters), 177 ml of 50% trichloroacetic acid is added. After the sample has stood for 10 minutes, the precipitated protein is collected by centrifugation at 20,000 *g* for 10 minutes and suspended in 330 ml of distilled water. The suspension is homogenized and the pH is adjusted to 7.3 with approximately 4.9 ml of 6 *N* KOH. After centrifugation at 20,000 *g* for 30 minutes, the slightly turbid supernatant solution is stored overnight at —20°. After thawing, a large, brown, gelatinous precipitate, devoid of activity, is occasionally noticed and is removed by centrifugation. More frequently, only a slight turbidity appears and may be ignored. To each 100 ml of solution 13.4 g of ammonium sulfate is added. The resulting precipitate is removed by centrifugation and discarded. To each 100 ml of supernatant solution 14.6 g of ammonium sulfate is added, and again the precipitate is removed by centrifugation and discarded.

[8] Determined by the procedure of K. Lohmann and L. Jendrassik, *Biochemistry* **2**, 178, 419 (1926), scaled down to 10 ml.

[9] M. E. Pullman and G. C. Monroy, *J. Biol. Chem.* **238**, 3762 (1963).

[10] D. E. Green, R. L. Lester, and D. M. Ziegler, *Biochim. Biophys. Acta* **23**, 516 (1957).

[11] Estimated by the biuret method as described by E. E. Jacobs, M. Jacob, D. R. Sanadi, and L. B. Bradley, *J. Biol. Chem.* **223**, 147 (1956), scaled down to a final volume of 0.3 ml and read at 540 mμ in a Zeiss spectrophotometer equipped with a microcell adapter.

To each 100 ml of the supernatant solution 15.9 g of ammonium sulfate is added; the resulting precipitate, which contains most of the activity, is dissolved in a final volume of 10 ml in 50 m*M* Tris-SO_4, pH 8.0, and the pH is adjusted to 8.0 with 0.5 *M* NH_4OH. The fractional removal of protein avoids the loss of activity by coprecipitation. The solution is dialyzed overnight against 2 liters of 5 m*M* ammonium sulfate, pH 8.0, with one change after 3 hours.

The dialyzed supernatant solution (15.6 ml) is placed on a column of DEAE-cellulose, type 20 (3×4 cm). The column is equilibrated with 5 m*M* ammonium sulfate, pH 7.4, at room temperature, and the same solution is used as eluent. Approximately one-half of the activity is eluted with 1.2 liters at a flow rate of 1.2 ml per minute. The entire effluent is lyophilized to dryness. The residue is dissolved in 41 ml of water and dialyzed overnight against 6 liters of water. The dialyzed solution is concentrated to 10 ml by lyophilization; it contains approximately 60 m*M* ammonium sulfate (conductivity measurement). A summary of the procedure is given in the table.

PURIFICATION OF ATPASE INHIBITOR

Fraction	Volume (ml)	Total protein[a] (g)	Total units	Specific activity (units/mg)	Yield (%)
Mitochondria	2,400	48.0	—	—	—
Alkaline extract	1,650	9.70	97,500	10	100
Eluate of trichloroacetic acid precipitate	365	5.40	90,000[b]	17[b]	90[b]
Ammonium sulfate precipitate	15.6	0.16	62,500	378	64
DEAE-cellulose eluent (lyophilized and dialyzed)	11.6	0.01	27,800	2,610	29

[a] Estimated by the biuret method as described by E. E. Jacobs, D. R. Sanadi, and L. B. Bradley [*J. Biol. Chem.* **223,** 147 (1956)], scaled down to a final volume of 0.3 ml and read at 540 mμ in a Zeiss spectrophotometer equipped with a microcell adapter.

[b] Approximate.

Properties of the Inhibitor

General Characteristics. The inhibitor retains its activity and can be redissolved following precipitation with 5% trichloroacetic acid. It is stable for *at least* 8 minutes at 90° and pH 7, for 60 minutes at 37° between pH 3.0 and 10.0, but is rapidly destroyed at 90° in 0.1 *N* HCl or 0.1 *N* NaOH. It is nondialyzable, soluble in 80% ethanol, and inactivated by trypsin and chymotrypsin. Neither pancreatic ribonuclease nor pancreatic deoxyribonuclease have any affect on its activity.

Stability during Storage. The purified inhibitor may be stored at —20° for several months without appreciable loss in activity.

Molecular Weight. The molecular weight of this protein is approximately 15,000 as estimated by sedimentation in a sucrose gradient.

Specificity of Inhibitor. The inhibitor is highly specific in its interaction with mitochondrial ATPase. No nonmitochondrial ATPase or phosphatase of those examined,[9] has been found to be sensitive to the inhibitor. ATPase of submitochondrial particles prepared from rat liver, beef heart, yeast, and kidney mitochondria are all inhibited. ATPase of intact mitochondria is not inhibited. The inhibitor prevents the hydrolysis not only of ATP, but also of ITP, UTP, and GTP, the other known substrates of mitochondrial ATPase.[12]

Effect of pH on Interaction between ATPase and Inhibitor. A fairly broad pH optimum between pH 5.8 and 7.0 exists for the interaction of the inhibitor and ATPase. Above pH 7.0 a sharp decline occurs.

Effect of Inhibitor on Oxidative Phosphorylation and the $^{32}P_i$*-ATP Exchange Reaction.* The inhibitor has no effect on oxidative phosphorylation catalyzed by the various submitochondrial particles thus far examined. The inhibitor–ATPase complex (see below), in which virtually all the hydrolytic activity of F_1 is masked, substitutes for F_1 in recoupling phosphorylation to oxidation in F_1-deficient particles.

The $^{32}P_i$-ATP exchange reaction catalyzed by F_1-deficient submitochondrial particles is dependent on the presence of F_1. If the inhibitor is added in excess to eliminate ATPase activity completely, the exchange is inhibited. ADP does not relieve the inhibition. On the other hand, the inhibitor has an apparent stimulatory effect on the exchange in some preparations of submitochondrial particles where an excess of ATPase is present. By sparing ATP, the total $^{32}P_i$ incorporated into ATP is increased severalfold, but no increase in the specific radioactivity of ATP occurs.[9]

Neither F_1 nor the F_1-inhibitor complex catalyzes an ADP-ATP exchange if precautions are taken to remove adenylate kinase from the inhibitor preparation. This may be accomplished by heating the purified inhibitor at 90° for 8 minutes at pH 7.4.[13]

The ATPase-Inhibitor Complex. An inhibitor–ATPase complex can

[12] M. E. Pullman, H. Penefsky, A. Datta, E. Racker, *J. Biol. Chem.* **235**, 3762 (1963).

[13] Recently it was found by Dr. B. Bulos that a 15-minute exposure at 100° was necessary to ensure removal of the last traces of adenylate kinase activity. Some loss of inhibitor activity is encountered under these conditions.

be separated from an incubation mixture containing ATPase and inhibitor by precipitation with 2 *M* ammonium sulfate.[9] The complex can be dissociated by heating for 2 minutes at 65° in the presence of 2 m*M* ATP. This procedure does not inactivate either the ATPase or the inhibitor. ATPase activity and coupling activity of F_1 is protected against cold inactivation by the inhibitor.

Section VI

Isolation and Assay of Coupling Factors for Oxidative Phosphorylation

Previously Published Articles from Methods in Enzymology Related to Section VI

Vol. II [97]. Mitochondrial ATPase. W. Wayne Kielley.
Vol. VI [39]. P_i^{32}-ATP Exchange Enzyme System. G. W. E. Plaut.

[81] The Isolation, Properties, and Assay of ATP Synthetase II

By ROBERT E. BEYER

ATP synthetase II restores phosphorylation coupled to the oxidation of both DPNH and succinate in submitochondrial particles with low phosphorylative capacity.[1] The enzyme appears to have little or no effect on phosphorylation during the oxidation of cytochrome *c* reduced by ascorbate. The major increase in phosphorylation induced by ATP synthetase II is at the second phosphorylation site between coenzyme Q and cytochrome *c* in the electron transfer sequence. The enzyme has also been called coupling factor II[1-3] and coupling factor 2.

Assay Method

The principal assay of the ATP synthetase II involves the measurement of the increase in phosphorylation (Δ P:O) of ETPH(EDTA-2) due to the presence of the soluble enzyme during the oxidation of succinate. The manometric assay employed and the preparation of the particle are described in this volume [34]. The enzyme may also be assayed by measuring the ATP-ADP exchange which it catalyzes.[3]

Preparation of the ATP Synthetase II

Mitochondrial Extract. Beef heart mitochondria (this volume [12]), obtained from the large-scale preparation, are separated into the light and heavy layers in 0.25 *M* sucrose–0.01 *M* Tris-HCl, pH 7.8. The heavy beef heart mitochondria are suspended in a solution 0.15 *M* in KCl, 0.01 *M* in Tris-HCl, pH 7.5, and 5 m*M* in EDTA to a concentration of 30 mg of protein per milliliter. The pH is adjusted to 7.5 with 1 *N* KOH, and 40-ml aliquots are subjected to sonic treatment for 60 seconds at full power in the Branson probe sonifier at between 7 and 8 amp output. The temperature is maintained below 5° with the use of a jacketed beaker and the circulation of ethylene glycol at between −5° and −10° through the jacket. The pH of the sonicated suspension is adjusted to 7.5 with 1 *N* KOH and centrifuged in the Spinco Model L ultracentrifuge (as are all centrifugations in this procedure) for 90 minutes in the No. 30 rotor at 30,000 rpm. The supernatant fluid (S-1) is carefully removed and is used for further fractionation.

[1] R. E. Beyer, *Biochem. Biophys. Res. Commun.* **16**, 460 (1964).
[2] R. E. Beyer, *Biochem. Biophys. Res. Commun.* **17**, 184 (1964).
[3] R. E. Beyer, *Biochem. Biophys. Res. Commun.* **17**, 764 (1964).

Ammonium Sulfate Precipitation. Solid ammonium sulfate is added to the S-1 to 45% saturation at 0° (25.6 g/100 ml), and the mixture is maintained between 0° and 5° overnight. The suspension is clarified by centrifugation for 5 minutes at 25,000 rpm in a No. 30 rotor and the clear solution (S-2) is adjusted to 70% saturation with solid ammonium sulfate (add 15.6 g/100 ml). The suspension is allowed to stand for 30 minutes at below 5°, and the resulting precipitate is collected by centrifugation for 5 minutes at 25,000 rpm in the No. 30 rotor. The supernatant fluid is discarded and the sediment is dissolved in a minimum volume of 0.05 *M* potassium bicarbonate which has been adjusted to pH 7.5 with HCl. This stage is termed the "45–70% SAS fraction." The next steps are determined by whether the preparation is to be chromatographed immediately or stored for further use. In the event storage is desired, the preparation is dialyzed for between 4 and 6 hours against 4 liters of 0.05 *M* potassium bicarbonate freshly adjusted to pH 7.5. The dialysis solution is changed once during the procedure, and the small flocculent precipitate which frequently is encountered is discarded by centrifugation. The soluble protein is lyophilized and may be stored as a dry powder for a period of at least several weeks at —20°. In the event the fresh 45–70% SAS fraction is to be chromatographed immediately, it is dialyzed for 4–6 hours against 4 liters of 5 m*M* Tris-H_2SO_4, pH 7.5, with one change of dialysis fluid during the procedure. The small flocculent precipitate is removed by centrifugation.

Chromatography on DEAE-Substituted Cellulose. DEAE-substituted cellulose[4] is settled from distilled water to remove fine particles and then washed successively with 1 *N* HCl, water, 1 *N* NaOH, water, 1 *N* HCl, water, 1 *M* Tris-H_2SO_4, pH 7.5, and stored in the latter solution. A 20×220 mm column of the cellulose is prepared and washed with approximately 100 ml of 5 m*M* Tris-H_2SO_4, pH 7.5. Up to 1 g of the dried 45–70% SAS fraction is dissolved in 50 m*M* Tris-H_2SO_4, pH 7.5, and passed through a column of Sephadex G-25M equilibrated with 5 m*M* Tris-H_2SO_4, pH 7.5, of sufficient volume to elute the protein in 5 m*M* Tris-H_2SO_4. The Sephadex eluate, or the fresh 45–70% SAS fraction dialyzed against 5 m*M* Tris-H_2SO_4, is applied to the DEAE-substituted cellulose and washed with sufficient 5 m*M* Tris-H_2SO_4, pH 7.5, to lower the optical density of the eluate to below 0.1 at 278 mμ. A linear gradient is applied between 0.005 and 0.15 *M* Tris-H_2SO_4, pH 7.5 (developed by using two chambers each containing 300 ml of solution). The ATP syn-

[4] BioRad Cellex-D, obtained from Calbiochem, P. O. Box 54282, Los Angeles 54, California, has given the most consistent resolution of this preparation.

thetase II is eluted as the second distinct peak at approximately 80 m*M* Tris-H_2SO_4, and is referred to as ATP synthetase II-D.

Recycling Molecular Sieve Chromatography. Several lyophilized preparations of ATP synthetase II-D are pooled, equilibrated with 50 m*M* Tris-H_2SO_4, pH 7.5, on a column of Sephadex G-25M, and applied to a 40 × 800 mm column of the polyacrylamide gel[5] P-200. The gel filtration is developed with the use of an apparatus such as the LKB ReCyChrom with an effluent recording device as described by Porath and Bennich.[6] The column is maintained at 2–4° by the use of a cooling jacket. The ATP synthetase is applied to the bottom of the column and the solvent is circulated in the same direction. Since pressure counteracts the gravity factor, compression is minimized and a good flow rate may be maintained. Three passes (corresponding to a bed height of 2.4 m) of the protein through the gel column are sufficient to separate the ATP synthetase II-D preparation into two discrete peaks. The second, slower moving, peak is collected and contains the activity. The table provides a summary of the purification procedure.

PURIFICATION OF ATP SYNTHETASE II

Stage	Volume (ml)	Protein (mg)	Saturation of particle[a] (μg/mg protein)	Δ P:O
HBHM	246	9470		
S-1	396	1410		
45–70% SAS	19	362	2000	+0.18
DEAE eluate[b]	4	10.2	60	+0.26
P-200 eluate[b]	4	6.4	28	+0.64

[a] Micrograms of protein of ATP synthetase per milligrams of particle protein required to give maximum increase in P:O ratio with succinate as substrate.
[b] Following concentration.

Properties

The enzyme appears to be specific in catalyzing phosphate transfer at the second phosphorylation site of ETPH(EDTA-2) in that the increases in P:O ratios (Δ P:O) during the oxidation of NADH and succinate are of the same order of magnitude while no increase is seen during the oxidation of cytochrome *c* reduced with ascorbate. The enzyme also accepts the terminal phosphate from ATP[2] and catalyzes an ATP–

[5] BioRad P-200, obtained from Calbiochem, gave very good resolution.
[6] J. Porath and H. Bennich, *Arch. Biochem. Biophys.*, Suppl. 1, 152 (1962).

ADP exchange reaction.[3] It is stable after both DEAE-cellulose and polyacrylamide elution in liquid N_2 for at least 3 weeks. The absorption spectrum shows a maximum at 278 mμ, a minimum at 250 mμ, and a shoulder at 290 mμ. It appears as a single component in the ultracentrifuge (mol. wt. = 124,000), but two minor bands appear in disc electrophoresis.

[82] Preparation and Properties of Mitochondrial ATPase (Coupling Factor 1)

By HARVEY S. PENEFSKY[1]

General Principles

The soluble ATPase of beef heart mitochondria also serves as a factor which couples the esterification of inorganic phosphate to the oxidation of substrates catalyzed by deficient submitochondrial particles.[1a, 2] The enzyme is solubilized when beef heart mitochondria are disrupted by shaking with glass beads in a high speed reciprocal shaker. This method is based on one already presented[3] and introduces changes which permit preparation of the enzyme on a larger scale.

Definition of Unit and Specific Activity. A unit of activity is defined as that amount of enzyme which catalyzes the turnover of 1 micromole of substrate per minute under the specified conditions of assay.[3] Specific activity is expressed as units per milligram of dry weight of protein.

Measurement of Protein Concentration. The protein concentration of the mitochondrial suspension and of the step 1 and 2 fractions is determined by a biuret method modified for mitochondria.[4] Soluble protein is measured in a Brice-Phoenix differential refractometer model BP-2000. The specific refractive index increment of bovine serum albumin, 188×10^{-5} *dl/g*,[5] is used to calculate protein concentration from the difference in refractive index between protein and solvent. Protein con-

[1] This manuscript was prepared while the author was at the Department of Biochemistry, The Public Health Research Institute of the City of New York, Inc., New York.

[1a] M. E. Pullman, H. S. Penefsky, A. Datta, and E. Racker, *J. Biol. Chem.* **235**, 3322 (1960).

[2] H. S. Penefsky, M. E. Pullman, A. Datta, and E. Racker, *J. Biol. Chem.* **235**, 3330 (1960).

[3] M. E. Pullman and H. S. Penefsky, Vol. VI [34].

[4] E. E. Jacobs, M. Jacob, D. R. Sanadi, and L. B. Bradley, *J. Biol. Chem.* **223**, 147 (1956).

[5] G. E. Perlmann and L. G. Longsworth, *J. Am. Chem. Soc.* **70**, 2719, (1948).

centration determined by refractometry agrees to within 5% with values based on the weight of dialyzed and dried enzyme. Values of protein concentration in step 3 and 4 fractions also may be determined by ultraviolet absorption[6] and are corrected to dry weight when multiplied by 1.87.[7]

Purification Procedure

Step 1. Disintegration of the Mitochondria. "Light layer" mitochondria (6.75 g) prepared according to the method of Green and Ziegler,[8] is suspended in ice cold 0.25 *M* sucrose containing 2 m*M* EDTA, pH 7.4, and 2 m*M* ATP to give a final volume of 225 ml. Fifteen milliliters of this suspension and 7 ml of glass beads[9] are placed in the stainless steel cup of a Nossal[10] type reciprocal shaker[11] and evacuated as described previously.[3] During the shaking period of 2 minutes, ice water is circulated around the cup and associated bearing and water at 25° is circulated around the main bearing. Following disruption of the mitochondria the contents of the cups are pooled and warmed to room temperature[12]; the supernatant solution is removed after allowing the beads to settle. The beads are washed once by resuspension with 70 ml of 0.25 *M* sucrose, 2 m*M* EDTA, pH 7.4, and 2 m*M* ATP. The extract and bead washings are combined and centrifuged at room temperature in the number 30 rotor of the Spinco Model L centrifuge. The packed brown pellet is discarded and the clear yellow supernatant solution is collected. The crude extract containing 1280 mg of protein is allowed to stand overnight at room temperature before the start of step 2.

Step 2. pH Fractionation. The pH of the crude extract is adjusted to pH 5.4 by addition of 1.9 ml of 1 *N* acetic acid at room temperature with mechanical stirring. The solution is centrifuged at room temperature for 5 minutes at 5000 *g*, and the precipitate is discarded. The pH of the supernatant solution is adjusted to pH 6.7 by addition of 0.7 ml of 2 *M* Tris-SO_4, pH 10.7.

Step 3. Protamine Fractionation. To prepare 0.5% protamine sulfate solution suspend 1.0 g protamine sulfate (Eli Lilly and Company) in 52

[6] See Vol. III [73].

[7] H. S. Penefsky and R. C. Warner, *J. Biol. Chem.* **240,** 4694 (1965).

[8] D. E. Green and D. M. Ziegler, Vol. VI [58].

[9] Purchased from Minnesota Mining and Manufacturing Co., Ridgefield, New Jersey, 3M "Superbrite," catalog number 090, 0.0110-inch diameter.

[10] P. M. Nossal, *Australian J. Exptl. Biol. Med. Sci.* **31,** 583 (1953).

[11] Model RS, manufactured by the Lourdes Instrument Corporation, Brooklyn, New York.

[12] All subsequent manipulations are carried out at room temperature (25°) unless otherwise indicated.

ml of distilled water. Adjust the pH to 7.0 with 0.1 *M* acetic acid, and bring the volume to 100 ml with water. Let stand in ice for 20 minutes. Centrifuge in the cold at 18,000 *g* for 5 minutes and discard the residue. To the supernatant solution add an equal volume of water, and allow the solution to warm to room temperature before use.

A 5.6-ml sample of 0.5% protamine sulfate solution is slowely added with mechanical stirring at room temperature to 224 ml of the step 2 fraction.[13] Stirring is continued for 15 minutes after the addition of protamine, and the light precipitate which forms is removed by centrifugation for 10 minutes at 15,000 *g*. All the ATPase activity remains in the clear supernatant solution. To the supernatant solution 73 ml of protamine sulfate solution is added as before. A heavy yellow precipitate forms and is collected by centrifugation. The supernatant solution is discarded, and the precipitate is dissolved by adding 6 ml of 0.25 *M* sucrose, 2 m*M* EDTA, 0.01 *M* Tris-SO_4, pH 7.4, 0.4 *M* $(NH_4)_2SO_4$, pH 7.4. The resulting slightly turbid, yellow solution is clarified by centrifugation, and the small amount of insoluble residue is discarded. To the clear yellow supernatant solution is added with gentle stirring an equal volume of saturated ammonium sulfate, pH 5.5. The suspension is kept in ice for 15 minutes to ensure complete precipitation of the enzyme and then is centrifuged. The precipitate is dissolved in 4 ml of 0.25 *M* sucrose–0.01 *M* Tris-SO_4 pH 7.4–2 m*M* EDTA[14] at room temperature. The dissolved enzyme is again precipitated by adding an equal volume of saturated ammonium sulfate solution. If desired the enzyme may be stored at this point at 4° before proceeding to the next step.

Step 4. Temperature Fractionation. The ammonium sulfate suspension of the enzyme from step 3 is centrifuged and dissolved in Sucrose–Tris–EDTA to give a final protein concentration of 20 mg/ml. This solution is made 8 m*M* in ATP and heated in 12-ml conical glass centrifuge tubes at 65° for 4 minutes. Immediately thereafter the tubes are cooled in a 25° bath and centrifuged to remove denatured protein. The enzyme is precipitated from the supernatant solution by the addition of an equal volume of saturated ammonium sulfate solution. The enzyme is dissolved in Sucrose–Tris–EDTA and once again precipitated

[13] Since the exact amount of protamine required varies slightly for different preparations, a preliminary fractionation with protamine is carried out on a 4-ml sample of the step 2 fraction and scaled up appropriately. Usually 0.1 ml of 0.5% protamine solution is sufficient to precipitate inactive protein and lipid without reducing enzymatic activity in the supernatant solution. The subsequent addition of approximately 1.3 ml of protamine to the 4-ml sample results in precipitation of the bulk of the enzyme.

[14] This solution will be referred to as Sucrose–Tris–EDTA.

PURIFICATION PROCEDURE FOR ATPASE

Fraction	Volume (ml)	Units	Protein (mg)	Specific activity (units/mg protein)	Yield (%)
Step 1. Crude extract	237	2760[a]	1280	2.2	100
		4950[b]		3.9	179
Step 2. pH fractionation	224	3860	620	6.2	140
Step 3. Protamine fractionation	10	3350	93.5	35.9	121
Step 4. Temperature fractionation	8	3240	58	74.8	157

[a] Activity measured on the day of preparation.
[b] Activity measured on the following day. The specific activity of the crude extract increased about 2-fold upon standing overnight.

with ammonium sulfate. The purified enzyme is stored at 4° as a suspension in ammonium sulfate. Samples of the enzyme may be prepared for assay by the centrifugation of desired aliquots of the ammonium sulfate suspension followed by removal of ammonium sulfate and solution of the pellet in appropriate buffers.

Properties

Stability. The enzyme is stable for several months when stored at 4° at a suspension in half-saturated ammonium sulfate solution. The activity of the dissolved enzyme decreases slowly over a period of several days at room temperature in Sucrose–Tris–EDTA, 4 mM ATP. The enzyme is stable for about 1 day in Sucrose–Tris–EDTA and in buffers which do not contain sucrose provided ATP is present. Aqueous solutions of the ATPase are stable for at least 5 days when stored at —70° if frozen quickly in liquid nitrogen and thawed rapidly at 30°.[7]

Cold Lability. ATPase activity and coupling factor activity of the purified enzyme are cold labile in aqueous buffers.[1a,2] The rate of cold inactivation is greater at low protein concentrations (the half-time at 0° is about 7 minutes at 0.05 mg/ml and about 21 minutes at 3 mg/ml), is accelerated by anions (the order of effectiveness of anions in accelerating cold inactivation is $I^- > NO_3^- > Br^- > Cl^- > SO_4^{--}$) and is more labile at 0° (half-time about 7 minutes) than at 10° (half-time about 30 minutes).[7] The enzyme is stabilized in the cold by factors derived from beef heart mitochondria such as a heat stable dialyzable inhibitor,[15] a heat stable protein inhibitor,[16] mitochondrial phospholipid,[15] and F_0, an insoluble protein containing phospholipid which confers sensitivity to

[15] E. Racker, *Federation Proc.* **21**, 54 (1962).
[16] M. E. Pullman and G. Monroy, *J. Biol. Chem.* **238**, 3762 (1963).

oligomycin on the ATPase.[17] The enzyme also is stabilized in the cold by ethylene glycol, methanol, glycerol, and ethanol.[7] The cold-inactivated enzyme may be partially reactivated by rewarming at concentrations greater than 0.1 mg/ml.

Physical Properties. At 25° in 20 mM P_i, pH 7.2, 2 mM EDTA, 4 mM ATP, 0.1 M KCl, $S^{\circ}_{20,w}$ is 12.9°, the molecular weight determined by equilibrium ultracentrifugation is 284,000, the intrinsic viscosity is 3.2 ml/g, and the partial specific volume is 0.74 ml/g.[7] When incubated in the cold, the enzyme dissociates forming an equilibrium mixture of 3.4 S, 9.2 S, and 11.9 S components. Rewarming of the cold-treated enzyme under conditions which restore enzymatic activity results in the formation of a single sedimenting species with a sedimentation coefficient identical with that of the native enzyme. The ATPase dissociates at 25° in 1% sodium dodecyl sulfate to form a single sedimenting species with an $S_{20,w}$ of 2.6 S and approximate molecular weight of 29,000.

Chemical Analysis. The minimum molecular weight based on the half-cystine content is 26,100.[18] Tryptophan, galactosamine, glucosamine, and fatty acids[19] were not detected. Sulfhydryl group analyses indicate 10 or 11 SH per mole. There is no evidence for disulfide bridges in the molecule.[7]

Other Properties. The specificity of the ATPase, activators, and inhibitors of the enzyme and the pH optima have been described.[3]

[17] E. Racker, *Biochem. Biophys. Res. Commun.* **10,** 435 (1963).
[18] The amino acid analyses were kindly performed by Dr. S. Moore.
[19] Fatty acid analyses were carried out by Dr. S. Fleischer and Dr. I. Mossbach.

[83] Preparation of ^{3}H-Acetyl-ATPase (Coupling Factor 1)

By YASUO KAGAWA[1]

General Principle

ATPase[1a] containing 5–6 moles of ^{3}H-acetyl groups per mole of enzyme is prepared by reaction with ^{3}H-acetic anhydride. The labeled preparations appear to retain all the properties of the native enzyme and are useful in studies of the binding of the soluble ATPase to insoluble mitochondrial fractions.

[1] See footnote 1, page 505.
[1a] See this volume [82].

Preparation

Reagents

^{3}H-Acetic anhydride, 1 mC/mg,[2] 25 mg/ml acetone
Buffer I, pH 7.4, 0.01 *M* $NaHCO_3$, 0.01 *M* sodium acetate, 1.0 m*M* ATP
Enzyme, 25 mg of step 4 fraction

Procedure. An appropriate volume of the ammonium sulfate suspension of the enzyme is centrifuged; after removal of the ammonium sulfate, the enzyme pellet is dissolved in 1.5 ml of 0.25 *M* sucrose, 0.01 *M* Tris-SO_4, pH 7.4, 0.001 *M* EDTA, 0.004 *M* ATP. Residual ammonium sulfate is removed by passage of the enzyme solution through a 1.8 × 9.2 cm column of Sephadex G-50[3] equilibrated with buffer I. The enzyme is collected in the 7–14 ml fraction of the column eluate.

To 5.5 ml of the enzyme solution obtained from the column (25.2 mg of protein) is added, with stirring at 25°, 5 μl of ^{3}H-acetic anhydride solution. After 10 minutes the enzyme is passed through a 1.6 × 14 cm column of Sephadex G-50 equilibrated with buffer I for removal of unreacted acetic anhydride. The enzyme is collected in the 10–18 ml fraction of the column eluate. This fraction contains 24.1 mg protein in 8 ml. To 6 ml of acetyl-ATPase is added 12 ml of saturated ammonium sulfate solution, pH 7.4, containing 1 m*M* ATP. The suspension is cooled in ice for 10 minutes then centrifuged in the cold at 18,000 *g* for 5 minutes. The enzyme pellet is resuspended in 4–5 ml of the supernatant solution and stored at 4°.

Properties

Stability. The stability of the acetylated enzyme appears to be equivalent to that of the native enzyme when stored at 4° as a suspension in ammonium sulfate.

Other Properties. Acetylation causes about 20% inactivation of ATPase activity but does not affect coupling factor activity, cold lability, oligomycin sensitivity when the enzyme is combined with F_0[4] (a protein

[2] Purchased from Nuclear-Chicago Corp., Des Plaines, Illinois. A sealed tube of 25 mC ^{3}H-acetic anhydride (specific radioactivity 6.7 mC/mg) is chilled in a dry ice-acetone bath for 1 hour before it is opened. Unlabeled acetic anhydride (21.3 mg dissolved in 1 ml of dry purified acetone) is added.

[3] Purchased from Pharmacia Fine Chemicals Inc., New Market, New Jersey.

[4] Y. Kagawa and E. Racker, *J. Biol. Chem.* **241**, 2467, 2475 (1966).

containing phospholipid), or inhibiton of ATPase activity by anti-ATPase antibody.[5]

A summary of the procedure and the stoichiometry of the acetylation reaction is presented in the accompanying table.

PROCEDURE FOR ACETYLATION OF THE ATPASE

Cpm[a] added	5.39×10^7
^{3}H-acetic anhydride (μg)	125
Specific radioactivity of acetic anhydride (cpm/μg)	4.31×10^5
ATPase recovered after acetylation (mg)	24.1
Cpm in ATPase fraction	9.98×10^6
Specific radioactivity (cpm/mg protein)	4.16×10^5
Moles acetyl/mole enzyme	5.2

[a] Counted in a Packard liquid scintillation counter.

[5] See this volume [107].

[84a] Coupling Factor 2 (F_2)[1]

By JUNE M. FESSENDEN,[1a] M. ANNE DANNENBERG,[1a] and EFRAIM RACKER[1a]

$$P_i + ADP \xrightarrow[F_1,\ F_4,\ F_2]{\text{A-particles}} ATP$$

Assay Method[2]

Principle. A-particles catalyze oxidations but require coupling factors for phosphorylation. In the presence of excess F_1 and F_3 or F_4 the stimulation of phosphorylation by F_2 is used as an assay of F_2 activity.

Purification Procedure

Step 1. Preparation of Mitochondrial Acetone Powder. Heavy layer beef heart mitochondria[3] (4 g of protein) are added to 500 ml of reagent-grade acetone at $-10°$ and blended in an explosion-proof Waring blendor for 30 seconds, then rapidly added to 2 liters of acetone at $-10°$. The suspension is stirred for 5 minutes, then centrifuged at 1100 g for 5 minutes. The pellet is resuspended in 500 ml of acetone ($-10°$), blended 30 seconds, and added to 2 liters of acetone. The suspension is mixed, for 5 minutes, then centrifuged at 1100 g for 5 minutes. The

[1] J. Fessenden and E. Racker, *J. Biol. Chem.* **241**, 2483 (1966).

[1a] This manuscript was prepared while the authors were at the Department of Biochemistry, The Public Health Research Institute of the City of New York, Inc., New York.

[2] See this volume [35].

[3] See Vol. VI [58].

pellet is resuspended in 500 ml of acetone (—10°), blended another 30 seconds, then filtered under suction through Whatman No. 1 filter paper. The cake is washed with 25 ml of anhydrous ether at —10°. To remove the remaining traces of solvents, the powder is placed in a desiccator and evacuated about 3 hours with an oil pump, then stored under vacuum at 4°. The yield is approximately 6 g of powder.

Step 2. Extraction. Six grams of the acetone powder are resuspended and homogenized in 80 ml of 0.1 *M* potassium phosphate pH 7.4 previously flushed with nitrogen. The suspension is immediately centrifuged at 18,000 *g* for 10 minutes. The supernatant fluid is discarded, and the pellet is homogenized in 300 ml of 0.06 *M* glycine pH 10.5 previously flushed with nitrogen. The suspension is kept in ice, and under nitrogen for 10 minutes, then centrifuged at 18,000 *g* for 10 minutes. The pellet is discarded. The supernatant solution is brought to pH 7.5 with 1 *N* acetic acid. At this stage 4 m*M* ATP is added and the preparation is stored overnight at —55°.

Step 3. Adsorption on Calcium Phosphate Gel. Calcium phosphate gel (3 mg per milligram of protein) is added to the glycine extract, homogenized, and kept at 0° with gentle stirring for 10 minutes. The suspension is centrifuged at 18,000 *g* for 10 minutes. The supernatant solution is discarded and the pellet is rinsed with cold glass-distilled water, then resuspended and homogenized in 25 ml of 0.3 *M* potassium phosphate pH 7.4. The suspension is gently stirred for 10 minutes at 0°, then centrifuged at 18,000 *g* for 10 minutes. The extraction with 0.3 *M* potassium phosphate is repeated twice as above with 15 ml of buffer. ATP (4 m*M*) is added to all fractions before storage. The fractions at this stage are virtually free of F_3 and suitable for studies. Only occasionally was the purification carried further.

Step 4. Chromatography of P-100 Gel. In the cold room a column (1×10 cm) of P-100 Gel, with a spacer gel of P-60 gel (1.5 cm) is set up and equilibrated with 0.3 *M* potassium phosphate pH 7.4. An aliquot from the active fraction of the previous step containing 5–8 mg of protein is put on the column. One-milliliter fractions are collected. The activity is eluted in the fractions immediately following the void volume. Before storage 4 m*M* ATP is added to the active fractions.

[84b] Coupling Factor 3 (F_3)[1]

By JUNE M. FESSENDEN[1a] and EFRAIM RACKER[1a]

$$\text{Succinate} + \text{ATP} + \text{DPN} \xrightarrow[\text{F}_1,\ \text{F}_3,\ \text{Mg}^{++}]{\text{A-particles}} \text{Fumarate} + \text{ADP} + \text{P}_i + \text{DPNH}$$

Assay Method

Principle. F_3 is measured with A-particles[2] by the stimulation of the ATP-linked reduction of DPN by succinate in the presence of F_1 or by the increase of the P:O ratio in the presence of F_1 and F_2.

Reagents

Tris-sulfate, 1 *M*, pH 7.5
$MgSO_4$, 0.2 *M*
Bovine serum albumin, 5% (crystalline, Armour Co.) dialyzed overnight against 100 volumes of 0.01 *M* Tris-SO_4, pH 7.5
Succinate, 0.25 *M*, succinic acid neutralized with 2 *M* Tris to pH 7.5
A-particles (5 mg/ml) (footnote 2)
F_1 (1 mg/ml) (footnote 3)
Na_2S, 0.3 *M*
DPN, 0.027 *M*, pH 6.8
ATP, 0.2 *M*, pH 7.4
F_3, 1–50 μl of the enzyme solution is used for assay. If necessary the enzyme may be diluted in 50 m*M* Tris-SO_4, pH 8, but the final concentration should not be below 3–5 mg of protein per milliliter. Protein is determined spectrophotometrically at 280 mμ, assuming that 1 mg/ml gives a density reading of 1.0.

Procedure. In a volume of 0.6 ml the following components are added to a 3-ml cuvette: 0.15 ml of Tris-SO_4, 0.05 ml of $MgSO_4$, 0.03 ml of bovine serum albumin, 0.06 ml of succinate, 0.05 ml of A-particles, 0.01 ml of F_1, F_3, and H_2O. After incubation for 10 minutes at room temperature, 2.33 ml of H_2O, 0.02 ml of Na_2S, and 0.02 ml of DPN are added, and the

[1] E. Racker, *Proc. Natl. Acad. Sci. U.S.* **48**, 1659 (1962). The current purification procedure of F_3 is a modification of the procedure described in the reference and results in the removal of pyrophosphatase activity.

[1a] This manuscript was prepared while the authors were at the Department of Biochemistry, The Public Health Research Institute of the City of New York, Inc., New York.

[2] See this volume [35].

[3] See this volume [82].

cuvette is placed in an Eppendorf fluorometer. After 1–2 minutes, 0.03 ml of ATP is added and the maximal rate of DPNH formation is recorded.

Purification Procedure

Step 1. Preparation of Crude Extract. Heavy layer beef heart mitochondria[4] (3 g of protein) are diluted to 150 ml in a final sucrose concentration of 0.15 *M*. The pH of the suspension is adjusted to 9.2 with freshly diluted ammonium hydroxide, care being taken that the ionic strength of the mixture is at a minimum. The suspension is exposed to sonic oscillation in 30-ml batches for 2 minutes in a Raytheon[5] sonic oscillator (250 watts, 10 kc) cooled by flowing ice water, then centrifuged at 104,000 *g* for 45 minutes. The pellet is discarded. The supernatant fluid is adjusted to pH 8 with 1 *N* acetic acid and recentrifuged at 104,000 *g* for 45 minutes. The supernatant solution is frozen at —55°.

Step 2. Chromatography on DEAE-cellulose. The supernatant fluid is thawed, and 100 ml is applied in the cold room to a DEAE-cellulose column (2.5 cm × 22 cm) that has been packed under light pressure and equilibrated with 20 m*M* Tris-SO_4, pH 8. The column is washed with 200–250 ml of 20 m*M* Tris-SO_4, pH 8, and the eluate is discarded. The buffer is changed to 0.1 *M* ammonium sulfate, 20 m*M* Tris-SO_4, pH 8, and 5-ml fractions are collected. Samples containing 200 μg of protein are assayed for activity. The active fractions are combined. The activity appears together with a yellow protein fraction, which is removed at step 3.

Step 3. Ammonium Sulfate Fractionation. The combined eluates in 0.1 *M* ammonium sulfate are brought to a final saturation of 33% by the slow addition of saturated ammonium sulfate, pH 8, and kept at 0° for 30 minutes. The solution is centrifuged at 18,000 *g* for 10 minutes. The supernatant fluid is collected (Sup I) and the precipitate is resuspended in 50 m*M* Tris-SO_4, pH 8, in one-tenth the volume of the combined eluates. Any insoluble material is removed by centrifugation at 18,000 *g* for 5 minutes (Sup II).

Sup I is brought to 50% saturation with a solution of saturated ammonium sulfate, pH 8.0. After 30 minutes at 0° the solution is centrifuged at 18,000 *g* for 10 minutes. The supernatant fluid is discarded, and the pellet is resuspended in 50 m*M* Tris-SO_4, pH 8 in one-tenth the volume of the combined eluates (Sup III). This fraction may be reprecipated at 30% ammonium sulfate saturation by the same treatment as above (Sup IV).

[4] See Vol. VI [58].

[5] Larger volumes can be exposed to sonic oscillation by means of a probe (exposure for 4 minutes is used with an MSE sonic oscillator).

The purification procedure is summarized in the table.

Purification Procedure for F_3

Fraction	Total (mg)	Total volume (ml)	Total units	Specific activity	Recovery (%)
Crude extract	940	100	34,800	37	100
DEAE-cellulose eluate	108	24	13,600	126	39
0–33% $(NH_4)_2SO_4$ (Sup II)	21	2.4	6,700	319	19.3
33–50% $(NH_4)_2SO_4$ (Sup III)	31	2.5	4,330	140	12.5
0–30% 2nd $(NH_4)_2SO_4$ (Sup IV)	12.5	2.5	3,120	250	9

Properties

Stability. In the frozen state F_3 (DEAE-cellulose eluate or Sup II) is stable for several months, but loses 80% of the activity in 10 minutes at 40°. Sup II is stable above 5 mg/ml but loses 40–60% of its activity when diluted to 1 mg/ml.

[85] Coupling Factor 4 (F_4)

By T. E. Conover and H. Zalkin[1]

It has been shown that A- and P-particles require the addition of coupling factor 4 (F_4) for respiratory chain phosphorylation[1a] and for all reactions associated with oxidative phosphorylation that involve a transphosphorylation step with ATP. These reactions include the ATP-dependent reduction of DPN,[2] the ATP-dependent reduction of TPN,[3] the $^{32}P_i$-ATP exchange,[1a] and the $H_2{}^{18}O$-P_i exchange.[4] On the other hand, the addition of F_4 is not required for the energy-linked reduction of DPN or TPN provided the energy is produced by oxidation of succinate rather than by ATP.[3]

Assay for Oxidative Phosphorylation

Principle. Submitochondrial particles that catalyze the oxidation of DPNH and succinate but that have lost the capacity for oxidative phosphorylation are used to assay F_4. Restoration of oxidative phosphoryla-

[1] This manuscript was prepared while the author was at the Department of Biochemistry, College of Physicians and Surgeons of Columbia University, New York.

[1a] T. E. Conover, R. L. Prairie, and E. Racker, *J. Biol. Chem.* **238**, 2831 (1963).

[2] R. L. Prairie, T. E. Conover, and E. Racker, *Biochem. Biophys. Res. Commun.* **10**, 422 (1963).

[3] E. Racker and T. E. Conover, *Federation Proc.* **22**, 1088 (1963).

[4] P. Hinkle, H. S. Penefsky, and E. Racker, unpublished experiments, 1964.

tion is obtained by addition of F_4 in the presence of an excess of coupling factor 1 (F_1).

Reagents

$MgSO_4$, 0.1 *M*

Sodium succinate, 0.1 *M*, pH 7.4

Potassium phosphate buffer, 0.1 *M*, pH 7.4

Stock mixture A contains 50 m*M* Tris-SO_4 buffer, pH 7.4; 20 m*M* $MgCl_2$; 10 m*M* ATP; 5 m*M* EDTA; 100 m*M* glucose; and 1% dialyzed bovine serum albumin.

Hexokinase, 100 units/ml. Commercial crystalline hexokinase may be used.[5] The ammonium sulfate suspension is dialyzed against 0.015 *M* sodium acetate, pH 5.4, containing 1.0% glucose.

P-particles.[1a,6] Preparation containing 20–30 mg protein per milliliter is used.

F_1 coupling factor.[7] Solution of 0.5–1.0 mg of protein per milliliter is prepared by sedimenting with centrifugation the required amount of protein from an ammonium sulfate suspension, decanting the supernatant fluid, and absorbing the residual traces of fluid with a piece of filter paper. The protein pellet is dissolved in a solution containing 0.25 *M* sucrose, 0.01 *M* Tris-SO_4, pH 7.4, 1 m*M* EDTA, and 4 m*M* ATP at room temperature.

F_4 coupling factor. Preparation should contain 10–20 mg of protein per milliliter freed of ammonium sulfate by dialysis against 0.2 *M* KCl, 0.02 *M* Tris, pH 7.4, and 1 m*M* EDTA.

Procedure. To reconstitute a phosphorylating system, 0.5–1.0 mg of P-particles is incubated with 50 μg of F_1, 4 micromoles of $MgSO_4$, and the preparation of dialyzed F_4 (0.5–1.0 mg) in a volume of 0.3 ml for 5 minutes at room temperature. A 0.2-ml aliquot of this mixture is added to the side arm of a 5-ml Warburg vessel which contains in the main compartment the following components: 0.05 ml of hexokinase (5–10 units); 0.05 ml of 0.1 *M* potassium phosphate buffer, pH 7.4; 0.1 ml of 0.1 *M* sodium succinate; and 0.1 ml of stock mixture A. The vessels are equilibrated for 6 minutes at 30°. The reaction is initiated by mixing the contents of the side arm with those of the main compartment.

[5] Boehringer Mannheim Corp., New York, New York, or Sigma Chemical Corp., St. Louis, Missouri.

[6] A-particles may also be used. The preparation of these particles is described in this volume [35].

[7] M. E. Pullman, H. S. Penefsky, A. Datta, and E. Racker, *J. Biol. Chem.* **235**, 3322 (1960). See Vol. VI [34] and this volume [82].

Alkali may be left out of the center well as there is no production of CO_2. The reaction is usually incubated for 20–30 minutes and terminated by removal of the vessels and immediate addition of 0.05 ml of 35% perchloric acid. Oxygen consumption is measured manometerically and phosphate esterification by the difference in the orthophosphate level in the reaction vessel and in a zero time control. The orthophosphate may be determined on the deproteinized reaction mixtures by a suitable method such as that of Lohmann and Jendrassik.[8]

Assay for $^{32}P_i$-ATP Exchange

Principle. In addition to oxidative phosphorylation, A- or P-particles have lost the capacity for $^{32}P_i$–ATP exchange. Restoration of $^{32}P_i$–ATP exchange is obtained by addition of F_4 in the presence of an excess of F_1. For routine assays the $^{32}P_i$–ATP exchange has the advantage of being rapid and giving satisfactory results even with crude preparations of F_4.

Reagents

$MgSO_4$, 0.1 *M*

Potassium phosphate-^{32}P, 0.2 *M*, pH 7.4, 10^{-3} cpm/micromole. Commercial $H_3{}^{32}PO_4$ is purified by ion exchange chromatography[9] or by recrystallization as $MgNH_4PO_4$.[1a]

Stock mixture B contains 40 m*M* $MgCl_2$; 40 m*M* ATP; and 5 m*M* EDTA, pH 7.4.

P-particles. Same as for oxidative phosphorylation.

F_1 coupling factor. Same as for oxidative phosphorylation.

F_4 coupling factor. Same as for oxidative phosphorylation.

Procedure. The P-particles and coupling factors are reconstituted as was described for the oxidative phosphorylation assay. As low concentrations of ammonium sulfate do not interfere with the $^{32}P_i$–ATP exchange, it is not necessary to dialyze the F_4 preparations. A 0.2-ml aliquot of the reconstituted preparation is added to 0.1 ml of 0.2 *M* potassium phosphate-^{32}P (10,000–20,000 cpm) and 0.2 ml of stock mixture B to initiate the reaction. The reaction is allowed to procede 15 minutes at 30° and is terminated by the addition of 0.05 ml of 35% perchloric acid.

The incorporation of $^{32}P_i$ into ATP is determined by extraction with

[8] K. Lohmann, and L. Jendrassik, *Biochem. Z.* **178,** 419 (1926). See also Vol. III [116].

[9] M. DeLuca, K. E. Ebner, D. E. Hultquist, G. Kreil, J. B. Peter, R. W. Moyer, and P. D. Boyer, *Biochem. Z.* **338,** 512 (1963).

molybdic acid and isobutanol–benzene.[10] An aliquot of the deproteinized reaction mixture is added to a tube containing 5 ml of 1% ammonium molybdate in 1.0 N $HClO_4$ and 5 ml of a 1:1 (by volume) mixture of isobutanol and benzene. The mixture is shaken vigorously for 20 seconds, and the upper organic phase is removed by aspiration as soon as the layers separate. To remove last traces of $^{32}P_i$, the aqueous phase is extracted a second time in the same manner with 5 ml of isobutanol saturated with H_2O. Generally 1 ml of the aqueous phase is removed for measurement of radioactivity. An alternative procedure is to measure the radioactivity in the nucleotides as adsorbed and washed on charcoal.[11] The breakdown of ATP by ATPase activity during the course of reaction is ignored for reasons of simplicity. It does not appear to effect the linearity of the reaction with time or concentration of protein.

Preparation of F_4 Coupling Factor

Beef heart mitochondria prepared as described by Green *et al.*[12] are used as the source of coupling factor F_4. The "light" layer mitochondria gave more satisfactory yields.

Alkali Extraction. Beef heart mitochondria, 6 g in 160 ml of 0.3 M KCl and 1.5 mM EDTA, are stirred with 80 ml of cold 1.2 M ammonium hydroxide for 5–10 minutes at 0°. The mixture is then centrifuged for 60 minutes at 78,000 g. The supernatant fluid is carefully decanted and recentrifuged for 60 minutes at 105,000 g. The clear supernatant solution is adjusted to pH 8.0 with 10 N acetic acid, and the small precipitate is removed by centrifugation at 105,000 g for 60 minutes.

First Ammonium Sulfate Precipitation. The protein concentration of the supernatant fluid is adjusted to 5–6 mg/ml, and for each 100 ml of solution, 67 ml of a saturated solution of ammoniacal ammonium sulfate, pH 8.0, are added. After centrifugation for 30 minutes at 18,000 g, the precipitate is separated and dissolved in 20 ml of 0.02 M Tris-SO_4, pH 8.0, containing 1 mM EDTA.

Second Ammonium Sulfate Precipitation. The ammonium sulfate concentration is determined by conductivity measurements, and ammoniacal ammonium sulfate is added to bring the final concentration to 0.48 M in a volume of 40 ml. This procedure results in a rather insoluble precipitate (fraction A), which is removed by centrifugation. This precipitate may be suspended in 0.25 M sucrose, 0.02 M Tris, pH 7.4, and

[10] O. Lindberg, and L. Ernster, *in* "Methods of Biochemical Analysis" (D. Glick, ed.), Vol. III, p. 1. Wiley (Interscience), New York, 1956. See also Vol. VI [39].

[11] P. D. Boyer, W. W. Luchinger, and A. B. Falcone, *J. Biol. Chem.* **223**, 405 (1956).

[12] D. E. Green, R. L. Lester, and D. M. Ziegler, *Biochim. Biophys. Acta* **23**, 516 (1957). See also Vol. VI [58] and this volume [12].

1 mM EDTA; it is active, but difficult to work with because of its insolubility. The ammonium sulfate concentration of the supernatant solution is raised to 1.4 M with solid ammonium sulfate, and the resulting precipitate (fraction B) is collected by centrifugation and dissolved in a small volume of 0.02 M Tris, pH 7.4, containing 1 mM EDTA. Fraction B may be dialyzed for 2–3 hours against 0.2 M KCl, 0.02 M Tris, pH 7.4, and 1 mM EDTA with little loss in activity. On dilution or dialysis with 0.02 M Tris, pH 7.4, fraction B forms a fibrous precipitate but remains fully active. These insoluble preparations may be suspended in 0.25 M sucrose–0.02 M Tris, pH 7.4, and sonicated for 2 minutes to give stable suspensions which do not sediment upon storage. For measurement of $^{32}P_i$–ATP exchange, preparations of F_4 may be used directly, but traces of ammonium sulfate must be removed by dialysis if oxidative phosphorylation is to be determined.

Properties of F_4

Stability. Preparations of F_4 retain full activity when stored frozen at −55° for several months. As a result of repeated freezing and thawing, particularly following dialysis, preparations of F_4 become fibrous and insoluble, but they are fully active. Sonication of F_4 in 0.25 M sucrose for 2 minutes yields suspensions that do not sediment upon storage.

F_4 Dependency for Reactions of Oxidative Phosphorylation. The F_4 requirement for reactions of oxidative phosphorylation catalyzed by P-particles or A-particles is shown in the table. All F_4 assays contain a saturating amount of F_1.

F_4 Requirement for Reactions of Oxidative Phosphorylation

Reaction	$-F_4$	$+F_4$
Oxidative phosphorylation[a] (P:O ratio)		
DPNH oxidation	0.03	0.65
Succinate oxidation	0.05	0.39
$^{32}P_i$–ATP exchange[a] (mμmoles/0.24 mg/min)	0	24.0
ATP-dependent reduction of DPN[a] (mμmoles/0.25 mg/min)	0.7	5.5
ATP-dependent reduction of TPN[b] (mμmoles/0.35 mg/min)	0.37	4.15

[a] Activity measured with P-particles [T. E. Conover, R. L. Prairie, and E. Racker, *J. Biol. Chem.* **238,** 2831 (1963); R. L. Prairie, T. E. Conover, and E. Racker, *Biochem. Biophys. Res. Commun.* **10,** 422 (1963); H. Zalkin and E. Racker, *J. Biol. Chem.* **240,** 4017 (1965)].

[b] Activity measured with A-particles [E. Racker and T. E. Conover, *Federation Proc.* **22,** 1088 (1963)].

Interaction with Cytochrome b, Phospholipids, and Uncouplers of Oxidative Phosphorylation.[13] At pH 7.4, approximately 2.6 millimicro-

[13] H. Zalkin and E. Racker, *J. Biol. Chem.* **240,** 4017 (1965).

moles of cytochrome *b* and 0.48 millimicromole of phospholipid (asolectin) are bound per milligram of F_4 as determined by direct analysis. Treatment of 2 mg of F_4 with desaspidin (0.002 mM), pentachlorophenol (0.5 mM), and trifluoromethoxycarbonyl cyanide phenylhydrazone (0.05 mM) results in complete loss of coupling factor activity. Dinitrophenol (1 mM) interacts with F_4 but is removed by washing with 0.25 M sucrose–0.01 M Tris-SO_4, pH 7.4. Sodium oleate (0.12 mM) does not interact with F_4.

Relationship between F_4 and Mitochondrial Structural Protein.[13] F_4 and mitochondrial structural protein[14] have similar specific activities for restoration of oxidative phosphorylation and the $^{32}P_i$–ATP exchange reaction catalyzed by P-particles. Both bind phospholipids, cytochrome *b*, and desaspidin. Both are insoluble at neutral pH and have similar heat inactivation curves. It has been suggested that F_4 may function to orient protein factors or phospholipids that are concerned with the phosphorylation process. Recent experiments,[15] however, revealed that preparations of F_4 contain coupling factors 2 and 3. As suggested previously,[1] the stimulation of phosphorylation may be due to bound protein factors rather than to the structural protein present in F_4 preparations.

[14] S. H. Richardson, H. O. Hultin, and S. Fleischer, *Arch. Biochem. Biophys.* **105**, 254 (1964).
[15] J. M. Fessenden and E. Racker, unpublished results, 1966.

[86] Extraction and Purification of an ATP–ADP Exchange Enzyme from Beef Liver Mitochondria

By Charles L. Wadkins[1] and Robert P. Glaze[1]

The ATP–ADP exchange reaction catalyzed by intact mitochondria obtained from beef and rat liver, rat brain, kidney, and heart tissues is inhibited by uncoupling agents of oxidative phosphorylation, such as 2,4-dinitrophenol, dicoumarol, and arsenate, and by coupling inhibitors, such as azide and oligomycin.[1a,2] These and other considerations led to a hypothesis that the ATP–ADP exchange reaction is a manifestation of a coupling step in which ATP is formed from ADP and an un-

[1] This manuscript was prepared while the authors were at the Department of Physiological Chemistry, The Johns Hopkins University School of Medicine, Baltimore, Maryland.
[1a] C. L. Wadkins, *J. Biol. Chem.* **236**, 221 (1961).
[2] C. L. Wadkins and A. L. Lehninger, *J. Biol. Chem.* **238**, 2555 (1963).

known phosphorylated intermediate generated during coupled respiration.[2]

Efforts to affect the solubilization and isolation of the enzyme responsible for that particular ATP-ADP exchange reaction have been handicapped because disruption of the mitochondrial structure results in the loss of oligomycin and dinitrophenol sensitivity of the ATP–ADP exchange reaction[2] and because the disruption also results in solubilization of adenylate kinase and other enzymes which also catalyze the exchange reaction.[3] Recent studies of the kinetic properties of the ATP–ADP exchange reaction catalyzed by soluble protein fractions obtained from beef liver mitochondria have demonstrated that the oligomycin sensitivity as well as certain kinetic effects by 2,4-dinitrophenol are observed when the solubilized exchange enzyme is assayed in the presence of low concentrations of reduced cytochrome *c*.[4] These kinetic effects have been utilized as an independent guide for the purification of an ATP–ADP exchange enzyme that is possibly related to the bound form present in intact mitochondria.

Assay Method

Principle. The ATP–ADP exchange reaction is assayed by incubation of the protein fraction with either ^{32}P-or ^{14}C-labeled ADP and unlabeled ATP at pH 6.8. The reaction results in the incorporation of radioisotope into ATP. The latter is measured by separation of the nucleotides by paper electrophoresis, quantitative elution, and determination of the specific activity of each nucleotide. The initial reaction rate is then calculated by use of the equation

$$X = -\frac{A \times B}{A + B} 2.303 \log (1 - F)$$

where X is the amount of ATP exchanged during the reaction period, A and B represent the amounts of ATP and ADP, respectively, in the total reaction system, and F is the ratio of the experimentally determined radioactivity of the ATP sample to the radioactivity of ATP expected at isotopic equilibrium.

Procedure

The ATP–ADP exchange reaction is carried out in a reaction system consisting of 9.0 mM ATP, 5.0 mM ADP containing approximately 20,000 cpm of ^{14}C-ADP,[5] 4.0 mM $MgCl_2$, 4.0 mM imidazole buffer, pH

[3] C. L. Wadkins and A. L. Lehninger, *Federation Proc.* **22,** 1092 (1963).

[4] R. P. Glaze and C. L. Wadkins, *Biochem. Biophys. Res. Commun.* **15,** 194 (1964).

[5] Obtained from Schwartz BioResearch Inc.

6.8, and enzyme in a final volume of 0.65 ml. The reaction is initiated by the addition of a standard reagent composed of the indicated quantities of ATP, ADP, and ^{14}C-ADP in a volume of 0.25 ml. The reaction system is incubated at 30° for 10 minutes and terminated by the addition of 0.25 ml of 2.5 *N* perchloric acid followed by 0.2 ml of 2.5 *N* KOH, and centrifuged to remove insoluble potassium perchlorate and protein.

Fifty microliters of the supernatant solution is then transferred to a sheet of Whatman 3 MM paper on a line approximately 10 cm from the end that will dip into the cathode vessel of the electrophoresis apparatus. Three or four supernatant samples can be run simultaneously on a single sheet. When the sample areas are dry, the paper sheet is dipped into the electrophoresis buffer and the excess buffer is removed by rapid blotting between two sheets of absorbent paper. The electrophoresis buffer used here was originally described by Sato *et al.*[6] and consists of 0.05 *M* citrate, pH 4.5, and 0.003 *M* zinc chloride. ATP, ADP, and AMP are easily separated in this system by application of a potential gradient of 15 volts/cm for 3 hours. Best resolution is obtained when the paper is compressed between glass sheets which are cooled by circulating tap water. After the paper has been air-dried at room temperature the nucleotides are located with an ultraviolet lamp and marked. For a typical separation with this system the leading edges of the spots corresponding to ATP, ADP, and AMP were 30 cm, 23 cm, and 5 cm, respectively, from the origin. The area corresponding to each nucleotide is then cut out and sectioned in 1 cm squares, immersed in 4 or 5 ml of water for 20 minutes with occasional agitation of the container. One milliliter is removed, placed in a planchet, evaporated to dryness and counted by means of a low background, gas-flow, thin-window counter. One milliliter is also transferred to a quartz cuvette, and the nucleotide concentration is estimated spectrophotometrically at 259 mμ using an extinction coefficient of 15.4×10^3 cm^2 per mole. The blank value for the absorbancy of water-soluble substances present in the paper is rarely more than 0.05 for the usual 1.0 ml aliquot, but the value should be determined and subtracted from that obtained for each nucleotide. Electrophoresis and elution of standardized reagents of the several nucleotides have shown that from 97–100% recovery can be expected by this procedure.

The effect of cytochrome *c* on the initial reaction rate of the purified ATP–ADP exchange enzyme has been described previously.[4] To summarize, the addition of 5–8 μ*M* cytochrome *c* (Sigma type III) and 5 m*M* ascorbic acid to the reaction system described above results in a two- to threefold stimulation of the rate of the exchange reaction. Addi-

[6] T. R. Sato, J. F. Thompson, and W. F. Danforth, *Anal. Biochem.* **5**, 542 (1963).

tion of 1 μg of oligomycin per milliliter will completely inhibit the stimulation caused by cytochrome *c*.

Preparation and Extraction of Beef Liver Mitochondria. Four thousand grams of fresh calf liver is chopped into small pieces and homogenized in 0.25 *M* sucrose (400 g tissue in 2 liters of sucrose) in a 1-gallon capacity Waring blendor controlled by a Variac rheostat. The blendor is operated at 60% of full voltage setting for 90 seconds at 4°. The resulting homogenate is strained through two thicknesses of cheesecloth and centrifuged at 700 *g* for 12 minutes in the No. 284 head of an International PR-2 centrifuge at 5°. The supernatant fraction is then centrifuged with the 3-RA rotor of the Lourdes AB centrifuge at 9500 *g* for 20 minutes. The supernatant layer together with the loosely packed fluffy layer of sediment are discarded, and the remaining mitochondrial pellet is suspended in 0.25 *M* sucrose at a concentration of 30 mg of protein per milliliter.

The mitochondrial suspension of about 4 liters is stirred slowly at 4° for 18 hours. The suspension is then centrifuged at 9500 *g* with the Lourdes centrifuge, using the 3 RA rotor, for 1 hour. The residue is discarded and the slightly turbid, reddish supernatant fraction is dialyzed against three changes of 0.001 *M* potassium phosphate, pH 7.5.

Ammonium Sulfate Fractionation. Ammonium sulfate is added slowly with stirring at 4° to the dialyzed extract. The final concentration should be 25 g per 100 ml of the extract. The pH of the solution is adjusted to 7.5, and the solution is stirred slowly for 30 minutes at 0°. The extract is then centrifuged at 9500 *g* for 20 minutes in the Lourdes 3 RA rotor, and the copious precipitate is discarded. To the supernatant fraction is then added at 0°, 40 g of ammonium sulfate per 100 ml of the supernatant fraction. The pH is again adjusted to 7.5 and the extract is stirred for 30 minutes at 0° and centrifuged as described above. The dark red pellet obtained is suspended in approximately 200 ml of 0.001 *M* potassium phosphate buffer, pH 7.5, and dialyzed for 24 hours against three changes of the buffer. Nearly all the protein dissolves during dialysis and any insoluble material is removed by centrifugation at 104,000 *g* in the Spinco Model E centrifuge and discarded.

Chromatography on Diethylaminoethyl-Cellulose. The supernatant fraction containing approximately 3500 mg of protein is then applied to a DEAE-cellulose column measuring 4.5 cm × 60 cm which was previously equilibrated with 0.001 *M* potassium phosphate, pH 7.5. The column is then eluted with the equilibrating buffer until the first protein peak is eluted. The tubes containing this protein are combined and the protein precipitated at 0° by addition of 65 g of ammonium sulfate per 100 ml of the combined eluate. The precipitate is dissolved in approximately

100 ml of 0.001 *M* potassium phosphate pH 7.5 and dialyzed for 24 hours against three changes of the buffer.

It is important to point out here that other protein fractions eluted from DEAE-cellulose by higher ionic strength buffers also catalyze an ATP–ADP exchange reaction, but these fractions do not respond to cytochrome *c* and oligomycin. The latter effects are associated only with that exchange activity associated with the first protein fraction eluted.

Carboxymethyl-Cellulose Chromatography. The dialyzed fraction from the above step is then applied to a 3.5 × 60 cm column of CM-cellulose which was preequilibrated with 0.001 *M* potassium phosphate, pH 6.5. The protein was eluted with a logarithmic gradient generated with 500 ml of 0.001 *M* potassium phosphate buffer pH 6.5 in a mixing chamber directly attached to the column and 500 ml of 0.05 *M* potassium phosphate, pH 6.5, in a reservoir. In a typical run, 10-ml fractions were collected, and the peak of ATP–ADP exchange activity was found in tubes 74–100. These fractions are then combined, the protein precipitated by addition of 65 g of ammonium sulfate per 100 ml of the combined fraction and centrifuged. The precipitate is then dissolved in approximately 50 ml of 0.001 *M* potassium phosphate buffer, pH 7.5 and dialyzed for 24 hours against the same buffer.

Chromatography on Hydroxylapatite. The column with a useful flow rate is prepared by mixing equal wet volumes of Biorad hydroxylapatite T and phosphate buffer-washed Johns Mansville Hyflo Super Cel Celite. A column of 3.5 × 5.0 cm is prepared and equilibrated with 0.05 *M* potassium phosphate buffer, pH 7.5. Approximately 50 ml of the active fraction from the preceding step is applied, and the column is eluted with a logarithmic gradient generated with 200 ml 0.05 *M* potassium phosphate buffer in the mixing vessel and 200 ml 0.5 *M* in a reservoir. Most of the protein is eluted between 0.05 and 0.2 *M* phosphate whereas the ATP–ADP exchange activity is eluted later at 0.25–0.3 *M* phosphate. The peak tubes are combined and concentrated to approximately 20 ml by use of a collodion membrane suspended in 0.001 *M* potassium phospate buffer in a closed container under vacuum.

Purification with Sephadex G-200. The phosphate concentration of the concentrated fraction from the preceding step is adjusted to 0.2 *M* and applied to a 2.5 × 30 cm column of Sephadex G-200 which had been equilibrated with 0.2 *M* potassium phosphate, pH 7.5, and eluted with the same buffer. The peak fractions which comprised approximately 80% of the exchange activity and 20% of the protein originally applied to the column are combined; they are concentrated to approximately 5 ml by use of the collodion membrane-vacuum apparatus described above.

The purification procedure and results for a typical preparation are

Purification Procedure

Purification step	Total protein	ATP–ADP exchange: Specific activity (μmoles min^{-1} mg^{-1} protein)	ATP–ADP exchange: Total activity (μmoles min^{-1})	Adenylate kinase: Specific activity (μmoles ADP formed min^{-1} mg^{-1} protein)	Adenylate kinase: Total activity (μmoles ADP formed min^{-1})
1. Sucrose extract	33,000	1	32,000	1.1	35,000
2. Ammonium sulfate	3,500	3	9,800	0.4	1,500
3. DEAE-cellulose	1,000	7	6,800	1.0	1,000
4. CM-cellulose	450	17	7,600	0.4	180
5. Hydroxylapatite	35	75	2,600	0.1	3
6. Sephadex G-200	7	220	1,500	<0.02	<0.1

summarized in the table. This method usually results in a yield of from 6–10 mg of enzyme protein from 4000 g of calf liver and a specific activity in excess of 200 micromoles of ATP exchanged per minute per milligram of protein. There is routinely little or no detectable adenylate kinase activity in the final fraction.

Properties of Purified Enzyme

The ATP–ADP exchange enzyme prepared by this method appears to be homogeneous with respect to sedimentation and diffusion studies and to electrophoresis on cellulose acetate. Sedimentation and diffusion studies suggest that the enzyme can exist in several states of aggregation. Freshly prepared samples usually possess molecular weight values of 30,000 or 70,000 suggesting the existence of a monomer-dimer relationship, but aged preparations possess molecular weights of 100,000 and in some instances of 170,000 suggesting even higher orders of association. The conditions specific for stabilizing one or the other of these states have not been determined.

The purified exchange enzyme also catalyzes nucleoside diphosphokinase activity but the pH optima for the two reactions are different. For the ATP–ADP exchange reaction the pH optima is at 6.8 and the exchange to nucleoside diphosphokinase activity there is approximately 12:1. The pH optimum for the nucleoside diphosphokinase reaction is at 8.3, and the ratio of the two activities at that pH is approximately 1.5:1.

Both reactions measured at their optimum pH values are activated to the same extent by Mg^{++} and Mn^{++}, both are inhibited by phenylmercuric acetate 50% at 10^{-5} M, and both are specifically inhibited by ADP but by no other nucleoside diphosphates. On the basis of these results it is tentatively concluded that both reactions are catalyzed by the same enzyme.

The extraction and purification procedure described here has been found to give reproducible results providing the mitochondria are not damaged by exposure to hypotonic sucrose solutions or to salt solution of ionic strength in excess of 0.01. Attempts to utilize hypotonic media, strong salt solutions, or freezing and thawing resulted in the extraction of large amounts of adenylate kinase activity which introduces complications for assay of the specific ATP–ADP exchange enzyme activity.

[87] Energy-Coupling in Nonphosphorylating Submitochondrial Particles

By Chuan-pu Lee[1] and Lars Ernster

Demonstration of Energy-Coupling

The energy-linked pyridine nucleotide transhydrogenase reaction[1a] (cf. this volume [113]) can be used as a tool for the demonstration[2] of energy-coupling in so-called "nonphosphorylating" submitochondrial electron transport particles. The approach is based on the concept that the energy-linked transhydrogenase reaction:

$$\text{NADH} + \text{NADP}^+ + \text{I} \sim \text{X} \rightarrow \text{NAD}^+ + \text{NADPH} + \text{I} + \text{X} \quad (1)$$

involves the utilization of a nonphosphorylated high-energy compound, denoted I ∼ X, which is an intermediate of the respiratory chain-linked oxidative phosphorylation system:

$$\text{I} + \text{X} \xrightarrow{\text{respiration}} \text{I} \sim \text{X} \quad (2)$$

$$\text{I} \sim \text{X} + \text{ADP} + \text{P}_i \rightleftharpoons \text{I} + \text{X} + \text{ATP} \quad (3)$$

(where I and X are hypothetic energy-transfer carriers). In phosphorylating submitochondrial particles, the energy-linked transhydrogenase reaction can be driven by either the respiratory chain, Reaction (2), or ATP, reversal of Reaction (3), at approximately equal capacities. In contrast, "nonphosphorylating" particles can derive energy for driving the energy-linked transhydrogenase reaction much less efficiently from

[1] See footnote 1, page 33.

[1a] L. Danielson and L. Ernster, *Biochem. Biophys. Res. Commun.* **10,** 91 (1963); *in* "Energy-Linked Functions of Mitochondria" (B. Chance, ed.), p. 157. Academic Press, New York, 1963; *Biochem. Z.* **338,** 188 (1963).

[2] C. P. Lee, G. F. Azzone, and L. Ernster, *Nature* **201,** 152 (1964).

ATP than from the respiratory chain. This is consistent with the conclusion that these particles still possess the capacity for generating $I \sim X$ by respiratory energy-coupling (Reaction 2), although their phosphorylating capacity, i.e., their ability to utilize $I \sim X$ for converting ADP and P_i into ATP (Reaction 3), is weakened or lost. It is also found[2] that in such particles oligomycin markedly stimulates, and may even be obligatory for, the respiratory chain-driven transhydrogenase reaction. Illustrative data are summarized in the table.

The data shown in this table compare two types of phosphorylating particles with three types of "nonphosphorylating" particles. "Heavy" beef-heart mitochondria, prepared according to Löw and Vallin,[3] and stored in 250 m*M* sucrose suspension at —10° for at least 3 days, were used as the starting material for the various preparations. The general procedure for the preparation of submitochondrial particles was as follows: The frozen mitochondrial suspension was thawed, and diluted with 250 m*M* sucrose to contain about 20–30 mg of protein per milliliter. Depending on the type of preparation to be investigated, various reagents were added as will be specified below. The suspension was then saturated with N_2 and subjected to sonic oscillation for 2 minutes at the maximal output in a 20-kc Raytheon sonicator cooled with running tap water (4–8°). The suspension was diluted with an equal volume of 250 m*M* sucrose and centrifuged at 12,000 *g* for 10 minutes. The supernatant fraction was decanted and centrifuged at 105,000 *g* for 40 minutes. The sediment was washed by homogenization with 10 volumes of 250 m*M* sucrose and centrifuged at 105,000 *g* for 40 minutes. The particles were finally suspended in 250 m*M* sucrose to give a protein concentration of about 20 mg/ml. A recovery of 20–30% of the mitochondrial protein was usually obtained.

In the case of the phosphorylating particles studied, the sonicating medium consisted of either 5 m*M* $MnCl_2$, 10 m*M* $MgSO_4$ and 1 m*M* sodium succinate, pH 7.5 (preparation described by Smith and Hansen[4]), or 15 m*M* $MgSO_4$ and 1 m*M* ATP, pH 7.5 (ETPH preparation of Linnane and Ziegler[5] as modified by Löw and Vallin[3]). "Nonphosphorylating" particles were prepared in a sonicating medium consisting of either 2 m*M* EDTA, pH 7.5 ("modified" ETPH of Linnane and Ziegler[5]), or 20 m*M* NH_4OH (A-particles of Conover *et al.*[6]). A third type of "nonphosphorylating" particle preparation was made by sonication of the

[3] H. Löw and I. Vallin, *Biochim. Biophys. Acta* **69,** 361 (1963).

[4] A. L. Smith and M. Hansen, *Biochem. Biophys. Res. Commun.* **8,** 33 (1962); *Biochim. Biophys. Acta* **81,** 214 (1964).

[5] A. W. Linnane and D. M. Ziegler, *Biochim. Biophys. Acta* **29,** 630 (1958).

[6] T. E. Conover, R. L. Prairie, and E. Racker, *J. Biol. Chem.* **238,** 2831 (1963).

COMPARISON OF ATP- AND SUCCINOXIDASE-SUPPORTED PYRIDINE NUCLEOTIDE TRANSHYDROGENASE ACTIVITIES OF VARIOUS SUBMITOCHONDRIAL PREPARATIONS FROM BEEF HEART[a,b]

			$NADH + NADP^+ \rightarrow NAD^+ + NADPH$ mμmoles/min/mg protein)			Ratio	
Type of preparation	Sonicating medium	P:O (succinate)	ATP	Succinate (−oligo)	Succinate (+oligo)	ATP: Succinate (+oligo)	Succinate (+oligo): Succinate (−oligo)
Phosphorylating	$Mn^{++} + Mg^{++}$ + succinate	1.0–1.5	46	46	46	1.0	1.0
	$ATP + Mg^{++}$	0.6–0.8	102	116	131	0.78	1.1
Nonphosphorylating	EDTA	<0.2	35	70	138	0.25	2.0
	NH_4OH	<0.2	8	13	55	0.14	4.2
	EDTA + P_i, treated with 2 *M* urea	<0.05	0	—	55	0	—

[a] From C. P. Lee, G. F. Azzone, and L. Ernster, *Nature* **201,** 152 (1964).

[b] Oxygen uptake was measured manometrically in the Warburg apparatus by the conventional techniques. The reaction mixture contained 125 m*M* sucrose, 25 m*M* Tris-acetate, pH 7.4, 4 m*M* $MgCl_2$, 10 m*M* $^{32}P_i$, pH 7.4, 10 m*M* succinate, 30 m*M* glucose, 1 m*M* ATP, 150 Kunitz-MacDonald units of yeast hexokinase, and 0.5–1 mg of submitochondrial particle protein, in a final volume of 2 ml. After 20 minutes of incubation at 30°, the samples were fixed with 0.3 ml 5 *M* H_2SO_4, and esterification of P_i was determined by the isotope distribution method as described by O. Lindberg and L. Ernster [*Methods Biochem. Anal.* **3,** 1 (1955)].

Pyridine nucleotide transhydrogenase activity was measured spectrophotometrically at 340 mμ with a Beckman DK-2 recording spectrophotometer. The reaction mixture contained 200 m*M* sucrose, 50 m*M* Tris-acetate pH 7.5, 10 m*M* $MgSO_4$, 0.33 m*M* NADH, 0.13 m*M* $NADP^+$, 0.66 m*M* oxidized glutathione, an amount of glutathione reductase capable of oxidizing 0.5 micromoles of NADPH per minute, and 0.5–1 mg of particle protein. When indicated, 5 μg of oligomycin ("oligo") was also added. Energy-linked transhydrogenase reaction was initiated by the addition of either 2 m*M* ATP or 3.3 m*M* succinate. The values given in the table have been corrected for the slight non-energy-linked transhydrogenase activities observed before the addition of ATP or succinate.

mitochondria in a medium containing 2 mM EDTA and 2 mM sodium-potassium phosphate, pH 7.5, and subsequent treatment of the particles so obtained with urea. One milliliter of particle suspension in 250 mM sucrose, containing 20 mg of protein per milliliter, was mixed with 1 ml of 4 M urea and incubated for 6 minutes at 20°. Five milliliters of cold distilled water was then added, and the incubation was continued for another 6 minutes at 0°. The reaction mixture was diluted with 5 ml 250 mM sucrose and centrifuged at 105,000 g for 40 minutes. The sediment was washed by homogenization with 10 volumes of 250 mM sucrose and recentrifuged at 105,000 g for 40 minutes. The particles were finally suspended in 250 mM sucrose.

The table indicates the phosphorylating efficiency of the various types of particles as measured with succinate as the substrate, as well as their energy-linked transhydrogenase activities with either succinate or ATP as the sucrose of energy, the former both in the absence and in the presence of oligomycin. The transhydrogenase reaction was assayed spectrophotometrically in the presence of oxidized glutathione and glutathione reductase as described in this volume [113].

Results similar to those shown in the table with the urea-treated EDTA particles have been reported by Haas[7] using a Keilin-Hartree heart muscle preparation.

Stimulation of Oxidative Phosphorylation and Its Reversal in "Nonphosphorylating" Particles by Oligomycin

The demonstration[8] of this effect of oligomycin has emerged from the observations that oligomycin stimulates the respiratory chain-driven pyridine nucleotides transhydrogenase reaction in "nonphosphorylating" particles[2] (cf. the table) and that the amount of oligomycin required for maximal stimulation is considerably smaller than that required for inhibition of the ATPase activity of the same particles.[8] Typical results obtained with EDTA particles are shown in Fig. 1. Forward oxidative phosphorylation is assayed with NADH as substrate, and the reverse process is assayed as ATP-supported reduction of NAD^+ by succinate. Maximal stimulation of both reactions occurs with an amount of oligomycin ranging between 0.2 and 0.3 μg per milligram of protein. Larger amounts of oligomycin are inhibitory. Mg^{++}, which is obligatory for both the forward and reverse reactions, in high concentrations diminishes the effect of oligomycin on the forward reaction but not on the reverse.

[7] D. W. Haas, *Biochim. Biophys. Acta* **89**, 543 (1964).

[8] C. P. Lee and L. Ernster, *Biochem. Biophys. Res. Commun.* **18**, 523 (1965); *Symp. Regulation Metabolic Processes in Mitochondria, Bari, 1965* Vol. 7, p. 218. B.B.A. Library, Elsevier, Amsterdam, 1966.

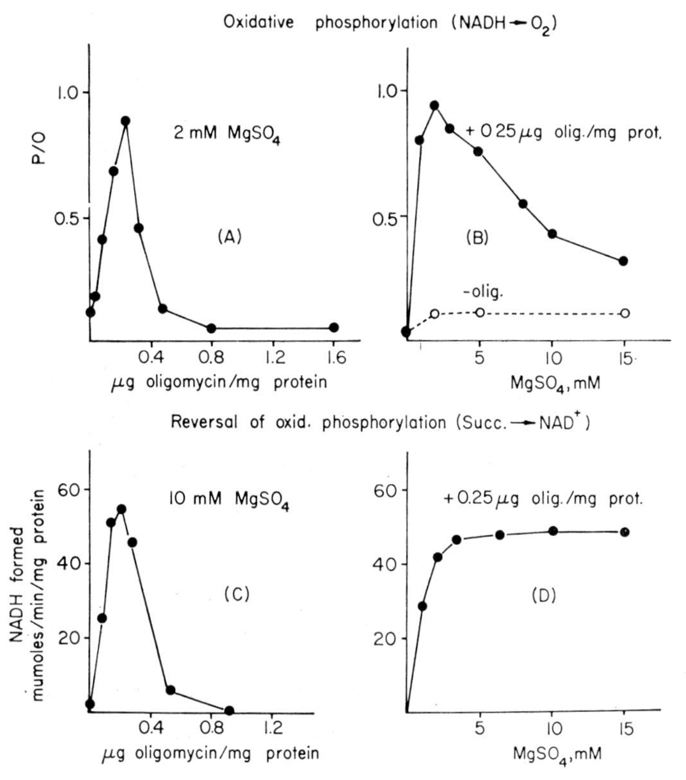

FIG. 1. Effects of oligomycin and Mg^{++} on oxidative phosphorylation and ATP-supported succinate-linked NAD^+ reduction. [From C. P. Lee and L. Ernster, *Symp. Regulation Metabolic Processes in Mitochondria, Bari, 1965* Vol. 7, p. 218. B.B.A. Library, Elsevier, Amsterdam, 1966.]

Oxidative phosphorylation was assayed in a reaction mixture consisting of 180 m*M* sucrose, 50 m*M* Tris-acetate buffer, pH 7.5, 2 m*M* ADP, 15 m*M* glucose, 75 Kunitz-MacDonald units of yeast hexokinase, 3 m*M* $^{32}P_i$ (1.2×10^6 cpm/micromole), particles (prepared in the presence of 2 m*M* EDTA, pH 8.6) containing 0.6 mg protein, and 1 m*M* NADH as the substrate. Other additions as indicated: panel A, 2 m*M* $MgSO_4$ and varying amounts of oligomycin; panel B, 0.25 μg oligomycin per milligram of protein and varying concentrations of $MgSO_4$. Final volume, 3.1 ml; temperature, 30°.

O_2 consumption was measured with a Clark oxygen electrode, and the esterification of P_i was determined by the isotope distribution method [O. Lindberg and L. Ernster, *Methods Biochem. Anal.* **3**, 1 (1955)]. The reaction mixture for the ATP-supported reduction of NAD^+ by succinate consisted of 180 m*M* sucrose, 50 m*M* Tris-acetate buffer, pH 7.5, 1.6 m*M* KCN, 0.2 m*M* NAD^+, 5 m*M* succinate, particles (prepared in the presence of 2 m*M* EDTA, pH 8.7) containing 0.6 mg protein; 3 m*M* ATP was added to start the reaction. Other additions as indicated: panel C, 10 m*M* $MgSO_4$ and varying amounts of oligomycin; panel D, 0.25 μg oligomycin per milligram of protein and varying concentrations of $MgSO_4$. Final volume, 3 ml; temperature 30°.

For obtaining the effects of oligomycin as described above, it is essential to maintain the pH of the EDTA-containing sonicating medium within the range 8.5–9.0. At pH values below 8.5, the particles may retain a substantial capacity for both oxidative phosphorylation and ATP-supported succinate-linked NAD^+ reduction even when assayed in the absence of oligomycin; at pH values above 9.0, the stimulating effect of oligomycin becomes less pronounced, especially as the stimulation of the forward oxidative phosphorylation is concerned. Particles prepared in the presence of ammonia,[8,9] as well as the Keilin-Hartree heart muscle preparation,[10] show a response to oligomycin similar to that of EDTA particles.

Results analogous to those shown in Fig. 1 are obtained with succinate or with ascorbate + TMPD (or PMS) rather than NADH as the substrate for oxidative phosphorylation, and with ascorbate + TMPD rather than succinate as the hydrogen donor for the ATP-supported NAD^+ reduction.[8] With NADH as substrate, in the absence of added Mg^{++}, oligomycin inhibits the respiration of EDTA particles by up to 80%, and the inhibition is relieved by phosphorylation-uncoupling concentrations of 2,4-dinitrophenol or dicoumarol. Oligomycin also stimulates the P_i–ATP exchange activity of EDTA particles by 4 to 5-fold.[11] The optimal oligomycin concentration for this effect is the same as for the maximal stimulation of oxdiative phosphorylation and its reversal. The effect of Mg^{++} is similar to that on the reversal of oxidative phosphorylation.

The optimal range of oligomycin concentration for all the above effects may vary with the batch of oligomycin used. A practical detail of great importance in performing titrations with oligomycin is to wash the reaction chambers very thoroughly with both water and ethanol between the single assays in order to remove adhering traces of oligomycin. Oligomycins A, B, C, and D (rutamycin) act in an essentially similar fashion; aurovertin does not replace oligomycin in any of the stimulating effects described above, and even abolishes these effects of oligomycin.[12]

Possible mechanisms involved in the above effects of oligomycin have been discussed by Lee and Ernster.[8] Fessenden and Racker[9] have studied the relationship between the phosphorylation stimulating effects of oligomycin and of soluble "coupling factors."

[9] J. Fessenden and E. Racker, *J. Biol. Chem.* **241**, 2483 (1966).

[10] K. van Dam and H. F. ter Welle, *Symp. Regulation of Metabolic Processes in Mitochondria, Bari, 1965* Vol. 7, p. 237. B.B.A. Library, Elsevier, Amsterdam, 1966.

[11] C. P. Lee and L. Ernster, unpublished results, 1965.

[12] O. Lindberg and L. Ernster, *Methods Biochem. Anal.* **3**, 1 (1955).

Section VII

Microsomal Electron Transport

[88] Microsomal DPNH-Cytochrome *c* Reductase

By BRUCE MACKLER

DPNH + ferricytochrome *c* ⇄ DPN^+ + ferrocytochrome *c*

Assay Method

Principle. Enzymatic activity at 38° was measured spectrophotometrically by following the rate of reduction of cytochrome *c* at 550 mμ.

Reagents

DPNH, 0.1% solution
Potassium phosphate buffer, 0.2 *M*, pH 7.5
Cytochrome *c* solution, 1%
Sodium azide solution, 0.1 *M*
Sodium cholate solution, 10%, pH 7.5
Enzyme diluted in 0.02 *M* potassium phosphate buffer, pH 7.5

Procedure. One-tenth milliliter of 0.1% DPNH, 0.2 ml of 0.2 *M* potassium phosphate buffer of pH 7.5, 0.1 ml of 1% cytochrome *c* solution, 0.1 ml of 0.1 *M* sodium azide solution, and sufficient water to make 1.0 ml are placed in a small test tube and brought to 38°. The solution is then added to a 1-ml optical cuvette having a 1-cm light path, and enzyme (2–4 μg) is added to start the reaction. Before assay it is necessary to preincubate the enzyme with sodium cholate (1 mg cholate per milligram of enzyme protein) to obtain maximal activity. The rate of reduction of cytochrome *c* is measured at 550 mμ in a spectrophotometer maintained at 38°. Optical density readings were obtained at 15-second intervals against a blank containing only cytochrome *c* and buffer in a total of 1 ml. Specific activity is expressed as micromoles of DPNH oxidized per minute per milligram of enzyme protein. Protein concentration is determined by the biuret method[1] with deoxycholate added to a final concentration of 1%.

Purification Procedure

Preparation of Microsomes. All procedures are carried out at 0–5°. Beef liver obtained fresh from the slaughterhouse is diced and ground in an electric meat grinder. To each 500 g of ground liver, 1500 ml of a solution containing dibasic potassium phosphate (0.01 *M*) and sucrose (0.25 *M*) are added along with 2 ml of 6 *N* potassium hydroxide. The

[1] B. Mackler and D. E. Green, *Biochim. Biophys. Acta* **21**, 1, 6 (1956).

mixture is blended at moderate speed for 20 seconds, then centrifuged for 15 minutes at 1300 *g*. The supernatant fluid (volume, 7–8 liters) is decanted through cheesecloth, and 2–3 liters of 0.25 *M* sucrose solution is added to the filtrate. The filtrate is centrifuged in the refrigerated Sharples centrifuge at full speed (flow rate of 300 ml/minute), and the effluent is collected and diluted with an equal volume of 0.9% potassium chloride solution. The mixture is again centrifuged at full speed in the Sharples centrifuge, but at a reduced flow rate of 100 ml/minute. The precipitate (microsomal fraction) is collected and suspended in an equal volume of a 0.25 *M* solution of sucrose.

Stage I Enzyme. Suspensions of microsomes are diluted with 0.2 *M* Tris buffer of pH 8.0 to a final protein concentration of 20 mg/ml. A neutral solution of deoxycholate (10%) is added in sufficient amount so that the final concentration of deoxycholate per milligram of protein is between 0.2 and 0.3 mg (the exact amount of deoxycholate for best yield of enzyme is determined for each preparation of microsomes). The mixture was centrifuged in the No. 30 rotor of the Spinco preparative ultracentrifuge for 90 minutes at 30,000 rpm, and the supernatant fluid is dialyzed for 24 hours against 30 volumes of 0.02 *M* phosphate buffer of pH 7.5. The dialyzed fraction is centrifuged in the No. 30 rotor of the Spinco ultracentrifuge for 90 minutes at 30,000 rpm, and the loosely packed residue (stage I) is suspended in a solution containing sucrose (0.25 *M*) and phosphate buffer (0.02 *M*, pH 7.5) to a protein concentration of about 20 mg/ml. This residue fraction (stage I) contains 70–80% of the total DPNH cytochrome *c* reductase activity of the original microsome fraction and is approximately 8 times as active per milligram of protein (cf. the table).

Stage II Enzyme. Suspensions of the particulate enzyme (stage I) are mixed with 0.25 volume of saturated ammonium sulfate(pH 7.0) and an equal volume of *n*-butanol. The mixture is quickly and thoroughly homogenized and then centrifuged in the No. 30 rotor of the Spinco ultracentrifuge. As soon as a speed of 30,000 rpm is attained with maximum acceleration, the centrifuge is turned off, and the water and butanol layers are poured off and discarded. The interface layer of solid material (between the butanol and water layers) is separated from the packed denatured material at the bottom of the tubes and suspended in approximately 200 ml of 0.02 *M* phosphate buffer of pH 7.5. The suspension is homogenized and centrifuged in the No. 30 rotor of the Spinco ultracentrifuge for 20 minutes at 30,000 rpm. The supernatant fluid (s) is set aside while the residue is again homogenized in 0.02 *M* phosphate buffer of pH 7.5 and then centrifuged in the No. 30 rotor at 4000 rpm for 15 minutes. The supernatant fluid is combined with (s) and the mixture is centrifuged in the No. 30 rotor for 65 minutes at 30,000. The supernatant is discarded, and the residue (stage II) is suspended in 0.02 *M* phosphate buffer of pH 7.5. The purification of the enzyme is shown in the table.

PURIFICATION OF MICROSOMAL DPNH-CYTOCHROME *c* REDUCTASE

Preparation	Recovery of activity (%)	Specific activity
Microsomes	100	0.35
Stage I enzyme	70	4.7
Stage II enzyme	40	23.0

Properties[2, 3]

The preparations slowly lose activity when stored at 0° or frozen at —20°, approximately 50% of the activity being lost over a period of 1–2 weeks. Stage II particle suspensions did not show TPNH-cytochrome *c* reductase, aldehyde oxidase, xanthine oxidase, or succinic dehydrogenase activity. The enzyme actively reduces ferricyanide as well as cytochrome *c* with DPNH as substrate. Preparations of the highest purity contain approximately 1 mole of flavin, 2 moles of cytochrome b_5 and 1 atom of nonheme iron per mole of enzyme. The enzyme has an approximate molecular weight of 1.5×10^6 including lipid, which constitutes 70% of the dry weight of the enzyme.

[2] N. Penn and B. Mackler, *Biochim. Biophys. Acta* **27**, 539 (1958).
[3] B. Mackler and W. B. Dandliker, *Biochim. Biophys. Acta* **30**, 639 (1958).

[89] Cytochrome b_5

By PHILIPP STRITTMATTER

Assay Method

This procedure is described under NADH-cytochrome b_5 reductase (see this volume [91]).

Purification Procedure

The procedure outlined below is a modification of that originally described for the enzyme obtained from rabbit liver[1] and applied to preparations from calf liver.[2] Preparations from calf liver appear to be homogeneous by sedimentation and diffusion, but on starch gel or cel-

[1] P. Strittmatter and S. F. Velick, *J. Biol. Chem.* **221**, 253 (1956).
[2] P. Strittmatter, *J. Biol. Chem.* **235**, 2492 (1960).

lulose acetate electrophoresis there are two major components of identical spectral and enzymatic properties that differ only in a dipeptide sequence at the carboxyl terminal end.[3] They can be separated by careful column chromatography,[3] but this is not necessary for normal enzymatic applications. The simpler procedure is therefore described here in which all steps are carried out at 0–5° unless otherwise specified.

Step 1. Isolation of Microsomes. Strips of calf liver, obtained fresh from the slaughterhouse, are ground in a meat grinder. Approximately 200 g of ground liver is homogenized with 2.5 volumes of cold sucrose medium (0.25 *M* sucrose, 0.001 *M* EDTA, pH 7.8) for 30 seconds. The homogenate is centrifuged for 7 minutes at 16,000 *g* and the supernatant suspension is decanted and saved. This step is repeated with a total of 3000 g of liver. The combined suspension (approximately 7 liters) is then brought to 0.4 saturation by the addition of saturated ammonium sulfate solution (720 g per liter, pH 7.2–7.5). The microsomal fraction obtained as a well packed precipitate, by centrifuging at 20,000 *g* for 20 minutes, is resuspended in 0.1 *M* Tris-acetate, 0.001 *M* EDTA buffer, pH 8.1, to yield a total of 2 liters. After dialysis against three 10-liter volumes of the same buffer, the suspension may be stored at —12° or used immediately.

Step 2. Incubation with Pancreatic Lipase. The dialyzed microsomal suspension is incubated at 37° for 4 hours with 200 mg soybean trypsin inhibitor and 500 ml of a partially purified lipase solution[4] (total volume, approximately 4 liters), then cooled in an ice bath 0–5°.

Step 3. Ammonium Sulfate Fractionation. The lipase-treated preparation is brought to 0.6 saturation by slowly adding solid ammonium sulfate and 1 *N* sodium hydroxide to maintain the pH between 7.0 and 7.5. After centrifuging this mixture at 20,000 *g* for 20 minutes, the supernatant fluid, containing some floating lipid rich precipitate, is decanted and filtered overnight through Whatman No. 1 fluted filter paper. This clear filtrate (approximately 3.3 liters) is brought to 0.9 saturation with solid ammonium sulfate and maintained at pH 7.0 by the intermittent addition of 1 *N* sodium hydroxide. This solution, which may become slightly turbid depending upon the cytochrome concentration, is stirred vigorously in an ice bath, and the pH is then lowered to 4.2 by the dropwise addition of 1 *N* HCl. The dark red precipitate is collected by cen-

[3] P. Strittmatter and J. Ozols, *J. Biol. Chem.* **241**, 4787 (1966).

[4] The pancreatic lipase was purified from the commercial "Steapsin" by using the fractional precipitation, adsorption, and elution steps of R. Willstätter and E. Waldschmidt-Leitz, ["Die Methoden der Fermentforschung" (K. Myrbäck and E. Bamann, eds.), p. 1560. Thieme, Leipzig, 1941].

trifugation at 20,000 g for 10 minutes and dissolved in approximately 50–100 ml of the pH 8.1 buffer.

Step 4. Reprecipitation at Low pH. The cytochrome b_5 solution from step 3 is again brought to 0.9 saturation by the addition of neutralized saturated ammonium sulfate solution. The small precipitate obtained by centrifuging at 20,000 g for 20 minutes is discarded. The heme protein is then precipitated and collected by lowering the pH to 4.2 with 1 N HCl as in step 3, dissolved in 50–100 ml of 0.1 M Tris-acetate, 0.001 M EDTA buffer, pH 8.1, and dialyzed three times against 500 ml of 0.02 M sodium phosphate buffer, pH 7.2.

Step 5. Column Fractionation. Approximately 20 ml of a 1% solution of crude cytochrome b_5 from step 4 is placed on a column 2 cm in diameter, containing a mixture of 8 g of Celite and 28 g of diethylaminoethyl cellulose that had been washed for 24 hours with 0.2 M sodium phosphate buffer, pH 7.2. The column containing the cytochrome b_5 is washed with 200 ml of the phosphate buffer, and the cytochrome is then eluted with 0.15 M sodium phosphate buffer, pH 7.2. The cytochrome b_5 (100–180 mg) is stored either in water, 0.1 M Tris-acetate buffer, pH 8.1, or 0.02 M sodium phosphate buffer, pH 7.2, at —12°.

Properties

Molecular Weight and Homogeneity. The calf liver cytochrome b_5 preparations appear to be homogeneous by sedimentation and in view of the fact that the protein contains 1 mole each of heme and tryptophan. Electrophoretically there are two distinguishable major components which differ in a single dipeptide sequence.[3] Since both components have identical spectra and reactivity in the cytochrome b_5 reductase system, the mixture of the two components need not be separated, but separation has been accomplished.[3] The molecular weight calculated from sedimentation velocity, diffusion, and partial specific volume measurements is 14,400 and is in good agreement with minimum molecular weight determinations, which yield a value of 12,900.

Stability. Cytochrome b_5 from both rabbit and calf liver is stable at 20–37° between pH 7.0 and 9.5 and can be stored indefinitely at —12°. In the acid pH region the heme protein becomes progressively more unstable. At 25° at pH 6.5, 4.3, and 3.5 the amounts of denaturation, as measured by the loss of cytochrome spectrum in 10 minutes, increases from 0 to 30 to 100%.

Effects of Various Reagents. Cytochrome b_5 is reduced by potassium borohydride, reduced indigo sulfonates, benzyl viologen, sodium hydrosulfite, and cysteine. The reduced heme protein is oxidized by potassium ferricyanide, ferric chloride, cytochrome *c*, various dyes of appropriate potential, mercuric chloride, and very slowly by oxygen. Carbon monoxide and potassium cyanide have no effect on the spectrum of the oxidized or reduced cytochrome. The enzymatic reduction is considered in this volume [91].

Resolution. Apocytochrome b_5 preparations have been obtained[2] which will recombine with heme to yield a heme protein indistinguishable from the original heme protein. The heme binding involves at least one specific imidazole group[2] and iron-protein interactions involving protein conformation changes.[5]

Distribution. Cytochrome b_5 has been isolated from rat, rabbit, calf, and pig liver and has been detected in preparations from kidney, pancreas, adrenal medulla, mammary gland, ovary, and intestinal mucosa.[6]

[5] J. Ozols and P. Strittmatter, *J. Biol. Chem.* **239,** 1018 (1964).
[6] P. Strittmatter, *in* "The Enzymes" (P. D. Boyer, H. Lardy, and K. Myrbäck, eds.), 2nd ed., Vol. VIII, p. 113. Academic Press, New York, 1963.

[90] Isolation of Cytochromes P-450 and P-420

By TSUNEO OMURA[1] and RYO SATO

Cytochrome P-450[1a] is sometimes called the microsomal carbon monoxide-binding pigment. It is present in the microsomal fraction of several animal tissues, and also in the mitochondrial fraction of adrenal cortex, but not in the mitochondria of other tissues. The function of P-450 as the oxygen activating enzyme for many mixed-function oxidases has recently been elucidated.[2] The carbon monoxide (CO) compound of reduced P-450 shows a unique absorption spectrum having a broad peak at 450 mμ.

P-450 is firmly associated with the microsomal membranes, and the solubilization of the pigment by treating microsomes with detergents or phospholipase is always accompanied by a drastic change in its spectral properties. The CO compound of the solubilized pigment has a strong absorption at 420 mμ and is called P-420[1] to distinguish it from the native membrane-bound form, P-450.

Starting from mammalian liver microsomes solubilized by phospholipase, P-420 may be purified severalfold. However, the solubilization

[1] See footnote 1, page 362.
[1a] T. Omura and R. Sato, *J. Biol. Chem.* **239,** 2370 (1964).
[2] T. Omura, R. Sato, D. Y. Cooper, O. Rosenthal, and R. W. Estabrook, *Federation Proc.* **24,** 1181 (1965).

and purification of this cytochrome in its P-450 form has never been achieved.

Assay Method

P-450. P-450 is identified by the formation of a yellow-colored CO compound of the reduced pigment with a broad absorption peak at 450 mμ. Since P-450 has a high affinity for CO, bubbling the dithionite-reduced sample with CO for 10–20 seconds is sufficient to give a fully developed spectrum of the CO compound of P-450. From the CO difference spectrum of the reduced sample, the concentration of P-450 can be calculated using the extinction coefficient-difference between 450 mμ (peak) and 490 mμ listed in Table I. Judging from the total heme content of microsomal preparations, P-450's of variable sources seem to have the same $\Delta\epsilon$. Presence of P-420 or hemoglobin does not interfere with the determination of P-450 by this method.

P-420. If P-450 is not present, the concentration of P-420 can be obtained by measuring the difference of the optical density between 420 mμ (peak) and 490 mμ of the CO-difference spectrum of dithionite-reduced samples (Table I). Since the reduced P-420 shows a hyperchromicity in alkaline solutions, the pH of the medium should be around 7.0–7.5. To prepare the CO compound of the sample, CO is first bubbled through the sample followed by the addition of dithionite. If dithionite is added to the sample prior to the addition of CO, the sample will give a erroneously low estimate of P-420.

TABLE I
EXTINCTION-DIFFERENCE ($\Delta\epsilon$) VALUES PER MILLIMOLE OF HEME FOR THE DIFFERENCE SPECTRA BETWEEN THE CO COMPOUND OF P-450 AND P-420 AND THE REDUCED PIGMENTS

Sample	$\Delta\epsilon_{heme}$ (CO compound minus reduced)
P-450	91 cm^{-1} mM^{-1} between 450 mμ and 490 mμ
P-450	−41 cm^{-1} mM^{-1} between 420 mμ and 490 mμ
P-420	111 cm^{-1} mM^{-1} between 420 mμ and 490 mμ

As the spectrum of the CO compound of P-420 is almost identical with that of hemoglobin, the sample for the determination of P-420 should be free from the blood pigment. However, since the microsomes prepared from animal tissues sometimes contain considerable amounts of absorbed hemoglobin, P-420 may be distinguished from hemoglobin by observing the differences in the spectra of the dithionite-reduced pigments. Reduced P-420 has a typical spectrum of *b*-type cytochromes (Table II).

TABLE II
POSITIONS OF PEAKS AND EXTINCTION COEFFICIENTS (ϵ) PER MILLIMOLE OF HEME IN THE ABSORPTION SPECTRA OF P-420 AT pH 7.5

Spectra	Wavelength (mμ)	ϵ (cm^{-1} mM^{-1})
Oxidized (Soret)	414	124
Reduced (Soret)	427	149
Reduced (β)	530	13
Reduced (α)	559	24
CO compound (Soret)	421	213

Even if P-420 and P-450 are present together in the sample, as in the case of steapsin-digested microsomes, the concentrations of these pigments may be calculated separately from the difference spectrum of the CO compounds of the reduced pigments. First, the concentration of P-450 is calculated from the optical density difference between 450 mμ and 490 mμ. Then the optical density difference (negative) between 420 mμ and 490 mμ, which should be caused by the presence of P-450, is calculated from the amount of P-450 and the $\Delta\epsilon$ value of Table I. By subtracting this value from the observed optical density difference between 420 mμ and 490 mμ, the contribution of P-420 to the spectrum will be obtained.

If the sample is pure enough to allow the observation of the absolute spectra, the concentration of P-420 may be calculated by using the extinction values of Table II. The ratio of optical densities for the oxidized form at 280 mμ and 414 mμ is a convenient index for the purity of samples. The best preparation obtained so far has a ratio of 1.9.

Purification Procedures

P-450-Containing Particles from Adrenal Cortex Mitochondria.[3] In the microsomal fractions obtained from animal tissues, P-450 is always accompanied by cytochrome b_5, which interferes with the observation of the spectral properties of P-450. In the case of liver microsomes, cytochrome b_5 can be selectively solubilized and removed from microsomal particles by digestion with steapsin.[4] However, by this treatment, about half of the original P-450 is converted to P-420, and the removal of P-420 is impossible. A particle preparation containing P-450 as the sole hemoprotein component may be prepared from adrenal cortex mitochondria.

[3] D. Y. Cooper, S. Narasimhulu, A. Slade, W. Raich, O. Foroff, and O. Rosenthal, *Life Sci.* **4**, 2109 (1965).

[4] T. Omura and R. Sato, *J. Biol. Chem.* **239**, 2379 (1964).

The mitochondrial fraction is prepared from a homogenate of beef adrenal cortex and washed once with 0.25 M sucrose.[5] The washed mitochondrial pellet is suspended in distilled water to give a final concentration of about 20 mg of protein per milliliter. The suspension is sonicated for 10 minutes (20 kc, 3 amp), and then centrifuged at 40,000 rpm for 30 minutes. The dark red pellet is discarded. The supernatant is further centrifuged at 50,000 rpm for 100 minutes, and the firmly packed dark red pellet is suspended in 0.1 M phosphate buffer, pH 7.5. The suspension is again sonicated for 5 minutes and then centrifuged at 50,000 rpm for 120 minutes. The pellet is finally suspended in 0.01 M phosphate buffer, pH 7.5.

The original concentration of P-450 in adrenal cortex mitochondria is 0.7–0.9 millimicromoles per milligram of protein, and the concentration in the final particle preparation is about the same. The recovery of P-450 from mitochondria is 10– 25%, most of the loss occurring in the first centrifugation of the sonicated mitochondrial suspension. The particle preparation contains a small amount of P-420, which increases gradually during the storage of the preparation, as a result of the conversion from P-450. However, neither detectable amounts of mitochondrial respiratory cytochromes, nor cytochrome b_5, are present in the final preparation. When kept frozen, the particle suspension may be stored for weeks without considerable change in the spectral properties of P-450.

P-420 from Rabbit Liver Microsomes.[4] Microsomes are first treated with crude pancreatic lipase, steapsin (Nutritional Biochemicals Corporation), to remove most of cytochrome b_5 from the particles. The digested microsomes are then solubilized by digestion with heat-treated snake venom. Lyophilized venom of the snake *Trimeresurus flavoviridis,* has been used.[4] Other snake venoms with strong phospholipase activity, such as *Naja naja* venom, are equally effective in solubilizing P-420. To prepare the heat-treated venom, a 1% solution of the snake venom in 0.1 M Tris buffer, pH 7.2, is heated in a boiling water bath for 5 minutes in order to inactivate most of the proteolytic activity of the crude venom. The precipitate is separated by centrifugation and discarded.

P-450 in intact microsomes in stable. However, after treatment with steapsin, P-450 and P-420 in the digested particles become labile to oxygen. Therefore, the digestion by steapsin as well as that by snake venom should be carried out under anaerobic conditions. When kept frozen in the absence of oxygen, both steapsin-digested microsomes and venom-solubilized preparations may be stored for weeks without considerable loss of P-450 and P-420. Purified P-420 is not sensitive to

[5] For the preparation of adrenal cortex mitochondria, see this volume [65].

oxygen. During the purification procedure, should the pH drop below 6, P-420 precipitates from the solution and can be redissolved only with difficulty.

Microsomes are prepared from the livers of 8 adult rabbits. To minimize the contamination with hemoglobin, the livers are well perfused with isotonic saline before homogenization with 4 volumes of a 0.15 *M* KCl solution. The homogenate is centrifuged at 10,000 rpm for 20 minutes, and the precipitate is discarded. The supernatant is further centrifuged at 30,000 rpm for 90 minutes to precipitate the microsomal fraction. The pellet of precipitated microsomes is suspended in 800 ml of 0.15 *M* KCl solution and recentrifuged at 30,000 rpm for 90 minutes.

The washed microsomes are suspended in 250 ml of 0.1 *M* phosphate buffer, pH 7.5 (10–15 mg of microsomal protein per milliliter), digested under nitrogen with 0.2% steapsin at 37° for 1 hour, and then centrifuged at 30,000 rpm for 2 hours. The supernatant is discarded, and the pellet is suspended in 200 ml of 0.1 *M* Tris buffer, pH 8.5.

Sodium deoxycholate and heat-treated snake venom, both at a final concentration of 0.1%, are added to the suspension, and the mixture is incubated under nitrogen at 0° for 36–40 hours. The mixture is then centrifuged at 30,000 rpm for 2 hours, and the precipitate and the fluffy layer are discarded. The clear red supernatant contains P-420.

The supernatant is fractionated by the addition of a saturated solution of ammonium sulfate, the pH of the supernatant being kept at 8.0–8.5 by the addition of a 1 *M* solution of Tris during the fractionation. The red precipitate formed between 25% and 45% saturation of ammonium sulfate is collected by centrifugation. The precipitate is dissolved in 50 ml of 0.1 *M* phosphate buffer, pH 7.5, and dialyzed overnight under nitrogen against 0.01 *M* phosphate buffer of the same pH.

The dialyzed solution is centrifuged at 10,000 rpm for 20 minutes, and the precipitate is discarded. Calcium phosphate gel[6] is added to the supernatant to a final concentration of 15–20 mg/ml, and the mixture is centrifuged. The sedimented gel is washed once with 0.05 *M*, pH 7.5, phosphate buffer, and then treated with 0.2 *M*, pH 7.5, phosphate buffer to elute P-420.

To the gel eluate, saturated ammonium sulfate solution is added to 25% saturation, the pH being maintained at 7.5. The red precipitate is collected by centrifugation and dissolved in 5 ml of 0.1 *M* phosphate buffer. The solution is applied to a Sephadex G-100 column (2 × 60 cm) which has been equilibrated with 0.1 *M* phosphate buffer. A red band of purified P-420 passes through the column without being appreciably retarded.

The content of P-420 in the final purified preparation is 5.0–6.5 milli-

[6] For the preparation of calcium phosphate gel, see S. P. Colowick, Vol. I [11].

micromoles of heme per milligram of protein which represents a 4- to 5-fold purification over microsomes. Recovery of P-420 is 10–20% of the microsomal P-450. The ratio of the optical densities at 280 mμ and 414 mμ of the oxidized sample is around 2.0. Cytochrome b_5 is almost completely removed by this purification procedure.

Properties

P-450. As the solubilization of P-450 is always accompanied by the conversion to P-420, little is known about the physicochemical properties of P-450. Absolute spectra have never been reported, and the reduced minus oxidized difference spectrum of P-450 is not yet well established.

Reduced P-450 combines reversibly with CO, and the dissociation constant of the CO compound of P-450 is about 2×10^{-7} M. P-450 is autoxidizable. A competition between oxygen and CO for reduced P-450 has been observed. Judging from spectral observations, cyanide does not combine with either the reduced or oxidized P-450.

Conversion to P-420 is the most unique property of P-450. Not only detergents and phospholipase, but also some sulfhydryl reagents (such as p-chloromercuribenzoate) and alcohols, convert P-450 quantitatively to P-420. The mechanism by which these various reagents cause such a drastic change of the absorption spectrum of P-450 is not yet known.

P-420. P-420 gives the typical spectra of a b-type cytochrome (Table II). The absorption spectra of oxidized P-420 and CO compound of reduced P-420 are little affected by the pH of the solution while the three absorption bands of the reduced pigment are considerably intensified in alkaline solution. In addition to the protoheme, which can be removed by an acid-acetone treatment, 1–2 atoms of nonheme iron per mole of heme are present in purified preparations.

Reduced P-420 combines reversibly with CO, but neither the reduced nor the oxidized pigment combines with cyanide. P-420 is autoxidizable. The oxidation reduction potential (E'_0) of purified P-420 is −20 millivolt at pH 7.0 and 20°.

[91] NADH-Cytochrome b_5 Reductase

By PHILIPP STRITTMATTER

$$\text{NADH} + 2 \text{ cytochrome } b_5 \text{ (Fe}^{+++}) \rightleftharpoons \text{NAD}^+ + \text{H}^+ + 2 \text{ cytochrome } b_5 \text{ (Fe}^{++})$$

Assay Method

Principle and Procedure. Cytochrome b_5 reductase is active with either cytochrome b_5 or any one of a number of artificial electron acceptors.[1] The spectrophotometric assay methods therefore simply utilize the absorbance changes which measure NADH oxidation or electron acceptor reduction. Normally potassium ferricyanide is used in the assay system since it yields the highest turnover number. When cytochrome b_5 is used, cytochrome *c* is added as the terminal electron acceptor to maintain the cytochrome b_5 largely in the oxidized state, and the increase in absorbance at 550 mμ is measured.

The standard assays[1] of catalytic activity are carried out at 25°, aerobically, in micro cells containing 0.025 micromoles of NADH, 0.05 micromole of potassium ferricyanide (added at zero time), and enzyme in 0.20 ml of 0.1 *M* Tris-acetate, 0.001 *M* EDTA buffer, pH 8.1. The oxidation of NADH is followed at 340 mμ at 15-second intervals for 2 minutes. These readings are corrected for the small contribution of potassium ferricyanide reduction to the 340 mμ absorbance change to express velocity in terms of micromoles substrate oxidized per minute.

Purification Procedure

The procedure described below is essentially the preparation that has been used for several years to obtain the enzyme from calf liver.[1,2] It does contain several minor modifications,[3] particularly in the snake venom extraction step and the insertion of a Sephadex gel filtration, and has consistently yielded homogeneous preparations of the highest yield. All operations are carried out at 0–5° unless otherwise specified.

Step 1. Isolation of Microsomes. A total of 3000 g of calf liver, fresh from the slaughterhouse, is ground in a meat grinder. Approximately 200-g portions are homogenized with 2.5 volumes of sucrose medium (0.25 *M* sucrose, 0.001 *M* EDTA, pH 7.8) in a Waring blendor for 30 seconds. The homogenate is centrifuged at 16,000 *g* for 7 minutes. The turbid supernatant fluid, containing the microsomes, is decanted and saved until all the tissue has been carried to this stage. The pH of the microsomal suspension (approximately 7.5 liters) is then adjusted to 5.3 (at 0–5°) by the dropwise addition of approximately 20–30 ml of 2 *N* acetic acid. The suspension is centrifuged for 8 minutes at 20,000 *g*, and the well packed precipitate is resuspended with 1500 ml of 0.1 *M* Tris-acetate, 0.001 *M* EDTA buffer, pH 8.1. The pH of this suspension is again lowered to pH 5.3 with 2 *N* acetic acid (approximately 30 ml), and the microsomes are recovered as a well packed precipitate by centrifugation at 20,000 *g* for 10 minutes. This reprecipitation is essential

[1] P. Strittmatter and S. F. Velick, *J. Biol. Chem.* **228**, 785 (1957).
[2] P. Strittmatter, *J. Biol. Chem.* **234**, 2661 (1959).
[3] P. Strittmatter and A. J. Wiksell, unpublished experiments, 1965.

for the removal of much of the absorbed hemoglobin from the microsomes. The packed microsomal precipitate is resuspended with 1100 ml of 0.05 *M* Tris-acetate 0.001 *M* EDTA buffer, pH 8.1, and either stored at −12° or used immediately.

Step 2. Snake Venom Extraction. A solution of 500 mg of *Naja naja* venom in 75 ml of 0.1 *M* Tris-acetate, 0.001 *M* EDTA buffer, pH 8.1, is added to the microsomal suspension while it is being stirred in an ice bath. The pH is carefully lowered to 6.45 (at 0–2°) with 2 *N* acetic acid. The suspension is then placed in a bath at 37° and stirred. After approximately 30 minutes, when the temperature is stable, the pH of the incubation mixture is adjusted to pH 5.9, if it is not at this value, and stirred for a total of 5 hours. The suspension is then rapidly cooled in an ice bath and centrifuged for 20 minutes at 20,000 *g*. The yellow-green supernatant fluid containing the reductase is decanted and brought to pH 8.3–8.5 with 1 *N* sodium hydroxide.

Step 3. Ammonium Sulfate Fractionation. Solid ammonium sulfate is slowly added to the enzyme solution from step 2 to 0.45 saturation, and the pH is maintained between 8.3 and 8.5 by the intermittent addition of 1 *N* sodium hydroxide. The suspension is then centrifuged for 10 minutes at 20,000 *g*, and the yellow supernatant fluid containing the enzyme is decanted. The reductase is precipitated by raising the ammonium sulfate concentration to 0.85 saturation, again with the intermittent addition of 1 *N* sodium hydroxide to maintain the pH between 8.3 and 8.5. The precipitate which forms in approximately 1 hour (or overnight if convenient) is collected by centrifuging at 20,000 *g* for 10 minutes. The enzyme is dissolved in a minimum amount of 0.1 *M* Tris-acetate, 0.001 *M* EDTA buffer, pH 8.1 (usually 75 ml). Some inactive material is usually present and is removed by centrifuging the solution briefly.

Step 4. Sephadex Gel Filtration. The enzyme solution from step 3 (approximately 100–150 ml) is placed on a Sephadex column, 4.5 by 150 cm, containing first 15 cm of Sephadex G-50, fine, and then 135 cm of Sephadex G-75, fine, previously equilibrated with Tris buffer, pH 8.1. When the column is developed with the same buffer, three protein fractions are obtained: a fast running inactive red band, a middle yellow band containing the reductase, and a slow moving red band. A fourth band containing free flavin moves extremely slowly. The fractions containing 80–90% of the enzyme, determined by assays, are pooled to yield approximately 250–300 ml of enzyme free of ammonium sulfate and most heme-containing impurities.

Step 5. Cγ Gel Adsorption. The crude reductase solution from step 4 is divided and placed on two Cγ gel columns, 2 cm in diameter, containing an even suspension of 20 g of Celite and 80 ml of Cγ gel (approximately 15.6 mg dry weight per milliliter) equilibrated with the 0.1 *M* Tris-acetate, 0.001 *M* EDTA buffer, pH 8.1. The columns are then washed with approximately 200 ml of 1 *M* Tris-acetate, 0.001 *M* EDTA buffer, pH 9.1, until the yellow band of enzyme has moved about half the distance of the column. The upper portion of the Cγ gel-Celite mixture is removed and discarded. The column is then washed with the 1 *M* buffer containing 0.01 saturated ammonium sulfate until the yellow protein band is within 2 cm of the bottom of the column. Again the gel above the yellow band is removed and discarded, and the enzyme is eluted with 1 *M* buffer containing 0.05 saturated ammonium sulfate. Saturated ammonium sulfate solution (pH 7.8, 5°) is then added to the eluate, and the precipitate obtained between 0.5 and 0.65 saturation contains enzyme of highest purity. As judged by the ultraviolet absorption spectra and specific activity, these preparations are only 60–80% pure; therefore, the ammonium sulfate is removed by gel filtration with Sephadex G-25, fine, and step 5 is repeated again. This yields approximately 40–60 mg of reductase, which appears to be 90–100% homogeneous, and a total recovery of 35–65% of the activity originally present in the crude snake venom extract.

Properties

Molecular Weight and Homogeneity. The reductase preparations appear to be homogeneous as judged by sedimentation and electrophoresis on cellulose acetate. The molecular weight of the enzyme calculated from sedimentation, diffusion, and partial specific volume data is 40,000. The protein contains 1 mole of FAD, has a turnover number of approximately 30,000 moles of NADH oxidized per minute per mole of enzyme with saturating concentrations of potassium ferricyanide or cytochrome b_5, and is characterized by a ratio of absorbance at 278 to 460 mμ of 6.9.[1]

Stability. The enzyme is stable from pH 7.0 to 9.0 at room temperature for hours, at 0–5° for days, and at —12° indefinitely. The reversible resolution of the holoenzyme at low pH in ammonium sulfate has been described.[4]

Specificity. NADH and analogs of this nucleotide are oxidized at various rates with either potassium ferricyanide, cytochrome b_5, or various dyes.[5] Because the K_m value for cytochrome b_5 is relatively high (2×10^{-5} *M*), it is difficult to obtain maximum velocities with the

[4] P. Strittmatter, *J. Biol. Chem.* **236**, 2329 (1961).

[5] P. Strittmatter, *J. Biol. Chem.* **234**, 2665 (1959).

heme protein. NADPH is oxidized extremely slowly and oxygen serves as a very poor electron acceptor.[5]

Equilibrium and pH Optimum. The oxidation of NADH proceeds virtually to completion, and product inhibition by NAD, at 50- to 100-fold molar excesses, is not discernible. The pH dependence of the reaction has not been examined thoroughly, but the optimum appears to be rather broad.[6]

Inhibitors. The reductase contains a reactive sulfhydryl group, which is essential for NADH interaction[7] and reacts rapidly with *p*-mercuribenzoate or with *n*-ethyl maleimide.[2]

[6] P. Strittmatter and S. F. Velick, *J. Biol. Chem.* **221**, 277 (1956).
[7] P. Strittmatter, *J. Biol. Chem.* **236**, 2336 (1961).

[92] The Preparation and Properties of Microsomal TPNH-Cytochrome *c* Reductase from Pig Liver[1]

By Bettie Sue Siler Masters, Charles H. Williams, Jr.,[1a] and Henry Kamin[1a]

TPNH-cytochrome *c* reductase from liver microsomes is a flavoprotein probably identical to an enzyme first prepared by Horecker[2] from whole liver acetone powder. It was purified from microsomes by Williams and Kamin[3] both as a partially purified microsomal subparticle and, with lipase treatment and fractionation, as a soluble flavoprotein essentially homogeneous in the ultracentrifuge. Phillips and Langdon[4] have independently reported purification of this enzyme by trypsin treatment of microsomes; their preparation, while apparently representing the same enzyme, differs from the Williams and Kamin[3] preparation in stability and sensitivity to sulfhydryl reagents. The procedures described

[1] The studies reported herein were supported in part by Grants No. AM 4662 and GM 11231 from the U.S.P.H.S., and a grant from the North Carolina Heart Association.
[1a] This manuscript was prepared while the authors were at the Duke University Medical Center, Durham, North Carolina.
[2] B. L. Horecker, *J. Biol. Chem.* **183**, 593 (1950).
[3] C. H. Williams, Jr., and H. Kamin, *J. Biol. Chem.* **237**, 587 (1962).
[4] A. H. Phillips and R. G. Langdon, *J. Biol. Chem.* **237**, 2652 (1962).

herein follow the methods of Williams and Kamin[3], as modified by Masters *et al.*[5]

This enzyme catalyzes the following reactions:

$$TPNH + H^+ + 2 \text{ cytochrome } c^{3+} \rightarrow TPN^+ + 2 \text{ cytochrome } c^{++} + 2 H^+ \quad (1)$$
$$TPNH + H^+ + \text{dye} \rightarrow \text{reduced dye} + TPN^+ \quad (2)$$
$$TPNH + H^+ + O_2 \rightarrow TPN^+ + H_2O_2 \quad (3)$$

The formation of hydrogen peroxide in reaction (3) is presumed but has not been directly demonstrated.

This enzyme, in its native milieu, may serve as a member of an electron transport chain directed toward hydroxylations. Orrenius and Ernster[6] have found parallel increases in activity of microsomal TPNH-cytochrome *c* reductase, cytochrome P-450 (but not cytochrome b_5), and an oxidative demethylase upon administration of phenobarbital to rats. Omura *et al.*[7] have fractionated what may be an analogous system from beef adrenal cortex particles, and they found it to contain a TPNH-oriented flavoprotein, a nonheme iron protein, and cytochrome P-450. The latter compound appears to serve as the terminal oxidase for steroid hydroxylation. The adrenal flavoprotein requires the nonheme iron protein for cytochrome *c* and P-450 reduction, but not for dye reduction.

Assay Method[3]

Cytochrome c Reduction

The assay depends upon measurement of the rate of cytochrome *c* reduction at 550 mμ. The following are added to a 1-ml spectrophotometer cuvette with buffer substituted for TPNH in the blank: enzyme in 0.05 *M* potassium phosphate buffer, pH 7.7 containing 1×10^{-4} *M* EDTA, 36 millimicromoles cytochrome *c* (Sigma type III) in the same buffer, and buffer to a volume of 1.0 ml. The reaction is initiated by addition of 0.1 ml of a fresh 1×10^{-3} *M* solution of TPNH (Sigma type I or II) in buffer (final volume = 1.1 ml), and the reaction is followed at 550 mμ during the linear portion of the reaction. When the reaction is followed in a recording spectrophotometer, such as the Cary 14, 0.03–0.3 units of enzyme represent a convenient amount; smaller quantities are convenient in the Beckman DU. Rates are proportional to enzyme concentration over a wide range.

[5] B. S. S. Masters, H. Kamin, Q. H. Gibson, and C. H. Williams, Jr., *J. Biol. Chem.* **240,** 921 (1965).

[6] S. Orrenius and L. Ernster, *Biochem. Biophys. Res. Commun.* **16,** 60 (1964).

[7] T. Omura, R. Sato, D. Cooper, O. Rosenthal, and R. W. Estabrook, *Federation Proc.* **24,** 1181 (1965).

Definition of Unit and Specific Activity. One unit is defined as an absorbance change of 1.0 per minute at 550 mμ at 25° in a 1-cm light path. This corresponds to reduction of 0.0476 micromole of cytochrome *c* per minute per milliliter of reaction mixture.

Neotetrazolium Reduction

Cytochrome *c* reductase activity in microsomes is associated with marked TPNH-neotetrazolium diaphorase activity.[8] This activity is disproportionately lost upon lipase treatment of microsomes, and its loss may serve as an indication of loss of an intermediate cofactor in the intact microsome, or of loss of a specific environmental or configurational state of the enzyme. Assay of this diaphorase activity is described in Williams and Kamin.[3]

Dichlorophenolindophenol (DCIP) Reduction

This activity is assayed by decrease in absorbance of DCIP measured at 600 mμ. The assay is identical to that for cytochrome *c* reduction, with substitution of 96 millimicromoles of DCIP (Sigma) for the cytochrome *c*. The molar absorptivity used for DCIP is $21 \times 10^3 M^{-1}$ cm^{-1}.

Purification Procedure

Preparation of Lipase. Since solubilization of enzyme from microsomes depends upon lipase treatment, the preparation of lipase will be described. This is a procedure of R. K. Crane[9] based on methods of Willstätter and Waldschmidt-Leitz.[10]

Mix 100 g of steapsin, U.S.P. (Nutritional Biochemicals Corporation) with 1 liter of 87% by volume aqueous glycerol. Maintain at 37° with stirring for 5 hours. Add 1500 ml of ice-cold glass-distilled water, and centrifuge at about 16,000 *g* for 20 minutes at 15° (Lourdes 3RA 3-liter rotor at maximum speed). To the supernatant, add 13 ml of 1 *N* acetic acid with stirring, followed by 250 ml of alumina cγ gel.[11] Stir for 30 minutes and centrifuge at 5000 *g* for 10 minutes at 15° in Lourdes 3RA rotor. Discard the supernatant fluid, and wash the gel twice with 500-ml portions of 30% by volume aqueous glycerol. Centrifuge after each wash at 5000 *g* for 10 minutes at 0–5° (Lourdes VRA 1.5-liter capacity rotor.)

[8] C. H. Williams, Jr., R. H. Gibbs, and H. Kamin, *Biochim. Biophys. Acta* **32**, 568 (1959).

[9] R. K. Crane, personal communication, 1960.

[10] R. M. Willstätter and E. Waldschmidt-Leitz, *in* "Die Methoden der Fermentforschung" (E. Bamann and K. Myrbäck, eds.), Vol. 1, p. 1560. Thieme, Leipzig, 1941.

[11] R. Willstätter, H. Kraut, and O. Erbacher, *Ber.* **58**, 2448 (1925).

Suspend the washed gel in 500 ml of 0.1 M diammonium phosphate, and allow to stir overnight at 0–3°. Centrifuge at 5000 g for 30 minutes at 0–5° in Lourdes VRA rotor. The supernatant is dialyzed against 3 changes of 10% by volume aqueous glycerol (8 hours per change), and centrifuged at about 20,000 g for 1 hour at 0–5° in Lourdes VRA rotor. Discard the sediment, and store the supernatant fluid in the deep freeze. It is stable indefinitely.

Preparation of Microsomes. Large-scale preparation of pork liver microsomes is performed by a modification of the method of Strittmatter and Velick.[12]

Whole livers, preferably from well bled animals, are obtained from the slaughterhouse immediately after killing and packed in plastic bags in crushed ice. Upon reaching the laboratory, the livers are perfused with 0.9% NaCl until the tissue appears pale: this perfusion helps minimize heme contamination of the final preparation. The liver is then sliced into strips, rinsed in cold 0.25 M sucrose containing 10^{-3} M EDTA (adjusted to pH 7.5 with 1 M acetic acid), and minced for 45 seconds at high speed in a 1 gallon capacity (Waring CB-5) blendor in 400 g batches with 1500 ml of cold sucrose per batch. Each minced batch is strained through two layers of cheesecloth into a 5-gallon carboy packed in ice.

Crude microsomes are first isolated by continuous flow centrifugation in the Lourdes LR centrifuge using the Lourdes CFR-1 rotor at 0–5°. The rotor speed is maintained at 12,000 rpm. Nuclei and cell debris are removed by pumping the suspension through the rotor at about 600 ml per minute (Sigmamotor Model OV22 kinetic clamp pump); the supernatant is collected in another ice-packed carboy. The mitochondria are then removed by pumping this supernatant through the rotor at 100–125 ml per minute. The liners of the continuous flow rotor should be changed every 5–6 liters to prevent overflow of mitochondria into the supernatant.

Then 1 N acetic acid is added to the supernatant to pH 5.35 ± 0.1 to precipitate the microsomes and some soluble protein; the suspension is allowed to stand in the cold at least an hour or until agglutination is evident. It is then pumped through the continuous flow rotor at 120 ml per minute. The liner is changed each time the supernatant becomes cloudy.

The contents of the rotor liners are emptied into a large chromatography jar and stirred overnight in a potassium phosphate buffer, pH 7.7, 1×10^{-4} M in EDTA, using about 150 ml of buffer for each 100 g of original liver. This procedure redissolves much of the nonmicrosomal protein precipitated in the previous step.

[12] P. Strittmatter and S. F. Velick, *J. Biol. Chem.* **228**, 785 (1957).

The viscous microsomal suspension is sedimented by 90 minutes centrifugation at 19,000 rpm in the 19 rotor of the Spinco Model L-2 ultracentrifuge, and the supernatant is discarded. The microsomes are resuspended in buffer, using a Ten Broeck homogenizer, to a total volume of 400–420 ml per liter of original viscous suspension. The thick microsomal suspension is stored at —20° in batches of about 400 ml; the batches are then processed individually. TPNH-cytochrome *c* reductase is stable for several months under these conditions.

The microsomes at this stage should be substantially free of mitochondria, as judged by succinate-neotetrazolium diaphorase activity. Most preparations have insufficient cytochrome oxidase activity to interfere with cytochrome *c* reductase assay, but they should be checked and, if necessary, cyanide can be added to the assay mixture. A small amount of mitochondrial contamination does not appear to interfere with subsequent purification.

Lipase Solubilization. Approximately 400 ml of the thawed microsomal suspension is brought to 37° and 10–20 ml[13] of lipase solution is added. The suspension is incubated for 15 minutes, chilled, and centrifuged for 1 hour at 30,000 rpm in the Spinco 30 rotor. The reddish yellow clear supernatant is taken directly through the subsequent steps, or it can be frozen.

pH Precipitation. The supernatant from the lipase solubilization is adjusted to pH 5.4 with 1 *N* acetic acid,[14] and the cloudy red suspension is centrifuged immediately at 9000 rpm (13,000 *g*) in the Lourdes VRA rotor for 10 minutes. The supernatant is adjusted to pH 4.6 with 1 *N* hydrochloric acid, and the resulting cloudy red suspension is again centrifuged. The precipitate is resuspended in about 70 ml of 0.05 *M* potassium phosphate, pH 7.7, 10^{-4} *M* EDTA. Although this step results in the removal of much of the heme protein, the enzyme solution still remains distinctly reddish.

Ammonium Sulfate Fractionation. Solid ammonium sulfate[15] is added to 45% saturation[16] over a 15–20 minute period to the enzyme solution

[13] Pilot experiments should be performed to establish the optimum amount for each new batch of lipase used. The quantity used should be sufficient to achieve nearly maximal solubilization of enzyme activity (in the Spinco supernatant of the lipase digest) in about 15 minutes. Prolonging lipase digestion, or increasing lipase concentration above the minimum required, does not materially affect the units of enzyme released, but can increase heme contamination in the final preparation.

[14] Broadening the pH limits (particularly the lower pH) increases heme contamination; the limits cited are followed rigorously, using a pH meter standardized with buffer at 5°.

[15] Analytical grade, recrystallized from 1×10^{-3} *M* EDTA.

[16] The amount of ammonium sulfate used to achieve the stated percentage of saturation is defined by the table of A. A. Green and W. L. Hughes. See Vol. I, p. 76.

maintained in an ice bath. The pH is kept at 7.7 with 1 *N* potassium hydroxide to minimize the loss of flavin which occurs in the presence of ammonium sulfate at pH 6 or below. The suspension is stirred for an additional 30 minutes and centrifuged at 9000 rpm (Lourdes 9-inch rotor) for 10 minutes. Sufficient ammonium sulfate is added to the supernatant to bring it to 65% saturation,[16] again maintaining the pH at 7.7. The suspension is again centrifuged, and the precipitate is resuspended in 20 ml of 0.05 *M* potassium phosphate buffer, pH 7.7,[17] 1×10^{-4} *M* EDTA. This clear yellow solution is stable at —20°.

Column Chromatography on Hydroxylapatite. A 1.9×10 cm hydroxylapatite column[18] (jacketed Beckman preparative column assembly) is packed in approximately 1-cm layers by alternately adding hydroxylapatite slurry and pumping 0.05 *M* potassium phosphate, pH 7.7, 1×10^{-4} *M* EDTA through the column at 16–18 ml per hour, using a Beckman Accu-flow pump.[19] The column is maintained at approximately 5°. The enzyme is added gently to the top of the column and nitrogen pressure (5–10 psi)is applied until the bottom of the meniscus touches the hydroxylapatite. Pressure is released, and 0.05 *M* potassium phosphate, pH 7.7, 1×10^{-4} *M* EDTA is added to fill the air space in the column assembly. The pump is connected via thin Tygon tubing to the linear gradient system, which consists of two connected 500 ml aspirator bottles. The bottle which delivers to the pump contains 200 ml of 0.05 *M* potassium phosphate, pH 7.7, 1×10^{-4} *M* EDTA, and is magnetically stirred. The second bottle, connected to the first at the same level with short rubber or tygon tubing, contains 200 ml of 0.2 *M* potassium phosphate, pH 7.7, 1×10^{-4} EDTA.

The pump is set to deliver 16–18 ml per hour to the fraction collector. An initial inactive reddish yellow peak is discarded; the first peak of

[17] If the enzyme solution appears reddish, the 65% ammonium sulfate step can be repeated, with overnight stirring of the suspension, and washing of the centrifuged pellet with 65% ammonium sulfate, pH 7.7, 1×10^{-4} *M* EDTA.

[18] The hydroxylapatite is prepared essentially according to A. Tiselius, S. Hjertén, and Ö. Levin, *Arch. Biochem. Biophys.* **65,** 132 (1956). The settling times are extended until all but the most highly dispersed material has settled. The resultant higher content of "fines" materially improves resolution. Commercial hydroxylapatite preparations tested have not yielded as high resolution as the hydroxylapatite prepared as described.

[19] Column chromatography under pressure has, in our hands, yielded better resolution than gravity flow. If the equipment cited is not available, pressure flow can be improvised using ball-jointed columns and clamps, and a nitrogen tank with pressure regulator. Gravity flow [see C. H. Williams, Jr. and H. Kamin, *J. Biol. Chem.* **237,** 587 (1962)] can yield satisfactory results, but often rechromatography of the final material is necessary.

enzyme activity is generally eluted after approximately one-fourth to one-third of the eluting buffer has passed through the column.

When chromatography is conducted under conditions of maximum resolution, enzyme activity generally appears in two peaks. Peak one usually contains about one-third (but sometimes more) of the total activity. Its spectrum always shows heme contamination. Peak two, emerging about 50 ml after the beginning of peak one, shows a typical flavin spectrum,[5] usually heme free. Both peaks yield identical kinetic and ultracentrifugal data; peak one specific activity may be 10% lower than peak two. Rechromatography of peak one yields both peak one and peak two, whereas rechromatography of peak two gives but a single peak. These observations suggest that the enzyme in the two peaks is identical, but that peak one contains small quantities of a contaminant which distorts the enzyme's chromatographic behavior.

Calcium Phosphate Gel Concentration. The enzyme solution from the column is dilute and at relatively high salt concentrations; the enzyme is relatively unstable under these conditions. Adsorption and elution with calcium phosphate gel is used primarily to concentrate the enzyme, but a small degree of purification is also usually achieved.

The activity peak from the hydroxylapatite column is dialyzed[20] (4 hours to overnight) against 0.001 *M* potassium phosphate, pH 7.7, 1×10^{-4} *M* EDTA. Approximately 1 ml of calcium phosphate gel[21] (0.08 g/ml) is added to the dialyzed solution for each 1000 units of enzyme activity. The suspension is stirred at 0–5° for 15 minutes and then centrifuged at 10,000 rpm in the Lourdes 9-inch rotor for 10 minutes. The supernatant is discarded, and the gel is resuspended in 0.4 *M* potassium phosphate buffer, pH 7.7, 10^{-4} *M* EDTA, (2 ml per 1000 units of activity), stirred for 30 minutes, and centrifuged as before. The supernatant is removed with a Pasteur pipette and dialyzed overnight against 1 liter of 0.05 *M* potassium phosphate buffer, pH 7.7, 10^{-4} *M* EDTA. The concentrated enzyme solution at this final stage is usually about 1.5×10^{-5} *M* in flavin, as determined by 450 minus 600 mμ absorbance using a molar absorptivity of 11.3×10^3 M^{-1} cm^{-1} for flavin,[22] and has a specific activity of 350–450 units per milligram of protein.[23]

[20] All dialysis tubing is first washed in distilled water, then brought to a boil in 1×10^{-3} *M* EDTA. It is stored at room temperature in the EDTA solution and is washed in distilled water before use.

[21] S. Swingle and A. Tiselius, *Biochem. J.* **48,** 171 (1951).

[22] P. Cerletti and N. Siliprandi, *Arch. Biochem. Biophys.* **76,** 214 (1958).

[23] Protein determined by the method of Murphy and Kies [J. B. Murphy and M. W. Kies, *Biochim. Biophys. Acta* **45,** 382 (1960)]. For this enzyme, this method gives results which are 10–20% higher than Kjeldahl analyses using a protein to nitrogen

If more extensive concentration of enzyme is required than that provided by the calcium phosphate gel procedure, the protein may be precipitated from the column eluate or the calcium phosphate gel eluate by addition of ammonium sulfate to 65% saturation at pH 7.7, collection of the precipitate, re-solution in a minimal volume of 0.05 *M* phosphate buffer, pH 7.7, 1×10^{-4} *M* EDTA, and dialysis against that buffer. This concentration procedure is usually accompanied by greater losses than the calcium phosphate gel procedure.

See the table for a summary of the purification procedure.

PURIFICATION OF MICROSOMAL TPNH-CYTOCHROME *c* REDUCTASE FROM PORK LIVER

Fraction	Total units	Units/ml	Protein (mg/ml)	Specific activity	Purification
Microsomes	8200	19	40	0.48	—
Lipase supernatant	7500	26	15	1.7	3.5
pH precipitate	6200	89	18	4.9	10
Ammonium sulfate precipitate	4200	209	10	21	43
Hydroxylapatite eluate	3400	78	0.2	390	810
Calcium phosphate gel concentrate	3000	402	0.9	446	930

Properties

Flavin Content and Molecular Weight. The prosthetic group of TPNH-cytochrome *c* reductase enzyme is flavin adenine dinucleotide,[2,3] but the apoenzyme can be reactivated by FMN as well as FAD.[2,3] The enzyme has an $S_{20,w}$ of 4.7 to 5.2×10^{-13}, consistent with a molecular weight range of 60,000–90,000.[3] The presence of a small (less than 10%) amount of a faster-sedimenting, higher molecular weight impurity has thus far prevented definitive establishment of molecular weight by sedimentation equilibrium. Most preparations have a FAD content indicating a minimal molecular weight of 40,000–60,000, and several analyses have fallen into the 35,000–40,000 range. These data indicate the presence of 2 FAD groups per mole. No metals appear to be present.

Specificity. The enzyme is specific for TPNH, but is relatively nonspecific for electron acceptor.[3] Cytochrome *c*, DCIP, ferricyanide, and neotetrazolium[3] react rapidly as electron acceptors, as does menadione[24,25]

factor of 7.1 established by amino acid analysis. Nevertheless, the Murphy and Kies method is convenient for assay purposes.

[24] H. Nishibayashi, T. Omura, and R. Sato, *Biochim. Biophys. Acta* **67**, 520 (1963).

[25] B. S. S. Masters, M. Bilimoria, H. Kamin, and Q. H. Gibson, *J. Biol. Chem.* **240**, 4081 (1965).

and several naphthoquinone derivatives.[26] Oxidase[3] and ferric ion reductase activity are low, about 0.1% and 3% of the rate of cytochrome *c* reduction, respectively. The enzyme can serve as an organic nitroreductase.[27] The enzyme purified by this procedure does not reduce cytochrome b_5[3,28] or coenzyme Q_{10}.[29]

Kinetics. TPNH-cytochrome *c* reductase has a pH optimum of 8.2, which appears to be independent of the buffer used.[3] The following K_m's have been established[2,3,25,30]: TPNH, $3.8 \times 10^{-6}\,M$; DCIP, $1.0 \times 10^{-5}\,M$; and menadione, 0.6 to $1.7 \times 10^{-5}\,M$. Lineweaver-Burk plots varying concentration of both TPNH and electron acceptor yield a series of parallel lines for both cytochrome *c* and DCIP.[5,25] These observations are consistent with a mechanism in which one substrate reacts with enzyme to form a product and modified enzyme, followed by reaction of modified enzyme with the second substrate.[31]

The V_{max} at "infinite" concentrations of both reactants corresponds to turnover numbers of about 1200 equivalents per minute using cytochrome *c* as acceptor[3,5] and of the same order for DCIP.[25]

Activators and Inhibitors. TPNH-cytochrome *c* reductase is markedly stimulated by low levels of *p*-chloromercuribenzoate[5]; maximal activation is reached at 2 moles of PCMB per mole of flavin, and higher concentrations inhibit.[3,5] TPN^+ is a competitive inhibitor,[2,3] with a K_I of $2.3 \times 10^{-6}\,M$[3]. The following compounds are without effect[3,5]: antimycin A, Amytal, quinacrine, 8-hydroxyquinoline, progesterone, arsenite, and *N*-ethyl maleimide.[5]

Mechanism. The mechanism of the electron transfers catalyzed by TPNH-cytochrome *c* reductase has been intensively studied.[5,25] The catalytic cycle consists of alternation of flavin between the fully reduced and half-reduced forms, and fully oxidized flavin does not appear to participate in catalysis. The same mechanism obtains for the reduction of both one- and two-electron acceptors.

[26] M. Bilimoria, B. S. S. Masters, and H. Kamin, unpublished observations, 1965.

[27] J. J. Kamm and J. R. Gillette [*Federation Proc.* **21,** 246 (1962)] have observed TPNH-organic nitroreductase activity in microsomes. Both Kamm at the National Institutes of Health and K. V. Rajagopalan of Duke University (personal communications) have shown that TPNH-cytochrome *c* reductase, prepared by the authors, exhibits this activity.

[28] J. Modirzadeh and H. Kamin, *Biochim. Biophys. Acta* **99,** 205 (1965).

[29] M. H. Bilimoria, B. S. S. Masters, and Henry Kamin, unpublished observations, 1965.

[30] The K_m value for DCIP published by Williams and Kamin was incorrectly calculated, and should read $1.5 \times 10^{-5}\,M$. The value given in the present text ($1.0 \times 10^{-5}\,M$) is based on the average of a number of subsequent experiments.

[31] R. A. Alberty, *Advan. Enzymol.* **17,** 1 (1956).

[92a] Microsomal Lipid Peroxidation

By LARS ERNSTER and KERSTIN NORDENBRAND

Enzymatically induced peroxidation of microsomal lipids was described in 1963 by Hochstein and Ernster,[1] who found that incubation of rat liver microsomes in the presence of oxygen, NADPH, and a nucleoside di- or triphosphate or inorganic pyrophosphate results in a rapid oxygen consumption, NADPH disappearance, and malonaldehyde formation. Shortly later it was found[2] that the presence of ferrous or ferric ions in conjunction with the nucleoside di- or triphosphate or pyrophosphate was essential for the reaction. Nucleoside diphosphate- and NADPH-induced O_2 uptake in microsomes was independently observed by Beloff-Chain *et al.*,[3] who subsequently confirmed[4] Hochstein and Ernster's demonstration of the concomitant lipid peroxide formation. The relationship of microsomal lipid peroxidation and drug hydroxylation has been studied by Orrenius *et al.*[5] and more recently by Gotto *et al.*[6] A detailed study of microsomal lipid peroxidation has been carried out by Ernster and Nordenbrand.[7]

Demonstration

Microsomal lipid peroxidation may be demonstrated by measuring (a) O_2 consumption, (b) NADPH disappearance, and (c) malonaldehyde (MA) formation. Results of an experiment involving the measurement of all three parameters are illustrated in Fig. 1. The experimental procedure is as follows:

Rat liver microsomes are prepared[8] by sedimenting the 10,000 *g* supernatant of a 0.25 *M* sucrose homogenate of rat liver at 105,000 *g* for 60 minutes. The surface of the tightly packed microsomal pellets is

[1] P. Hochstein and L. Ernster, *Biochem. Biophys. Res. Commun.* **12**, 388 (1963); *Ciba Found. Symp. Cellular Injury, 1963,* p. 123. Little, Brown, Boston, Massachusetts, 1964.

[2] P. Hochstein, K. Nordenbrand, and L. Ernster, *Biochem. Biophys. Res. Commun.* **14**, 323 (1964).

[3] A. Beloff-Chain, R. Catanzaro, and G. Serlupi-Crescenzi, *Nature* **198**, 351 (1963).

[4] A. Beloff-Chain, G. Serlupi-Crescenzi, R. Catanzaro, D. Venetacci, and M. Balliano, *Biochim. Biophys. Acta* **97**, 416 (1965).

[5] S. Orrenius, G. Dallner, and L. Ernster, *Biochem. Biophys. Res. Commun.* **14**, 329 (1964).

[6] A. M. Gotto, R. M. Hutson, A. W. Meikle, and O. Touster, *Biochem. Pharmacol.* **14**, 989 (1965).

[7] L. Ernster and K. Nordenbrand, in preparation.

[8] L. Ernster, P. Siekevitz, and G. E. Palade, *J. Cell Biol.* **15**, 541 (1962).

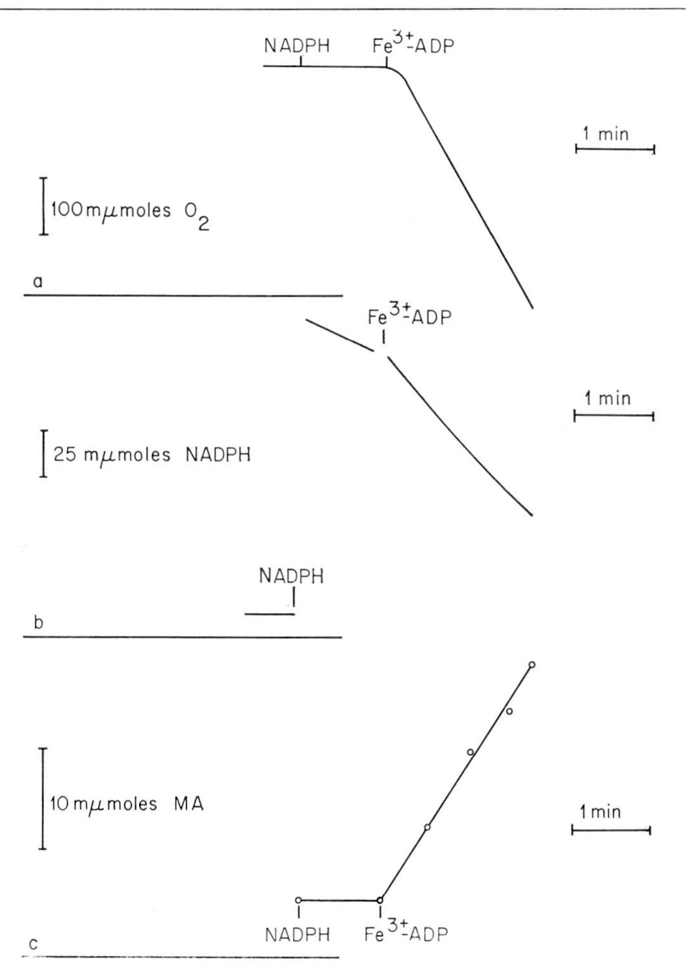

FIG. 1. NADPH-linked lipid peroxidation in rat liver microsomes.

thoroughly rinsed with 0.15 *M* KCl in order to remove the bulk of adhering sucrose, which interferes with the malonaldehyde determination. The microsomes are suspended in 0.15 *M* KCl to an approximate final concentration of 20 mg of protein per milliliter (approximately 1 g of liver per milliliter). Protein is determined by the biuret method.[9]

The reaction mixture consists of 0.15 *M* KCl, 0.025 *M* Tris-HCl buffer, pH 7.5, and microsomes containing 2–3 mg protein, in a final volume of 3 ml.

O_2 consumption (Fig. 1*a*) is measured with a Clark O_2 electrode.

[9] A. G. Gornall, C. J. Bardawill, and M. M. David, *J. Biol. Chem.* **177**, 751 (1949).

After short thermoequilibration of the reaction mixture in the electrode chamber (indicated by a horizontal trace on the recorder), 10 μl of 15 mM NADPH is added. Very little or no O_2 consumption (downward deflection of the trace) is observed. Upon the further addition of 5 μl of a solution containing 0.6 M ADP and 6 mM $FeCl_3$ (ADP-Fe), a high rate of O_2 uptake is observed.

NADPH disappearance (Fig. 1*b*) is monitored fluorometrically[10] at 450 mμ with an excitation wavelength of 365 mμ. An Eppendorf photometer with fluorometer attachment is a suitable instrument for this purpose. The instrument is coupled to an extended-scale recorder and standardized with a known solution of NADPH; arrangement for simultaneous recording of NADPH disappearance and O_2 consumption may be made by mounting the O_2 electrode in the fluorometer cuvette.

Addition of NADPH to the microsomal suspension is reflected in an instantaneous increase in fluorescense, followed by a constant, slow decrease due to NADPH oxidation. Upon the addition of ADP-Fe, the rate of NADPH oxidation increases 3 to 4-fold.

MA formed (Fig. 1*c*) is measured colorimetrically with the thiobarbituric acid (TBA) reaction.[11] The incubation is interrupted at suitable intervals by the addition of 0.1 ml of 100% trichloroacetic acid (TCA). Two milliliters of the fixed reaction mixture is transferred into a centrifuge tube and mixed with 2 ml of 30% TCA, 0.2 ml of 5 M HCl (cf. Hunter *et al.*[12]), and 2 ml of 0.75% TBA. The mixture is put in a boiling water bath for 15 minutes; during this time the red color of the TBA–MA complex develops. After centrifugation the color of the clear solution is measured at 535 mμ. An extinction coefficient of $1.56 \times 10^5 M^{-1}$ cm^{-1} is used[13] for calculating the amount of MA.

O_2 consumption may alternatively be measured manometrically in the Warburg apparatus. Measurement of NADPH disappearance spectrophotometrically at 340 mμ is complicated by the turbidity of the microsomal suspension which, in addition, may change during the incubation.

General Features

The maximal rate of NADPH-linked lipid peroxidation at 30° is approximately 160 millimicromoles of O_2 consumed per minute per milligram of microsomal protein. The reaction is strongly temperature dependent, its rate increasing by about 4-fold from 20° to 35°. The

[10] R. W. Estabrook and P. K. Maitra, *Anal. Biochem.* **3**, 369 (1962).

[11] F. Bernheim, M. L. Bernheim, and K. M. Wilbur, *J. Biol. Chem.* **174**, 257 (1948).

[12] F. E. Hunter, J. M. Gebicki, P. E. Hoffsten, J. Weinstein, and A. Scott, *J. Biol. Chem.* **238**, 828 (1963).

[13] R. O. Sinnhuber, T. C. Yo, and Te Chang Yo, *Food. Res.* **23**, 626 (1958).

NADPH disappearance accompanying microsomal lipid peroxidation ranges between one-third and one-fifth mole of NADPH per mole of O_2 consumed. The molar ratio O_2:MA varies between 16 at 20° and 23 at 35°. Hydrogen peroxide is formed as an oxidation product of NADPH, which can be demonstrated with catalase and methanol as described by Gillette *et al.*[14] The maximal extent of lipid peroxidation is approximately 1 micromole of O_2 consumed per milligram of microsomal protein.

Enzymatic lipid peroxidation in microsomes requires specifically NADPH as the electron donor; NADH is without effect. The K_m for NADPH is 0.55 μM. Added NADPH may be replaced by $NADP^+$ and an NADPH-generating system. This is necessary in long-time experiments in order to ensure a constant supply of NADPH. Either glucose 6-phosphate (G-6-P) + G-6-P dehydrogenase or isocitrate + isocitrate dehydrogenase may be used as a NADPH-generating system. In the former case it is important to consider that liver microsomes contain a highly active glucose 6-phosphatase (approximately 0.25 micromoles of G-6-P split per minute per milligram of protein), and thus, that a relatively high concentration of G-6-P ought to be used. In the case of the isocitrate system, the concentration of Mn^{++} (which is an activator of isocitrate dehydrogenase) should be relatively low, $\leqslant 10^{-5} M$, since higher concentrations of Mn^{++} inhibit lipid peroxidation. With either NADPH-generating system it is necessary to add 50 mM nicotinamide to the incubation mixture in order to minimize the breakdown of $NADP^+$ by the microsomal NADPase.

ADP may be replaced by any nucleoside 5′-di- or tri- (but not mono-) phosphate. Maximal activity with all these nucleotides is reached at a concentration of 0.5–1.0 mM. Inorganic pyrophosphate also activates lipid peroxidation. The optical concentration in this case is in the range of 0.01–0.02 mM above which the reaction is inhibited. Likewise, oxalate activates lipid peroxidation, maximal activation being reached at a concentration of 1.0 mM. In all cases, the presence of iron is necessary, Fe^{++} and Fe^{3+} being equally active. The optimal concentration of iron is 10–40 μM. In the absence of these complexing agents Fe^{3+} is completely inactive, whereas Fe^{++} gives a transient lipid peroxidation, the rate and extent of which are dependent on the amount of Fe^{++} added but independent of the presence of NADPH. The EDTA, cyanide, and phenanthroline complexes of iron, as well as hematin compounds, are inactive in inducing microsomal lipid peroxidation.

The antioxidant, diphenyl-*p*-phenylenediamine (DPPD), is a potent

[14] J. R. Gillette, B. B. Brodie, and B. N. LaDu, *J. Pharmacol. Exptl. Therap.* **119**, 532 (1957).

inhibitor of microsomal lipid peroxidation. In a concentration of 5×10^{-8} *M* it completely inhibits MA formation and suppresses the rates of O_2 uptake and NADPH disappearance to the levels occurring in the absence of iron complex. A similar, though less efficient, inhibition is obtained with the vitamin E metabolite,[15] 2-(3-hydroxy-3-methylcarboxypentyl)-3,5,6-trimethylbenzoquinone (complete inhibition of MA formation at 5×10^{-6} *M*), whereas α-tocopherol itself is not inhibitory up to a concentration of 10^{-4} *M*. A strong inhibitor of lipid peroxidation also is vitamin A,[16] which abolishes MA formation completely at a concentration of 10^{-6} *M*. The inhibitory concentration is increased with microsomes from vitamin A deficient rats, which in addition show a higher rate of lipid peroxidation than those from normal rats. With both DPPD and vitamin A the inhibition of lipid peroxidation may be transient when low concentrations of the compounds are used, probably because of a conversion of the compounds into a noninhibitory form during the incubation.

Drugs undergoing NADPH-linked hydroxylation in liver microsomes, such as codeine and aminopyrine, inhibit lipid peroxidation,[5] probably by competition for the NADPH oxidizing flavoprotein, NADPH-cytochrome *c* reductase. The inhibitor of drug hydroxylation, SKF-525A, suppresses lipid peroxidation as well (by about 80% at a concentration of 5×10^{-4} *M*). Carbon monoxide, which is a characteristic inhibitor of drug hydroxylation, does not affect lipid peroxidation and even relieves the inhibition of the latter by drugs. *p*-Hydroxymercuribenzoate, 5×10^{-5} *M*, completely inhibits NADPH-linked lipid peroxidation and drug hydroxylation.

There are some similarities between the NADPH-linked, enzymic, lipid peroxidation and the nonenzymatic peroxidation of lipids known[17-19] to occur upon incubation of subcellular particles with ascorbate. Thus the ascorbate-linked lipid peroxidation is activated by the iron complexes of nucleoside di- or triphosphates and of oxalate, and inhibited by DPPD and other antioxidants. In contrast, however, the ascorbate-linked reaction is also activated by Fe^{++} (but not Fe^{3+}) alone, as well as by iron malonate. Furthermore, inorganic pyrophosphate, which in conjunction

[15] E. J. Simon, A. Eisengart, L. Sundheim, and A. T. Milhorat, *J. Biol. Chem.* **221,** 807 (1956).

[16] S. Orrenius, K. Nordenbrand, and L. Ernster, in preparation.

[17] A. Ottolenghi, *Arch. Biochem. Biophys.* **79,** 355 (1959).

[18] E. H. Thiele and J. W. Huff, *Arch. Biochem. Biophys.* **88,** 203 (1960).

[19] F. E. Hunter, A. Scott, P. E. Hoffsten, F. Guerra, J. Weinstein, A. Schneider, B. Schutz, J. Fink, L. Ford, and E. Smith, *J. Biol. Chem.* **239,** 604 (1964). See also R. C. McKnight, F. E. Hunter, and W. H. Oehlert, *J. Biol. Chem.* **240,** 3439 (1965).

with iron is a potent activator of the NADPH-linked lipid peroxidation, fails to activate the ascorbate-linked reaction and even inhibits the latter as activated by other iron complexes. *p*-Hydroxymercuribenzoate, which blocks NADPH-linked lipid peroxidation, has no effect on the ascorbate-linked, nonenzymatic process. Finally, whereas the NADPH-linked lipid peroxidation is an exclusively microsomal process, the ascorbate-linked reaction occurs also in mitochondria (and presumably with other subcellular fractions as well).

Both the NADPH-linked and the ascorbate-linked lipid peroxidations cause a striking lysis of the microsomes, as revealed by a decrease of light-scattering of the microsomal suspension and a release of microsomal protein. The lysis is accompanied by characteristic changes in the activity of various microsomal enzymes, similar to those following the exposure of microsomes to treatment with detergents, e.g., deoxycholate.[8,20] Among these changes are a decrease of the glucose 6-phosphatase and an increase of the nucleoside diphosphatase activities.

Mechanism

The reactions involved in microsomal lipid peroxidation may be described by the following equations:

$$\text{NADPH} + \text{H}^+ + \text{O}_2 \rightarrow \text{NADP}^+ + \text{H}_2\text{O}_2 \quad (1)$$

$$\text{Lipid} + \text{O}_2 \rightarrow \text{lipid peroxide} \quad (2)$$

$$\text{Lipid peroxide} \rightarrow \text{malonaldehyde} \quad (3)$$

Reaction 1 probably involves the flavoprotein, NADPH-cytochrome *c* reductase, as well as an exogenous iron complex, with the formation of an iron peroxide or hydroperoxide free radical as a possible intermediate (cf. Nilsson *et al.*[21]). Reaction 2 may proceed by a mechanism similar to that proposed for nonenzymatic peroxidation of polyunsaturated fatty acids, or for fatty acid peroxidation catalyzed by the enzyme lipoxidase; in a simplified form, this mechanism may be written as follows (for details, see references cited in footnotes 22–26):

[20] L. Ernster and L. C. Jones, *J. Cell Biol.* **15**, 563 (1962).
[21] R. Nilsson, S. Orrenius, and L. Ernster, *Biochem. Biophys. Res. Commun.* **17**, 303 (1964).
[22] S. Bergström and R. T. Holman, *Advan. Enzymol.* **8**, 425 (1948).
[23] A. M. Siddiqi and A. L. Tappel, *J. Am. Oil Chemists' Soc.* **34**, 529 (1957).
[24] H. S. Mason, *Advan. Enzymol.* **19**, 79 (1957).
[25] W. Franke and H. Freshe, *in* "Handbuch der Pflanzenphysiologie," Vol. III, p. 137. Springer, Berlin, 1957.
[26] M. Hayano, *in* "Oxygenases" (O. Hayaishi, ed.), p. 181. Academic Press, New York, 1962.

$$\begin{array}{ccccccc} & \mathrm{H} & \mathrm{H} & \mathrm{H} & \mathrm{H} & \mathrm{H} & \\ -\mathrm{CH_2}- & \mathrm{C}= & \mathrm{C}- & \mathrm{C}- & \mathrm{C}= & \mathrm{C}- & \mathrm{CH_2}- \\ & & & \mathrm{H} & & & \end{array}$$

$$\downarrow -\mathrm{H}^{\cdot} \qquad (2a)$$

$$\begin{array}{ccccccc} & \mathrm{H} & \mathrm{H} & \mathrm{H} & \mathrm{H} & \mathrm{H} & \\ -\mathrm{CH_2}- & \mathrm{C}- & \mathrm{C}= & \mathrm{C}- & \mathrm{C}= & \mathrm{C}- & \mathrm{CH_2}- \\ & \cdot & & & & & \end{array}$$

$$\downarrow +\mathrm{O_2} \qquad (2b)$$

$$\begin{array}{ccccccc} & \mathrm{H} & \mathrm{H} & \mathrm{H} & \mathrm{H} & \mathrm{H} & \\ -\mathrm{CH_2}- & \mathrm{C}- & \mathrm{C}= & \mathrm{C}- & \mathrm{C}= & \mathrm{C}- & \mathrm{CH_2}- \\ & \mathrm{O} & & & & & \\ & \mathrm{O} & & & & & \\ & \cdot & & & & & \end{array}$$

$$\downarrow +\mathrm{H}^{\cdot} \qquad (2c)$$

$$\begin{array}{ccccccc} & \mathrm{H} & \mathrm{H} & \mathrm{H} & \mathrm{H} & \mathrm{H} & \\ -\mathrm{CH_2}- & \mathrm{C}- & \mathrm{C}= & \mathrm{C}- & \mathrm{C}= & \mathrm{C}- & \mathrm{CH_2}- \\ & \mathrm{O} & & & & & \\ & \mathrm{O} & & & & & \\ & \mathrm{H} & & & & & \end{array}$$

In Reaction (2a), a hydrogen atom is extracted from an unsaturated fatty acid, with the formation of a fatty acid free radical; in reaction (2b), this free radical reacts with molecular oxygen, yielding a fatty acid peroxide free radical; and in reaction (2c), the fatty acid peroxide free radical reacts with a hydrogen atom to form the fatty acid peroxide. In the NADPH-linked lipid peroxidation, reaction (1) initiates reaction (2a), probably by way of the iron peroxide or hydroperoxide free radical formed in the course of reaction (1). Once the fatty acid peroxide free radical has been formed, it can initiate the peroxidation of an adjacent unsaturated fatty acid, and thus the process can continue as a chain reaction as long as adjacent unsaturated fatty acids are available. This explains the lack of stoichiometry between reactions (1) and (2). There is also no stoichiometric relationship between reactions (2) and (3), since only certain types of fatty acid peroxides can undergo cleavage into malonaldehyde.[27]

[27] L. K. Dahle, E. G. Hill, and R. T. Holman, *Arch. Biochem. Biophys.* **98,** 253 (1962).

Section VIII

Special Assays and Techniques

Previously Published Articles from Methods in Enzymology Related to Section VIII

Vol. IV [12]. Techniques for the Assay of Respiratory Enzymes. Britton Chance.
Vol. IV [19]. Electron Microscopy of Cellular Constituents. Fritiof S. Sjöstrand.

[93] Diffuse Reflectance Spectrophotometry

By GRAHAM PALMER

In biochemical studies it is frequently necessary to obtain spectroscopic information from material which is partially or completely opaque to visible light. For systems in which a finite amount of light is transmitted the elegant double-beam techniques developed by Chance[1] are convenient, although the resultant information is in the form of a difference spectrum. Similar but not so precise experiments can be performed in a manual spectrophotometer if provision is made for positioning the cuvettes close to the photodetector, thus minimizing the amount of light lost by scattering.

However, when the amount of light transmitted by a sample is vanishingly small it becomes more profitable to use reflection methods; such a situation arises, for example, with samples prepared for low-temperature electron paramagnetic resonance (EPR) spectroscopy.

There are at least two techniques which can be used to obtain reflectance spectra; they are the integrating sphere method and the opal-glass reflection method. We will only consider the former method in this chapter; the latter method has been discussed by Shibata.[2]

Characteristics of Reflected Light

When an opaque body is illuminated, light may be reflected from it in two ways: (1) *specularly,* or directly, from the surface of the material; this is characteristic of the quality of the reflecting surface and, if the illumination is polarized, the specular reflection is also polarized; (2) *diffusely;* this is a consequence of multiple internal reflections in the sample and is influenced by the absorbancy characteristics of the components of the sample. As a consequence of this scattering the diffusely reflected light changes its state of polarization. It is to be anticipated that the light reflected from most samples will have both specular- and diffuse-reflection components.

In reflectance spectroscopy one compares the intensity of light reflected from the sample (R) with that reflected from the reference (R_0). The per cent reflectance is defined as 100 R/R_0 and is analagous to per cent transmittance 100 I/I_0. Thus reflectance measurements are most conveniently made on transmittance scales of spectrophotometers.

[1] For a detailed description of these methods see Volume IV [12].

[2] K. Shibata, "Methods of Biochemical Analysis" (D. Glick, ed.), Vol. 9, p. 217. Wiley (Interscience), New York, 1962.

Quantitation

Instead of Beer's law so familiar from absorbtion spectroscopy, the basic equation of reflectance spectroscopy is due to Kubelka and Munk,[3,4] viz.

$$F(R_\infty) \equiv \frac{(1 - R_\infty)^2}{2R_\infty} = \frac{K}{S}$$

where $F(R_\infty)$ is called the remission function, R_∞ is the diffuse reflectance of an infinitely thick powder referred to a nonabsorbing standard (e.g., MgO), S is a scattering coefficient, and K is the absorption coefficient of the Bouger-Lambert law. K is related to the familiar molar absorbancy A_m by

$$K = 2.3 \times c \times A_m$$

where c is the concentration of the absorbing material. From the experimentally determined reflectance, K/S [or $F(R_\infty)$] can be obtained most readily from tables that present K/S as a function of R_∞ (cf. the accompanying table). A plot of $F(R_\infty)$ against K should yield a straight line and does so over a limited range the extent of which is rather critically dependent on the physical characteristics of the sample. There are, however, restrictions in the use of this relationship. These are that (1) the absorbing material should have the same refractive index as the medium, (2) K and S should be independent of thickness, (3) the absorbing particles should be randomly oriented, (4) the incident light must be perfectly diffuse. This ensures that the top layer of the material also received diffuse light, a condition of the mathematical analysis.

Provided S is independent of wavelength, $F(R_\infty)$ should vary as K, i.e., the reflectance spectrum should mimic the transmittance spectrum. However, as can be seen by reference to Fig. 1, the spectrum of cytochrome c is grossly distorted at shorter wavelengths, the Soret band being only 50% more intense than the α-band (after conversion to transmittance values). While this attenuation with decreasing wavelength may, in part, be due to increased scattering losses at the sample face there is another affect that will contribute to this distortion. The specular reflectivity (R_s) of a sample is related to its refractive index (μ) by Fresnel's Law. Thus if μ were to change, it is to be anticipated R_s will change in a like manner. From an analysis of the reflectivity of single crystals of barium oxide, Jahoda concluded that K and μ (and there-

[3] P. Kubelka and F. Munk, *Z. Techn. Physik.* **12,** 593 (1931).

[4] P. Kubelka, *J. Opt. Soc. Am.* **38,** 448 (1948).

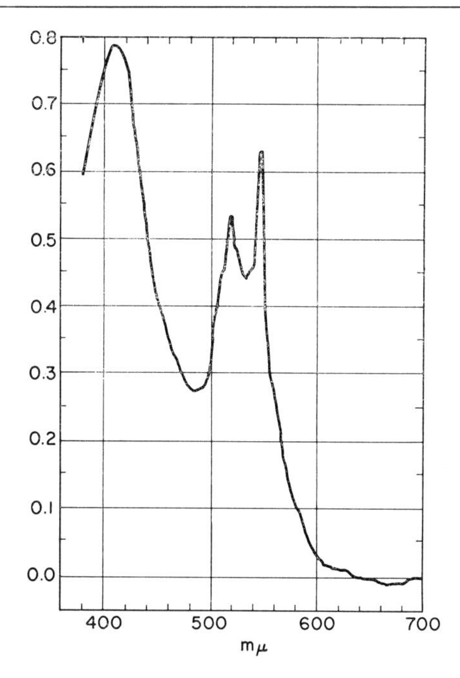

FIG. 1. Reflectance spectrum of partially reduced cytochrome *c* presented on an absorbancy scale.

fore R_s) changed simultaneously.[5] Similar results have been obtained with single crystals of potassium chloride.[6] Consequently we find that the specular reflection, which contributes most to the total reflected light at high absorbancies when diffuse reflection is low, increases in the region of high absorption and hence depresses the reflectance spectrum.

The error introduced by variation on the specular component can be eliminated rather simply.[7] If the illuminating light is polarized and observation is made with an analyzer crossed against the polarizer, only the diffuse reflectance will be observed (see above).

Integrating Sphere

Ideally, all the light which is reflected from the sample should fall on the photodetector, and the most convenient way of achieving this end is through the use of an integrating sphere. In essence this is a large

[5] F. C. Jahoda, *Phys. Rev.* **107,** 1261 (1957).

[6] P. L. Hartmann, J. R. Nelson, and J. G. Siegfried, *Phys. Rev.,* **105,** (1957) 183.

[7] A. S. Makas, *J. Opt. Soc. Am.* **52,** 43 (1962).

TABLE I

RATIO OF ABSORPTION COEFFICIENT TO SCATTERING COEFFICIENT (K/S) AS A FUNCTION OF REFLECTIVITY IN PERCENT ($100R_\infty$)[a]

$100R_\infty$	K/S	$100R_\infty$	K/S	$100R_\infty$	K/S	$100R_\infty$	K/S
0.0		5.0	9.02	10.0	4.050	15.0	2.408
.1	449.0	.1	8.83	.1	4.001	.1	2.387
.2	249.0	.2	8.64	.2	3.953	.2	2.365
.3	165.7	.3	8.46	.3	3.906	.3	2.344
.4	124.0	.4	8.29	.4	3.860	.4	2.324
.5	99.0	.5	8.12	.5	3.814	.5	2.303
.6	82.3	.6	7.957	.6	3.770	.6	2.283
.7	70.4	.7	7.800	.7	3.726	.7	2.263
.8	61.5	.8	7.650	.8	3.684	.8	2.244
.9	54.6	.9	7.504	.9	3.642	.9	2.224
1.0	49.0	6.0	7.363	11.0	3.600	16.0	2.205
.1	44.5	.1	7.227	.1	3.560	.1	2.186
.2	40.7	.2	7.096	.2	3.520	.2	2.167
.3	37.5	.3	6.968	.3	3.481	.3	2.149
.4	34.7	.4	6.844	.4	3.443	.4	2.131
.5	32.3	.5	6.725	.5	3.405	.5	2.113
.6	30.3	.6	6.609	.6	3.368	.6	2.095
.7	28.4	.7	6.496	.7	3.332	.7	2.078
.8	26.79	.8	6.387	.8	3.296	.8	2.060
.9	25.33	.9	6.281	.9	3.261	.9	2.043
2.0	24.01	7.0	6.178	12.0	3.227	17.0	2.026
.1	22.82	.1	6.078	.1	3.193	.1	2.009
.2	21.74	.2	5.980	.2	3.159	.2	1.9930
.3	20.75	.3	5.886	.3	3.127	.3	1.9767
.4	19.85	.4	5.794	.4	3.094	.4	1.9606
.5	19.01	.5	5.704	.5	3.062	.5	1.9446
.6	18.24	.6	5.617	.6	3.031	.6	1.9289
.7	17.53	.7	5.532	.7	3.001	.7	1.9134
.8	16.87	.8	5.449	.8	2.970	.8	1.8980
.9	16.26	.9	5.369	.9	2.940	.9	1.8828
3.0	15.68	8.0	5.290	13.0	2.911	18.0	1.8678
.1	15.14	.1	5.213	.1	2.882	.1	1.8529
.2	14.64	.2	5.139	.2	2.854	.2	1.8382
.3	14.17	.3	5.066	.3	2.826	.3	1.8237
.4	13.72	.4	4.994	.4	2.798	.4	1.8094
.5	13.30	.5	4.925	.5	2.771	.5	1.7952
.6	12.91	.6	4.857	.6	2.744	.6	1.7812
.7	12.53	.7	4.791	.7	2.718	.7	1.7673
.8	12.18	.8	4.726	.8	2.692	.8	1.7536
.9	11.84	.9	4.662	.9	2.667	.9	1.7400
4.0	11.52	9.0	4.601	14.0	2.641	19.0	1.7266
.1	11.22	.1	4.540	.1	2.617	.1	1.7133
.2	10.93	.2	4.481	.2	2.592	.2	1.7002
.3	10.65	.3	4.423	.3	2.568	.3	1.6872
.4	10.39	.4	4.366	.4	2.544	.4	1.6743
.5	10.13	.5	4.311	.5	2.521	.5	1.6616
.6	9.89	.6	4.256	.6	2.498	.6	1.6490
.7	9.66	.7	4.203	.7	2.475	.7	1.6366
.8	9.44	.8	4.151	.8	2.452	.8	1.6242
.9	9.23	.9	4.100	.9	2.430	.9	1.6121

[a] Reprinted with permission from D. B. Judd, "Color in Business, Science, and Industry," pp. 258–262. Wiley, New York, 1952.

TABLE I (*Continued*)

$100R_\infty$	K/S	$100R_\infty$	K/S	$100R_\infty$	K/S	$100R_\infty$	K/S
20.0	1.6000	25.0	1.1250	30.0	0.8167	35.0	0.6036
.1	1.5881	.1	1.1175	.1	.8116	.1	.6000
.2	1.5763	.2	1.1101	.2	.8066	.2	.5964
.3	1.5646	.3	1.1028	.3	.8017	.3	.5929
.4	1.5530	.4	1.0955	.4	.7967	.4	.5894
.5	1.5415	.5	1.0883	.5	.7918	.5	.5860
.6	1.5302	.6	1.0811	.6	.7870	.6	.5825
.7	1.5190	.7	1.0740	.7	.7822	.7	.5791
.8	1.5078	.8	1.0670	.8	.7774	.8	.5756
.9	1.4968	.9	1.0600	.9	.7726	.9	.5723
21.0	1.4860	26.0	1.0531	31.0	0.7679	36.0	0.5689
.1	1.4752	.1	1.0462	.1	.7632	.1	.5655
.2	1.4645	.2	1.0394	.2	.7586	.2	.5622
.3	1.4539	.3	1.0326	.3	.7539	.3	.5589
.4	1.4434	.4	1.0259	.4	.7494	.4	.5556
.5	1.4331	.5	1.0193	.5	.7448	.5	.5524
.6	1.4228	.6	1.0127	.6	.7403	.6	.5491
.7	1.4126	.7	1.0062	.7	.7358	.7	.5459
.8	1.4026	.8	.9997	.8	.7313	.8	.5427
.9	1.3926	.9	.9933	.9	.7269	.9	.5395
22.0	1.3827	27.0	0.9868	32.0	0.7225	37.0	0.5364
.1	1.3729	.1	.9805	.1	.7181	.1	.5332
.2	1.3632	.2	.9742	.2	.7138	.2	.5301
.3	1.3536	.3	.9680	.3	.7095	.3	.5270
.4	1.3441	.4	.9618	.4	.7052	.4	.5239
.5	1.3347	.5	.9557	.5	.7010	.5	.5208
.6	1.3254	.6	.9496	.6	.6967	.6	.5178
.7	1.3161	.7	.9436	.7	.6926	.7	.5148
.8	1.3070	.8	.9376	.8	.6884	.8	.5118
.9	1.2979	.9	.9316	.9	.6843	.9	.5088
23.0	1.2889	28.0	.9257	33.0	0.6802	38.0	0.5058
.1	1.2800	.1	.9199	.1	.6761	.1	.5028
.2	1.2712	.2	.9140	.2	.6720	.2	.4999
.3	1.2624	.3	.9083	.3	.6680	.3	.4970
.4	1.2538	.4	.9026	.4	.6640	.4	.4941
.5	1.2452	.5	.8969	.5	.6600	.5	.4912
.6	1.2366	.6	.8912	.6	.6561	.6	.4883
.7	1.2282	.7	.8857	.7	.6521	.7	.4855
.8	1.2198	.8	.8801	.8	.6483	.8	.4827
.9	1.2116	.9	.8746	.9	.6444	.9	.4798
24.0	1.2033	29.0	.8691	34.0	0.6406	39.0	0.4770
.1	1.1952	.1	.8637	.1	.6368	.1	.4743
.2	1.1871	.2	.8583	.2	.6330	.2	.4715
.3	1.1791	.3	.8530	.3	.6292	.3	.4688
.4	1.1712	.4	.8477	.4	.6255	.4	.4660
.5	1.1633	.5	.8424	.5	.6218	.5	.4633
.6	1.1555	.6	.8372	.6	.6181	.6	.4606
.7	1.1478	.7	.8320	.7	.6144	.7	.4580
.8	1.1401	.8	.8268	.8	.6108	.8	.4553
.9	1.1325	.9	.8217	.9	.6072	.9	.4526

(*Continued*)

TABLE I (*Continued*)

$100R_\infty$	K/S	$100R_\infty$	K/S	$100R_\infty$	K/S	$100R_\infty$	K/S
40.0	0.4500	45.0	0.3361	50.0	0.25000	55.0	0.18409
.1	.4474	.1	.3342	.1	.24850	.1	.18294
.2	.4448	.2	.3322	.2	.24702	.2	.18180
.3	.4422	.3	.3302	.3	.24554	.3	.18066
.4	.4396	.4	.3283	.4	.24406	.4	.17953
.5	.4371	.5	.3264	.5	.24260	.5	.17840
.6	.4345	.6	.3245	.6	.24114	.6	.17728
.7	.4320	.7	.3226	.7	.23969	.7	.17617
.8	.4295	.8	.3207	.8	.23825	.8	.17506
.9	.4270	.9	.3188	.9	.23682	.9	.17396
41.0	0.4245	46.0	0.3170	51.0	0.23539	56.0	0.17286
.1	.4220	.1	.3151	.1	.23397	.1	.17177
.2	.4196	.2	.3132	.2	.23256	.2	.17068
.3	.4172	.3	.3114	.3	.23116	.3	.16960
.4	.4147	.4	.3096	.4	.22976	.4	.16852
.5	.4123	.5	.3078	.5	.22837	.5	.16746
.6	.4099	.6	.3060	.6	.22699	.6	.16639
.7	.4075	.7	.3042	.7	.22562	.7	.16533
.8	.4052	.8	.3024	.8	.22425	.8	.16428
.9	.4028	.9	.3006	.9	.22289	.9	.16323
42.0	0.4005	47.0	0.2988	52.0	0.22154	57.0	0.16219
.1	.3982	.1	.2971	.1	.22019	.1	.16116
.2	.3958	.2	.2953	.2	.21885	.2	.16013
.3	.3935	.3	.2936	.3	.21752	.3	.15910
.4	.3912	.4	.2918	.4	.21620	.4	.15808
.5	.3890	.5	.2901	.5	.21488	.5	.15707
.6	.3867	.6	.2884	.6	.21357	.6	.15606
.7	.3845	.7	.2867	.7	.21227	.7	.15505
.8	.3822	.8	.2850	.8	.21097	.8	.15405
.9	.3800	.9	.2833	.9	.20968	.9	.15306
43.0	0.3778	48.0	0.2817	53.0	0.20840	58.0	0.15207
.1	.3756	.1	.2800	.1	.20712	.1	.15109
.2	.3734	.2	.2783	.2	.20585	.2	.15011
.3	.3712	.3	.2767	.3	.20459	.3	.14913
.4	.3691	.4	.2751	.4	.20333	.4	.14816
.5	.3669	.5	.2734	.5	.20208	.5	.14720
.6	.3648	.6	.2718	.6	.20084	.6	.14624
.7	.3627	.7	.2702	.7	.19960	.7	.14529
.8	.3606	.8	.2686	.8	.19837	.8	.14434
.9	.3584	.9	.2670	.9	.19714	.9	.14340
44.0	0.3564	49.0	0.26541	54.0	0.19593	59.0	0.14246
.1	.3543	.1	.26383	.1	.19471	.1	.14152
.2	.3522	.2	.26226	.2	.19351	.2	.14059
.3	.3502	.3	.26070	.3	.19231	.3	.13967
.4	.3481	.4	.25915	.4	.19112	.4	.13875
.5	.3461	.5	.25760	.5	.18993	.5	.13784
.6	.3441	.6	.25606	.6	.18875	.6	.13693
.7	.3421	.7	.25454	.7	.18758	.7	.13602
.8	.3401	.8	.25302	.8	.18641	.8	.13512
.9	.3381	.9	.25150	.9	.18525	.9	.13422

TABLE I (*Continued*)

$100R_\infty$	K/S	$100R_\infty$	K/S	$100R_\infty$	K/S	$100R_\infty$	K/S
60.0	0.13333	65.0	0.09423	70.0	0.06429	75.0	0.04167
.1	.13245	.1	.09355	.1	.06377	.1	.04128
.2	.13156	.2	.09287	.2	.06325	.2	.04089
.3	.13069	.3	.09220	.3	.06274	.3	.04051
.4	.12981	.4	.09153	.4	.06223	.4	.04013
.5	.12895	.5	.09086	.5	.06172	.5	.03975
.6	.12808	.6	.09020	.6	.06122	.6	.03938
.7	.12722	.7	.08954	.7	.06071	.7	.03900
.8	.12637	.8	.08888	.8	.06021	.8	.03863
.9	.12552	.9	.08823	.9	.05972	.9	.03826
61.0	0.12467	66.0	0.08758	71.0	0.05923	76.0	0.03789
.1	.12383	.1	.08693	.1	.05873	.1	.03753
.2	.12299	.2	.08629	.2	.05825	.2	.03717
.3	.12216	.3	.08565	.3	.05776	.3	.03681
.4	.12133	.4	.08501	.4	.05728	.4	.03645
.5	.12051	.5	.08438	.5	.05680	.5	.03609
.6	.11969	.6	.08375	.6	.05632	.6	.03574
.7	.11887	.7	.08313	.7	.05585	.7	.03539
.8	.11806	.8	.08250	.8	.05538	.8	.03504
.9	.11725	.9	.08188	.9	.05491	.9	.03470
62.0	0.11645	67.0	0.08127	72.0	0.05444	77.0	0.03435
.1	.11565	.1	.08066	.1	.05398	.1	.03401
.2	.11486	.2	.08005	.2	.05352	.2	.03367
.3	.11407	.3	.07944	.3	.05306	.3	.03333
.4	.11328	.4	.07884	.4	.05261	.4	.03299
.5	.11250	.5	.07824	.5	.05216	.5	.03266
.6	.11172	.6	.07764	.6	.05171	.6	.03233
.7	.11095	.7	.07705	.7	.05126	.7	.03200
.8	.11018	.8	.07646	.8	.05081	.8	.03167
.9	.10941	.9	.07588	.9	.05037	.9	.03135
63.0	0.10865	68.0	0.07529	73.0	0.04993	78.0	0.03103
.1	.10789	.1	.07471	.1	.04949	.1	.03070
.2	.10714	.2	.07414	.2	.04906	.2	.03039
.3	.10639	.3	.07356	.3	.04863	.3	.03007
.4	.10564	.4	.07299	.4	.04820	.4	.02976
.5	.10490	.5	.07243	.5	.04777	.5	.02944
.6	.10416	.6	.07186	.6	.04735	.6	.02913
.7	.10343	.7	.07130	.7	.04693	.7	.02882
.8	.10270	.8	.07074	.8	.04651	.8	.02852
.9	.10197	.9	.07019	.9	.04609	.9	.02821
64.0	0.10125	69.0	0.06964	74.0	0.04568	79.0	0.02791
.1	.10053	.1	.06909	.1	.04526	.1	.02761
.2	.09982	.2	.06854	.2	.04485	.2	.02731
.3	.09910	.3	.06800	.3	.04445	.3	.02702
.4	.09840	.4	.06746	.4	.04404	.4	.02672
.5	.09769	.5	.06692	.5	.04364	.5	.02643
.6	.09699	.6	.06639	.6	.04324	.6	.02614
.7	.09630	.7	.06586	.7	.04284	.7	.02585
.8	.09560	.8	.06533	.8	.04245	.8	.02557
.9	.09492	.9	.06481	.9	.04206	.9	.02528

(*Continued*)

TABLE I (*Continued*)

$100R_\infty$	K/S	$100R_\infty$	K/S	$100R_\infty$	K/S	$100R_\infty$	K/S
80.0	0.02500	85.0	0.01324	90.0	0.005556	95.0	0.001316
.1	.02472	.1	.01304	.1	.005439	.1	.001262
.2	.02444	.2	.01285	.2	.005324	.2	.001210
.3	.02417	.3	.01267	.3	.005210	.3	.001159
.4	.02389	.4	.01248	.4	.005097	.4	.001109
.5	.02362	.5	.01230	.5	.004986	.5	.001060
.6	.02335	.6	.01211	.6	.004876	.6	.001013
.7	.02308	.7	.01193	.7	.004768	.7	.000966
.8	.02281	.8	.01175	.8	.004661	.8	.000921
.9	.02255	.9	.01157	.9	.004555	.9	.000876
81.0	0.02228	86.0	0.01140	91.0	0.004451	96.0	0.000833
.1	.02202	.1	.01122	.1	.004347	.1	.000791
.2	.02176	.2	.01105	.2	.004246	.2	.000751
.3	.02151	.3	.01087	.3	.004145	.3	.000711
.4	.02125	.4	.01070	.4	.004046	.4	.000672
.5	.02100	.5	.01054	.5	.003948	.5	.000635
.6	.02075	.6	.01037	.6	.003852	.6	.000598
.7	.02050	.7	.01020	.7	.003756	.7	.000563
.8	.02025	.8	.01004	.8	.003662	.8	.000529
.9	.02000	.9	.00987	.9	.003570	.9	.000496
82.0	0.01976	87.0	0.009713	92.0	0.003478	97.0	0.000464
.1	.01951	.1	.009553	.1	.003388	.1	.000433
.2	.01927	.2	.009394	.2	.003299	.2	.000403
.3	.01903	.3	.009238	.3	.003212	.3	.000375
.4	.01880	.4	.009082	.4	.003126	.4	.000347
.5	.01856	.5	.008929	.5	.003041	.5	.000321
.6	.01833	.6	.008776	.6	.002957	.6	.000295
.7	.01809	.7	.008625	.7	.002874	.7	.000271
.8	.01786	.8	.008476	.8	.002793	.8	.000247
.9	.01764	.9	.008328	.9	.002713	.9	.000225
83.0	0.01741	88.0	0.008182	93.0	0.002634	98.0	0.000204
.1	.01718	.1	.008037	.1	.002557	.1	.000184
.2	.01696	.2	.007893	.2	.002481	.2	.000165
.3	.01674	.3	.007751	.3	.002406	.3	.000147
.4	.01652	.4	.007611	.4	.002332	.4	.000130
.5	.01630	.5	.007472	.5	.002259	.5	.000114
.6	.01609	.6	.007334	.6	.002188	.6	.0000994
.7	.01587	.7	.007198	.7	.002118	.7	.0000856
.8	.01566	.8	.007063	.8	.002049	.8	.0000729
.9	.01545	.9	.006930	.9	.001981	.9	.0000612
84.0	0.01524	89.0	0.006798	94.0	0.001915	99.0	0.0000505
.1	.01503	.1	.006667	.1	.001850	.1	.0000409
.2	.01482	.2	.006538	.2	.001786	.2	.0000323
.3	.01462	.3	.006410	.3	.001723	.3	.0000247
.4	.01442	.4	.006284	.4	.001661	.4	.0000181
.5	.01422	.5	.006159	.5	.001601	.5	.0000126
.6	.01402	.6	.006036	.6	.001541	.6	.0000080
.7	.01382	.7	.005914	.7	.001483	.7	.0000045
.8	.01362	.8	.005793	.8	.001426	.8	.0000020
.9	.01343	.9	.005674	.9	.001370	.9	.0000005
						100.0	0.0000000

sphere with sample and reference ports and apertures for sample and reference beams and photomultiplier. The sphere is coated on the inside by a shell of magnesium oxide which is very highly reflecting (coefficient of reflectance greater than 0.98 throughout the visible spectrum).[8] Thus any light reflected from the sample or reference undergoes multiple internal reflections and eventually finds its way to the photomultiplier. The system is shown in essence in Fig. 2. In this arrangement the sample and reference are illuminated alternately with chopped monochromatic light and the photomultiplier, and allied circuitry detects and compares the signals from the sample and reference beam. With the exception of a few accessories constructed for a specific application[9] this system is commercially available.[10]

The theory of the integrating sphere has been considered *in extenso*.[11] Jacquez and Kuppenheim showed that to minimize errors the sphere should be as large as possible with small apertures with the exception of that for the photodetector, which should be as large as possible (ideally the photomultiplier should extend over the whole wall of the sphere). However, for high efficiency a small sphere is demanded, and most commercial designs compromise between the two opposing requirements.

Jacquez and Kuppenheim further showed that the comparison method as outlined above is substantially better than the substitution method, in which measurements are made by presenting the sample and reference in turn at a single observation port. In the comparison method the sphere efficiency is identical for sample and reference; however, in

[8] The spheres are generally coated with magnesium oxide by igniting magnesium ribbon and directing the fumes of the oxide onto the walls of the sphere. The shell formed in this way is very fragile and should be inspected regularly for flaws. In the event that a permanent sphere coating is convenient, the following method can be used [W. E. T. Middleton and C. C. Sanders, *Illum. Engr.* **48** (5), May 1953].

Remove all the old magnesium oxide coating and ensure that the surface to be painted is smooth, clean, dry, and lint free. Treat this surface with Alumiprep (Neilson Chemical Co., 2300 Gainsboro, Detroit, Michigan) and cover the rest of the sphere with masking tape. Prepare the paint, which is a 44% (w/v) suspension of barium sulfate in water. Apply the paint with a spray gun using smooth even strokes from a distance of 2 feet. The gun is driven from a cylinder of high quality nitrogen. Apply about 10 coats allowing each coat to dry before applying the next: take care not to apply too much paint at a time. The finished coating is smooth and white. Application of too many coats results in a granular texture; the excess may be wiped off with lint-free tissues.

[9] G. Palmer and H. Beinert, *Anal. Biochem.* **8**, 95 (1964).

[10] Cary Model 14 with model 1411 diffuse reflectance accessory: Applied Physics, Palo Alto, California.

[11] J. A. Jacquez and H. F. Kuppenheim, *J. Opt. Soc. Am.* **45**, 460 (1955).

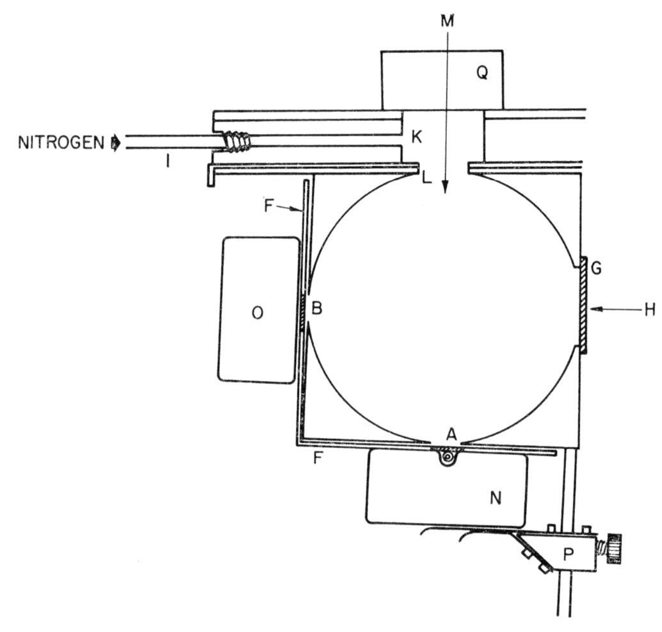

FIG. 2. Schematic drawing showing a cross section of the integrating sphere and accessories used for mounting EPR sample tubes. The drawing shows: the integrating sphere in top view, seen from photomultiplier position with port *L* for entrance of sample beam *M*; port *G* with quartz window for entrance of reference beam *H*; sample port *A*, reference port *B*; metal plate *F* containing quartz windows for *A* and *B*; filter holder *I* with drilled-in hole for introducing dry gaseous nitrogen through port masking cavity *K*; lens *Q*, which closes the system toward the body of the spectrophotometer; magnesium carbonate block *O* covering the reference port; and clamp *P* which holds block *N*.

the comparison method, the act of replacing one segment of the sphere wall (the sample) with a different material (the reference) changes the sphere efficiency.

Flat cells which can be made to sit recessed into the sphere walls would be the ideal sample walls. In this way light losses by internal scattering along the walls of the sample container can be avoided.

Further, although most work in the visible is not energy limited, good resolution of sharp bands (cf. Fig. 1) necessitates the use of the high-intensity light sources, e.g., quartz-iodide or projection lamps which are offered by spectrophotometer manufacturers.

An Example of the Adaptation of Diffuse Reflectance Spectroscopy to a Specific Need

The EPR spectroscopy of enzymes is frequently concerned with those systems which also show characteristic changes in their optical properties during catalysis. For reasons detailed elsewhere,[12] EPR spectroscopy is most commonly performed in cylindrical tubes at liquid nitrogen temperature; i.e., the samples are frozen. To be most meaningful the optical and EPR spectroscopy must be performed on samples in the same state for certain properties of paramagnetic systems are dramatically temperature dependent.[12] Thus we have resorted to low-temperature diffuse reflectance spectroscopy as a means of obtaining this information.[9,13]

This consisted of modifying the Cary 1411 diffuse reflectance attachment to allow mounting the EPR tubes and of providing a suitable means of refrigerating the sample *in situ*. Because this refrigeration was

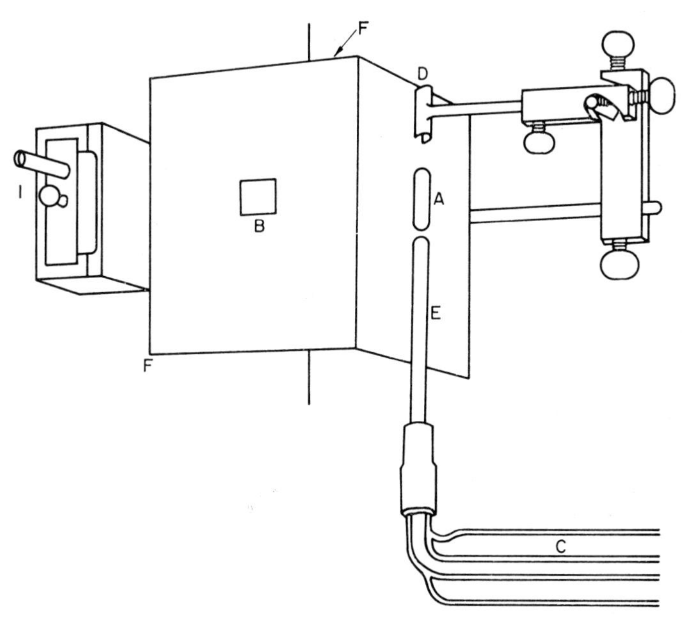

FIG. 3. Schematic drawing of accessory to Cary model 14 M recording spectrophotometer modified for measuring diffuse reflectance of frozen samples. The drawing shows the right front end of the spectrophotometer where the integrating sphere is mounted behind the metal plate *F*. The drawing shows: sample part *A* and reference port *B* covered by quartz windows of corresponding shape which are mounted in the metal plate *F*; filter holder *I* with hole for introducing dry gaseous nitrogen; collar *D* for positioning sample tube, which is not shown; end of Dewar system *C*, which conducts cold nitrogen gas (−100 to −110°) to copper tube *E*. The sample tube rests in the upper end of *E*.

[12] G. Palmer, this volume [94].

[13] G. Palmer and H. Beinert, *in* "Rapid Mixing and Sampling Techniques in Biochemistry" (B. Chance, R. H. Eisenhardt, Q. H. Gibson, and K. K. Lonberg-Holm, eds.), p. 205. Academic Press, New York, 1965.

achieved by flowing cold gas up and over the sample tube, it was necessary to seal the integrating sphere with quartz windows to prevent the cold gas from entering the sphere (Fig. 3). Furthermore the sphere was purged with dry gaseous nitrogen which was introduced via the filter holder and presumably escaped through various gaps in the mechanical construction. The sample tube was secured at the sample port by a metal collar (D) and rested on a copper tube (E), the tip of which was crimped for this purpose. The gaseous refrigerant—generated by boiling liquid nitrogen–was conducted to the sample via this copper tube. To minimize scattering losses the sample tube was backed by a block of magnesium carbonate in which was cut a groove just large enough to contain the 4 mm i.d. tube. (This groove is most satisfactorily made using a clean wooden applicator stick mounted in a hand drill which is laid carefully on the surface of the magnesium carbonate block.) The surface of this block and that at the reference port must be kept extremely clean: this is best achieved by routinely exposing a new layer of magnesium carbonate using the surface of a third block as a grinding agent. Further details of this arrangement are given in Figs. 2 and 3 and in the original paper.[9] With this system a linear relationship between concentration and reflectance has never been observed, presumably as a result of the affects described above, and to difficulty in reproducibly freezing samples. However, with the use of suitable calibration curves made from standards prepared under the appropriate conditions, results have been obtained which are surprisingly good.[8,13] With the recent advances in our understanding of the reflectance phenomena, it is to be hoped that more precise measurements can be made in the future, for reflectance spectrophotometry is potentially a very useful technique for the biologist in his efforts to comprehend the structure and function of macromolecular systems.

[94] Electron Paramagnetic Resonance

By Graham Palmer

The phenomenon and theory of electron paramagnetic resonance (EPR) has been amply described elsewhere, the article by Jardetsky and Jardetsky[1] being particularly useful for those with a limited training in the physical sciences. A review of the application of EPR to enzyme

[1] O. Jardetsky and C. D. Jardetsky, *in* "Methods of Biochemical Analysis" (D. Glick, ed.), Vol. 9, p. 235. Wiley (Interscience), New York, 1962.

systems is available.[2] Thus, this chapter starts from the premise that the interested investigator is familiar with the rudiments of the resonance phenomenon, and it will be devoted entirely to the methodology of EPR with special emphasis on the peculiar requirements of the biological scientist.

Although EPR is considerably more sensitive than bulk susceptibility measurements, the biochemist is still faced with the problem of working with materials that exhibit feeble resonance signals, for even with pure enzymes it is rare that one has enough material to prepare a working solution of 10^{-3} M in the potentially paramagnetic species. Even if this is possible, one is faced with the possibility that only a fraction of the material will exhibit a signal. Furthermore, at these concentrations, the working material is often quite viscous and offers severe problems of addition and mixing of reagents, achievement of anaerobic conditions, etc. Thus, the biochemist frequently resorts to measurements in the frozen (amorphous) state as a means of enhancing sensitivity. This is achieved in two obvious ways.

1. Measurements in aqueous solution are hampered by the severe dielectric losses of water—the solvent *par excellence* for biological materials. Ice, on the other hand, does not have this drawback, and by working with frozen solutions it is possible to use much larger effective sample sizes (about 120 μl with a 4 mm i.d. capillary compared with 20 μl with the 0.2 mm i.d. aqueous flat cell).

2. EPR spectra—usually those of transition metals—are often sharpened by lowering the temperature; this is due to a decrease in efficiency of certain relaxation processes. This does not mean that the total absorption is greater, only that it occurs over a smaller range of field, with the consequence that the amplitude is increased, the half-width decreases, and the spectrum is intensified.

Unfortunately, at the high sensitivity necessary in many biological experiments, the experimenter is bothered by the presence of signals due to the contamination of the equipment with extraneous resonances. The most common source of these impurities is the quartz used in the construction of the accessories and sample tubes: fortunately, there are at least two companies who supply quartz free of significant amounts of troublesome trace metals.[3] However, having ensured that the quartz is free of contamination, the investigator must apply himself to ensure

[2] H. Beinert and G. Palmer, *Advan. Enzymol.* **27**, 105.

[3] Engelhardt Industries, Inc., 685 Ramsey Avenue, Hillside, New Jersey; and Thermal American Fused Quartz, Route 202 and Changebridge Road, Montville, New Jersey.

that both the sample tubes and dewar are kept scrupulously clean. Clearly, for this, prevention is better than a cure and attention should be paid at all times to the manipulating and storage of the equipment with this in mind. In the event that contamination does occur, the offending item may be immersed in hot concentrated acid for several hours, thoroughly rinsed with water, and dried. It will be appreciated that introducing acid into a large number of narrow bore tubes is both tedious and messy—another reason for encouraging clean habits. In the unhappy event that the cavity itself becomes contaminated, one's only hope is to dismantle it and try washing with mild detergents and careful swabbing with solvents, or to gently polish the walls and side plates with a mild abrasive using one's finger (toothpaste followed by flour is an efficient, if somewhat exotic, combination). However, the dismantling and reassembly of a cavity should not be undertaken lightly, and there is good reason to return the unit to the manufacturer for his attention. (In particular, the reassembly of a cavity appears to be a black art, the amount of torque applied on the various retaining screws affecting significantly the performance of the reassembled cavity, due in large part to shaking of the modulation coils when insecure.)

EPR Measurements at Low Temperatures

The easiest way to perform low temperature measurements is to insert a dewar containing liquid nitrogen into the cavity. Unfortunately, at the high instrumental settings routinely employed, the boiling of the liquid nitrogen generates a considerable amount of noise and the automatic frequency control is often overpowered. The variable-temperature devices, which work by cooling the sample with a stream of prechilled gaseous nitrogen are much more satisfactory in this regard and are highly recommended. Their main disadvantages, namely, the inability to attain quite the same temperature as liquid nitrogen and the difficulty in maintaining a prescribed temperature (this latter requiring constant attention from the operator), have recently been obviated by the development of generators of cold nitrogen which rely on boiling of liquid nitrogen by an electric heater. The cold nitrogen gas so produced is further chilled by passage through a standard heat exchanger immersed in liquid nitrogen and then is led to the sample situated in the cavity. The temperature is controlled by adjusting the rate of boiling of the liquid nitrogen, and the whole system can be made completely automatic by using suitable temperature or pressure-sensing devices as active control elements. A recent version of this system can reach temperatures of —194° and maintain temperatures of —180° to within a degree for several hours.[3a]

[3a] R. E. Hansen and H. Beinert, personal communication, 1965.

With this variable-temperature cooling system, much shorter sample tubes are permissible. To maximize signal size, a suitable tube would have 4 mm i.d. and 0.5 mm wall thickness; a convenient length is 12 cm. Note that with this large internal diameter thin-wall tubing is essential because of the limited space available in the variable temperature cavity dewar. In fact, the commercially available dewars do not allow maximum use of the space available in the cavity, and it is necessary to fabricate a larger version with 10 mm i.d., 0.5 mm wall thickness and 6 mm i.d., 0.5 mm wall thickness tubing for the outer and inner walls of the dewar, respectively.

The introduction of one's sample into these tubes is facilitated by the use of plastic tubing and syringes; for this purpose, Intramedic polyethylene tubing which comes in a variety of sizes, is highly recommended. Although suitable gauge hypodermic needles are adequate for adapting the hose to the syringe, they have a tendency to prick holes in the tubing, and Luer Stub Adapters[4] are more convenient for this purpose.

The Intramedic tubing has various nominal bores and can easily be calibrated in terms of volumes. The larger-bore tubing, e.g., PE 160, can be used for loading the enzyme into the sample tube; the narrow-bore tubing is suitable for making the small additions of substrate, inhibitor, etc. When additions are made to the sample *in situ,* mixing can be achieved using a suitable length of nichrome wire with a small loop at the tip. This can be rotated between the fingers and simultaneously moved up and down to accomplish mixing in a few seconds, or, alternatively, the wire can be mounted in the chuck of a top drive stirrer. The samples can be frozen by immersing them slowly into liquid nitrogen taking 10–15 seconds to accomplish this. Thawing is best achieved by plunging the sample tube into a beaker of water. If the solutions contain a high salt content, particularly ammonium sulfate, frequent breakages will be encountered; the remedy is obvious. Faster freezing of samples can be obtained by plunging the EPR tube into isopentane chilled to −150°. Under these conditions, the freezing process appears to be complete in about a second.

Because many of the enzymes that it is of interest to study by EPR are oxidation-reduction systems, it is natural that accessories have been developed for performing anaerobic experiments; a convenient design for an anaerobic EPR tube is shown in Fig. 1. The enzyme in one side arm and any additional reagent in the second side arm are readily out-gassed and flushed with pure nitrogen, the tube finally being left under vacuum. It is then easy to transfer the contents of either side arm to the capillary

[4] Clay-Adams, Inc., 141 East 25th Street, New York, New York.

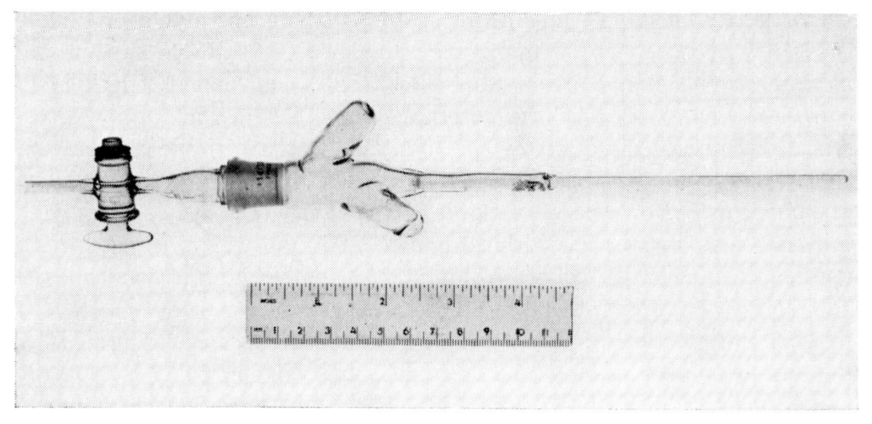

FIG. 1. Photograph of a sample cell suitable for anaerobic EPR work at low temperature.

or to mix these materials prior to examination. An alternative design has been described by Beinert and Sands.[5]

Preparation of Samples for Solution Work

Because of the lossy nature of aqueous media several devices are used to minimize the coupling of the electric field of the microwave power to the solvent. The most popular of these is the so-called flat cell by means of which the sample is constrained as a thin (approximately 0.2 mm) layer and positioned in the cavity in the region of maximum H-field and minimum E-field. This positioning is quite critical and can prove to be rather tricky, and the amount of application involved appears to vary from instrument to instrument, some operators finding that very little readjustment is needed in between samples (in the same cell), whereas others find that some effort is required.

Inasmuch as various flat cells appear to differ in thickness (internal diameters as large as 0.4 mm have been found) quite large readjustments are usually necessary if different flat cells are used in the same experiment. However, it is technically, if not economically feasible to purchase a number of these flat cells and select matched sets. Filling the flat cell presents no problems; however, manipulating the solution *in situ* is very difficult unless the contents are kept under vacuum (see below).

The standard taper joints fitted to the tubular portion of this device permit the attachment of numerous accessories. One which has proved

[5] H. Beinert and R. H. Sands, *in* "Free Radicals in Biological Systems" (M. S. Blois, Jr., H. W. Brown, R. M. Lemmon, R. O. Lindblom, and M. Weissbluth, eds.), p. 17. Academic Press, New York, 1961.

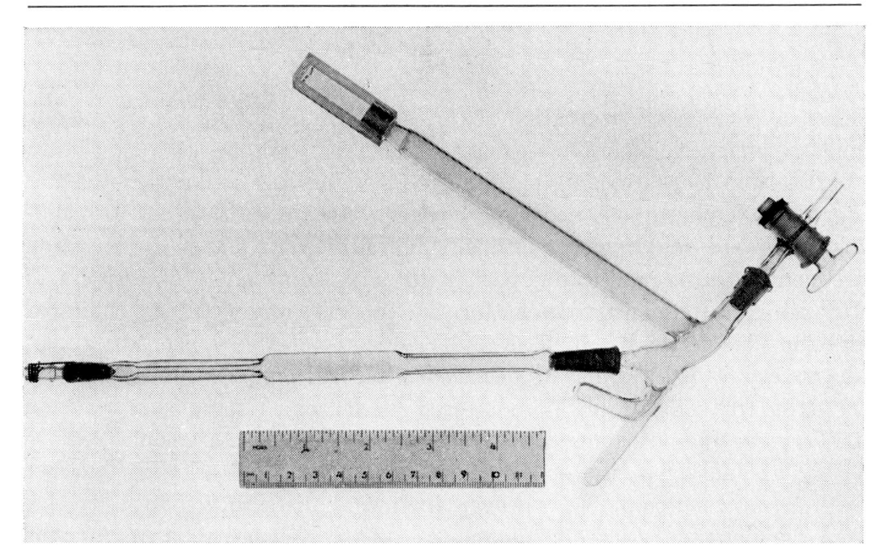

FIG. 2. Photograph of a sample cell suitable for combined anaerobic EPR and optical spectroscopy (courtesy of Dr. V. Massey).

to be exceptionally useful is for anaerobic experiments; in addition, it has a spectrophotometer cuvette attached, thus permitting optical and EPR spectra on the same sample. This device is shown in Fig. 2. As the cell contents are kept *in vacuo* it is possible to transfer material to and from the flat portion simply by warming the lower portion of the apparatus with one's hand. Any air bubbles trapped in the sensitive part of the flat cell can be removed by cautiously waving the apparatus and thus centrifuging the bubbles to the surface.

One unexpected bonus is the facility to use the flat cell directly as a spectrophotometer cuvette. This demands that one is studying an intensely colored sample, but because of the concentration requirements of the EPR spectrophotometer, often this condition is not difficult to meet. It is an easy task to design an adapter for inserting the cell into a spectrophotometer.

Obtaining Experimental Parameters

g values

The g value is obtained from the relationship

$$h\nu = g\beta H$$

where h = Planck's constant, β = the Bohr magneton, and ν and H are the frequency and magnetic field at resonance. Rearranging,

$$g = \frac{h}{\beta} \cdot \frac{\nu}{H} = 714 \cdot 443 \frac{\nu}{H}$$

Thus, a g-value determination requires determining both the frequency and the magnetic field.

Precision measurements require rather elaborate auxiliary equipment. For most biological applications, however, the resonance lines are sufficiently broad that high precision is not necessary. Thus, the klystron frequency can be read to one part in 10^4 with an accuracy of one part in 10^3 using a frequency meter[6] for the modest outlay of two hundred dollars. This device is a tunable cavity which is inserted in the waveguide; the procedure is to adjust its dimensions until it is in resonance with the klystron frequency. This is monitored by searching for a sharp dip on the crystal bias (leakage) meter. More precise measurements require either a transfer oscillator[7] or a frequency multiplier chain,[8] together with a frequency counter. However, the most satisfactory solution to this problem is the recent development of an X-band plug-in for the Hewlett-Packard Model 5245L electronic frequency counter, which will allow a direct monitoring of the klystron frequency during operation with accuracies much higher than needed. Inasmuch as the counter is necessary for monitoring the magnetic field (see below), this is clearly an optimal solution.

Although magnetic fields can be read directly from field-regulating devices such as the Fieldial,[9] the accuracy of such measurements is not very high, approximately 0.3%. Thus, most measurements of magnetic field are made with an NMR or proton probe,[10] a marginal oscillator which is tuned to give proton resonances at the magnetic field in question and the oscillator frequency measured with an electronic counter. From the relationship that the proton precision frequency is 4.5286 kc/gauss, the magnetic field is easily calculated.

While it is possible to measure the field and klystron frequency with a large degree of accuracy, there is a distinct limitation in that the position of the g value is not always known exactly, and one often resorts to measuring the positions of arbitrary points, e.g., the point of maximum slope (abbreviated as S_m). Furthermore, with broad resonances there is

[6] Hewlett-Packard Model X532B, Hewlett-Packard, 1501 Page Mill Road, Palo Alto, California.

[7] Hewlett-Packard Model 540B.

[8] MicroNow Instrument Co., 6124 N. Pulaski, Chicago, Illinois.

[9] Varian Associates.

[10] D. J. E. Ingram, "Free Radicals As Studied by Electron Spin Resonance." Butterworth, London, 1958.

a certain subjectivity in deciding when the magnetic field is at the g value; second derivative presentation is very useful in reducing this uncertainty, although this is limited by the reduced signal-to-noise level produced by the additional audiomodulation of the magnetic field. Thus, with a frequency counter and Fieldial, g values can be determined to a few parts in a thousand. For an additional five thousand dollars they can be determined to a part in 10^5.

Idealized spectra for $S = ½$ in spherical, axial, and rhombic spectra are shown in Fig. 3, together with the position of the g values: it is

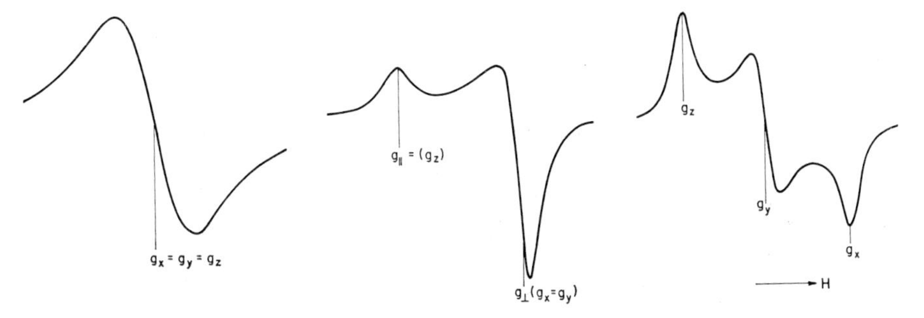

FIG. 3. Idealized spectra for a single unpaired electron in the frozen state in (left) spherical symmetry, (middle) axial symmetry, (right) rhombic symmetry.

interesting to note that for axial symmetry, g lies between the points of maximum absorption and maximum slope and can be identified only by computer simulation of the spectrum.

Linewidths

Another parameter of interest is the linewidth, which for radicals means the peak-to-peak distance in gauss. Alternatively, one can use different units, remembering that one gauss equals 2.8 Mc and 30,000 Mc equals 1 cm^{-1}. Thus, a 20 gauss separation can also be reported as 56 Mc or 0.0019 cm^{-1}. If less than 100 gauss, these are most easily determined using the incremental field potentiometer of the Fieldial with an accuracy of better than 100 milligauss. Higher accuracies can be obtained by accurate g-value measurements of the appropriate points of interest. Note that when the position of the line, *viz* the g value is required, then overmodulation is desirable in that it produces the largest signal without affecting the position of the crossover point. However, the lineshape is altered by too high a modulation, and one can use the rough rule that the modulation amplitude should not be greater than one-tenth the linewidth for really accurate lineshapes, but that it can be as high as one-third the linewidth for a good approximation to the real spectrum. It is important to realize that the modulation amplitude control on most spectrometers has only an arbitrary calibration. To convert these numbers into field, it is necessary to monitor the voltage generated in a single loop of wire situated in the cavity in the maximum H-field, i.e., move the loop to obtain the maximum voltage. From the relation

$$B = \frac{10^4 E}{2\pi A \omega}$$

(where B is the modulation field in peak-to-peak gauss when E is the peak-to-peak voltage detected, ω = modulation frequency, and A is the area of the wire loop in meters^2), one obtains the amplitude of the field modulation corresponding to the particular setting of the modulation amplitude control.

To determine whether or not a resonance line is Lorentzian or Gaussian in shape, measure the width of the spectrum at the position where the height is 10% of maximum. If the lineshape is Gaussian, the width will be 1.4 times the linewidth; if the spectrum is Lorentzian the width will be 2.5 times the linewidth. Intermediate results are common.

For spectra exhibiting an anisotropy in g, it is not possible to measure the natural linewidth directly, although the values can be obtained empirically by curve-fitting processes using digital computers. Inasmuch as the spectra frequently broaden with increasing temperature, it is clearly important to specify the temperature when reporting the linewidth. It must be borne in mind that linewidths observed in the frozen microcrystalline state are frequently broadened as a result of anisotropic dipole-dipole interactions, and the apparent linewidth so observed will be an upper limit on the true linewidth.

Power

Although for routine work it is sound practice to keep the microwave power incident on the cavity below the level at which saturation occurs, it is frequently of interest to study the signal intensity as a function of microwave power, i.e., saturation studies.

To do this successfully requires changing the power over a wide range from a few tenths of a milliwatt to a few hundred milliwatts; most commercial bridges are unsatisfactory for this purpose, being designed for either high or low power operation, but not both. Furthermore, under these conditions the magic-T becomes inefficient as only one-half of the power is incident on the cavity: a circulator[11] is to be preferred. A sketch

[11] Microwave Development Laboratory, Model 90 CR 16-9E X-band Circulator, 87 Crescent Road, Needham Heights, Massachusetts.

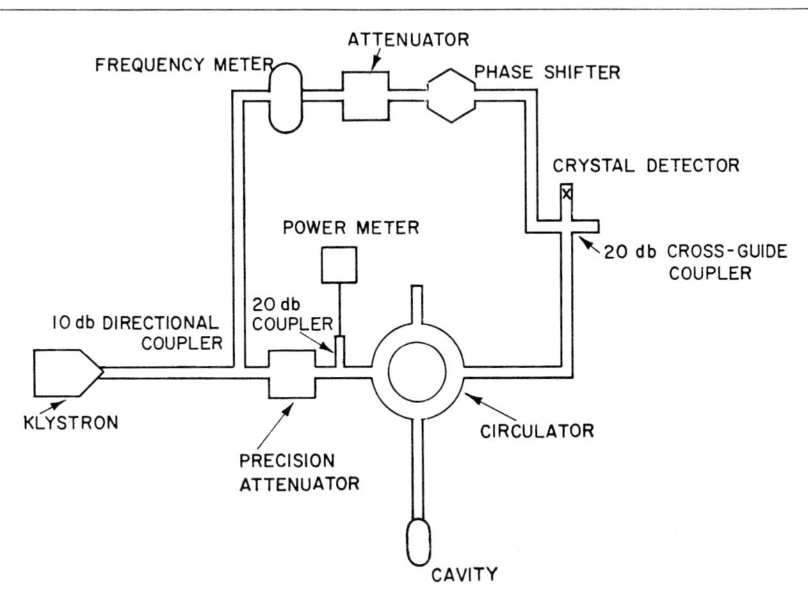

FIG. 4. Block diagram of a microwave bridge suitable for operation over a wide-range of microwave power incident on the cavity.

of a desirable arrangement is shown in Fig. 4. Crystal bias is provided via the reference arm and hence very low powers are obtainable with a standard high power X-band klystron (e.g., Varian 153C).

Quantitation

Because the area delimited by an absorption curve is directly proportional to the number of unpaired electrons, it would appear that quantitative estimates of the paramagnetic species would be a relatively simple problem. Unfortunately, such is not the case. Typically, these estimations involve comparison of the intensity of a standard with that of an unknown—the standard having been prepared by conventional analytical means (e.g., weighing, spectroscopic standardization). Ideally, the standard would have the same g value, line shape relaxation time, and spin state as the unknown—but this is never achieved and many compromises have to be made. Furthermore, the sample and unknown should be run in the same solvent with the same sample geometry, temperature, field modulation, and at levels of microwave power such that saturation of the resonances do not occur. In solution work the difficulties of reproducing the exact orientation of the flat cell is an additional limiting factor.

In this latter respect low temperature work is more favorable and

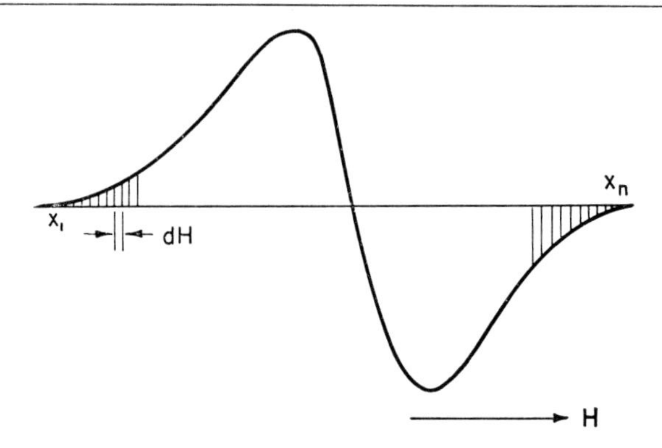

FIG. 5. Double integration.

signal intensities of duplicate standards can be reproduced to $\pm 1\%$ providing due allowance is made for any possible variation in the diameter of the sample tubes. Furthermore, the graphical processes of estimation—whether double integration or first moments—appear to be quite reproducible to within a few per cent. Similarly, mechanization of the process by either analog[12] or digital[13] computer gives good reproducibility.

Double Integration (*Fig. 5*). The spectrum is divided into a series of η vertical strips of equal width dH and the height of each strip X_η is measured. The X_η are added and a record of the accumulating subtotal S_η is made, i.e., $S_1 = X_1$, $S_2 = X_1 + X_2$, $S_3 = X_1 + X_2 + X_3$. . . . After passing the point of maximum absorption, the subsequent X_η should be subtracted so that S_η should decrease, ideally to zero. S_η is then plotted with constant horizontal increment (not necessarily dH) to obtain the absorption curve. The second integration is readily performed by cutting out the absorption curve and weighing it. (It is advisable to use good quality paper to avoid errors due to variations in paper thickness.) Because the area is distributed equally around the point of maximum absorption, S_η should return to zero, and reasonable adjustments in the baseline are permissible to promote this. However, it is frequently found that the absorption curve still does not return exactly to zero and the wings of the curve should be joined by a straight line.

First Moments (*Fig. 6*)

$$\text{First moment} = \sum_{1}^{\eta} (H_\eta - H_0) \cdot X_\eta$$

[12] R. E. Hansen and H. Beinert, personal communication, 1965.
[13] R. Estabrook, personal communication, 1965.

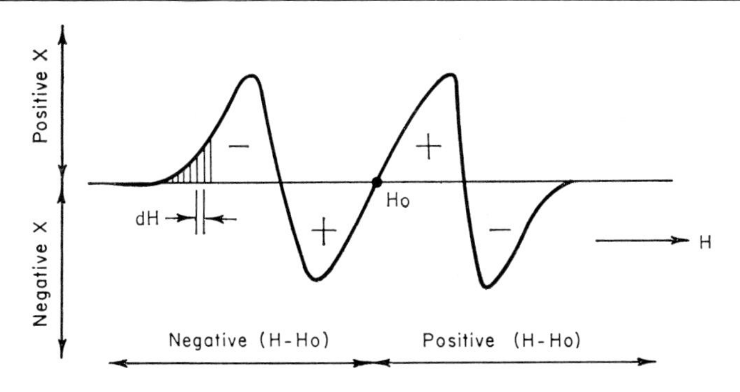

FIG. 6. First moments.

The spectrum is divided up into a series of η vertical strips of equal width dH. The height of each strip (X_η) is multiplied by the distance of this strip from some arbitrary reference point H_0 $(H - H_0)$, with due attention being paid to the algebraic sign. The η products are then added to yield the desired first moment.

Wyard[13a] has recently shown that if the first moment is taken about the middle of the spectrum, errors due to constant baseline drift are eliminated.

In our hands, double integration has proved to be more convenient than first moments: however, the latter estimation is uninfluenced by modulation broadening, and highly structured spectra can be rendered more convenient for estimation by deliberate overmodulation. It has been found empirically that double integration is also uninfluenced by a moderate amount of overmodulation.[14]

It is very desirable that the spectra unknown and standard be run under identical conditions. In particular the incident microwave power must be attenuated sufficiently so that saturation of resonance is not occurring (remember that the signal amplitude should double for every 6 db of power, in the absence of saturation), and the field modulation should be kept constant. However, when necessary the following corrections should be remembered: (1) The signal intensity is directly proportional to modulation amplitude in the absence of modulation broadening. However, the modulation amplitude control is nonlinear and must be calibrated in terms of gauss (cf. linewidths). (2) In the absence of saturation, the signal intensity is proportional to the square

[13a] S. J. Wyard, *J. Sci. Instr.* **42**, 709 (1965).

[14] M. L. Randolph, *in* "Free Radicals in Biological Systems," (M. S. Blois, Jr., H. W. Brown, R. M. Lemmon, R. O. Lindblom, and M. Weissbluth, eds.), p. 249. Academic Press, New York, 1961; *Rev. Sci. Instr.* **31**, 949 (1960).

root of the microwave power. (3) The signal intensity is proportional to the square of the sweep rate; as the latter is influenced by several factors, it is most satisfactorily recorded as gauss per centimeter of chart paper.

A final correction which is necessary with metals is due to possible g value differences between standard and unknown, the correction being $(g_{\perp}, \text{sample}/g_{\perp}, \text{standard})^2$ when the ratio $g_{\perp}/g_{\parallel}$ are similar for sample and unknown; otherwise more complicated expressions are necessary.[15]

It would appear that the major source of error resides in the choice of standard.[16] In some cases, e.g., with copper compounds which give relatively well defined spectra and where reasonable standards are available, errors are put as low as $\pm 5\%$; conversely, with broad signals often found with iron compounds, the estimation can easily be in error by a factor of three. These are extreme samples, but they indicate the nature of the problem and substantially detract from the merit of this feature of EPR spectroscopy.

Suitable standards appear to be, for metals, 1 mM Cu^{II} in 10 mM EDTA; for radicals, nitrosyl disulfonate,[17] A_M (545 mμ) $= 21.6$; diphenyl picryl hydrazyl,[18] A_M (525 mμ) $= 11.9 \times 10^3$ (recrystallize from carbon disulfide); and quinhydrone.[19]

Temperature

It is essential at this point that we consider the temperature at which EPR measurements are made. Ideally, we would like to make as many measurements as possible at room temperature, for this approximates more nearly physiological condition, allows simple kinetic and optical observation, and avoids artifacts of freezing. These latter include shifts of spin state with decrease in temperature, possible changes in equilibria, and other temperature-dependent processes. One artifact, which is probably not as serious for protein-bound constituents (which, by virtue of the bulk of the protein, tend to remain magnetically dilute), is dipolar broadening of spectra due to solute segregation during the freezing process.[20]

The low temperature measurements are in general more convenient to make (apart from the tedium of assembling the refrigeration system).

[15] R. Aasa and T. Vanngard, *Proc. 7th Intern. Conf. Coordination Chem.* p. 137. Almquist & Wiksell, Stockholm, 1962.

[16] H. Beinert and B. Kok, *Biochim. Biophys. Acta* **88**, 278 (1964).

[17] Alpha Inorganics, 8 Congress Street, Beverly, Massachusetts.

[18] Aldrich Chemical Company, Milwaukee, Wisconsin.

[19] Hydroquinone, 5 mM in degassed 0.1 M phosphate pH 7.4 yields 6.4 μM unpaired spins [G. Narini, H. S. Mason, and I. Yamazaki, *Anal. Chem.* **38**, 367 (1966)].

[20] R. T. Ross, *J. Chem. Phys.* **42**, (1965) 3919.

Sample changing is easier, signals are sharper—hence, easier to quantitate—signal-to-noise is improved, and species of short relaxation times may become observable. However, one cannot extrapolate from this and aim to work routinely at liquid helium temperature; because operation at these low temperatures requires quite specialized techniques and because saturation occurs rather readily, provision for extremely low power measurements is obligatory. Even those enzymes that cannot be saturated at 77°K with 0.2 watt (e.g., ceruloplasmin, and the $g = 1.94$ iron species) are readily saturated with a few microwatts at 4°K.

The temperature can be monitored by a thermocouple inserted into the gas stream just before it enters the sensitive part of the cavity; the temperature of the sample has been found empirically to be about 1° warmer than indicated by the thermocouple.

Miscellaneous

Crystal Diodes

X-band. For routine work at 100 kc/s modulation the IN23F, IN23G, and M403A diodes manufactured by Microwave Associates[21] are excellent. However, the Philco L4154[22] is only slightly inferior at 100 kc/s and markedly superior at 400 c/s and hence is a better choice for investigators who will use both high and low frequency modulation (e.g., as in second derivative presentation).

K-band. The Microwave Associates IN53B is superior to the Sylvania IN53C.

K-band (35 Gc) vs. X-band (9 Gc)

K-band measurements are extremely useful in that greater resolution of g value anisotropy is obtained. However, it appears that hyperfine interactions clearly seen at X-band may not be resolved at K-band. Thus, there is not a clear-cut answer as to the absolute merits; ideally a laboratory should have facilities for working at both frequencies. However, the X-band system appears to be much more satisfactory for routine operation, although low temperature measurements at K-band are just as feasible as at X-band. Solution work at K-band appears to be forbiddingly difficult.

Concentration of Samples

It is almost invariably true that investigators with no experience of EPR attempt to obtain results with enzyme solutions which are far too

[21] Microwave Associates, Burlington, Massachusetts.
[22] Philco Corporation, P. O. Box 115, N. Wales, Pennsylvania.

dilute. Several methods exist for concentrating these solutions. They may either be dialyzed against high molecular weight polymers (e.g., polyvinylpyrollidone,[23] Aquacide[24]) or concentrated by ultrafiltration in collodion sacs, for which convenient and simple apparatus is available commercially.[25] For most applications the author favors the latter, it being quick and clean.

Auxiliary Techniques

Spectroscopic Measurements

The feasibility of obtaining optical spectra on samples prepared for room temperature EPR has already been mentioned. In addition, it can be mentioned that with suitable masking, visible spectra can also be obtained in the 4 mm i.d. cylindrical tubes; linearity has been observed with absorbancies up to 2.0. However, as these tubes are predominantly employed for low temperature work, this is not very useful for the reasons outlined earlier.

For spectra on frozen samples there appear to be two solutions. The first, which has found a certain amount of use, is the application of reflectance spectrophotometry. This is discussed elsewhere in this volume. An alternative method is to employ the scattered transmission technique.[26] Whether or not this is fundamentally different from the reflectance method is difficult to judge, for the frozen samples are quite opaque and the photomultiplier is presumably observing light scattered from the surface of the sample.

Kinetic Studies

Inasmuch as the preponderance of the systems studied by EPR are components of enzymes, there is an obvious need for kinetic measurements on these species; for if these species are real participants in the mechanism of action of the enzyme, the rates of their appearance or disappearance cannot be slower than the overall activity of the enzyme.

At one time it was feared that the high enzyme concentration necessary for these experiments might modify the activity of the system. However, in the two systems which have been examined this has not been true.[27,28]

[23] "Plasdone," General Aniline and Film Corp., 435 Hudson St., New York, New York.

[24] Calbiochem, 3625 Medford Street, Los Angeles, California.

[25] Sleicher and Schuell, Keene, New Hampshire.

[26] Y. Morita and H. S. Mason, *J. Biol. Chem.* **240**, 2654 (1965).

[27] H. Beinert, G. Palmer, T. Cremona, and T. P. Singer, *J. Biol. Chem.* **240**, 475 (1965).

The most obvious way of reforming these measurements is by suitable modification of existing conventional flow techniques. While this is the most persuasive method, it has three serious disadvantages: (1) large consumption of material, (2) poor signal-to-noise ratio—a consequence of the large spectrometer bandwidth necessary to follow the rapid events, and (3) that the method just cannot be applied to those systems with short relaxation times, for they can be observed only at cryogenic temperatures.

A solution to this has been devised by Bray and is known as the rapid-freezing technique. It is basically a quenching method in which the reacting solution is squirted as a fine jet into a large volume of cold (—140°) isopentane, where the reaction mixture is frozen very rapidly.

These techniques have recently been discussed *in extenso*.[29]

Recently Borg[29a] has achieved excellent results with a continuous flow device for a K-band spectrometer, obtaining both substantial fluid economy and improved time resolution.

Postscript

Although the components of magnetic resonance equipment are manufactured by several companies,[30] the products of Varian Associates are most generally used. Although it is difficult to give an evaluation of the merits of the other instruments available, from *published specifications* there is not a great deal of difference between them.

Because advanced experimentation is limited by the characteristics of the magnet—physical size as well as magnetic field capability—considerable thought should be given to the purchase of this component. For most biological work a 9-inch magnet with a 4-inch air gap gives excellent field homogeneity with plenty of working space for any unusual accessories the creative investigator might wish to utilize.

---

[28] H. Beinert and G. Palmer, *J. Biol. Chem.* **239,** 1221 (1964). See also Q. H. Gibson, C. Greenwood, D. C. Wharton, and G. Palmer, *J. Biol. Chem.* **240,** 888 (1965).

[29] "Rapid Mixing and Sampling Techniques in Biochemistry," (B. Chance, R. H. Eisenhardt, Q. H. Gibson, and K. K. Lonberg-Holm, eds.), Academic Press, New York, 1965. In particular, the papers by L. H. Piette, D. C. Borg, R. C. Bray, and G. Palmer and H. Beinert.

[29a] D. C. Borg, personal communication, 1966.

[30] Alpha Scientific, 940 Dwight Way, Berkeley, California; JEOLCO (USA) Inc., 461 Riverside Avenue, Medford, Massachusetts; Magnion Inc., 144 Middlesex Turnpike, Burlington, Massachusetts; Perkin-Elmer Corp., 723 Main Avenue, Norwalk, Connecticut; Spectromagnetic Industries, 25377 Huntingdon, Hayward 3, California; Varian Associates, 611 Hansen Way, Palo Alto, California.

[95] The Preparation of Cytochrome-Deficient Mutants of Yeast

By FRED SHERMAN

Aerobically grown bakers' yeast (*Saccharomyces cerevisiae*) has a classical cytochrome system which is comparable to mammalian systems and which contains cytochromes a, a_3, b, c, and c_1. By a variety of techniques, it is possible to obtain mutants of yeast that are partially or completely deficient in one or more of the cytochromes. Some of these mutants have been employed in many diverse and unrelated studies concerning mitochondrial structure and function, radiobiology, enzyme regulation, active transport, etc.

The types of mutants can be considered in two *genetic* classes: (1) the cytoplasmic or ρ^- mutants (vegetative "petities")[1] and (2) the chromosomal mutants[2,3] ("segregational" mutants), which can be further classified[4] as p or cy. The ρ^- mutants are (i) completely deficient in cytochromes a and a_3; (ii) either completely or almost completely deficient in cytochromes b and c_1; and (iii) have normal or slightly higher amounts of cytochrome c. Different chromosomal mutants can have a variety of partial or complete deficiencies. The symbol p is used to denote single-gene mutants which fail to grow on nonfermentable carbon sources, such as lactate, ethanol, glycerol, acetate, but which may or may not have altered cytochrome spectra. The symbol cy denotes single-gene mutants which have altered absorption spectra, but which can utilize nonfermentable carbon sources for growth. The symbols P, CY, and ρ^+, respectively, will be used to describe the normal counterparts of the mutant determinants p, cy, and ρ^-.

It is also possible to prepare mutant strains having more than one genetic defect, and thereby increase the types of cytochrome deficiencies. For example, chromosomal mutants (p and cy) usually can be obtained both as ρ^+ and ρ^-. Also haploid strains can be constructed by genetic recombination to have two or more mutant genes.

It also should be mentioned that it is possible to obtain yeast mutants

[1] B. Ephrussi, "Nucleo-cytoplasmic Relations in Microorganisms." Oxford Univ. Press (Clarendon), London and New York, 1953.

[2] S. Y. Chen, B. Ephrussi, and H. Hottinguer, *Heredity* **4**, 337 (1950).

[3] F. Sherman, *Genetics* **48**, 375 (1963).

[4] F. Sherman, *in* "Mécanismes de Régulation des Activités Cellulaires chez les Microorganismes", p. 465. C.N.R.S., Paris, 1965.

which have defects in the tricarboxylic acid cycle,[5] and mutants which have high concentrations of porphyrins and metalloporphyrins.[4,6]

The Preparation of ρ^- Strains

Since ρ^- mutants occur spontaneously at relatively high frequencies, they can be easily isolated from all strains of bakers' yeast. Several techniques have been successfully employed for distinguishing between ρ^+ (normal) and ρ^- (mutant) colonies. One simple method is to plate approximately 100–200 cells on a YPDG (1% Bacto-yeast extract, 2% Bacto-peptone, 0.1% dextrose, 3% (v/v) glycerol, and 2% agar) dish which is then incubated at 30° for 2–3 days. At this time the ρ^+ colonies are several times larger than the ρ^- colonies, owing to the fact that the ρ^- mutants cannot utilize glycerol for growth and that glucose was in growth-limiting amounts. The ρ^- colonies can be isolated and tested for their cytochrome content and respiratory capacity in order to verify the phenotype. In growing populations of most laboratory strains of *Saccharomyces cerevisiae* there are over 0.5% of ρ^- cells. However, if the proportion of mutant cells is extremely low, the strain can be grown in the presence of acriflavin, which induces high frequencies of ρ^- cells.[7] Also, for rapid and routine isolations one can first grow a high density of cells on the surface of acriflavine medium[8] for 1 day and then streak diluted cells on YPDG plates for single colonies. In most instances, the majority of colonies would be ρ^-. It should be mentioned that a variety of physical and other chemical agents are also known to induce ρ^- cells.[9]

Other methods used for distinguishing between ρ^+ and ρ^- strains are (1) overlaying with 2,3,5-triphenyltetrazolium chloride[10,11]; (2) replica-plating or streaking on YPG (1% Bacto-yeast extract, 2% Bacto-peptone, and 3% glycerol) or similar media[3,12]; (3) by employing ad_1 or ad_2 strains which produce a red pigment that is effected by the respiratory genotype[13]; and (4) the inclusion of various dyes in the

[5] M. Ogur, A. Roshanmanesh, and S. Ogur, *Science* **147,** 1590 (1965).

[6] T. Pretlow and F. Sherman, in preparation.

[7] B. Ephrussi, H. Hottinguer, and A.-M. Chimenes, *Ann. Inst. Pasteur* **76,** 351 (1949).

[8] The exact concentration of acriflavin will depend on the strain in question and the type of nutrient medium. An effective medium consists of: 1% Bacto-yeast extract; 2% Bacto-peptone; 2% glucose; 2% agar; and 0.025% acriflavin.

[9] S. Nagai, N. Yanagishima, and H. Nagai, *Bacteriol. Rev.* **25,** 404 (1961).

[10] M. Ogur, R. St. John, and S. Nagai, *Science* **125,** 928 (1957).

[11] A dry mix of tetrazolium, glucose, and agar (TTC Overlay Agar) is commercially available from the Fisher Scientific Company (New York).

[12] M. Ogur and R. St. John, *J. Bacteriol.* **72,** 500 (1956).

[13] J. Tavlitzki, *Rev. Can. Biol.* **10,** 48 (1951).

growth medium.[14-16] The tetrazolium overlay method has been conveniently used in studies where either the colony size was modified by other agents, or the detection of sectored colonies was desirable.

In view of the consistent results of many previous investigations, genetics tests are not usually considered necessary to verify the cytoplasmic nature of these mutants.

The Preparation of Chromosomal Mutants

It should be emphasized that chromosomal mutants are generally found at low frequencies, even after mutagenic treatments, and are therefore much more difficult to isolate in comparison to ρ^- mutants. In addition, most initial tests do not readily distinguish the many types of chromosomal mutants from the ρ^- mutants, which comprise the major class of mutants in a cell population. However, successful methods are now available for preparing chromosomal mutants which have cytochrome and related deficiencies. The initial step is to treat a haploid culture with a mutagenic agent in order to increase the frequencies of mutants. The second step consists of scoring the colonies that were derived from the treated cells for possible cytochrome mutants. Scanning procedures, which have proved to be effective, depend primarily on three connected but different properties: (1) failure to grow on nonfermentable carbon and energy sources; (2) an altered absorption spectrum as determined by low-temperature spectroscopy; and (3) a negative benzidine reaction due to a decreased cytochrome content.

Mutagenic Treatments

Any mutagenic treatment which is used to produce general types of chromosomal mutants, also can be employed to obtain cytochrome mutants. Ultraviolet irradiation and the chemical mutagens nitrous acid and nitrosoguanidine (*N*-methyl-*N'*-nitro-*N*-nitrosoguanidine) have been found to be effective.

A simple procedure for producing mutants with nitrous acid[17] is to suspend freshly grown haploid cells in 0.5 *M* sodium acetate buffer (pH 4.6, 30°) to a density of approximately 2×10^7 cells/ml. $NaNO_2$ is quickly dissolved in sterile water to give a concentration of approximately 0.3 *M*. Equal volumes of the $NaNO_2$ solution and the buffered cell suspension are mixed and kept at 30° for 10 minutes. After this

[14] S. Nagai, *J. Bacteriol.* **86**, 299 (1963).
[15] S. Nagai, *Stain Technol.* **40**, 147 (1965).
[16] S. Nagai, *J. Bacteriol.* **90**, 220 (1965).
[17] F. Sherman, J. Parker, J. Stewart, E. Margoliash, and W. Campbell, in preparation.

desired time of treatment, the cells are quickly diluted 1:10 in a solution of 1% yeast extract and 2.7% $Na_2HPO_4 \cdot 7\ H_2O$, and approximately 150 viable cells per plate are streaked on YPD medium (1% Bacto-yeast extract, 2% Bacto-peptone, 2% dextrose, and 2% agar). The exact concentration of $NaNO_2$ (around 0.3 M) and the time of treatment (approximately 10 minutes) should be adjusted to produce approximately 0.1% survival. The plates are then incubated at 30° for 4–7 days.

Isolation Procedures

Growth on Nonfermentable Substrates.[18] A convenient method of testing for a functional electron transport system is to determine whether the mutant in question can utilize nonfermentable carbon sources (e.g., glycerol, lactate, acetate, ethanol, etc.) for growth. $P\,\rho^-$, $p\,\rho^+$, and $p\,\rho^-$ strains all fail to grow on nonfermentable carbon sources. However, p genes segregate in a normal Mendelian fashion,[2,3] while the ρ^- determinant exhibits cytoplasmic inheritance.[1] In addition, $p\,\rho^+$ strains can be rapidly distinguished from $P\,\rho^-$ mutants (vegetative "petites") by crossing to known $P\,\rho^-$ strains and determining whether hybrids from the mating mixtures can grow on nonfermentable substrates.

Plates containing the colonies that were derived from the HNO_2-treated cells are replicated on YPG medium in order to determine which colonies are unable to utilize nonfermentable substrates for growth. Portions of each mutant colony are taken from the master plates and streaked on a nutrient medium (YPD); second portions are mixed with a haploid $P\,\rho^-$ strain of opposite mating type on the same YPD plate. After 1 day of incubation the plates are replica-plated on YPG plates. Strains which do not grow on YPG medium, but which grow after crossing with $P\,\rho^-$ strains, are probably $p\,\rho^+$. Single-gene segregation should be verified by genetic analysis. It should be mentioned that certain p mutants have not been observed to be complemented by ρ^- strains.[3] Therefore, one would not expect to find this class of mutants by the isolation technique.

p mutants can also be obtained by testing pedigrees of yeast stocks. During the course of extensive genetic studies involving large numbers of nutritional mutants, many p genes were uncovered after testing segregants for growth on glycerol medium.[3,19,20] Some of the p mutants concurrently had requirements for lysine and glutamate.[3,5]

[18] D. Pittman, J. M. Webb, A. Roshanmanesh, and L. E. Coker, *Genetics* **45**, 1023 (1960).
[19] D. C. Hawthorne and R. K. Mortimer, *Genetics* **45**, 1085 (1960).
[20] D. C. Hawthorne and R. K. Mortimer, unpublished experiment, 1965.

Almost all *p* mutants isolated by the above methods have been found to be deficient in respiration and in one or more of the cytochromes.[21] Some of the *p* ρ^+ strains had different cytochrome deficiencies than *P* ρ^- strains. However, some strains are known which respire and have all the cytochromes, but nevertheless do not grow on nonfermentable substrates.

Altered Absorption Spectra.[22] A technique has been described for rapidly examining the absorption spectra of large numbers of strains. Each colony is reinoculated on a square plastic petri dish[23] containing YPD medium. After 3 days' incubation, the square plates containing 36 clones per dish are frozen in liquid nitrogen and placed on a plastic block mounted under a Hartree microspectroscope.[24] The plastic block contains a number of grooves in order that the square dish can be quickly moved by hand to center each strain in the light path. In this manner, it is possible to examine over 2000 strains a day. After a mutant is detected, the plate is defrosted and the strain reinoculated on a slant culture.

Many diverse types of mutants have been isolated by this technique: mutants which were deficient in some of the cytochromes and which failed to grow on media containing nonfermentable carbon sources (*p* mutants); mutants which were partially deficient in the various cytochromes but still retained the ability to grow on at least most of the nonfermentable substrates (*cy* mutants); and mutants having high concentrations of abnormal pigments. It would be difficult to isolate these latter two types of mutants by screening on medium containing nonfermentable carbon sources. However, the spectroscopic method could not be used for isolating *p* mutants having a normal cytochrome spectrum. With this method it would be tedious to isolate *p* mutants having a phenotype similar to ρ^- strains, since crosses would have to be tested.

Benzidine Test.[17] The benzidine reaction is known to be effective with hemes and hemoproteins, and therefore the degree of coloration would be expected to be correlated with the cytochrome content. However, the degree of the benzidine reaction differs only slightly from the normal and the vegetative mutant (ρ^-), and both strains are stained to a dark blue color.

The essential steps of staining yeast colonies with the benzidine reagents consists of (1) overlaying the plates with approximately 15 ml of the benzidine reagent (1 g of benzidine dihydrochloride, 20 ml of glacial acetic acid, 30 ml of distilled water, and 50 ml of 95% etha-

[21] F. Sherman and P. P. Slonimski, *Biochim. Biophys. Acta* **90,** 1 (1964).
[22] F. Sherman, *Genetics* **49,** 39 (1964).
[23] Obtained from Falcon Plastics, Los Angeles, California.
[24] Obtained from Beck Limited, London, England.

nol)[25,26]; (2) pouring the reagent off after approximately 15 seconds of contact; (3) overlaying the plate with a 5% solution of hydrogen peroxide; (4) pouring off the peroxide solution after approximately 1 minute when the colonies developed a slightly blue color; (5) allowing the color to intensify for an additional few minutes; and finally (6) transferring the benzidine-negative clones to slants with a sterile loop. If the experiments are hampered by the breakage and smearing of the colonies during the addition and removal of solutions, the plates can be first sprayed with hair spray (several commercial products are all equally effective).

The benzidine method is more restricted to the types of mutants that can be isolated. One major class of benzidine-negative mutants obtained from a normal strain was deficient in all cytochromes. This would be expected if lesions occurred in the biosynthetic pathway of the porphyrins. However, if one started with a ρ^- strain which lacks cytochromes *a* and *b*, then many cytochrome *c* deficient mutants were isolated by checking the benzidine-negative strains with known *cy* strains.[17] The main advantage of the benzidine method lies in the large numbers of colonies that can be easily examined.

Types of Mutants

The various types of mutants that have been isolated by the three techniques are summarized in the table. Examples are known of mutants which failed to grow on nonfermentable carbon sources (*p* mutants), but nevertheless were either normal or deficient with respect to respiration, the benzidine reaction, and components of their cytochrome system.

THE PROPERTIES OF VARIOUS TYPES OF CHROMOSOMAL MUTANTS OBTAINED BY THE THREE ISOLATION TECHNIQUES

	Isolation technique		
	Non-fermentable substrates	Spectroscopic examination	Benzidine test
Genotypes	*p*	*p* and *cy*	*p* and *cy*
Growth on nonfermentable substrates	−	+ and −	+ and −
Normal absorption spectrum	+ and −	−	−
Benzidine reaction	+ and −	+ and −	−
Ability to respire	+ and −	+ and −	+ and −

[25] H. Wu, *J. Biochem.* (*Tokyo*) **2,** 189 (1923).
[26] R. H. Deibel and J. B. Evans, *J. Bacteriol.* **79,** 356 (1960).

Likewise, mutants obtained by spectroscopic examinations all have, by definition, abnormal absorption spectra and yet either could or could not (1) grow on nonfermentable substrates, (2) respire, or (3) give rise to a positive benzidine reaction. Moreover, all mutants, isolated as benzidine negative, were at least partially deficient in their cytochrome system, but could have various other properties. Some of the yeast mutants that have been described were deficient in cytochromes $a + a_3$ (strain p_5)[21]; cytochrome b (strain p_{10})[27]; cytochrome c (strains cy_1, cy_2, cy_3, etc.)[17,22]; and more than one cytochrome.[21,27]

[27] C. Reilly and F. Sherman, *Biochim. Biophys. Acta* **95**, 640 (1965).

[96] Methods of Determining the Photochemical Action Spectrum

By Otto Rosenthal and David Y. Cooper

If the degree of promotion or inhibition of a biological reaction by bands of monochromatic light of equal quantum intensity is plotted as a function of wavelength of irradiating light, a photochemical action spectrum results which depicts the light absorption spectrum of the biocatalyst responsible for the light sensitivity of the reaction. The method was developed by Otto Warburg[1-3] to demonstrate the hemoprotein nature of the enzyme responsible for the light-reversible CO inhibition of cell respiration. It was subsequently used[4,5] to identify the CO-combining microsomal pigment P-450 as the terminal oxidase of several mixed function oxidase systems which were inhibited by CO and reactivated by illumination with white light. Finally this method resulted in the discovery and characterization of phytochrome, the light-sensing pigment controlling flowering and development of many plants,[6,7] and in the resolution of the multipigment systems participating in photosyn-

[1] O. Warburg and E. Negelein, *Biochem. Z.* **202**, 202 (1928).
[2] O. Warburg and E. Negelein, *Biochem. Z.* **214**, 64 (1929).
[3] O. Warburg, "Heavy Metal Prosthetic Groups and Enzyme Action," Chapters XII and XIII. Oxford Univ. Press. (Clarendon), London and New York, 1949.
[4] R. W. Estabrook, D. Y. Cooper, and O. Rosenthal, *Biochem. Z.* **338**, 741 (1963).
[5] D. Y. Cooper, S. Levin, S. Narasimhulu, O. Rosenthal, and R. W. Estabrook, *Science* **147**, 400 (1965).
[6] M. W. Parker, S. B. Hendricks, H. A. Borthwick, and N. J. Scully, *Botan. Gaz.* **108**, 1 (1946).
[7] H. W. Siegelman and S. B. Hendricks, *Advan. Enzymol.* **26**, 1 (1964).

thesis.[8,9] The essential technical requirements for this method are a light source from which monochromatic bands of sufficient intensity can be isolated, and a radiometer for the accurate measurement of the quantum energy of the bands.

Light Sources[10]

Carbon arcs, high pressure xenon lamps, projector type metal filament lamps, and high pressure metal vapor lamps (Hg, Cd, HgCd, etc.) have been most frequently used.[11] The first three types furnish continuous spectra convenient for scanning a broad spectral range with the same light source and for resolving details of the action spectrum. Xenon lamps and tungsten filament lamps are presently the preferred sources of continuous spectra, since they do not produce smoke and do not require intensity readjustment during operation. Both types yield high radiation intensities from the blue to the near infrared. The luminescence of tungsten filaments drops steeply below 450 mμ whereas the useful range of xenon lamps extends to approximately 280 mμ.

The line spectra of the metal vapor lamps have the advantage of yielding narrow monochromatic light bands of high intensity and purity between 260 and 650 mμ. The position of the bands depends on the nature of the metals. These lamps are mainly used to provide reliable reference points for the photochemical action spectrum. They do not permit the resolution of every detail of the spectrum.

The isolation of the monochromatic bands of light from the different light sources discussed has been accomplished with monochromators (prisms),[2,6] diffraction gratings,[8] or interference filters,[5,12] usually in combination with colored glasses and dye solutions. Narrow-band interference filters with half-band widths of 12–15 mμ are commercially available for the range of 340–800 mμ. Interference filters with half-band widths of 2–8 mμ are available on special order according to specifications by several manufacturers.[13] The filters transmit 50–70% of the

[8] F. T. Haxo and L. R. Blinks, *J. Gen. Physiol.* **33**, 389 (1950).

[9] O. Warburg, G. Krippahl, and W. Schröder, *Z. Naturforsch.* **10b**, 631 (1955).

[10] F. Weigert, "Optische Methoden der Chemie," Chapter II. Akad. Verlagsges. Leipzig, 1927. An extensive description of design and optical properties of different light sources.

[11] High pressure xenon and mercury vapor lamps are manufactured by Osram G.m.b.H., Berlin and Munich, West Germany. Distributor in U.S.A. (Macbeth Sales Corporation, P.O. Box 950, Newburgh, New York) will give information on the type of power supply required for operating the lamps.

[12] W. Bladergroen, "Problems in Photosynthesis." Thomas, Springfield, Illinois, 1960.

[13] Thin Film Products Corporation, Cambridge, Massachusetts.

specified wavelength in the range of 450–600 mμ. Below and above this range the transmittance decreases to 35–40%.

Figure 1 shows schematically the irradiation arrangement for studying the light reversal of the CO inhibition of various mixed function oxidase reactions in order to determine the relative action spectrum of the CO components of these oxygenases between 400 and 500 mμ. The arrangement is a simplified adaptation of the technique developed in O. Warburg's laboratory.[12] The latter arrangement includes a running-water heat filter; adjustable circular and rectangular diaphragms for modifying the light intensity and for adapting the dimension of the light beam to reaction vessels of different shape; and a reflecting mirror moving synchronously with a horizontally shaken reaction vessel. The determination of the action spectrum of the growth and flowering regulating system in higher plants has been described in detail by Parker *et al.*[6]

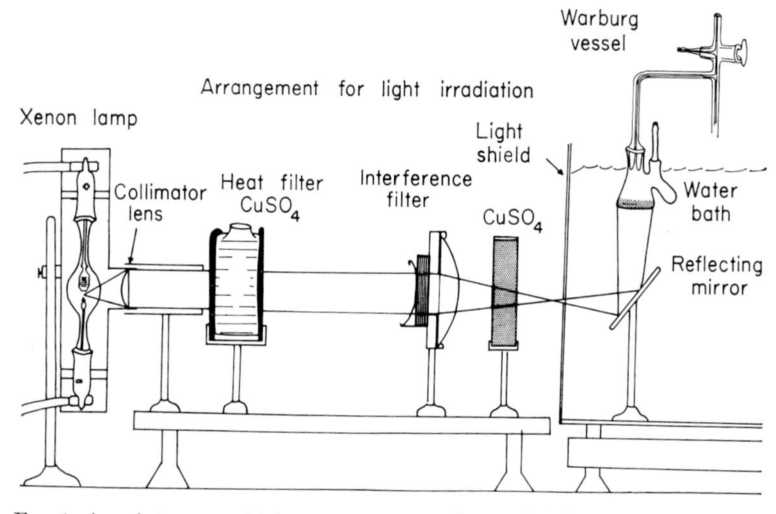

FIG. 1. A point source high pressure xenon lamp (XBO/1600 watt) is mounted in a water-jacketed metal housing, through the cavity of which air from an electric fan (not shown) is circulated. The power is derived from a 230-volt, 10 amp Christie Silicon Rectifier. The light beam passes through a collimator lens of Pyrex glass; heat filter (5 cm of 7% $CuSO_4$ solution); interference filters and, if needed, neutral density filters, focusing lens, and 2.5 cm layer of $CuSO_4$ solution for absorption of the second-order spectrum, enters the light-shielded glass-walled water bath through an opening in the shielding and is reflected by the 45° mirror onto the bottom of the Warburg vessel, which is shaken within the area of the beam at 130 oscillations per minute. Reproduced from D. Y. Cooper, S. Levin, S. Narasimhulu, O. Rosenthal, and R. W. Estabrook, *Science* **147,** 400 (1965).

Radiometry

The energy of the irradiating light is usually measured with thermopiles or bolometers.[14] The Lummer-Kurlbaum bolometer (Fig. 2), used in O. Warburg's and the writer's laboratories, has the advantage that, if a beam of light is shifted over the surface of its platinum strips, the electrical response remains constant. This is not the case with thermopiles.

Thermopiles and bolometers are standardized against a Total Radiation Standard, a carbon filament lamp, calibrated by or according to the specifications of the U.S. National Bureau of Standards.[15] The radiant flux density of this light source at specified distance and operating conditions is given in terms of watts/cm^2 (1 watt $= 1 \times 10^{-7}$ erg/sec $= 1$ joule/sec $= 0.239$ g cal/sec $= 14.34$ g cal/min).

The energy of 1 mole quanta (1 einstein), the photochemical equivalent, is defined by Eq. (1).

$$E = Nh\nu = Nh\frac{c}{\lambda} = \frac{1.197 \times 10^8}{\lambda}\ \text{erg} = \frac{2.86}{\lambda}\ \text{cal} \tag{1}$$

where N is Avogadro's number (6.02×10^{23}); h, Planck's constant $(6.626 \times 10^{-27}$ erg sec$^{-1})$; ν, the frequency of radiation (sec^{-1}); c, the velocity of light $(2.998 \times 10^{10}$ cm $\times$ sec$^{-1})$; and λ, the wavelength in centimeters. If the latter is expressed in millimicrons, E has to be multiplied by 10^7. A radiant flux density of 1 gram calorie per square centimeter per minute is then equivalent to $\lambda \times 3.5 \times 10^{-2}$ micro einstein per square centimeter per minute—a convenient unit for expressing the intensity of the radiation applied to the reaction vessels.

The photochemical action spectrum is usually determined at infinitely small light absorption of the reaction system in order to ensure that every enzyme molecule along the light path is exposed to the same intensity of radiation. Hence, it is necessary to ascertain whether this condition is fulfilled. In the arrangement shown in Fig. 1, the bolometer was placed in the empty water bath at the position of the 45° mirror, and the intensity of the different beams of monochromatic light selected for studying the action spectrum was determined. The intensities were then approximately equalized by means of neutral density filters. After the water bath was filled, the intensity of the light beam emerging

[14] Cf. F. Weigert, "Optische Methoden der Chemie," Chapter X. Akad. Verlagsges., Leipzig, 1927.

[15] U. S. Comm.-*Natl. Bur. Std.* (*U.S.*)-DC, Feb. 24, 1960. Reprints of the instructions as well as calibrated total radiation standards can be obtained from the Eppley Laboratory, Inc., Newport, Rhode Island.

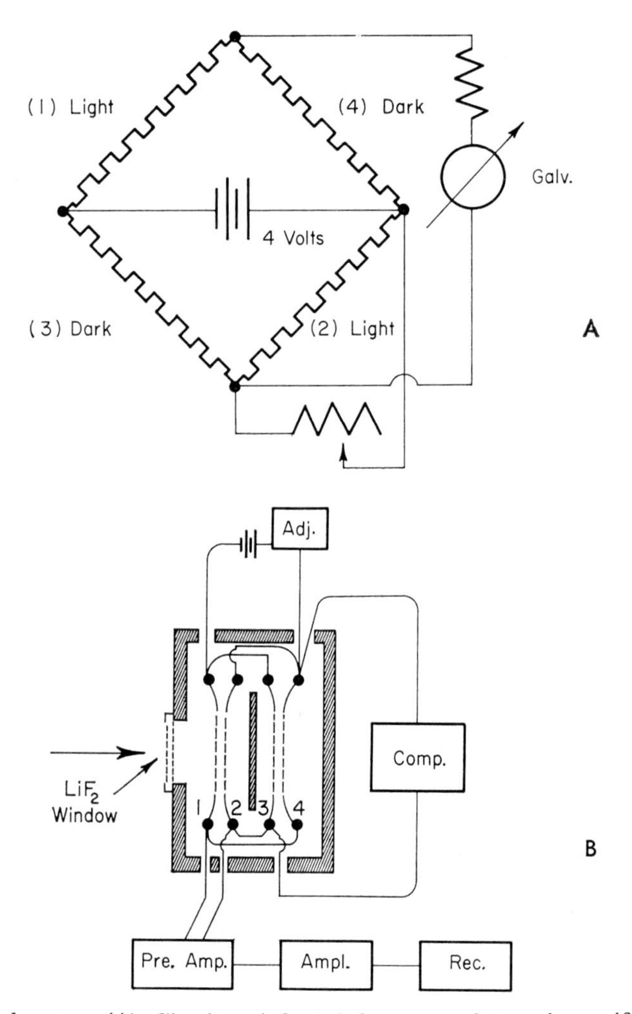

FIG. 2. Bolometer. (A) Circuitry (adapted from manufacturer's specifications). Each branch of the Wheatstone bridge consists of two sets of 10 strips ($1.8 \times 36 \times 0.001$ mm) of blackened platinum foil. Illuminated and darkened sets alternate. The resistance of each set is 38 ohm. Four volt storage battery, galvanometer (recommended 2×10^{-8} amp sensitivity, 60 ohm internal resistance), and variable compensating resistance (100–700 ohm) complete the circuit. (B) Cross section of the bolometer box, showing the partition into a "light" and "dark" compartment and the arrangement of the frames supporting the platinum strips. The interstices between the strips of frame *1* are alined with the strips of frame *2*. The total light-exposable platinum foil area is 10 cm^2. It can be diminished in steps of 2 cm^2 by inserting diaphragms behind the window. This permits checking the utilizable area, that is, the area of linear electric responses to variations of light intensity. To the circuitry shown in Fig. 2A, a voltage adjuster has been added and the galvanometer has been replaced by a strip chart recorder permitting permanent records of the measurements and the performance of the instrument (Grossflächenbolometer nach Kurlbaum. Manufacturer: Ing. H. Röhrig, Brentanostr. 17, Berlin-Steglitz, Germany).

from an open cuvette immersed to the same depth as the manometer vessel, and having the same bottom area and depth of enzyme solution as the latter, was then measured by means of a photocell previously calibrated against the bolometer at the position of the 45° mirror. The two types of estimate agreed within 1%.

The procedure of estimating the incident light, introduced by O. Warburg,[12] consists of reflecting the vertical light beam emerging from the water bath onto the bolometer by means of a mirror-lens system after removal of the reaction vessel. For the precise measurement of the light absorbed by enzyme preparations or cell suspensions, Ulbricht's sphere[16] should be used.

The Determination of the Photochemical Action Spectrum of CO Derivatives of Hematin Enzymes and Nonheme Iron Enzymes

O. Warburg's determination of the action spectrum of the CO component of the respiration enzyme[17] is based on the assumption that the distribution of the enzyme between O_2 and CO is analogous to that of hemoglobin as represented by the partition equation (2).

$$\frac{FeO_2}{FeCO} \cdot \frac{CO}{O_2} = K \tag{2}$$

If $FeO_2 + FeCO$ is equal to the total amount of enzyme present, and if the reaction rate in the absence of CO is taken as an estimate of this amount, then n, the ratio of the rate in the presence of CO to the rate in the absence of CO, is a measure of the term FeO_2 in Eq. (2), and the partition constant K can be readily determined experimentally by Eq. (3).

$$\frac{n}{1-n} \cdot \frac{CO}{O_2} = K \tag{3}$$

The photodissociation of FeCO results in an increase of the partition constant from the dark value, K_d, to the light value, K_h. The relative displacement of K at constant partial pressures of O_2 and CO is computed by Eq. (4).

$$\frac{\Delta K}{K_d} = \frac{K_h - K_d}{K_d} = \left[\frac{n_h}{1-n_h} - \frac{n_d}{1-n_d}\right] \Big/ \left[\frac{n_d}{1-n_d}\right] \tag{4}$$

This quotient is a linear function of the light intensity, i, for any given wavelength. Thus

[16] O. Warburg and G. Krippahl, *Z. Naturforsch.* **9b,** 181 (1954). Cf. also the reference cited in footnote 12.

[17] Cf. references cited in footnotes 1 and 2 and, for a summary, reference cited in footnote 3.

$$\frac{\Delta K}{K_d} = Li \tag{5a}$$

$$L = \frac{1}{i}\frac{\Delta K}{K_d} \tag{5b}$$

The correlation factor L, termed light sensitivity, is the reciprocal of the quantum intensity that doubles the partition constant. If L is determined for a suitable reference wavelength, i_{ref}—usually the light band that produces maximal or nearly maximal reversal of the CO inhibition—and if the quotients L_x/L_{ref} are plotted versus the wavelength of light, the relative light absorption spectrum of the CO derivative of the enzyme is obtained.

In regard to the physical significance of the constants K and L, it should be noted that the partition constant K is the ratio of 4 velocity constants which govern the equilibrium (Eq. 6).

$$\frac{n}{1-n}\frac{p_{\text{CO}}}{p_{\text{O}_2}} = \frac{zB}{bZ'} = K \tag{6}$$

The equation is mathematically identical with the equilibrium equation for hemoglobin. As in the latter, the rate constants b and z refer, respectively, to the formation and decomposition of the CO compound. For the enzyme partition, however, B is the rate constant for the formation of the ferric from the ferrous form—equivalent to the rate constant for O_2 utilization—while Z' is the rate constant for the conversion of the ferric to the ferrous form.

If the four rate constants are substituted into the equilibrium Eq. (4), b, B, and Z' cancel out and one obtains Eq. (7).

$$\frac{\Delta K}{K_d} = \frac{K_h - K_d}{K_d} = \frac{z_h - z_d}{z_d} = \frac{z_i}{z_d} = K \tag{7}$$

where z_h and z_d connote the rate constants for the decomposition of the CO compound during illumination and darkness, respectively. The difference, z_i, between these two constants is termed the light decomposition constant. It is equal to $i\ \varphi\ \beta$, i.e., the product of the light intensity, the photochemical yield[18] and the light absorption coefficient[19] of the CO compound at a given wavelength. Substituting into Eq. (5b) we obtain

$$L = \frac{1}{i}\frac{z_i}{z_d} = \frac{\varphi\beta}{z_d} \tag{8}$$

The light sensitivity is thus inversely proportional to the dark decomposition constant of the CO component of the enzyme. Since z_d

[18] φ = (moles dissociated by irradiation)/(mole quanta absorbed).

[19] $\beta = \text{cm}^2/\text{mole}$.

decreases with decreasing temperature, the light sensitivity should increase with decreasing temperature, a principle utilized by Warburg in determining L.

If the change of n as a function of time can be measured during the transition from light to darkness and darkness to light, z_d can be determined. Since the transition times are usually too short for precise measurements of the time course of O_2 uptake, this measurement can be accomplished by the intermittent exposure of the enzyme systems to different periods of darkness and light. The experimental arrangement and the method of calculating z_d have been described in detail by Warburg.[20]

If z_d is known and if φ can be estimated, the light absorption coefficient β is obtained and the molar absorption spectrum of the CO component of the hematin enzyme can be constructed from the photochemical action spectrum.

While the action spectrum method has been originally designed for the terminal oxidase of cell respiration in conjunction with continuous manometric assay of the oxygen uptake, the method has proved to be equally applicable to CO-sensitive hematin-catalyzed oxygenase reactions if the rate of incorporation of oxygen into the specific oxygenase substrate, measured by a discontinuous assay of the reaction product, is substituted for the rate of oxygen consumption.[4,5] In any case, however, it is essential to ascertain whether the assumptions upon which the derivation of Eqs. (2) to (4) is based, are valid for the reaction under examination. Particularly, the partial pressures of O_2 and CO have to be kept constant during the transitions from darkness to light and the enzyme system has to be kept saturated with respect to electron donors and, in the case of mixed function oxidases, also with respect to the specific substrate of oxygen incorporation.

Zero-order reaction kinetics in the presence and absence of CO; constant K values at varying $CO:O_2$ ratios, as well as at equal ratios and varying partial pressures of the two gases; and a linear relationship between $\Delta K/K_d$ and light intensity are indications that the basic assumptions are valid for the enzyme system under study. In addition, it is necessary to determine the reaction rate in the absence of CO at the same oxygen pressure as in the presence of CO and to check whether the reaction rate in the absence of CO is affected by the light intensities needed for the reversal of the CO inhibition.

Warburg (3) found K values of about 10 for the respiration of animal tissues and values up to 30 for yeast cells, depending on cell strain and

[20] O. Warburg and E. Negelein, *Biochem. Z.* **202**, 202 (1928).

age of the culture. K-values for mixed function oxidases are around 1 (range 0.5–2.0).[5] At 0–10°, the light sensitivity factor L for the respiration of acetobacter, yeast, and hemoglobin was of the order of 10^8–10^9, while L for microsomal mixed function oxidases at 25° was two orders of magnitude lower.[5] One thus needs comparatively high light intensities to reverse the CO inhibition of mixed function oxidations.

Warburg and Negelein[21] observed that, with certain strains of the yeast *Torula utilis* for which the partition Eq. (3) had proved to be valid at 32° and 40°, the equation took at 10° the form

$$\frac{n}{1-n}\left(\frac{CO}{O_2}\right)^{\beta} = K \tag{9}$$

where β is greater than one. The exponent β can be computed from

$$\beta = \log\left[\left(\frac{n}{1-n}\right)_1 \Big/ \left(\frac{n}{1-n}\right)_2\right] \Big/ \log\left[\left(\frac{CO}{O_2}\right)_2 \Big/ \left(\frac{CO}{O_2}\right)_1\right] \tag{10}$$

where the subscripts 1 and 2 refer to the ratios $n/(1\text{-}n)$ determined at the corresponding ratios of $CO{:}O_2$. The exponent β for *Torula utilis* at 10° was approximately 1.5. If β had been taken as one, K would have decreased from 10 to 4.55 when the ratio of $CO{:}O_2$ was raised from 3.7 to 19.8.[22]

There is as yet no satisfactory explanation of this phenomenon. Warburg[3,21] mentioned the possibility that the ratio p_{CO}/p_{O_2} inside the cell might differ from the outside ratio. At the same time he emphasized that it is immaterial for the determination of the relative photochemical action spectrum whether or not β is greater than one. The term $CO{:}O_2$ does not appear in Eq. (4). As far as we are aware, however, no determinations of the action spectrum of the respiration enzyme have been under conditions where β was greater than one.

We have observed[23] decreasing K values with increasing $CO{:}O_2$ ratios at constant oxygen pressure when studying the CO inhibition of steroid 11-β hydroxylase preparations from bovine adrenocortical mitochondria. With this system, however, the K values tended to become constant when the system was reinforced with nonheme iron protein

[21] O. Warburg and E. Negelein, *Biochem. Z.* **193,** 334 (1928).

[22] Decreasing K values have also been observed when yeast cells were not saturated with oxidizable substrates. Under these conditions the first term of the partition equation changes to $\epsilon n/(1-\epsilon n)$ where ϵ is the ratio of respiration at nonsaturating substrate concentration to respiration at saturating concentration [O. Warburg, *Biochem. Z.* **189,** 354 (1927)].

[23] D. Y. Cooper and O. Rosenthal, *Federation Proc.* **25,** 765 (1966).

and the specific flavoprotein, the components of the electron transport system from TPNH to cytochrome P-450.[24]

The application of the partition equation, derived from hemoglobin, to hematin enzymes of unknown concentration presumes that the rate of oxygen utilization, in respiration or mixed function oxidation, is a measure of the total amount of enzyme present. The rate-limiting step is the reduction of Fe^{3+} to Fe^{2+}, since the velocity constant B for the oxidation of Fe^{2+} is much greater than Z', the velocity constant for the reduction of Fe^{3+}. If the electron transfer capacity of the reducing system is small in comparison to the amount of Fe^{3+} present, only a fraction of the latter will function in the measured electron transfer to O_2 while the remaining fraction of "free" Fe^{3+} will remain undetected.

If then part of the Fe^{2+} combines with added CO, reducing equivalents become available for the partition of "free" Fe^{3+} between CO and O_2, since the dark decomposition constant, z_d, of the CO complex is low. That is, the value of n, the ratio of the rate of reaction in the presence of CO to the rate in its absence, will thus be higher than the true value for the partition equilibrium. With increasing $CO:O_2$ ratios, the "free" Fe^{3+} becomes saturated and K decreases in an exponential fashion. Conversely, if the CO compound is decomposed by light, the ratio of $\Delta K:K_d$ will not be a linear function of the light intensity, since the increase in the reaction rate is limited by the reduction rate of Fe^{3+}. The result will be a distortion of the photochemical action spectrum.

Warburg[25] pointed out that it is probably incorrect to treat Z' as a constant at all the possible changing values of the Fe^{3+} concentration. Nevertheless, the simplifying assumption of unchanging velocity constants has proved to be applicable to the respiration of most cells, as well as to a variety of mixed function oxidations by subcellular preparations. It is important, however, to realize that there are exceptions and to take this into consideration when applying the action spectrum method to aerobic hydroxylation reactions catalyzed by various subcellular tissue preparations in order to decide whether the incorporation of molecular oxygen into a given substrate is mediated by a hematin enzyme such as cytochrome P-450.

It has also to be taken into consideration that carbon monoxide may disappear from the reaction system through side reactions.[26] To mention are: (a) hydration to formate by the KOH used for absorption of CO_2

[24] T. Omura, R. Sato, D. Y. Cooper, O. Rosenthal, and R. W. Estabrook, *Federation Proc.* **24**, 1181 (1965).

[25] Cf. page 107 of Warburg (cited in footnote 3).

[26] Cf. pages 75-77 of Warburg (cited in footnote 3).

in respiration measurements; (b) oxidation to CO_2 by certain hemins present in the medium; and (c) enzymatic hydration or oxidation within cells. It is the latter type of reaction, for which some experimental evidence is available, that provides the basis for Warburg's suggestion of different $CO:O_2$ ratios inside and outside the cell. The disappearance of CO can be determined manometrically by appropriate control experiments.

The action spectrum method has also been applied to CO-inhibitable, light-reversible enzyme reactions that appear to be catalyzed by non-heme iron proteins. The equations for computing the relative and the molar absorption spectrum of the CO compounds of these enzymes are in principle the same as those for the hematin enzymes. In the case of the anaerobic fermentation of glucose to butyric acid, CO_2 and H_2 by *Clostridium butyricum*,[27] the partition equation reverts to the simpler form (Eq. 11).

$$[n/(1 - n)]\, p_{CO} = K \tag{11}$$

Recently a similar, but O_2-activating enzyme, involved in the photosynthesis by *Chlorella*, has been reported from O. Warburg's laboratory.[28] The light absorption of the CO compounds of the two enzymes starts in the blue and extends into and increases continuously toward the ultraviolet. No distinct absorption bands have as yet been observed.

Determination of Photochemical Activation and Inactivation Spectra of Various Biocatalysts

Irreversible Reactions. The irreversible photoinactivation of the sulfhydryl enzyme urease by ultraviolet light has been studied in Warburg's laboratory.[29] The velocity constant, z_i, of the destruction of the enzyme can be computed from Eq. (12).

$$z_i = (1/t) \ln (W_0/W) \tag{12}$$

where W_0 and W refer to the enzyme activity before and after irradiation, t is the time of irradiation, and z_i has the physical significance explained in connection with Eq. (8). Since z_i/i is equal to $\varphi\beta$, a plot of $(z_i/i)_{\lambda x}/(z_i/i)_{\lambda\,\mathrm{ref}}$ versus the wavelength of the irradiating light repre-

[27] W. Kempner, *Biochem. Z.* **257,** 41 (1933); W. Kempner and F. Kubowitz, *ibid.* 265, 285 (1934); F. Kubowitz, *ibid.* 274, 285 (1934). Cf. also Chapters IX and XVIII of Warburg (cited in footnote 3).

[28] H. S. Gewitz and W. Völker, *Z. Naturforsch.* **18b,** 649 (1963). See also Warburg, *Ann. Rev. Biochem.* **33,** 1 (1964).

[29] F. Kubowitz and E. Haas, *Biochem. Z.* **257,** 337 (1933); cf. also page 100 of Warburg (cited in footnote 3).

sents the relative action spectrum or the relative light absorption spectrum of the photodecomposed biocatalyst. If its molecular weight is known, φ and β can be computed. The measurement should be performed at infinitely small light absorption—low enzyme concentration, short light path—so that the intensity, i of the incident light can be taken as a measure of the photochemical efficiency.

Reversible Reactions. Haxo and Blinks[8] determined the action of monochromatic light bands of equal quantum intensity on the photosynthesis of marine algae of different colors. The consumption and production of oxygen during darkness and illumination were measured polarographically while the light absorbed by the algae was determined by means of Ulbricht's sphere. Using a common reference point for the relative action spectra and relative light absorption spectra, the authors obtained simultaneous plots of both types of spectra which permitted their comparison with the known absorption spectra of the pigments present in the algae and evaluation of the function of individual pigments, such as phycoerythrin, in photosynthesis.

Warburg, Krippahl, and Schröder[9] determined the action spectrum of the enzyme responsible for the catalytic action of low intensities of blue-green light on the photosynthetic efficiency of *Chlorella* suspensions exposed to high intensities of red light. For this purpose the quantum intensities, I, were determined that produced equal effects with different colors of light. The relative photochemical action, W, of two wavelengths of monochromatic light is thus defined by

$$W_1/W_2 = I_2/I_1 \tag{13}$$

During the passage of a light beam through a *Chlorella* suspension of sufficient density for manometric determination of the gas exchange, the quantum intensity to which the enzyme in the algae is exposed, decreases with increasing distance from the point of light entrance as a consequence of dispersion of the light by the *Chlorella* cells and absorption by the sum total of *Chlorella* pigments. The decrease in intensity with increasing distance, dx, from entrance point was calculated from Eq. 14.

$$-\Delta I = \bar{I}\beta c dx \tag{14}$$

where $\bar{I}$ is $\frac{1}{2}$ $(I_0 + I)$, the arithmetic mean of the intensities of the incident and the emerging light measured with Ulbricht's sphere; β the molar absorption coefficient of the enzyme; and c the enzyme concentration.

If the quantum intensities producing equal effects are determined for the two wavelengths 1 and 2 upon the same cell suspension, $-\Delta I_1$; $-\Delta I_2$; c_1; c_2, dx_1, and dx_2 can be taken as equal and the relative action spectrum and hence the relative absorption spectrum of the enzyme can be computed from Eq. (15).

$$\beta_1/\beta_2 = \bar{I}_2/\bar{I}_1 \tag{15}$$

In regard to the significance of this equation, Warburg *et al.*[9] emphasize two points: (a) only the absorption spectrum of the sum total of the *Chlorella* pigments can be directly measured by means of Ulbricht's sphere, whereas the absorption spectrum of the enzyme can be deduced only from the action spectrum; and (b) the simplifying assumption that dx_1 equals dx_2 is not correct, since the dispersion of light by the algal cells increases with decreasing wavelengths of light. The absorption coefficients at the shorter wavelengths are thus somewhat overestimated. The authors indicate that this error could be eliminated by determining the light absorption of a *Chlorella* suspension and an equivalent methanol extract of the *Chlorella* pigments under identical conditions. Such measurements were carried out for the wavelengths 645 mμ and 546 mμ. At 645 mμ the absorptions of cells and dissolved pigments were not significantly different, while at 546 mμ the absorption of the cell suspension was 1.57 times that of the methanol extract.

The determination of the action spectrum of phytochrome, that is the biocatalyst of the reversible photoreaction that controls germination of seeds of some higher plants and of fern spores, photoperiodicity of flowering, coloration of some fruits, and etiolation has been described by Parker *et al.*[6] and reviewed by Siegelman and Hendricks.[7] The enzyme has two forms with absorption maxima at 660 mμ (P-660) and 730 mμ (P-730), respectively. Irradiation with light of 660 mμ wavelength converts P-660 into P-730, while irradiation with the wavelength 730 mμ produces the reverse reaction. The action spectrum is obtained as in Eq. (13). Regarding quantitation of the different physiological responses associated with the conversion reaction, the publications cited in footnotes 6 and 7 should be consulted. The effective form of the pigment appears to be P-730.

By expressing the physiological response as a function of the fractional conversion of the pigment into its effective form estimates of the product of the absorption coefficients and the quantum efficiencies for the conversion of the two forms can be obtained.[30] The method is based on the fact that the photoreaction follows first-order kinetics with respect to energy in both directions. The computations are similar to those employed by Warburg (20) for deriving $\varphi\beta$ from the reversible photodissociation of the CO compound of the respiration enzyme.

[30] S. B. Hendricks, H. A. Borthwick, and R. J. Downs, *Proc. Natl. Acad. Sci. U.S.* **42,** 19 (1956).

The fractional conversion of the pigment[7] is given by Eq. 16.

$$\log(\text{P-660}) = -0.43\, I_\lambda S_\lambda (\alpha_\lambda \varphi_\lambda) t + C_1 \quad (16)$$

where I is the quantum intensity, α the molar absorption coefficient (M^{-1} cm^{-1}), t the time of irradiation, C an integration constant, and S the fraction of the incident light energy reaching the pigment; i.e., the fraction not "screened" by dispersion and absorption by other pigments. An analogous equation holds for the mole fraction P-730. If $I_\lambda t$ is determined for two degrees of physiological response to irradiation at 660 mμ as well as at 730 mμ, the mole fractions of P-660 and P-730 (= 1 − P-660) that correspond to the physiological responses observed, are obtained and $\alpha\ \varphi\ S$ can be calculated for the two absorption maxima. This method has formed the basis for the characterization and purification of phytochrome.

[97] Determination of Oxygen Affinity

By FREDERICK J. SCHINDLER

The "determination of oxygen affinity" may be defined as the measurement of the rate of oxygen utilization as a function of oxygen concentration (O_2) over a range of oxygen concentrations including the concentration at which the rate is half-maximal. The latter concentration is generally referred to as the "apparent K_m" for the system. The present paper discusses the main methods for the determination of the K_m of respiratory systems for oxygen and gives a detailed procedure for use of the polarographic method.

Manometric Method

In this method, known oxygen concentrations are introduced in the gas phase of a manometric vessel, and the rate of oxygen utilization is measured in the classical manner. The basic technical problem is maintenance of oxygen concentration equilibration between gas and liquid phases at measurable rates of oxygen consumption. This problem sets a lower limit to the apparent K_m's which can be measured using manometry. The limit may be estimated using the data of Roughton[1] or Matsen and Myers.[2] Matsen and Myers found that at 25° with 120 shakes per minute, 2.5 cm stroke, 3 × 5 × 1 cm rectangular vessels, and 10 ml fluid

[1] F. J. W. Roughton, *J. Biol. Chem.* **141,** 129 (1941).
[2] F. Matsen and J. Myers, *Arch. Biochem. Biophys.* **55,** 373 (1955).

volume, the concentration of oxygen in the fluid phase was less than the equilibrium concentration by $(R/7.7)$ μl/ml, where R is the oxygen uptake in microliters per minute. Thus when R equals 0.1 μl/minute the oxygen concentration in the fluid phase is less than the equilibrium concentration by 0.013 μl/ml or 0.58 μM. Clearly this technique will be unsuitable for measurement of apparent K_m's lower than 5 μM oxygen. Extension of the manometric method to lower apparent K_m's has been achieved but is very difficult.[3,4]

Froese Method[5]

Froese used a manometer vessel containing a polarographic probe to measure oxygen in the fluid phase. He was thus able to measure the deviation from the equilibrium oxygen concentration which, as noted above, is proportional to the rate of oxygen utilization. A basic technical problem of this method is maintenance of a constant rate of gas transfer between phases over extended periods of time. Drift problems in polarographic sensitivity and residual current are magnified in the Froese method, but on the other hand all other technical problems of the "polarographic method" (see below) are minimized or completely eliminated.

Polarographic Method

In this method an attempt is made to obtain a continuous record of oxygen concentration in the fluid phase as a function of time over the range $(O_2) >$ apparent K_m to $(O_2) <$ apparent K_m, all changes in (O_2) being due to the reaction under consideration. The basic technical problems are achievement of (1) a rigorously closed system with respect to oxygen in the fluid phase (2) a precise (noise-free) polarographic trace, (3) a negligible, or at least a stable and oxygen-independent residual current due to other electrode reactions, (4) negligible oxygen-dependent interference of the enzyme system with the diffusion field at the electrode, and (5) negligible time lag in the response of electrode current to oxygen concentration in the bulk of the fluid.

Longmuir[6,7] extended the rapidly moving solid cathode polarographic technique[8-11] to measurements of very low oxygen concentrations. After

[3] O. Warburg and F. Kubowitz, *Biochem. Z.* **214**, 18 (1929).
[4] A. Baender and M. Kiese, *Arch. Exptl. Pathol. Pharmakol.* **224**, 312 (1955).
[5] G. Froese, *Biochim. Biophys. Acta* **57**, 509 (1962).
[6] I. S. Longmuir, *Biochem. J.* **57**, 81 (1954).
[7] I. S. Longmuir, *Biochem. J.* **65**, 378 (1957).
[8] H. A. Laitinen and I. M. Kolthoff, *J. Phys. Chem.* **45**, 1061 (1941).
[9] F. D. Harris and A. J. Lindsey, *Nature* **162**, 413 (1948).
[10] B. Chance, *Science* **120**, 767 (1954).

empirical trial of many glasses, he succeeded in making gold to glass seals with rather large gold diameters. After pretreating the cathode at high polarizing voltages he found residual currents corresponding to less than 0.01 μM oxygen at pH 7 or greater, using —0.8 volt polarizing voltage with respect to a saturated calomel electrode. By rotating the 1.5 mm diameter cathode very reproducibly at 1300 rpm in an arc of several centimeters, he obtained extremely high current sensitivity (0.6 μamperes (μa)/μM oxygen) with negligible noise. Technical problem (1) was met by using a glass vessel with no cavities for gas bubble trapping and with provision for positive displacement and replacement of the reaction mixture via syringes attached to the vessel.

Investigators who intend to determine oxygen affinities on a long-term basis would do well to copy Longmuir's technique. The much simpler mechanical arrangement of Hagihara[12] for rotating the cathode may be substituted. Also, as indicated below, residual currents of the same order of magnitude may be obtained with platinum cathodes by the use of a lower polarizing voltage.

The following procedure[13] is recommended for investigators who do not wish to invest time and effort in a system as complex mechanically as that of Longmuir. The residual currents obtained are in the range 4 to 8×10^{-3} μa/μM oxygen. Apparent K_m's as low as 0.03 μM can be measured. The apparatus is inferior to that of Longmuir in the lower precision of the polarographic trace (peak-to-peak noise about 1.5% of signal) due to irregularities in fluid flow past the electrode surface and in the less rigorous achievement of a closed system in the fluid phase.

An apparatus of the type illustrated in Fig. 1 has been routinely used with success. The apparatus is constructed as follows: fit a flat-bottomed cylindrical bottle (about 5 cm diameter and over 10 cm height) with a 1-inch magnetic stirring bar and a rubber stopper. Insert through the stopper (a) the platinum cathode (see below), (b) a large size fiber-type liquid junction calomel reference electrode with resistance less than 300 ohms, (c) a No. 19 syringe needle long enough to reach well into the bottle, and (d) a short syringe needle reaching just beyond the stopper. Introduce the reaction mixture, complete except for enzyme, leaving only enough gas space to accommodate subsequent additions. Prepurified nitrogen (less than 10 ppm oxygen) is led into the reaction vessel through

[11] P. W. Davies, *in* "Physical Techniques in Biological Research" (W. L. Nastuk, ed.), Vol. IV, p. 137. Academic Press, New York, 1961.

[12] B. Hagihara, *Biochim. Biophys. Acta* **46**, 134 (1961).

[13] Work done with Dr. H. Ikuma.

FIG. 1. The reaction vessel and electrode arrangement for the polarographic measurement of oxygen affinity.

a short length of tubing (preferably nylon, glass, or metal) and a Luer slip fitting to the long needle. Oxygen is purged from the system as determined by monitoring the polarographic current. The short needle in the stopper allows gas to escape. After establishing the residual current level, stop the gas flow and insert by syringe known volumes of reaction mixture containing known quantities of dissolved oxygen, thus calibrating the polarographic current over the span of oxygen concentrations to be used in the determination of the oxygen affinity. Perform the calibration rapidly so that the small gas volume remains essentially oxygen-free nitrogen. Next, record the rate of loss of oxygen to the gas phase (slight due to the small interface area:volume ratio). Finally, insert the enzyme system under study and record polarographic current as a function of time as the enzyme system removes the oxygen. Add more oxygen-containing reaction mixture, recalibrating in the presence of enzyme, and repeat the attainment of anaerobiosis. It will be possible to adjust the enzyme concentration so that both oxygen losses to the gas phase and electrode time lags are negligible.

The preparation of the oxygen cathode is critical for the success of the method. As the area of the cathode is increased, the sensitivity of the polarographic current to changes in the fluid flow rate increases. On the other hand, the residual current in terms of equivalent oxygen concentration decreases as 1/cathode diameter.[14] To attain the cathode characteristics mentioned above, use about 0.5 mm diameter cathodes. These are prepared as follows: melt 8 mil platinum wire at one end so as to form a ball about 1 mm in diameter. Seal in a soft glass tube long enough to reach through the stopper to just above the stirring bar in the reaction vessel. Grind to expose the desired area of platinum, and polish with very fine carborundum spread on paper. It is best to make about ten electrodes at one time and empirically select the best with respect to residual current and insensitivity to fluctuations in stirring rate. Cracks or air bubbles at the platinum-glass interface must be completely absent.

It is necessary to apply the polarizing voltage from a high quality potentiometer which has only the necessary voltage (0.6 volt) across it. Solder all connections well. When working at low oxygen and correspondingly high load resistors, small transient changes in polarizing voltage give significant artifactual signals due to the large capacitance at the platinum-solution interface. Minimize the residual current to current sensitivity ratio by adjusting the polarizing voltage in the range 0.4–0.6 volt with respect to the saturated calomel electrode. To achieve a sensitivity of about 0.1 μM oxygen full scale on a chart recorder, use a millivoltmeter on the 0.1-mv scale with a 300,000 ohm load resistor for the polarographic current. To reduce the sensitivity, switch to less sensitive scales on the amplifier, but use lower load resistors above 10 μM oxygen so that the voltage drop across the load resistor is never greater than 10 mv.

Bacterial Luminescence Method[15]

In this method the oxygen indicator in the fluid phase is simply bacterial luminescence instead of polarographic current. With the vessel described above (in a light-tight box) viewed by a photomultiplier, oxygen could be measured down to 0.0004 μM with *Photobacterium fischeri* concentrations which caused negligible oxygen uptake with respect to the enzyme being studied. Luminescence is linearly related

[14] I. S. Longmuir, personal communication, 1963.

[15] F. J. Schindler, Ph.D. Dissertation, Univ. of Pennsylvania (1964), University Microfilms, Inc., Ann Arbor, Michigan.

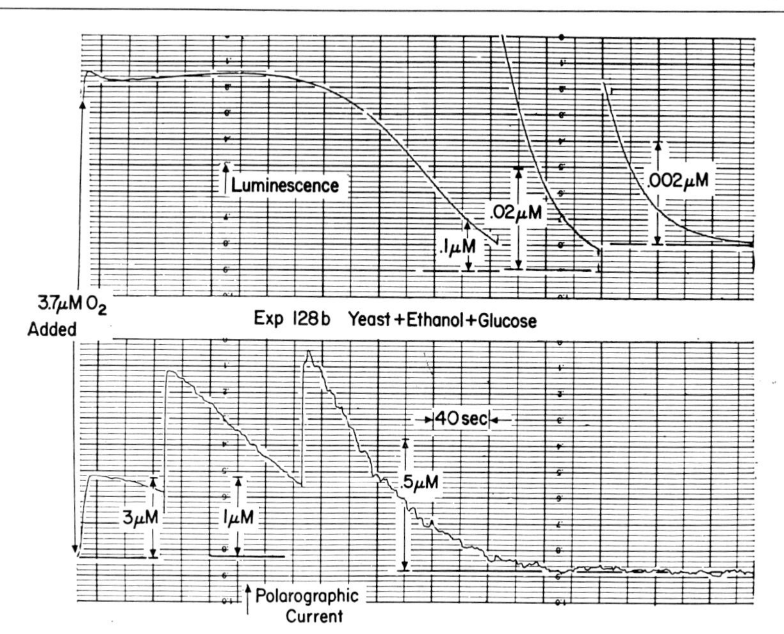

FIG. 2. Comparison of the luminescence and polarographic method for determining the oxygen affinity of yeast respiration.

to oxygen concentration below about 0.1 μM oxygen. Sugar or salt must be present in the reaction medium to provide an osmolarity of at least 0.6 M. Temperature should be between 10° and 30°. This method becomes useful for determination of apparent K_m's which are too low for determination by the polarographic method or for measurement of oxygen utilization in the region (O_2) much less than apparent K_m, particularly in the case of cytochrome oxidase systems. An example of the comparative characteristics of the luminescence and polarographic methods for determining the oxygen affinity of yeast is illustrated in Fig. 2.

[98] Determination of Acid-Labile Sulfide and Sulfhydryl Groups

By TSOO E. KING and ROY O. MORRIS

Acid-Labile Sulfide

Principle

"Acid-labile sulfide," "inorganic sulfide" or "sulfide" recently reported in enzymes and ferredoxins (see, for example, Handler *et al.*[1] and

[1] P. Handler, K. V. Rajagopalan, and V. Aleman, *Federation Proc.* **23**, 30 (1964).

San Pietro[2]) is operationally defined by the method of determination according to Fogo and Popowsky.[3] In the original method, hydrogen sulfide is absorbed in a suspension containing sodium hydroxide and zinc acetate. The zinc sulfide formed is coupled with *p*-aminodimethylaniline (*N,N*-dimethyl-*p*-phenylenediamine) in acidic solution in the presence of ferric chloride to give, presumably, methylene blue, which is then determined spectrophotometrically. For biological material, the sample is directly treated with NaOH–$Zn(CH_3COO)_2$ mixture and then coupled with *p*-aminodimethylaniline as in the original procedure. The protein precipitated is removed by centrifugation, and the reading at 670 mμ is taken in the clear, blue supernatant solution.

Assay Method

The following is an adaptation[4] of the method by Fogo and Popowsky and has been used in this laboratory for some time. The concentration of the stock zinc acetate solution is increased so that the test may accommodate dilute enzyme samples.

Reagents

Zinc acetate, 2.6% in water

Sodium hydroxide, 6% in water

N,N-Dimethyl-*p*-phenylenediamine monohydrochloride (Eastman No. 492), 0.1% in 5 *N* HCl

Ferric chloride, 0.0115 *M* in 0.6 *N* HCl

Standard sodium sulfide, 2×10^{-2} *M* in oxygen-free water. The sodium sulfide crystals should be first thoroughly washed with water to remove oxidation products such as sulfite and some unknown inhibitory compounds on the surface. It is standardized by the conventional iodometric method. For the use in the colorimetry, it is carefully diluted to 1×10^{-4} *M* with oxygen-free water. Solutions of sodium sulfide are not stable as a result of inevitable contact with oxygen and should be prepared frequently.

"Oxygen-free" water. Water used throughout the procedure is redistilled from a glass apparatus and also passed through ion exchangers. "Oxygen-free" water used in the preparation of sulfide solutions is made by first boiling and then cooling with simultaneous bubbling of an inert gas such as "prepurified" nitrogen and helium.

[2] A. San Pietro, *Proc. 6th Intern. Congr. Biochem. New York, 1964,* Vol. 5, p. 369.

[3] J. K. Fogo and M. Popowsky, *Anal. Chem.* **21**, 732 (1949).

[4] T. E. King, *Biochem. Biophys. Res. Commun.* **6**, 511 (1964).

Procedure.[5] In a glass-stoppered test tube of 13 × 100 mm (such as No. 60818, Van Waters and Rogers, Inc., Seattle) are placed 0.5 ml of zinc acetate, 0.1 ml of NaOH, and 0.7 ml of sample. The tube is stoppered with a small drop of water touched on the stopper as a seal. The tube is shaken on a Vortex test tube shaker for 1 minute. Open the stopper. Rapidly add 0.25 ml of the diamine reagent, close the stopper, gently swirl (the content now becomes clear), then open the stopper, and quickly add 0.1 ml ferric chloride. The tube is immediately stoppered and shaken on a Vortex for 1 minute.

Allow the mixture to stand at room temperature for 30 minutes, then add 0.85 ml of water. Mix well. The precipitate is removed by centrifugation at about 10,000 *g* for 10 minutes. The reading is taken at 670 mμ against a reagent blank.

Usually, four levels of standard are made at 5, 10, 20, and 40 millimicromoles of sulfide per tube corresponding to 0.05, 0.1, 0.2, and 0.4 ml of the dilute standard solution. Within these concentrations, the absorbance is linear with sulfide concentration. It is advisable to use three levels of the sample with the reading of the highest not more than 0.4 absorbance unit.

Calculation

Once a standard curve is established it may be used for all determinations when the conditions are not changed. Nonetheless, it is recommended to check from time to time. Under the conditions described, 1 micromole sulfide gives an absorbance of 10.6 (10–11.1) at 670 mμ. Values of 11.4[6] and 10.7[3] have been reported.

Comments

The blue color formed, presumably methylene blue, shows absorption maxima[3,4,7] at 668–669 and 746 mμ, a shoulder at 620 mμ, and a trough at 707 mμ. The maximum at 746 mμ may also be used for the spectrophotometric reading, especially for those solutions with very slight turbidity due to some fine protein precipitate which cannot be easily centrifuged down.

When sulfide content is high in some samples, such as ferredoxins, the centrifugation for removal of protein may not be necessary. On the

[5] Advice received in our initial work from Dr. Bob B. Buchanan, Dr. A. San Pietro, and Professor H. Freund is gratefully acknowledged.

[6] W. Lovenberg, B. B. Buchanan, and J. C. Rabinowitz, *J. Biol. Chem.* **238**, 3899 (1963).

[7] T. E. King, *in* "Non-heme Iron Proteins" (A. San Pietro, ed.), p. 413. Antioch Press, Yellow Spring, Ohio, 1965.

other hand, certain protein precipitates especially in bulk amount adsorb the blue compound formed in the colorimetry. Care must be taken on this aspect in interpreting the results. Actually this method is not satisfactory for those samples with low sulfide content and with activity to significantly adsorb methylene blue.

The coupling reaction is inhibited[5] by cations of many heavy metals, e.g., mercury, nickel, and cadmium. A few tenths of millimicromoles of Mersalyl, α-[(3-hydroxymercuri-2-methoxypropyl)carbamyl] phenoxyacetate, significantly inhibits the color development. Another factor of lower results can be attributed to the loss of sulfide to the gas phase after the addition of the diamine reagent. The amount that escapes from the colorimetry is a function of the solubility and total content of sulfide present.

The configuration or nature of sulfide in enzyme is not known, although several schemes have been proposed (see, for example, Blomstrom[8]). The existence of sulfide as such in protein (but not an operational artifact) has also been questioned.[9] However, cysteine, glutathione, or bovine serum albumin does not show the color reaction under the conditions tested.

Sulfhydryl Groups

Principle

The method described here is the Thomson and Martin[10] modification of the procedure used by Benesch *et al.*[11] The protein is titrated in solution with silver ions which form silver mercaptides as shown in Eq. 1.

$$RSH + Ag^+ \rightarrow RSAg + H^+ \quad (1)$$

Here R is the protein moiety. The excess Ag^+ over the stoichiometric amount required is detected amperometrically.

Where it is desired to determine the total sulfhydryl content the titration should be performed in 8 M urea in order to expose "masked" sulfhydryl groups. The "masked" sulfhydryl groups normally react slowly, if at all, with silver ions.

[8] D. C. Blomstrom, E. Knight, W. D. Phillips, and J. F. Weiher, *Proc. Natl. Acad. Sci. U.S.* **51**, 1085 (1964).

[9] W. Bayer, W. Parr, and B. Kazmaier, *Arch. Pharm.*, **298**, 196 (1965). From further experimental findings, this view has been challenged [see H. Matsubara, IEG-1-580 and R. Malkin and J. C. Rabinowitz, *Biochemistry* **5**, 1262 (1966)].

[10] C. G. Thomson and H. Martin, *in* "Glutathione," *Biochem. Soc. Symp.* **17**, 17 (1959).

[11] R. E. Benesch, H. A. Lardy, and R. Benesch, *J. Biol. Chem.* **216**, 663 (1955).

Assay Method

Apparatus. The apparatus for the titration is shown diagrammatically in Fig. 1. It consists of a platinum sensing electrode (*A*) in a beaker (*B*) containing the titration medium, a micrometer syringe burette (*C*) for addition of silver nitrate, and a mercury-mercuric oxide reference electrode (*D*) connected to the beaker by means of the potassium chloride bridge (*E*). During the titration, the beaker is rotated at constant speed by means of a synchronous motor. The diffusion current, due to excess silver ions, is measured by the galvanometer (*G*). It is advisable to perform the titration in the absence of air by blowing a gentle stream of nitrogen over the solution.

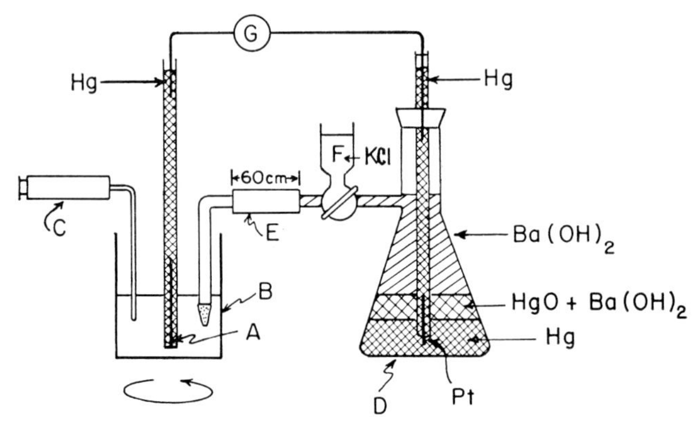

FIG. 1. Diagram of the titration apparatus for the determination of sulfhydryls (not in scale). See text for explanation.

Construction details are not critical, but the following points may prove helpful. The platinum electrode, similar to that used in the polarographic determination of oxygen, is constructed by sealing a 2 cm length of 26 gauge platinum wire into the end of a 6 mm (OD) glass tube. The wire is sealed so that it is perpendicular to the length of the tube. Contact with the external circuit is made by filling the tube with mercury. If necessary, the diffusion current, and thereby the sensitivity, may be increased by increasing the area of the electrode.

The usual level of thiol assayed by this technique is about 0.1 micromole. Since this is conveniently estimated in about 15 ml of titration medium, the beaker is of 50 ml capacity. Although the rate of rotation of the beaker is not critical, it must be held constant during the titration and should not be less than 70 rpm.[12] But the rate of the rotation should

[12] We use a phonograph motor of 78 rpm; a gear is changed to make actual rotation at about 100 rpm.

not be high enough to cause the electrode to break through the liquid surface; turbulence must be avoided.

The connection between the titration vessel and the reference electrode is accomplished by a saturated potassium chloride salt bridge. The bridge (*E*) is 60 cm in length and is conveniently made from Tygon tubing. At the electrode vessel it is sealed by a tapered glass tube containing a plug of 2% agar gel in saturated potassium chloride solution. If, between periods of use, the plug is immersed in saturated KCl solution, its useful life is indefinite.

The reference electrode is constructed from a 125 ml Erlenmeyer flask to which a three-way stopcock is sealed. The upper arm of the stopcock ends in the funnel (*F*) and is used to fill the bridge with saturated potassium chloride solution. A layer of mercury to a depth of at least 1 cm is placed in the flask and is overlaid with a slurry of mercuric oxide in saturated barium hydroxide solution. The flask and side arm, up to the stopcock, are then filled with saturated barium hydroxide solution. An electrical contact is established by a platinum wire immersed in the mercury layer. If protected from the light, this reference electrode remains stable for many months.

The galvanometer (*G*) should have a sensitivity of at least 0.006 μa/mm.

Reagents. For the success of the method all reagents should be the purest available. The technique is especially sensitive to heavy metals, and all reagents must be made up in deionized and glass-redistilled water. "Oxygen-free" water is recommended for precious work. Normally, reagent grade urea contains sufficient ammonium cyanate to render the method useless. Therefore, wherever this reagent is used, it should be recrystallized. The recrystallization is done by first treating a saturated urea solution at 50° with ion exchanger, Amberlite MB1, for 10 minutes.

The titration medium has the following composition: Tris, 1 *M*, free base, 27 ml; potassium chloride, 1 *M*, 2 ml; gelatin, 0.1% w/v, 6 ml; urea, solid, 48 g.

The volume of this mixture is made up to 100 ml and the pH is adjusted to 7.2 with nitric acid. Gelatin is necessary only when nonprotein samples are titrated. Urea is not essential when the sulfhydryl groups are reactive, as they are in low molecular weight compounds or "exposed" on the protein surface.

The silver nitrate for titration is prepared by direct weighing of the analytically pure compound. A stock solution 0.1 *M* in concentration is stable over long periods in the dark and is diluted to 0.001 *M* for use.

Procedure. The protein (containing an equivalent sulfhydryl content of 0.1–0.4 micromole) in a 1 ml volume is added to 15 ml of titration medium at room temperature. The titration is performed by adding 0.001 *M* silver nitrate in 0.01 ml aliquots through a micrometer syringe burette and reading the current 2 minutes after each addition. Prior to the end point, the current rises very slowly with each addition and a base line can be established. As the end point is approached, the magnitude of each rise becomes greater, and finally when silver is in excess the current rises very rapidly in a linear manner. The end point is taken as the intersection of the excess Ag^+ line with the base line. A typical titration curve is illustrated in Fig. 2. Since some masked sulfhydryl groups react very sluggishly even in 8 *M* urea, it is often necessary to add silver nitrate solution at intervals greater than 2 minutes.

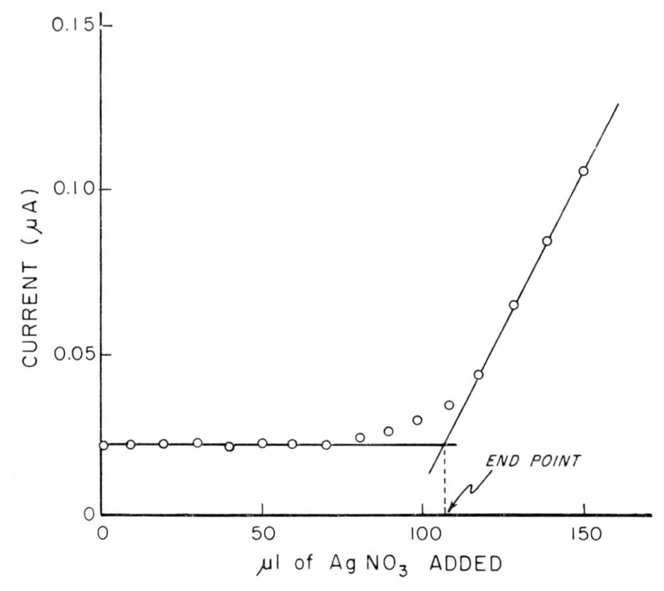

FIG. 2. Titration of glutathione (0.11 micromole) with 0.001 *M* $AgNO_3$.

Before satisfactory results can be obtained, it is necessary to condition the platinum electrode. The electrode is first cleaned in strong nitric acid and subsequently in water. A series of titrations are then carried out using a low molecular weight compound such as glutathione. It is found that with each successive titration, the base line current gradually falls, finally reaching a constant, low value.[13] When this occurs, the electrode is ready for use, and up to fifty titrations may be performed without loss of sensitivity. However, with time the response

[13] The absolute value in term of amperage is dependent on a number of factors, such as the area of platinum electrode (cf. Fig. 2).

becomes sluggish and it becomes necessary to clean and recondition the electrode as described above.

Calculation

Since there is a 1:1 stoichiometry involved in the reaction, the sulfhydryl content of the sample is given by the Eq. (2).

$$\text{Sulfhydryl content (micromoles)} = V \times M \times 1000 \qquad (2)$$

Where V is the volume in milliliters of silver nitrate of molarity M required to reach the end point.

Comments

The sulfhydryl content thus determined should not be considered as the absolute value. There are two factors, among others, of opposite directions. One is the accessibility of Ag^+ to —SH group; even in 8 M urea not all may be exposed. Second, other species may react with Ag^+. This is especially true for the recently described acid-labile sulfide in several enzymes. Likewise, when an amino group is proximate to the sulfhydryl, a compound of the type $(RSAg)_2Ag$ may be formed (cf. Kolthoff and Strichs[14]).

Acknowledgment

The experimental work was supported by grants from the National Science Foundation, the National Institutes of Health, the American Heart Association, and the Life Insurance Medical Research Fund.

[14] I. M. Kolthoff and W. Strichs, *J. Am. Chem. Soc.* **72,** 1952 (1950).

[99] Sensitive Measurements of Changes of Hydrogen Ion Concentration

By B. Chance and M. Nishimura

Biological reactions involved in the production or consumption of acid or base can be studied easily by measuring pH. This method has been applied, among others, in studies of organic acid production[1,2] and phosphorylation.[3] For quantitative calculations, it is necessary to know pK's of all the reactants and products, pH where the reaction is followed, and the buffering capacity of the reaction system (see Calibration).

[1] P. Rona and R. Ammon, *Biochem. Z.* **249,** 446 (1932).
[2] B. Chance, *Science* **120,** 767 (1954).
[3] S. P. Colowick and H. M. Kalckar, *J. Biol. Chem.* **148,** 117 (1943).

As an example, the application of this method to the measurement of photophosphorylation in bacterial chromatophores is shown.[4] In this case, electron flow is cyclic, and no net oxidation or reduction products are accumulated. Therefore, only the concentrations of P_i, ADP, and ATP (and their magnesium complexes) are taken into consideration.

In the simplified stoichiometry of the reaction,

$$ADP + P_i + nH^+ \rightarrow ATP + H_2O \tag{1}$$

the value of n ($= \Delta H^+/\Delta P_i$) is calculated from pK's of the reactants. This value of n and the buffering capacity ($=$ added $H^+/\Delta pH$) determined experimentally by titration are used for the calculation of reaction rates. In the pH range of 7–9, the above-mentioned reaction can be rewritten as follows:

$$\begin{pmatrix} ADP^{-3} \\ \rightleftharpoons \\ ADP^{-2} \end{pmatrix} + \begin{pmatrix} P_i^{-2} \\ \rightleftharpoons \\ P_i^{-1} \end{pmatrix} + nH^+ \rightarrow \begin{pmatrix} ATP^{-4} \\ \rightleftharpoons \\ ATP^{-3} \end{pmatrix} + H_2O \tag{2}$$

Or, if we take the magnesium complexes of ADP and ATP into consideration, Eq. (2) becomes

$$\begin{pmatrix} ADP{\cdot}Mg^{-1} \\ \rightleftharpoons \\ ADP{\cdot}Mg \end{pmatrix} + \begin{pmatrix} P_i^{-2} \\ \rightleftharpoons \\ P_i^{-1} \end{pmatrix} + nH^+ \rightarrow \begin{pmatrix} ATP{\cdot}Mg^{-2} \\ \rightleftharpoons \\ ATP{\cdot}Mg^{-1} \end{pmatrix} + H_2O \tag{3}$$

The large values of apparent stability constants of the magnesium complexes of ADP and ATP[5-7] suggest that Eq. (3) takes place in the presence of magnesium ion. These reactants, P_i, ADP, ATP and their magnesium complexes, can be regarded as weak acids, and the following relationship can be assumed:

$$pH = pK_a + \log \frac{\gamma_{A^-}}{\gamma_{HA}} + \log \frac{[A^-]}{[HA]} = pK'_a + \log \frac{[A^-]}{[HA]} \tag{4}$$

where γ_{A^-} and γ_{HA} are the activity coefficients for the conjugate basic and acidic forms, respectively. The values of pK's used for the calculation of n are shown in the table. The result of the calculation appears in Fig. 1. In this figure, n is expressed as a function of pH. It is seen that, between pH 7 and 9, the n-value is close to, but lower than, unity. The values of n calculated with and without assumption of magnesium complex formation differed slightly. In Fig. 1, the value of n experimentally obtained from the comparison of the two independent methods (P_i determination and pH recording) is also indicated by a heavy vertical line.

[4] M. Nishimura, T. Ito, and B. Chance, *Biochim. Biophys. Acta* **59**, 177 (1962).
[5] K. Burton and H. A. Krebs, *Biochem. J.* **54**, 94 (1953).
[6] R. M. Smith and R. A. Alberty, *J. Am. Chem. Soc.* **78**, 2376 (1956).
[7] R. A. Alberty, R. M. Smith, and R. M. Bock, *J. Biol. Chem.* **193**, 425 (1951).

VALUES OF pK'_a USED FOR CALCULATION OF n[a]

Acid	pK'_a	Acid	pK'_a
ATP^{-3}	6.48	ATP Mg^{-1}	4.97
ADP^{-2}	6.26	ADP Mg	5.13
P_i^{-1}	6.64		

[a] Based on data of R. M. Smith and R. A. Alberty [*J. Am. Chem. Soc.* **78,** 2376 (1956)] and of R. A. Alberty, R. M. Smith, and R. M. Bock [*J. Biol. Chem.* **193,** 425 (1951)].

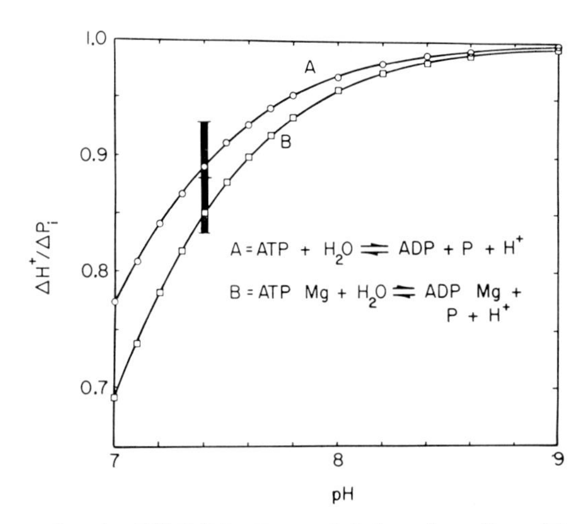

FIG. 1. Values of n $(=\Delta H^+/\Delta P_i)$. Curve A is based on Eqs. (2) and (4); curve B is derived from Eqs. (3) and (4). Heavy vertical line corresponds to the experimentally determined value of n at pH 7.40 from the comparison of the phosphorylating rates obtained by two independent methods (pH measurement and P_i determination).

Typical examples of pH recordings in the photophosphorylation of chromatophores isolated from purple bacterium *Rhodospirillum rubrum* are shown in Fig. 2. Light-induced pH rise is clearly demonstrated in these pH traces. This rise is P_i-, ADP-, and magnesium-dependent and shows a good agreement with other methods of determination of phosphorylation. However, under certain conditions, biological particulate systems, including mitochondria, chloroplasts, and chromatophores, show marked pH changes which are not directly correlated to the concentration changes of substrates or metabolites.

Calibration

In the measurement of pH changes in biological reaction system, the amount of H^+ produced or consumed, as well as the actual pH change,

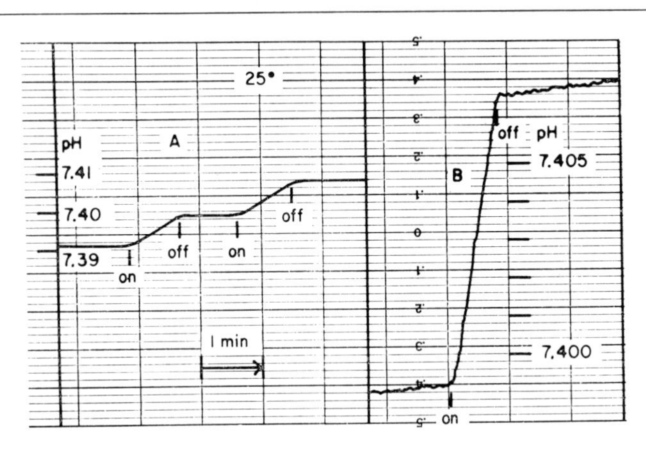

FIG. 2. Recordings of the pH change by illumination of *Rhodospirillum rubrum* chromatophores. Reaction mixture contained 10 micromoles of KH_2PO_4, 30 micromoles of $MgCl_2$, 1 micromole of sodium succinate, 1 micromole of ADP, chromatophores 7.78 millimicromoles of bacteriochlorophyll, in 2.34 ml.

is often necessary. For the calibration of $\Delta H^+/\Delta pH$ (H^+ produced or consumed/ΔpH), a small volume of dilute strong acid or alkali (e.g., HCl or NaOH) of known concentration is added to the reaction mixture and the deflection of recorder is measured. This calibration must be carried out at or close to the pH where the reaction is studied, because the buffering capacity and $\Delta H^+/\Delta$ substrate are dependent on pH. Calibration of this type can be conveniently done by an automatic titrating device or pH-stat, though this is slower and more sluggish in response than the direct pH recording when used for the measurement of time-course of H^+ production or consumption.

Stirring of Reaction Mixture

Adequate stirring is necessary for the uniform mixing of components and for the rapid response of electrodes. Stirring of the reaction mixture is usually done mechanically, i.e., by a vibrating rod, a magnetically driven "flea" or a slow rotation (e.g., 60 rpm) of the reaction cuvette by a micromotor. Too rapid stirring may give an artifactual electrode potential.

Some Possible Causes of Errors

Besides the common causes of errors in pH determination (liquid junction potential, presence of high concentration of salts, high protein concentration, high flow rate, etc.), the following are often encountered as possible causes of errors in studies of pH change with a high sensitivity.

(1) Dissolution of volatile acid, etc., especially carbon dioxide. A proper design of the reaction cell will minimize this effect. (2) Electrostatic disturbance by charged objects (clothing, etc.). A metal shield of the glass electrode assembly reduces the external electrostatic effect. (3) Illumination of electrode. When pH recording is applied with photosynthetic systems or in a system with a high illumination for fluorescence excitation, etc., this effect at times must be taken into consideration. Some of the electrodes, especially the colored ones, were found to be more sensitive to light than others. When trouble is encountered in this respect, the use of cut-off filters to remove the light of shorter wavelengths is often helpful.

The glass electrode is also subject to the so-called suspension error[8] so that the actual value of the pH may be in error by several tenths of a unit. This may make little difference in a mitochondrial suspension or protein solution where the pH is initially set by a buffer and only changes of pH are to be recorded. The changes appear to be reliable even though the protein concentration may be 10 and 20 mg/ml.

Limitations of Sensitivity

A number of authors have measured pH changes down to the region of a 10^{-4} pH,[9] and it is not difficult to measure liberation or uptake of H^+ of a few millimicromoles. In the presence of fairly high concentrated protein solutions (for example, 10 mg/ml mitochondrial protein[4,10-12]). Often the limitation of the stability of the glass electrode for pH measurement may be set by the stability of the reference electrode.

Other electrodes (e.g., quinhydrone) might well be suitable for sensitive measurements; however, these have not been tested in detail in this laboratory with biological preparations.

Meters

Many meters can be satisfactory for recording pH changes if this is the only quantity being recorded. However, more complicated devices currently in use where oxygen, potassium, and sodium as well as hydrogen ions are measured[13] lead to interaction between the electrode which

[8] W. Bartley and R. E. Davies, *Biochem. J.* **57**, 37–49 (1954).

[9] B. Chance, *Harvey Lectures Ser.* **49**, 145 (1955).

[10] See, for example, B. Chance, G. R. Williams, and G. Hollunger, *J. Biol. Chem.* **278,** 439 (1963).

[11] See, for example, P. Mitchell and J. Moyle, *Nature* **208,** 147 (1965).

[12] See, for example, B. Chance, T. Ito, P. K. Maitra, and R. Oshino, *J. Biol. Chem.* **238,** 1516 (1963).

[13] See, for example, B. Pressman, this volume [111].

is best controlled by allowing each electrode system to "float," isolated from ground and not connected to ground even at the recording end. In this way, the most convenient system for interconnection of the electrode grounds can be selected. It should be noted that some meters show a large amount of 60-cycle ripple in their output and are therefore not suitable for the registration of rapid pH changes. We have used extensively a simple battery-operated electrometer circuit not greatly different from the original DuBridge electrometer,[14] and more recently chopper-type circuits such as the Radiometer Model 22 pH meter or the Leeds and Northrup pH meter. The radiometer device gives a large voltage output, whereas that of the Leeds and Northrup may be as small as 10 millivolts full scale. Thus, the recording of a thousandth of a pH unit requires a highly sensitive amplifier. However, many pH meters are now being made with an expanded scale and a higher voltage output. The Radiometer gives about 0.7 volt/pH and is satisfactory in this respect.

Vibrating reed electrometers manufactured by the London Co.[15] give outstanding performance which is scarcely necessary for that of the ordinary size electrode but may be extremely useful in applications where the electrode resistance is very high.

A large number of electrodes have been found suitable for precise pH measurements, but not all have shown rapid response speed. The larger electrodes have been used in the flow apparatus,[16] and the Leeds and Northrup type 124138 have been used in some relatively rapid experiments in cuvettes. Certain of the "combination electrodes" are also available, such as the Radiometer type GK 2024C, the Arthur H. Thomas 4858-L15 and the Instrumentation Laboratory type 14040.

Indicator Methods

Colorimetric pH indicators (for a summary of types, see Chance[17]) have the advantages that they respond extremely rapidly (in the order of microseconds), that, with appropriate spectrophotometric equipment, extremely sensitive pH measurements may be made, and that they are usable in extremely small volumes. Their disadvantages are that their light absorption may interfere with other optical observations, that they may react chemically in oxidation-reduction reactions, and that they

[14] B. Chance and V. Legallais, *Rev. Sci. Instr.* **22**, 627 (1951).

[15] B. Chance, *in* "Waveforms" (B. Chance, V. Hughes, E. F. MacNichol, D. Sayre, and F. C. Williams, eds.), MIT Radiation Laboratory Series, **19** (2nd ed.), 389 (1964).

[16] See W. Love and F. J. W. Roughton, *in* "Rapid Reactions, Technique of Organic Chemistry" (S. L. Friess and A. Weissberger, eds.), 1st ed., Vol. 8, p. 733. Wiley (Interscience), New York, 1953.

[17] B. Chance, *J. Biol. Chem.* **240**, 2729 (1965).

may be bound to constituents of the reaction mixture with an alteration in the chemical properties. One advantage may accrue from an apparent disadvantage: the indicators may be bound on the "inside" of organelle membranes so that pH changes in localized volumes may be detected. (An example of this is afforded by the employment of phenol red in yeast cells.[18])

Exploitation of the advantages of the indicators with respect to sensitivity requires the use of sophisticated optical apparatus, particularly the double beam spectrophotometer or equivalent devices which are capable of measuring 1×10^{-4} in optical density at moderate response speeds or somewhat larger absorbance changes with response speeds suitable for employment with flow apparatuses. Under these conditions, micromolar amounts of the change of hydrogen ions may be detected in time of fractions of milliseconds. It is appropriate to point out that the combination of such apparatuses is fairly cumbersome compared to that of a simple pH meter. Nevertheless, the increased range of performance is striking. In order to avoid interference with the absorption bands of components under study, two types of indicator have been found valuable. For example, in studies on preparations containing cytochromes, indicators absorbing in the region of 620 mμ, such as bromothymol blue, have been useful. Alternatively, fluorescence indicators, such as 4-methylumbelliferon can be excited at 366 mμ, giving emission in the region around 420 mμ.

The protein error of indicators has been discussed by Clark[19] and has been investigated in detail, for example, in the binding of bromothymol blue to bovine serum albumin. The chief effect is a shift of the indicator pK several tenths of a pH unit in the alkaline direction. If, however, the binding constant for the indicator is high, and constant protein concentrations are used throughout the experiment, pH changes are readily measured following suitable calibrations of the glass electrode.

Binding of bromothymol blue to intact mitochondria, submitochondrial particles, chloroplasts and chromatophores has become of recent interest, since it is apparent that the indicator binds to the hydrophobic portion of the membrane involved, and in each case, appears to indicate changes of hydrogen ion concentration of the opposite sense to those shown by an external pH indicator, for example, the glass electrode. Under such circumstances, the binding of the indicator to the membrane system involves a shift of pH which makes the determination of absolute

[18] K. M. Brandt, *Acta Physiol. Scand.* **10**, Suppl. 30 (1945).

[19] W. M. Clark, "The Determination of Hydrogen Ions," 3rd ed. Ballière, London, 1928.

pH inaccurate, but allows the determination of changes of H^+ concentration. The indicator can be calibrated by titrations with acids and bases in the presence of the glass electrode. We include below a typical experiment where performance of the indicator method may be compared with that of the glass electrode.

Comparison of Performance of Glass Electrode with Indicator Methods. Figure 3 compares the glass electrode and color indicator methods for measuring a fast light-induced pH change in a suspension of chromatophores prepared from *R. rubrum.*[20] This record is useful, since it emphasizes strengths and weaknesses of the two methods. A chromatophore suspension illuminated by infrared light at an intensity

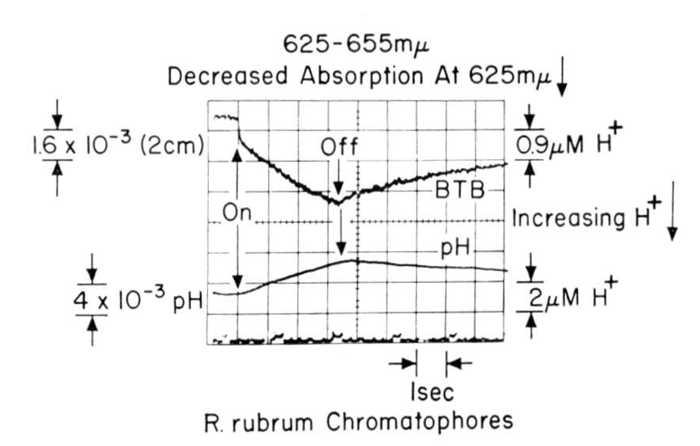

FIG. 3. Comparison of bromothymol blue and glass electrode responses to an illumination of *R. rubrum* chromatophores. The traces from top to bottom are spectrophotometric trace, pH, and time markers. Chlorophyll concentration 37 μM; 0.03 M NaCl; 2 mM glycylglycine buffer, pH 7.0; 8 μM bromothymol blue.

of 4 nEinsteins cm^2/second exhibits an external alkalinization and an internal acidification due to ion movements or a proton pump in the membrane of the vesicle. The glass electrode indicates by its upward deflection the time course of pH change (2 μM H^+) in 7 seconds. This illustrates the great sensitivity of recording in unbuffered suspensions of biological material (2 mM glycylglycine was present). The pH change corresponding to 2 μM hydrogen ion was 4×10^{-3}. The slow response of the glass electrode is indicated in the beginning of the curve; when the light is flashed on, the electron transport system responds in 50 msec, there is no pH change, and the maximum rate is obtained only after a half-second illumination.

A color indicator, bromothymol blue, has been found to indicate complementary acidification within the vesicular structure of the chromato-

[20] B. Chance, M. Nishimura, M. Avron, and M. Baltscheffsky, *Arch. Biochem. Biophys.* **117,** 158 (1966).

phore to which it is bound.[20] Here a linear decrease of absorbancy corresponding to acidification in the structure is observed from the moment of turning on the light. The absorbancy change corresponding to 0.9 μM H^+ change is 1.6×10^{-3} for a 2-cm optical path. Some noise is indicated on the trace, but actually the response speed of the electronic circuit for the optical recording was 50 msec compared to 250 msec for the glass electrode circuit. Thus one can say for equal electrical response speeds, the signal to noise ratio is about the same. It should be pointed out, however, that the color indicator, while sensitive inside the vesicular membrane, will continue to give sensitive indications over a wide range of buffer strengths, in contradistinction to the glass electrode.

The weakness of the color indicator is that other absorbancy changes in the biological material will interfere with the indication. This is shown by the jump in the trace at the point of turning on the light, and suitable calibrations of the amount of this jump had to be made in the absence of the indicator.

In summary, the glass electrode has the advantage of specificity, while the color indicator has the advantage of response speed. The glass electrode suffers from slow response speed; the color indicator technique suffers from confusion of the absorbancy change with those of other pigments in the biological material. The fact that the indicator can be bound on either side of the membrane has great fundamental advantage but requires a suitable control in order to determine where the indicator is bound and under what experimental conditions.

Flow Apparatuses. The glass electrode has been used to measure the kinetics of rapid reactions on several time scales; for example, Steinhardt and Zaiser adapted the MacInnes-Belcher tubular glass electrode (now manufactured by Cambridge Instruments) in a flow apparatus simply by attaching a mixer to it. They achieved a flow rate of 0.34 ml/second and thus made pH measurements at a time of less than 3 seconds after mixing. They claimed an accuracy of 0.005% per pH.

A large glass electrode with a small resistance was an important feature of their apparatus. They also used a flowing junction for the reference electrode. The authors state that pH measurements made in stationary and flowing standard solutions of acid were the same. Thus, there is no "flow artifact."[21]

Love[16] has used the stopped-flow method to achieve a time resolution of 60 msec which was limited by the speed of response of the glass electrodes. In this case glass syringes drove the reactants through a mixing chamber and into a chamber specially constructed for a glass electrode

[21] J. Steinhardt and E. M. Zaiser, *J. Biol. Chem.* **190,** 197 (1951).

in which jets projected streams of the mixed reactants onto the electrode surface. This modification greatly decreased the "washout" time for the apparatus. The particular electrodes employed had a resistance of 200 megohms and a capacitance of 40 $\mu\mu$farads which set the time resolution of the system.

Flow Apparatus for Indicators. The spectrophotometric method for determination of small changes of hydrogen ion concentration is admirably adapted to use with flow apparatuses, and here shows the great advantages over the glass electrode method, particularly with respect to response speed. A number of flow apparatuses are available for this purpose, and have been described in a recent symposium.[22] The double-beam spectrophotometer is particularly suitable for use with the flow apparatuses because of its insensitivity to nonspecific absorbance changes in the optical path.

[22] B. Chance, R. H. Eisenhardt, Q. H. Gibson, and K. K. Lonberg-Holm, eds., "Rapid Mixing and Sampling Techniques in Biochemistry." Academic Press, New York, 1965.

[100] $^{14}CO_2$ Measurements during Mitochondrial Respiration

By S. C. Stuart and G. R. Williams

It is widely acknowledged that any full understanding of the respiration of mitochondria must include a knowledge of the flux of substrate through those metabolic cycles, the dehydrogenases of which supply reducing equivalents to the respiratory chain. One potentially useful method of obtaining this knowledge is to follow the production of $^{14}CO_2$ from specificially labeled substrates by respiring mitochondria.[1]

Methods based on the collection of $^{14}CO_2$ during Warburg respirometry cannot provide detailed kinetic information for comparison with spectrophotometric and polarographic data, although by substituting a trap containing ethanolic hyamine hydroxide for the conventional alkali-soaked filter paper Wang and his colleagues[2] have been able to sample respiratory $^{14}CO_2$ at intervals as short as 10 minutes, a sampling rate which enabled them to follow adequately the evolution of $^{14}CO_2$ by intact organisms.

[1] G. R. Williams, *Can. J. Biochem.* **43**, 603 (1965).

[2] C. H. Wang, I. J. Stern, C. M. Gilmour, S. Klungsoyr, D. J. Reed, J. J. Bialy, B. E. Christensen, and V. H. Cheldelin, *J. Bacteriol.* **76**, 207 (1958).

Method

The method of Wang,[2] referred to by him as radiorespirometry, has been modified for mitochondrial studies. First, hyamine hydroxide has been replaced by phenylethylamine as suggested by Woeller.[3] Secondly, the CO_2 adsorbent has been distributed on a solid support and packed into small columns which gives ease of handling and permits the use of a simple manifold for frequent sampling.

Materials

- 2-Phenylethylamine (phenethylamine, Matheson, Coleman and Bell is used without further purification)
- Acid-washed diatomaceous silica (Chromosorb W 60/80 or 80/100, Johns Manville)
- A manifold with a capillary three-way stopcock to each side arm. The side arms carry male 10/30 standard taper joints.
- Small tubes about 7 cm in length and 0.5 cm in diameter with female 10/30 standard taper joints at one end and a narrower tube at the outlet end.

Procedure. Phenethylamine, 3.0 ml, is mixed with 7.0 g of Chromosorb W. The amine is taken up readily in these proportions to give a mixture that is still free running and can be easily poured into the small tubes. The tubes are plugged with glass wool and are packed by tapping on a hard surface without tamping. A second plug of glass wool is inserted to hold the packing in place when the tube is inverted. Each tube contains about 0.5 g of the mixture in a column which fills the 7-cm tube. The traps are placed on the greased joints of the manifold, which is in turn joined to the vessel in which $^{14}CO_2$ evolution is taking place (Fig. 1). A simple pipette-like rocking chamber has been used,[1] but the mixing of additions to this system is not entirely adequate and reaction vessels that permitted more efficient mixing could doubtless be devised. Using a manifold with five side arms it is a relatively simple matter to divert the outflowing gas stream through each trap in turn, and sample times as low as 10 seconds can easily be achieved and maintained, subject only to the supply of tubes and the patience of the investigator. The contents of the trap are tapped out into a standard vial containing 10 ml of a conventional solution for liquid scintillation counting (4 g 2,5-diphenyloxazole, 100 mg 1,4-bis-2-(5 phenyloxazolyl)benzene in 1.0 liter of toluene). The vial is agitated gently by hand to ensure dissolution of

[3] F. A. Woeller, *Anal. Biochem.* **2**, 508 (1961).

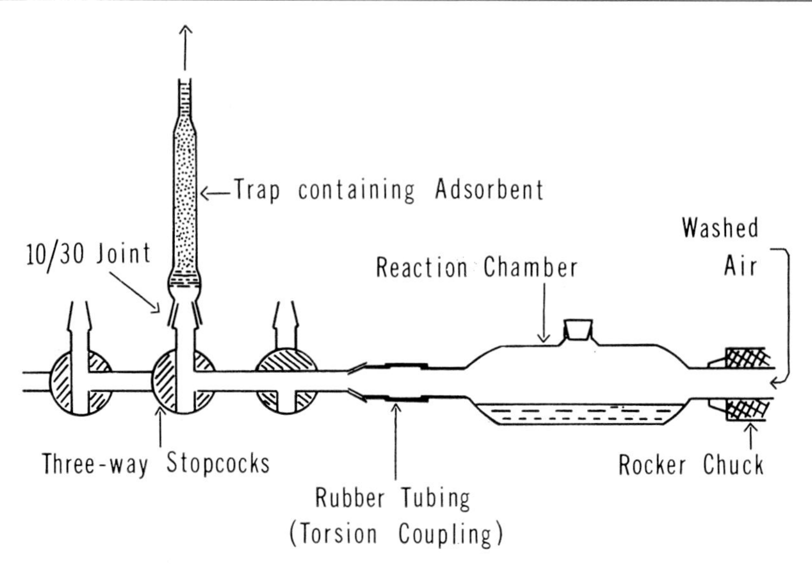

FIG. 1. Block diagram of the apparatus used.

the carbamate; the Chromosorb and glass wool settle rapidly to the bottom of the vial. The contents may then be counted in the usual manner. The efficiency of counting as judged by a channels ratio technique was around 65%, indicating that some quenching occurs.

Recovery. The percentage recovery of $^{14}CO_2$ on the traps is excellent. An $NaH^{14}CO_3$ solution was added in microliter amounts to phosphate buffer pH 5.9 in the reaction vessel, and the evolved $^{14}CO_2$ was collected for 15 minutes. The contents were counted as described above. Similar additions were made, using the same radioactive solution and micropipettes, to the standard counting system to which 1.0 ml of methanol had been added. The radioactive bicarbonate was held in solution under these conditions. All counts were corrected to 100% efficiency; the results are compared in the table. Similar high efficiencies of trapping (~100%) are obtained in metabolic experiments.

RECOVERY OF $^{14}CO_2$ FROM $NaH^{14}CO_3$

Volume of $NaH^{14}CO_3$ solution (μl)	Dpm by addition to vial	Dpm trapped as $^{14}CO_2$	Recovery (%)
10	96,300	94,000	98
15	145,700	142,000	97
20	184,900	184,000	99

Limitations of $^{14}CO_2$ Monitoring and Sampling Techniques

In experiments involving intact higher organisms the ionization chamber has been used to follow continuously the excretion of $^{14}CO_2$ after administration of labeled compounds.[4-6] Recently the attempt has been made to use such a system to follow $^{14}CO_2$ appearance in suspensions of unicellular organisms[7] and mitochondria.[1] In rapidly metabolizing sys-

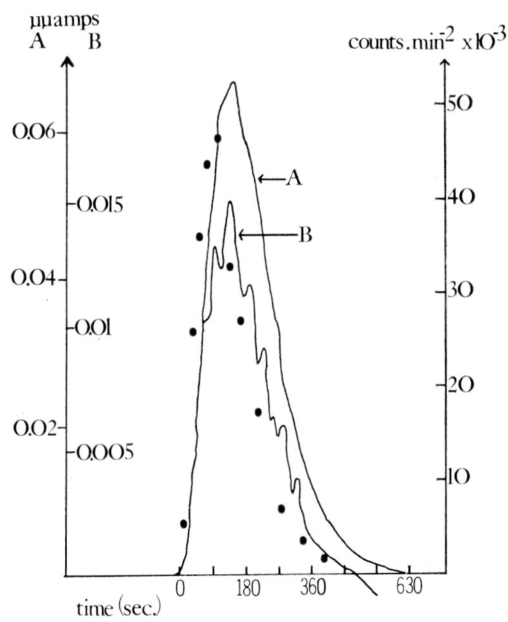

FIG. 2. $^{14}CO_2$ expulsion curves from isolated rat heart mitochondria incubated in a medium of the following composition: 0.050 *M* Tris-HCl, 0.010 *M* Na_2HPO_4, 0.050 *M* KCl, 0.005 *M* $MgSO_4$, 0.15 *M* sucrose, 0.010 *M* sodium pyruvate, 0.0003 *M* adenosine diphosphate, 1.0 g/l carbonic anhydrase per liter, pH = 7.4. At zero time on the abscissa 0.1 μC of succinate-1,4-^{14}C was added ([succinate] = 28 μM). The reaction was carried out in a rocking vessel (Fig. 1). Curve *A* was obtained at an air flow rate of 275 ml/minute using a 275 ml spherical ionization chamber (Applied Physics Corporation 3295200). Curve *B* was obtained at an air flow rate of 100 ml/minute using a 5 ml cylindrical ionization chamber (Nuclear Chicago DCF-5). The points (●) were obtained by the sampling technique described here.

[4] B. M. Tolbert, M. Kirk, and E. Baker, *Am. J. Physiol.* **185,** 269 (1956).
[5] B. M. Tolbert, M. Kirk, and F. Upham, *Rev. Sci. Instr.* **30,** 116 (1959).
[6] L. Levenbook, and M. L. Dinamarea, *Anal. Biochem.* **11,** 391 (1965).
[7] J. A. Bassham and M. Kirk, *in* "Rapid Mixing and Sampling Techniques in Biochemistry" (B. Chance, R. H. Eisenhardt, Q. H. Gibson, and K. K. Lonberg-Holm, eds.), p. 319. Academic Press, New York, 1965.

tems, the rate of change of the ionization current may not reflect accurately the rate of production of $^{14}CO_2$. The factors responsible for such distortion have been listed by one of us.[1] Some of these factors become less important in the rapid sampling method here and a comparison of the results obtained in a typical metabolic experiment is presented in Fig. 2. However, the simple sampling technique lends itself to an even more critical evaluation of the discrepancy between the rate of metabolic production of $^{14}CO_2$ and the rate of its appearance in the gas phase. If the enzymatic reactions are terminated by injection of perchloric acid at a given time, and sampling of the gas stream is continued, it is possible to estimate the amount of $^{14}CO_2$ which had been produced but not released from the suspension. By repeating this experiment but varying the time of addition of acid, it is possible to construct composite curves comparing the true rate of $^{14}CO_2$ production with that indicated by sampling the gas stream. The results are shown in Fig. 3.

It is clear that analysis of the gas phase for $^{14}CO_2$ is not an adequate method for following the initial stages of moderately rapid metabolic reactions. One of us has discussed the parameters which may be used to

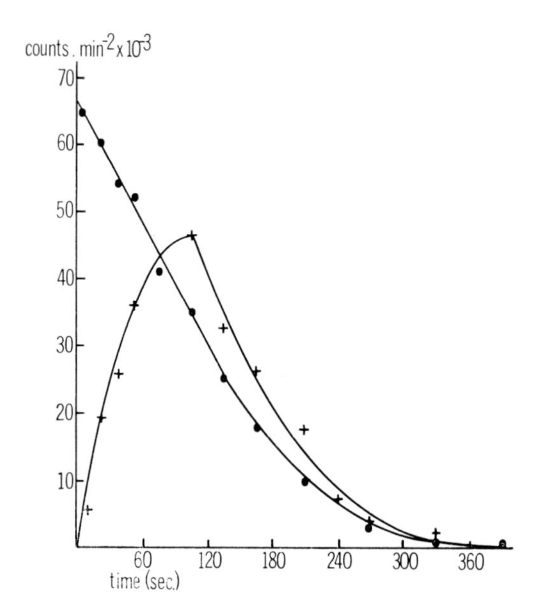

FIG. 3. Correction of $^{14}CO_2$ expulsion curve for $^{14}CO_2$ retained in the medium. The points (+) are taken from Fig. 1 (sampling technique). A correction is applied from a standard curve relating the total $^{14}CO_2$ obtainable after stopping the reaction with perchloric acid to that obtained before acidification. The points (●) have been corrected in this fashion.

define $^{14}CO_2$ expulsion curves as obtained by the ionization chamber technique.[1] These parameters varied according to the substrate added, and presumably the variations do reflect differences in the kinetic constants governing the metabolism of those substrates. However, Fig. 3 suggests that precise interpretation of variations in these parameters will be very hazardous since the form of the $^{14}CO_2$ expulsion curves is related only in an indirect fashion to the rate of production of $^{14}CO_2$ in the metabolizing system.

[101] New Developments in Electron Microscopy As Applied to Mitochondria

By D. F. PARSONS[1]

Introduction

The recently developed electron microscope preparation technique of negative staining gives new information about the localization and arrangement of phosphorylation enzymes and electron transport carriers on the mitochondrial membranes. The method serves as a valuable adjunct to the classical electron microscope method of ultrathin sectioning[2] of fixed and embedded mitochondria. In ultrathin sections the structures are made visible in the electron microscope as a result of their impregnation by heavy metals (osmium in combination with lead or uranyl salts). Needless to say, metal impregnation is destructive to proteins, etc., and the thin section method cannot be expected to give much detail at the level of macromolecular structure.[3] More recently, a method of negative staining was introduced for studying the fine structure of viruses.[4-8] In this method a suspension of virus is mixed with a solution

[1] Work supported by Canadian M.R.C. Grant MA-1611, Canadian N.R.C. Grant A-1723 and a Grant from the National Cancer Institute of Canada. The manuscript was prepared while the author was at the Department of Medical Biophysics, University of Toronto, Toronto, Ontario, Canada.

[2] D. C. Pease, "Histological Techniques for Electron Microscopy," 2nd ed. Academic Press, New York, 1964.

[3] D. F. Parsons, *Symp. Quant. Electron Microscopy, Washington, 1964. Lab. Invest.* **14** (Part 2), 431 (1965).

[4] J. L. Farrant, *Biochim. Biophys. Acta* **13**, 569 (1954).

[5] C. E. Hall, *J. Biophys. Biochem. Cytol.* **1**, 1 (1955).

[6] H. E. Huxley, *Proc. 1st European Regional Conf. Electron Microscopy, Stockholm, 1956* p. 260. Academic Press, New York, 1957.

[7] R. W. Horne and S. Brenner., *Proc. 4th Intern. Conf. Electron Microscopy, Berlin, 1958* Vol. 1, p. 625. Springer, Berlin, 1960.

[8] S. Brenner and R. W. Horne, *Biochim. Biophys. Acta* **34**, 103 (1959).

of potassium phosphotungstate and a thin film of the mixture dried down onto the electron microscope specimen grid.

The same method can be applied to suspensions of isolated cell components[9,10] and even to whole cells[11,12] or frozen sections of cells.[13,14] At or near neutral pH potassium phosphotungstate has little affinity for proteins, or lipids, so cell structures are unstained and remain transparent to electrons while the space between the structures is filled with a thin film of electron opaque phosphotungstate. Other heavy metal salts[15,16] can be used as negative stains for special purposes, but none is as generally useful as phosphotungstate. The negatively stained preparation has certain advantages over positively stained thin sections, although there are also severe disadvantages (particularly the fact that cells and other large structures have to be broken open to obtain a thin specimen). However, the high contrast of the preparation and its stability to heat damage by the electron beam are important advantages that allow the specimens to be examined at high magnification (40,000–80,000 times) and enables a resolution of fine specimen detail of about 12 Å to be achieved.

Negative Staining Technique

Potassium Phosphotungstate (3%) Negative Staining Solution (KPTA). It is essential that no turbidity should develop during preparation of the negative stain, or the negatively stained material will show background granularity (even if the solution is filtered). Of reagent grade phosphotungstic acid ($P_2O_5 \cdot 24\ WO_3 \cdot nH_2O$), 3 g is dissolved slowly with rapid stirring in about 80 ml of glass-distilled water. The very acid solution is slowly neutralized with freshly prepared 10% KOH (about 4.2 ml is required) to give a pH of 6.8, and the volume is made up to 100 ml. The solution is stored in the cold room to minimize growth of mold or bacteria, but otherwise the phosphotungstate forms a very stable buffer solution. The pH of 6.8 allows for a slight shift in pH to-

[9] R. W. Horne and V. P. Whittaker, *Z. Zellforsch. Mikroskop. Anat.* **58,** 1 (1962).
[10] H. Fernández-Morán, *Circulation* **26,** 1039 (1962).
[11] D. F. Parsons, *J. Cell Biol.* **16,** 626 (1963).
[12] D. F. Parsons, *in* "Subcellular Pathology" (J. W. Steiner, H. Z. Movat, and A. C. Ritchie, eds.), Vol. I. Harper (Hoeber), New York, in press.
[13] J. D. Almeida and A. F. Howatson, *J. Cell Biol.* **16,** 616 (1963).
[14] W. Bernhard, *J. Appl. Phys.* **35,** 616 (1964). (Abstract.)
[15] R. W. Horne, *in* "Techniques for Electron Microscopy" (D. Kay, ed.), p. 158. Thomas, Springfield, Illinois, 1961.
[16] R. C. Valentine and R. W. Horne. *Interpretation Ultrastruct. Symp. Intern. Soc. Cell Biol.* **1,** p. 263. Academic Press, New York, 1962.

ward alkalinity which occurs on drying (Parsons, unpublished, 1966; see also Bradley[17]). Some applications of phosphotungstate staining have been reviewed recently.[18-20]

Negative Staining of Mitochondria by the Surface Spreading Method. Mitochondria are isolated by one of the described methods (see Section II of this volume). The mitochondria may be prepared for negative staining in different ways. Mitochondria can be sampled direct from the isotonic isolating media (sucrose or KCl), after swelling in dilute phosphate buffer (1 mM, pH 7.0), or after other treatments such as fragmentation to submitochondrial particles.[21-26] In addition it is advisable, following the initial pretreatment of the mitochondria, to fix some of the material before negative staining. This is necessary because the membrane fragments may be distorted or broken to some extent if negatively stained without fixation. However, it must also be remembered that the fixatives may destroy some parts of the structure rather than preserve them. Ideally, mitochondria should be negatively stained after several kinds of pretreatments. The best fixation procedure is to mix the suspension of mitochondria (in sucrose or dilute phosphate buffer) with an equal volume of Palade's 1% osmium tetroxide solution[2] and to fix at 0° for 30 minutes. Tris buffer must not be present since it interferes with the fixation by osmium tetroxide.

The surface spreading method[11,27] is the most convenient way to negatively stain mitochondria. This method gives a fairly uniform distribution of membrane on the specimen grid and has the advantage that moderate concentrations of sucrose, salts, or fixative do not interfere. These diffuse away from the surface film. Three per cent KPTA is placed in a small dish (Fig. 1) about 1 cm in diameter (disposable plastic tube tops are convenient). A steel sewing needle 1 mm thick is cleaned by moistening it with a drop of KPTA and rubbing the needle with lens

[17] D. E. Bradley, *J. Gen. Microbiol.* **29**, 503 (1962).
[18] R. W. Horne and P. Wildy, *Advan. Virus Res.* **10**, 102 (1963).
[19] Ö. Levin, *Acta Univ. Upsaliensis* (Abstract of Uppsala Dissertations in Science) No. 24. Almqvist & Wiksell, Uppsala, 1963.
[20] A. P. Waterson, *in* "Techniques in Experimental Virology" (R. J. C. Harris, ed.), p. 359. Academic Press, New York, 1964.
[21] C. Cooper and A. L. Lehninger, *J. Biol. Chem.* **219**, 489, 519 (1956).
[22] T. M. Devlin and A. L. Lehninger, *J. Biol. Chem.* **219**, 507 (1956).
[23] D. Ziegler, R. Lester, and D. E. Green, *Biochim. Biophys. Acta* **21**, 80 (1956).
[24] W. W. Kielley and J. R. Bronk, *J. Biol. Chem.* **230**, 521 (1958).
[25] W. C. McMurray and H. A. Lardy, *J. Biol. Chem.* **233**, 754 (1958).
[26] H. S. Penefsky, M. E. Pullman, A. Datta, and E. Racker, *J. Biol. Chem.* **235**, 3330 (1960).
[27] H. Fernández-Morán, *Arkiv Zool.* **40A**, 1 (1948).

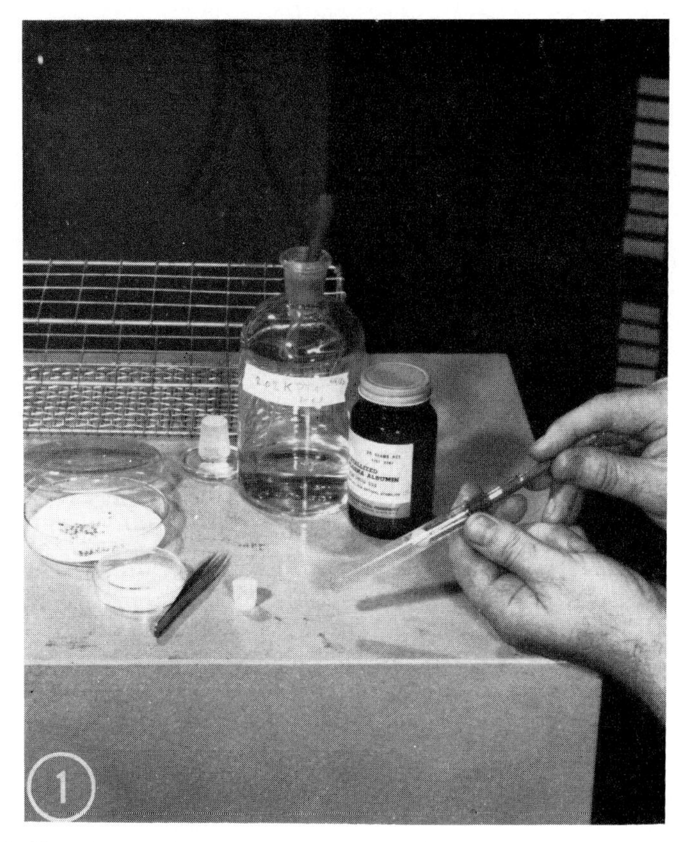

FIG. 1. Negative staining of a suspension of mitochondria by the surface spreading technique. The mitochondria in isolating medium or 1 mM phosphate are diluted to a moderate turbidity (1–5 mg protein per milliliter). A steel sewing needle is dipped 1 cm into the suspension. A small dish is filled with 2 or 3% potassium phosphotungstate (pH 6.8) containing about 0.01% of bovine plasma (or serum) albumin.

paper. The needle is dipped about 1 cm into the suspension of fixed or unfixed mitochondrial membranes and then dipped slowly into the KPTA in the dish (Fig. 2). A very thin surface film forms on the KPTA. The distribution of membranes is controlled by adjusting the concentration of the original suspension (1–5 mg of mitochondrial protein per milliliter). Electron microscope specimen grids (400 mesh with a Formvar film coated with a gray film of carbon) are floated briefly, film side down on the surface (Fig. 3) and then picked up again. The excess KPTA is absorbed off with the torn edge of a piece of filter paper (Fig. 4), and the grid is allowed to dry in air. Unless the surface of the grids is exception-

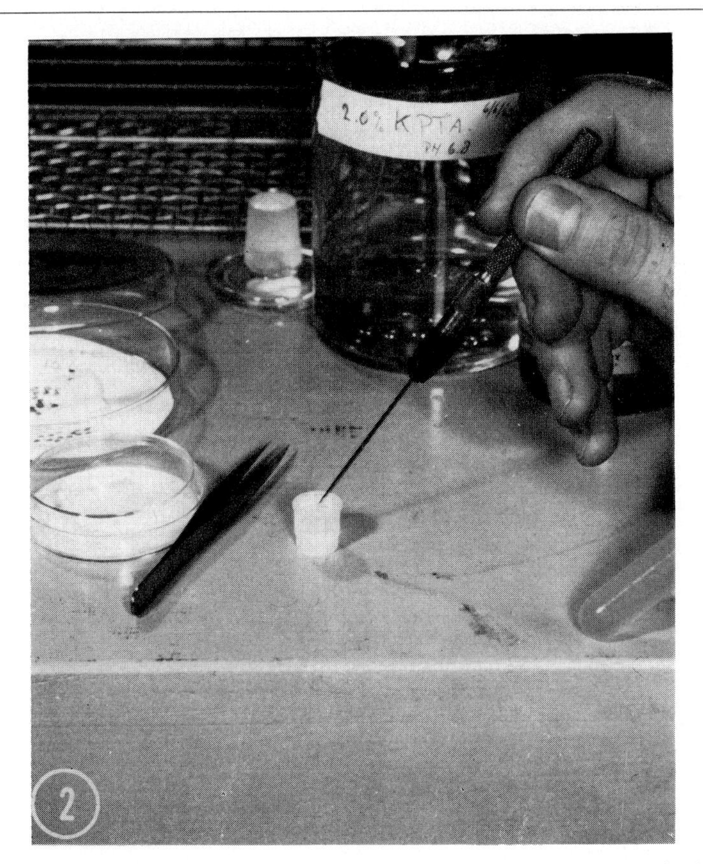

FIG. 2. The needle is dipped slowly, while moving it from side to side, into the phosphotungstate.

ally free from grease it will be necessary to add a wetting agent to the KPTA. Approximately 0.01% bovine serum albumin is added to a small volume of the stock KPTA just before use. The wetting effect lasts only for 1 day. Fortunately, the serum albumin appears to be denatured into a form which makes it unrecognizable in the background of the preparation. Unless adequate wetting is obtained, the membranes will be unevenly distributed on the grid, the KPTA will form an uneven layer, and the Formvar carbon film will be partially broken.

The Appearance of Negatively Stained Mitochondria after Different Pretreatments

Prefixation of Whole Mitochondria for Particle Counting, or Estimation of Fragmentation and Contamination by Other Membranes. When

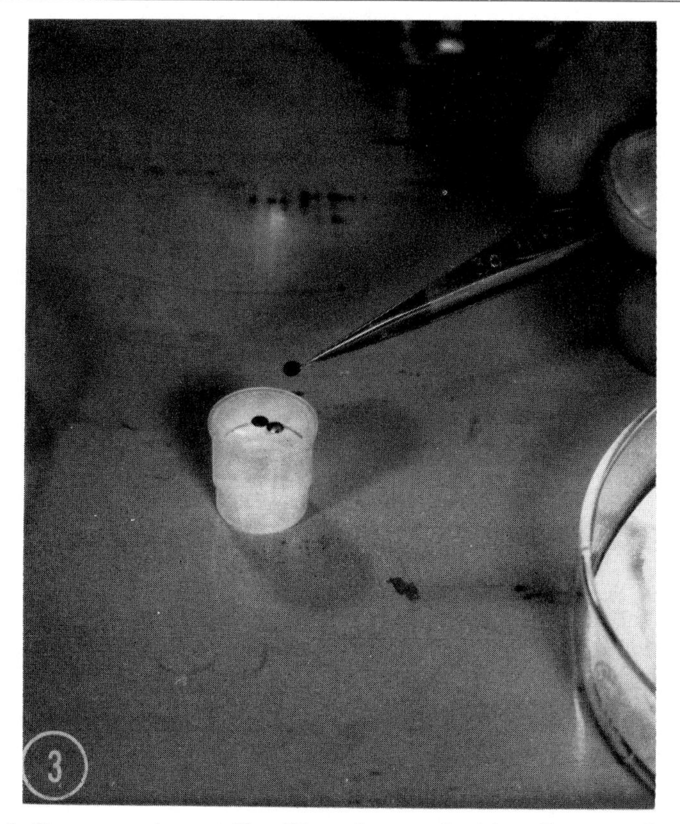

FIG. 3. Copper specimen grids, 400 mesh, coated with a Formvar film plus an evaporated carbon film, are placed, film side downward, on the surface. Nothing is visible on the surface, but the fixed position of the floating grids indicates that a film is present.

fixed, as described, with osmic acid the mitochondria appear as dense spheres or cylinders. This method provides the biochemist with a convenient method of checking a mitochondrial preparation for contamination. In many cases it is sufficient to surface spread the mitochondria direct from the isolative medium (Fig. 5), but some of the mitochondria are fragmented by the negative staining process. Vesicles and tubes of endoplasmic reticulum are readily recognized as white structures with rounded edges (Fig. 6) even though the ribosomes are poorly preserved and barely visible. In estimating the degree of fragmentation of a mitochondrial preparation, it should be remembered that even osmium-tetroxide-fixed mitochondria fragment slightly (5–15%) under negative staining conditions. It appears feasible to obtain particle counts of

FIG. 4. The grids are immediately picked up again, and the excess phosphotungstate is absorbed off with filter paper. Sufficient plasma albumin has been added to give good wetting of the grid if a thin wet film remains on the grid after the maximum amount of phosphotungstate has been absorbed off with filter paper. The grids dry in about 1 minute and can then be examined in the electron microscope. The grids are stable for at least 1 week and can be examined repeatedly.

mitochondria by the surface spreading method. Direct mixing gives more reproducible counts, but the isotonic medium gives a large background of sucrose or salts.

Preswelling of Mitochondria before Negative Staining. In order to examine the structure of the inner and outer membranes at high resolution it is necessary to break open the mitochondria and to spread out the membranes to form only single or double sheets. One way to do this is to swell the mitochondria before surface spreading. Swelling weakens the membrane so that surface tension during surface spreading causes the membranes to rupture and spread out (Fig. 7). This method reveals striking fine detail on the inner membranes (including the cristae). The

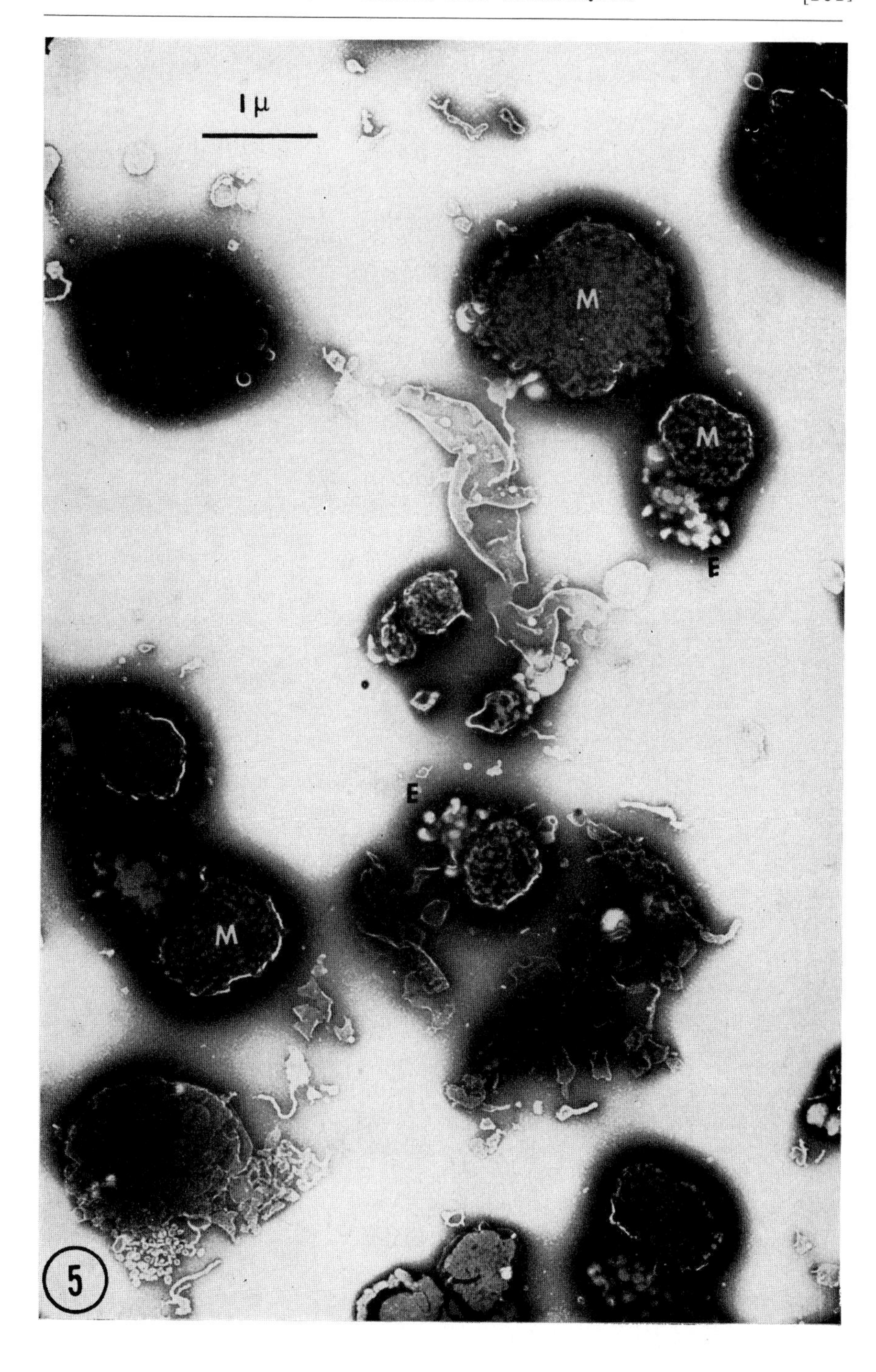
1 μ
M
M
E
E
M
5

most obvious features are projecting 90 Å subunits (elementary particles[10,28]; inner membrane subunits, *IMS*[29]). Such subunits are visible on the inner membranes of many, if not all, types of mammalian cell,[28-31] plant cell,[32,33] and insect cell[34] mitochondria. Similar subunits appear to be present on chloroplast membranes,[33,35,36] chromatophores,[37] and bacterial plasma membranes.[38] Although it was originally supposed that the subunits contained cytochromes,[10,28,39] no significant loss of cytochromes or interference with electron transport was found to occur in preparations stripped of inner membrane subunits,[40] and the subunits had the same appearance in mitochondria deficient in certain cytochromes.[41] On the other hand, stripping of the subunits was found to be associated with loss of ATPase activity.[42] The heads of the subunits appeared morphologically similar to individual molecules of purified mitochondria ATPase. Recent experiments on partial restoration of oxidative phosphorylation by recombination of the ATPase and other factors with subunit stripped submitochondrial particles also suggested an identification of the subunit head with mitochondrial ATPase.[43]

[28] H. Fernández-Moran, T. Oda, P. V. Blair, and D. E. Green, *J. Cell Biol.* **22,** 63 (1964).

[29] D. F. Parsons, *Science* **140,** 985 (1963).

[30] W. Stoeckenius, *J. Cell Biol.* **16,** 483 (1963).

[31] J. T. Stasny and F. L. Crane, *J. Cell Biol.* **22,** 49 (1964).

[32] M. J. Nadakavukaren, *J. Cell Biol.* **23,** 193 (1964).

[33] D. F. Parsons, W. D. Bonner, Jr., and J. G. Verboon, *Can. J. Botany* **43,** 647 (1965).

[34] D. S. Smith, *J. Cell Biol.* **19,** 115 (1963).

[35] E. N. Moudrianikis, *J. Cell Biol.* **23,** 63A (1964).

[36] T. Oda and H. Huzisige, *Exptl. Cell Res.* **37,** 481 (1965).

[37] H. Löw and B. Afzelius, *Exptl. Cell Res.* **35,** 1431 (1964).

[38] D. Abram, *J. Bacteriol.* **89,** 855 (1965).

[39] D. E. Green, P. V. Blair, and T. Oda, *Science* **140,** 382 (1963).

[40] B. Chance, D. F. Parsons, and G. R. Williams, *Science* **143,** 136 (1964).

[41] B. Chance and D. F. Parsons, *Science* **140,** 985 (1963).

[42] E. Racker, D. D. Tyler, R. W. Estabrook, T. E. Conover, D. F. Parsons, and B. Chance, *in* "Oxidases and Related Systems," Proc. Intern. Symp. Oxidases and Related Oxidation-Reduction Systems (T. E. King, H. S. Mason, and M. Morrison, eds.), Vol. 2, p. 1077. Wiley, New York, 1965.

[43] E. Racker, B. Chance, and D. F. Parsons, *Federation Proc.* **23,** 431 (1964).

FIG. 5. A negatively stained, surface spread preparation of rat kidney mitochondria. The needle was dipped into a suspension of the mitochondria in isotonic sucrose. Most of the mitochondria remained intact although there is more breakage than occurs if the mitochondria are prefixed with osmic acid before negative staining. Particularly with fixed preparations, relative counts of broken and unbroken mitochondria and of contaminating nonmitochondrial membranes can be made. Endoplasmic reticulum (*E*) has a characteristic white vesicular appearance. *M*, mitochondria. No. 9085-4.

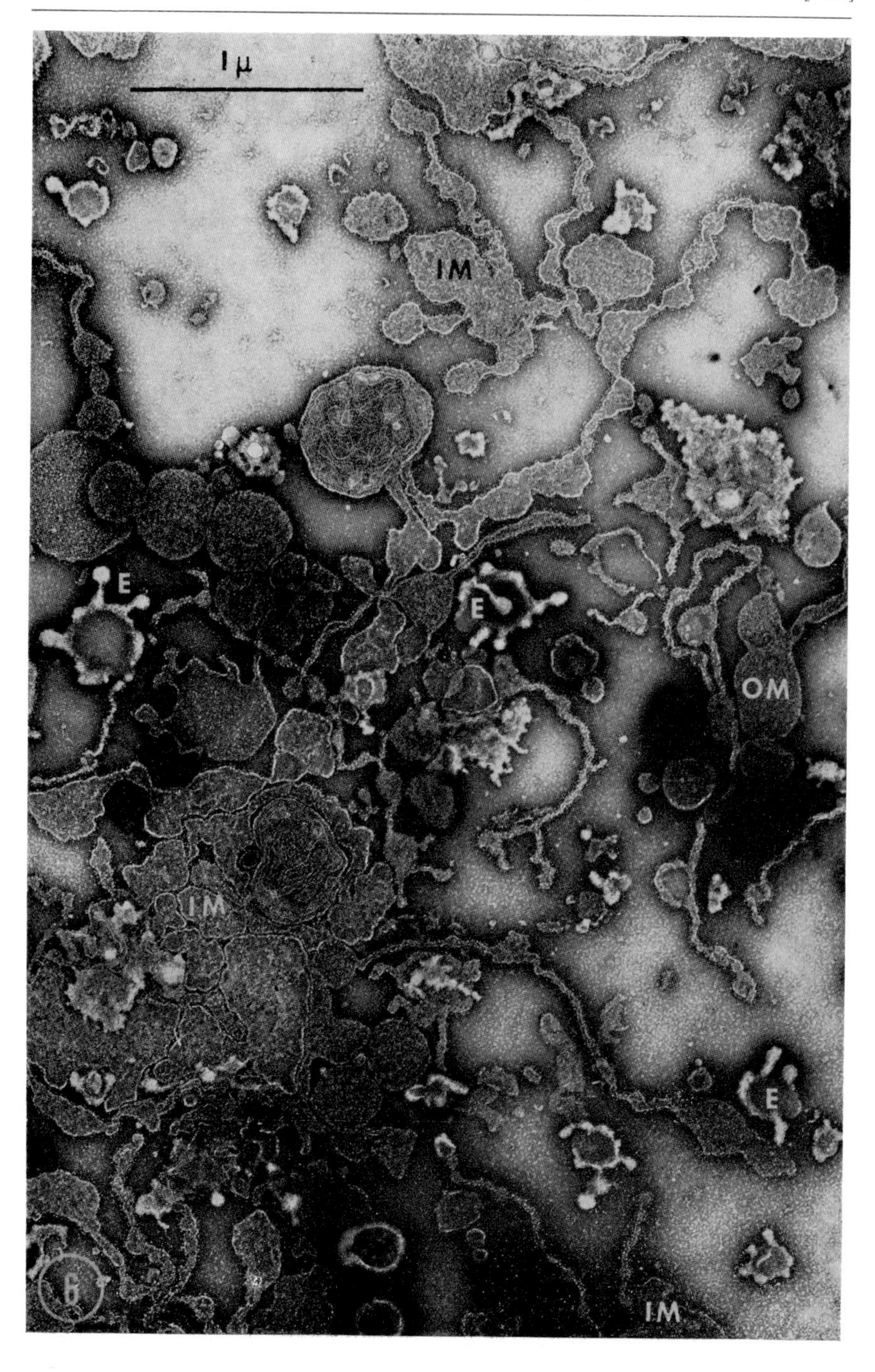
1 μ
IM
E
E
OM
IM
E
IM
6

In addition, pretreatment by swelling causes the cristae to assume a tubular rather than a flattened vesicular form. Sjöstrand *et al.*[44] suggested that the tubular cristae are myelin forms indicating a reorganization of membrane lipids. They suggest that the inner membrane subunits are artifacts associated with myelin degeneration of the membrane. However, there are several reasons why this suggestion is unlikely to be correct. The tubular cristae of swollen mitochondria are visible also in thin sections in association with a broken and partly detached outer membrane.[45,46] It appears that the tubular cristae are formed as a result of evagination of the cristae surface to form the expanded surface of the swollen inner membrane "ghost" rather than by myelin degeneration. In addition the subunits appear to be partly preserved in thin sections in certain insect mitochondria[47] and in mammalian cell mitochondria if the normal fixation process is modified.[45]

Regardless of the final answer about the chemical nature and natural form of the inner membrane subunits, they represent useful morphological markers for identifying the inner membrane of mitochondria in cell fractions.[46] Negative staining thus provides a convenient rapid method for checking very small quantities of cell fractions for mitochondrial contamination.

Negatively Stained Submitochondrial Particles. Phosphorylating submitochondrial particles are in a very convenient form for negative staining studies, and should be studied more extensively. The particles are usually of relatively small (several thousand Angstroms) diameter and consist of small vesicles. Apparently, the originally flat membrane fragments unite at the edges to form spherical vesicles during preparation, in a similar manner to the formation of vesicles in a microsome preparation as a result of fragmentation of endoplasmic reticulum. The vesicular shape is seen in thin section or negatively stained preparations of fixed submitochondrial particles. However, when negatively stained without

[44] F. S. Sjöstrand, E. A. Cedergren, and V. Karlsson, *Nature* **202**, 1075 (1964).

[45] D. F. Parsons, *Intern. Rev. Exptl. Pathol.* **4**, 1 (1965).

[46] D. F. Parsons, G. R. Williams, and B. Chance, "Symposium on Cell Membrane Problems: Recent Advances." *N.Y. Acad. Sci.* in press (1966).

[47] D. E. Ashurst, *J. Cell Biol.* **24**, 500 (1965).

FIG. 6. Nonfixed, negatively stained fragmented rat liver mitochondrial fraction to show the three main types of membrane present. Some irregular pieces of membrane show numerous projecting 90 Å subunits at the edges (inner membrane or cristae, *IM*). Some round pieces of membrane show no projecting subunits (outer membrane, *OM*). The white pieces with rounded edges are contaminating endoplasmic reticulum (*E*). No. 747. Magnification: ×32,000.

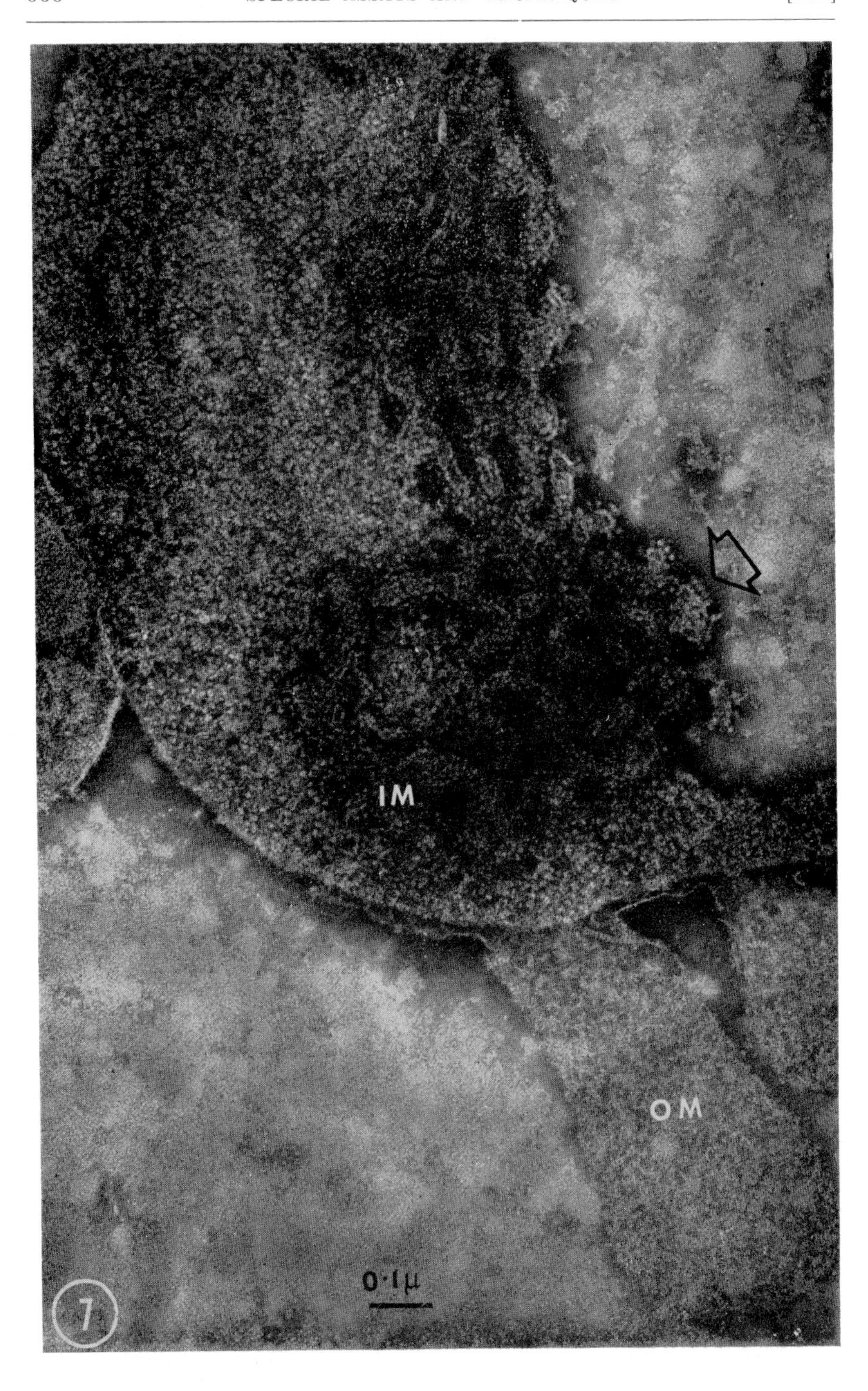
IM
OM
0·1μ
7

fixation the vesicles break open and flatten out (see Fig. 19 in Parsons[45]). Their surface is covered with inner membrane subunits, and their structure can be observed during treatment with specific reagents which modify electron transport or oxidative phosphorylation.

Conclusions

Because of difficulties in interpretation, the negative staining method must be regarded as supplementary to the thin section method, not as a substitute for it. In any study relating the structure of mitochondria to function the examination of thin sections is the first step. If this step is omitted grosser changes in the mitochondria are liable to be overlooked. Thin sections will clearly demonstrate such changes as differences in volume, breakage of the outer membrane, change of shape of cristae, change in matrix granules. If the established methods of thin sectioning had been used more extensively in the past in studies of swelling, compartmentation, ion transfer, and ion accumulation, such misconceptions as the reversibility of structural changes in mitochondria during large amplitude swelling could have been avoided. There appear to be two main indications for negative staining in mitochondrial studies. First, fixed mitochondria negatively stained by the surface spreading method give a rapid method of checking and counting a mitochondrial preparation for fragmentation and contamination by other cellular structures. The improved resolution available over phase contrast light microscopy outweighs the disadvantage of using chemical fixation. Secondly, negative staining of swollen or fragmented mitochondria can be expected to provide useful indications about the nature of the interaction of specific reagents affecting electron transport, or oxidative phosphorylation. No extensive and systematic studies of this type have yet been carried out. The subunits of mitochondria are so distinctive that even small pieces of membrane can be recognized as being of mitochondrial origin. Hence in cell fractionation studies examinations of fractions suspended in dilute phosphate buffer readily reveals the presence of contaminating mitochondria.

FIG. 7. Mitochondrion from a 15091A mouse tumor negatively stained and surface spread after swelling of the mitochondria in 1 m*M* phosphate (pH 7.0). The outer membrane (*OM*) is partly detached and shows a characteristic pitted appearance in the unfixed material. The inner membrane (*IM*) is completely covered with projecting 90 Å subunits. As a result of the swelling, the cristae have been transformed into narrow tubules (arrow) covered with 90 Å subunits. No. 5597-4. From D. F. Parsons, *in* "Subcellular Pathology" (J. W. Steiner, H. Z. Movat, and A. C. Ritchie, eds.). Harper (Hoeber), New York, in press. (Reprinted by permission.)

[102] Methods for Studying Cytochrome-Mitochondrial Structural Protein Interaction

By RICHARD S. CRIDDLE

Introduction

Throughout this discussion, the fraction of protein first isolated from beef heart mitochondria by Green *et al.*[1] and termed mitochondrial structural protein will be considered to be a major constitutent of the membrane systems of mitochondria. It will further be considered to be a single protein species which plays a major structural role in the organization and stabilization of the membrane-bound components of the mitochondrion to ultimately aid in the systematic functioning of the overall complex system.[2] This implies of course a specific interaction or binding of mitochondrial components with structural protein. No further elaboration of details on specific location and function of this material can be unequivocally proposed at this time although recent evidence regarding the specific nature of the interaction[3] and its effect on mitochondrial enzyme activity[4] may in the near future provide some answers to these aspects of the problem.

The original information obtained on binding of cytochromes comes from determination of the relative ease of extractability of the various cytochromes from the membrane system. It was noted for example that cytochrome *c* could be readily removed from the insoluble matrix by extraction with dilute aqueous saline solutions. The other cytochromes of the electron transport chain were much more difficult to extract however and required the addition of detergents to break up their major interactions. This type of observation has so far been of little aid in the understanding of structural protein–cytochrome interaction due to the overall complexity of the membrane material involved. It does, however, offer some guidelines in the study of more simplified systems and serves as a standard of comparison for answers to the question of how closely the observed cytochrome-structural protein reactions compare with what is noted in the intact system. It also serves as a reminder

[1] D. E. Green, H. Tisdale, R. S. Criddle, P. Y. Chen, and R. M. Bock, *Biochem. Biophys. Res. Commun.* **2**, 109 (1961).

[2] R. S. Criddle, R. M. Bock, D. E. Green, and H. Tisdale, *Biochemistry* **1**, 827 (1962).

[3] D. L. Edwards and R. S. Criddle, *Biochemistry* **5**, 583 (1966).

[4] D. O. Woodward, *Proc. Natl. Acad. Sci. U.S.* **55**, 872 (1966).

that the ultimate understanding of cytochrome-structural protein interaction will have to be placed in terms of its role in a system containing numerous additional proteins and small molecules as well as a large compliment of phospholipids.

Materials

The starting material for preparation of structural protein has usually been beef heart mitochondria obtained by methods such as those described by Green and Ziegler.[5] However, other mitochondrial sources have also been employed. Structural protein may be isolated and purified as described by Richardson *et al.*[6] and subsequently solubilized using the methods outlined by Criddle *et al.*[2] As the methods normally required for solubilization of structural protein initially employ the action of rather extreme values of pH (i.e. > 10) or of surface active agents, consideration of the solvent effects must be superimposed upon the cytochrome binding results for interpretation of strength, specificity, and stoichiometry of the interactions. Care should therefore be taken to keep the method of preparation and solubilization in the foreground for the interpretation of experimental observations.

Preparation of Structural Protein. A suspension of mitochondria (20 mg/ml) in pH 7.5, 0.01 *M* phosphate, 0.25 *M* sucrose is mixed with one-third its volume of 0.9% KCl and homogenized. The homogenate is frozen at $-20°$ and thawed. It is then centrifuged for 20 minutes at 78,000 *g*, and the supernatant fraction is discarded. The pellet is suspended and washed in 0.9% KCl three more times to remove soluble proteins and yield mitochondrial fragments. The sodium salts of cholate and deoxycholate are added (pH 7.5) with stirring to bring their concentrations to 1.0 and 2.0 mg per milligram of protein, respectively. A small amount of insoluble material is removed by centrifugation at 20,000 *g* for 20 minutes. Two micrograms of sodium dithionite are added per milligram of protein, and the solution is adjusted to 12% saturation with a neutralized solution of saturated ammonium sulfate. The solution is stirred in the cold (4°) during the addition of ammonium sulfate and for 15 minutes thereafter. If precipitation does not start immediately, an additional small amount of ammonium sulfate is added. The precipitate is collected by centrifugation and washed twice in Tris-sucrose buffer pH 8.0. The protein is then washed with excess acetone–water (90:10% by volume) to remove lipid and detergent.

[5] D. E. Green and D. M. Ziegler, Vol. VI [58].

[6] S. H. Richardson, H. O. Hultin, and D. E. Green, *Proc. Natl. Acad. Sci. U.S.* **46**, 1470 (1963).

Binding Studies by Mutual Solubilization

Both the structural protein and the cytochromes, with the exception of cytochrome *c*, have a marked tendency to form large aggregated particles in solution. Under certain conditions this aggregation phenomenon can be controlled into formation of specific-sized *n*-mers such as the hexameric form of cytochrome c_1[7] or the reported pentameric form of cytochrome *a*.[8] In many instances, however, it is difficult to obtain specific-sized polymers, and either a general aggregation, resulting in a distribution of high polymeric forms, or precipitation is observed.

A simple, direct method of observation of a complex formation between the structural protein and cytochromes results from a mutual solubilization phenomenon. The effect of complex formation on the solubilization of structural protein is particularly marked since its solubility in neutral aqueous solution is negligible while it may be very high in the presence of cytochromes. The reaction with cytochrome c_1 can be used to best illustrate this solubilization. To cytochrome c_1 (hexamer) in solution pH 7.5, 0.5 *M* Tris buffer is added an essentially unbuffered aqueous solution of solubilized structural protein at pH 11. At the resultant pH of 7.8 a clear solution is obtained which, upon analysis, shows the presence of a structural protein–cytochrome c_1, complex (1:1). The hexameric cytochrome c_1 polymer has been depolymerized by the structural protein while the structural protein which would normally form insoluble aggregates under these conditions has been monomerized by the cytochrome.

Mutual solubilization of structural protein and cytochromes *a* and *b* have also been noted and a solubilization of structural protein with myoglobin and cytochrome *c* have also been studied.[2,3] A summary of conditions which have been used to obtain cytochrome–structural protein complexes is shown in the table.

Ultracentrifugation Measurements

Both sedimentation velocity determination and molecular weight measurements by equilibrium or approach to equilibrium methods have been used to follow the formation of complexes between structural protein and cytochromes. Sedimentation velocity experiments are particularly adapted to following the mutual solubilization of aggregates of the interacting species. Using the schlieren optical system it is possible to observe directly the sedimentation of the heme proteins because of their bright colors and to correlate this with migration of the boundaries.

[7] R. S. Criddle, R. M. Bock, D. E. Green, and H. Tisdale, *Biochem. Biophys. Res. Commun.* **5**, 75 (1961).

[8] S. Takemori, I. Sekuzu, and K. Okunuki, *Nature* **188**, 593 (1960).

METHODS USED TO PREPARE STRUCTURAL PROTEIN–HEME PROTEIN COMPLEXES[a]

Heme protein	Source or method of preparation	Solution conditions for complex formation		
		Heme protein	Structural protein	Comments
Cytochrome c_1	*c*	5 mg/ml protein dissolved in 0.025 *M* Tris-HCl, pH 7.6	5 mg/ml dissolved in pH 10.0; 0.1% SDS[b] Dialyze vs. pH 10.0; 0.05 *M* NaCl; no buffer	Equal molar quantities mixed. Final pH = 7.8
Cytochrome *c*	*d*	Soluble in pH 8.5, 0.005 *M* Tris-HCl buffer	As above except for omission of NaCl	—
Cytochrome *a*	*e*	0.05% SDS, pH 10.0, 0.01 *M* phosphate	Same as solution for cytochrome *a*	Mix and allow to stand 24 hours. Dialyze vs. pH 8, Tris-HCl, 0.005 *M* NaCl
Cytochrome *b*	*f*	Adjust 5 mg/ml solution to pH 11 with KOH	Dissolve at pH 11.0 in KOH	Mix and allow to stand 4–5 hours. Adjust slowly to pH 7.4 with 0.1 *M* phosphate
Myoglobin	*g*	0.01 *M* phosphate pH 7–9	As for cytochrome c_1	Mix equal molar quantities

[a] The methods outlined are not particularly critical to the formation of complex and may be varied considerably with no noticeable effects except for the complex formation with cytochrome *b*. In this case, the method employed was the only one found to work satisfactorily.

[b] SDS, sodium dodecyl sulfate.

[c] R. Bornstein, R. Goldberger, and H. Tisdale, *Biochim. Biophys. Acta* **50,** 527 (1961).

[d] Horse heart, type III, Sigma Chemical Co.

[e] K. S. Ambe and A. Venkataraman, *Biochem. Biophys. Res. Commun.* **1,** 133 (1959).

[f] R. Goldberger, A. Pumphrey, and A. Smith, *Biochim. Biophys. Acta* **58,** 307 (1962).

[g] Horse heart, Mann Research Laboratories, Inc.

The stoichiometry involved in complex formation may be followed using molecular weight determinations in the centrifuge. This method, however, suffers from the inability to work routinely over wide ranges of protein concentration and of varying solvent conditions. Protein concentrations in the range of approximately 2–10 mg/ml of each reactant are normally employed for these runs. A supporting electrolyte such as 0.05–0.1 *M* NaCl is generally added to the solution to reduce electrostatic effects on the molecular weight determinations. While such salt concentrations have no observed effect on the binding of other cytochromes, this ionic strength is sufficiently high to prevent the binding of cytochrome *c*. This led to the erroneous initial conclusion that structural protein would not complex with cytochrome *c*. In light of this observation it is interesting to recall that cytochrome *c* is readily extracted from mitochondria with dilute salt solutions.

Fluorescence Measurements

Discussion. Structural protein, when excited with 280 mμ light, exhibits a normal protein fluorescence emission spectrum with a maximum near 340 mμ which can be observed using standard fluorescence methods.[9] On the other hand, heme containing proteins such as the cytochromes have a very low fluorescence yield, as first noted by Weber and Teal,[10] due to the transfer of energy from the excited aromatic amino acids to the porphyrin residues. As this transfer of the excitation energy can take place over a distance of 40–50 Å in proteins, it is possible to follow the complex formation between the nonheme containing structural protein and a heme protein by observing the net decrease in the fluorescence emission from structural protein upon complex formation.

This procedure offers several advantages over the previously described methods for following structural protein–cytochrome interaction when it is applicable. The sensitivity of the fluorescence method permits measurements on submilligram quantities. Changes in binding due to solvent conditions, temperature, and the addition of specific reagents can readily be followed and quantitated to yield information concerning the nature and strength of the binding. The major limitation of the procedure is the requirement that both reactants and products be completely soluble through out the course of the experiment to minimize the effects of scattered light. While the sensitivity of the method allows measurements in dilute solution and thus favors solubility, the marked tendency for most mitochondrial proteins to aggregate into large polymers must be taken into consideration at all times.

Method of Fluorescence Titration. Solubilized structural protein with known concentration in the range of 0.05–0.10 mg/ml is placed in a 3-ml, thermostatted fluorescence cell and incubated to allow temperature equilibration. The solution containing heme protein (about 10^{-5} *M*) is then added with the aid of a micrometer syringe, and the titration is followed by observation of the decrease of fluorescence emission following the

[9] S. Udenfriend, "Fluorescence Assay in Biology and Medicine." Academic Press, New York, 1962.

[10] G. Weber and F. W. J. Teal, *Discussions Faraday Soc.* **27**, 134 (1959).

addition of 0.01-ml aliquots. The sample is stirred following successive additions of titrant and then allowed to equilibrate for a period of a few seconds up to a few minutes, as required, prior to recording the change in fluorescence. The observed fluorescence may be plotted as a function of concentration of added heme protein to yield curves of the type shown in Fig. 1 for the binding of cytochrome *c* to structural protein.

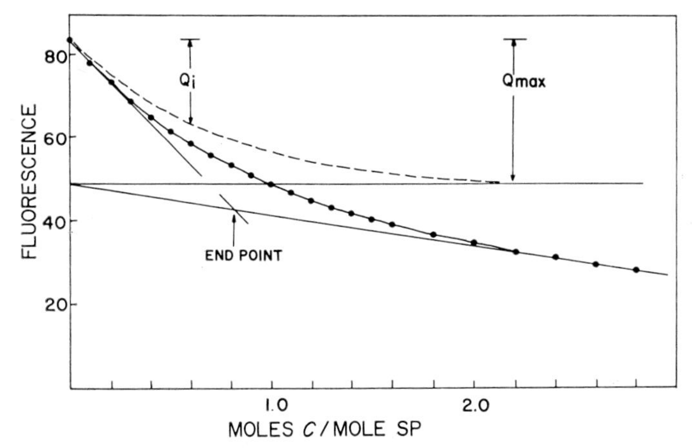

FIG. 1. Fluorescence titration curve illustrating the quenching of structural protein fluorescence upon complexing with cytochrome *c*. Q_{max} is the total fluorescence quenching observed. Q_i is the amount of quenching observed after *i* additions of cytochrome *c* (0.01 ml per addition) and is determined using the corrected curve shown by the dashed line (- - -). The end point of the titration, and therefore the stoichiometry of the reaction can be determined from the intersection of the initial and final extrapolated slopes.

The dissociation constant for the cytochrome–structural protein complex may be determined from the fluorescence quenching curve using the method suggested by Velick *et al.*[11] The total fluorescence quenching (Q_{max}) and the quenching at each experimental point (Q_i) are determined from the corrected curve and related to the amount of free structural protein, free cytochrome c_1, and of complex for calculation of the dissociation constant.[12] In general, the reaction stoichiometry may be determined from the end point of the titration curve by inspection or by mathematical analysis of the titration curve. However, limitations in the experimental method may prevent stoichiometry determination in some instances. For a discussion of such situations see Edwards and Criddle.[3]

[11] S. Velick, C. W. Parker, and H. N. Eisen, *Proc. Natl. Acad. Sci. U.S.* **46,** 1470 (1960).

[12] H. N. Eisen and G. W. Siskind, *Biochemistry* **3,** 996 (1964).

It is mandatory in experiments of this type that adequate controls be run to ensure that the observed decrease in fluorescence intensity is the result of quenching due to specific complex formation and not merely an effect of beam attenuation resulting from addition of a strongly absorbing heme protein or the result of a nonspecific protein interaction. One should also ensure by means of concentration variation that a readily reversible equilibrium is being monitored by this procedure if a quantitative interpretation of the binding results is to be made.

Fluorescence polarization methods have not yet been applied in these studies, however the system should be particularly well suited to their use. By use of this technique one should be able to correlate the information on stoichiometry and the equilibrium constants with a direct measurement of molecular size.[13]

The Use of Gel Filtration to Follow Binding

Passage of structural protein–cytochrome solutions through columns of Sephadex G-200 has been used to a limited extent to look at complex formation. Using a carefully calibrated column for measurements, one can use this procedure to make accurate estimates of particle size and composition and thus follow cytochrome-structural protein interaction directly.[14] The major additional advantage offered by Sephadex studies compared to the previously described methods is the possibility of following the relative affinities of different cytochromes for structural protein, through investigation of possible competition between cytochromes for the same binding site or region on the structural protein molecule.

An example which will serve to illustrate the method is the observed displacement of cytochrome *c* from the structural protein by myoglobin. Equal molar concentrations of cytochrome *c* and structural protein were mixed to form a complex and were passed through a Sephadex column (1×20 cm) by elution with 0.005 M Tris-HCl buffer pH 8.5. The effluent was monitored by measuring light absorption at 280 and 550 mμ which allowed isolation of the complex and also calculation of the amount of cytochrome bound per milligram of protein. This calculation can be made using the values: $E^{1\%}_{\text{cm}}$ for reduced cytochrome *c* at 550 mμ equals 22.6 and at 280 mμ equals 17.2 and for structural protein at 280 mμ equals 11.0. The structural protein–cytochrome *c* complex was then incubated with a twofold molar excess of myoglobin and again passed through the Sephadex column. Eluent peaks of cytochrome *c*, and myoglobin (overlapping) and of myoglobin–structural protein complex were

[13] D. J. R. Laurence, Vol. IV, p. 174.

[14] P. Andrews, *Biochem. J.* **91**, 222 (1964).

isolated and identified by their spectral characteristics. The binding ratio structural protein:myoglobin was determined from the ratios of optical density at 280 and 630 mμ; this ratio being 7.25 for metmyoglobin and 12.0 for the 1:1 complex of structural protein and myoglobin.

The displacement of cytochrome *c* by myoglobin suggests that both may be bound at or near the same site on the structural protein molecule and that myoglobin is bound more tightly to this site than is cytochrome *c*. This type of study can be used to further study the sites and nature of other cytochrome or mitochondrial enzyme interactions with structural protein and can be used to check the relative values determined for the different cytochrome binding constants determined by other means.

Effect of Chemical Reagents on Binding

The mitochondrial structural protein has been shown to undergo oxidation of a single sulfhydryl group per molecule with a simultaneous increase in average molecular weight to a value approaching that of a structural protein dimer. A simultaneous decrease in the cytochrome binding capacity of the structural protein preparation with increased dimer formation is noted. The effect of disulfide reducing reagents on the binding can be followed directly using the previously described methods. Dithioerythritol added to a concentration of 0.001 *M* is capable of reversing the dimerization and restoring binding. Other reducing reagents have not been investigated. Iodoacetamide and *p*-chloromercuribenzoate added at concentrations of 0.001 *M* in the fluorescence titration experiments, cause a reduction of free —SH titer of the protein but have no appreciable effect on binding.

Effects of other added reagents on the binding can also be readily followed using the methods outlined. As noted earlier, sensitivity of cytochrome *c* binding to ionic strength has been followed using fluorescence quenching. Other reagents selected for their ability to interfere with the normal types of chemical bonds stabilizing protein complexes or for their ability to cause modification of chemical groups on either the cytochromes or the structural protein may be added and their effects followed in an analogous manner.

Binding Studies by Altered Activity or Physical Properties

If structural protein functions as postulated in ordering and stabilizing complex mitochondrial structure, it may be expected that structural protein binding may affect the overall activity of some or all of the functional constituents of the complex. One such example of this has been demonstrated for cytochrome *b* binding and others are coming to light for interaction with other mitochondrial components.

It has been shown by Goldberger *et al.*[15] that the change in oxidation-reduction potential of cytochrome *b* upon complex formation can be used to both follow structural protein binding and to illustrate a functional role for this material. Recently it has also been shown that mitochondrial structural protein from *Neurospora* has a marked effect on the kinetic parameters K_m and V_{max} upon binding with the mitochondrial enzymes malic dehydrogenase and fumarase.[4] In experiments of this type, solubilized preparations of structural protein are preincubated with the enzyme in approximately equal molar ratios and the kinetics followed using the usual enzyme assay procedures. Binding is then indicated by the effect on the enzyme kinetics together with determinations as a function of concentration to indicate stoichiometry. It is expected that examples of this type will become increasingly apparent as studies of the functional role of structural protein progress.

[15] R. Goldberger, A. Pumphrey, and A. Smith, *Biochim. Biophys. Acta* **58**, 307 (1962).

[103] Characterization of Respiratory Chain Components by Polyacrylamide Gel Electrophoresis

By KUNI TAKAYAMA[1] and CLINTON D. STONER[1]

The entire functional electron transfer chain of the mitochondria appears to be composed of subunit proteins of low molecular weight. This concept has been supported by the isolation of specific proteins such as cytochrome *b*, cytochrome c_1, nonheme iron protein, and copper protein. Classical fractionation methods have been applied in isolating these components, which are characterized by their extreme insolubility in water. These methods include the use of bile salts, detergents, and organic solvents in various combinations followed by fractionation with ammonium sulfate or ammonium acetate. The objective in each case has been to isolate a specific component, all other components being regarded as impurities. A large amount of starting material is usually required for the preparation of the final purified fraction. Moreover, the preparative procedures are usually difficult and time consuming.

A simpler method for the fractionation of complex protein mixtures is by zone electrophoresis in gels. High resolution and sensitivity of the method has allowed successful characterization of water-soluble proteins. Recently, successful application of zone electrophoresis to water-insoluble proteins has been reported. Waller[1a] solubilized ribosomal proteins

[1] This manuscript was prepared while the authors were at the Institute for Enzyme Research, The University of Wisconsin, Madison, Wisconsin.

[1a] J. P. Waller, *J. Mol. Biol.* **10**, 319 (1964).

from *Escherichia coli* and separated them by starch-gel electrophoresis in a system of 0.05 M sodium acetate, pH 5.6 containing 6 M urea. Maizel[2] used polyacrylamide gel electrophoresis in 0.1 M acetic acid and 8 M urea in separating the poliovirus proteins. Using the phenol–acetic acid solvent, Work[3] solubilized virus proteins and fractionated them by electrophoresis on polyacrylamide gel.

We have utilized zone electrophoresis in fractionating the hydrophobic proteins from the mitochondrial electron transfer chain.[4] The protein can be completely solubilized in (a) phenol–acetic acid–water (2:1:1, w/v/v) and (b) 35% acetic acid–5 M urea. Polyacrylamide gel electrophoresis of the proteins solubilized in these solvent systems allows rapid separation of the components into sharp and discrete bands.

Method

Reagents

Phosphate, 0.01 M, pH 7.5
Acetone
Phenol–acetic acid–water (2:1:1, w/v/v)
Urea
Stock A: acrylamide, 6 g; *N,N'*-methylene bisacrylamide (BIS), 0.16 g; urea 12 g; glacial acetic acid 28 ml; water to make 60 ml
Stock B (prepared fresh daily): ammonium persulfate, 0.30 g; urea, 12 g; water to make 20 ml.
N,N,N',N'-Tetramethylethylenediamine (TEMED)
Acetic acid, 7%, 10%, and 75%
Amido Black solution: 1% Amido Black 10B in 7% acetic acid
Bathophenanthroline solution: 0.1% 4,7-diphenyl-1,10-phenanthroline sulfonate and 1% hydroquinone in 0.1 M acetate, pH 4.5

Solubilization of Mitochondrial Proteins. The insoluble mitochondrial protein is suspended in 0.01 M phosphate buffer, and the protein concentration is adjusted to 20 mg/ml. This suspension is added slowly with stirring to the ice-cold aqueous acetone solution (18 volumes of acetone and 1 volume of water) and allowed to stand for 15 minutes in an ice bath. Acetone extraction removes the bile salts and lipids from

[2] J. V. Maizel, Jr., *Biochem. Biophys. Res. Commun.* **13,** 483 (1963).
[3] T. S. Work, *J. Mol. Biol.* **10,** 544 (1964).
[4] K. Takayama, D. H. MacLennan, A. Tzagoloff, and C. D. Stoner, *Arch. Biochem. Biophys.* **114,** 223 (1966).

the protein.[5] The suspension is centrifuged and the residue is washed twice with water. The protein is dissolved in the phenol–acetic acid–water solution, and the protein concentration is adjusted to 10 mg/ml. Solid urea is added to give a final concentration of 2 *M*.

Polyacrylamide Gel Electrophoresis. A conventional disc electrophoresis apparatus is used.[6] A "working solution" consisting of 7½% acrylamide, 35% acetic acid, and 5 *M* urea is prepared by mixing the following: Stock A–Stock B–TEMED (3:1:0.02, v/v).

One milliliter of the "working solution" is placed in a clean glass tube (5 mm inner diameter and 65 mm long) which is covered at one end with a triple layer of parafilm; 75% acetic acid is carefully layered over the acrylamide solution and the tube is incubated at 50° for 15 minutes to polymerize the acrylamide. The tube containing the gel is rinsed with 75% acetic acid and filled with the same solution. About 10–15 μl of the protein sample (0.10–0.15 mg) is carefully layered between the acetic acid solution and the gel. Both upper and lower reservoirs of the electrophoresis apparatus are filled with 10% acetic acid. The lower electrode serves as the cathode. Electrophoresis is carried out at room temperature for 1 hour at a constant current of 5 ma per tube.

Staining. After electrophoresis, the gel is removed from the glass tube by carefully trimming the gel away from the surface of the glass tube under water with a wire or needle. The gel can be stained by soaking it in the Amido Black solution for 1 hour. The gel is then rinsed several times in 7% acetic acid solution and destained electrophoretically in the same solution at a constant current of 10 ma per tube. Excessively long destaining is to be avoided since the stained protein bands will begin to fade out.

If the protein originally contains iron, this iron dissociates in the phenol–acetic acid solution and then becomes nonspecifically bound to all the protein species. This allows an alternative staining method. The gel is soaked in a bathophenanthroline solution overnight at room temperature to reveal the proteins as red bands.

Scope

This procedure was applied to complex I (DPNH–coenzyme Q reductase), complex II (succinic–coenzyme Q reductase), complex III (reduced coenzyme Q–cytochrome *c* reductase), complex IV (cytochrome oxidase), and soluble succinic dehydrogenase.[4] In each case the complexes were resolved into a number of protein bands as shown in Fig. 1. The protein

[5] R. L. Lester and S. Fleischer, *Biochim. Biophys. Acta* **47**, 358 (1961).
[6] B. J. Davis, *Ann. N.Y. Acad. Sci.* **121**, 404 (1964).

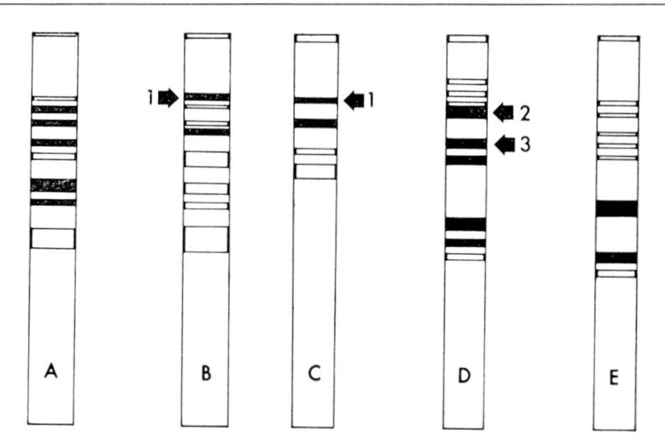

FIG. 1. Schematic representation of the electrophoretic patterns of the four mitochondrial complexes and succinic dehydrogenase: (A) complex I; (B) complex II; (C) succinic dehydrogenase; (D) complex III; (E) complex IV; (*1*) fluorescent flavoprotein band; (*2*) cytochrome c_1 band; (*3*) protein band associated with nonheme iron protein. Cathode was at the bottom.

bands of the flavoprotein in complex II and soluble succinic dehydrogenase as well as the cytochrome c_1 and nonheme iron protein in complex III were tentatively identified.

The system probably would be effective in rapidly separating the component proteins of other hydrophobic protein complexes. Intermolecular disulfide bonds should be reduced prior to solubilizing the protein. Mitochondrial proteins that are treated with thioglycolic acid (or with 2-mercaptoethanol) to reduce the disulfide bonds followed by alkylation with iodoacetate in the presence of 8 M urea cannot be stained effectively with Amido Black. Electrophoretic destaining removes the dye completely. However, destaining by washing with 7% acetic acid or 2% trichloroacetic acid does allow the protein bands to become lightly stained.

Phenol appears to be the most effective solubilizing agent for these hydrophobic mitochondrial proteins. Although the incorporation of phenol into the acrylamide gel would give a more effective system, this is not directly possible since phenol prevents the polymerization of the acrylamide. Phenol can be incorporated into the gel by soaking but this causes the gel to expand considerably and become fragile. Thus the 35% acetic acid–5 M urea in the gel is a compromise system which still allows the proteins to remain in solution. The pattern of the protein bands obtained from the mitochondrial electron transfer complexes are very similar whether phenol impregnated gel or the system described in this text is used. Acrylamide gel concentrations of 5 and 2½% acrylamide were tried on cytochrome oxidase. With decrease in the gel concentration, the protein bands showed greater mobility but the pattern of the protein bands remained the same.

There are certain limitations in the method. In all cases, some protein remains at the origin which is either nonmobile or unable to penetrate the gel. There is also the question of the degree of dissociation achieved by the solubilizing system. The phenol–acetic acid–water solution of the complexes studied are completely transparent, and the protein remains suspended even after centrifugation at 100,000 *g* for 30 minutes. It is possible that the phenol–acetic acid system functions by minimizing the hydrophobic interactions. This solvent system is admittedly drastic. All enzymatic properties are probably destroyed and most of the prosthetic groups are either dissociated, altered, or destroyed. Thus only the primary structure of the native protein could be assured of remaining intact.

[104] Means of Terminating Reactions

By MARTIN KLINGENBERG and ERICH PFAFF

Investigations on mitochondrial metabolism include measurements of the instantaneous level of metabolites or the functional state of cofactors. For this purpose the reactions have to be instantaneously terminated ("quenched") by rapid mixing with a "quenching" agent. The quenched extract is then available for further analysis of the metabolites. This extraction can be combined with simultaneous continuous recording of spectrophotometric absorption of metabolic components and continuous recording of other functions such as respiration, pH changes, etc.

The requirements for the appropriate quenching method are instantaneous termination of the reaction, which can be followed by an extraction that preserves the metabolites or the cofactors in their functional state.

Acid Extraction[1]

The most widely used quenching reaction is the addition of acid, mostly perchloric acid, which gives an effective inactivation of the enzymes and destruction of the mitochondria. Thus the cofactors and substances bound in the mitochondria are effectively released. The lability of some α-keto acids and phosphate compounds containing phos-

[1] M. Klingenberg, *in* "Methoden der enzymatischen Analyse" (H.-U. Bergmeyer, ed.), p. 528. Verlag Chemie, Weinheim/Bergstrasse, 1962.

phate in anhydride or amide bond (creatine-arginine-phosphate) requires that the acid extracts are rapidly neutralized to avoid degradation. The reduced pyridine nucleotides are at the low pH rapidly degraded to the nicotinamide-deficient moiety. The acid extracts are, after neutralization, well suited for enzymatic tests[1] or for chromatographic analysis.[2]

The procedure is to add 3 *M* $HClO_3$ to the mitochondrial suspension to give a final concentration of 0.5 *M* $HClO_4$. The acid is added under rapid mixing. Immediate quenching is effective under a broad temperature range from about 0° to 40°.

Alkaline Extraction[1]

For the extraction of DPNH and TPNH the quenching reagent has to be alkaline; 2 *M* KOH in ethanol is added to the mitochondrial suspension to give a final concentration of 0.5 *M* KOH. The enzymatic reactions are immediately terminated in the alkaline extract. In some determination procedures a complete destruction of DPN and TPN in the alkaline extract is required for the further determination of the DPNH and TPNH. To achieve this and the complete irreversible inactivation of the enzymes, the alkaline extract must be kept at 25° for 30 minutes before it is neutralized. It has to be taken into account that after neutralization the reduced coenzymes can be slowly reoxidized by nonenzymatic reaction catalyzed by the mitochondrial extract.

Lipophilic Extraction[3]

For the extraction of a lipophilic component, such as ubiquinone, the following procedure is recommended; it preserves the ubiquinone at its steady state reduction oxidation level. The mitochondrial suspension is rapidly mixed with double its volume of 70% methanol and 30% petroleum ether. The mixture is well shaken for 30 seconds. The upper petroleum ether phase contains the oxidized and reduced ubiquinone. The extraction is instantaneous since the protein is precipitated by the petroleum ether and the ubiquinone is rapidly removed from its environment in mitochondria by the mediation of methanol.

The extraction can be modified to allow simultaneous measurement of hydrophilic metabolites, such as pyridine nucleotides. For this purpose it must be combined with an acid quenching in order to inactivate completely the DPNH oxidase, ATPase, etc. The extraction mixture contains 0.2 *M* $HClO_4$ in 70% methanol and 30% petroleum ether (composed of 70 ml of methanol, 30 ml of petroleum ether, 1.5 ml of 70% $HClO_4$).

[2] H. W. Heldt, M. Klingenberg, and K. Papenberg, *Biochem. Z.* **342,** 508 (1965).
[3] A. Kröger and M. Klingenberg, *Biochem. Z.* **344,** 317 (1966).

After extraction the lower phase contains the acid-extractable, hydrophilic components such as DPN and adenine nucleotides; the upper phase contains the ubiquinone. The various components can be studied in these two extracts.[3]

Separate Measurements of Intra- and Extramitochondrial Metabolic States

For the separate analysis of sediment and medium the method of sedimentation and separation by a sucrose gradient has been frequently used.[4] By this method the mitochondria do not stay in equilibrium with the medium in the separation layer. The metabolic state will change further in the sediment since the mitochondria are not immediately inactivated. The sucrose density layer should be appropriately replaced by a dextrane density layer in order to avoid too strong an increase of the osmolarity, which would also influence the mitochondria. This layer filtration method can be used only in cases where the internal substances of the mitochondria do not leak out during the washing process or change in the sedimented state.

Termination of the reaction combined with instantaneous fixation can be accomplished with another type of centrifugal filtration. This method—in principle first described by Werkheiser and Bartley[5]—separates the suspension and incubation zone of the mitochondria from the acid fixation layer by a silicone layer. The layers must be arranged in a suitable density gradient. During centrifugation, the mitochondria pass the silicone layer and become stripped of their surrounding medium. The difficulties encountered at first[6] have been overcome after a detailed investigation of the filtration and sedimentation process on the basis of which the method was newly developed.[7,8] During centrifugal sedimentation on the phase limit medium/silicone, portions of the mitochondria temporarily accumulate. From here minute droplets ($\emptyset \approx 0.2$ mm) sediment very fast (in far less than 3 seconds) through the silicone.[9]

[4] B. C. Pressman, *J. Biol. Chem.* **232**, 967 (1958).

[5] W. C. Werkheiser and W. Bartley, *Biochem. J.* **66**, 79 (1957).

[6] J. E. Amoore, *Biochem. J.* **70**, 718 (1958).

[7] E. Pfaff, Diplomarbeit, University of Marburg, 1962.

[8] E. Pfaff, Doctoral thesis, University of Marburg, 1965.

[9] At the entrance to the acid layer, mixtures of deproteinized mitochondria can form that are at first less dense than the silicone. If the amount of mitochondria exceeds a certain limit, less dense droplets may migrate back in the centrifugal field and cause contamination by acid of the upper suspension. The reverse migration of acid is avoided when the concentration of mitochondria is small enough. Furthermore, it is important also that the density of the acid sufficiently exceeds that of the silicone. On the other hand, retarded inactivation can be caused by a too high density of the acid.

The following experimental procedure has been the one principally used by the authors. Siliconized centrifuge tubes made of Pyrex glass (Ø ≈ 6–8 mm) are prepared for the swinging-bucket rotor SW-39, Spinco-Beckman Co. The centrifuge tubes are placed in a polyethylene adapter containing water. They are run up to 22,000 rpm (35,000 g) and then stopped. The following densities can be recommended for suspension of mitochondria in 0.25 M sucrose ($\rho = 1.037$): acid layer containing 1.6 M $HClO_4$, $\rho = 1.08$ g cm^{-3}; silicone layer, $\rho = 1.05$ to 1.06 g cm^{-3}. The experiment is started by the addition of the mitochondria to the incubation layer; in other cases the suspension may be layered on top of the silicone.

Control experiments proved that the intramitochondrial metabolic state is preserved through the silicone until the inactivation. The method is therefore proved to yield the instantaneous metabolic pattern of mitochondria. The extramitochondrial fluid which accompanies the mitochondrial droplets can be corrected for by the use of labeled high molecular glucose (^{14}C-etherated globular polyglucose or ^{14}C-dextrane). The filtered amount is composed of approximately equal parts of mitochondrial and extramitochondrial volume.

The shortest time attainable for the incubation and separation of the mitochondria is about 50 seconds after the start of incubation. This includes the time for preparing the rotor and acceleration of the centrifuge until it reaches 8000 rpm, when the mitochondria start to penetrate the silicone layer.

For kinetic experiments requiring shorter times for the separation of mitochondria, the method can be modified to include an additional incubation layer (centrifugal-layer filtration)[8,10] (cf. Fig. 1). Here the mitochondria are first present in an upper layer under which is a layer containing the reactive substances. The layers are isolated by a small amount of isolating medium. The density of the three layers is increased by the addition of dextran. During centrifugation the mitochondria migrate for only a short time through the reaction layer, the time of which can be calibrated. Incubation times as low as 10 seconds can be obtained. The lag time at the phase-limit medium-silicone can be excluded by placing an additional washing layer before the silicone.

Termination of Reaction by Sieve Filtration

Another approach for separating intra- and extramitochondrial metabolites is sieve filtration. Here the mitochondrial suspension is pressed

[10] M. Klingenberg, E. Pfaff, and A. Kröger, *in* "Rapid Mixing and Sampling Techniques in Biochemistry" (B. Chance, R. H. Eisenhardt, Q. H. Gibson, and K. K. Lonberg-Holm, eds.), p. 333. Academic Press, New York, 1965.

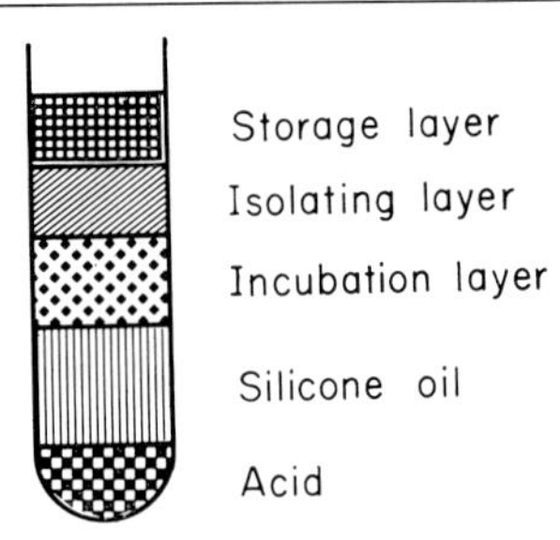

FIG. 1. Scheme of centrifugal layer filtration. The layers have the following composition: the storage layer contains the mitochondria in 0.25 *M* sucrose medium (density $\rho = 1.037$ g/ml), the density of the 0.25 *M* sucrose medium in the isolating and the incubation layer is increased by the addition of 15.0 mg and 22.5 mg of dextran per milliliter, respectively. The silicone oil has a density of 1.065 and the acid (1.6 *M* $HClO_4$) of 1.08.

through a filter which may consist of a Celite layer,[11] Millipore filter,[8] or glass fiber filter.[12] In this case the filtrates can be collected in acid and thus immediately deproteinized. This allows determination of the true metabolic state of the extramitochondrial space. However, the intramitochondrial metabolites may have greatly changed, since collecting the mitochondria on the filter surface separates them from their substrate and oxygen supply. Furthermore it is difficult to account reliably for the amount of medium trapped in the mitochondrial layer on the filter by high molecular compounds since these are also concentrated with the mitochondria on the filter surface.[8]

The following method has been applied by the authors. Filters of the Millipore Corporation ($\emptyset \approx 0.65\,\mu$) have been used in combination with a stainless steel pressure filter holder (Millipore Corporation). An air pressure of up to 6 atmospheres is used. The filter holder was modified by a stirring device which allowed starting of the incubation by mixing within the filter chamber and immediate filtration in less than 2 seconds. A suspension of 4 ml of mitochondria with a total of 10 mg of protein can be filtered within less than a second.

[11] G. F. Azzone and L. Ernster, *J. Biol. Chem.* **236,** 1501 (1961).
[12] G. Brierley and R. L. O'Brien, *J. Biol. Chem.* **240,** 4532 (1965).

[105a] Energy-Linked Low Amplitude Mitochondrial Swelling

By LESTER PACKER

Reversible swelling-shrinkage associated with metabolic states is an energy-linked function of mitochondria.[1] Similar to other energy-dependent processes, volume changes are affected by reagents which interact with the energy transfer pathway.[2] Thus, mitochondrial swelling is supported by electron flow through the respiratory chain and requires the presence of an anion-like phosphate. Moreover, the process is inhibited by ADP, which causes shrinkage presumably by competing for energy. Another distinguishing feature of this energy-linked process is the action of oligomycin, which does not inhibit swelling, but abolishes the shrinkage effect of ADP.[3] This swelling process results in a rapid but small volume change, and therefore direct gravimetric methods for measuring water content of mitochondria are difficult to employ to quantitatively assess its extent. Optical methods have been widely adopted because they afford a rapid and empirical measure of volume.[4]

The high mitochondrial concentration used in gravimetric studies depletes the reaction mixture of O_2, causing anaerobiosis which alters the metabolic, and thereby, the structural state. However, packed volume (or mitocrit) determinations have been employed to detect swelling successfully by pretreating the reaction mixture with O_2 and performing the experiment rapidly,[5] or by redesigning centrifuge tubes to contain a well in the bottom, where mitochondria are collected from the bulk solution and their volume determined.[6] The Coulter counter technique is also valuable for measuring particle volumes, and it has been used to detect high amplitude mitochondrial swelling[7] but is difficult to use for low amplitude studies.[8] The types of methods needed to assess the

[1] L. Packer, *J. Biol. Chem.* **235,** 242 (1960); **236,** 214 (1961); **237,** 1327 (1962).

[2] For properties of other energy-linked functions, see Vol. VIII [9].

[3] L. Packer, E. Corriden, and R. Marchant, *Biochim. Biophys. Acta* **78,** 534 (1963).

[4] This arises from the large differences in refractive index between mitochondria and the medium in which they usually are suspended.

[5] L. Packer, *J. Cell Biol.* **18,** 487 (1963); **18,** 495 (1963).

[6] W. Bartley and M. Enser, *Biochem. J.* **93,** 322 (1964).

[7] J. Gebicki and F. Hunter, Jr., *J. Biol. Chem.* **239,** 631 (1964).

[8] However, the energy-linked volume changes of spinach chloroplasts *in vitro* are easily measured by the Coulter counter because the initial volume of chloroplasts is larger than mitochondria [cf. L. Packer, P. A. Siegenthaler, and P. S. Nobel, *J. Cell Biol.* **26,** 593 (1965)].

occurrence of energy-linked volume changes are those providing great sensitivity at low mitochondrial concentration. The most convenient and simplest to use—light scattering and transmission (or absorbancy) determinations—are described below.

Techniques

Test System. Rabbit heart mitochondria are used as an example of a system manifesting energy-linked swelling. The mitochondria are isolated[9] in sucrose (0.32 M) EDTA (0.001 M, pH 7.5) and tested at 25° in the following basic medium: Tris (0.02 M, pH 7.5), sucrose (0.05 M), KCl (0.020 M); where indicated substrate (0.005 M), sodium phosphate (0.005 M, pH 7.5), and ADP (250 μM) were added to the basic medium. [For liver mitochondria, a medium containing Tris (0.010 M, pH 7.5), sucrose (0.05 M), $MgCl_2$ (0.005), and KCl (0.010 M) may be employed.]

Optical Measurements at 0° and 90°. Volume changes are measured optically, at a wavelength where respiratory pigments do not appreciably absorb the incident light. A Brice-Phoenix light-scattering photometer, containing a dual photomultiplier system and adapted for time recordings, was used.[10] For light scattering (90°) or transmission (0°) measurements at 546 mμ, the initial level (one full-scale width on the recorder chart) is set, and changes, obtained with various mitochondrial concentrations or by the addition of reagents to the system, are recorded. Scattering decreases and transmission increases reflect either swelling or a decrease of the mitochondrial concentration.[5,6] The use of an ordinary laboratory spectrophotometer for the measurement of energy-linked swelling at 0° requires a dual beam instrument which subtracts the reading in the reference from the sample cuvette.

Procedure

Light-Scattering Photometer. The use of light scattering at 90° and transmission (or absorbancy) at 0° for measuring energy-linked mitochondrial swelling and its reversal by ADP are shown by simultaneous recordings in Fig. 1. The mitochondrial suspension is added to the basic reaction medium to a final concentration of 1 mg protein per milliliter and then transferred to the cuvette. The initial light intensity is adjusted for full-scale width of the recorder (i.e., relative setting of 100, or 100%). Electron flow is initiated by the simultaneous addition of substrate + phosphate on the tip of a stirring rod as a concentrated solution to permit

[9] Procedures for isolating heart mitochondria are described in Vol. VIII [2].

[10] A DC amplifier (Houston Instruments-M 10) and a recording milliammeter (Texas Instruments, Rectiriter 1 mA full-scale) are the components employed.

rapid mixing and negligible dilution. Electron flow causes a prompt decrease of light scattering and increase in transmission (or absorbancy decrease) which corresponds to swelling. The volume change is completed in about 2 minutes and a steady state is established. The extent of the optical change resulting from swelling is usually 20–40%. Swelling requires both substrate and phosphate; the addition of either alone will not result in swelling if mitochondria are depleted of endogenous reserves of these substances.[1]

An inhibition of mitochondrial swelling is brought about by certain reagents that interact with the energy transfer pathway to compete for available energy. Hence, initiation of phosphorylation by addition of ADP (Fig. 1) results in a partial reversal of the swelling process; when

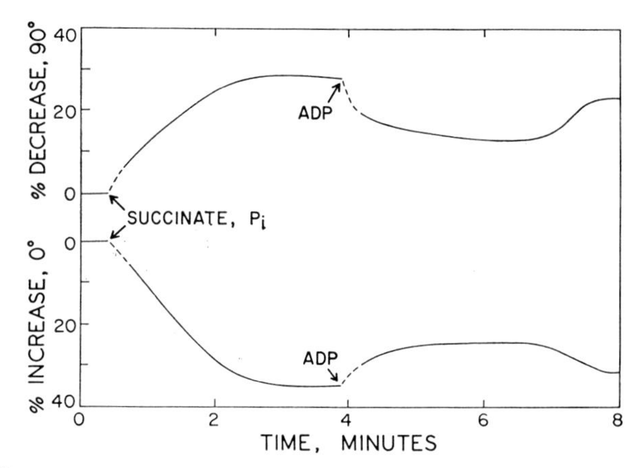

FIG. 1. Time course of energy-linked mitochondrial swelling. Measurements were recorded simultaneously at 0° and 90° with initial readings set at one full-scale deflection (zero point in figure). The per cent changes caused by the indicated additions were recorded on the charts.

ADP is converted to ATP, the mitochondria again swell to the previous level. Besides ADP, other substances (uncouplers, cations, and inhibitors) may result in shrinkage, unless tested at inordinately high concentration, where other effects are introduced that result in high-amplitude swelling.[1,5] The extent of swelling is dependent upon the system of electron flow, being frequently larger with substances, such as succinate and ascorbate + tetramethylphenylenediamine, which support both forward and reverse pathways of electron flow.[1] Oligomycin, which inhibits ATP synthesis, but does not generally affect other energy-linked functions of mitochondria, will not inhibit swelling, but it prevents ADP from causing shrinkage,[3] and therefore this reagent is useful in identifying the presence of an energy-linked swelling process. The extent of swelling will probably vary in different mitochondria because of the existence of multiple energy-linked functions and unknown quantities of endogenous reserves. The concentration of reagents and mitochondria employed are important. The action of substrate plus phosphate, or ADP, on swelling and shrinkage, respectively, follows closely upon their effect on respiration; 1 m*M* phosphate and 20 μM ADP give about half-maximal volume changes. To ensure the full swelling or shrinkage response, it is necessary to add these reagents at concentrations which fully activate the reactions involved.

The mitochondrial concentration, the type of optical technique used, and the osmolarity are also important factors in evaluating the extent of swelling.[5] Figures 2C and D show the variation in light scattering or

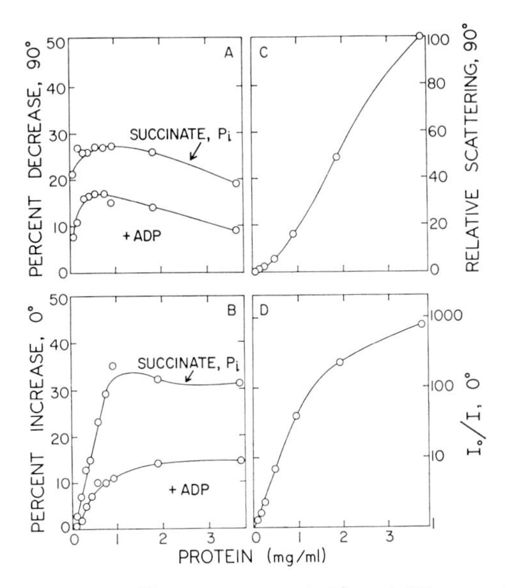

FIG. 2. Dependence of swelling responses at 0° and 90° on mitochondrial concentration. (A and B) Conditions as in Fig. 1 at indicated mitochondrial concentrations. (C and D) Relative light intensity at 0° readings recalculated as I_0/I and 90° of mitochondria suspended in basic reaction medium at indicated protein concentration.

transmission at a range of mitochondrial concentrations up to 6 mg protein per milliliter. The scattering method is probably best to use at low mitochondrial concentrations, where slight changes of concentration have only a small effect on relative scattering intensity, but an enormous effect upon light transmission (these values are divided into the medium values and plotted logarithmically). Figures 2A and B also show the effect of mitochondrial concentration upon the extent of the swelling response. The responses are not independent of mitochondrial concentration in the low concentration range especially at 0°; however, at 0.25 mg protein per milliliter, the 90° readings become independent of concentration. Hence, 90° data are a more sensitive measure of the response than 0° readings in the low concentration range.[11]

Dual-Beam Spectrophotometer. The data in Fig. 2B indicate that the most useful range of concentration for heart mitochondria is between 1 and 4 mg protein per milliliter for 0° measurements. However Fig. 2D shows that this range of concentration spans several units of transmission or absorbancy, so the measurement of swelling with a laboratory spectrophotometer at 0° requires the presence of a reference compartment to subtract the high initial optical opacity of the mitochondrial suspension. A typical experiment is made in a Beckman DB spectrophotometer set at 546 mμ with the manual slit in the wide open position. Mitochondria (1–4 mg protein per milliliter) are mixed with reaction mixture and distributed equally into reference and sample cuvettes. The reading is adjusted to 100% transmission (or zero absorbance) with the balancing adjust knob, and then substrate + P_i is added to the reference cuvette. The kinetics of the reaction are followed either visually or on a suitable strip chart recorder (cf. Fig. 1). The results obtained by this procedure closely parallel those shown in Fig. 2B both with regard to concentration dependence and extent of absorbancy change.

[11] The slight depression of the response at higher concentrations results from increasing osmolarity (from mixing mitochondria in 0.3 *M* sucrose with the 0.05 *M* sucrose of medium). Energy-linked volume changes occur over a wide range of osmolarity (cf. footnote 1 references).

[105b] Measurement of Mitochondrial Swelling and Shrinking—High Amplitude

By F. Edmund Hunter, Jr., and E. E. Smith

When isolated mitochondria are treated with a variety of agents the optical density or turbidity of dilute suspensions undergoes relatively large changes. Since early studies indicated a correlation with swelling, these changes became known as high amplitude swelling and shrinking.[1–3] The term was really derived from the changes in optical density, as the

[1] A. L. Lehninger, *Physiol. Rev.* **42,** 467 (1962).
[2] J. B. Chappell and G. D. Greville, *Biochem. Soc. Symp. (Cambridge, Engl.)* **23,** 39 (1963).
[3] L. Packer, *J. Cell Biol.* **18,** 487 (1963).

mitochondrial volume changes were unknown in most cases. High amplitude swelling is reversible under certain circumstances. However, incomplete reversibility and the slow rate of swelling in comparison with low amplitude changes has led some investigators to consider it a change that goes beyond physiological processes, being characteristic of mitochondria under *in vitro* conditions and possibly of pathological conditions *in vivo*. The distinction between low and high amplitude changes and physiological versus pathological changes is not entirely clear.

High amplitude swelling has been widely studied for the following reasons. (1) Induction of swelling by many agents, especially by hormones such as thyroxine, seemed to offer clues to control of membrane permeability. (2) Swelling *in vivo* was reported for both physiological and pathological states. (3) Important implications in mitochondrial organization and control seemed highly probable because swelling with many agents was dependent on electron transport and intermediates of oxidative phosphorylation. (4) Some cases of high amplitude swelling seemed to be related to active transport of the ions Ca^{++} and K^{+}.

Methods for Measurement of High Amplitude Swelling

Optical Density Changes

Principle. Dilute suspensions of mitochondria are turbid and absorb and scatter visible light. When the mitochondria swell the light scattering decreases and the optical density of the suspension decreases. Most investigators have employed this method, using light at 520 mμ. The theoretical basis of light scattering measurements have been considered by several workers.[4–6]

Equipment and Procedure. (a) DILUTE SUSPENSIONS OF MITOCHONDRIA. Suspensions of isolated mitochondria are diluted, usually in 0.125–0.175 *M* KCl or 0.25–0.33 *M* sucrose, to give a protein concentration of 100–200 μg/ml, or an optical density between 0.4 and 0.9. The exact reading will depend on the basic medium and constituents added.[2] The experiment is carried out by adding the substances under study, mixing, and following the optical density either by frequent manual readings or in a recording instrument. The changes will vary from 10 to 90% of the initial optical density. Visual checks are essential to detect possible aggregation.

Almost any colorimeter or spectrophotometer can be used to follow

[4] M. Bier, Vol. IV, p. 147.

[5] A. L. Koch, *Biochim. Biophys. Acta* **51**, 429 (1961).

[6] H. Tedeschi and D. L. Harris, *Biochim. Biophys. Acta* **28**, 392 (1958).

the changes in turbidity by measuring the transmitted light. Both rectangular and round cuvettes of about 10-mm light path are used. A set of selected test tubes or matched cuvettes permits a series of tests with readings every 3–5 minutes. Each tube serves as its own control for the initial reading. Initial readings should be taken as quickly as possible after mixing (usually 3–5 seconds) if the experiment is initiated by addition of mitochondria.

The apparent optical density will vary between instruments because of differences in geometry of design. Most colorimeters and spectrophotometers, and especially those using round cuvettes, record some scattered light as well as the direct beam. In machines with sensitive phototubes and adequate amplification the distinction between light with a low angle of scatter and the directly transmitted beam can be sharpened by the insertion of appropriate slits.

(b) CONCENTRATED SUSPENSIONS OF MITOCHONDRIA. It is highly desirable at times to study more concentrated suspensions, for mitochondria do undergo changes such as rapid loss of respiratory control when diluted sufficiently to use a 10-mm light path. Two techniques have been used: one uses cuvettes with a 1- or 2-mm light path; the other, spectrophotometers and colorimeters with the ability to amplify the signal from the small amount of transmitted light at high optical densities.

90° Light Scattering

Principle. When large amplitude swelling occurs, the change in 90° light scattering will usually be greater than that in total light scattering as measured by optical density.[3,7] Thus, 90° light scattering has been used as a sensitive method to follow large amplitude swelling. However, whether one is measuring large amplitude or small amplitude swelling quickly becomes a matter of semantics. Gotterer *et al.*[8] used light scattering at different angles as a sensitive method to detect small changes and loss of protein.

Instruments and Procedures. 90° light scattering is measured on mitochondrial suspensions similar to those used for optical density changes. Commercial light scattering photometers are used, but less elaborate instruments such as fluorometers with appropriate slits serve the same purpose.

Gravimetric Measurements

The wet weight of the mitochondrial pellet obtained by standard centrifugation in tared tubes gives an estimate of the mitochondrial

[7] B. Chance and L. Packer, *Biochem. J.* **68**, 295 (1958).

volume.[9] The method requires fairly large amounts of mitochondria in relatively concentrated suspensions. It has been widely used as a check on optical methods. However, since the extramitochondrial water in such pellets can vary severalfold, accurate estimates of mitochondrial volume require simultaneous estimation of extramitochondrial space through the use of some nonpenetrating substance such as ^{14}C-carboxypolyglucose.[10]

The wet weight of mitochondria relative to other constituents should be a good measure of swelling. In practice, however, dry weight of mitochondrial material is impossible to obtain because of KCl, sucrose, or other constituents present. A more satisfactory reference is mitochondrial protein, but this is subject to error if there is protein loss.

Mitocrit (Packed Wet Volume)

This measurement correlates fairly well with the volume occupied by the mitochondria, although exact values for absolute volume require measurements of extramitochondrial space.[11] The method has been used to determine whether optical methods were giving values related to real volume changes.

Procedure. Any standardized high speed centrifugation procedure in tubes calibrated to read the volume of the pellet will do. Since the pellet volume is not large, concentrated suspensions or tubes especially designed for the purpose give more reliable readings. Small-diameter tubes similar to those for hematocrits have been used.[12] Wet weight of the pellet probably gives a more reliable figure for the volume than that obtained by trying to read volume directly if tubes of special design are not used.

Viscosity

Changes in mitochondrial volume in a suspension should result in viscosity changes, and Packer[3,13] has obtained results consistent with other measurements. In practice, however, even with more concentrated suspensions the technical problem is to get enough viscosity change to measure with accuracy.

[8] G. S. Gotterer, T. E. Thompson, and A. L. Lehninger, *J. Biophys. Biochem. Cytol.* **10,** 15 (1961).

[9] M. G. MacFarland and A. G. Spencer, *Biochem. J.* **54,** 569 (1953).

[10] W. C. Werkheiser and W. Bartley, *Biochem. J.* **66,** 79 (1957).

[11] S. Malamed and R. O. Recknagel, *Proc. Soc. Exptl. Biol. Med.* **98,** 139 (1958).

[12] J. W. Harman and M. T. O'Hegarty, *Exptl. Mol. Pathol.* **1,** 573 (1962).

Electronic Counting and Sizing of Mitochondria

Principle. This method[14] is based on the high electrical resistance of biological membranes, which effectively exclude the volume within the membrane in the conduction of direct current. Electronic particle counters measure this volume by the momentary increase in resistance when the particle is drawn through a small aperture, displacing electrolyte solution conducting the current.

Advantages and Disadvantages. The method is based on electrical properties of the membrane, which remain unchanged after considerable swelling.[15] The values are proportional to particle volume regardless of shape.

One disadvantage is that the mitochondrial suspension must be diluted to extreme degrees in an ionic medium. Aggregation of mitochondria due to the salt does not seem to be a major problem at these dilutions, but it is less certain that rapid initial changes in the mitochondria do not occur. After dilution the count and size distribution of normal mitochondria are remarkably stable. Other disadvantages include the probable necessity for selection of larger mitochondria for study because of the limits of the machine. If smaller apertures are used in order to cover the full range of normal mitochondria, technical problems increase greatly. Preventing clogging of the aperture by particles of dirt or debris becomes extremely difficult.

Preparation of Solutions. Mitochondrial suspensions contain debris, but the major problem in practice stems from the diluting medium. All glassware must be given final rinses with dust-free water. Water and dilution media must be repeatedly filtered through Millipore filters with 0.22 μ or smaller pores.[14]

Procedure. Most measurements have been made with mitochondria diluted to 100–200 μg protein per milliliter. A 50-μl aliquot of this suspension is added to 8 ml of isotonic KCl and mixed thoroughly by inversion. The mitochondria in different size ranges are counted at different settings of the machine. Close attention by an experienced operator is required to handle clogging of the aperture.

Removal of Dirt from the Aperture. Simple tapping of the aperture tube will very frequently dislodge dirt particles and permit prompt resumption of counting. Release of suction, a brief "burn out" with high current, or brushing of the aperture with a camel hair brush may remove dirt particles. If these fail, alcohol, acetone, filtered detergent, or a brief

[14] J. M. Gebicki and F. E. Hunter, Jr., *J. Biol. Chem.* **239**, 631 (1964).

[15] H. Pauly, L. Packer, and H. P. Schwan, *J. Biophys. Biochem. Cytol.* **7**, 589, 603 (1960).

treatment with chromic acid cleaning solution may be required, but this means starting a new size distribution curve. Chromic acid may remove the cement holding the aperture plate.

Interpretations. Polystyrene spheres can be used to calibrate the machine for absolute volumes. Signals are proportional to volumes. With swelling, previously undetected small mitochondria may enter the counting range. Partial disintegration of swollen mitochondria could increase the count of smaller particles.

Phase Microscopy

Visual and photographic examination of mitochondria by this means provided the earliest information on swelling changes in mitochondria.[16] It is used by most workers for preliminary examinations and for general confirmation of observations made by other techniques. A few have carried out systematic studies.[12,15] Experience is required, since continuous movement and problems with plane of focus make evaluations difficult. Mitochondria which do not move are probably fixed on the glass surface and do not show normal swelling-shrinking behavior.

Electron Microscopy

On theoretical grounds this method should tell us most about the changes which mitochondria undergo with swelling. A considerable measure of success has been achieved with normal mitochondria. Some workers have used the technique to check on the changes in swelling experiments, and most agree that swelling induced with the milder agents involves separation of the membranes of the cristae and a marked decrease in the number of cristae, with much of the material becoming part of the enclosing membrane structures.[1,17–19]

The application of electron microscopy to swollen mitochondria has the difficulty of greater fragility and susceptibility to artifacts during fixation. Malamed *et al.*[20] have pointed out that some electron microscope fixatives can cause swelling of isolated mitochondria. However, the relatively normal appearance of control mitochondria[19] when these chemicals are used to fix pellets for electron microscopy indicates that the dangers, although real, do not obviate electron microscope work if care and judgment are used.

Sjöstrand presented a detailed discussion of the electron microscopic

[16] J. W. Harman and M. Feigelson, *Exptl. Cell Res.* **3**, 509 (1952).

[17] F. E. Hunter, Jr., R. Malison, W. F. Bridgers, B. Schutz, and A. Atchison, *J. Biol. Chem.* **234**, 693 (1959).

[18] E. C. Weinbach, *Proc. Natl. Acad. Sci. U.S.* **50**, 561 (1963).

[19] S. Malamed, *Z. Zellforsch. Microskop. Anat.* **65**, 10 (1965).

[20] S. Malamed, J. Weissman, and W. Chenitz, *Federation Proc.* **23**, 519 (1964).

approach.[21] Since success with isolated mitochondria requires specialized skills, the interested investigator is advised to consult detailed studies with the most recent techniques.[18,19,22] The need to sample many layers in the pellet and the fact that sectioning gives a wide variety of cross sections of mitochondria pose serious problems with respect to the conclusions that can be drawn. The average maximum diameter of all the mitochondria in the section may be the most useful measure representative of the population being examined. Published pictures usually represent attempts to select a typical field after examination of hundreds.

Degree of Correlation between Measurements Made by Different Methods

Since each method entails possible errors, the degree of correlation between techniques is of prime importance. Early workers established the general validity of the optical methods by direct examination with the phase microscope, packed wet volume, and wet weight of the pellet. While predictable light scattering relationships consistent with theory were found for changes in tonicity,[5,6,11,23] the complexities in size and shape have caused most investigators to consider only gross relative changes. Even in the early work there were some discrepancies between packed volume, phase microscopy, and optical density.[1,2] More recently others[12,24-26] have seriously questioned the correlations with optical density changes and tried to define the nature of the changes measured by each of the several methods. These studies have led to the suggestion that changes in internal structures such as the cristae may be more responsible for the light scattering changes than changes in the overall volume, although both can change simultaneously. Light scattering changes usually are large with phenomena related to electron transport in membranes and membrane fragments. Simple osmotic swelling and structural or colloid osmotic swelling in cristae have been suggested.[11,12,19] Final conclusions are still not possible, but certainly multiple changes are occurring. In summary, the full definition of the exact changes can be made only by applying several different methods.

Reversibility of High Amplitude Swelling

Under many circumstances the optical density changes can be reversed to a large degree. Lehninger[1] demonstrated the prime role for

[21] F. S. Sjöstrand, Vol. IV, p. 391.
[22] S. Malamed, *J. Cell Biol.* **18,** 696 (1963).
[23] L. Packer, *J. Biol. Chem.* **235,** 242 (1960).
[24] R. L. Klein and R. J. Neff, *Exptl. Cell Res.* **19,** 133 (1960).
[25] S. I. Honda and A. M. Muenster, *Arch. Biochem. Biophys.* **88,** 118 (1960).
[26] W. Bartley and M. D. Enser, *Biochem. J.* **93,** 322 (1964).

ATP in the process, with BSA, Mg^{++}, Mn^{++}, or other substances required under specific circumstances. ATP may restore the membrane to its original state. The extrusion of water and solute suggests an additional contractile process. Other methods also indicate a shrinking or reduction in size. Electron microscopy indicates a significant but incomplete return toward normal structure.[18, 26]

[106] Preparation and Use of Antisera to Respiratory Chain Components[1]

By STARKEY D. DAVIS,[2] THOMAS D. MEHL, RALPH J. WEDGWOOD, and BRUCE MACKLER[3]

Enzymes are proteins and are potentially antigenic. Antisera to enzymes have been used to study enzyme antigens and structure, enzyme evolution, embryology, genetics, localization, enzyme interrelationships, and the nature of the enzyme–substrate association at the active site. An extensive survey of the immunochemistry of enzymes has been published.[4]

Antisera and enzymes may form insoluble complexes and precipitate in solution or in agar. The use of immunodiffusion to study antigens has been reviewed briefly in this series[5] and more completely elsewhere by Ouchterlony.[6] Antisera may also inhibit enzyme activity. Some enzymes are completely inhibited, others are partly inhibited, and some not at all. Inhibition apparently results from steric hindrance of substrate or acceptor binding by antiserum bound near the active sites.[4]

Antisera to respiratory enzymes have been used in both immunologic and enzymatic tests to study the structural and functional relationship between respiratory enzymes.[7, 8]

[1] Supported in part by grant H-5457 from the National Institutes of Health, and by a grant from the Initiative 171 fund, University of Washington to R. J. W.

[2] Special fellowship awardee of the National Institutes of Health.

[3] Research Career Development Awardee of the National Institutes of Health.

[4] B. Cinader, Ed., *Ann. N.Y. Acad. Sci.* **103**, 495 (1963).

[5] See Vol. VI [119].

[6] Ö. Ouchterlony, *Progr. Allergy* **6**, 30 (1962).

[7] H. R. Mahler, B. Mackler, P. P. Slonimski, and S. Grandchamp, *Biochemistry* **3**, 677 (1964).

[8] B. Mackler, R. J. Erickson, S. D. Davis, T. D. Mehl, and R. J. Wedgwood, to be published.

Preparation of Antisera

Enzymes prepared in the usual manner[8,9] are suspended or dissolved in 0.1 *M* phosphate buffer, pH 7.4, to contain approximately 5 mg of enzyme per milliliter. Antiserum is produced in rabbits by multiple injections of enzyme mixed with equal volumes of complete Freund's adjuvant (Difco). Control serum is obtained from each rabbit before beginning the injections. Up to 30 ml of blood can be obtained from the marginal ear vein of rabbits, and this volume will usually be sufficient. To facilitate emulsification, the Freund's adjuvant is warmed to 37° in a water bath, the enzyme solution is added, and the mixture is thoroughly agitated, with care not to denature the enzyme. Weekly intramuscular injections of 5 mg of enzyme in adjuvant for 8–12 weeks will usually yield a satisfactory antiserum. Sites of injections should be varied to prevent abscess formation. Trial bleedings from an ear vein are tested for potency after 8 weeks.

Potency of antisera is tested by immunodiffusion as follows: For each liter of solution, add 10 g of Bacto Noble agar (Difco) and 8.5 g NaCl to distilled water. Dissolve by bringing the solution to boiling with frequent agitation to prevent charring. Allow the agar to cool to 50° and pour 10-ml aliquots into petri dishes. Allow to harden by cooling. The agar dishes will keep for months in the refrigerator if sealed to prevent drying. Bacterial contamination is reduced by the addition to the agar solution of a preservative, such as 200 mg of Merthiolate. Wells are cut in the agar dishes using a cork borer or other tubing about 3 mm in diameter. The wells are placed about 5 mm apart. Agar is aspirated from the wells with Pasteur pipettes attached to suction. To perform the test, fill a well with the enzyme solution or suspension used as antigen and fill an adjacent well with the serum to be tested. Cover the dish, seal to prevent drying, and leave overnight at 37°. A potent antiserum will produce a heavy precipitin line between the two wells within 24 hours.

When a satisfactory antiserum is obtained, the rabbits are exsanguinated by cardiac puncture. The blood is allowed to clot for several hours at room temperature, and the serum is separated by centrifugation. The serum is stored frozen (—20° to —70°) without the addition of preservatives. In our experience, antiserum stored at —64° has been stable for several years.

We have found that longer periods of injections (over 8 weeks) may be needed to produce a potent antiserum to particulate enzymes than

[9] H. R. Mahler, B. Mackler, S. Grandchamp, and P. P. Slonimski, *Biochemistry* 3, 668 (1964).

to soluble enzymes. Also, as in the preparation of any antiserum, an occasional animal may fail to develop a potent antiserum despite an extended series of immunizations. For this reason, several animals should be injected with each enzyme. For a more detailed discussion of the injection and bleeding of animals, see the text by Campbell *et al.*[10]

Control serum may sometimes be inhibitory to enzyme activity. Fractionation of the serum with $(NH_4)_2SO_4$ yields a preparation which is not inhibitory and contains γ-globulin in a partially purified form. It is therefore advantageous to purify the antisera and control sera as follows: To serum in an ice bath add a saturated (at 0°) solution of $(NH_4)_2SO_4$, pH 7.0, to 50% final saturation. Stir for 10 minutes and centrifuge at 0–5°. Decant and discard the supernatant. Take up the residue in a minimum volume of distilled water, dialyze the solution against 0.15 *M* NaCl for 24 hours, and store frozen until used.

Applications of Antisera to the Study of Respiratory Enzymes

Examples of the techniques involved in the use of antisera in studies of respiratory enzymes are described below, using antisera to beef heart DPNH oxidase and DPNH dehydrogenase.

Immunologic Studies.[8] Antisera and enzymes were studied by double diffusion in agar as described above. As illustrated in Fig. 1, each anti-

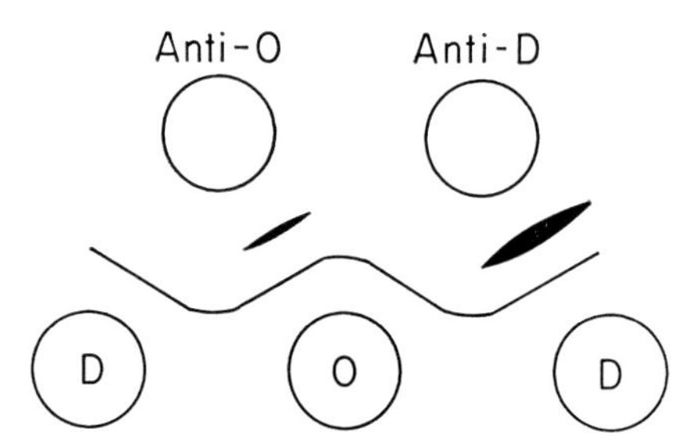

FIG. 1. Immunodiffusion tests of antisera to beef heart DPNH oxidase and DPNH dehydrogenase. *Anti-O,* antiserum to oxidase; *Anti-D,* antiserum to DPNH dehydrogenase; *O,* DPNH oxidase; *D,* DPNH dehydrogenase.

serum produced two precipitin lines against its homologous enzyme preparation. Each antiserum produced one precipitin line against the heterologous enzyme which fused in a reaction of identity with one of the lines against the homologous enzyme preparation. This demonstrated that each enzyme had a specific antigen not shared with the other, and both enzyme preparations had a common second antigen.

Enzyme Inhibition Studies.[7,8] For studies of antiserum inhibition of

[10] D. H. Campbell, J. S. Garvey, N. E. Cremer, and D. H. Sussdorf, "Methods in Immunology." Benjamin, New York, 1964.

enzyme activity, enzyme, antiserum, and the appropriate assay buffer were preincubated in cuvettes at the reaction temperature for 5 minutes. DPNH, and where called for NaCN and EDTA were then added, and the reaction was started by addition of the appropriate electron acceptor. In studies of DPNH oxidase activity the reactions were begun by addition of DPNH. Control serum and NaCl, in volumes present in the preparations of antiserum, were substituted for antiserum in control experiments. Preincubation of the enzymes with antiserum for periods of 10 minutes resulted in greater loss of enzyme activity, with no significant increase in inhibition of enzymatic activity by the antisera.

[107] Antibody[1] against F_1[2]

By JUNE M. FESSENDEN[2a] and EFRAIM RACKER[2a]

In a multi-enzyme system which is associated with membrane structure, such as the system of oxidative phosphorylation, it is often difficult to obtain satisfactory resolution of the participating catalysts. With the availability of a specific antibody against a single component of the pathway, its catalytic function can be evaluated even under conditions of incomplete resolution. For example, mitochondrial ATPase (F_1) which has been obtained in homogeneous form[2] stimulates oxidative phosphorylation in F_1-deficient submitochondrial particles.[3] Yet such particles usually contain residual ATPase, sometimes in considerable quantities, and energy-producing reactions can be readily demonstrated under suitable conditions even without added coupling factors.[4,5] Specific antibodies seem to be particularly suited for the evaluation of the role of residual coupling factors in deficient particles.[6]

Assay Method

Principle. The incorporation of $^{32}P_i$ into ATP, catalyzed by submitochondrial particles is inhibited by an antiserum against F_1. The degree

[1] Unpublished method. We are very indebted to Dr. F. Adler for his assistance and advice with the immunization procedure and evaluation of the data.

[2] See this volume [82].

[2a] This manuscript was prepared while the authors were at the Department of Biochemistry, The Public Health Research Institute of the City of New York, Inc., New York.

[3] H. S. Penefsky, M. E. Pullman, A. Datta, and E. Racker, *J. Biol. Chem.* **235**, 3330 (1960).

[4] E. Racker and G. Monroy, *Abstr. 6th Intern. Congr. Biochem., New York, 1964,* p. 760. Federation Am. Soc. Exptl. Biol.

[5] L. Danielson and L. Ernster, *Biochem. Z.* **338**, 188 (1963). See also C. P. Lee and L. Ernster, *Biochem. Biophys. Res. Commun.* **18**, 523 (1965).

[6] J. M. Fessenden and E. Racker, *J. Biol. Chem.* **241**, 2483 (1966).

of inhibition can be used in titration experiments to evaluate the potency of the antiserum.

Reagents

Potassium phosphate, 0.2 *M*, pH 7.4
ATP, 0.2 *M*, pH 7.4
$MgSO_4$, 0.2 *M*
Recrystallized $^{32}P_i$ (10^6 cpm/ml)[7]
SMP (10 mg/ml)[8]
Normal serum
Antiserum against F_1
Trichloroacetic acid, 50%

Procedure. Into a volume of 0.1 ml, 0.05 ml of SMP, 0.01–0.05 ml of antiserum or normal serum are pipetted into test tubes and incubated for 10 minutes at 23°. To this mixture into a final volume of 0.5 ml, 0.1 ml of potassium phosphate, 0.05 ml of $MgSO_4$, 0.05 ml of ATP, and 0.05 ml of $^{32}P_i$ are added. After incubation at 30° with shaking for 10 minutes, the reaction is stopped by the addition of 0.05 ml of trichloroacetic acid. After centrifugation, 0.2 ml of the supernatant fluid is removed and extracted with isobutanol-benzene.[9] One milliliter of the water layer is plated, dried, and counted in a Nuclear-Chicago flow counter.

Definition of Unit and Specific Activity. One unit equals that amount of serum necessary to inhibit the $^{32}P_i$-ATP exchange reaction by 50%. Specific activity is expressed as units per milligram of protein.

Immunization of Chicken. An equal volume of Freund's complete adjuvant (Difco) is added dropwise from a syringe into the antigen solution of F_1 (2 mg/ml) and the suspension is stirred vigorously for 2 hours at 23°. One milliliter of the mixture is injected into the breast muscle of the chicken. About 4 weeks later an injection of 1 ml of the antigen (1 mg/ml) in isotonic saline is made; 5–7 days later, 10–15 ml of blood are removed from the wing vein. The injections and bleedings are repeated every 2 weeks until maximal titers are reached. The antibody response in chicken is very variable and usually only one out of three gives a satisfactory serum. Once the titer has reached its peak, further injections of antigen cause a rapid drop in the level of antibodies.

Both normal and immune chicken sera contain some substance which interferes with respiration and for studies of oxidative phosphorylation,

[7] See this volume [4].

[8] See this volume [79]. See also E. Racker, *Proc. Natl. Acad. Sci. U.S.* **48**, 1659 (1962).

[9] See this volume [9].

the serum is precipitated with $2 M$ ammonium sulfate, dissolved in $0.01 M$ Tris-sulfate, pH 7.4–$0.15 M$ KCl, and dialyzed for 16 hours at 4° against large volumes of the same buffer to remove the salt.

Immunization of WRA Mice.[10] The antigen F_1 (1 mg/ml) is mixed with Freund's complete adjuvant as above. Of this mixture, 0.4 ml is injected intraperitoneally into each of approximately 20 mice. Four weeks later 0.2 ml of the antigen (500 μg/ml) in isotonic saline is injected as above. After 5–7 days the mice are bled from the ophthalmic plexus with a disposable pipette with a broken tip. This schedule of injections and bleedings is repeated weekly until the titer is at a satisfactory level, at this time the animals are bled out by decapitation.

Purification of γ-Globulins from Antiserum

All steps are carried out at 23°, at room temperature unless specified. As a control, a sample of normal serum is carried through the same procedure.

Step 1. Four milliliters of serum (50 mg/ml) are mixed with 14 ml of a 0.4% aqueous Rivanol[11] (6,9-diamino-2-ethoxyacridine lactate) solution. Without removal of the precipitate the pH is adjusted to 8.5 with $1 N$ NaOH. The heavy flocculent precipitate is removed by centrifugation, and acid-washed Norit A charcoal (0.5 g) is added to remove excess Rivanol. The solution is centrifuged, filtered, and lyophilized. The lyophilized residue is resuspended in 2.0 ml of isotonic saline.

Step 2. To this solution, 360 mg of solid sodium sulfate[12] is slowly added with constant stirring. After 10 minutes, the precipitate is collected by centrifugation at 18,000 g for 10 minutes, dissolved in 2.0 ml of $0.15 M$ KCl–$0.01 M$ Tris-SO_4, pH 7.4, and dialyzed overnight at 4° against 500 volumes of the same buffer. This fraction contains γ-globulin and some β-globulin. The table shows the yield and purification achieved by this procedure.

Purification of an Antibody Against F_1

Fraction	Volume	Mg/ml	Units/ ml	Total units	Specific activity	Per cent recovery
Serum	4.0	50	95	380	1.9	100
Rivanol-treated preparation	1.6	12	50	80	4.2	21
Na_2SO_4 precipitate	1.5	4	50	75	12.5	20

[10] The WRAIR albino mouse strain was supplied by Dr. Edward L. Buescher, and a colony developed by Mr. Walter Sapanski.

[11] J. Hořepí and R. Smetana, *Acta Med. Scand.* **155**, 65 (1956).

[12] R. A. Kekwick, *Biochem. J.* **34**, 1248 (1940).

Properties

The antibody from both chicken and mice is specific for F_1. Using double diffusion analysis,[13] a single precipitin band appears with F_1, but nothing appears with any of the other coupling factors. The ATPase activity of F_1 as well as oxidative phosphorylation in submitochondrial particles is inhibited by the antibody.

[13] Ö. Ouchterlony, *Acta Pathol. Microbiol. Scand.* **32**, 231 (1953).

[108] Synthesis of γ-^{32}P-ATP

By Harvey S. Penefsky[1]

General Principle

ATP labeled only in the terminal phosphate group is synthesized from ADP and $^{32}P_i$ by coupling the reactions catalyzed by glyceraldehyde 3-phosphate dehydrogenase and 3-phosphoglyceric acid kinase.[1a]

Method

Reagents

Triethanolamine, 0.5 *M*, pH 8.0 (50 m*M*)
DPN (4 m*M*), 0.02 *M*
ADP (3 m*M*), 0.1 *M*
$MgCl_2$ (5 m*M*) 0.1 *M*
DL-Glyceraldehyde 3-phosphate[2] (3 m*M*), 0.03 *M*
Sodium pyruvate[3] (4 m*M*), 0.1 *M*
$^{32}P_i$, 13.3×10^6 cpm/micromole[4] (0.3 m*M*)
Lactate dehydrogenase[5] (10 μg)

[1] See footnote 1, page 522.

[1a] H. S. Penefsky, M. E. Pullman, A. Datta, and E. Racker, *J. Biol. Chem.* **235**, 3330 (1960). A somewhat similar procedure has been described by I. M. Glynn and J. B. Chappell, *Biochem. J.* **90**, 147 (1964).

[2] Purchased from Schwarz BioResearch, Inc. as the barium salt of DL-glyceraldehyde-3-P diacetal and converted to glyceraldehyde-3-P by the method of E. Racker, V. Klybas, and M. Schramm, *J. Biol. Chem.* **234**, 2510 (1959).

[3] Pyruvate and lactate dehydrogenase is necessary to prevent the decline in the oxidation of G-3-P caused by DPNH (cf. reference of footnote 2).

[4] $^{32}P_i$ was purchased from Oak Ridge National Laboratories and purified before use according to the method of T. E. Conover, R. L. Prairie, and E. Racker, *J. Biol. Chem.* **238**, 2831 (1963).

[5] All enzymes used are crystalline preparations obtained from C. F. Boehringer und Soehne, Mannheim, Germany.

3-Phosphoglyceric acid kinase (10 μg)
Glyceraldehyde 3-phosphate dehydrogenase (200 μg)

The numbers in parentheses indicate the final concentration of each component in the reaction mixture. Glyceraldehyde 3-phosphate is the final addition made in a total volume of 2.0 ml. The reaction is allowed to proceed at room temperature for 20 minutes. At the end of the incubation period 4 micromoles of carrier ATP is added and the entire reaction mixture is transferred without deproteinization to a 1 × 3 cm column of Dowex 1-chloride. ATP is eluted from the column according to the following schedule[6]: 20 ml of H_2O; 20 ml of 0.003 *N* HCl; 35 ml of 0.01 *N* HCl, 0.02 *M* NaCl; 25 ml of 0.01 *N* HCl, 0.2 *M* NaCl. The ATP is recovered in the final fraction.

The yield, expressed in terms of counts per minute recovered in the ATP fraction, has been greater than 90%. More than 99.5% of the radioactivity present in the ATP fraction is labile to hydrolysis in 1 *N* HCl for 7 minutes at 100° but stable to hydrolysis if pretreated with hexokinase, glucose, and Mg^{++}.

[6] W. E. Cohn and C. E. Carter, *J. Am. Chem. Soc.* **72,** 4273 (1950).

[109] Methods for the Elevation of Hepatic Microsomal Mixed Function Oxidase Levels and Cytochrome P-450

By HERBERT REMMER, HELMUT GREIM, JOHN B. SCHENKMAN, and RONALD W. ESTABROOK

Liver microsomes contain a mixed function oxidase system (monooxygenase) which is responsible for the oxidation of a wide variety of lipid-soluble drugs and aromatic compounds.[1-3] Reactions catalyzed by this system include *N*-dealkylations (e.g., aminopyrine), *O*-dealkylations (e.g., codeine), side-chain oxidations (e.g., barbiturates), aromatic ring hydroxylations (e.g., aniline, hydrocarbons), and sulfoxidation (e.g., chlorpromazine) among other reactions.[4]

Many substances have been shown to elevate the content of the mixed function oxidase activity of liver microsomes; these range from bar-

[1] J. Axelrod, *J. Pharmacol. Exptl. Therap.* **114,** 430 (1955).
[2] B. B. Brodie, J. Axelrod, J. R. Cooper, L. Gaudette, B. N. LaDu, C. Mitoma, and S. Udenfriend, *Science* **12,** 603 (1955).
[3] C. Mitoma, H. S. Posner, A. C. Reitz, and S. Udenfriend, *Arch. Biochem. Biophys.* **61,** 431 (1956).
[4] J. Gillette, *Proc. 1st Intern. Pharmacol. Meeting, Stockholm, 1961,* Vol. 6, p. 13. Macmillan, New York, 1962.

biturates to tranquilizers, insecticides, and polycyclic hydrocarbons such as 3,4-benzpyrene and 3-methylcholanthrene (for a listing see Conney[5]). The increase in oxidative activity is associated with an increase in the microsomal content of cytochrome P-450.[6,7] Experiments with the inhibitors of protein synthesis (puromycin and actinomycin D)[8,9] have indicated that the increase in enzyme activity is due to the synthesis of new enzyme protein in the microsomes.

Best known is the induction of the mixed function oxidase after treatment of rats with phenobarbital. The increase in the level of enzyme activity for the oxidation of all the different substrates known to be metabolized by liver microsomes, and to require NADPH and molecular oxygen, precedes an increase in the activity of a number of other enzymes, such as esterases, reductase, and transferases associated with the microsomes. An enhancement of intracellular structure can also be observed with the electron microscope, i.e., an increase of the smooth membranes of the endoplasmic reticulum of the liver cells. At the same time the weight of the liver related to the body weight increases. This may be regarded as a real hypertrophy of the liver.

The time necessary to achieve maximal levels of microsomal mixed function oxidase activities varies with the inducer compound used. Methylcholanthrene or benzpyrene treatment exerts a maximal effect within 24 hours.[7,10] Why polycyclic hydrocarbons increase hydroxylating enzyme activity toward only certain substances[5] is a question which cannot be answered at the present time. After one injection of DDT, the maximal activity is achieved after 1 or 2 weeks. When phenobarbital is used, maximal enzyme activity toward all substrates is reached after 3–5 days of treatment.

Three procedures which increase the level of liver microsomal mixed function oxidase activity and cytochrome P-450 are described below. One method uses phenobarbital as the inducer; a second uses 3-methylcholanthrene or 3,4-benzpyrene; and the third method employs DDT.

Induction with Phenobarbital

The barbiturate of choice to elevate the level of microsomal mixed function oxidase is phenobarbital. Experiments by Remmer[11] have shown

[5] A. H. Conney, *Proc. 2nd Intern. Pharmacol. Meeting Prague, 1963,* Vol. 4, p. 277. Macmillan, New York, 1965.
[6] H. Remmer and H. J. Merker, *Ann. N.Y. Acad. Sci.* **123,** 79 (1965).
[7] S. Orrenius, *J. Cell Biol.* **26,** 725 (1965).
[8] H. V. Gelboin and N. Blackburn, *Biochim. Biophys. Acta* **72,** 657 (1963).
[9] A. H. Conney, *J. Biol. Chem.* **228,** 753 (1957).
[10] A. H. Conney and A. Gilman, *J. Biol. Chem.* **238,** 3682 (1963).
[11] H. Remmer, *Arch. Exptl. Pathol. Pharmakol.* **237,** 296 (1959).

that the long-acting barbiturates are more effective than shorter-acting ones for this purpose. The dosages and frequency of treatment employed by various investigators have varied widely, although the response is dose dependent (cf. Orrenius[7]).

The procedure of choice requires single daily intraperitoneal injections of 100 mg of phenobarbital, sodium, in 0.15 *M* NaCl, per kilogram of body weight, for 5 days. Male rats weighing about 200 g are generally used, since control animals have a level of enzyme activity somewhat above that of immature male and female or adult female rats,[12] and adult male rats can better tolerate this level of phenobarbital. With immature male and female and adult female rats, a lower dosage of phenobarbital (about 80 mg/kg) is suggested. Higher levels of phenobarbital and longer periods of treatment do not evoke a faster or greater response in elevation of cytochrome P-450 content or enzyme activity in rat liver microsomes. The parallel increase in microsomal cytochrome P-450 and hexobarbital (Evipan) oxidase activity in microsomes from

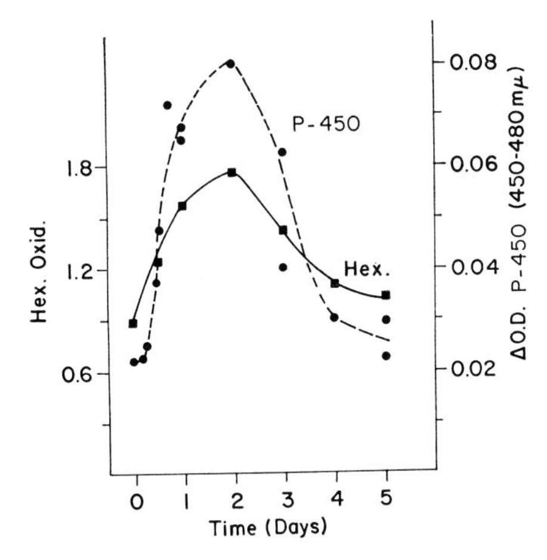

FIG. 1. Time course of induction after a single intraperitoneal injection of 80 mg/kg of phenobarbital, sodium, to immature (50–60 g) female rats. Scale on left is hexobarbital oxidation (HEX) expressed as millimicromoles of hexobarbital oxidized per milligram of microsomal protein per minute, in a medium containing 9000 *g* supernatant from 0.5 g of liver (about 5 mg microsomal protein), 0.2 m*M* TPN, 5 m*M* nicotinamide, 0.4 m*M* HEX, 25 m*M* $MgCl_2$, 0.1 *M* phosphate buffer, pH 7.4. Total volume 5 ml. The abscissa is days after injection of phenobarbital. Scale on right is the difference in absorption at 450 mμ (P-450) relative to 480 mμ of a medium containing supernatant from 0.25 g of liver, 0.6 m*M* TPNH, in 3 ml of 0.1 *M* phosphate buffer, pH 7.4. The Δ absorbance was determined between one cuvette with the above suspension gassed with N_2, and an identical solution gassed with CO, and was expressed as Δ O.D. per milligram of microsomal protein.

[12] R. Kato, E. Chesara, and G. Frontino, *Biochem. Pharmacol.* **11**, 221 (1962).

immature female rats, pretreated with a single injection of 80 mg/kg of phenobarbital, sodium, is shown in Fig. 1.

Preparation of Microsomes

The rats, starved overnight, are sacrificed about 48 hours after the last injection of phenobarbital, and the livers are rapidly removed and put into ice-cold 0.25 *M* sucrose to chill. The animals are starved overnight prior to being killed in order to remove glycogen from the liver. Microsomes are prepared in essentially the manner of Hogeboom.[13]

The chilled livers are rinsed free of excess blood and minced into fresh, ice-cold, 0.25 *M* sucrose to make a 10% suspension (1 g liver: 9 ml sucrose). The tissue is disrupted, in a glass homogenizer with a motor-driven Teflon pestle, by using a moderate speed for about 1 minute or until homogeneous.

Cell debris, unbroken cells, and nuclei are removed by centrifuging at 600 *g* for 10 minutes at 0° in a refrigerated centrifuge (Servall).[14] The supernatant is then centrifuged consecutively for 10-minute periods at 12,000 *g* to remove mitochondria and at 18,000 *g* to remove remaining light mitochondria; the latter step also removes some heavier microsomes.[15]

The supernatant of the 18,000 *g* centrifugation is next centrifuged at 105,000 *g* (measured from the bottom of the centrifuge tube) for 60 minutes in a refrigerated preparative ultracentrifuge (Spinco Model L). The microsomal fraction is sedimented at the bottom of the centrifuge tube and is contaminated with hemoglobin. The supernatant is decanted, the pellet resuspended in about half the previous volume using ice cold 0.15 *M* KCl, and the suspension is centrifuged at 105,000 *g* for 30 minutes. This washing procedure removes most of the adventitious hemoglobin. The pinkish supernatant is decanted and about 5 ml of fresh, ice-cold 0.15 *M* KCl, containing 50 m*M*, pH 7.5 Tris-HCl buffer, is added to the pellet. The microsomal pellet is poured into a chilled homogenizing vessel for suspension to a convenient protein concentration (about 20 mg/ml).

The yield of microsomes by this procedure is about 15–20 mg microsomal protein per gram of rat liver for uninduced animals, and about 25 mg of protein per gram of liver for phenobarbital-induced rats. An alternative method for preparation of microsomes is given by Siekevitz.[16]

[13] G. H. Hogeboom, Vol. I, p. 16.

[14] H. Remmer, J. Schenkman, R. W. Estabrook, H. Sasame, J. Gillette, S. Narasimhulu, D. Y. Cooper, and O. Rosenthal, *Mol. Pharmacol.* **2**, 187 (1966).

[15] V. R. Potter, R. O. Recknagel, and R. B. Hurlbert, *Federation Proc.* **10**, 646 (1951).

[16] P. Siekevitz, Vol. V [5].

Induction with 3-Methylcholanthrene or 3,4-Benzpyrene

These two hydrocarbons are equally effective in causing an elevation of certain hepatic microsomal mixed function oxidase activities.[10] A single intraperitoneal injection of 20 mg/kg of 3,4-benzpyrene in 0.25 ml of corn oil (Mazola oil) to rats causes a maximal increase in cytochrome P-450 and enzyme activity within 18 hours. Maximal enzyme activity is maintained for at least 24 hours and declines thereafter to reach control levels by 6 days. Smaller amounts of benzpyrene cause lesser increases in enzyme levels,[10] while repeated injections of benzpyrene (20 mg/kg) do not cause further elevation. Livers of treated rats are excised 24 hours after injection with benzpyrene and put into ice-cold 0.25 *M* sucrose to chill. Microsomes are prepared as described above.

Induction with DDT

Pretreatment of rats with a single intraperitoneal injection of DDT [2,2-bis(*p*-chlorophenyl)-1,1,1-trichloroethane] causes a slow rise in the microsomal content of cytochrome P-450 and aminopyrine and hexobarbital oxidative activities, until about 10 days. The elevated levels are

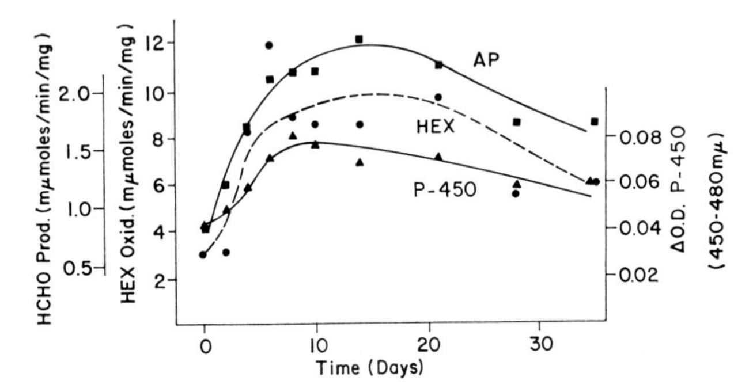

FIG. 2. Time course of induction after a single injection of 200 mg/kg DDT in corn oil to 140-g female rats. The scale on the right is the difference in absorption at 450 mμ (P-450) relative to 480 mμ of a medium containing 2 mg of microsomal protein per milliliter, in 50 m*M* Tris buffer, pH 7.5; difference spectra were obtained between a cuvette containing the above suspension plus 0.3 m*M* TPNH and an identical solution which had been gassed with CO for 1 minute; Δ O.D. values are expressed as per milligrams of microsomal protein. Scale on inner left ordinate is rate of hexobarbital oxidation (HEX), while that on far left is rate of aminopyrine (AP) demethylation, as measured by formaldehyde (HCHO) production; both are expressed as millimicromoles per minute per milligram of microsomal protein. Assay medium contained 50 m*M* Tris, pH 7.5; 8 m*M* isocitrate; 0.33 m*M* TPN; 15 μg/ml Sigma type IV isocitric dehydrogenase; 5 m*M* $MgCl_2$, and either 8 m*M* aminopyrine or 5 m*M* hexobarbital as substrate. Assay time 10 minutes at 37°.

maintained for about another 10 days, and thereafter decrease over a further 15–20 day period (Fig. 2). Because of the very long time period required to elevate the mixed function oxidase system, this procedure is not a recommended one; it is added only to lend a cautionary note concerning the use of apparent control animals.

Comments

In view of the rapidity with which the polycyclic hydrocarbons act to elevate the level of microsomal mixed-function oxidase activities, it would be expected that this is the method of choice for increasing this enzyme system. However, because of the apparent selective increase in enzyme activities stimulated by these compounds, and the fact that one can still detect benzpyrene-like fluorescence in the liver microsomes 24 hours after pretreatment of rats with benzpyrene, caution should be used in the interpretation of enzymatic data obtained with liver microsomes of rats pretreated with these hydrocarbons.

In a comparison of the extent to which certain substances elevate the level of liver microsomal mixed function oxidase, the magnitude of the increase can be disappointingly low if the control animals have inadvertently been exposed to insecticides[17] like DDT. As shown in Fig. 2, even isolating animals for several weeks will not restore the microsomal mixed function oxidase or cytochrome P-450 content to true control levels, when a long-acting insecticide is the cause of the induction.

[17] J. R. Fouts, *Ann. N.Y. Acad. Sci.* **104,** 875 (1963).

[110] Thermal Measurements during Oxidative Reactions[1]

By M. Poe,[2] H. Gutfreund,[3] and R. W. Estabrook[4]

The thermodynamic quantities which describe a biological process express in concise form all the macroscopic physical information that is known about the process. This information can resolve the energy balance of the process, show directly the changes in organization of the participants in the process, and predict what the equilibrium state of the process will be.

[1] Supported in part by a U.S. Public Health Service Research Grant GM 12202.
[2] Predoctoral fellow of the National Science Foundation.
[3] During the preparation of this manuscript, this author was at the National Institute for Research in Dairying, Shinfield, Reading, England.
[4] This work was carried out during the tenure of a U.S. Public Health Service Research Career Development Award No. GM-K3-4111.

Concomitant with the oxidative metabolism of substrates by mitochondria is the liberation of energy; energy which may be conserved in the formation of ATP or energy which may be expended as heat. One means of determining the balance of energy during oxidative reactions is to measure the rate of heat production associated with oxygen utilization. The present paper describes the design and operating characteristics of a differential calorimeter suitable for measurement of enthalpy changes during metabolism by fragile subcellular particulates such as mitochondria.

Description of Apparatus

A differential calorimeter has been constructed, consisting of two dewar flasks, a sample and a reference, whose temperatures are compared with a thermopile. The 50 ml dewar flasks are wrapped in aluminum foil and placed in hollow aluminum cylinders which are closed at one end. The cylinders are then immersed in a large constant temperature water bath. The dewars may be easily removed for cleaning and change of contents.

A diagrammatic representation of the sample dewar is presented in Fig. 1. Immersed in the 30 ml of reaction medium is a 5 or 10 couple

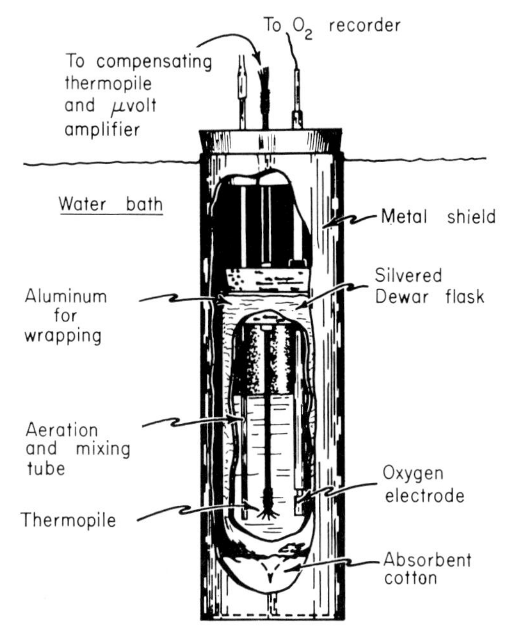

Fig. 1. Diagrammatic representation of the sample dewar with associated thermopile, oxygen electrode, and gassing tube.

copper–constantan thermopile, an oxygen electrode mounted on a vibrator, and a tube for aeration and mixing. Large corks are cut to fit the top of the dewars and the aluminum metal shields. The reference dewar, which contains the other end of the thermopile, is the same as the sample dewar except for the presence of an oxygen electrode. The use of two dewar flasks whose temperatures are compared eliminates any significant contribution from drifts in temperature of the environment, allowing the experimenter to monitor only those enthalpy changes due to the processes taking place in the sample dewar.

Electronic Circuitry

A block diagram of the electronic circuitry associated with the calorimeter is given in Fig. 2. The thermopile voltage is amplified by a Leeds and Northrup DC microvolt amplifier (Model 8739-A). The thermopile output, per copper-constantan thermocouple, is 40 microvolts per degree temperature difference between the two dewars (at 16°).[5]

The thermopile consists of 5 or 10 copper–constantan thermocouples, made of Leeds and Northrup No. 30 wire, joined in series. The thermocouples are joined by spot welding, sealed into thin-walled plastic tubing

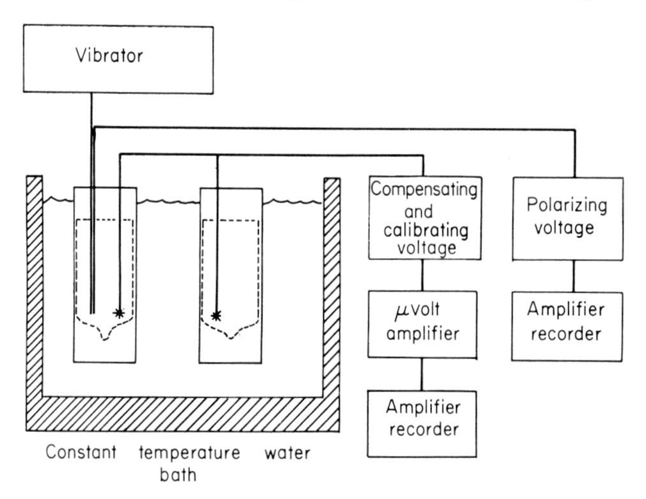

FIG. 2. Schematic representation of the electronic components utilized with the differential calorimeter. The compensating and calibration unit for thermal measurements has been described by Roughton.[6] A Leeds and Northrup microvolt amplifier, together with a Varian G-10 amplifier recorder, were used to measure small changes in voltage generated during the heat measurements. The polarizing voltage unit for the oxygen electrode has been described (see this volume [7]).

[5] National Bureau of Standards, Circular 508, May 7, 1951.

about 3 inches long, and these tubes are then firmly tied together with nylon thread. The external resistance of a thermopile consisting of 5 copper–constantan thermocouples was about 35 ohms. The plastic tube covering of the thermocouples (Intramedic polyethylene tubing, type PE 160) gives bulk strength to the thin thermocouple wires and protects the wires from attack by material in the dewar flasks. One junction of the thermopile is immersed in the sample dewar, and the other is immersed in the reference dewar. The two ends of the thermopile are connected to a compensating and calibrating voltage unit designed similarly to that described by Roughton.[6] This bucking voltage circuit can be used to null out the thermopile signal due to a steady state temperature difference between the two dewars. In general, if a temperature difference existed between the contents of the sample and reference dewars at the beginning of an experiment, it was possible to use the 1 mm diameter aerating tube to cool the contents of one of the dewars at a slow uniform rate until the temperatures of the sample and reference dewars balanced exactly. As shown in Fig. 2, the output of the calibrating and compensating voltage unit was connected to the Leeds and Northrup microvolt amplifier and then to a Varian G-10 recorder with variable attenuation.

The oxygen electrode[7] consists of a silver wire and platinum wire sealed in soft glass suspended from the vibrating surface of a Brown chopper. The polarizing voltage circuit has been described by Estabrook.[7] The output from the oxygen electrode was also connected to a Varian G-10 amplifier-recorder set to run simultaneously with the recorder used for heat measurements. When the sample dewar's temperature was made equal to that of the reference dewar, the temperature difference would remain constant for long periods of time. The rate of heating, due to the vibrating motion of the oxygen electrode, was immeasurably small.

Calibration of Heat Measurements

The calibration of the specific heat of the sample dewar can be measured by two different methods: (a) ohmic heating with a precision resistor, and (b) the neutralization of a known amount of a strong acid by a strong base. For this reaction of method b, the amount of heat liberated is 13.34 kcal per mole of acid neutralized.[8]

The specific heat of the dewar was measured to be slightly larger

[6] F. J. W. Roughton, *in* "Technique of Organic Chemistry" (S. L. Friess and A. Weissberger, eds.), Vol. VIII, Investigation of Rates and Mechanisms of Reactions, p. 669. Wiley (Interscience), New York, 1953.

[7] R. W. Estabrook, this volume [7].

[8] "International Critical Tables," Vol. V, p. 212. McGraw-Hill, New York, 1926.

than the specific heat of the liquid contents of the dewar. There was a small contribution due to the thermopile, oxygen electrode, and gassing tube.

In the course of calibration of the specific heat of the sample dewar, the response time and heat transfer function of the thermopile were also investigated. With only the vibrating oxygen electrode as stirrer, the time constant of the thermopile's response was 5–7 seconds, mostly due to the finite rate of thermal diffusion in the sample dewar's contents.

The thermopile was found to have a heat transfer function that was not independent of the heating rate of the sample. At heating rates usually encountered in metabolic systems (less than 2 calories per minute) the thermopile followed the temperature of the dewar quite faithfully, but at higher rates of heating the thermopile would tend to pick up heat faster than the rest of the sample dewar, especially when heated with a solid heating element. This upper limit on the heating rate seems to be imposed by the thermal relaxation time of water.

The oxygen content of the reaction media can be calibrated as described in a preceding section.[7]

Sample Application

An example of the use of the calorimeter is seen in Fig. 3. In this experiment the nonphosphorylating heart muscle preparation, labeled ETP,[9] with NADH as substrate, is observed to consume oxygen and produce heat. The sample dewar contained 29.0 ml of a phosphate buffer–KCl solution, while the reference dewar contained 30 ml of the same solution. The two dewars were gassed with oxygen until oxygen saturation was indicated by the oxygen electrode. The rate of gassing was adjusted until the temperature of the contents of the two dewars was nearly equal. After the gassing by oxygen was stopped and the rate of temperature change due to thermal drift determined, a 1.0 ml aliquot of ETP was added by pipette. The sample of ETP was previously equilibrated to the temperature of the external water bath; it was introduced into the sample dewar through the gassing tube, followed by a brief period (10 seconds) of gassing with oxygen to ensure adequate mixing. After determining that thermal equality of the sample and reference dewars was not disturbed, an aliquot of NADH was introduced with a brief period of gassing for mixing. The rate of heat production and oxygen utilization was then recorded as indicated. Upon attaining anaerobiosis, the production of heat ceases, and the thermal drift of the calo-

[9] S. Minakami, F. J. Schindler, and R. W. Estabrook, *J. Biol. Chem.* **239**, 2042 (1964).

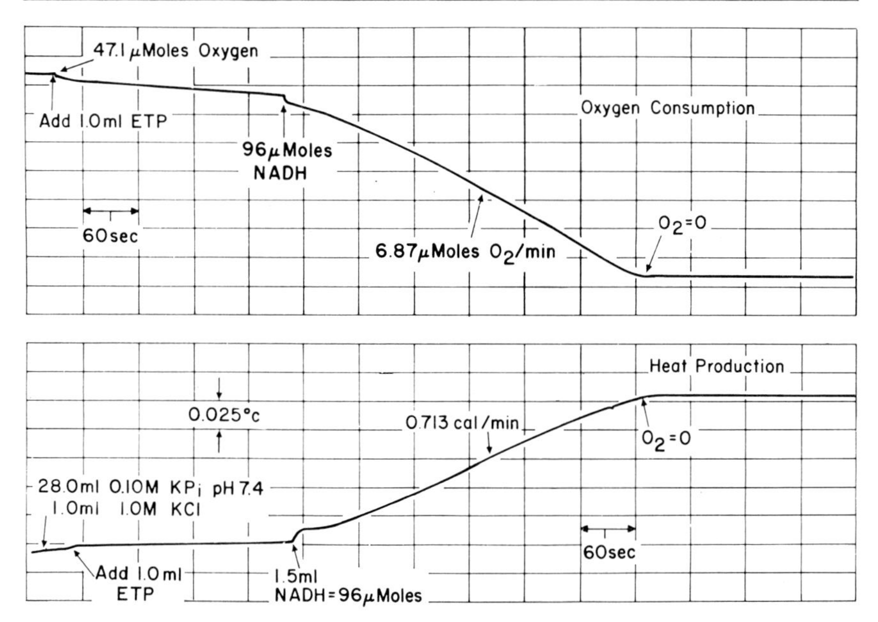

FIG. 3. The simultaneous measurement of oxygen utilization and heat production during NADH oxidation by ETP. Conditions as described in the text.

rimeter is seen to be both minimal and about the same as before the start of heat production. This indicates that the diffusion of oxygen-containing air into the dewar is relatively slow, and that the heat loss of the dewar contents to the water bath during the course of the experiment was minimal. By measuring the average rate of oxygen utilization (dO_2/dt) from the slope of the oxygen electrode recording, and dividing it into the average rate of net heat production (dH/dt) as determined from the thermopile measurements, an accurate and quite reproducible value for the enthalpy of oxidation of NADH was obtained. In six different experiments, ΔH was determined to be -54.3 ± 1.0 kcal/mole. With heart muscle preparations of this type, the affinity of the respiratory chain for NADH is high,[10] and it was possible to add limiting amounts of NADH to titrate the amount of heat produced. The experimental ΔH obtained in this manner agreed with the value determined from the steady state rates.

Recently[11] this calorimetric method has been applied to a study of the energy balance during oxidative phosphorylation by liver mitochondria. In addition the technique can be applied to the study of yeast and bacterial metabolism.

[10] R. W. Estabrook and B. Mackler, *J. Biol. Chem.* **229**, 1091 (1957).

[11] R. W. Estabrook and H. Gutfreund, *Federation Proc.* **19**, 39 (1960).

A description of the use of enthalpy measurements to obtain free energy measurements has been described by Benzinger.[12] The classical antecedents of these measurements are the extensive heat measurements during muscle contraction that were carried out by A. V. Hill.[13]

[12] "Temperature: Its Control and Measurement," Vol. III, Part 3, p. 43. Reinhold, New York, 1963.

[13] A. V. Hill, *Proc. Roy. Soc.* **B136**, 195, 211, 220 (1949).

[111] Biological Applications of Ion-Specific Glass Electrodes

By BERTON C. PRESSMAN

Glass membrane pH electrodes are standard equipment in virtually all biological laboratories. The slight sensitivity of such electrodes to excessive concentrations of Na^+ at high pH has usually been regarded chiefly as an annoyance, but the recent commercial availability of electrodes especially designed to respond to various monovalent cations has opened up new areas of experimentation of particular concern for the biological sciences. This article will stress certain practical applications of these electrodes. Those desiring a more comprehensive treatment of the technological and theoretical aspects of ion-specific glass electrodes are referred to the recent review by Eisenman.[1]

Theory

Introduction of appropriate amounts of trivalent oxides such as Al_2O_3 into an alkali glass confers on the latter in effect a permeability toward monovalent cations. If a membrane of such a glass separates two solutions of a permeant cation M^+ of activities $(M^+)_a$ and $(M^+)_b$, a potential develops across the membrane:

$$E = E_0 + \frac{RT}{F} \ln \frac{(M^+)_a}{(M^+)_b} \tag{1}$$

where E_0 is a term arising from the asymmetry of the two sides of the membrane plus various junction potentials contributed by the measuring system. The latter term is the Nernst diffusion potential arising from the permeant ion and, under ordinary laboratory conditions, reduces to $0.058 \log [(M^+)_a/(M^+)_b]$.

The usual form of the electrode is a bulb filled with a permanent reference solution. If, during the course of the experiment, the external

[1] G. Eisenman, *Advan. Anal. Chem. Instr.* **10**, 213 (1965).

$(M^+)_a$ changes from $(M^+)_1$ to $(M^+)_2$, the change in observed potential is given as:

$$\Delta E = 0.058 \log \frac{(M^+)_1}{(M^+)_2} \tag{2}$$

Thus, for the differential measurements of primary interest to the biologist, the reference $(M^+)_b$ within the bulb and the E_0 term remain constant and cancel out. A useful form of this expression which holds approximately when $\Delta(M^+) \ll (M^+)$ is:

$$\Delta E = 0.058 \frac{\Delta(M^+)}{(M^+)_1 + \frac{1}{2}\Delta(M^+)} \tag{3}$$

The complete measuring circuit usually employs an Ag/AgCl reference electrode within the ion electrode assembly in contact with the reference solution, and a reference electrode, either Ag/AgCl or Hg/Hg_2Cl_2, in contact with the test solution via a salt bridge. The complete system of junctions may be depicted as:

$$\underbrace{\text{Ag} \mid \text{AgCl} \mid \text{KCl}}_{\text{Reference solution}} \parallel \underbrace{(M^+)_a}_{\text{Test solution}} \parallel \underbrace{\text{glass} \mid (M^+)_b \mid \text{AgCl} \mid \text{Ag}}_{\text{Ion electrode}}$$

E may then be measured by connecting the two silver elements to a high impedance potential measuring device such as a vacuum tube pH meter or a vibrating reed electrometer. The ion permeability of the glass arises from the atomic configuration within the glass lattice, and a given glass composition exhibits spectrum of permeabilities not restricted to those ions present within the lattice. Since it is accordingly not possible to construct an electrode absolutely specific to a single ionic species, we must consider the behavior of an electrode toward several permeant ions; for two permeant ions this is described by the equation:

$$E = E_0 + 0.058 \log [(M^+) + K_{MN}(N^+)] \tag{4}$$

where K_{MN} represents the "relative sensitivity" of a given glass composition toward ions M^+ and N^+ and is a function of their relative abilities to permeate the glass. For differential measurements Eq. (4) takes the form:

$$\Delta E = 0.058 \log \left[\frac{(M^+)_1 + K_{MN}(N^+)_1}{(M^+)_2 + K_{MN}(N^+)_2}\right] \tag{5}$$

or the approximation:

$$\Delta E = 0.058 \frac{\Delta(M^+) + K_{MN}\Delta(N^+)}{(M^+) + \frac{1}{2}\Delta(M^+) + K_{MN}[(N^+) + \frac{1}{2}K_{MN}\Delta(N^+)]} \tag{6}$$

When (N^+) remains constant the terms containing $\Delta(N^+)$ drop out and effect of (N^+) is to reduce the ΔE obtained for a given $\Delta(M^+)$ below that predicted by Eq. (3).

Specificity of Electrodes toward H^+, Na^+, and K^+

In biological systems the permeant ions most likely to be encountered by the electrodes are H^+, Na^+, and K^+. By omitting Al_2O_3 from the glass, electrodes virtually specific for H^+ at the usual biological pH ranges are easily constructed. Na^+ sensing electrodes are highly specific, Na^+/K^+ sensitivity typically exceeding 100:1, i.e., $K_{NaK} < 0.01$. These electrodes have a small but appreciable pH dependence, particularly at the lower ranges of biological pH, and for experiments in which pH and (Na^+) change simultaneously, some corrections for the pH are necessary. This interference can also be held to a minimum by appropriate buffering to hold down the pH. K^+ sensing electrodes, on the other hand, although relatively insensitive to pH, possess comparatively lower K^+/Na^+ sensitivities ranging from 3 to 20. With these electrodes, if concomitant K^+ and Na^+ changes are encountered, corrections should be made for the Na^+ interference according to Eq. (5) or (6).

Despite the lack of total specificity of the cation electrodes, simultaneous measurement of changes in H^+, Na^+, and K^+ can be made by monitoring the test solution with all three of the electrode types described and solving simultaneous equations derived by calibrating the electrodes with standards of each of the three ionic species. Since the E obtained from the standard pH electrode may be evaluated without correction, it may then be used to correct the E obtained from the Na^+ selective electrode, and the corrected value of Na^+ may be used in turn to correct the E of the K^+ selective electrode. Procedures have been described whereby the signals obtained simultaneously from Na^+ and K^+ selective electrodes are coupled in such a manner as to provide a record of the potentials produced by changes in (K^+) continuously corrected for the concomitant changes in (Na^+).[2,3] If it has been determined that (Na^+) remains essentially constant during an experiment, both H^+ and K^+ records can be evaluated without any need for correction for the lack of complete specificity of the electrodes.

Sensitivity of Electrodes toward Other Ions

The commercially available glass electrodes respond to NH_4^+ much as they do to K^+, although glass compositions have been found which,

[2] F. Gotch, Y. Tazaki, K. Hamaguchi, and J. S. Meyer, *J. Neurochem.* **9**, 81 (1962).
[3] S. M. Friedman and F. K. Bowers, *Anal. Biochem.* **5**, 471 (1963).

in conjunction with commercially available K^+ sensitive electrodes, would permit good discrimination between simultaneous changes in (NH_4^+) and (K^+).[1] Evaluation of the E arising from (NH_4^+) may have to take into account the pH dependent equilibrium of NH_4^+ with NH_3.

Other basic nitrogen compounds that occasionally occur in biological systems react appreciably with cation sensitive electrodes including alkyl amines, quaternary amines, hydrazonium and hydroxylammonium and, to a lesser extent, guanidium and amino acids.[1] We have observed no marked responses of Beckman electrode No. 39047 to Tris buffer or choline chloride, which we occasionally include in our experimental systems.

The electrodes also exhibit sensitivity toward the other alkali ions which, although not normally occurring in biological systems, are subject to transport by biological membranes. In principle, all that is needed to follow simultaneous changes in any number of ions is a combination of the same number of electrodes of differing ion specificity. The characteristics of enough glass compositions have been determined to permit selection of an optimal combination for any given pair of ions.[1] This would permit continuous monitoring of, for example, a K^+ for Rb^+ exchange across a biological membrane. For membranes which do not discriminate appreciably between K^+ and Rb^+, such as occur in mitochondria,[4] this would yield a continual record of an exchange process presently subject to study by isotope techniques alone.

Cation electrodes also show good sensitivity toward nonalkali ions such as Tl^+ and Ag^+. Glass membrane electrodes are usually considerably more sensitive to Ag^+ than to any other ion, which suggests their application in biology to the Ag^+ titration of protein sulfhydryl groups. Glass electrodes have also been reported which have sensitivity to Ca^{++} and other divalent ions, but they require rather stringent preconditioning and have limited specificity and slow responses.[5,6]

Calibration of Electrodes

For many biological applications it is more convenient to calibrate electrode responses empirically by addition of standards to the test solution than compute ion concentrations by Eq. (1) from precise measurements of the electrode potentials. The latter procedure would also require knowledge of the activities of all ions which could modify the

[4] C. Moore and B. C. Pressman, *Biochem. Biophys. Res. Commun.* **15,** 562 (1964).

[5] R. M. Garrels, M. Sato, M. E. Thompson, and A. H. Truesdell, *Science* **135,** 1045 (1962).

[6] A. H. Truesdell and A. M. Pommer, *Science* **142,** 1292 (1963).

response of the electrode to the ion being followed, as well as the appropriate K values in Eq. (5). In the case of the Na^+ selective electrode, pH appreciably affects the potential obtained at a given Na^+ activity, while the K^+ selective electrode response could be influenced by Na^+, NH_4^+ as well as various amino compounds arising from the biological test material, added as substrate (e.g., glutamate) or as buffer (e.g., Tris).

It must also be borne in mind that Eqs. (1–6) deal with ion *activities* rather than absolute *concentrations*. Determination of net ion concentrations in a test solution by means of auxiliary assays, i.e., flame photometry, is therefore inadequate for permitting precise evaluation of electrode signals via Eqs. (1–6). The concentration $[M^+]$ and activity (M^+), of a given ion M^+ are related by the activity coefficient, γ, according to the following equation:

$$(M^+) = \gamma[M^+] \tag{7}$$

At concentrations around 100 mM, often attained in biological fluids, γ, for K^+ and Na^+, is about 0.75, and, at 10 mM, γ is 0.90. The discrepancy between the concentration and activity of an ion in solution would be further exaggerated if the ion associated with a soluble component of the system. The application of the electrodes we have been most concerned with is the measurement of net ion movements in and out of a biological material as correlated with changes in metabolic states. In this case evaluation of the electrode responses by addition of standards directly to the test system obviates the need for precise knowledge of the actual activities of the ions in the system.

Determining the ΔH^+ is perhaps a special case, as the ΔpH corresponding to a given increase or decrease of H^+ can be controlled by varying the buffer concentration, and the standard pH electrodes are almost unaffected by other ions. If the ΔpH can be held thus to 0.1, the ΔH^+ can be regarded as virtually proportional to the measured ΔpH. Calibration is best accomplished by determining the amount of acid or base required to be added to the test solution to return the electrode to its original potential.

Calibration of the cation electrodes by restoration of the initial potential is possible only when the activity of the ion being followed has been reduced during the experiment. Otherwise it is advisable to calibrate with a relatively small amount of standard or, in a separate operation, to determine the amount of a standard required to produce the same potential change observed under experimental conditions.

A second technique for calibration, particularly applicable for relatively large changes of an ion during an experiment, consists of the addition of identical small calibration doses of standard before and after the experimental shift in ion activities. From Eq. (3) the expression follows:

$$(M^+)_f = (M^+)_i \cdot \frac{\Delta E_i}{\Delta E_f} \tag{8}$$

where $\Delta E_i/\Delta E_f$ is the ratio of the potential changes by calibrating dose of M^+. $(M^+)_f$ and $(M^+)_i$ are the final and initial activities of M^+. This expression is particularly useful when the initial activity can be calculated from the known components added to the system, and the sum total of ions other than M^+ to which the electrode is sensitive is not significant.

A third mode of employing the ion-specific electrodes is to measure the amount of ion addition required to maintain the electrode potential constant by means of a commercially available pH stat, substituting a cation sensitive electrode for the normally employed pH electrode. Recordings of the amount of ion addition are accurately proportional to the amount of ion removed from the system, undistorted by the log function governing potential changes in the non-statted measuring systems (cf. Eq. 1). A method of linearizing the relationship between the recorded signal and ion concentrations by means of an analog computer has also been described.[3]

Apparatus

We have found the Radiometer Model 22 and Leeds and Northrup expanded scale Model 7405 pH meters equally satisfactory for initial amplification of the electrode potentials. The Radiometer pH meter does not have enough offset potential available to balance out the potentials obtained from the commercial electrodes. This can be corrected by the introduction of an external offset potential between the reference electrode input of the instrument. Both makes of meters cited have outputs suitable for driving a standard 10 mv potentiometer recorder directly with sufficient sensitivity.

When several glass electrodes are used to monitor the same test solution, a common reference electrode is mandatory to minimize cross interaction. Economy of space is achieved by measuring pH with a concentric combination electrode such as the A. H. Thomas 4858-L15, utilizing its Ag/AgCl electrode as the common reference. The cation ion electrodes which we have employed most extensively are the Beckman 39046 Na^+ sensitive electrode and the Beckman 39047 K^+ sensitive electrode (so called "cation" electrode). Electrodes of equivalent ion specificity are available from Electronic Industries Limited (England), designated as

BH 68 and BH 69. Glass for the custom fabrication of cation-sensitive electrodes can be obtained from the Research and Development Division of the Corning Glass Works. We have been able to test out experimental Corning ion selective electrodes and find them to be comparable to their Beckman equivalents. These electrodes should be commercially available in early 1966.

The transport of ions by mitochondria is accompanied by other mitochondrial changes which the apparatus developed in our laboratory monitors continuously by appropriate techniques. Oxygen consumption, an index of mitochondrial energy production, is measured polarographically (cf. article by R. W. Estabrook, this volume [7]). The simplest way to incorporate an oxygen-sensing device into the reaction vessel is to insert a Clark-type electrode assembly such as the Radiometer E-5044. The fact that its electrode elements are isolated from the reaction medium and cation electrode system by means of a plastic membrane minimizes problems of electrode interaction. Differentiation of the oxygen electrode output, as indicated in Fig. 1, serves to make changes in mitochondrial respiration more evident.

In addition to the aforementioned electrodes for monitoring H^+, Na^+, K^+, and O_2, our apparatus includes optical devices for measuring light scattering and fluorescence. Simultaneous presentation of all data on a single record is best obtained by means of a light writing recording oscillograph such as the Consolidated Electrodynamic 5-124 or the Honeywell Visicorder. These instruments have galvanometers of about 30 Ω impedance, and it is not possible to drive them with sufficient sensitivity directly from the pH meters. For this purpose special driving amplifiers have been designed in our laboratory which, to minimize interactions between the electrodes, are chopper modulated and "floating," i.e., not grounded to the chassis. These amplifiers include variable damping controls, step-switched gain controls, input balance controls, and an output offset for rapidly controlling the placement of each tracing on the record. A block diagram of the electronic components is presented in Fig. 1, and Fig. 2 illustrates the configuration of the optical components of our apparatus. Figure 3 gives a representative recording with the apparatus. Details about the construction of such amplifiers can be obtained from their designer, Mr. Dieter Mayer, Johnson Foundation, University of Pennsylvania, Philadelphia, Pennsylvania.

Special Characteristics of Electrodes

The Beckman Instrument Company makes relatively modest claims of about 3–4 for the K^+/Na^+ specificity of their 39047 "cationic" electrode.

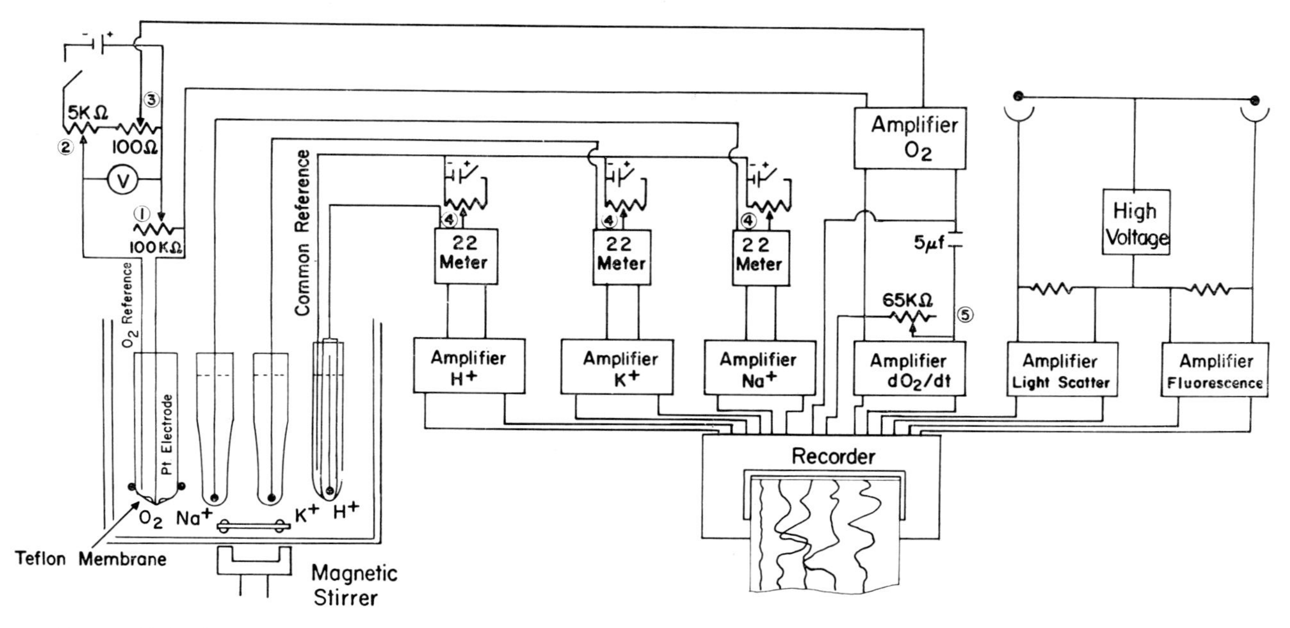

FIG. 1. Block diagram of electronic components of ion uptake apparatus. Each amplifier is equipped with variable time constant selectors, stepped gain controls, and an offset voltage across the output to control the light pen positions. Other controls not integral parts of the amplifier include: (*1–3*) oxygen electrode controls, (*1*) variable sensitivity, (*2*) variable polarizing voltage and (*3*) variable offset; (*4*) external balancing offset controls for pH meters; (*5*) variable time constant selector for oxygen electrode differentiator.

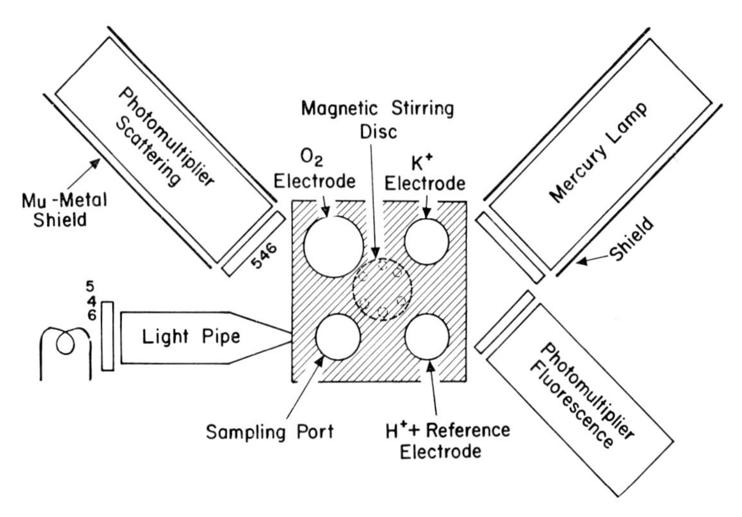

FIG. 2. Overhead view of ion uptake apparatus showing orientation of optical components. Filtered light (546 mμ) from an incandescent lamp is led through an aluminized light pipe butting perpendicularly against the reaction cuvette. The cuvette was constructed of optical glass and has inside dimensions of 2.4 cm $\times$ 2.4 cm $\times$ 4.0 cm. The pipe is tapered to a rectangular slit 8 mm high and 2 mm across. Scattered light appearing at the same surface passes through a second 546 mμ interference filter into the photomultiplier detector, which is shielded against pick-up from the magnetic stirrer by mu-metal. Exciting light from the phosphor coated mercury lamp (G.E. F4T4BL) is filtered to isolate the 366 mμ line before entering the cuvette. Emitted light is detected from the same surface, exciting light blocked by a Wratten 2A filter, and scattering light by a Wratten 24 filter. Placement of the electrodes in the black Lucite cell cover is also indicated. Temperature is controlled by means of water jackets surrounding the cuvette compartment and both lamp housings.

However, according to Eisenman, the selectivity improves progressively for over a month as the glass becomes hydrated during immersion in water. He reported K^+/Na^+ selectivity of 10:1 for most K^+ electrodes, including the Beckman 39047.[1] On checking three well aged Beckman electrodes, we found K^+/Na^+ selectivity ratios of 13, 16, and 17:1. Another electrode gave ratios of 3:1, but its $\Delta E/\Delta(K^+)$ was noticeably lower than predicted by Eq. (1) and inspection revealed it to be badly scratched.

Both types of Beckman electrodes typically show a response time in our apparatus of 2.5 seconds for $t_{1/2}$. However, while the K^+ selective electrode potential stabilizes rapidly after a standard addition, the Na^+ selective electrode exhibits a slow tailing off, lasting over 2 minutes. The intrinsic response time of the K^+ selective electrode is considerably less,

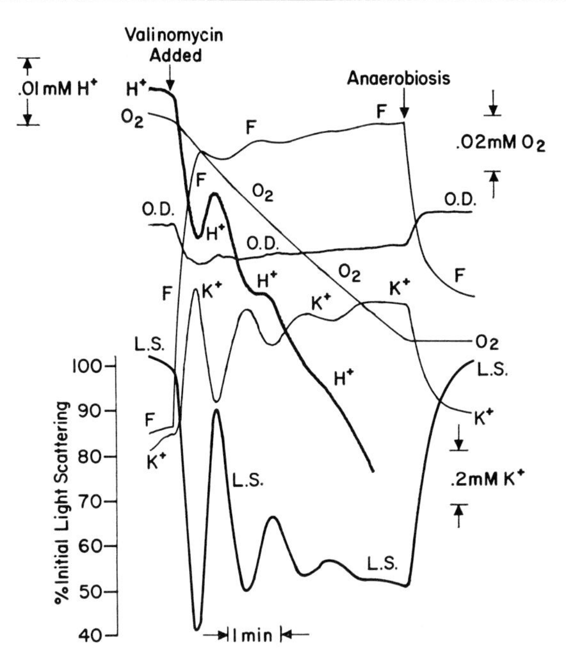

FIG. 3. Representative experimental record obtained from ion uptake apparatus. The tracings are identified as: K^+, potassium level in medium (up signifies decrease in medium, i.e., increased concentration in mitochondria; H^+, pH in medium (down signifies decrease); O_2, level of dissolved O_2 in medium; *O.D.*, rate of change of oxygen consumption ("oxygen derivative"); *F*, fluorescence (up signifies decrease, equivalent to the oxidation of mitochondrial pyridine nucleotides); *L.S.*, light scattering (down signifies decrease, equivalent to an increase in mitochondrial volume). The experimental conditions consisted of: rat liver mitochondria (45 mg protein equivalent); KCl, 10 m*M*; Tris-chloride, 20 m*M*; Tris-phosphate, 5 m*M*; sucrose, 250 m*M*; final pH, 7.2; final volume, 10.0 ml temperature, 24°. The medium was saturated with 100% O_2 before addition of the mitochondria. After equilibration of the electrodes, the transport of K^+ and attendant changes were initiated by the addition of 0.5 μg valinomycin in alcohol. Since, under these conditions, mitochondria do not concentrate Na^+, tracings with a Na^+-sensitive electrode are not included. In this particular medium, the induced movements of K^+ show an oscillation which is reflected in all other parameters measured. Note that the oscillations are not apparent in the direct oxygen tracing, but can be discerned in the more sensitive oxygen derivative (O.D.) tracing.

and, by reducing the amplifier damping, a $t_{1/2}$ of 0.8 second has been obtained, part of which must still be ascribed to amplifier response and the time required for homogeneous mixing. The Na^+ selective electrode also shows a transient change in response to K^+ which rapidly decays, the potential eventually returning to its original value. These departures from ideal behavior of the Beckman 39046 electrode appear to be a general characteristic of all Na^+ selective glass compositions.

The electrodes exhibit streaming potentials when in contact with moving fluid, and in order to limit the noise arising from this source, stirring must be as smooth as possible. Noise levels appear lower in the actual biological test systems than in the presence of pure electrolytes, perhaps because of the beneficial effects of increased viscosity on the smoothness of stirring. In our apparatus, stirring is accomplished by means of a magnetic flea driven by a rotating magnet below the reaction chamber.

Taking the noise level as the practical limit of sensitivity, both the Na^+ and K^+ selective electrodes can detect changes of ion concentrations as little as 0.2% over a range of at least 10^{-1} *M* to 10^{-4} *M*. If one sacrifices response time by additional damping, even greater precision of measurement is possible in the case of slow reactions. Since the same noise patterns often show up on each of the glass electrode tracings, a major source of the noise apparently lies in a common component, e.g. the reference electrode, possibly at the frit junction. Utilization of low noise reference junction techniques might further improve precision. The optimal precision available thus exceeds considerably that of flame photometry at comparable sensitivity.

Applications

Variation in contact potentials, produced by dipping electrodes into a series of solutions, limits their precision for routine analytical purposes to 2–3%. The ultimate potentialities of the electrode are realized only when they can be immersed in a test solution long enough to develop stable potentials and their response to a subsequent perturbation recorded continuously. The most interesting biological applications fall into the latter category.

Ion-specific glass electrodes have been employed extensively in our laboratory for measurement of the translocation of K^+, Na^+, and H^+ across the mitochondrial membrane.[4,7,8] After the equilibration period, rapid ion movement is initiated by addition of minute amounts (10^{-8} to 10^{-10} *M*) of certain antibiotics, e.g., valinomycin, to the isolated mitochondria in an appropriate medium. Although the total ion concentration of the system as a whole remains constant, concentration of an ionic species within the intramitochondrial phase lowers its activity in the extramitochondrial phase which is the phase sensed by the electrodes.

[7] E. J. Harris, R. Cockrell, and B. C. Pressman, *Biochem. J.* **99**, 200 (1966).

[8] B. C. Pressman, *Proc. Natl. Acad. Sci. U.S.* **53**, 1076 (1965).

Thus, it has been concluded that valinomycin induces an energy dependent uptake of K^+ almost exclusively, while gramicidin induces accumulation of both K^+ and Na^+ by mitochondria.[8] The electrodes are particularly valuable for studying this process, since the intra-extramitochondrial distribution of ions is radically altered by the packing produced while separating the intra-extramitochondrial phases centrifugally prior to chemical analysis.

In principle, redistribution of ions between a soluble phase sensed by a glass electrode and a second phase associated with any microdispersed material may be followed by the techniques described here. Such an ion translocation could be brought about either by movement of ions across a membrane into a closed vesicle, or ion exchange directly with the membrane. We have also measured ion movements in and out of intact yeast and bacteria as initiated by substrate addition, intact ascites cells as initiated by valinomycin, and photosynthetic bacteria as initiated by light. Other cell fractions, such as rat liver nuclei or microsomes, exhibit non-energy-dependent ion uptakes, probably representing ion exchange processes. Dilley and Vernon have similarly followed light induced ion movements in chloroplasts.[9]

Other interesting applications of the electrodes include the simultaneous monitoring *in vivo* of brain surface, cortex, and blood.[2] Friedman, in a review of the application of ion-specific electrodes, has described in detail the use of metal connected flow through electrodes for the continuous monitoring of Na^+ and K^+ in arterial plasma.[10] The basis for the routine clinical determination of alkali ions in blood,[11] as well as urine and cerebrospinal fluid,[12] has also been presented. Hinke has described the construction and application of ion specific microelectrodes to the determination of intracellular ion activities in crab and lobster muscle[13] and squid axion.[14] Lev has made similar measurements within the frog muscle.[15] Khuri *et al.*[16] have also employed glass microelectrodes to examine the *in vivo* ionic composition within a single kidney tubule. An interesting generalization which emerges from these studies is that Na^+ appears to be bound by biological materials to a much greater extent

[9] R. A. Dilley and L. P. Vernon, *Arch. Biochem. Biophys.* **111**, 365 (1965).
[10] S. M. Friedman, *Methods Biochem. Anal.* **10**, 71 (1962).
[11] S. M. Friedman, S. L. Wong, and J. H. Walton, *J. Appl. Physiol.* **18**, 950 (1963).
[12] E. W. Moore and D. W. Wilson, *J. Clin. Invest.* **42**, 293 (1963).
[13] J. A. M. Hinke, *Nature* **184**, 1257 (1959).
[14] J. A. M. Hinke, *J. Physiol.* (*London*) **156**, 314 (1961).
[15] A. A. Lev, *Nature* **201**, 1132 (1964).
[16] R. N. Khuri, D. A. Goldstein, D. L. Maude, C. Edmonds, and A. K. Solomon, *Am. J. Physiol.* **204**, 743 (1963).

than K^+,[12,14,15] and in particular the variable degrees of Na^+ binding by blood plasma might prove to be diagnostically significant.[12]

As the capabilities of the ion-specific glass electrodes become more widely appreciated, and improvements are made in the performance and available selection of the electrodes and apparatuses employing them, we may expect to see their biological applications extended considerably. This would seem particularly true for the area of *in vivo* and *in vitro* transport by biological membranes and particulate preparations derived from them.

A recent promising development is the silicone membrane "Pungor" electrode, which is reported to exhibit highly specific concentration potentials to a wide variety of anions and cations, and behaves analogously to the glass membrane electrode. Pungor electrodes sensitive to iodide, sulfate, phosphate, chloride, and nickel are currently being offered by National Instrument Laboratories, Inc.

Section IX

Measurement of Energy-Linked and Associated Reactions

[112] Energy-Linked Reduction of NAD^+ by Succinate

By LARS ERNSTER and CHUAN-PU LEE[1]

Historical

In 1957 Chance and Hollunger[1a] briefly reported that the addition of succinate to liver mitochondria in the controlled respiratory state results in a rapid and extensive reduction of the endogenous pyridine nucleotide. The reaction was abolished by amytal and by uncouplers of oxidative phosphorylation. They interpreted the phenomenon as an energy-dependent transfer of hydrogen from succinate via succinate dehydrogenase and the respiratory chain to NAD^+, by a process that involved a reversal of oxidative phosphorylation. Similar observations were shortly reported by Bücher and Klingenberg[2] with insect flight-muscle mitochondria and another flavin-linked substrate, glycerol 1-phosphate. During 1959–1961 both laboratories published a series of extensive investigations of this flavosubstrate-linked pyridine nucleotide reduction,[3-9] all of which supported the original interpretation of a reversal of oxidative phosphorylation. This interpretation was criticized by Krebs and others,[10-13] who claimed that the observed reduction of the mitochondrial pyridine nucleotide in the presence of succinate might involve endogenous substrates or malate derived from succinate, rather than succinate itself, as the

[1] See footnote 1, page 33.

[1a] B. Chance and G. Hollunger, *Federation Proc.* **16**, 163 (1957).

[2] T. Bücher and M. Klingenberg, *Angew. Chem.* **70**, 552 (1958).

[3] M. Klingenberg and W. Slenczka, *Biochem. Z.* **331**, 486 (1959).

[4] M. Klingenberg, W. Slenczka, and E. Ritt, *Biochem. Z.* **332**, 47 (1959).

[5] B. Chance and G. Hollunger, *Nature* **185**, 666 (1960); *J. Biol. Chem.* **236**, 1534, 1555, 1562 (1961).

[6] B. Chance, *Biol. Struct. Function, Proc. 1st IUB/IUBS Intern. Symp. Stockholm, 1960* p. 119. Academic Press, New York, 1961. See also *J. Biol. Chem.* **236**, 1544, 1569 (1961). See also B. Chance and B. Hagihara, *Proc. 5th Intern. Congr. Biochem., Moscow, 1961* Vol. 5, p. 3. Pergamon, Oxford, 1963.

[7] M. Klingenberg, *Collog. Ges. Physiol. Chem.* **11**, 82 (1961). See also *Biol. Struct. Proc. 1st IUB/IUBS Intern. Symp. Stockholm, 1960* p. 227. Academic Press, New York, 1961. See also *Biochem. Z.* **335**, 263 (1961).

[8] M. Klingenberg and P. Schollmeyer, *Biochem. Z.* **333**, 335 (1960); **335**, 231, 243 (1961). See also *Proc. 5th Inter. Congr. Biochem., Moscow, 1961,* Vol. 5, p. 46. Pergamon, Oxford, 1963.

[9] M. Klingenberg and T. Bücher, *Biochem. Z.* **334**, 1 (1961).

[10] R. G. Kulka, H. A. Krebs, and L. V. Eggleston, *Biochem. J.* **78**, 95 (1961).

[11] H. A. Krebs, L. V. Eggleston, and A. d'Alessandro, *Biochem. J.* **79**, 537 (1961).

[12] H. A. Krebs, *Biochem. J.* **80**, 275 (1961).

[13] H. A. Krebs and L. V. Eggleston, *Biochem. J.* **82**, 134 (1962).

direct source of reducing equivalents, and may thus not proceed via the respiratory chain but by way of a simple reversal of a pyridine nucleotide-linked dehydrogenase reaction.

In 1960 Ernster and associates[14] demonstrated an amytal- and uncoupler-sensitive reduction of acetoacetate by succinate in tightly-coupled rat liver mitochondria. Evidence was presented[14–18] that the reaction proceeded via succinate dehydrogenase, the respiratory chain-linked NADH dehydrogenase, mitochondrial NAD^+, and β-hydroxybutyrate dehydrogenase. The process was thus analogous to that studied by Chance and Hollunger[1a] except that the mitochondrial NADH was continuously reoxidized by acetoacetate. The amount of acetoacetate that could be reduced in this manner greatly exceeded the amount of endogenous substrates conceivably present in the mitochondria, thus eliminating the possibility that endogenous substrates could be the source of the reducing equivalents. Furthermore, malate did not replace succinate under the conditions employed, and, in addition, as later shown by Klingenberg and von Häfen,[19] the amount of acetoacetate reduced was equivalent to the amount of succinate oxidized. Moreover, in 1961, Löw and associates[20] demonstrated an energy-dependent reduction of externally added NAD^+ by succinate catalyzed by phosphorylating submitochondrial particles from beef heart, and presented evidence[20–23] that the reaction involved a reversal of oxidative phosphorylation over the NAD^+-flavoprotein region of the respiratory chain. More recently these conclusions have further been substantiated with the aid of tritiated succinate[24] and NAD^+.[25]

[14] L. Ernster, *Biol. Struct. Function, Proc 1st IUB/IUBS Intern. Symp., Stockholm, 1960* p. 139. Academic Press, New York, 1961.

[15] L. Ernster, *Proc. 5th Intern. Congr. Biochem., Moscow, 1961,* Vol. 5, p. 115. Pergamon, Oxford, 1963. See also *Symp. Funktionelle und morphologische Organisation der Zelle, Rottach-Egern, 1962,* p. 98. Springer, Berlin, 1963. See also *Nature* **193,** 1050 (1962).

[16] L. Ernster, G. Dallner, and G. F. Azzone, *J. Biol. Chem.* **238,** 1124 (1963).

[17] G. F. Azzone, L. Ernster, and E. C. Weinbach, *J. Biol. Chem.* **238,** 1825 (1963).

[18] L. Ernster, G. F. Azzone, L. Danielson, and E. C. Weinbach, *J. Biol. Chem.* **238,** 1834 (1963).

[19] M. Klingenberg and H. von Häfen, *Biochem. Z.* **337,** 120 (1963).

[20] H. Löw, H. Krueger, and D. M. Ziegler, *Biochem. Biophys. Res. Commun.* **5,** 231 (1961).

[21] H. Löw and I. Vallin, *Biochim. Biophys. Acta* **69,** 361 (1963).

[22] H. Löw, I. Vallin, and B. Alm, *in* "Energy-Linked Functions of Mitochondria" (B. Chance, ed.), p. 5. Academic Press, New York, 1963.

[23] I. Vallin and H. Löw, *Biochim. Biophys. Acta* **92,** 446 (1964).

[24] H. D. Hoberman, L. Prosky, P. G. Hempstead, and H. W. Afrin, *Biochem. Biophys. Res. Commun.* **17,** 490 (1964).

Over the past few years, studies of the succinate-linked pyridine nucleotide reduction have proved to be a most valuable source of information regarding the mechanisms of conservation and transfer of energy connected with the respiratory chain, at both the mitochondrial and submitochondrial levels. Two major concepts have evolved from these studies: the concept of the general reversibility of oxidative phosphorylation in the respiratory chain; and the concept of direct transfer of energy among the individual coupling sites of the respiratory chain via nonphosphorylated high-energy intermediates, without the involvement of the phosphorylating enzyme system. Surveys of the pertinent literature are found in recent review articles.[26, 27]

General Features

Possible reaction pathways involved in the succinate-linked reduction of NAD^+ and related systems are illustrated schematically in Fig. 1. Succinate-linked NAD^+ reduction involves a reversal of electron trans-

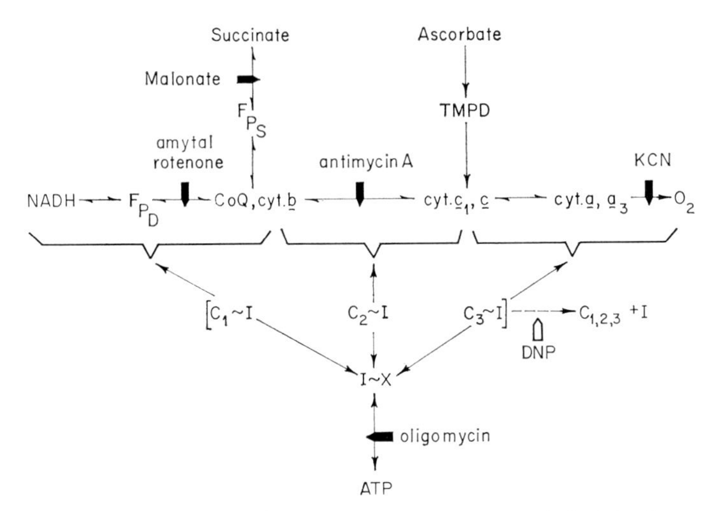

FIG. 1. Hypothetic scheme of the energy-transfer system linked to the respiratory chain. C_1, C_2, C_3 denote electron carriers at the energy-coupling sites of the respiratory chain; I and X denote hypothetic energy transfer carriers [for further definitions, see Ernster and C. P. Lee, *Ann. Rev. Biochem.* **33,** 729 (1964)]. Solid bars indicate site of action of inhibitors; open bar indicates site of stimulation by 2,4-dinitrophenol (DNP).

[25] C. P. Lee, N. Simard-Duquesne, L. Ernster, and H. D. Hoberman, *Biochim. Biophys. Acta* **105,** 397 (1965).

[26] L. Ernster and C. P. Lee, *Ann. Rev. Biochem.* **33,** 729 (1964).

[27] D. R. Sanadi, *Ann. Rev. Biochem.* **34,** 21 (1965).

port through coupling site 1 of the respiratory chain, with the expenditure of one high-energy bond per molecule of NAD^+ reduced by succinate. The net reaction may be described by the equation:

$$\text{Succinate} + NAD^+ + I \sim X \rightleftharpoons \text{fumarate} + NADH + I + X \quad (1)$$

where $I \sim X$ is a hypothetical high-energy intermediate of oxidative phosphorylation. As can be deduced from Fig. 1, the energy for succinate-linked NAD^+ reaction can be derived either from ATP:

$$ATP + I + X \rightleftharpoons ADP + P_i + I \sim X \quad (2)$$

or directly from coupling site 2 or 3 of the respiratory chain:

$$C_{2,3} \sim I + X \rightleftharpoons C_{2,3} + I \sim X \quad (3)$$

Suitable energy-generating systems of the latter type are (cf. Ernster and Lee[26]): aerobic oxidation of succinate (coupling sites 2 and 3); aerobic oxidation of ascorbate plus TMPD or PMS (coupling site 3); anaerobic oxidation of succinate by ferricyanide (coupling site 2).

Amytal, rotenone, and malonate, which block Reaction (1), are diagnostic inhibitors of succinate-linked NAD^+ reduction under all conditions. Likewise, uncouplers of oxidative phosphorylation such as 2,4-dinitrophenol, dicoumarol, etc., which probably cause a splitting of the $C \sim I$ compounds, are general inhibitors of succinate-linked NAD^+ reduction. Respiratory inhibitors acting in the cytochrome region, such as antimycin A and cyanide, do not inhibit the succinate-linked NAD^+ reduction when ATP serves as the source of energy, but may of course inhibit the respiratory chain-driven process.

The energy-transfer inhibitor, oligomycin, characteristically inhibits succinate-linked NAD^+ reduction when this is driven by ATP,[19,21] but not when it is driven by energy derived directly from the respiratory chain.[14] Oligomycin also restores the respiratory chain-driven succinate-linked NAD^+ reduction as abolished by P_i plus ADP[14,15] or by arsenate (but not by dinitrophenol).[15,28] Furthermore, oligomycin in appropriate concentrations can restore ATP-driven succinate-linked NAD^+ reduction in certain types of "nonphosphorylating" electron transport particles[29] (cf. this volume [87]). Another inhibitor of oxidative phosphorylation, aurovertin, inhibits the ATP-driven succinate-linked NAD^+ reduction only at levels much higher than those needed for blocking oxidative phosphorylation.[30] The electron- and energy-transfer inhibitor, octylguani-

[28] A. M. Snoswell, *Biochim. Biophys. Acta* **60,** 143 (1962).

[29] C. P. Lee and L. Ernster, *Biochem. Biophys. Res. Commun.* **18,** 523 (1965). See also *Symp. Regulation of Metabolic Processes in Mitochondria, Bari, 1965,* B.B.A. Library, **7,** 218. Elsevier, Amsterdam, 1966.

[30] G. Lenaz, *Biochem. Biophys. Res. Commun.* **21,** 170 (1965).

dine, abolishes succinate-linked NAD^+-reduction under all conditions[29,31] Finally Mg^{++} inhibits the ATP-driven succinate-linked NAD^+ reduction in intact mitochondria or submitochondrial particles prepared with digitonin,[6,8,19] whereas it is obligatory for the same reaction in submitochondrial particles prepared by sonic oscillation.[21]

Assay

Succinate-linked NAD^+ reduction may be assayed with different reaction systems, depending on whether the assay is performed (a) with mitochondria or submitochondrial particles; (b) with energy derived from electron transport through the respiratory chain or from ATP; and (c) by measuring the NADH formed directly or by transferring hydrogen from the NADH formed to a terminal acceptor and measuring the disappearance of the acceptor or the appearance of the reduced product.

Three illustrative assay systems are described below and discussed with regard to the possible variations in conditions.

Example 1. Reduction of Endogenous NAD⁺ by Succinate in Intact Mitochondria with ATP as the Source of Energy. This system (see Fig. 2) consists of intact mitochondria incubated in an isotonic buffered, Mg^{++}-free medium with succinate under nonrespiring conditions (e.g., in the presence of Na_2S). Reduction of the pyridine nucleotide is monitored

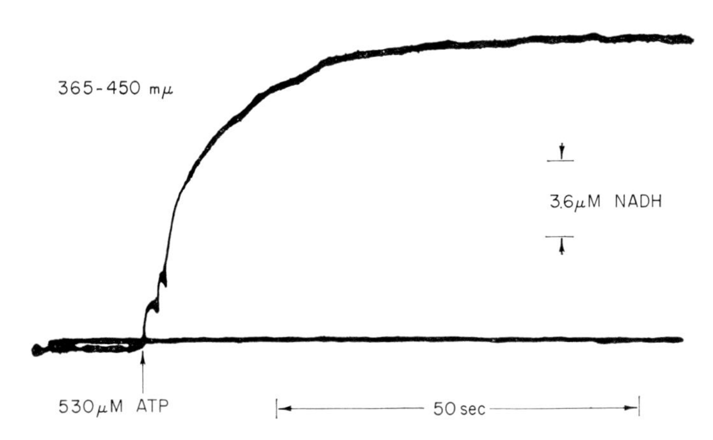

FIG. 2. Reduction of endogenous NAD^+ by succinate in intact mitochondria. Pigeon heart mitochondria were suspended in mannitol-sucrose-Tris medium, pH 7.4, in the presence of 4 m*M* succinate and 360 μ*M* Na_2S. Reduction of pyridine nucleotide was initiated by the addition of 530 μ*M* ATP as indicated. Temperature, 26°. Upward deflection of trace indicates increase in fluorescence. From B. Chance, *J. Biol. Chem.* **236**, 1544 (1961).

[31] F. A. Hommes, *Biochim. Biophys. Acta* **77**, 183 (1963).

fluorometrically at 450 mμ with an excitation at 365 mμ. The ATP-induced reduction of pyridine nucleotide is indicated by the increase in fluorescence.

Alternatively, the pyridine nucleotide reduction can be monitored spectrophotometrically at 340 mμ, provided that a sensitive spectrophotometer is available, preferably a double-beam spectrophotometer, which minimizes errors due to unspecific light absorption. When using the fluorometric method, caution must be exercised in employing chemicals that may interfere either by exhibiting fluorescence themselves (e.g., aurovertin) or by quenching the fluorescence of reduced pyridine nucleotide (e.g., 2,4-dinitrophenol).

It is also important to be aware of the fact that both the fluorometric and spectrophotometric procedures are not specific for NADH but measure NADPH as well. This circumstance is particularly critical when working with mitochondria, such as those from rat liver, which contain relatively much $NADP^+$. In such a case it may be essential to determine NADH and NADPH separately. This can be done by specific enzyme procedures[32] after extraction of the reduced pyridine nucleotides with alkali. Such a determination may be necessary even when no $NADP^+$ is present, if the assay is performed fluorometrically, since the fluorescence intensity of NADH may vary with the state of binding of the pyridine nucleotide to the mitochondria.[33-36]

High-energy intermediates generated by the respiratory chain may be used instead of ATP as the source of energy in the above system, providing that the mitochondria are devoid of endogenous NAD^+-linked substrates; this can be achieved by aeration of the mitochondria in the cold for a suitable length of time prior to the assay. Respiratory inhibitor is omitted from the incubation medium, and the pyridine nucleotide reduction occuring upon the addition of succinate is monitored. Succinate here serves two purposes: it provides high-energy intermediates by aerobic oxidation, and it furnishes reducing equivalent to NAD^+.

Succinate-linked reduction of endogenous NAD^+ can also be demonstrated with submitochondrial particles derived from mitochondria by treatment with digitonin (but not with those prepared by sonic oscilla-

[32] M. Klingenberg, *in* "Methoden der enzymatischen Analyse" (H.-U. Bergmeyer, ed.), p. 531, 537. Verlag Chemie, Weinheim, 1962.

[33] R. W. Estabrook, *Anal. Biochem.* **4**, 231 (1962).

[34] Y. Avi-Dor, J. M. Olson, M. D. Doharty, and N. O. Kaplan, *J. Biol. Chem.* **237**, 2377 (1962).

[35] R. W. Estabrook and S. P. Nissley, *Symp. Funktionelle und morphologische Organisation der Zelle, Rottach-Egern, 1962,* p. 119. Springer, Berlin, 1963.

[36] R. W. Estabrook, J. Gonze, and S. P. Nissley, *Federation Proc.* **22,** 1071 (1963).

tion, which are devoid of bound NAD^+). The assay procedure is the same as that for intact mitochondria.

Example 2. Succinate-Linked Reduction of Exogenous NAD^+ in Submitochondrial Particles with ATP as the Source of Energy. This assay (Fig. 3) involves incubation of phosphorylating submitochondrial particles in a buffered Mg^{++}-containing medium in the presence of NAD^+, succinate and a terminal electron transport inhibitor, e.g., KCN. Reduction of NAD^+ is initiated by the addition of ATP. The reaction can be monitored by measuring the light absorbancy at 340 mμ with a commercial spectrophotometer. Alternatively it may be monitored fluorometrically as described above in example 1.

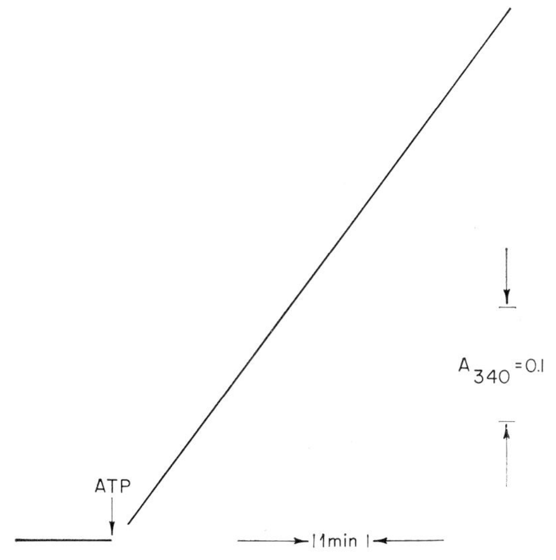

FIG. 3. ATP-supported succinate-linked NAD^+ reduction catalyzed by phosphorylating electron-transport particles from beef heart. The reaction mixture consisted of 180 m*M* sucrose, 50 m*M* Tris-acetate, pH 7.5, 6 m*M* $MgSO_4$, 1.6 m*M* KCN, 1 m*M* NAD^+, 5 m*M* succinate, particles (ETPH) containing 0.6 mg protein; 3 m*M* ATP was added to start the reaction. Final volume, 3 ml; temperature, 30°.

NAD^+ may be replaced by NAD^+-analogs[20,21]; in that case the reaction is monitored at the wavelength of maximal absorption of the respective reduced analog. $NADP^+$ does not replace NAD^+. On the other hand, the succinate-linked NAD^+ reduction may be coupled to the energy-linked reduction of $NADP^+$ by NADH, catalyzed by the same particles.[37] The conditions for the coupled reaction are analogous to

[37] L. Danielson and L. Ernster, *Biochem. Biophys. Res. Commun.* **10**, 91 (1963). See also *in* "Energy-Linked Functions of Mitochondria" (B. Chance, ed.), p. 157. Academic Press, New York, 1963. See also *Biochem. Z.* **338**, 188 (1963).

those for the noncoupled one, except that both NAD^+ and $NADP^+$ are included in the reaction mixture; the former may be added in a catalytic concentration.

Aerobic oxidation of ascorbate plus TMPD or PMS can substitute for ATP as the source of energy of or the succinate-linked NAD^+ reduction.[23] This system is assayed in the presence of antimycin A to block the aerobic oxidation of succinate and NADH. The amount of antimycin A is critical in that too large an excess of the amount needed for blocking the NADH and succinate oxidase may inhibit the succinate-linked NAD^+ reduction per se.[23,38]

ATP-supported succinate-linked NAD^+ reduction in "nonphosphorylating" submitochondrial particles may be stimulated by the addition of suitable "coupling factors"[39-41] or of an appropriate amount of oligomycin.[29]

A "reversal" of ATP-supported succinate-linked NAD^+ reduction, i.e., a phosphorylating oxidation of NADH by fumarate, has been demonstrated with phosphorylating submitochondrial particles.[42]

Example 3. Succinate-Linked Reduction of Acetoacetate in Rat Liver Mitochondria with Aerobically Generated High-Energy Intermediates as the Source of Energy (see the table). Tightly coupled mitochondria from rat liver are incubated aerobically in an isotonic buffered medium containing succinate and acetoacetate. Aliquots are removed at successive time intervals and fixed with perchloric acid. Acetoacetate is determined, e.g., by the method of Walker.[43] Parallel determinations of β-hydroxybutyrate, succinate, and malate plus fumarate may also be performed as described by Klingenberg and von Häfen.[19]

Various controls may be run, involving, for example, omission of succinate, replacement of succinate by malate, or addition of amytal (or rotenone). In all three cases, the disappearance of acetoacetate is greatly diminished. An inhibition of the succinate-linked acetoacetate reduction will also occur upon the addition of agents that diminish the steady state level of high-energy intermediates supplied by the respiratory chain, such as P_i + P-acceptor,[14,15] dinitrophenol,[14] or arsenate.[15,28] Oligomycin,

[38] D. R. Sanadi, T. E. Andreoli, R. L. Pharo, and S. R. Vyas, *in* "Energy-Linked Functions of Mitochondria" (B. Chance, ed.), p. 26. Academic Press, New York, 1963.

[39] D. R. Sanadi, A. L. Fluharty, and T. E. Andreoli, *Biochem. Biophys. Res. Commun.* **8**, 200 (1962).

[40] F. A. Hommes, *Biochim. Biophys. Acta* **71**, 595 (1963).

[41] T. E. Conover, R. L. Prairie, and E. Racker, *J. Biol. Chem.* **238**, 2831 (1963).

[42] D. R. Sanadi and A. L. Fluharty, *Biochemistry* **2**, 523 (1963).

[43] P. G. Walker, *Biochem. J.* **58**, 699 (1954).

SUCCINATE-LINKED ACETOACETATE REDUCTION IN RAT LIVER MITOCHONDRIA[a,b]

Expt. No.	Additions	Acetoacetate removed (micromoles)
1	None	0.3
	Succinate	7.2
	Succinate, amytal	1.3
	Malate	0.5
2	Succinate	5.5
	Succinate, dinitrophenol	0.4
	Succinate, dicoumarol	0.3

[a] From G. F. Azzone, L. Ernster, and E. C. Weinbach, *J. Biol. Chem.* **238,** 1825 (1963).

[b] Each vessel contained, in a final volume of 2 ml: mitochondria from 0.5 g (Expt. 1) or 0.3 g (Expt. 2) rat liver (ca. 20 mg mitochondrial protein per gram of liver); acetoacetate, 4.2 m*M*; $MgCl_2$, 8 m*M*; glycylglycine buffer, pH 7.5, 20 m*M*; KCl, 50 m*M*; sucrose, 62 m*M* (Expt. 1), or 50 m*M* (Expt. 2); P_i, 12.5 m*M* (Expt. 2 only); and, when indicated, succinate, 10 m*M*; L-malate, 10 m*M*; amytal, 2 m*M*; 2,4-dinitrophenol, 0.05 m*M*; dicoumarol, 0.02 m*M*. Incubation was made at 30° for 20 minutes.

which does not inhibit the reaction, even abolishes the inhibition produced by P_i plus P-acceptor, or by arsenate, but not that produced by dinitrophenol. P_i is not essential for the reaction and may even inhibit it when present in high concentrations (higher than 10 m*M*).[15] This effect is counteracted by Mg^{++} or Mn^{++}. In the absence of added P_i, these cations neither stimulate nor inhibit the aerobically driven succinate-linked acetoacetate reduction. Externally added ATP can serve only as a poor source of energy for succinate-linked acetoacetate reduction as compared with the aerobically generated high-energy intermediates.[15] The ATP-driven reaction is inhibited by Mg^{++} and stimulated by EDTA.[19]

Succinate-linked NAD^+ reduction in liver mitochondria has also been studied with α-ketoglutarate plus NH_4^+, rather than acetoacetate, as the oxidant for the mitochondrial NADH formed, with glutamate as the product.[44-47] Arsenite is added to inhibit the oxidation of α-ketoglutarate.

[44] E. C. Slater and J. M. Tager, *in* "Energy-Linked Functions of Mitochondria" (B. Chance, ed.), p. 97. Academic Press, New York, 1963. See also *Biochim. Biophys. Acta* **77,** 276 (1963).

[45] J. M. Tager and E. C. Slater, *Biochim. Biophys. Acta* **77,** 227 (1963).

[46] J. M. Tager, *Biochim. Biophys. Acta* **77,** 258 (1963).

[47] J. M. Tager, J. L. Howland, and E. C. Slater, *Biochim. Biophys. Acta* **77,** 266 (1963).

The reaction sequence taking place is somewhat more complex than in the case of the succinate-linked acetoacetate reduction, in that the glutamate formed interacts with oxaloacetate derived from succinate by way of transaminase, yielding aspartate. Thus, the reducing equivalents mediated by the mitochondrial NAD^+ originate from both succinate and malate. Moreover, since glutamate dehydrogenase reacts with both NADH and NADPH, part of the reaction—and presumably a major part of it—proceeds in the mitochondria via NADPH. The latter may be generated from NADH by way of the energy-linked transhydrogenase reaction. The process may be followed by measuring the appearance of glutamate + aspartate. In contrast to the succinate-linked acetoacetate reduction, the glutamate synthesis is stimulated by P_i. Furthermore it is not inhibited by P_i plus P-acceptor, and the presence of α-ketoglutarate plus NH_4^+ even causes a lowering of the P:O ratio with succinate as substrate. Ascorbate plus TMPD can replace succinate as the energy-yielding substrate. The reaction can also be driven under nonrespiring conditions with added ATP as the source of energy.

[113] Energy-Linked Pyridine Nucleotide Transhydrogenase

By LARS ERNSTER and CHUAN-PU LEE[1]

Klingenberg and Slenczka[1a] found in 1959 that incubation of rat liver mitochondria with NAD^+-specific substrates or succinate in the absence of phosphate acceptor resulted in a rapid and almost complete reduction of the mitochondrial $NADP^+$. From this and related observations,[2] they postulated that the pyridine nucleotide transhydrogenase reaction earlier described by Kaplan *et al.*[3] can be controlled by energy derived from the respiratory chain with a consequent shift of its equilibrium toward the formation of NAD^+ and NADPH. Similar observations were later reported by Estabrook and Nissley.[4]

In 1963 Danielson and Ernster[5,6] demonstrated that submitochondrial

[1] See footnote 1, page 33.

[1a] M. Klingenberg and W. Slenczka, *Biochem. Z.* **331**, 486 (1959).

[2] M. Klingenberg, *Collog. Ges. Physiol. Chem.* **11**, 82 (1961). See also M. Klingenberg and P. Schollmeyer, *Proc. 6th Intern. Congr. Biochem., Moscow, 1961,* Vol. 5, p. 46. Pergamon Press, Oxford, 1963.

[3] N. O. Kaplan, S. P. Colowick, and E. F. Neufeld, *J. Biol. Chem.* **205**, 1 (1953).

[4] R. W. Estabrook and S. P. Nissley, *Symp. Funktionelle und morphologische Organisation der Zelle, Rottach-Egern, 1962,* p. 119. Springer, Berlin, 1963.

[5] L. Danielson and L. Ernster, *Biochem. Biophys. Res. Commun.* **10**, 91 (1963).

[6] L. Danielson and L. Ernster, *in* "Energy-Linked Functions of Mitochondria" (B. Chance, ed.), p. 157. Academic Press, New York, 1963. p. 157. See also *Biochem. Z.* **338**, 188 (1963).

particles from rat liver and beef heart catalyze an energy-dependent reduction of $NADP^+$, coupled to the energy-transfer system of the respiratory chain. Various aspects of this reaction have subsequently been investigated in great detail both with submitochondrial particles[7-19] and intact mitochondria,[19-24] and the pertinent literature has been reviewed by Ernster and Lee.[25]

General Features

The energy-linked reduction of $NADP^+$ by NADH catalyzed by submitochondrial particles may be described by the following net equation

[7] F. A. Hommes and R. W. Estabrook, *Biochem. Biophys. Res. Commun.* **11,** 1 (1963).
[8] F. A. Hommes, *in* "Energy-Linked Functions of Mitochondria" (B. Chance, ed.), p. 39. Academic Press, New York, 1963.
[9] R. W. Estabrook, F. A. Hommes, and J. Gonze, *in* "Energy-Linked Functions of Mitochondria" (B. Chance, ed.), p. 143. Academic Press, New York, 1963.
[10] G. McGuire, L. Pesch, and H. Fanning, *Nature* **200,** 71 (1963).
[11] C. P. Lee, G. F. Azzone, and L. Ernster, *Nature* **201,** 152 (1964).
[12] C. P. Lee and L. Ernster, *Biochim. Biophys. Acta* **81,** 187 (1964).
[13] T. Kawasaki, K. Satoh, and N. O. Kaplan, *Biochim. Biophys. Res. Commun.* **17,** 648 (1964).
[14] C. P. Lee and L. Ernster, *Biochem. Biophys. Res. Commun.* **18,** 523 (1965).
[15] D. W. Haas, *Biochim. Biophys. Acta* **82,** 200 (1964); **89,** 543 (1964).
[16] T. E. Andreoli, R. L. Pharo, and D. R. Sanadi, *Biochim. Biophys. Acta* **90,** 16 (1964).
[17] C. P. Lee, N. Simard-Duquesne, L. Ernster, and H. D. Hoberman, *Biochim. Biophys. Acta* **105,** 397 (1965).
[18] C. P. Lee and L. Ernster, *Symp. Regulation of Metabolic Processes in Mitochondria, Bari, 1965,* B.B.A. Library, **7,** 218. Elsevier, Amsterdam, 1966. See also *Biochem. Biophys. Res. Commun.* **23,** 176 (1966). See also L. Ernster, C. P. Lee, and S. Janda, *Proc. 3rd Meeting Fed. Europ. Biochem. Soc., Warsaw, 1966.* Academic Press, London, in press. See also C. P. Lee *in* "Round Table Discussion on Mitochondrial Structure and Compartmentation" (E. Quagliariello, S. Papa, E. C. Slater, and J. M. Tager, eds.), Adriatica Editrice, Bari, in press.
[19] K. van Dam and H. F. ter Welle, *Symp. Regulation of Metabolic Processes in Mitochondria, Bari, 1963,* B.B.A. Library, **7,** 237. Elsevier, Amsterdam, 1966. K. van Dam, Ph.D. Thesis, University of Amsterdam, 1966.
[20] E. C. Slater and J. M. Tager, *in* "Energy-Linked Functions of Mitochondria" (B. Chance, ed.), p. 97. Academic Press, New York, 1963. See also *Biochim. Biophys. Acta* **77,** 276 (1963).
[21] J. M. Tager and E. C. Slater, *Biochim. Biophys. Acta* **77,** 227 (1963).
[22] J. M. Tager, *Biochim. Biophys. Acta* **77,** 258 (1963).
[23] J. M. Tager, J. L. Howland, and E. C. Slater, *Biochim. Biophys. Acta* **77,** 266 (1963).
[24] M. Klingenberg, *in* "Energy-Linked Functions of Mitochondria" (B. Chance, ed.), p. 121. Academic Press, New York, 1963. See also *Biochem. Z.* **343,** 479 (1965). See also M. Klingenberg and H. von Häfen, *Biochem. Z.* **343,** 452 (1965).
[25] L. Ernster and C. P. Lee, *Ann. Rev. Biochem.* **33,** 729 (1964).

$$NADH + NADP^+ + I \sim X \rightarrow NAD^+ + NADPH + I + X \quad (1)$$

where $I \sim X$ is a high-energy intermediate of oxidative phosphorylation which can be generated either by the three energy-coupling sites of the respiratory chain:

$$C_{1,2,3} + I \xrightarrow{\text{respiration}} C_{1,2,3} \sim I \quad (2)$$

$$C_{1,2,3} \sim I + X \rightleftharpoons C_{1,2,3} + I \sim X \quad (3)$$

or by ATP

$$ATP + I + X \rightleftharpoons ADP + P_i + I \sim X \quad (4)$$

$C_{1,2,3}$ denoting electron-transfer carriers at the respective energy-coupling sites of the respiratory chain; and I and X, hypothetical energy-transfer carriers (cf. Ernster and Lee[25]).

Reaction (1) differs from the pyridine nucleotide transhydrogenase reaction earlier described by Kaplan *et al*[3]:

$$NADH + NADP^+ \rightleftharpoons NAD^+ + NADPH \quad (5)$$

in that it involves the expenditure of one high-energy bond per molecule of $NADP^+$ reduced by NADH; this stoichiometry has been deduced from several lines of evidence.[6,15,18] Its apparent equilibrium constant

$$K' = \frac{[NAD^+][NADPH]}{[NADH][NADP^+]}$$

has been estimated[12] to be 480 ($\Delta E'_0 = 83$ mvolt; $-\Delta F'_0 = 3.8$ kcal), as compared with the equilibrium constant

$$K = \frac{[NAD^+][NADPH]}{[NADH][NADP^+]}$$

of the non-energy-linked transhydrogenase reaction of Kaplan *et al.*[3] catalyzed by the same particles, which is 0.79 ($\Delta E'_0 = 4$ mvolt, $-\Delta F'_0 = 0.19$ kcal). The activation energy of both reactions has been calculated[7,9] to be approximately 20 kcal/mole.

Both the energy-linked and the non-energy-linked transhydrogenase reactions involve the 4*A*-hydrogen atom of NADH and the 4*B*-hydrogen atom of NADPH, as revealed by studies with tritiated pyridine nucleotides.[17] The hydrogen transfer between the two pyridine nucleotides proceeds directly, i.e., without an exchange of hydrogen atoms between these and water. These features strongly suggest that the two reactions are catalyzed by the same enzyme, the non-energy-linked reaction possibly representing an "uncoupled" form of the energy-linked one. The identity of the enzymes involved in the two types of transhydrogenase reaction is further suggested the finding[13] that antibodies produced against the purified (non-energy-linked) transhydrogenase inhibit the energy-linked transhydrogenase reaction catalyzed by submitochondrial particles. On the other hand, the stereospecificity of the transhydrogenase reactions clearly distinguishes them from the reaction catalyzed by DT diaphorase (4*A* specific with respect to both NADH and NADPH[17]) as well as from the respiratory chain-linked NADH dehydrogenase reaction (4*B*-specific[17,26]).

Adenine nucleotides inhibit the non-energy-linked transhydrogenase reaction but are without effect on the energy-linked one.[18] The same holds for Mg^{++}, except that it is even obligatory for the energy-linked reaction when this is driven by ATP as the source of energy.[7,8,12,25] 3,5,3′-Triiodothyronine inhibits both types of reaction, as well as the succinate-linked reduction of NAD^+[7,9] and the ATPase reaction catalyzed by the same particles.[27] 2,4-Dinitrophenol and related uncoupling agents inhibit the energy-linked transhydrogenase reaction driven by either the respiratory chain or ATP as the source of energy, but the concentration of uncoupler needed for complete inhibition is higher in the latter than in the former case.[6] Oligomycin is without effect on the transhydrogenase reaction per se, but inhibits the energy-linked reaction when this is driven by ATP.[5,6] In "nonphosphorylating" submitochondrial particles, oligomycin stimulates, and may even be obligatory for, the energy-linked transhydrogenase reaction driven by either the respiratory chain or ATP.[11,18] In the case of ATP, however, this effect of oligomycin occurs only in a narrow range of concentration, beyond which oligomycin inhibits. Octylguanidine,[18] phenethyl biguanide,[18] aurovertin,[28] atractyloside,[6] and arsenate[6] have been reported to be without effect on the energy-linked transhydrogenase reaction.

Assay

The energy-linked transhydrogenase reaction is assayed by incubating submitochondrial particles in a buffered medium in the presence of NADH, $NADP^+$, and a suitable source of energy. The assay may be performed principally in three ways: (1) by linking the reaction to the alcohol dehydrogenase system, to regenerate NADH from NAD^+, and following the reduction of $NADP^+$; (2) by linking the reaction to the glutathione reductase system to reoxidize the NADPH formed to $NADP^+$, and following the oxidation of NADH; (3) without the use of an acces-

[26] L. Ernster, H. D. Hoberman, R. L. Howard, T. E. King, C. P. Lee, B. Mackler, and G. Sottocasa, *Nature* **207**, 940 (1965).

[27] O. Lindberg, H. Löw, T. E. Conover, and L. Ernster, *Biol. Struct. Function, Proc. 1st IUB/IUBS Intern. Symp. Stockholm, 1960*, p. 3. Academic Press, New York, 1961.

[28] G. Lenaz, *Biochem. Biophys. Res. Commun.* **21**, 170 (1965).

sory dehydrogenase system, by fixing aliquots of the reaction mixture at suitable time intervals and determining the concentrations of the individual pyridine nucleotides (NADH and NADPH and/or NAD^+ and $NADP^+$) by specific enzymatic methods.[29] In all cases, the measurement of the reduced pyridine nucleotides can be made either spectrophotometrically at 340 mμ, or fluorometrically at 450 mμ with an excitation at 365 mμ. Examples of the three types of assay are shown in Figs. 1–3.

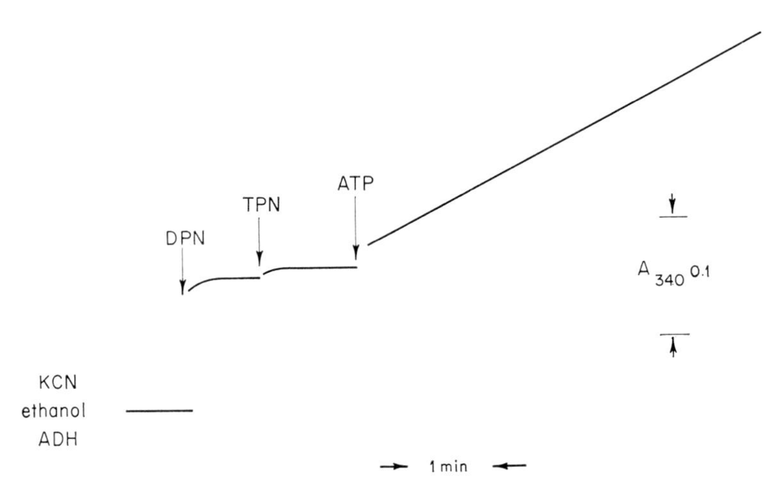

FIG. 1. ATP-dependent reduction of $NADP^+$ by NADH. The reaction mixture consisted of 50 mM Tris buffer, pH 8.0, 6 mM $MgCl_2$, 250 mM sucrose, 1 mM KCN, 57 mM ethanol, 0.25 mg of alcohol dehydrogenase, and rat liver particles containing 0.65 mg of protein. Further additions were: NAD^+, 0.0167 mM; $NADPH^+$, 0.2 mM; ATP, 2 mM; oligomycin, 3 μg. Final volume, 3 ml; temperature, 30°. From L. Danielson and L. Ernster, *Biochem. Z.* **338**, 188 (1963).

The source of energy to be preferred varies with the type of particles used. With phosphorylating particles, aerobically generated high-energy intermediates and ATP are equally efficient in driving the transhydrogenase reaction (cf. Fig. 2); oligomycin has no effect on the aerobically driven reaction, but inhibits the ATP-driven one. With "nonphosphorylating" particles capable of energy-linked transhydrogenation (cf. Lee *et al.*[11] and this volume [87]), aerobically generated high-energy intermediates are preferred to ATP as the source of energy; low concentrations of oligomycin stimulate both types of reaction, whereas high concentrations of oligomycin stimulate the aerobically driven, and inhibit the ATP-driven, reaction.

[29] M. Klingenberg, *in* "Methoden der enzymatischen Analyse" (H.-U. Bergmeyer, ed.), pp. 528, 531, 535, 537. Verlag Chemie, Weinheim, 1962.

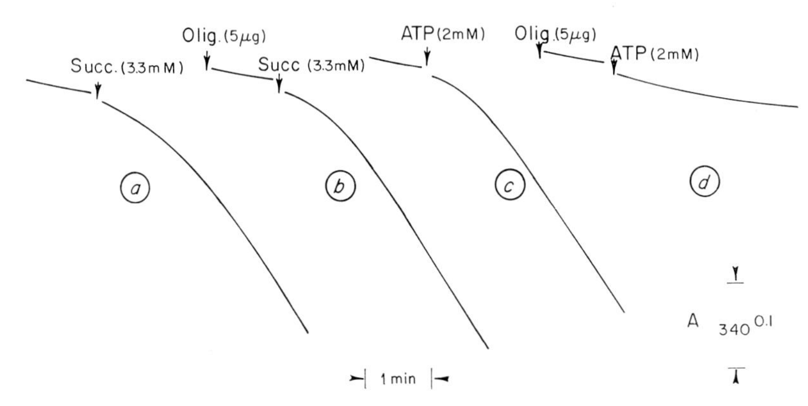

FIG. 2. Energy-dependent reduction of $NADP^+$ by NADH. The reaction mixture consisted of 200 mM sucrose, 50 mM Tris-acetate buffer, pH 7.5, 10 mM $MgSO_4$, 3.3 μM rotenone, 400 μM NADH, 130 μM $NADP^+$, 0.66 mM glutathione, and an amount of glutathione reductase capable of oxidizing 0.5 micromoles of NADPH per minute, particles (ETPH) containing 0.63 mg of protein. Further addition(s) as indicated. Final volume, 3 ml; temperature, 30°. From C. P. Lee, G. F. Azzone, and L. Ernster, *Nature* **201,** 152 (1964).

Succinate and ascorbate + TMPD are the most suitable substrates for the aerobically driven transhydrogenase reaction.[18] With particles from rat liver, NADH may also serve as the energy-yielding substrate,[5,6] provided that sufficient alcohol dehydrogenase is added to maintain a level of NADH necessary for the transhydrogenase reaction. In the case of beef-heart particles, the NADH oxidase activity usually is too high to allow the maintenance of an appreciable level of NADH even by the use of massive amounts of alcohol dehydrogenase. The assay system for these particles must therefore contain an inhibitor of NADH oxidase. Rotenone is the best agent for this purpose, since it does not interfere with the oxidation of succinate or ascorbate plus TMPD, or with the respiratory chain-linked energy-transfer system. Amytal may suppress succinate oxidation and also exhibit an uncoupling effect.[25] Antimycin A may be used in the case of ascorbate plus TMPD, but not, of course, in the case of succinate as the energy-yielding substrate. KCN is unsuitable in both cases. When ATP is used as the source of energy, rotenone, antimycin A, and KCN are equally suitable as inhibitors of the NADH oxidase.

ITP, UTP, GTP, and CTP can replace ATP in driving the energy-linked transhydrogenase reaction,[6] but the rates are only about 40% that with ATP in the case of ITP and UTP, and about 15% in the case of GTP and CTP.

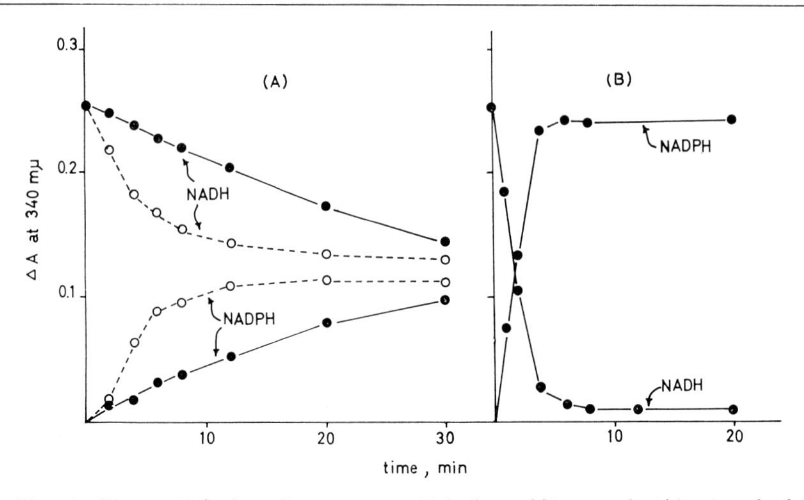

FIG. 3. Energy-linked and non-energy-linked pyridine nucleotide transhydrogenase reactions catalyzed by submitochondrial particles from beef heart. From C. P. Lee and L. Ernster, *Biochim. Biophys. Acta* **81**, 187 (1964).

(A) The reaction mixture consisted of 200 m*M* sucrose, 50 m*M* Tris-acetate buffer, pH 7.5, 3 μ*M* rotenone, 88 μ*M* NADH, 86 μ*M* $NADP^+$. Particles containing 5.1 mg of protein were added to start the reaction. Final volume, 20 ml; temperature, 30°. Samples of 2 ml were removed at time intervals as indicated, and the reaction was stopped by the addition of 0.1 ml of 2 *M* KOH. The samples were then diluted with 2 ml of a solution containing 200 m*M* sucrose and 50 m*M* Tris buffer, pH 7.5, and neutralized by the addition of 1 *M* acetic acid (final pH 7.5–8.0). NADH and NADPH were estimated by measuring the decrease of absorbancy at 340 mμ in a Beckman DK-2 spectrophotometer. To 3 ml of the above neutralized solution were added first 1 m*M* pyruvate + 8 μg of lactate dehydrogenase, to estimate the content of NADH, and then 1 m*M* α-ketoglutarate, 3 m*M* NH_4Cl, and 20 μg of glutamate dehydrogenase to estimate the content of NADPH. ○---○, without Mg^{++}: ●——●, with 10 m*M* Mg^{++}.

(B) The reaction mixture had essentially the same composition as that in (A) except that 85 μ*M* NADH (instead of 88 μ*M*), 10 m*M* $MgSO_4$, and 5 m*M* succinate were present. Assay systems for NADH and NADPH were the same as described for panel (A).

In order to evaluate the rate of the energy-linked transhydrogenase reaction, it is necessary to determine the rate of the non-energy-linked reaction occurring under the prevailing conditions. The contribution of the latter can be lowered by using a relatively high concentration of Mg^{++} (see Fig. 3). It may be further decreased by the addition of adenine nucleotides.

[114] Energy-Linked Transport of Ca^{++}, Phosphate, and Adenine Nucleotides

By ERNESTO CARAFOLI[1] and ALBERT L. LEHNINGER

Mitochondria isolated from a variety of tissues can accumulate certain ions from the suspending medium in a process linked to electron transport; among these are the cations Ca^{++}, Sr^{++}, Mn^{++}, and Mg^{++} and the anions phosphate, sulfate, and acetate. When accumulation of divalent cations occurs, it replaces oxidative phosphorylation; these processes are therefore alternative and each is stoichiometric with electron transport. During uptake of Ca^{++}, large amounts of ADP or ATP may also be accumulated.

The accumulation of Ca^{++} and phosphate, the most intensively studied ions, can be observed under two different sets of conditions. In the first, termed *massive loading,* mitochondria are exposed to relatively high concentrations of Ca^{++} ($\sim$3 m*M*) and they accumulate amounts of Ca^{++} up to 2.5 micromoles per milligram of mitochondrial protein; phosphate is required in this case and it is also accumulated.[1a, 2] Although such massive accumulation of Ca^{++} and phosphate is stoichiometric with electron transport, the excess Ca^{++} in the medium causes loss of acceptor control and of capacity for oxidative phosphorylation. In the second condition, termed *limited loading,* mitochondria are presented with small amounts of divalent cations ($\sim$0.1 m*M*, or $\sim$100 micromoles Ca^{++} per milligram of protein); phosphate is not required, but if it is present, it is also accumulated. Under these conditions there is no damage to oxidative phosphorylation or respiratory control. The dynamics and stoichiometry of ion uptake are best studied in limited-loading experiments.[3-5]

Massive Loading with Ca^{++}

Massive accumulation of Ca^{++} and P_i can be supported either by electron transport or by ATP hydrolysis.[1a, 2] When it is supported by oxidation of a respiratory substrate, ATP is also required in the medium for the mitochondria to retain the accumulated calcium phosphate. In the ATP-driven system, no respiratory substrate is needed, but high

[1] This manuscript was prepared while the author was at the Department of Physiological Chemistry, The Johns Hopkins University School of Medicine, Baltimore, Maryland.

[1a] F. D. Vasington and J. V. Murphy, *J. Biol. Chem.* **237**, 2670 (1962).

[2] C. S. Rossi and A. L. Lehninger, *Biochem. Z.* **338**, 698 (1963).

[3] C. S. Rossi and A. L. Lehninger, *J. Biol. Chem.* **239**, 3971 (1964).

[4] J. B. Chappell, M. Cohn, and G. D. Greville, *in* "Energy-Linked Functions of Mitochondria" (B. Chance, ed.), p. 253. Academic Press, New York, 1963.

[5] B. Chance, *J. Biol. Chem.* **240**, 2729 (1965).

concentrations of ATP (~15 m*M*) are required.

Reagents and Reaction Media. The components are given on the basis of a 3.0 ml total volume; the systems may be scaled down or up as required.

RESPIRATION-SUPPORTED SYSTEM

Stock solutions	Volume (ml)	Final concentration (m*M*)
0.1 *M* Tris-chloride, pH 7.2	0.3	10
0.1 *M* Na-succinate or DL-β-hydroxybutyrate, pH 7.2	0.3	10
0.1 *M* $MgCl_2$	0.3	10
0.4 *M* Na-K phosphate buffer, pH 7.2	0.3	4.0
0.8 *M* NaCl	0.3	80
0.15 *M* Na-ATP, pH 7.2	0.06	3.0
0.03 *M* $^{45}CaCl_2$ (0.05 μC per μmole)	0.3	3.0
Mitochondrial suspension (protein, 10 mg/ml in 0.25 *M* sucrose)	0.3	1.0 mg protein per ml
H_2O	0.84	
Total volume:	3.0	

ATP-SUPPORTED SYSTEM. Same as above, with respiratory substrate replaced by water and the final ATP concentration raised to 15 m*M*.

Procedure. Mitochondria are added last to start the reaction, which is usually carried out for 10 minutes with shaking in air at 30° in beakers or large centrifuge tubes. The tubes are then rapidly cooled in cracked ice, and centrifuged in the cold at 20,000 *g* for 4 minutes to separate the "loaded" mitochondria from the suspending medium.

Determination of the Uptake of Ca^{++} and Inorganic Phosphate

Reagents

Sucrose, 0.25 *M*
Sodium lauryl sulfate, 0.2%
Trichloroacetic acid 8%

Procedure. After sedimentation of the mitochondria, the supernatant reaction media are saved for analysis. The sedimented mitochondria are washed twice with 5.0-ml portions of cold 0.25 *M* sucrose. Prior to the last centrifugation, the suspensions of washed mitochondria are divided into two equal portions. After centrifugation, one pellet is extracted with 5.0 ml of cold trichloroacetic acid and aliquots of the clarified extract are used for P_i determination by the method of Gomori.[6] The other pellet is

[6] G. Gomori, *J. Lab. Clin. Med.* **27**, 955 (1942).

dissolved in 5.0 ml sodium lauryl sulfate. Aliquots (0.3 ml) of the sodium lauryl sulfate extracts of the pellets, and of the original clarified reaction media, are plated on aluminum planchets (2.0 cm diameter), evaporated to dryness to yield "thin" samples, and counted in a gas-flow Geiger tube. The distribution of radioactivity between washed pellets and the incubation medium is used to calculate the uptake of Ca^{++}. The preexisting intramitochondrial Ca^{++} (10 millimicromoles per milligram of protein) is very low compared to the large amounts of Ca^{++} taken up (up to 2500 millimicromoles Ca^{++} per milligram of protein); the Ca^{++} uptake data therefore need not be corrected for exchange. Since P_i uptake is stoichiometric with that of Ca^{++}, measurement of accumulated P_i in the mitochondrial trichloroacetic acid extract will suffice as a simpler means of following accumulation of Ca^{++} and phosphate.

Properties of the Massive Loading Reaction

For each pair of electrons passing through each energy-conserving site of the respiratory chain, about 1.67 molecules of Ca^{++} and 1.0 molecule of P_i are accumulated; Ca^{++} and P_i are thus accumulated in a molar ratio of 1.67, which is that of hydroxylapatite.[2] The ATP-supported system is less efficient. Mg^{++} is essential in the incubation medium, and some of it is accumulated with the Ca^{++} and phosphate.[7] Antimycin A and cyanide block Ca^{++} uptake in systems supported by electron transport, whereas oligomycin has no effect. Oligomycin, on the other hand, greatly inhibits the ATP-supported system, which is otherwise only slightly inhibited by antimycin A or cyanide. Ion uptake in either type of system is inhibited by true uncoupling agents such as 2,4-dinitrophenol.[1a, 2]

Massive uptake of $^{85}Sr^{++}$ can be studied under the same conditions described above;[8] the uptake of $^{85}Sr^{++}$ is determined by counting aliquots of the clarified reaction medium or of the trichloroacetic acid extracts of the washed mitochondria in a scintillation counter.

Uptake of ATP or ADP during Massive Uptake of Ca^{++} and P_i

Accumulation of ATP and ADP may be measured in the same medium used for study of the uptake of Ca^{++} and P_i. In this case, ATP-8-^{14}C (0.05–0.1 μC) is employed, and the Ca^{++} added is unlabeled. AMP of the medium is not accumulated under these conditions.[9]

[7] E. Carafoli, C. S. Rossi, and A. L. Lehninger, *J. Biol. Chem.* **239**, 2055 (1964).
[8] E. Carafoli, S. Weiland, and A. L. Lehninger, *Biochim. Biophys. Acta,* **97**, 88 (1965).
[9] E. Carafoli, C. S. Rossi, and A. L. Lehninger, *J. Biol. Chem.* **240**, 2254 (1965).

Procedure. At the end of the incubation, the mitochondria are centrifuged and washed twice with 5.0-ml portions of 0.25 *M* sucrose. Before the last centrifugation, the suspensions are divided into two aliquots. One of the resulting pellets is dissolved in 0.2% sodium lauryl sulfate; aliquots of the solutions are plated and counted in a low-background gas flow counter. The other pellet is extracted with 6% cold perchloric acid. Aliquots of the perchloric acid extract may be applied to paper, and the labeled ATP and ADP separated and identified by paper electrophoresis.[9]

Limited Loading with Ca^{++}

In the limited loading system,[2] the mitochondria are usually incubated in cuvettes of the Clark oxygen polarograph linked to a strip chart recorder.[10] A known amount of Ca^{++} is added to mitochondria respiring in state 4, and the rate and amount of extra oxygen uptake is recorded, up to the point when oxygen uptake returns to the initial resting rate. Ca^{++} uptake is then measured in an aliquot of the medium.[11]

Reagents and Reaction Medium. This is equilibrated with air.

Reagent	Volume (ml)	Final concentration (m*M*)
0.1 *M* Tris-chloride, pH 7.2	0.19	10
0.8 *M* NaCl	0.19	80
0.1 *M* Na-succinate, pH 7.2	0.09	10
H_2O	1.33	
Total volume:	1.80	

Procedure. After thermoequilibration of the above medium at 25° in the polarograph cuvette, 0.10 ml of mitochondrial suspension (5.0 mg protein) is added and the course of the state 4 oxygen uptake is followed. The contents of the cuvette are continuously stirred by a magnetic bar. Sixty seconds after the addition of the mitochondria, $^{45}CaCl_2$ (200–600 millimicromoles) is added with a microsyringe (6.7–20 μl). When the respiration has returned to the resting rate, a 1.0-ml aliquot of the suspension is removed with a 1.0-ml syringe equipped with a needle. The needle is quickly replaced with a Swinny adapter containing a Millipore filter (0.8 μ pore size) and 0.3–0.4 ml of clear filtrate are collected by exerting moderate pressure on the plunger of the syringe.[12] The entire

[10] W. W. Kielley and J. R. Bronk, *J. Biol. Chem.* **230,** 521 (1958).
[11] E. Carafoli, *Biochim. Biophys. Acta* **97,** 107 (1965).
[12] H. Rasmussen, A. Waldorf, D. D. Dziewiatowski, and H. F. DeLuca, *Biochim. Biophys. Acta* **75,** 250 (1963).

operation requires approximately 10 seconds. An aliquot of the clear filtrate (0.1–0.2 ml) is plated and counted for $^{45}Ca^{++}$.

The Ca^{++}:0 accumulation ratio is the ratio of millimicromoles Ca^{++} accumulated to millimicroatoms of extra oxygen uptake stimulated by the addition of Ca^{++}, as recorded by the polarograph trace. Its value will depend on the respiratory substrate; it is approximately 5–6 for β-hydroxybutyrate, 3.6–4.0 for succinate.[2] The baseline state 4 respiration is also capable of coupling to Ca^{++} accumulation, but the efficiency is low.

The Ca^{++}:0 activation ratio is the ratio (millimicromoles Ca^{++} added): (millimicroatoms extra oxygen taken up); it is usually slightly larger than the Ca^{++}:0 accumulation ratio.

Properties of the Reaction. Usually not more than 1% of the added Ca^{++} is left in the supernatant 30 seconds after the respiration has returned to the resting rate. Mg^{++} ions and ATP are not necessary for simple uptake of Ca^{++} in the absence of P_i. When P_i is added, it is also accumulated if the system contains 1.0 m*M* ATP plus 10 m*M* $MgCl_2$, or 1.0 μg of oligomycin.[2]

Limited uptake of Sr^{++}[11] and Mn^{++}[4] can be studied under the same conditions, by employing $^{85}Sr^{++}$ or $^{54}Mn^{++}$. With these cations, respiration returns to the resting rate in the presence of inorganic phosphate; ATP is not required.

[114a] Fatty Acid Activation and Oxidation by Mitochondria

By SIMON G. VAN DEN BERGH

In animal cells fatty acid activation occurs both in the mitochondria and in the microsomes, but the mitochondria are the only site of fatty acid oxidation.

Fatty acids are different from other mitochondrial respiratory substrates in three important respects: (a) they need activation with formation of thiol esters of coenzyme A prior to their oxidation; (b) they are potent uncoupling agents; and (c) they are powerful mitochondrial swelling agents. These properties all have specific consequences and require special precautions.

Although fatty acid breakdown has been studied in mitochondria isolated from a variety of animal and plant tissues, we shall in this chapter mainly refer to rat liver mitochondria and only in the last paragraph consider some of the properties of fatty acid breakdown in other types of mitochondria.

Assay Conditions

Protection against Swelling. Failure to protect the mitochondria against the swelling action of fatty acids has given rise to the erroneous contention that fatty acids are inhibitory to mitochondrial respiration. If proper care is taken to avoid swelling of the mitochondria, no such inhibition can be observed. The easiest way to prevent fatty acid-induced mitochondrial swelling is by adding a low concentration (2 m*M*) of EDTA to the reaction medium. Bovine serum albumin (1–2%) can also be used,[1] but this has other effects as well (see below).

Reaction Medium. It has been stated[2] that isolated mitochondria do not oxidize fatty acids unless ATP, cytochrome *c* and, under certain conditions, a Krebs-cycle intermediate are added to the reaction medium. However, these additions are unnecessary if the reaction is started by the addition of freshly isolated mitochondria to an otherwise complete reaction medium. A normal reaction medium for mitochondrial phosphorylating oxidations can be used, provided that EDTA is included (see above). Even a glucose-hexokinase trapping system for ATP can be added; evidently, the intramitochondrial fatty acid-activating system can compete successfully with the extramitochondrial hexokinase trap for the ATP generated inside the mitochondria. In studies of the energetics of fatty acid oxidation it is advisable to use AMP as phosphate acceptor; added ADP may cause complications, since it can activate fatty acids through the combined action of myokinase and fatty acyl-CoA synthetase.

Substrates. Straight-chain, saturated fatty acids with up to nine carbon atoms are added as 0.1 *M* neutralized aqueous solutions[2]; longer-chain acids are added as alcoholic solutions of the free acids.[3] Although the latter may precipitate in the aqueous reaction medium, this does not interfere with their oxidation by the mitochondria.

General Properties

Products. In liver mitochondria, fatty acid oxidation proceeds only partly via the Krebs cycle with CO_2 as end product. Part of the fatty acid is converted into acetoacetate, which may, under conditions of a high phosphate potential, be partly reduced to β-hydroxybutyrate. The extent to which these reactions proceed may be estimated by measuring both oxygen uptake and ketone-body formation during fatty acid oxidation. For the determination of β-hydroxybutyrate and acetoacetate,

[1] P. Björntorp, H. A. Ells, and R. H. Bradford, *J. Biol. Chem.* **239,** 339 (1964).

[2] See Vol. I [89].

[3] The final concentration of ethanol in the reaction medium should not exceed 2%.

enzymatic methods[4] should be preferred over colorimetric methods. From these determinations it can be calculated that with even-numbered fatty acids under normal conditions 25–35% of the oxygen uptake is connected with Krebs-cycle activity. Odd-numbered fatty acids give rise preferentially to CO_2 rather than to acetoacetate. If Krebs-cycle intermediates are added (e.g., 10 m*M* L-malate), an almost complete conversion to CO_2 occurs.

Uncoupling Effect of Fatty Acids. Since fatty acids are uncouplers of oxidative phosphorylation, the phosphorylation efficiency will decrease with increasing concentrations of the fatty acid. The rate of oxidation is unaffected by the increased uncoupling as long as some phosphorylation occurs. But as soon as the P:O ratio drops to zero, the oxidation of the fatty acid is inhibited by lack of ATP, necessary for the activation of the fatty acid.[5] The inhibitory concentration of the fatty acid is not an absolute figure, but should be related to the amount of mitochondria present.[6] For most fatty acids, therefore, a critical ratio of fatty acid to mitochondrial protein exists, above which the oxidation of the fatty acid is inhibited.[5] For rat liver mitochondria a number of these ratios is shown in the table.

CRITICAL RATIOS OF FATTY ACID TO RAT LIVER MITOCHONDRIAL PROTEIN

Fatty acid	Critical ratio (micromoles/mg protein)
Heptanoic acid (C_7)	5.5
Octanoic acid (C_8)	1.9
Pelargonic acid (C_9)	0.55
Capric acid (C_{10})	0.26
Undecanoic acid (C_{11})	0.09
Lauric acid (C_{12})	0.06
Tridecanoic acid (C_{13})	0.05
Myristic acid (C_{14})	0.07
Palmitic acid (C_{16})	∞ [a]
Oleic acid ($C_{18:1}$)	0.04

[a] With normal amounts of mitochondria, the uncoupling concentration of palmitic acid and the longer-chain fatty acids cannot be reached because of the low solubility of these acids in the aqueous reaction medium. By greatly lowering the concentration of the mitochondria, it is sometimes possible to pass the critical ratio.

[4] D. H. Williamson, J. Mellanby, and H. A. Krebs, *Biochem. J.* **82**, 90 (1962).

[5] S. G. Van den Bergh, *in* "Regulation of Metabolic Processes in Mitochondria" (J. M. Tager, E. Quagliariello, S. Papa, and E. C. Slater, eds.), p. 125. Elsevier, Amsterdam, 1966.

[6] P. Borst, J. A. Loos, E. J. Christ, and E. C. Slater, *Biochim. Biophys. Acta*, **62**, 509 (1962).

Unsaturated fatty acids are more powerful uncouplers than the corresponding saturated fatty acids.[6] Therefore, their critical ratios are lower.

Serum albumin prevents the uncoupling effect of fatty acids. Consequently, addition of 1–2% purified[7] serum albumin to the reaction medium allows the oxidation of fatty acids at concentrations above the critical ratio.

Effect of Inhibitors and Uncouplers.[8] When added in concentrations that completely abolish respiratory-chain phosphorylation, uncouplers inhibit the mitochondrial oxidation of fatty acids. Oligomycin also completely inhibits fatty acid oxidation; evidently, high-energy intermediates of oxidative phosphorylation are unable to activate fatty acids.[5] Arsenate does not affect fatty acid oxidation.[5] Arsenite (1 mM) and malonate (5–10 mM) inhibit the oxygen uptake by about 40% in the absence of added malate; in the presence of malate, these inhibitors have almost no effect.[5] Fluoride (5–10 mM) inhibits all fatty acid oxidation.

Atractyloside does not inhibit the activation of fatty acids by ATP generated in the respiratory chain, but it makes fatty acid oxidation dependent on α-oxoglutarate-linked substrate-level phosphorylation, since the AMP formed in the activation reaction can no longer be phosphorylated by ATP (myokinase is located outside the atractyloside barrier), but has to be phosphorylated by GTP.[9]

GTP-Linked Fatty Acid Activation

In addition to the well known ATP-linked fatty acid-activating enzymes, catalyzing reaction (1)

$$\text{Fatty acid} + \text{ATP} + \text{CoA} \rightleftharpoons \text{fatty acyl-CoA} + \text{AMP} + \text{PP}_i \qquad (1)$$

liver mitochondria contain a GTP-specific fatty acid-activating enzyme, which catalyzes reaction (2)[10]

$$\text{Fatty acid} + \text{GTP} + \text{CoA} \rightleftharpoons \text{fatty acyl-CoA} + \text{GDP} + \text{P}_i \qquad (2)$$

The latter enzyme is strongly inhibited by inorganic phosphate.

Thus, in the absence of added inorganic phosphate liver mitochondria can activate fatty acids via reaction (2) and fatty acid oxidation can

[7] Commercially available serum albumin contains a considerable amount of fatty acids. When it is used in studies on fatty acid metabolism it should, therefore, be extensively purified, e.g., according to D. S. Goodman, *Science* **125,** 1296 (1957).

[8] See this volume [8].

[9] S. G. Van den Bergh, *in* "Mitochondrial Structure and Compartmentation" (J. M. Tager, S. Papa, E. Quagliariello, and E. C. Slater, eds.). Adriatica Editrice, Bari, in press.

[10] C. R. Rossi and D. M. Gibson, *J. Biol. Chem.* **239,** 1694 (1964).

occur even when respiratory-chain phosphorylation is completely uncoupled.

Under the following sets of conditions fatty acid oxidation occurs at an appreciable rate and is completely dependent on activation via reaction (2) by GTP, generated in α-oxoglutarate oxidation: (a) at fatty acid concentrations below the critical ratio, in the presence of 0.1 mM dinitrophenol and in the absence of inorganic phosphate; (b) at fatty acid concentrations above the critical ratio, in the absence of inorganic phosphate; (c) with some fatty acids at concentrations far above the critical ratio, both in the presence and in the absence of inorganic phosphate (e.g., oleic acid at concentrations of 0.3 micromole per milligram of mitochondrial protein).[5] In all these cases, fatty acid oxidation is insensitive to uncouplers of respiratory-chain phosphorylation, to oligomycin, and to atractyloside. It is completely inhibited by arsenite (both in the presence and in the absence of added malate) and by fluoride, and in the first two cases by phosphate and by arsenate.

Activation of Fatty Acids by Added ATP

In order to study activation of fatty acids by added ATP it is necessary to inhibit or prevent all endogenous formation of ATP. This can be done in two ways:

(a) By addition of a respiratory inhibitor. In this case fatty acid oxidation does not occur and the formation of fatty acyl-CoA has to be measured directly. This can be done either with added coenzyme A or hydroxylamine as acyl-acceptor[11] or by measuring the endogenous acyl-CoA of the mitochondria.[12]

(b) By addition of an uncoupling agent or by using fatty acid concentrations above the critical ratio (20–30 mM orthophosphate must be present to inhibit GTP-linked fatty acid activation). Added ATP, however, cannot support fatty acid oxidation under these conditions unless either oligomycin or 1 mM L-carnitine (or 2 mM DL-carnitine) are also added. Whereas with added carnitine, concentrations of ATP as low as 0.1 mM may be used, with added oligomycin at least 2 mM ATP must be added to effect optimal rates of fatty acid oxidation.[9]

Fatty acid oxidation, driven by added ATP, is insensitive to uncouplers and to oligomycin. It is inhibited by fluoride and by atractyloside. On the other hand, when catalytic amounts of coenzyme A (10–100 μM) are present in addition to carnitine, the process is insensitive to atractyloside.[9]

[11] See Vol. III [137].

[12] P. B. Garland, D. Shepherd, and D. W. Yates, *Biochem. J.* **97**, 587 (1965).

Oxidation of Activated Fatty Acids

When an activated form of the fatty acids is used as respiratory substrate, the mitochondrial oxidation of fatty acids can be studied irrespective of the activation process. Added coenzyme A esters of fatty acids are, however, poorly oxidized since they cannot penetrate toward their intramitochondrial sites of oxidation. On the other hand, carnitine esters of fatty acids, which also represent an activated form of fatty acids, are readily oxidized.[13] Evidently, the barrier in the mitochondria which is impermeable to fatty acyl-CoA, is freely permeable to fatty acyl-carnitine. Since the mitochondria contain enzymes, which catalyze the reversible reaction (3)[14]

$$\text{Fatty acyl-CoA} + \text{carnitine} \rightleftharpoons \text{fatty acyl-carnitine} + \text{CoA} \quad (3)$$

they will oxidize added fatty acyl-CoA if catalytic amounts of carnitine are present in the reaction medium.

Carnitine esters of fatty acids can be synthesized nonenzymatically.[13] Long-chain acyl-carnitines are potent detergents, and in concentrations in which they are used as respiratory substrates in manometric experiments, they completely lyse the mitochondria within a few seconds. Addition of 1–2% bovine serum albumin will protect the mitochondria against this damaging action of the long-chain acyl-carnitines.

When added in an activated form, fatty acids behave in every way as normal mitochondrial respiratory substrates, with one exception: fluoride completely inhibits their oxidation.

Fatty Acid Breakdown in Other Types of Animal Mitochondria[15]

With the exception of mitochondria from carbohydrate-utilizing insects,[16] all types of isolated mitochondria readily oxidize carnitine esters of fatty acids.[17] The enzymes of fatty acid oxidation must, therefore, be present. Yet, most types of isolated mitochondria do not oxidize free fatty acids. This inability obviously resides in the activation process. Mitochondria from kidney and heart oxidize only free fatty acids with up to twelve carbon atoms. Longer-chain acids are oxidized only in the presence of carnitine and in the absence of a glucose-hexokinase trapping system for ATP. The same holds true for mitochondria from calf diaphragm and from locust flight muscle, but not for mitochondria from

[13] J. Bremer, *J. Biol. Chem.* **237**, 3628 (1962).

[14] S. Friedman and G. Fraenkel, *Arch. Biochem. Biophys.* **59**, 491 (1955); K. R. Norum, *Biochim. Biophys. Acta* **89**, 95 (1964).

[15] For plant mitochondria, see Vol. I [90].

[16] See this volume [22].

[17] C. Bode and M. Klingenberg, *Biochem. Z.* **341**, 271 (1965).

other skeletal muscles.[17]

GTP-linked activation has to date been demonstrated only in liver mitochondria. It is definitely absent from kidney and heart mitochondria.

In most types of mitochondria ketone-body formation does not occur and CO_2 is the only end product of fatty acid oxidation. In those mitochondria fatty acid oxidation is completely dependent on the formation of oxalacetate in the Krebs cycle. In heart mitochondria, for instance, fatty acid oxidation is completely inhibited by malonate and by arsenite; this inhibition can be released by addition of malate.

[115] Measurement of Protein Synthesis in Mitochondria

By MELVIN V. SIMPSON,[1] MAURILLE J. FOURNIER, JR.,[1] and DOROTHY M. SKINNER[1]

Since the demonstration that mitochondria could effect the incorporation of amino acids into protein,[1a] the problem has been pursued in a number of laboratories (for methods, see footnotes 2–7). The recent discovery of the presence of DNA in mitochondria, work on the origin of these organelles, and the realization of the close structure-function analogy between mitochondria and chloroplasts, all have combined to make this area a major field of investigation. We will be concerned here only with the system as employed for amino acid incorporation.

Homogeneity of Mitochondrial Preparation

Amino acid incorporation into mitochondrial protein has been shown to be dependent upon ATP, generated in the rat liver or rat heart system either by oxidative phosphorylation or by an external ATP generating system[1a, 3] or, in the case of the calf heart system, solely by oxidative phosphorylation.[8, 9]

[1] During the preparation of this manuscript M. V. Simpson and M. J. Fournier, Jr. were at the Department of Biochemistry, Dartmouth Medical School, Hanover, New Hampshire; D. M. Skinner was at the Department of Physiology, New York University School of Medicine, New York.

[1a] J. R. McLean, G. L. Cohn, I. K. Brandt, and M. V. Simpson, *J. Biol. Chem.* **233**, 657 (1958).

[2] D. B. Roodyn, P. J. Reis, and T. S. Work, *Biochem. J.* **80**, 9 (1961).

[3] G. F. Kalf, *Arch. Biochem. Biophys.* **101**, 350 (1963).

[4] A. M. Kroon, *Biochim. Biophys. Acta* **91**, 145 (1964).

[5] D. E. S. Truman and A. Korner, *Biochem. J.* **83**, 588 (1962).

[6] M. V. Simpson, *Ann. Rev. Biochem.* **31**, 333 (1962).

[7] D. B. Roodyn, *Biochem. J.* **97**, 782 (1965).

[8] G. F. Kalf and M. V. Simpson, *J. Biol. Chem.* **234**, 2943 (1959).

[9] D. M. Skinner and M. V. Simpson, unpublished data.

Various kinds of evidence have been employed to show that the incorporation does not result from contaminating microsomes or ribosomes; the most telling observation is the resistance of the incorporating system to RNase, even when this enzyme is added at extremely high levels.[1a,2] Even if microsomes or ribosomes were resistant to this enzyme, it would not be expected that cytoplasmic sRNA (on whose addition such systems are usually dependent) would be. Further evidence for the absence of microsomes comes from the failure to detect microsomal enzymes in well washed mitochondrial preparations active in amino acid incorporation.[2,5,10,11] Also, experiments have been performed in which mitochondria have been washed free of added highly labeled microsomes.[1a,2] Moreover, *in vivo* studies on skeletal muscle have shown mitochondria to be more active in amino acid incorporation (per milligram of protein) than are muscle microsomes[12]; isolated ribosomes obtained from this tissue have also been found to be rather inactive.[13] Hence, contamination of mitochondrial preparations would need to occur to a very high and easily detetable extent to account for the amino acid incorporation activities of these preparations. However, no such large contamination has in fact been observed.[9]

The absence of intact nuclei in a well prepared liver mitochondrial preparation can be demonstrated by light or electron microscopy; however, not much attention has been given to the possible presence of nuclear fragments. From what is known about protein synthesis in nuclei, however, it may be concluded that the system is similar to the cytoplasmic one and, normally, should be sensitive to RNase.

Because the rates of amino acid incorporation are fairly low in mitochondrial systems, contamination by any more than minimal numbers of bacteria represents an ever present threat, particularly when amino acids of high specific radioactivity are used. The evidence against any significant contribution by bacteria to amino acid incorporation in mitochondrial systems where such studies have been done, may be considered conclusive, and is summarized as follows: (a) The time curve of incorporation of amino acids into mitochondrial protein of liver generally shows an initial linear region, then a progressive falling off, until, in an hour or less, incorporation ceases.[2,5] Such kinetics are not characteristic of bacterial growth. (b) Assay of the bacterial level in various mitochondrial incorporating systems showed that the incorporating activity

[10] A. M. Kroon, *Biochim. Biophys. Acta* **72,** 391 (1963).
[11] D. B. Roodyn, K. B. Freeman, and J. R. Tata, *Biochem. J.* **94,** 628 (1965).
[12] M. V. Simpson and J. R. McLean, *Biochim. Biophys. Acta* **18,** 573 (1955).
[13] J. R. Florini and C. B. Breuer, *Biochemistry* **4,** 253 (1965).

was independent of the level of bacterial contamination.[2,3] In addition, in rat liver mitochondrial systems, the number of viable organisms decreases by a factor of 50% or greater during the 1-hour incubation period.[14] (c) Sucrose density gradient centrifugation of the incubation mixture after amino acid incorporation, and subsequent analysis of the four fractions obtained, showed most of the radioactivity to be in the mitochondrial region.[11] (d) In more recent studies, density gradient runs of higher resolution than those described above, showed no radioactivity in the bacterial region; almost all the isotope was found in the rather narrow mitochondrial band.[14] However, we have, as have Roodyn *et al.*,[11] found that from 20% to 35% of the total bacteria in the gradient remain in the mitochondrial band. We have therefore turned to germ-free animals, aseptic operative techniques, and sterile solutions, to produce final incubation media with very little bacterial contamination. In a series of such experiments, independent of one another, contamination was reduced from a range of 60,000–100,000 cells per milliliter at zero incubation time to a range of 40–1000 per milliliter, without a sign of any loss in amino acid incorporating activity. No anaerobic organisms were detectable.

If germ-free animals and sterile solutions are not used but exceptional care is taken during the operative procedure, levels of contamination no higher than 1000 cells per milliliter can be attained routinely. We have now modified our preparative procedure to include these conditions which are described below.

Procedure for Rat Liver Mitochondria

There are a number of procedures available for studying incorporation in the rat liver system; we describe the one now in use in our laboratory. The incubation medium resembles that originated by Roodyn *et al.*[2]; one obtains the same initial incorporation rates as in the medium originally described by McLean *et al.*,[1a] but the total period of incorporation is somewhat longer. The reader is referred to McLean *et al.*[1a] for the use of an external ATP generating system.

Reagents

Medium A: 0.25 M sucrose, 0.02 M K_2HPO_4, 2.0 mM disodium EDTA, 0.03 M nicotinamide, adjusted to pH 6.8 at 25°

Medium B: 0.1 M sucrose, 0.04 M KCl, 1.3 mM disodium EDTA, 0.01 M dipotassium succinate, 0.016 M KH_2PO_4, 4.0 mM AMP, 0.05 mM DPN, 8.0 mM $MgSO_4$, adjusted to pH 7.0 at 30°

[14] M. J. Fournier and M. V. Simpson, unpublished data.

L-Amino acid mixture: 5.0 micromoles each of arginine, asparagine, aspartic acid, cysteine, glutamic acid, glutamine, glycine, histidine, hydroxyproline, isoleucine, lysine, methionine, phenylalanine, proline, serine, threonine, tryptophan, tyrosine, and valine adjusted to pH 7.0, in a final volume of 1.0 ml

DL-Leucine-1-^{14}C (20 mC/mmole), 20 μC/ml (Under the conditions of counting used here, 1 μC = 1.0×10^6 cpm.)

Trichloroacetic acid, 10%

Trichloroacetic acid, 5%

Ethanol, 95%

Ethanol, 70%

Ethyl ether

The first four solutions are stored at $-20°$.

Preparation of Mitochondria

Male Sprague-Dawley or Wistar rats weighing between 100 and 150 g are stunned by a blow on the head followed by decapitation. The anterior portion of the body is scrubbed with liberal amounts of 70% ethanol, and is then flushed with this reagent. A rectangle of skin about 5 cm by 8 cm is removed with sterile instruments, care being taken to avoid penetration into the abdominal cavity. The exposed area is then flushed with 70% ethanol; the liver is then removed, again with sterile instruments. The liver is then rapidly weighed, cut into several pieces, and immersed in a large volume of ice cold medium A. All subsequent operations are carried out at 0°–4°.

About 6 g of liver at a time are homogenized in about 3–4 volumes of medium A with the aid of a fairly loose-fitting all-glass homogenizer (pestle diameter, 0.740 inch; tube inside diameter, 0.758 inch) rotating at about 1500 rpm. Homogenization is performed in 15–20 seconds with about 6–8 slow vertical passes of the pestle. The homogenate is made up to a final ratio of 9 volumes of medium A per gram of liver and the resulting suspension is sedimented twice at 2600 rpm (600 *g*, average radius) for 10 minutes in the SS-34 rotor of the RC 2 Sorvall centrifuge, each time leaving behind a few milliliters of supernatant fluid so as not to include the loosely packed layer of sediment. The pellets are discarded and the supernatant solution is recentrifuged at 8300 rpm (6100 *g*, average radius) for 10 minutes in the same centrifuge.[15] The pellets from no more than 20 g of liver are combined and resuspended in 15–18 ml of

[15] This value should be reduced by about 25% if the No. 30 rotor of the Spinco ultracentrifuge is used in order to compensate for the shorter path length in this rotor.

medium A by means of 3 or 4 gentle strokes in a smaller normal fitting glass homogenizer (pestle diameter, 0.618 inch; tube inside diameter, 0.622 inch) used manually. The homogenate is diluted to about 45 ml and recentrifuged at 8300 rpm for 10 minutes. The supernatant fluid is discarded along with any fluffy layer. This washing procedure is repeated 4 additional times. The supernatant fluid becomes completely clear at the fourth wash. For the sixth and last wash, all sediments are combined into a single centrifuge tube and suspended and sedimented as described above. The mitochondria are then resuspended in a few milliliters of medium A to give a final concentration of about 100 mg of mitochondrial protein per milliliters (biuret method).

Incubation Conditions

The reaction mixture consists of 0.75 ml of medium B, 0.1 ml of L-amino acid mixture, 0.05 ml of DL-leucine and 0.05 ml of H_2O in a final volume of 0.95 ml. To the reaction mixture in 13 × 100 mm test tubes is added 0.05 ml of mitochondrial suspension, and the vessels are then incubated in air in a shaking water bath at 30°. For incubation volumes in excess of 2 ml, it is advisable to use Erlenmeyer flasks to ensure adequate aeration. At approriate time intervals, 0.1-ml samples are removed and transferred to paper disks to be washed and counted by the slightly modified method[16] of Mans and Novelli.[17]

Results

In an average experiment, incorporation of leucine is linear for at least 20 minutes and falls off during the succeeding hour to an extent which varies from preparation to preparation. Under the conditions described here, the specific activity of the mitochondrial protein at the end of the first hour of incubation is about 200 $\mu\mu$C (200 cpm) per milligram of protein. A typical time course follows: 5 minutes, 24 $\mu\mu$C; 10 minutes, 48 $\mu\mu$C; 15 minutes, 76 $\mu\mu$C; 20 minutes, 92 $\mu\mu$C; 30 minutes, 132 $\mu\mu$C; 45 minutes, 168 $\mu\mu$C; 60 minutes, 194 $\mu\mu$C. While such rates seem low by comparison with ribosomes, if the basis of the computation of the specific activity is changed to the parameter milligrams of mitochondrial RNA, then the specific activity becomes 20,000 $\mu\mu$C (20,000 cpm) per milligram of RNA.

Reaction Product

Evidence that the incorporation of labeled amino acids into mitochondrial protein reflects the actual biosynthesis of a protein molecule

[16] G. W. Dietz, B. R. Reid, and M. V. Simpson, *Biochemistry* **4,** 2340 (1965).

[17] R. J. Mans and G. D. Novelli, *Arch. Biochem. Biophys.* **94,** 48 (1961).

is strong but not yet conclusive. Both N-terminal[18] and C-terminal[3] analyses show that the labeled amino acid has been incorporated into the interior of the protein molecule (but see anomalous results reported by Suttie[18]). Partial hydrolysis of the leucine-labeled protein indicates that labeled leucine is found in a multitude of different peptides in association with a variety of neighboring amino acids.

What specific proteins are formed is not clear. Present evidence indicates that soluble proteins such as cytochrome *c*[19,20] or malic dehydrogenase[20] do not become labeled. On the other hand, appreciable amounts of radioactivity are found in lipoprotein fractions.[21-24]

Procedure for Calf Heart Mitochondria

Several laboratories have devised procedures for studying protein synthesis in heart mitochondria.[3,8-10,19,22] These mitochondria possess the advantage over liver mitochondria of a longer lived amino acid incorporating period; incorporation may be linear for at least an hour, and may continue for as long as 4 hours. Too many variables exist in these two systems, including isotope dilution by endogenous amino acids, to permit a direct comparison of their incorporating activities. We describe here the procedure used in our laboratory which is particularly useful for preparing active heart mitochondria in large amounts. It and the incubation medium were derived from an oxidative phosphorylation system developed by Slater and Cleland.[25]

Reagents

Medium C: 0.32 *M* sucrose, 0.005 *M* sodium EDTA, pH 7.2

Medium D: 0.30 *M* KH_2PO_4, 1.2 m*M* AMP, 1.2 m*M* ADP, 0.010 *M* $MgCl_2$, 0.004 *M* sodium EDTA, 0.2 *M* sucrose, adjusted to pH 7.0

L-Amino acid mixture: 10 micromoles each of arginine, asparagine, aspartic acid, cysteine,[26] glutamic acid, glutamine, glycine, histi-

[18] J. W. Suttie, *Biochem. J.* **84,** 382 (1962).

[19] M. V. Simpson, D. M. Skinner, and J. M. Lucas, *J. Biol. Chem.* **236,** P. C. 81 (1961).

[20] D. B. Roodyn, J. W. Suttie, and T. S. Work, *Biochem. J.* **83,** 29 (1962).

[21] D. B. Roodyn, *Biochem. J.* **85,** 177 (1962).

[22] D. E. S. Truman, *Biochem. J.* **91,** 59 (1964).

[23] J. B. Marsh, *J. Biol. Chem.* **238,** 1752 (1963).

[24] G. F. Kalf and M. A. Grece, *Biochem. Biophys. Res. Commun.* **17,** 674 (1964).

[25] E. C. Slater and K. W. Cleland, *Biochem. J.* **53,** 557 (1953).

[26] At the pH of this solution, cysteine is oxidized to cystine fairly rapidly, resulting in a decreased rate of amino acid incorporation. It is therefore advisable to prepare the mixture without cysteine and to neutralize and add this amino acid immediately before use.

dine, isoleucine, leucine, lysine, methionine, phenylalanine, proline, serine, threonine, tryptophan, and tyrosine, adjusted to pH 7.0 in a final volume of 1.0 ml
α-Ketoglutarate, 0.10 *M*, adjusted to pH 7.0
DL-Valine-1-^{14}C (1.0 mC/millimole), 20 μC/ml (under the conditions of counting used here, 1 μC = 1.0×10^6 cpm)
Penicillin G
Streptomycin sulfate
Sucrose, 0.2 *M*
Trichloroacetic acid, 10%
Trichloroacetic acid, 5%
Ethanol, 95%
Ethyl ether

The first five reagents are stored at —20°.

Preparation of Mitochondria

Calf hearts weighing 150–200 g are cut into several pieces, wrapped in aluminum foil, and packed in ice immediately after the animals are killed, for the trip back to the laboratory. All further preparative procedures are conducted in the cold. The hearts are trimmed of fat, connective tissue, and blood vessels, are weighed and minced with scissors. Then 100-g portions are homogenized in 3 volumes of medium C in a Waring blendor (1000 ml size) for 5 seconds. An additional 250 ml of medium C is added, and the homogenization is continued for an additional 7 seconds. Medium C is added to the homogenate to bring it to a final concentration of 1:9, tissue to medium.

The homogenate is centrifuged for 12 minutes at 1000 *g* (average radius) in the 840a angle rotor of the PR2 International Centrifuge. Fat and other floating contaminants are removed by filtering the supernatant fluid through 4 layers of gauze. The filtrate is then centrifuged at 7500 *g* (average radius) for 7.5 minutes in the No. 30 rotor of the Spinco ultracentrifuge. (In dealing with large volumes, the material can be centrifuged in the VRA-V rotor of the Lourdes Vacufuge at 1300 *g* for the low speed spin and 10,000 *g* for the high speed spin.) The mitochondrial pellets are resuspended in medium C with the aid of a loose fitting homogenizer and washed 3 times in this medium under the same conditions of sedimentation. After the third washing, pellets are combined and resuspended in 0.3 *M* sucrose at a concentration of 30 mg of mitochondrial protein per milliliter.

Incubation Conditions

The reaction mixture consists of 0.5 ml of medium D, 0.05 ml of α-ketoglutarate, 0.1 ml of L-amino acid mixture, 0.05 ml of valine- 1-^{14}C, 0.1 ml of sucrose, 40 μg of streptomycin sulfate, and 250 units of penicillin G, in a final volume of 0.8 ml. To the reaction mixture in 10-ml Erlenmeyer flasks is added 0.1 ml of mitochondrial suspension, and the vessels are incubated in a shaking water bath at 37°. Samples are removed at appropriate intervals and treated as described earlier in the rat liver system.

Results

Under the conditions described here, levels of incorporation routinely reach 45–60 $\mu\mu$C (45–60 cpm) per milligram of protein at the first hour of incubation, at which time the incorporation is still linear. A typical time course follows: 1 hour, 48 $\mu\mu$C; 2 hours, 98 $\mu\mu$C; 3 hours, 120 $\mu\mu$C; 4 hours, 130 $\mu\mu$C; 6 hours, 135 $\mu\mu$C. The period of linearity varies from preparation to preparation but is never less than 1 hour. It should be noted that the specific activity of the valine-1-^{14}C used in these experiments was rather low.

[116] Sodium and Potassium-Stimulated ATPase

By R. L. POST and A. K. SEN

The ATPase activity which requires Na^+, K^+, and Mg^{++} together and which is inhibited by cardiac glycosides is a part of the enzyme system for the stoichiometric transport of Na^+ outward and K^+ inward across cell membranes. It is widely distributed in animal tissues and species.[1] Organs which transport Na^+ and K^+ to energize electrical activity or secretion show much activity. For example, the electric organ of the electric eel has yielded 3 units/g,[2] brain 11 units/g,[3] kidney 17 units/g,[4] and the salt gland of the sea gull 40 units/g.[5] The units give the activity of homogenates per gram of wet tissue. The activity has been found also in

[1] J. C. Skou, *Physiol. Rev.* **45**, 596 (1965).

[2] I. M. Glynn, C. W. Slayman, J. Eichberg, and R. M. C. Dawson, *Biochem. J.* **94**, 692 (1965).

[3] P. D. Swanson, H. F. Bradford, and H. McIlwain, *Biochem. J.* **92**, 235 (1964).

[4] A. K. Sen, unpublished data, 1965.

[5] S. L. Bonting, L. L. Caravaggio, M. R. Canady, and N. M. Hawkins, *Arch. Biochem. Biophys.* **106**, 49 (1964).

preparations of isolated membranes of erythrocytes,[6] liver,[7] brain,[8] and squid giant axon.[9]

Assay Method

Principle. From the rate of release of inorganic phosphate from ATP in the presence of Na^+, K^+, and Mg^{++} is subtracted the rate of release in the presence of Mg^{++} and a cardiac glycoside inhibitor, such as ouabain.

Procedure. Into three 16×125 mm thick-walled test tubes in an ice bath is placed 1.0 ml of a solution of 4.5 mM Na_2ATP, 7.5 mM $MgCl_2$, 0.75 mM H_2Na_2EDTA, 30 mM imidazole, and 30 mM glycylglycine. Then is added 0.1 ml of a *homogeneous* enzyme suspension which will release 0.2–1 micromole of inorganic phosphate during incubation with added ($Na^+ + K^+$). To the first tube is added 0.1 ml of a solution of 1.5 M NaCl and 0.3 M KCl. To the second tube is added 0.1 ml of 2.5 mM ouabain. To all tubes is added enough water to bring the volume to 1.5 ml. The first two tubes are incubated at 37° for 20 minutes, with shaking if the enzyme suspension tends to settle out. The last tube is kept on ice. The reaction is stopped by returning the heated tubes to the ice bath.

Released inorganic phosphate may be measured simply by adding 2.5 ml of 0.48 M $HClO_4$, mixing, filtering and taking a 2.0-ml aliquot for assay by the method of Fiske and SubbaRow,[10] or by the modification of Nakamura and Mori.[11] It is necessary to read each of the heated tubes simultaneously with the unheated control to correct for an increase in all the readings with time. This increase is due to the instability of ATP in the presence of acid molybdate. The following more accurate method separates phosphomolybdate from ATP by extraction into butyl acetate. It is derived from that of Wahler and Wollenberger.[12]

Reagents

$HClO_4$, 1.2 M, (0.5 ml) containing 80 g of silicotungstic acid per liter

Solution (1.5 ml) of 60 mM Na_2MoO_4 and 2.25 M NaCl

[6] R. L. Post, C. R. Merritt, C. R. Kinsolving, and C. D. Albright, *J. Biol. Chem.* **235,** 1796 (1960).

[7] P. Emmelot, C. J. Bos, E. L. Benedetti, and P. Rümke, *Biochim. Biophys. Acta* **90,** 126 (1964).

[8] R. J. A. Hosie, *Biochem. J.* **96,** 404 (1965).

[9] P. F. Baker and T. I. Shaw, *J. Physiol.* (*London*) **180,** 424 (1965).

[10] C. H. Fiske and Y. SubbaRow, *J. Biol. Chem.* **66,** 375 (1925). See also Vol. III [115].

[11] M. Nakamura and K. Mori, *Nature* **182,** 1441 (1958).

[12] B. E. Wahler and A. Wollenberger, *Biochem. Z.* **329,** 508 (1958).

Butyl acetate, (4.00 ml)
Isopropanol (4.00 ml) containing 0.06 mM $CuCl_2 \cdot 2\ H_2O$ (dissolved first) and 0.27 M H_2SO_4
2-Mercaptoethanol (5 μl)

Procedure. If the reaction mixture contains less than 0.1 mg of protein, it need not be removed beforehand. The perchloric acid mixture is added to each 1.5 ml sample in the ice bath. The molybdate solution is added, and the contents are swirled briefly. Then butyl acetate is added, the outside of the tube is wiped dry, and the tube is agitated vigorously for at least 20 seconds at room temperature on a vortex mixer (e.g., Cyclo-Mixer, Clay-Adams, Inc., New York 10, New York) so that the contents are completely emulsified. The tube is centrifuged for a minute to clear the emulsion, the upper layer is swirled gently by hand, and 2.00 ml of the upper layer is added to the isopropanol solution in a second tube. The mercaptoethanol is added, and the contents are mixed on the vortex. The resulting blue color is stable after 10 minutes for at least 4 hours. The absorbancy is read in a colorimeter with a red filter or in a spectrophotometer at 625 mμ. If the absorbancy is inconveniently high, 2.00 ml of the unknown are diluted with 2.00 ml of butyl acetate and 4.00 ml of isopropanol mixture. The protein concentration of the enzyme suspension is measured by the method of Lowry *et al.*[13] The activity in the presence of ($Na^+ + K^+ + Mg^{++}$) less that in the presence of (Mg^{++} + ouabain) is expressed as micromoles of P_i $(min)^{-1}$ (milligrams of protein)$^{-1}$ at 37°. One unit of activity splits 1 micromole of ATP per minute at 37° and corresponds to about 10^{-10} mole.[14] The sensitivity of the preparation is the ratio of the ($Na^+ + K^+$)-sensitive activity to the total activity.

Comments. The butyl acetate and isopropanol must be spectroquality, electronic grade, or redistilled. If the addition of one-third volume of concentrated H_2SO_4 to either solvent produces a yellow color, the readings will not be stable. The reading is insensitive to the sample volume in the range from 1.0 to 1.5 ml. Substitution of arsenate or silicate for phosphate does not produce any color. This method fails if a precipitate or a blue color appears in the aqueous phase. A precipitate may be formed between silicotungstic acid and some buffers, for instance. Silicotungstic acid minimizes the adsorption of phosphomolybdate to denatured protein. If silicotungstic acid is omitted, protein should be removed before molybdate is added. A blue color is due to reduction of molybdate. It may be prevented by addition of an oxidant, for instance NaClO as 5 μl of a household bleach such as Clorox.

[13] O. H. Lowry, N. J. Rosebrough, A. L. Farr, and R. J. Randall, *J. Biol. Chem.* **193**, 265 (1951). See also Vol. III [73].
[14] R. L. Post, A. K. Sen, and A. S. Rosenthal, *J. Biol. Chem.* **240**, 1437 (1965).

For accurate kinetic studies inorganic cations must be absent except as added. Contamination enters through adsorption of ions on membranes and in reagents. Contamination can be estimated by analysis of the complete reaction mixture for alkali metal ions by flame photometry and for NH_4^+.[15] Na_2ATP can be converted to the Tris form by passage through a Dowex 50 column in the Tris form. A 1×4 cm column in the H^+ form is washed with 1 *N* HCl until the effluent is free of Na^+ by flame photometry. It is then washed with 1 *M* Tris until the effluent is alkaline and finally with water until the effluent is neutral. One mmole of Na_2ATP in 2 ml of water is placed on the column and run slowly through with water. All water is distilled and deionized or redistilled. H_4EDTA can be made from a solution of H_2Na_2EDTA by precipitating with HCl and washing with water. It can be redissolved with an equivalent amount of Tris.

Purification Procedure

The general principle is to obtain a "microsomal" fraction by differential centrifugation of a sucrose homogenate and to form dispersed and activated particles with a detergent, urea, or a concentrated iodide solution.[16] The results reported for four procedures are given in the table.

The following procedure is designed for a centrifuge which can spin 8×40 ml at 35,000 *g*. Four guinea pigs are killed by a blow on the neck and the kidneys are removed. During all subsequent procedures the preparation is kept close to 2° and neutral pH. The kidneys are sliced in half lengthwise and the cortex is dissected off. The combined cortices are forced through holes (1 mm in diameter) in a tissue press (e.g., from Harvard Apparatus Co., Dover, Massachusetts), and the resulting paste is divided into three equal portions of about 6 g each. Each portion is homogenized in a Potter-Elvehjem homogenizer with a Teflon pestle (e.g., No. 4288-B, size 55 ml, clearance 0.15–0.23 mm from A. H. Thomas Co., Philadelphia). Homogenization is at 1000–1500 rpm with 34 ml of a solution of 0.25 *M* sucrose, 0.02 *M* NaCl, 5 m*M* H_2Na_2EDTA, 1 m*M* $MgCl_2$, and 10 m*M* imidazole. The pestle is passed up and down the tube 12 times. The homogenate is centrifuged at 300 *g* for 10 minutes and the supernatant is saved. The same homogenization procedure is repeated on the precipitate twice. The final precipitate is discarded and the three supernatants are combined, diluted to 320 ml with the homogenizing solution, mixed, and centrifuged at 35,000 *g* for 30 minutes. All subsequent

[15] R. H. Brown, G. D. Duda, S. Korkes, and P. Handler, *Arch. Biochem. Biophys.* **66**, 301 (1957).

[16] T. Nakao, Y. Tashima, K. Nagano, and M. Nakao, *Biochem. Biophys. Res. Commun.* **19**, 755 (1965).

CHARACTERISTICS OF $(Na^+ + K^+)$-ATPASE PREPARATIONS

Animal	Tissue	Dispersing agent	Specific activity units, (mg protein)$^{-1}$	Yield, units (g tissue)$^{-1}$	Sensitivity (%)
Rabbit[a]	Brain or kidney cortex	NaI	3.1	3.4	97
Guinea pig[b]	Kidney cortex	Urea	2.5	3.9	90
Guinea pig[c]	Brain cortex	Polyoxy-ethylene	1.2	10	83
Electric eel[d]	Electric organ	Urea	1.0[e]	2.5	100

[a] T. Nakao, Y. Tashima, K. Nagano, and M. Nakao, *Biochem. Biophys. Res. Commun.* **19,** 755 (1965).

[b] A. K. Sen, unpublished data, 1965.

[c] P. D. Swanson, H. F. Bradford, and H. McIlwain, *Biochem. J.* **92,** 235 (1964).

[d] I. M. Glynn, C. W. Slayman, J. Eichberg, and R. M. C. Dawson, *Biochem. J.* **94,** 692 (1965).

[e] Assayed at 25°. According to R. W. Albers (personal communication) $(Na^+ + K^+)$-ATPase from this tissue is relatively unstable, but activities of 4 to 5 are routinely possible in fresh preparations. See also S. Fahn, G. J. Koval, and R. W. Albers, *J. Biol. Chem.* **241,** 1882 (1966).

centrifugations will be at this volume, force, and duration also. The supernatant is discarded and the sediment is homogenized again in the same volume of a solution of 0.25 *M* sucrose, 2 m*M* H_2Na_2EDTA, 0.1 m*M* $MgCl_2$, 4 m*M* imidazole and 0.02% (w/v) sodium heparin.[17] Homogenization is in the same homogenizer at 700–1000 rpm for 10 passes of the pestle. After centrifugation the sediment consists of a translucent light yellow upper layer of membranes and an opaque brown lower layer of mitochondria. The supernatant is discarded cautiously by aspiration to avoid losing any of the loose upper layer. About 2 ml of a solution of 2 *M* urea, 6 m*M* $(NH_4)_2SO_4$, 2 m*M* H_2Na_2EDTA, 0.1 m*M* $MgCl_2$ and 12 m*M* imidazole is added to each precipitate and swirled over the surface by hand. This supernatant is decanted and saved. The procedure is repeated and the compact part of the yellow layer and the underlying brown layer of the precipitate are discarded. The combined supernatants are diluted to 60 ml with more of the urea solution, mixed, and allowed to stand overnight. The next day any sediment is discarded by decantation and 100 ml of a

[17] W. L. Stahl, J. C. Smith, L. M. Napolitano, and R. E. Basford, *J. Cell Biol.* **19,** 293 (1963).

solution of 15 mM NaCl, 1 mM H_2Na_2EDTA, and 3 mM imidazole is added and the suspension is centrifuged. The supernatant is discarded and the upper half of each precipitate is suspended, combined, and diluted to 10 ml in a solution of 10 mM imidazole, 0.1 mM H_4EDTA, and 5 mM HCl. The lower part of the precipitate contains about 70% as much activity as the upper part, and its specific activity is about half as high. It is usually not used. The enzyme can be washed free from traces of inorganic ions by suspension in, and centrifugation from, a solution of 25 mM imidazole, 13 mM histidine-HCl, and 0.1 mM H_4EDTA. This procedure produces about 70 units of activity from 18 g of cortex with the characteristics shown in the table.

Properties

Stability. The activity decreases about 1% per day at 2°. Traces of Ca^{++} slowly inactivate the enzyme, and it is best to store it in the presence of EDTA. It is stable for months or longer when frozen or lyophilized. The presence of sucrose or buffer appears to be necessary for stability during lyophilization, however. Suitable heating has increased the sensitivity to 100% with only a 17% loss of activity.[18]

Activators and Inhibitors. Aging, freezing and thawing, and mild detergents (e.g., 0.1% sodium deoxycholate) may improve the activity or sensitivity of fresh homogenates or "microsomal" fractions. Lecithin partially restored activity lost by solubilization with 0.33% sodium deoxycholate.[19] The enzyme is inhibited rapidly by Hg^{++} and *p*-chloromercuribenzoate, and slowly by *N*-ethyl maleimide and diisopropyl fluorophosphate. It is inhibited by fluoride, oligomycin, erythrophleum alkaloids,[20] and cardiac glycosides. Sensitivity to particular glycosides varies with the species, e.g., rat enzyme is more sensitive to scillaren A than to ouabain. The aglycons of the glycosides are also effective and can be removed by washing more easily. (Digitonin is not a glycoside but a saponin.)

Kinetics, Stoichiometry, and Localization. The K_m for ATP is about 0.3 mM with Mg^{++} in slight excess. As the amount of Mg^{++} is reduced with maximal ATP, the activity decreases but the sensitivity increases. The K_m for Na^+ is about 1.5 mM and for K^+ about 0.4 mM. Each of these monovalent cations is a competitive inhibitor of the activating effect of the other. The apparent affinity of the Na^+-site for K^+ is about

[18] J. Somogyi, *Biochim. Biophys. Acta* **92**, 615 (1964).

[19] R. Tanaka and K. P. Strickland, *Arch. Biochem. Biophys.* **111**, 583 (1965).

[20] S. L. Bonting, N. M. Hawkins, and M. R. Canady, *Biochem. Pharmacol.* **13**, 13 (1964).

7-fold less than for Na^+ whereas the apparent affinity of the K^+-site for Na^+ is about 160-fold less than for K^+. The activation energy is about 17,000 cal $(°)^{-1}$ $(mole)^{-1}$ and higher than that of the insensitive ATPase. The optimum pH is 7.5 and broad. In intact human erythrocytes the stoichiometry is 3 Na^+ outward, 2 K^+ inward, and one terminal phosphate bond of ATP split in the presence of Mg^{++} on the intracellular surface of the membrane.[21] In squid axons ouabain acts on the extracellular surface. There are partial competitive effects between cardiac glycosides and K^+ at certain concentrations.

Specificity. NH_4^+, Rb^+, Cs^+, and Li^+ can substitute for K^+. Mn^{++} but not Ca^{++} can substitute for Mg^{++}. CTP can substitute for ATP.

Associated Activities. $(K^+ + Mg^{++})$-dependent and ouabain-sensitive *p*-nitrophenyl phosphatase and acetyl phosphatase activities are present. Traces of adenylate kinase may be present.

[21] R. Whittam and M. E. Ager, *Biochem. J.* **97**, 214 (1965).

[117] ^{32}P-Labeling of Mitochondrial Protein and Lipid Fractions

By P. D. Boyer and L. L. Bieber

General Considerations

The incorporation of ^{32}P into mitochondrial phosphoproteins has usually been measured by two different approaches. One is by precipitation of the protein, generally by acid, followed by suitable washing or extraction to remove nonprotein ^{32}P components. The other consists of dissolving or suspending the protein in a suitable medium, followed by removal of nonprotein ^{32}P by ion exchange column or extraction procedures.

Only two types of covalently linked phosphoryl groups have been demonstrated in mitochondrial proteins, namely the *O*-phosphorylserine and the 3-phosphorylhistidine derivatives. Assays for ^{32}P bound to proteins may also include apparently noncovalently linked P_i, such as the binding of P_i by structural protein, and an unidentified phosphate that passes through anion exchange columns with protein and lipid.[1,2] The functions of mitochondrial phosphoserine are unknown. The phospho-

[1] A. W. Norman, L. L. Bieber, O. Lindberg, and P. D. Boyer, *Biochem. Biophys. Res. Commun.* **17**, 108 (1964).

[2] L. L. Bieber and P. D. Boyer, *J. Biol. Chem.* **241**, 5375 (1966).

histidine appears to be closely associated with succinate thiokinase[3,4] and nucleoside diphosphokinase.[5]

Mitochondrial phospholipids are complex, with composition similar to that of the parent organ,[6] mainly lecithin, phosphatidylethanolamine, phosphotidylserine, sphingomyelin, phosphatidylinositols, and phosphatidic acid. The predominant rapidly labeled lipid in rat liver mitochondria is phosphatidyldiphosphoinositol.[7,8] The phosphatidylinositol phosphates may be quite difficult to remove from protein, and the water solubility of univalent salts of triphosphoinositide[9] could result in the loss of this component during isolation procedures.

The procedures described herein suffice for measurement of total ^{32}P in protein and lipid fractions, with subdivision of the protein fraction into the phosphohistidine and phosphoserine components. In incubations with $^{32}P_i$ and AT^{32}P of several seconds' duration, only phosphohistidine is present in measurable amounts.[2,10] Phosphoserine and phospholipid fractions are labeled much more slowly, and appear to be labeled only from ^{32}P-nucleoside triphosphate, not from $^{32}P_i$. Several alternative procedures are described; they are short and may serve for different objectives. The procedures have been used chiefly with rat liver mitochondria, but also with bovine heart and liver mitochondria. They probably can be applied directly to mitochondria from various sources under various incubation conditions.

Purification of $^{32}P_i$ and "Zero-Time" Measurements

Commercial preparations of $^{32}P_i$ may contain impurities that bind tenaciously to protein–lipid fractions. Such impurities may reappear upon storage of purified samples, perhaps by radiation-induced reactions. These impurities, as well as occluded or noncovalently bound P_i or ATP, may contribute considerably to the apparent protein or lipid ^{32}P. The investigator should test his $^{32}P_i$ or AT^{32}P in a suitable "zero-time" sample to measure the amount of apparent ^{32}P-protein or lipid present

[3] R. A. Mitchell, L. G. Butler, and P. D. Boyer, *Biochem. Biophys. Res. Commun.* **16**, 545 (1964).

[4] G. Kreil and P. D. Boyer, *Biochem. Biophys. Res. Commun.* **16**, 551 (1964).

[5] A. W. Norman, R. T. Wedding, and K. Black, *Biochem. Biophys. Res. Commun.* **20**, 703 (1965).

[6] F. D. Collins and V. L. Shotlander, *Biochem. J.* **79**, 316, 321 (1961).

[7] T. Galliard and J. N. Hawthorne, *Biochim. Biophys. Acta* **70**, 479 (1963).

[8] J. Garbus, H. F. DeLuca, M. E. Loomans, and F. M. Strong, *J. Biol. Chem.* **238**, 59 (1963).

[9] R. M. C. Dawson, *Biochem. J.* **97**, 134 (1965).

[10] O. Lindberg, J. J. Duffy, A. W. Norman, and P. D. Boyer, *J. Biol. Chem.* **240**, 2850 (1965).

as compared to incubated samples. For procedures where metabolic reactions are stopped by phenol or trichloroacetic acid, addition of the $^{32}P_i$ or $AT^{32}P$ with the stopping agent gives a fairly satisfactory measure of contamination, occlusion, and noncovalent binding.

If $^{32}P_i$ purification proves desirable, a simple empirical procedure that has been found useful in the author's laboratory is as follows:

A 1-10 mC sample of $^{32}P_i$ for purification, containing about 0.1 micromole of P_i is made neutral or slightly alkaline by use of 0.2 *N* NaOH in small drops, wide-range pH test paper, and a disposable, fine-tip glass rod. Addition of P_i to carrier-free preparations lessens losses by adsorption on glass and may decrease production of radiation-induced impurities. The 1–5 ml sample is adsorbed on an approximately 0.5×1.5 cm column of weakly basic anion exchange resin, with use of a fine-tipped disposable dropper. The resin may be conveniently suspended in a medicine dropper stopped with a glass bead or glass wool. The authors have used Rohm and Haas CG-4B resin, but other similar resin would likely suffice. The adsorbed sample and original container is washed with several small portions (1–2 ml total volume) of distilled water, using the same dropper as for transfer of the original sample. The $^{32}P_i$ is eluted with 0.2 *N* NaOH. Appearance of the ^{32}P in the eluate is conveniently measured by "focusing" an approximately 2-mm hole in a 1-cm thick lead shield on the column outlet. A Geiger tube mounted behind the shield and suitable count-rate meter and recorder serve to monitor the elution. Impurities appear to be retained near the top of the column. If desired, the pH of the eluate is adjusted to near neutral by 0.1 *N* HCl addition.

The purification is made with appropriate precautions in a separate room, with considerable care to avoid contamination of equipment or personnel.

Phenol Extraction Procedure

Principle. Mitochondrial protein and lipid are extracted into phenol. $^{32}P_i$ and/or $AT^{32}P$ are removed by careful washing. ^{32}P in the phenol extract gives a measure of total covalently bound protein and lipid ^{32}P. For elucidation of the nature of the phosphoprotein and the lipid phosphate, the protein and lipid are precipitated from the phenol by acetone addition. The aqueous-insoluble lipid phosphate is extracted from the precipitate by organic solvents, the phosphate released from phosphohistidine by acid hydrolysis, and the phosphorserine measured as the acid-stable fraction.

Reagents

Buffered phenol: 88 g of phenol plus 12 ml of 0.01 *M* Na_2HPO_4, pH 8.0–8.5

Washing buffer: a solution of 0.01 *M* Na_2HPO_4, 0.01 *M* Na_2EDTA, and 0.01 *M* $Na_4P_2O_7$ (or their equivalent) brought to pH 8–8.3

NH_4OH, Conc.

7.5 *M* Urea-0.3 *M* NH_4OH: 455 g of urea and 20 ml of conc. NH_4OH made to 1 liter

Acetone, chloroform, and methanol; reagent grades

Procedure. To a 1–5 ml incubation mixture containing 3–50 mg of mitochondrial protein and about 10^6 to 10^7 cpm of ^{32}P as P_i or ATP is added to 10 ml of the buffered phenol solution, with rapid mixing by a magnetic stirrer. The suspension in a 40 ml heavy-walled centrifuge tube is thoroughly mixed with 25 ml of the washing buffer, a small magnetic stirring bar being used. The solution is centrifuged at about 2500 rpm in a swinging-bucket centrifuge (about 18 cm radius to the end of the centrifuge tube), and the upper aqueous layer is removed by careful aspiration. Any interfacial material is dispersed in the phenol by addition of concentrated NH_4OH or 7.5 *M* urea-0.3 *M* NH_4OH, or both as required. Wash buffer is added, and the washing procedure repeated 4–7 times. If the final phenol volume is greater than 5 ml, it is desirable to reduce the volume to less than 5 ml by distilled water wash before acetone precipitation.

Plating of a 1–2 ml aliquot of the above phenol extract, corrected for a "zero-time" sample, gives a measure of the total protein and lipid phosphate present. In short incubations under conditions as used by Bieber and Boyer[2] nearly all the ^{32}P is present as phosphohistidine, and the procedure will give a reasonable estimation of the phosphohistidine content. For measurements of the phosphoserine and lipid phosphate fractions, the sample is treated as follows:

The ^{32}P containing proteins and acetone-insoluble lipids are precipitated by rapid addition of about 5 volumes of cold acetone with rapid mixing. The mixture is stored at $-20°$ for several hours or, preferably, overnight. The precipitate is collected by centrifugation, and the lipids are removed by successive extraction with about 5 ml of chloroform–methanol (1:1, v:v), 5 ml of chloroform–methanol–H_2O (20:10:1, v:v:v), and 5 ml of methanol. The residue is separated from the organic solvents by centrifugation. The counts in an aliquot of the combined solvent extracts can be used as a measure of the total counts in the rapidly labeled lipid fraction.

The protein residue is treated with about 5 ml of 0.3 N trichloroacetic acid–0.001 M P_i for 3 minutes in a boiling water bath. The mixture is cooled, protein removed by centrifugation, and an aliquot of the supernatant solution is used as a measure of the acid-labile phosphoprotein phosphate (phosphohistidine). The protein residue is washed twice with about 10 ml of phosphate buffer, and the counts remaining with the precipitate used as a measure of the acid-stable phosphoprotein (phosphoserine).

Acid Precipitation and Column Procedure

Principle. Proteins are precipitated and freed of most $^{32}P_i$ and/or $AT^{32}P$ by trichloroacetic acid near 0°. Remaining P_i or ATP is removed by dispersing the protein in urea-NH_3 and passage over an anion exchange resin. Phosphoprotein is precipitated by acetone. Phosphate is liberated from phosphohistidine by acid hydrolysis, and counted.

Reagents

Trichloroacetic acid, 0.45 M
Dowex 1, 100–200 mesh 8X-(OH^-). Remove fines and wash with H_2O
Urea-NH_3 solution: 445 g of urea and 20 ml of concentrated NH_3 made to 1000 ml
Acetone

Procedure. To a 1–2 ml incubation mixture containing about 10^6 to 10^7 cpm of ^{32}P is added two volumes of cold 0.45 M trichloroacetic acid, and the mixture is rapidly chilled to near 0°. The protein is removed by rapid centrifugation near 0°, the tubes are drained in the cold and their mouths are carefully wiped with tissue, and the protein is immediately dispersed in 1–3 ml of cold urea-NH_3 solution. The acid precipitation aids in removing $^{32}P_i$ and ^{32}P-contaminants. Alternatively 5 ml of urea-NH_3 solution can be added directly to 1 ml incubation solutions.

The dispersed solution is passed through a 1×10 cm column of Dowex 1 resin, and the column is washed with 7–10 ml of urea-NH_3 solution or with H_2O. A 1–2 ml aliquot of the mixed column eluate and wash may be plated and counted as a measure of total phosphoprotein and lipid present. Difficulty is frequently encountered in "creeping" of the urea-NH_3 samples during drying. As an alternative, the protein may be precipitated from the column eluate by the addition of 5 or more volumes of cold acetone, and the determinations of acid-stable and acid-labile phosphate be performed as described above.

With short-term incubations under conditions as used by Lindberg *et al.*[10] and by Bieber and Boyer,[2] nearly all the initially labeled material in the protein-lipid fraction will be phosphohistidine.

Comments

Various modifications of the above procedures may prove useful. For example, combination of urea-NH_3-column assay with the phenol assay showed the presence of an unidentified phosphate fraction in the column eluate.[1,2]

The procedures cannot be used for unambiguous identification of phosphorylated components. For this purpose, isolation and comparison of the phosphorylated product with authentic material is recommended.

For alternate methods of extraction of lipid fractions rapidly labeled by ^{32}P and their characterization, the reader is referred to references cited in footnotes 7, 8, 9, and 11.

[11] A. K. Hajra, A. B. Seiffert, and B. W. Agranoff, *Biochem. Biophys. Res. Commun.* **20,** 199 (1965).

[118] ^{32}P-Labeling of a ($Na^+ + K^+$)-ATPase Intermediate

By R. L. Post and A. K. Sen

$$[\gamma\text{-}^{32}\text{P}]\ \text{ATP} + \text{enzyme} \xrightarrow[\text{Mg}^{++}]{\text{Na}^+} [^{32}\text{P}]\ \text{enzyme-P} + \text{ADP}$$

$$[^{32}\text{P}]\ \text{enzyme-P} \xrightarrow{\text{K}^+} \text{enzyme} + [^{32}\text{P}]\ \text{P}_i$$

Preparation of [γ-^{32}P] ATP

The procedure is modified from that of Glynn and Chappell.[1]

Reaction Mixture

(1) Carrier-free [^{32}P] P_i in dilute HCl, 15 mC
(2) Solid Tris, 1.5 micromoles for each micromole of HCl supplied with [^{32}P] P_i
(3) 1.0 ml of a cofactor solution containing 2 mM H_2Na_2EDTA, 2.5 mM Na_2ATP, 0.5 mM Na_2ADP, 2.5 mM 3-phosphoglyceric acid (tricyclohexylamine salt), and 0.1 mM NAD
(4) 0.5 ml of 0.5 M Tris and 0.5 M Tris-Cl, pH 8.1
(5) 2 μl of mercaptoethanol
(6) 10 μg of phosphoglyceric kinase (Boehringer)

[1] I. M. Glynn and J. B. Chappell, *Biochem. J.* **90,** 147 (1964).

(7) 30 μg of phosphoglyceraldehyde dehydrogenase (Boehringer)
(8) 0.1 ml of 0.05 *M* $MgCl_2$

Procedure. The reaction is carried out at about 23° in the lead-glass bottle in which the ^{32}P comes. The reagents are added in the sequence given. Neutralization at step 2 is tested by adding a tiny drop to pH paper. Any pH between 4 and 8 is satisfactory at this point. The cofactor solution (3) may be stored frozen. During the addition of the $MgCl_2$ (step 8), the reaction mixture is swirled gently in order to minimize the possibility of precipitating [^{32}P] $MgNH_4PO_4$. Formation of this precipitate delays labeling of ATP for about 10 hours. The enzyme preparations may provide enough P_i and NH_4^+ to form the precipitate if the procedure given above is modified.

After 10 minutes of incubation, 1 μl or less of the mixture is added to 1 ml of 0.5 m*M* KH_2PO_4 and tested for incorporation of ^{32}P into ATP as follows.[2] Five milliliters of 5% (w/v) trichloroacetic acid is added and an aliquot is counted; 0.8 g of dry Norit A charcoal is added, and the tube is shaken for 20 minutes at room temperature. After the charcoal is filtered out, a similar aliquot is counted. The ratio of counts in the second aliquot to those in the first is the fraction of ^{32}P not incorporated into ATP. If this is 0.5 or less, labeling reactions are stopped after 30 minutes of incubation by heating at 100° for 3 minutes and cooling in air. (Ice water will break the lead glass container.)

The radioactive ATP is separated by column chromatography with gradient elution as follows. The reaction mixture is filtered through Whatman No. 541 paper, and most of the radioactivity is washed through the paper with water. The mixture is diluted to about 50 ml and passed through a 1 × 3 to 4 cm column of 2 g of Dowex 1 anion exchange resin in the chloride form at about 1 ml per minute. (Sharper separations are obtained if the column is first washed in sequence with 20 ml each of water, acetone, water, 1 *M* NaOH, water, acetone, water, 1 *M* HCl, and water, followed by 2 ml of 1 *M* Tris and water until the effluent is no longer alkaline to pH paper.) The column is washed with 20 ml of water and then eluted at pH 2.5 with 3 m*M* HCl containing a continually rising gradient of Tris-Cl from a 4-chambered Varigrad (e.g., from Buchler Instruments, Inc., Fort Lee, New Jersey). Each chamber contains 35 ml with Tris-Cl concentrations as follows: No. 1, 10 m*M*; No. 2, 30 m*M*; and Nos. 3 and 4, each 300 m*M*. Then 3.5–4 ml fractions are collected. A monitor of the ultraviolet absorbance of the column effluent helps in evaluating the performance of the column. The contents of the

[2] R. K. Crane and F. Lipmann, *J. Biol. Chem.* **201,** 235 (1953).

tubes with significant amounts of ATP are combined, neutralized by the addition of 20 micromoles of solid Tris per milliliter, and stored frozen. Incorporation of radioactivity is 80–85%, giving a specific activity of 5 mC/micromole. A preparation is usable for 2–3 months. Contact with molybdate glass may reduce the stability of the ATP. In one case the ATP transferred its radioactivity to the walls of a plastic bottle. Polycarbonate containers have been satisfactory.

Precautions against Radiation Injury. A portable Geiger-Müller detector is helpful. It is desirable to wear glasses to protect against the possibility of formation of an opacity in the eye and to work behind a sheet of ordinary window glass or a sheet of Lucite 2 cm thick.

Preparation of Phosphorylated Intermediate

The procedure is that of Post *et al.*[3]

Reaction Mixture

(1) Imidazole glycylglycine, 10 micromoles
(2) $MgCl_2$, 2 micromoles
(3) NaCl, 20 micromoles
(4) ($Na^+ + K^+$)-ATPase, 3 units[4] with about 1.2 mg of protein
(5) [γ-^{32}P] ATP, 4×10^{-8} mole, 10^6 cpm

The volume is 1.0 ml.

Procedure. The reaction is carried out at 0° with rapid stirring. The reagents are added in the sequence given. Five seconds after the addition of the radioactive ATP, the reaction is stopped by the addition of 35 ml of 0.3 *M* $HClO_4$ or trichloroacetic acid (fresh) containing 0.6 m*M* unlabeled ATP. The acid is at 0°, and the ATP is added to it less than 1 hour before the reaction is carried out. The suspension of denatured protein is centrifuged at 12,000 *g* for 20 minutes at 4°, and the supernatant is decanted. The precipitate is homogenized with a glass rod in the residual supernatant and diluted with 35 ml of the acid at 0° (now without added ATP). Centrifugation and decantation of the supernatant are repeated as before. The washing cycle is repeated once more. The centrifugation may also be done at 1500 *g* at room temperature with perhaps 30% less recovery. About 20% of the protein is lost in the procedure. About 10% of the radioactivity in the precipitate is side products. To estimate the amount of side products the procedure is con-

[3] R. L. Post, A. K. Sen, and A. S. Rosenthal, *J. Biol. Chem.* **240,** 1437 (1965).
[4] See this volume [116].

ducted in parallel in a reaction mixture which is identical except that sodium ion is replaced by potassium ion. Only side products appear in this precipitate. The procedure yields 0.24×10^{-7} moles of intermediate containing 6000 cpm, or 0.6% of those added.

Comments. The rate of labeling is limited by the rate of mixing. The turnover time of the intermediate under these conditions is about 12 seconds. Formation of the intermediate is poorly reversible by ADP so that greater efficiency of incorporation of radioactivity could probably be obtained if the amount of carrier ATP were reduced (with a concomitant increase in the specific activity of the ATP). The limiting factor here is the simultaneous splitting of ATP by the $(Na^+ + K^+)$-insensitive ATPase, which contaminates most preparations. Because of a difference in the temperature coefficients of the two kinds of ATPase, the activity of the independent ATPase may be much greater than that of the dependent ATPase at 0°. In case of uncertainty it is best to estimate the splitting of ATP during the labeling reaction by analyzing the first acid supernatant for ATP by adsorption of radioactivity on charcoal.[2]

Properties

The chemical homogeneity of the material is not completely established.

Stability of the Phosphate Bond. Between pH 2 and pH 6 at 4° hydrolysis is about 1½% per hour. Outside this range the rate is more rapid, and hydrolysis is very rapid above pH 9.[5,6,7] The temperature coefficient of hydrolysis is greater at pH 6 than at pH 2.[6] Hydrolysis is accelerated by molybdate and hydroxyl amine.[5,6,7] In the presence of methanol or ethanol, methyl or ethyl phosphate is formed.[7,8] The bond is probably a carboxyl phosphate.

Properties of the Material to Which the Phosphate is Attached. Peptic digestion solubilizes the intermediate.[6,7,9] The first soluble material is converted by further digestion into two radioactive subfragments which have a molecular weight of about 2000. The subfragments can be acetylated and/or oxidized with performic acid.[7]

[5] K. Nagano, T. Kanazawa, N. Mizuno, Y. Tashima, T. Nakao, and M. Nakao, *Biochem. Biophys. Res. Commun.* **19,** 759 (1965).

[6] L. E. Hokin, P. S. Sastry, P. R. Galsworthy, and A. Yoda, *Proc. Natl. Acad. Sci. U.S.* **54,** 177 (1965).

[7] H. Bader, A. K. Sen, and R. L. Post, *Biochim. Biophys. Acta* **118,** 106 (1966).

[8] D. A. Hems and R. Rodnight, *Biochem. J.* **96,** 57P (1965).

[9] R. W. Albers, S. Fahn, and G. J. Koval, *Proc. Natl. Acad. Sci. U.S.* **50,** 474 (1963).

[119] Analysis of Phosphoproteins

By K. Ahmed and J. D. Judah

There are two methods for the analysis of "phosphoprotein." First, alkaline hydrolysis at 38° yields inorganic orthophosphate (P_i) quantitatively. Secondly, partial acid hydrolysis (2 *N* HCl, 100° in sealed tubes for 8 hours) yields *O*-phosphorylserine (P-serine).

Both methods have been applied to the determination of tissue phosphoproteins, in particular for radioactivity measurements. Alkaline hydrolysis lacks specificity and is unable to distinguish between P_i liberated from the protein and that which may be present as a contaminant. However, it is quick and useful for many purposes. The second method gives a poor yield of P-serine and is not useful for quantitation. But the identification of P-serine is positive. The possibility exists that P-serine is formed by the migration of phosphoryl groups from some other amino acid (histidine?) during the isolation procedure. The point is that isolation of P-serine, radioactive or otherwise, is a firm indication of the presence of a phosphoprotein in the material hydrolyzed.

Assay Methods

Principle. Alkaline hydrolysis. Tissue phosphoproteins are washed free of acid-soluble components, lipids, and nucleic acids using the Schmidt and Thanhauser procedure.[1] They are then treated with dilute alkali to liberate P_i, which is estimated by the method of Berenblum and Chain.[2] It may also be analysed by the method of Fiske and SubbaRow,[3] providing that the P_i is first separated (e.g., as the magnesium-ammonium salt) since the products of hydrolysis interfere with the analysis.

Acid hydrolysis. The protein material obtained after the Schmidt and Thanhauser[1] procedure is subjected to acid hydrolysis in 2 *N* HCl. *O*-Phosphorylserine is liberated from the phosphoprotein[4] and is separated by paper chromatography, by column chromatography or by electrophoresis.

Reagents

Trichloroacetic acid, 5% (w/v)
Trichloroacetic acid, 20% (w/v)

[1] G. Schmidt and S. J. Thanhauser, *J. Biol. Chem.* **161,** 83 (1945).
[2] I. Berenblum and E. Chain, *Biochem. J.* **32,** 295 (1938).
[3] C. H. Fiske and Y. SubbaRow, *J. Biol. Chem.* **66,** 375 (1925).
[4] E. P. Kennedy and S. W. Smith, *J. Biol. Chem.* **207,** 153 (1954).

Ethanol–ethyl ether (3:1 v/v)
Ethanol
HCl, conc.
HCl, 2 *N*
KOH, 1 *N*
H_2SO_4, 10 *N*
Ammonium molybdate, 10%
$SnCl_2$ reagent prepared by dissolving 10 g $SnCl_2$ in conc. HCl to a final volume of 25 ml. This keeps for a month or two. Before use it is diluted 1:200 in *N* H_2SO_4
Acetone
Benzene–isobutanol (1:1) saturated with water
Cation exchanger AG-50 (Dowex 50 analytical grade), H^+ form X4–8, 200–400 mesh, can be obtained from Bio-Rad laboratories, California.

Procedure

ALKALINE HYDROLYSIS. Tissue (approximately 100–200 mg) is deproteinized with a final concentration of 5% trichloroacetic acid and is thoroughly homogenized (5–6 ml, is a convenient volume). The homogenized material is centrifuged for 5–10 minutes in a clinical centrifuge. Stronger centrifugation may be needed with certain tissues and preparations.[5] The residue is washed in this manner 5 or 6 times with thorough suspension each time. If the phosphoproteins are labeled with ^{32}P, necessary precautions should be taken to avoid cross-contamination of various samples. In this case it is advisable to use 5% trichloroacetic acid containing 0.1 *M* sodium phosphate as carrier, for the first three washes. The acid-washed protein residue is washed once or twice with 5 ml distilled water. This step is quite important, as in the presence of trichloroacetic acid the phosphoproteins tend to dissolve in the organic solvents used to extract lipids.[5-7] The residue is then suspended in 1 ml H_2O followed by 4 ml ethanol. It is stirred for 5 minutes and centrifuged. The residue is washed twice more with ethanol–ethyl ether (5 ml each time). The residue so obtained is suspended in 3 ml of 5% trichloroacetic acid and placed in a water bath maintained at 96° for 15 minutes. The tubes are occasionally stirred. They are then allowed to cool. The suspension is centrifuged after the volume is made to 5 ml with 5% trichloroacetic acid. The residue is washed once more with 5 ml of 5% trichloroacetic

[5] K. Ahmed and J. D. Judah, *Biochim. Biophys. Acta* **104,** 112 (1965).
[6] H. N. Munro and E. D. Downie, *Arch. Biochem. Biophys.* **106,** 516 (1964).
[7] J. D. Judah, K. Ahmed, and A. E. M. McLean, *Biochim. Biophys. Acta* **65,** 472 (1962).

and decanted carefully to remove trichloroacetic acid. This step should remove nucleic acid. The residue is suspended in 0.5 ml *N* KOH, and the volume is made to 1.0 ml. The tubes are covered and allowed to stand for 16–18 hours at 38°. At the end of this period, add 0.1 ml of concentrated. HCl and enough 20% trichloroacetic acid to a final volume of 3.0 ml and mix well. Place the tubes on crushed ice for 20 minutes to ensure complete precipitation. Centrifuge to obtain a clear supernatant. A suitable aliquot (1 ml) of the supernatant is taken for analysis of the P_i liberated. This can be done by a modification of the method of Berenblum and Chain.[2] Place the sample (1 ml) in a graduated, glass-stoppered centrifuge tube of 15-ml volume (or a small separating funnel) and add 0.5 ml of 10 *N* H_2SO_4 and 1.0 ml of 10% ammonium molybdate. Make the final volume to 3.0 ml with water, add 1.0 ml of acetone, and mix. Add 3.0 ml of benzene–isobutanol solution and shake vigorously. Allow the layers to separate and shake again. Remove the lower aqueous layer and add 3 ml of *N* H_2SO_4; shake well again. Remove the aqueous layer and repeat this washing twice more. Vigorous washing of the benzene–isobutanol layer is important to remove all the acetone. Add 2 ml of diluted $SnCl_2$ reagent (diluted 1:200 with *N* H_2SO_4), and gently mix with the benzene–isobutanol layer. Avoid vigorous shaking since the blue color tends to pass into the aqueous layer. Remove the aqueous layer and make up the volume of the benzene–isobutanol layer to 4 ml (or as required, depending upon the density of the color) with ethanol. Mix well and estimate the density of the blue color at 660 mμ against a reagent blank.

Measurements of radioactivity are easily made on the final solution, after its optical density has been obtained. Evaporate an aliquot on a suitable planchet and use any conventional end-window counter.

PHOSPHOPROTEIN ASSAY BY P-SERINE ANALYSIS. The protein residue is washed as described above. It is advisable to use a larger sample of material as about 80% or more of *O*-phosphorylserine is lost during the acid hydrolysis. The tissue is suspended in 2 *N* HCl (about ten times the volume of final residue), and 5 micromoles of carrier P-serine is added. If an estimate of specific activity is desired, the carrier must not be added. The tubes are sealed (hard-glass Pyrex tubes must be used) and placed at 104° for 8 hours. At the end of hydrolysis, centrifuge off any small amounts of residue and lyophilize the hydrolyzate to remove HCl. The dried material is dissolved in a small volume of 0.05 *N* HCl and placed on a Dowex 50 column prepared as follows: Dowex 50, 200–400 mesh, X4-8, analytical grade is used. It should be washed and recycled at least once, then washed several times with 0.05 *N* HCl. The resin suspension in 0.05 *N* HCl is poured into a column and allowed to settle. The authors have used columns 32 × 1 cm. Elution is with 0.05 *N* HCl.

Flow is adjusted to approximately 0.5 ml/minute. Fractions of 1.5 ml are collected using any convenient fraction collector. P_i moves with the solvent front and is soon gone (tubes 12–20 in the above example). P-serine emerges from about tube 35. The peak should be symmetrical. An appearance of lag is often due to the presence of small amounts of *O*-phosphorylthreonine which appears soon after P-serine. If radioactivity measurements are desired, these tubes may be counted immediately and the specific activity determined after the phosphoserine is completely hydrolyzed to liberate P_i. Alternately the tubes containing P-serine may be pooled and counted. If further characterization is desired, the material in the pooled tubes may be lyophilized and subjected to paper chromatography or paper electrophoresis. The details of the paper chromatography are described by Kennedy and Smith[4] and of electrophoresis by Dawson.[8] The chromatogram or electrogram can be radioautographed (if radioactive material is studied) and/or stained to reveal phosphate and phosphate esters by the method of Wade and Morgan.[9] The ninhydrin reaction may also be used. It may be pointed out that the yield of P-serine during acid hydrolysis under similar conditions appears to vary from tissue to tissue.[5,7]

[8] R. M. C. Dawson, *Biochem. J.* **75,** 45 (1960).

[9] H. E. Wade and D. M. Morgan, *Nature* **171,** 529 (1953).

Author Index

The numbers in parentheses are footnote numbers and are inserted to enable the reader to locate a cross reference when the author's name does not appear at the point of reference in the text.

A

B

C

H

M

N

T

Y

Z

Subject Index

A

E

F

G

N

O

P

Due